AF594964

The Tight Junction and Its Proteins

The Tight Junction and Its Proteins: More Than Just a Barrier

Volume 1

Editors

Michael Fromm
Susanne M. Krug

MDPI • Basel • Beijing • Wuhan • Barcelona • Belgrade • Manchester • Tokyo • Cluj • Tianjin

Editors
Michael Fromm
Charité—Universitätsmedizin Berlin
Germany

Susanne M. Krug
Charité—Universitätsmedizin Berlin
Germany

Editorial Office
MDPI
St. Alban-Anlage 66
4052 Basel, Switzerland

This is a reprint of articles from the Special Issue published online in the open access journal *International Journal of Molecular Sciences* (ISSN 1422-0067) (available at: https://www.mdpi.com/journal/ijms/special_issues/Tight_Junction).

For citation purposes, cite each article independently as indicated on the article page online and as indicated below:

LastName, A.A.; LastName, B.B.; LastName, C.C. Article Title. *Journal Name* **Year**, *Article Number*, Page Range.

Volume 1
ISBN 978-3-03943-224-0 (Pbk)
ISBN 978-3-03943-225-7 (PDF)

Volume 1-2
ISBN 978-3-03943-302-5 (Pbk)
ISBN 978-3-03943-303-2 (PDF)

Cover image courtesy of Susanne M. Krug.

Contents

About the Editors

Michael Fromm is a senior professor and former Head of the Institute of Clinical Physiology at the Charité—Universitätsmedizin Berlin, Germany, where he still works. He has published more than 230 research articles, resulting in an h-index of 61. Through this, he has made seminal contributions to transport mechanisms and barrier functions of intestinal and renal epithelia in health and disease. For some claudins, he has discovered that they form channels selective for ions and/or water. Starting in 2006, he was the coordinator of the Deutsche Forschungsgemeinschaft (DFG) Research Unit "Molecular structure and function of the tight junction", which paved the way to a currently running DFG graduate school focusing on tight junction research. Within this, his lab focuses on protein prerequisites of tricellular tight junction water permeability.

Susanne M. Krug has studied Biochemistry and is now a Group Leader at the Institute of Clinical Physiology at the Charité—Universitätsmedizin Berlin, Germany. She has made remarkable contributions to the understanding of the tricellular tight junction as a regulated passage site for macromolecules, especially in inflamed intestinal epithelium. Presently, she holds three grants of the Deutsche Forschungsgemeinschaft (DFG) and, based on >60 publications, has reached an h-index of 25. Her current research interests still deal with tricellular tight junctions, but also with interaction of immune cells and tight junction proteins in inflammatory bowel diseases.

Preface to "The Tight Junction and Its Proteins: More Than Just a Barrier"

Most accredited FAO statistics predict that in 30 years, the world's population will have reached 9 billion people. In order to satisfy the nutritional needs of humans, the demand for raw materials, especially protein sources, will increase. It has been estimated that by 2050, the production of meat will have increased by 50%, while the demand for fish, milk, and eggs will have grown by 75%. An increase in animal products requires an increase in farmed animals, and this will be accompanied by a significant intensification in livestock farming (higher animal densities and production units, more concentrated feed, pharmaceuticals, and vaccinations, etc.). A large number of animals, farmed in relatively small areas, will result in a larger demand for protein and energy sources on which to feed them and in the deposition of large amounts of excreta, containing nitrogen, phosphorus, organic matter, and fecal microbes, in the water, with a consequent contamination of water systems globally, which will include surface water eutrophication and groundwater nitrate enrichment. Thus, the livestock sector is an important user of natural resources and has a great influence on air, soil, and water quality, the global climate, and biodiversity maintenance. Our research proposes innovative ideas to control the environmental damage through the management of animal nutrition. At the same time, the perception of animals as sentient beings capable of feeling emotions, like joy and pain, will increase in prevalence in the future. Thus, it will be increasingly important to adopt nutritional strategies and breeding techniques capable of increasing animal welfare and at the same time to reduce the use of pharmacological treatments in full respect of the environment, animal health, and food safety.

Michael Fromm, Susanne M. Krug
Editors

International Journal of
Molecular Sciences

Editorial

Special Issue on "The Tight Junction and Its Proteins: More than Just a Barrier"

Susanne M. Krug and Michael Fromm *

Institute of Clinical Physiology/Nutritional Medicine, Department of Gastroenterology, Rheumatology and Infectious Diseases, Charité—Universitätsmedizin Berlin, Campus Benjamin Franklin, 12203 Berlin, Germany; susanne.m.krug@charite.de
* Correspondence: michael.fromm@charite.de

Received: 22 June 2020; Accepted: 26 June 2020; Published: 29 June 2020

Abstract: For a long time, the tight junction (TJ) was known to form and regulate the paracellular barrier between epithelia and endothelial cell sheets. Starting shortly after the discovery of the proteins forming the TJ—mainly, the two families of claudins and TAMPs—several other functions have been discovered, a striking one being the surprising finding that some claudins form paracellular channels for small ions and/or water. This Special Issue covers numerous dedicated topics including pathogens affecting the TJ barrier, TJ regulation via immune cells, the TJ as a therapeutic target, TJ and cell polarity, the function of and regulation by proteins of the tricellular TJ, the TJ as a regulator of cellular processes, organ- and tissue-specific functions, TJs as sensors and reactors to environmental conditions, and last, but not least, TJ proteins and cancer. It is not surprising that due to this diversity of topics and functions, the still-young field of TJ research is growing fast. This Editorial gives an introduction to all 43 papers of the Special Issue in a structured topical order.

Keywords: tight junction; claudin; occludin; tricellulin; angulin; paracellular channels and barriers; cell polarity; pathogens; immune cells; environmental sensors; cancer

---

1. Introduction

The aim of the Special Issue "The Tight Junction and Its Proteins: More Than Just a Barrier" is to collect original studies and reviews on the functions of the tight junction (TJ) and its proteins, which extend beyond the classical, and therefore name-giving, task of sealing the paracellular pathway of epithelia. Terms like "tight" junction, occludin and claudin (both derived from Latin words indicating closure) are suggestive of a sole barrier function for these structures. However, there are many other functions, including those in gene expression, morphogenesis, cell proliferation, adhesion, signaling, immune responses, reaction to bacteria and viruses, cancer, and the enhancement of drug uptake. All 43 articles of this Special Issue are briefly introduced in Section 2 to Section 6 of this editorial review.

2. Tight Junction Barriers and Structures

As briefly delineated by the first section, the TJ is mainly known as being structure forming and regulating the paracellular barrier. However, besides the dynamic and complex regulation of the paracellular pathways, many other aspects have been discovered where TJs and their proteins are involved, indicating that there are many functions beyond this.

In this Special Issue, we would like to introduce and cover several of the new arising functions. However, even the classical function as a paracellular transport-regulating protein complex is still not completely understood, and regulation in health and disease, as well as the molecular and structural aspects of TJs and their components, need to be fully elucidated.

In the following sections, a short overview of the articles within this Special Issue is given to indicate the different facets of TJ protein functions that may extend beyond the typical paracellular barrier aspects.

2.1. Pathogens Affecting the TJ and Regulation via Immune Cells

The barrier function of the TJ helps to protect the organisms against the entry of pathogens from the outer environment and thus is often an initial target of pathogens, leading to barrier impairment and subsequent inflammation processes in disease statuses. Many studies have shown direct interactions of pathogens with TJ proteins or indirect influence due to their secreted toxins aiming to break down the barrier. The impairment of the intestinal barrier leads to osmotic imbalances and diarrhea as the main symptoms and can be caused by many pathogenic bacteria. However, many pathogenic bacteria do not only act on TJs directly but also induce apoptosis and thereby the loss of epithelial integrity. For example, *Klebsiella oxytoca*-induced antibiotic-associated hemorrhagic colitis and diarrhea occur mainly to the secreted cytotoxins tilivalline and tilimycin, which induce apoptosis, while in addition, independently of these, *Klebsiella oxytoca* also affects claudin-5 and -8 and thus paracellular barrier integrity [1]. Toxins like *Clostridium perfringens* enterotoxins (CPEs) are also known to directly interact with claudins and have come into focus for use as tools targeting claudins specifically [2,3].

Inflammatory processes, which often go together with barrier impairment, are linked to activated immune cells reacting upon enhanced antigen uptake or other stimuli. Cytokines that are secreted not only further stimulate immune cells but also change the regulation and expression of TJ proteins, affecting barrier properties. In airway epithelia, the secretion of interleukin-13 (IL-13) induces the ubiquitination of TJ proteins and leads to proteasomal aggregation via the JAK/STAT-dependent expression of ubiquitin-conjugating E2 enzyme, which finally results in barrier disruption and lung damage [4]. Immune cells, furthermore, usually balance barrier-disturbing and -protective cytokine secretion; for example, IL-4 inhibits the barrier-enhancing effects of IL-17A, and an imbalance of such cytokine profiles has been shown in atopic dermatitis [5]. Furthermore, in other pathologies, changes in the cytokine patterns of immune cells seem to be key factors for barrier defects. In celiac disease, peripheral monocytes possess elevated pro-inflammatory cytokine levels, altering the TJ's protein composition and thus the barrier function [6].

2.2. The TJ and Its Regulators as Therapeutic Targets

Barrier restoration and improvement are the main aim when targeting TJs and their proteins. Here, on the one hand, natural substances—which are already often used to treat pathological conditions in traditional medicine—are in focus; on the other hand, targeted modifications of proteins and peptides are used to induce TJ modulation.

Among the many studied cytokines that influence the barrier during inflammation, tumor necrosis factor-α (TNF) has been studied the most and is also linked to regulatory factors like myosin light chain kinase (MLCK) [7]: activated MLCK triggers the reorganization of the intestinal TJ via perijunctional actomyosin ring contraction, which further leads to TJ protein endocytosis and thus barrier changes. MLCK inhibition leads to toxic effects, as other essential enzymatic effects are inhibited too, but a small molecule—divertin—prevents only MLCK1 recruitment to the pathologic antibody-mediated rejection (PAMR), making it a potential therapeutic approach.

The effects of natural compounds are analyzed regarding their modes of action and signaling mechanisms beyond. For example, quercetin, a very common flavonoid, is known to improve the intestinal TJ barrier and also to reduce the occurrence of kidney stone formation by the reduction of ion and water reabsorption due to inducing PI3K/AKT-regulated claudin expression changes [8]. Curcumin, another polyphenol, shows protective effects in pathological conditions, which can be induced by infections with *Campylobacter jejuni*. This bacterial strain affects the TJ barrier via immune cell-mediated cytokine effects, which can be alleviated by curcumin [9].

Besides barrier enhancement and thus often protective effects, food compounds can also weaken or open the paracellular barrier. Caprate, a fatty acid, can reversibly remove claudin-5 and tricellulin from the TJ and, by this, modulate the intestinal barrier's function. This also is observed in the immune-relevant Peyer's patch follicle-associated epithelium, suggesting that caprate could trigger paracellular antigen uptake and might be connected to food allergy development [10]. However, the targeted opening of the TJ can be used to allow not only antigen uptake, as a more unwanted effect, but also enhanced drug uptake, which might be of special interest for the blood–brain barrier (BBB) or for treating tumors.

Numerous approaches have been taken to modify binding peptides originating from bacterial toxins. Such toxins can bind to specific TJ proteins; it is known that CPE can bind to members of the claudin family and that the *Clostridium perfringens* iota-toxin can bind to other TJ proteins such as the angulins. Peptide fragments that still possess binding capacities can be either used based on their direct binding to their original targets or can be mutated to gain binding specificity for other TJ proteins [3]. In detail, the mutation of the claudin-binding domain of CPE (cCPE) leads to its binding to the usually-not-targeted claudin-1, which is involved, for example, as a co-receptor for hepatitis C virus (HCV) infections or is a main component of the epidermal barrier limiting drug delivery. Such new interactions created by specifying the cCPE might thus allow its usage for therapeutic claudin-1 targeting or drug delivery [2].

2.3. Molecular Structure of TJ Networks and Proteins

To achieve the specific modification of binding peptides and to understand the basic principles of TJ protein interaction and the formation of barriers and channels, knowledge of the molecular structure of TJ proteins is essential. Technical approaches to elucidate the structural features of TJs and the interaction of TJ proteins within the TJ network include super-resolution imaging, which resolves strand formation and appearance at levels already comparable to those achieved with freeze-fracture electron microscopy. Here, different aspects are in development, and technical challenges still have to be taken into account [11].

A higher resolution is achieved by crystal structure analyses, which, for a long time, were difficult for TJ proteins. It took a long time from the first published structures of JAM-a [12] and the cytosolic C-terminus of occludin [13] for the first complete claudin structure to be published [14]. In this Special Issue, an overview is given over the molecular structures of TJ proteins published to date, focusing also on structural motifs and the types of interactions between TJ proteins and their ligands involving these motifs [15]. The resolution of such structural features of TJ proteins may further be used to model interactions and to suggest the organization of these proteins to form TJ strands with dynamic barrier- and channel-forming properties [16]. The understanding of such interactions and motifs may furthermore allow the specification and development of new compounds targeting TJs. For claudin-5, the predominant claudin of the BBB, *in silico* modulations have been performed to understand how this claudin may be targeted and have also suggested interface formations, which, however, need to be confirmed by experimental studies [17].

3. Tight Junctions as Regulators of Cellular Processes

3.1. TJ and Cell Polarity

The interactions of TJ proteins with each other not only build paracellular barriers but also maintain the organization and separation of apically and basolaterally located membrane proteins and lipids via a fence function. Thus, they are essential for polarity and thereby the organization of epithelial functionality. Vice versa, polarity complex proteins influence TJ network formation and the resulting paracellular barrier. Polarity and the associated TJ barriers are also essential for transport processes, depending on solute gradients and/or electrical potentials. For the paracellular transport of phosphatidylcholine in the biliary tract, such negative potentials are essential and generated by cystic fibrosis transmembrane conductance regulator (CFTR) and the anion exchange protein 2 (AE2) [18].

The regulation of polarity, which is not only based on the TJ and polarity complex proteins but also involves the adherens junction (AJ), depends on several regulatory proteins. One of these is the adenosine monophosphate (AMP)-activated protein kinase (AMPK), which is well-known as a central regulator of energy metabolism. However, besides that, it is an upstream regulator of the AJ in response to junctional tension and affects TJ function and apicobasal cell polarization. Although still not completely understood, the functions and regulation of and via AMPK are reviewed in this Special Issue and may give further indications for AMPK as a potential target in barrier and polarity regulation [19].

Polarity complexes are also a target of pathogens, as their interruption also leads to barrier disturbance. Enteropathogenic *Escherichia coli* (EPEC) secrete the effector protein EspF, which leads to the disruption of the scaffold protein PAR (partitioning-defective) complex, instability of the actin cytoskeleton, and perturbation of the TJ via aPKCζ [20].

3.2. TJ and Cell Proliferation

Closely connected to cell polarity is also the ability of cells to proliferate, and the crosstalk between both fundamental cellular processes is known. The control of epithelial and endothelial cell proliferation is not only linked to the adherens junction (AJ); the involvement of the TJ is increasingly being elucidated. TJ proteins may act as critical cell cycle modulators by binding to and regulating nuclear access of other factors. Furthermore, they may regulate transcription and proliferation by being located at cellular organelles instead of the apicolateral cell membranes. In this Special Issue, a review on TJs and cell proliferation explains the control of cell proliferation and differentiation by TJs, which is required for the formation and also maintenance of tissue barriers [21]. In this context, TJ-associated proteins that are not anchored to the membrane like the ZO proteins or cingulins seem to be especially important mediators, as they connect the cytoskeleton to the TJ and thereby have a huge impact on cytoarchitecture and regulation. One example of a protein involved as a key regulator in gene expression, cell proliferation, and cell size is ZO-2 [22].

3.3. Regulatory Functions of Tricellular TJ Proteins

The tricellular tight junction (tTJ) is mainly involved in the regulation of paracellular macromolecule passage, and in tight epithelia, it also significantly contributes to ion permeability. Besides such classical functions of TJ proteins, tricellulin turned out to be able to directly regulate paracellular water transport in tight epithelial cells [23], and it was found that the tTJ is also of importance for cell and tissue differentiation. For example, during apoptosis, the tTJ is dissolved as a consequence of tricellulin fragmentation and the resulting instability of the tTJ [24]. In the differentiation of cells and tissues, the expression and localization of tTJ proteins is also of significance. For endometrial tissue, the menstrual cycle-dependent expression and localization of tricellulin and angulin-1 have been shown, and changes in these lead to dysfunctional barriers, promoting cancer progression and metastasis [25].

4. Organ- and Tissue-Specific Functions

The expression profile of TJ proteins within different tissues and organs is very specific and does not only have general barrier or regulatory functionality for endothelia and epithelia. Within an organ, this profile is essential for the whole organ's function, which can be very complex and thus easily and severely affected by small changes in the TJ protein composition or functionality of single proteins. To study the function and impact of single TJ proteins in such complexity, often, mouse models are generated that can provide information on human-associated disorders. Such models may be knock-outs, knock-downs, or knock-ins of the respective proteins [26], and the findings can be related to observed pathologies in patients. However, interspecific differences need to be taken in account, which, for example, can be observed in intestinal preservation injury, where progression at the tissue and molecular levels differs between rats, pigs, and humans [27].

From pathological conditions, conclusions are also drawn as to the actual physiological function within a complex organ or tissue. Some organs and the function of specific TJ proteins within are focus of this Special Issue.

4.1. Sensory Organs

In the inner ear, for example, the maintenance of ion gradients and therefore strictly regulated barriers are necessary for hearing. Several loss-of-function mutations in TJ proteins have been linked to deafness. A novel variant of claudin-14, besides nine others, is linked to the phenotypic expansion of hereditary hearing loss, DFNB2 [28].

Another sensory organ is the eye, where, unlike for other epithelia, the apical surface of the retinal pigment epithelium (RPE) is in direct contact with neural tissue. It is involved in the daily phagocytosis of the effete tips of photoreceptor cells, and besides the well-studied transcellular trafficking of these, here, the role of TJs is still under investigation. However, the molecular composition of the TJ of the RPE is affected in several retinopathies, underlying an important function of this barrier [29].

4.2. Barriers of the Brain and Nerves

The disturbance of the epithelial and endothelial barriers within the brain may be caused by different pathologies. One example is cerebral cavernous malformation (CCM), a disease characterized by mulberry-shaped clusters of dilated microvessels leading to seizures, headaches, and strokes from brain bleeding. A group of genes (CCM genes) that are involved in cellular signaling and barrier function may possess mutations leading to loss of function and thereby resulting in this pathology, but they may be also involved in others [30]. As the brain is an organ where epithelial, endothelial, and glial barriers are involved in functionality, high complexity in barrier formation and maintenance is essential. While the endothelial blood–brain barrier (BBB) and the epithelial blood–cerebrospinal fluid barrier (BCSFB) have been often studied, less is known about the epithelial and mesothelial arachnoid barrier and the glia limitans. Besides being physical barriers, they seem to also be of importance in immune compartmentation and immune surveillance [31].

Additionally, the nervous system possesses specialized barriers as, for example, the dorsal root ganglion (DRG) consists of neuron-rich and fiber-rich regions, and their putative epineurium/perineurium. The blood–DRG barrier (BDB) properties in these physiologically distinct regions are altered after nerve injury by changes in specific spatial TJ protein expression. For example, in the neuron-rich regions, claudin-5 was specifically downregulated, leading to the higher permeability of this barrier, while the other regions were not affected [32].

4.3. Gastrointestinal Organs

In the liver, TJ proteins maintain tissue homeostasis and regulate paracellular diffusion between adherent hepatocytes or cholangiocytes at the blood–biliary barrier (BBIB). Non-junctional localization has regulatory cell functions such as those in differentiation, proliferation, and migration. In addition, TJ proteins are hepatocyte entry factors for the hepatitis C virus (HCV), and impaired TJ protein expression has been reported in several hepatobiliary diseases [33]. TJs are essential for the maintenance of negative potentials as driving forces for the paracellular transport of phosphatidylcholine in the biliary tract needed for mucus production [18].

In the intestine, in classic studies, barrier functions and transport have been investigated as the main functions of TJs. However, interactions with immune cells, mainly via their secreted cytokines in inflammatory processes, have been also described in a barrier-regulating context (see Section 2.1). The interaction of TJs and transcellular transporters do not only affect transport and typical barrier properties but also play a role in intestinal epithelial repair. TJ junction complex components are linked to ion channels and transporters that respond to loss of the homeostasis within the intestine during acute injury and finally induce TJ closure [34].

4.4. The Kidney

Within the kidney, nephrons are the functional units regulating the re-uptake of nutrients, ions, and water as well as the excretion of unwanted substances and excess ions and water. Here, it is a pre-requisite that the different segments are well regulated in their function and that the TJ highly influences this. Claudins in the renal collecting duct differ in their expression and function [35] compared to those of the proximal segments. For example, in the distal regions, claudin-7 modulates salt homeostasis and WNK4 expression [36], and claudin-19 is finely adjusted in its function depending on the osmolality [37].

In the proximal tubule, claudin-12 has been shown to be important for permeability to calcium, as the proximal tubules of claudin-12 knock-out (KO) mice had different permeabilities. However, a decreased permeability to calcium in the proximal tubule was compensated by the downregulation of claudin-14 in the thick ascending limb of the Henle's loop [38].

Claudin-10a is also expressed in the proximal tubules of the kidney but facilitates anion permeability. Defects in claudin-10a thus affect the proximal tubule, while claudin-10b is expressed in the thick ascending limb of the Henle's loop but also in many other organs, such as the skin, salivary glands, sweat glands, brain, lung, and pancreas; pathological changes can be observed in these as well [39]. Such effects can be very different in the specific organs, which is also known for other claudins. Claudin-2, which is another channel-forming TJ protein, not only regulates paracellular cation and water transport within organs such as the proximal tubules, small intestine, liver, gall bladder, and pancreas but is also involved in proliferation, migration, and cell fate determination by the alteration of key signaling pathways [40].

5. Tight Junctions Sense and React to Environmental Conditions

Environmental factors also belong to influences that affect barrier integrity and thereby lead to pathologies. Ultraviolet (UV) radiation, like UVB, and oxidative stress in skin epithelial cells (HaCaT) lead to the dephosphorylation of claudin-1, which then is removed from TJs, leading to their disruption. This effect can be rescued by an extract of Brazilian green propolis [41].

In euryhaline fish, the change between fresh and saltwater conditions leads to adaptive changes to allow the handling of the imbibed seawater [42]. In the fish intestine, transcellular aquaporin-dependent water transport is downregulated, and paracellular transport via the two paralogs of claudin-15 present in fish, cldn15a and cldn15b, plays a major role.

The environment can be comprised of the outer environmental condition, but also the microenvironment in a tissue or a cell. Already, under physiological conditions, the TJ forming the paracellular barrier is dynamically regulated. Microenvironmental factors change the tissue's properties and lead to adaptation to the new conditions. Such factors may be osmolality changes, which can be huge and are of significance, especially in the kidney. In the inner medullary collecting duct, claudin-19 expression and localization in TJs or vesicles depend on osmolality, which leads to transformation from tight to leaky epithelia and vice versa, due to changes in ion transport and ion selectivity [37]. Osmolality and hydrostatic pressure furthermore affect not only transepithelial transport but also cell proliferation and the cytoskeletal architecture. Changes in these parameters may be sensed by TJ proteins and also play a role in carcinogenesis [43].

Aging, in addition, results in changes in microenvironmental conditions. For example, age-dependent changes in the immune system can be linked to pathologies affecting the TJs of the blood–brain barrier (BBB), altering the microvasculature and, by this, possibly impairing cognitive skills and abilities, which can be observed in several age-related disorders like Alzheimer's or Parkinson's disease [44]. The TJ attains increasing focus, as barrier dysfunction and microenvironmental changes appear to be common denominators in neurological disorders.

6. TJ Proteins and Cancer

Facing the fact that tumor and cancer development is usually based on various disturbed cell processes, it is obvious that many different factors can be involved. Besides the clearly regulatory factors of proliferation, the TJ and its proteins are also affected. Besides their role in the typically connected barrier integrity and paracellular transport regulation, TJ proteins are involved in many other aspects. This makes TJ proteins also important for such processes that are strictly regulated but out of control in cancer. Many types of cancer exhibit changes in the expression of TJ proteins. Some of them are also used as markers for cancer subtypes.

In general, it is assumed that the barrier impairment and loss of polarity due to TJ dysregulation are responsible for tumor progression and metastasis. However, the role of distinct TJ proteins within tumorigeneses is less understood, and some claudins can be upregulated in one kind of cancer while being downregulated in another. Nevertheless, claudins are emerging targets for therapeutic approaches and the elucidation of cancer malignancy, as they have been shown to be causally linked to mesenchymal transition [45]. Especially, on the example of claudin-1, it becomes clear that it can be involved in different ways in cancer development and that the generation of therapeutics relies on further investigations that explore the involvement of the respective claudins in the different types of cancer [46].

7. Conclusions

Knowledge about the TJ and its proteins has evolved from the early findings of its role in sealing the paracellular cleft of epithelia and endothelia, establishing barriers, and maintaining gradients towards a more differentiated view that these barriers are selective in their properties—as some TJ proteins form paracellular channels through the bTJ—and that different paracellular pathways exist for small ions, small uncharged solutes, water, and macromolecular solutes. In addition, the paracellular pathways regulated by the TJ are of a dynamic nature and are well-regulated under physiological conditions but are also affected under pathological states.

Besides such properties that influence the formation and regulation of paracellular barriers, an increasing number of TJ proteins involved in various cellular processes and regulations are identified, indicating the diverse and dynamic functionality of these proteins. This Special Issue gives an overview of several of those mechanisms besides the well-established barrier function. The different aspects, covered by a total of 43 articles, demonstrate that the field of TJ research is about to explode.

Author Contributions: Conceptualization, M.F. and S.M.K. (equally contributing); writing, S.M.K. and M.F.; funding acquisition, M.F. and S.M.K. All authors have read and agreed to the published version of the manuscript.

Funding: The research of the authors was funded by the Deutsche Forschungsgemeinschaft (DFG), DFG Research Training Group GRK 2318, DFG Transregio Collaborative Research Center TRR 241, and the Open Access Publication Fund of the Charité–Universitätsmedizin Berlin.

Conflicts of Interest: The authors declare no conflict of interest.

Abbreviations

AQP	Aquaporin
CPE	*Clostridium perfringens* enterotoxin
KD	Knock-down
KO	Knock-out
SGLT	Sodium-glucose-linked transporter
TAMP	Tight junction-associated MARVEL protein
TER	Transepithelial resistance
TJ	Tight junction
bTJ	Bicellular tight junction
tTJ	Tricellular tight junction
TM	Transmembrane
TNF-α	Tumor necrosis factor-α

References

1. Hering, N.A.; Fromm, A.; Bucker, R.; Gorkiewicz, G.; Zechner, E.; Hogenauer, C.; Fromm, M.; Schulzke, J.D.; Troeger, H. Tilivalline- and tilimycin-independent effects of klebsiella oxytoca on tight junction-mediated intestinal barrier impairment. *Int. J. Mol. Sci.* **2019**, *20*, 5595. [CrossRef] [PubMed]
2. Beier, L.S.; Rossa, J.; Woodhouse, S.; Bergmann, S.; Kramer, H.B.; Protze, J.; Eichner, M.; Piontek, A.; Vidal, Y.S.S.; Brandner, J.M.; et al. Use of modified clostridium perfringens enterotoxin fragments for claudin targeting in liver and skin cells. *Int. J. Mol. Sci.* **2019**, *20*, 4774. [CrossRef] [PubMed]
3. Hashimoto, Y.; Tachibana, K.; Krug, S.M.; Kunisawa, J.; Fromm, M.; Kondoh, M. Potential for tight junction protein-directed drug development using claudin binders and angubindin-1. *Int. J. Mol. Sci.* **2019**, *20*, 4016. [CrossRef] [PubMed]
4. Schmidt, H.; Braubach, P.; Schilpp, C.; Lochbaum, R.; Neuland, K.; Thompson, K.; Jonigk, D.; Frick, M.; Dietl, P.; Wittekindt, O.H. Il-13 impairs tight junctions in airway epithelia. *Int. J. Mol. Sci.* **2019**, *20*, 3222. [CrossRef]
5. Brewer, M.G.; Yoshida, T.; Kuo, F.I.; Fridy, S.; Beck, L.A.; De Benedetto, A. Antagonistic effects of il-4 on il-17a-mediated enhancement of epidermal tight junction function. *Int. J. Mol. Sci.* **2019**, *20*, 4070. [CrossRef]
6. Delbue, D.; Cardoso-Silva, D.; Branchi, F.; Itzlinger, A.; Letizia, M.; Siegmund, B.; Schumann, M. Celiac disease monocytes induce a barrier defect in intestinal epithelial cells. *Int. J. Mol. Sci.* **2019**, *20*, 5597. [CrossRef]
7. He, W.-Q.; Wang, J.; Sheng, J.-Y.; Zha, J.-M.; Graham, W.V.; Turner, J.R. Contributions of myosin light chain kinase to regulation of epithelial paracellular permeability and mucosal homeostasis. *Int. J. Mol. Sci.* **2020**, *21*, 993. [CrossRef]
8. Gamero-Estevez, E.; Andonian, S.; Jean-Claude, B.; Gupta, I.; Ryan, A.K. Temporal effects of quercetin on tight junction barrier properties and claudin expression and localization in mdck ii cells. *Int. J. Mol. Sci.* **2019**, *20*, 4889. [CrossRef]
9. Lobo de Sa, F.D.; Butkevych, E.; Nattramilarasu, P.K.; Fromm, A.; Mousavi, S.; Moos, V.; Golz, J.C.; Stingl, K.; Kittler, S.; Seinige, D.; et al. Curcumin mitigates immune-induced epithelial barrier dysfunction by campylobacter jejuni. *Int. J. Mol. Sci.* **2019**, *20*, 4830. [CrossRef]
10. Radloff, J.; Cornelius, V.; Markov, A.G.; Amasheh, S. Caprate modulates intestinal barrier function in porcine peyer's patch follicle-associated epithelium. *Int. J. Mol. Sci.* **2019**, *20*, 1418. [CrossRef]
11. Gonschior, H.; Haucke, V.; Lehmann, M. Super-resolution imaging of tight and adherens junctions: Challenges and open questions. *Int. J. Mol. Sci.* **2020**, *21*, 744. [CrossRef] [PubMed]
12. Kostrewa, D.; Brockhaus, M.; D'Arcy, A.; Dale, G.E.; Nelboeck, P.; Schmid, G.; Mueller, F.; Bazzoni, G.; Dejana, E.; Bartfai, T.; et al. X-ray structure of junctional adhesion molecule: Structural basis for homophilic adhesion via a novel dimerization motif. *Embo J.* **2001**, *20*, 4391–4398. [CrossRef] [PubMed]
13. Li, Y.; Fanning, A.S.; Anderson, J.M.; Lavie, A. Structure of the conserved cytoplasmic c-terminal domain of occludin: Identification of the zo-1 binding surface. *J. Mol. Biol.* **2005**, *352*, 151–164. [CrossRef]
14. Suzuki, H.; Nishizawa, T.; Tani, K.; Yamazaki, Y.; Tamura, A.; Ishitani, R.; Dohmae, N.; Tsukita, S.; Nureki, O.; Fujiyoshi, Y. Crystal structure of a claudin provides insight into the architecture of tight junctions. *Science* **2014**, *344*, 304–307. [CrossRef]
15. Heinemann, U.; Schuetz, A. Structural features of tight-junction proteins. *Int. J. Mol. Sci.* **2019**, *20*, 6020. [CrossRef]
16. Fuladi, S.; Jannat, R.-W.; Shen, L.; Weber, C.R.; Khalili-Araghi, F. Computational modeling of claudin structure and function. *Int. J. Mol. Sci.* **2020**, *21*, 742. [CrossRef] [PubMed]
17. Rajagopal, N.; Irudayanathan, F.J.; Nangia, S. Computational nanoscopy of tight junctions at the blood-brain barrier interface. *Int. J. Mol. Sci.* **2019**, *20*, 5583. [CrossRef]
18. Stremmel, W.; Staffer, S.; Weiskirchen, R. Phosphatidylcholine passes by paracellular transport to the apical side of the polarized biliary tumor cell line mz-cha-1. *Int. J. Mol. Sci.* **2019**, *20*, 4034. [CrossRef]
19. Tsukita, K.; Yano, T.; Tamura, A.; Tsukita, S. Reciprocal association between the apical junctional complex and ampk: A promising therapeutic target for epithelial/endothelial barrier function? *Int. J. Mol. Sci.* **2019**, *20*, 6012. [CrossRef]

20. Tapia, R.; Kralicek, S.E.; Hecht, G.A. Enteropathogenic escherichia coli (epec) recruitment of par polarity protein atypical pkcζ to pedestals and cell–cell contacts precedes disruption of tight junctions in intestinal epithelial cells. *Int. J. Mol. Sci.* **2020**, *21*, 527. [CrossRef]
21. Díaz-Coránguez, M.; Liu, X.; Antonetti, D.A. Tight junctions in cell proliferation. *Int. J. Mol. Sci.* **2019**, *20*, 5972. [CrossRef]
22. González-Mariscal, L.; Gallego-Gutiérrez, H.; González-González, L.; Hernández-Guzmán, C. Zo-2 is a master regulator of gene expression, cell proliferation, cytoarchitecture, and cell size. *Int. J. Mol. Sci.* **2019**, *20*, 4128. [CrossRef]
23. Ayala-Torres, C.; Krug, S.M.; Schulzke, J.D.; Rosenthal, R.; Fromm, M. Tricellulin effect on paracellular water transport. *Int. J. Mol. Sci.* **2019**, *20*, 5700. [CrossRef]
24. Janke, S.; Mittag, S.; Reiche, J.; Huber, O. Apoptotic fragmentation of tricellulin. *Int. J. Mol. Sci.* **2019**, *20*, 4882. [CrossRef] [PubMed]
25. Kohno, T.; Konno, T.; Kojima, T. Role of tricellular tight junction protein lipolysis-stimulated lipoprotein receptor (lsr) in cancer cells. *Int. J. Mol. Sci.* **2019**, *20*, 3555. [CrossRef] [PubMed]
26. Seker, M.; Fernández-Rodríguez, C.; Martínez-Cruz, L.A.; Müller, D. Mouse models of human claudin-associated disorders: Benefits and limitations. *Int. J. Mol. Sci.* **2019**, *20*, 5504. [CrossRef]
27. Søfteland, J.M.; Casselbrant, A.; Biglarnia, A.-R.; Linders, J.; Hellström, M.; Pesce, A.; Padma, A.M.; Jiga, L.P.; Hoinoiu, B.; Ionac, M.; et al. Intestinal preservation injury: A comparison between rat, porcine and human intestines. *Int. J. Mol. Sci.* **2019**, *20*, 3135. [CrossRef] [PubMed]
28. Kitano, T.; Kitajiri, S.-i.; Nishio, S.-y.; Usami, S.-i. Detailed clinical features of deafness caused by a claudin-14 variant. *Int. J. Mol. Sci.* **2019**, *20*, 4579. [CrossRef]
29. Naylor, A.; Hopkins, A.; Hudson, N.; Campbell, M. Tight junctions of the outer blood retina barrier. *Int. J. Mol. Sci.* **2019**, *21*, 211. [CrossRef]
30. Wei, S.; Li, Y.; Polster, S.P.; Weber, C.R.; Awad, I.A.; Shen, L. Cerebral cavernous malformation proteins in barrier maintenance and regulation. *Int. J. Mol. Sci.* **2020**, *21*, 675. [CrossRef]
31. Castro Dias, M.; Mapunda, J.A.; Vladymyrov, M.; Engelhardt, B. Structure and junctional complexes of endothelial, epithelial and glial brain barriers. *Int. J. Mol. Sci.* **2019**, *20*, 5372. [CrossRef] [PubMed]
32. Lux, T.J.; Hu, X.; Ben-Kraiem, A.; Blum, R.; Chen, J.T.-C.; Rittner, H.L. Regional differences in tight junction protein expression in the blood–drg barrier and their alterations after nerve traumatic injury in rats. *Int. J. Mol. Sci.* **2019**, *21*, 270. [CrossRef] [PubMed]
33. Roehlen, N.; Roca Suarez, A.A.; El Saghire, H.; Saviano, A.; Schuster, C.; Lupberger, J.; Baumert, T.F. Tight junction proteins and the biology of hepatobiliary disease. *Int. J. Mol. Sci.* **2020**, *21*, 825. [CrossRef] [PubMed]
34. Slifer, Z.M.; Blikslager, A.T. The integral role of tight junction proteins in the repair of injured intestinal epithelium. *Int. J. Mol. Sci.* **2020**, *21*, 972. [CrossRef]
35. Leiz, J.; Schmidt-Ott, K.M. Claudins in the renal collecting duct. *Int. J. Mol. Sci.* **2019**, *21*, 221. [CrossRef]
36. Fan, J.; Tatum, R.; Hoggard, J.; Chen, Y.-H. Claudin-7 modulates cl− and na+ homeostasis and wnk4 expression in renal collecting duct cells. *Int. J. Mol. Sci.* **2019**, *20*, 3798. [CrossRef]
37. Ziemens, A.; Sonntag, S.R.; Wulfmeyer, V.C.; Edemir, B.; Bleich, M.; Himmerkus, N. Claudin 19 is regulated by extracellular osmolality in rat kidney inner medullary collecting duct cells. *Int. J. Mol. Sci.* **2019**, *20*, 4401. [CrossRef]
38. Plain, A.; Pan, W.; O'Neill, D.; Ure, M.; Beggs, M.R.; Farhan, M.; Dimke, H.; Cordat, E.; Alexander, R.T. Claudin-12 knockout mice demonstrate reduced proximal tubule calcium permeability. *Int. J. Mol. Sci.* **2020**, *21*, 2074. [CrossRef]
39. Milatz, S. A novel claudinopathy based on claudin-10 mutations. *Int. J. Mol. Sci.* **2019**, *20*, 5396. [CrossRef]
40. Venugopal, S.; Anwer, S.; Szászi, K. Claudin-2: Roles beyond permeability functions. *Int. J. Mol. Sci.* **2019**, *20*, 5655. [CrossRef]
41. Marunaka, K.; Kobayashi, M.; Shu, S.; Matsunaga, T.; Ikari, A. Brazilian green propolis rescues oxidative stress-induced mislocalization of claudin-1 in human keratinocyte-derived hacat cells. *Int. J. Mol. Sci.* **2019**, *20*, 3869. [CrossRef]
42. Tipsmark, C.K.; Nielsen, A.M.; Bossus, M.C.; Ellis, L.V.; Baun, C.; Andersen, T.L.; Dreier, J.; Brewer, J.R.; Madsen, S.S. Drinking and water handling in the medaka intestine: A possible role of claudin-15 in paracellular absorption? *Int. J. Mol. Sci.* **2020**, *21*, 1853. [CrossRef] [PubMed]

43. Tokuda, S.; Yu, A.S.L. Regulation of epithelial cell functions by the osmolality and hydrostatic pressure gradients: A possible role of the tight junction as a sensor. *Int. J. Mol. Sci.* **2019**, *20*, 3513. [CrossRef] [PubMed]
44. Costea, L.; Meszaros, A.; Bauer, H.; Bauer, H.C.; Traweger, A.; Wilhelm, I.; Farkas, A.E.; Krizbai, I.A. The blood-brain barrier and its intercellular junctions in age-related brain disorders. *Int. J. Mol. Sci.* **2019**, *20*, 5472. [CrossRef] [PubMed]
45. Gowrikumar, S.; Singh, A.B.; Dhawan, P. Role of claudin proteins in regulating cancer stem cells and chemoresistance-potential implication in disease prognosis and therapy. *Int. J. Mol. Sci.* **2019**, *21*, 53. [CrossRef]
46. Bhat, A.A.; Syed, N.; Therachiyil, L.; Nisar, S.; Hashem, S.; Macha, M.A.; Yadav, S.K.; Krishnankutty, R.; Muralitharan, S.; Al-Naemi, H.; et al. Claudin-1, a double-edged sword in cancer. *Int. J. Mol. Sci.* **2020**, *21*, 569. [CrossRef]

International Journal of
Molecular Sciences

Article

Tilivalline- and Tilimycin-Independent Effects of *Klebsiella oxytoca* on Tight Junction-Mediated Intestinal Barrier Impairment

Nina A. Hering [1,*], Anja Fromm [2], Roland Bücker [2], Gregor Gorkiewicz [3], Ellen Zechner [4], Christoph Högenauer [5], Michael Fromm [2], Jörg-Dieter Schulzke [2] and Hanno Troeger [6]

1. Medical Department of General, Visceral and Vascular Surgery, Charité – Universitätsmedizin Berlin, 12203 Berlin, Germany
2. Institute of Clinical Physiology/Nutritional Medicine, Medical Department, Division of Gastroenterology, Infectiology and Rheumatology, Charité – Universitätsmedizin Berlin, 12203 Berlin, Germany; anja.fromm@charite.de (A.F.); roland-felix.buecker@charite.de (R.B.); michael.fromm@charite.de (M.F.); joerg.schulzke@charite.de (J.-D.S.)
3. Institute of Pathology, Medical University of Graz, A-8036 Graz, Austria; gregor.gorkiewicz@medunigraz.at
4. BioTechMed-Graz, Institute of Molecular Biosciences, University of Graz, A-8010 Graz, Austria; ellen.zechner@uni-graz.at
5. Division of Gastroenterology and Hepatology, Medical University of Graz, A-8036 Graz, Austria; christoph.hoegenauer@medunigraz.at
6. Medical Department, Division of Gastroenterology, Infectiology and Rheumatology, Charité – Universitätsmedizin Berlin, 12203 Berlin, Germany; hanno.troeger@charite.de

* Correspondence: nina.hering@charite.de

Received: 4 September 2019; Accepted: 7 November 2019; Published: 8 November 2019

Abstract: *Klebsiella oxytoca* causes antibiotic-associated hemorrhagic colitis and diarrhea. This was attributed largely to its secreted cytotoxins tilivalline and tilimycin, inductors of epithelial apoptosis. To study whether *Klebsiella oxytoca* exerts further barrier effects, T84 monolayers were challenged with bacterial supernatants derived from tilivalline/tilimycin-producing AHC6 or its isogeneic tilivalline/tilimycin-deficient strain Mut-89. Both preparations decreased transepithelial resistance, enhanced fluorescein and FITC-dextran-4kDa permeabilities, and reduced expression of barrier-forming tight junction proteins claudin-5 and -8. Laser scanning microscopy indicated redistribution of both claudins off the tight junction region in T84 monolayers as well as in colon crypts of mice infected with AHC6 or Mut-89, indicating that these effects are tilivalline/tilimycin-independent. Furthermore, claudin-1 was affected, but only in a tilivalline/tilimycin-dependent manner. In conclusion, *Klebsiella oxytoca* induced intestinal barrier impairment by two mechanisms: the tilivalline/tilimycin-dependent one, acting by increasing cellular apoptosis and a tilivalline/tilimycin-independent one, acting by weakening the paracellular pathway through the tight junction proteins claudin-5 and -8.

Keywords: antibiotic-associated hemorrhagic colitis; tight junction; claudin; apoptosis; *Klebsiella oxytoca*

---

1. Introduction

A vast number of microorganisms colonize the intestinal mucosa, forming a complex ecosystem. The importance of this diverse microbiota for intestinal health and function becomes evident once the balance of this ecosystem is disturbed. This can occur by overgrowth of a single species after antibiotic medication, as it is well-known from the toxinogenic strains of *Clostridium difficile* causing antibiotic-associated diarrhea and pseudomembranous colitis [1]. Within the last years, *Klebsiella oxytoca* (*K. oxytoca*) was recognized to cause antibiotic-associated hemorrhagic colitis (AAHC), a distinct form of antibiotic-associated colitis, accompanied by bloody diarrhea and abdominal cramps [2]. *K. oxytoca* is a Gram-negative bacterium that is ubiquitous in the environment, but also colonizes the healthy intestinal mucosa [2–4] and is associated with nosocomial infections [5]. Most clinical isolates are resistant to amino- and carboxypenicillins [6], favoring selective intestinal overgrowth of *K. oxytoca* in antibiotic-treated patients.

K. oxytoca secretes cytotoxic compounds into the external milieu in vitro and in the host intestine [7–10]. Schneditz and colleagues identified a cluster of cytotoxin synthesis genes that is shared by toxin-positive *K. oxytoca* strains. They showed the pentacyclic pyrrolobenzodiazepine tilivalline to be responsible for inducing epithelial apoptosis in Hep-2 cells in vitro, and that the biosynthetic gene cluster was required for disease in a mouse model of AAHC. Subsequent studies showed that the gene cluster produces a second pyrrolobenzodiazepine toxin, tilimycin, which also induces apoptosis in epithelial cells and contributes to the development of colitis [9–12]. Knock-out of the *npsB* gene, encoding a nonribosomal peptide synthetase within the cluster, exhibited a toxin-negative phenotype and was incapable of inducing apoptosis [8].

The "leaky gut concept" is defined by an impaired paracellular barrier, resulting in increased flux across that barrier [13]. Consistent with that concept, stimulated epithelial apoptosis was demonstrated to have a direct leak effect [14]. Tilivalline causes a decrease in transepithelial resistance in epithelial T84 monolayers, indicating an impaired barrier function. Interestingly, this tilivalline-induced decrease could be abolished when apoptosis was inhibited pharmacologically [8]. Epithelial apoptosis is also a feature of barrier dysfunction induced by other bacterial enteropathogens, for example, *Campylobacter spp.* [15,16] or *Arcobacter buzleri* [17]. Other studies showed epithelial apoptosis can induce cleavage of cell adhesion molecules [18–20].

The epithelial tight junction forms a sealing structure between the lateral cell membranes of adjacent epithelial cells and is built by different types of transmembranal proteins. The two types capable of barrier formation are the claudin family, with 27 members in mammals [21], and the family of TJ-associated MARVEL proteins (MAL and related proteins for vesicle trafficking and membrane link), comprising occludin, tricellulin, and marvel-D3 [22]. Extracellular loops of these proteins interact with those from the neighboring cells, and by this, build up the paracellular barrier. However, some of the claudins form paracellular channels for small cations, anions, or water. Under pathological conditions, upregulation of channel-forming claudins or downregulation of barrier-forming claudins lead to barrier dysfunction [23]. Because of this key function for intestinal barrier function, the present study aimed to clarify how *K. oxytoca* affects tight junction protein function.

2. Results

After infection of T84 monolayers with vital tilivalline/tilimycin-producing *K. oxytoca* strains AHC6 or #204, we observed a strong drop in transepithelial resistance (TER) within 24 h (Figure 1a). For comparison, we infected the monolayers with the toxin-negative AHC6 mutant strain Mut-89, which is incapable of producing tilivalline and tilimycin. Interestingly, the Mut-89 strain decreased TER as well. To assess the role of tilivalline and tilimycin, we prepared supernatants of broth cultures of the cytotoxin-producing strain AHC6 and the mutant strain Mut-89. Different doses of supernatant preparations (10, 50, 100, and 150 µL) of AHC6 or Mut-89 were assessed after 24 and 48 h. As expected, the toxin-positive strain reduced TER in T84 monolayers in a dose- and time-dependent manner. However, we also observed a reduction in TER by the mutant strain, although AHC6 was more effective than Mut-89. Since the strongest effect was observed with 150 µL at 48 h (Figure 1b), this setup was used for all following cell culture experiments. To further differentiate the nature of the supernatants derived from – AHC6 and Mut89, thermal stability was tested by heating. TER effects of supernatants of AHC6 and Mut-89 were abolished by heat treatment at 95 °C, but not at 60 °C. Supernatants treated at 60 °C were as effective as native preparations (Figure 1c). TER (or epithelial resistance, R^{epi}) consists of a paracellular (R^{para}) and a transcellular (R^{trans}) component. While the effects of AHC6 supernatant on R^{epi} were due to a severe reduction of both R^{para} and R^{trans}, the R^{epi} drop caused by supernatant of the mutant strain Mut-89 was caused solely by a decrease in R^{para} (Figure 1d).

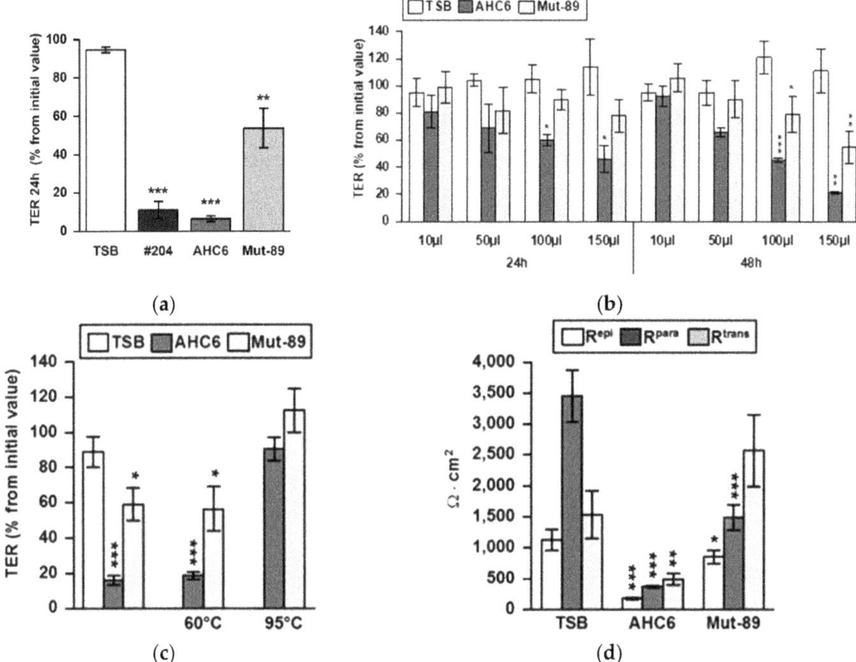

Figure 1. Tilivalline/tilimycin-dependent and -independent effects on transepithelial resistance. (**a**) T84 monolayers were infected with vital tilivalline/tilimycin-producing *K. oxytoca* strains #204 and AHC6 or the AHC6 isogenic mutant strain Mut-89. TER (transepithelial resistance) levels were measured at 24 h after infection and were compared with initial values (set 100%) and untreated controls (n = 5–7; ** $p > 0.01$; *** $p < 0.001$ versus control). (**b**) Dose- and time-dependent TER effects of preparations of bacterial culture supernatants from *K. oxytoca* strains AHC6 and from tilivalline/tilimycin knock-out Mut-89 were assessed on T84 monolayers. Controls were treated with equal amounts of TSB culture medium (n = 4–5; * $p < 0.5$; ** $p < 0.01$; *** $p < 0.001$ versus TSB). (**c**) Supernatants of ACH6 or Mut-89 were heated to 95 or 60 °C prior to T84 challenge and TER measurements (n = 5–10; $p < 0.05$, $p < 0.001$ versus TSB). (**d**) Discrimination of paracellular (R^{para}) and transcellular resistance (R^{trans}) by two-path impedance spectroscopy was performed 48 h after challenging cellular monolayers with supernatant of AHC6 or Mut-89 (n = 4–13; ** $p < 0.01$; *** $p < 0.001$ versus TSB control).

These findings were in accordance with permeability changes of the paracellular marker molecules fluorescein (332 Da) and FITC-dextran-4000 (FD4, 4 kDa). Here, AHC6 enhanced the permeability for fluorescein about 14-fold and for FD4 about 4-fold compared with TSB-treated monolayers (Table 1). Preparations of Mut-89 also increased permeability, but the effect was less pronounced than with AHC6. Fluorescein permeability was increased about 2-fold and FD4 permeability about 1.6-fold versus TSB control (Table 1).

Table 1. Measurements of permeability to fluorescein (332 Da, $P_{Fluorescein}$) or FITC-dextran 4000 (4 kDa, P_{FD4}) and corresponding transepithelial resistance (TER) (mean ± SEM (n), * $p < 0.05$; ** $p < 0.01$; *** $p < 0.001$).

	$P_{Fluorescein}$ (10^{-6} cm·s^{-1})	P_{FD4} (10^{-6} cm·s^{-1})	TER (Ω·cm^2)
TSB	0.10 ± 0.01 (7)	0.02 ± 0.00 (8)	1754 ± 179 (7)
AHC6	1.46 ± 0.31 (7) **	0.09 ± 0.01 (8) ***	231 ± 53 (7) ***
Mut-89	0.21 ± 0.02 (6) *	0.03 ± 0.00 (8) *	814 ± 174 (6)

To assess the impact of tilivalline/tilimycin-dependent epithelial apoptosis on TER and tight junction changes, T84 monolayers were treated with Q-VD-OPh, a potent inhibitor of caspase-3 cleavage, prior to challenging with *K. oxytoca* supernatants (Figure 2). TER in monolayers treated with Q-VD-OPh and AHC6 supernatant was not significantly different from that challenged with Mut-89 supernatant (without Q-VD-OPh), but was lower than in TSB controls, pointing to an apoptosis-independent mechanism. Interestingly, Q-VD-OPh treatment abolished the AHC6-induced TER decrease only by about 15.5% but inhibited caspase-3 cleavage completely (Figure 2).

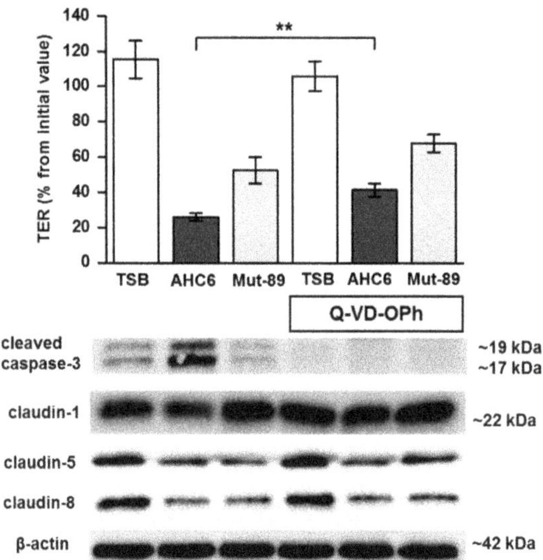

Figure 2. Impact of apoptosis inhibition on transepithelial resistance and tight junction protein expression. T84 monolayers were pretreated with apoptosis inhibitor Q-VD-OPh before challenging with supernatants from AHC6 or Mut-89. Control monolayers were treated with TSB. TER measurement was performed over 48 h (n = 7; ** $p < 0.01$), followed by analysis of protein expression. Blocking of Caspase-3 cleavage served as positive control for apoptosis inhibition by the pan-caspase inhibitor Q-VD-OPh (bars on the right). A representative Western blot shows tilivalline/tilimycin-dependent and -independent tight junction changes (n = 5–6).

We next examined the expression profile of distinct barrier-relevant tight junction proteins by Western blotting and subsequent densitometry (Table 2). Challenging for 48 h with AHC6 or Mut-89 supernatants reduced protein levels of claudin-5 and -8 to a similar extent, when compared with TSB controls. In addition, supernatant from AHC6, but not from Mut-89, reduced claudin-1 protein level, which was prevented by pretreatment with Q-VD-OPh. The expression of claudin-2, -4, or tricellulin was neither affected by AHC6 nor by Mut-89 supernatants (Table 2). In Q-VD-OPh-treated monolayers, claudin-1 protein expression did not differ from that in TSB controls, while protein levels of claudin-5 and -8 remained significantly different (Table 2, and signal intensity in Figure 2).

Table 2. Densitometric quantification of tight junction protein expression (mean ± SEM (n), values are given in % of TSB-treated control, which was set to 100%, * $p < 0.05$; ** $p < 0.01$; *** $p < 0.001$ versus TSB control). QV (Q-VD-OPh).

Tight Junction Protein	Expression (%)				
	AHC6	Mut-89	TSB QV	AHC6 QV	Mut-89 QV
claudin-1	65 ± 6 (9) ***	88 ± 11 (9)	81 ± 12 (5)	87 ± 11 (6)	95 ± 11 (6)
claudin-5	47 ± 10 (8) **	48 ± 8 (8) ***	90 ± 6 (5)	38 ± 12 (5) *	47 ± 14 (5) *
claudin-8	33 ± 5 (9) ***	43 ± 6 (8) ***	102 ± 5 (5)	36 ± 10 (6) **	45 ± 10 (6) **
claudin-2	91 ± 11 (6)	78 ± 11 (6)			
claudin-4	91 ± 8 (6)	89 ± 9 (6)			
tricellulin-a	83 ± 9 (6)	101 ± 7 (6)			

Beside protein expression, the proper functional localization of tight junction proteins was investigated by immunostaining and subsequent laser scanning microscopy (LSM) on T84 monolayers. The zonula occludens protein 1 (ZO-1) served as a tight junction marker. We monitored the localization and signal intensity of claudin-1, -5 and -8 by z-stack imaging and intensity–distance plots of representative areas. Effects of AHC6 supernatants were investigated without and with Q-VD-OPh apoptosis inhibition and were compared with effects of Mut-89 supernatants and TSB-treated control. Monolayers challenged with AHC6 supernatant showed claudin-1 to be redistributed out of the tight junction domain of the cells, especially within areas of apoptotic foci (Figure 3a, second row). In contrast, intensity–distance plots of control monolayers showed a clear merging of red (ZO-1) and green (claudin-1) peaks with high intensities (amplitudes) at regions of cell contact (tight junctions), and signal intensity in intracellular regions was very low (background intensity). In AHC6-treated monolayers, the intensities of claudin-1 and ZO-1 signals were reduced and more diffuse (Figure 3a, second row). While ZO-1 localization seemed unaffected, intensity–distance plots showed irregular and flattened curves, pointing to reduced protein expression and redistribution of claudin-1 off the tight junction domain of the cell. Claudin-1 changes in signal intensity and localization were relieved by Q-VD-OPh apoptosis inhibition, while ZO-1 intensity reduction was unaffected by Q-VD-OPh (Figure 3a, fourth row). Monolayers treated with Mut-89 supernatant looked similar to those of the TSB-controls (Figure 3a, third row).

In contrast to claudin-1, signal intensity and focal localization of claudin-5 (Figure 3b) and claudin-8 (Figure 3c) were affected by both AHC6 and Mut-89 supernatants. Overall, signal intensity was reduced in challenged monolayers compared with control monolayers (Figure 3b,c, second and third rows). The intensity–distance plots of controls showed single matching peaks of red and green curves and only slight background signal. However, plots of monolayers treated with *K. oxytoca* supernatants revealed smaller amplitudes of green (claudin-5 or -8) and red (ZO-1) curves. Focally enhanced intracellular claudin signals were indicated by irregular curves, due to enhanced background or additional peaks in intracellular regions. Q-VD-OPh treatment did not prevent reduction of claudin-5 and -8 and also single foci of delocalization were still evident (Figure 3b and c, fourth and fifth rows). Compared with its respective controls, the signal intensity of the tight junction anchor ZO-1 was diminished in all AHC6- or Mut-89-challenged monolayers (Figure 3a–c).

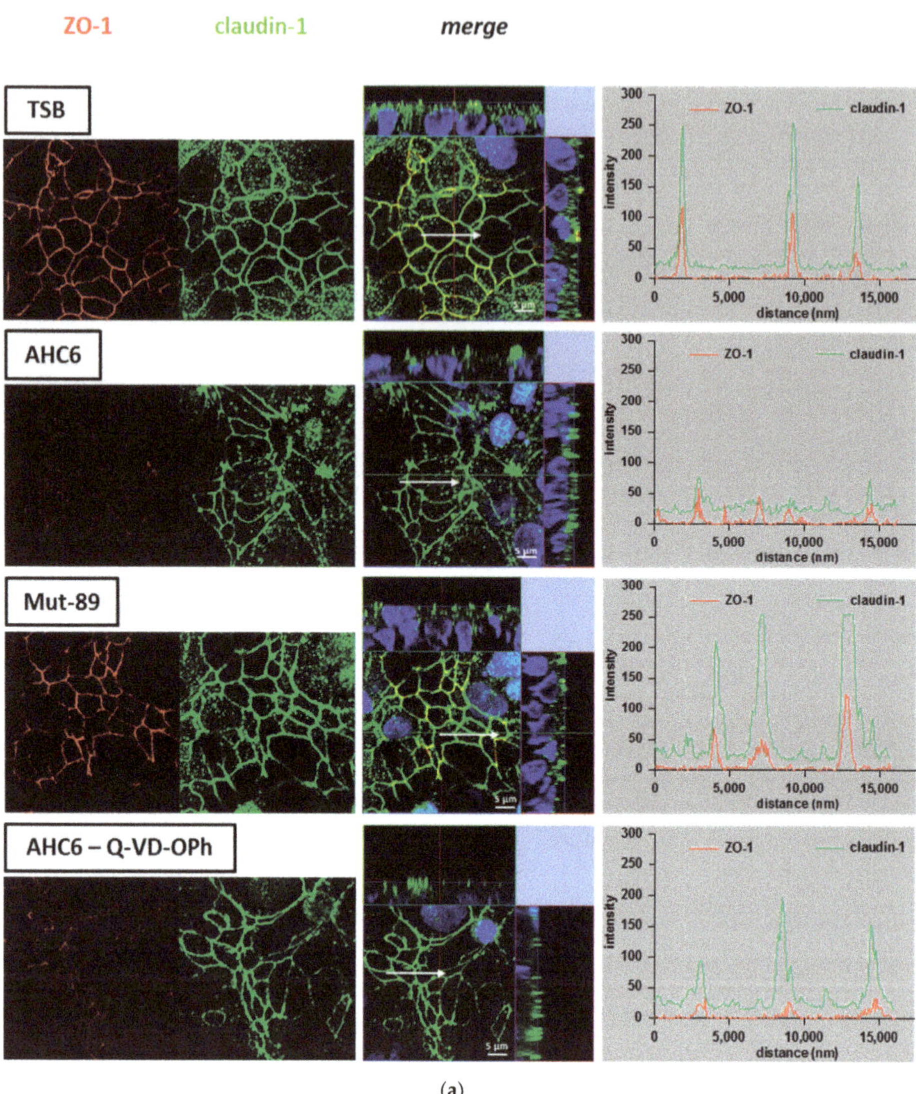

(a)

Figure 3. *Cont.*

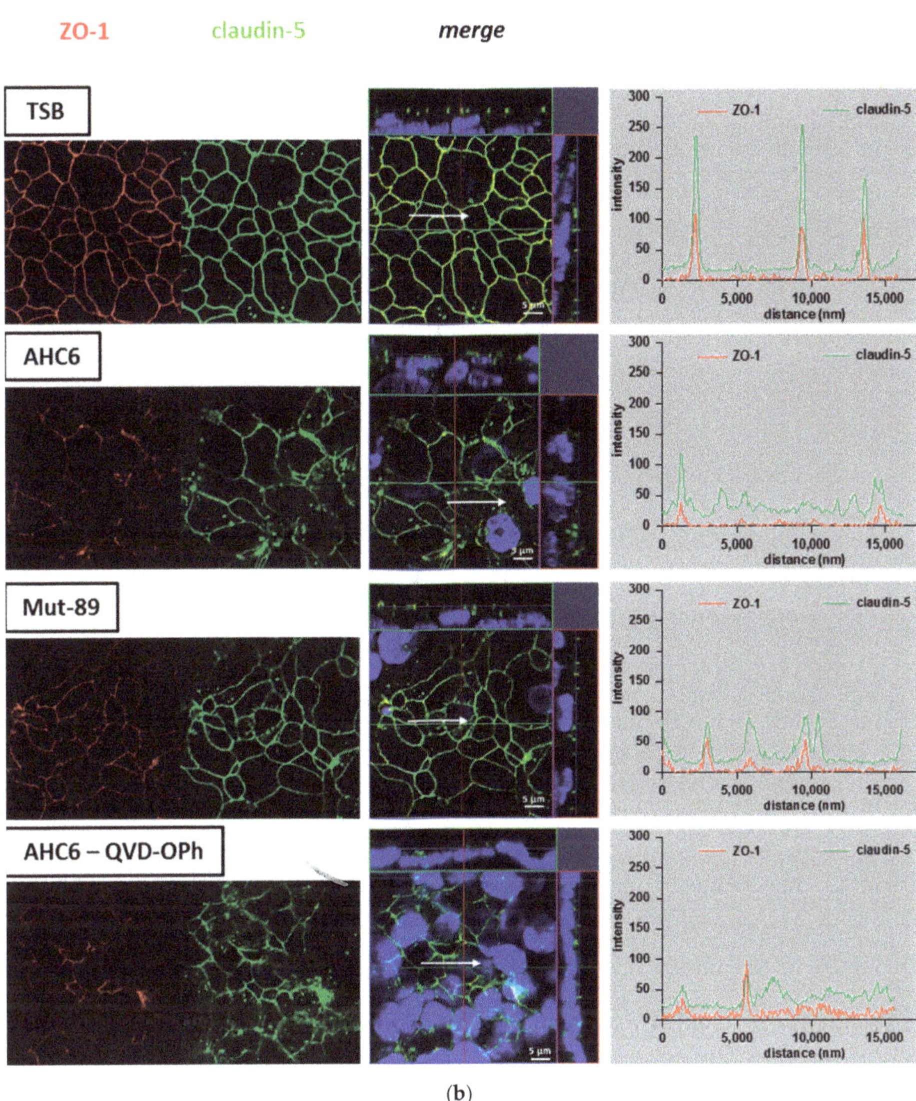

(**b**)

Figure 3. *Cont.*

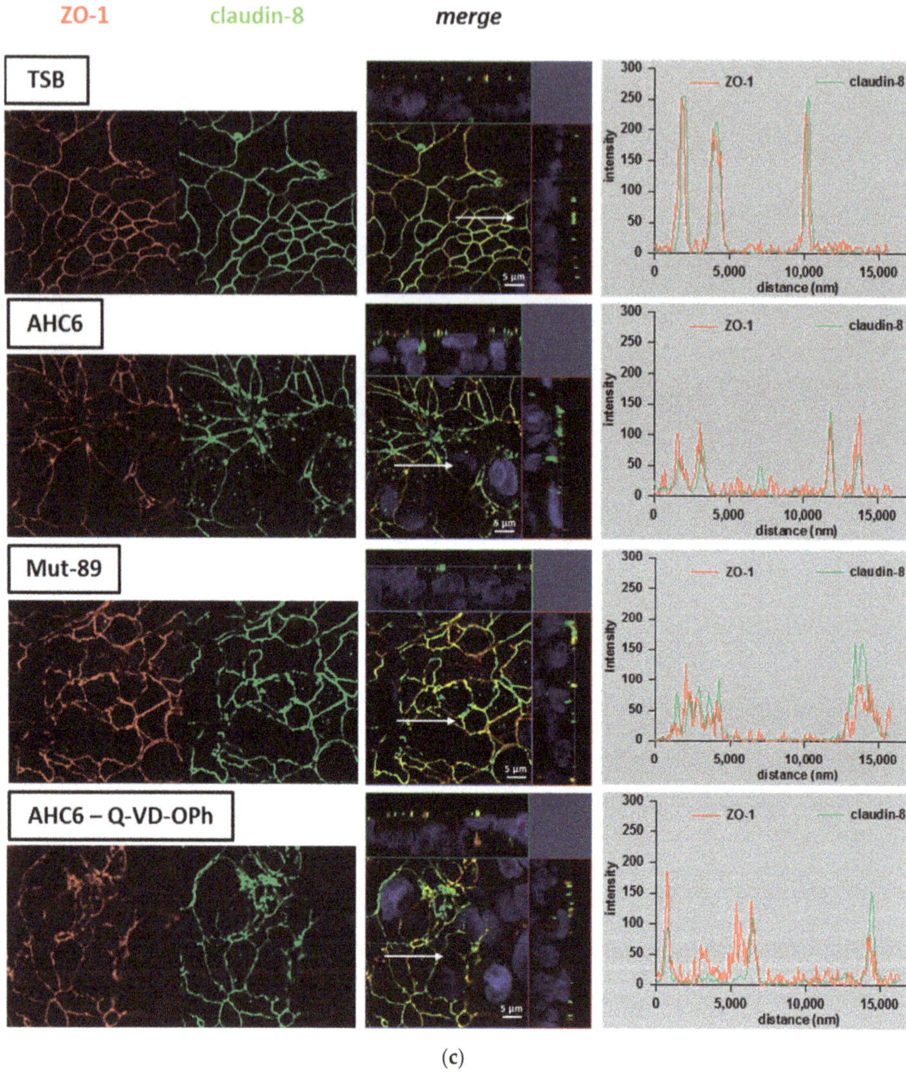

(c)

Figure 3. Redistribution of tight junction proteins in T84 monolayers. T84 monolayers were immunostained after 48 h of challenging with supernatants from AHC6, with or without previous apoptosis inhibition (Q-VD-OPh) or supernatants from Mut-89. TSB-treated monolayers served as control. Immunofluorescence staining of (**a**) claudin-1, (**b**) claudin-5, and **c**) claudin-8 (green) were analyzed by confocal laser scanning microscopy ($n = 3$ each). ZO-1 (red) served as tight junction marker. Distribution of tight junction proteins was considered by z-stack imaging. Intensity–distance plots show the signal intensity and merging of ZO-1 and claudin signals in representative areas. Arrows were located using Zeiss Zen software and mark the range and direction of the intensity–distance plot. Nuclei were DAPI-stained (4′,6-Diamidin-2-phenylindol; blue).

These findings were corroborated by immunostaining of claudin-5 and -8 on colon samples, derived from a mouse model of AAHC (Figure 4). Occludin was used as a tight junction marker, and tight junction staining was analyzed in colon crypts of untreated controls and AHC6- or Mut-89-infected mice. In comparison with controls, signal intensities of claudin-5 (Figure 4a), claudin-8 (Figure 4b), and occludin (Figure 4a,b) were reduced in AHC6- and Mut-89-infected mice. Intensity–distance plots of representative areas showed reduced amplitudes of green (claudin) and red (occludin) curves. In case of claudin-8, signals were more diffuse in infected animals, and intensity–distance plots displayed a wide-ranging and flat-topped curve (Figure 4b).

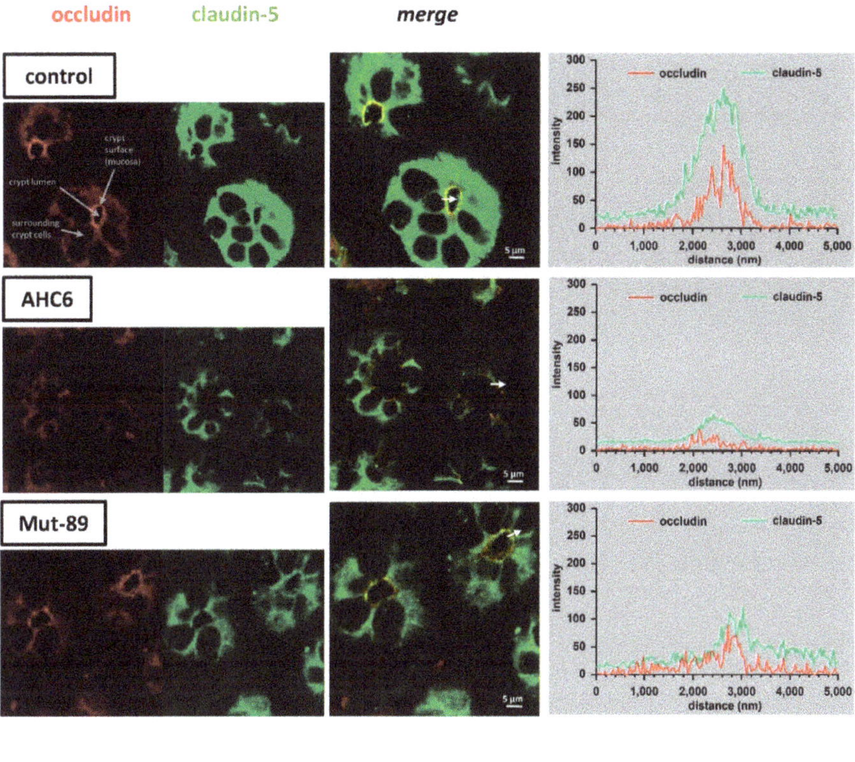

(a)

Figure 4. *Cont.*

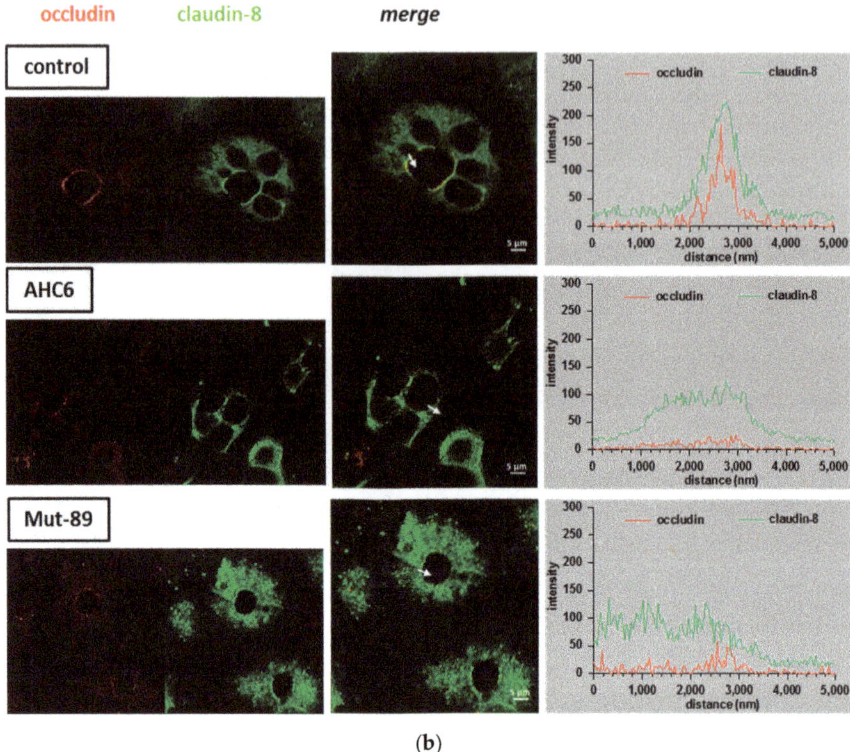

(b)

Figure 4. Immunofluorescence staining of claudins in colon crypts from an AAHC (antibiotic-associated hemorrhagic colitis) **mouse model.** Representative micrographs show immunostaining of occludin (red) and (**a**) claudin-5 or (**b**) claudin-8 (green) in mouse colon crypts of control, AHC6-infected, and Mut-89-infected mice (n = 4 each group; claudin-5 and -8). Signal intensity and localization of representative areas are shown in intensity–distance plots. Arrows were located using Zeiss Zen software and mark the range and direction of the intensity–distance plot.

3. Discussion

The pathomechanisms of *K. oxytoca*-induced antibiotic-associated hemorrhagic colitis are not completely understood. Recently, the cytotoxins tilivalline and tilimycin were identified and shown to cause epithelial apoptosis [8–10]. However, so far it was not clear if epithelial apoptosis was the only mechanism of *K. oxytoca*-caused barrier perturbation. Within the present study, we showed that *K. oxytoca*-induced barrier impairment involves not only epithelial apoptosis, but also tilivalline/tilimycin-dependent and -independent changes to the expression and localization of distinct tight junction proteins.

As expected, vital tilivalline/tilimycin-producing *K. oxytoca* isolates AHC6, and #204 induced a severe TER drop in T84 colon epithelial monolayers. Barrier perturbation in the host intestine was so far attributed to the cytotoxins tilivalline and tilimycin [8,10]. However, our study showed that vital bacteria and a cell-free preparation of bacterial culture supernatant of the mutant strain Mut-89, which is incapable of producing tilivalline/tilimycin, was also able to induce barrier dysfunction in T84 monolayers, indicated by a TER decrease and an increase in permeability for fluorescein and FD4. The biosynthesis of *K. oxytoca*-secreted compounds tilivalline/tilimycin and also culdesacin are dependent on the nonribosomal peptide synthetase operon including *npsB* [11]. Therefore, they cannot be responsible for the barrier defects caused by Mut-89, which is based on a *npsB* knock-out [8].

In a study on oral epithelial wound recovery, De Ryck and colleagues already pointed out that *K. oxytoca* must secret other effector molecules beyond tilivalline and tilimycin. They hypothesized that *K. oxytoca*-produced quorum-sensing molecules might contribute to its inhibitory impact on wound healing [24]. Former studies showed the toxicity of culture supernatant of *K. oxytoca* derived from patients with AAHC was not inactivated by 60 °C treatment, but was sensitive to 100 °C [25]. Interestingly, Mut89 supernatant preparations were also stable at 60 °C, equivalent to the wild type. Although, the particular nature of the virulence factor affecting claudin-5 and -8 remains elusive, the present study revealed a distinct pathomechanism beyond tilivalline/tilimycin-induced apoptosis.

While the AHC6 supernatant preparation reduced R^{trans} and R^{para}, preparations of the tilivalline/tilimycin knock-out strain affected only R^{para}, which reflects tight junction integrity. In fact, we found *K. oxytoca*-induced tight junction changes in the T84 cell culture model as well as in colon tissue derived from mice of an AAHC model. The reduction of claudin-1 and its proper localization within the tight junction turned out to be functionally dependent on the cytotoxins tilivalline and tilimycin, as the supernatant of Mut-89 did not induce any changes to this tight junction protein in vitro. In a very recent study that predominantly examined the impact of microbial balance in the gut on the microbiota–gut–brain axis, treatment of mice with ampicillin induced an imbalance of gut microbiota, with a clear prevalence of *K. oxytoca*. Interestingly, the treatment with ampicillin or with *K. oxytoca* itself suppressed claudin-1 in the colon [26]. This observation fits to our findings of a reduced claudin-1 level. Claudin-5 and -8 are barrier-forming tight junction proteins [27]. Perturbation of these proteins can be associated with barrier loss, as already found in other studies. In a mouse model of *Yersinia enterocolitca* infection, barrier defects involved a reduced expression of claudin-8 and redistribution of claudin-5 [28]. Claudin-5 and -8 are also reduced and redistributed in Crohn's disease, which is characterized by intestinal inflammation and diarrhea [29]. Within the present study, the reduction and redistribution of these proteins occurred independently from tilivalline/tilimycin in vitro and in vivo. The observations from the animal model of AAHC supported the findings from our T84 cell cultures. Analyses of crypts of AHC6- and Mut-89-infected mice revealed that claudin-5 and -8 as well as occludin were reduced and distributed off the tight junction. In case of claudin-8, a redistribution into subapical cytoplasmatic compartments might be assumed from Figure 4b. The tight junction defects caused by *K. oxytoca* are very similar to effects observed with other enterobacterial pathogens. For example, the pore-forming toxin aerolysin, from *Aeromonas hydrophila* induces tight junction protein redistribution [30]. The *Clostridium perfringens* enterotoxin uses claudin-3, -4, and -7 as receptors to damage the epithelial cells [31]. Samples of the mucosa obtained from patients with *Camplylocbacter jejuni* infection revealed reduced expression of claudin-3, -4, -5, and -8 and redistribution of claudin-1, -5, and -8 off the tight junction [16].

Interestingly, occludin and the scaffold protein ZO-1 that serves as a tight junction anchor were also reduced by *K. oxytoca*. Both molecules have diverse regulatory functions, especially for Rho-kinase-dependent actin remodeling [32] or myosin light chain kinase-dependent trafficking [33]. ZO-1 and occludin interactions play a crucial role for proper epithelial cell polarization, as ZO-1 knock down is associated with delayed occludin recruitment and tight junction assembly [34]. It is tempting to speculate that *K. oxytoca* causes epithelial cell depolarization similar to enteropathogenic *E. coli* [35]. In a very recent study, *K. oxytoca* was found to be increased and associated with intestinal dysfunction in cachectic mice. The authors pointed out that these effects were independent from tilivalline/tilimycin, as the *K. oxytoca* isolated from mouse feces revealed no cytotoxic activity. Here the mRNA level of occludin and immune relevant barrier factors were reduced [36]. Thus, the loss of ZO-1 or occludin might be also relevant for the barrier destabilization in *K. oxytoca* infection.

In conclusion, *K. oxytoca*-induced intestinal barrier impairment involves two pathomechanisms acting in parallel. One is known to act via tilivalline/tilimycin and causes a barrier defect by increased apoptosis. As an additional indirect effect, apoptosis appears to downregulate and to delocalize claudin-1. As a further mechanism, *K. oxytoca* causes via tilivalline/tilimycin-independent action a direct weakening of the tight junction barrier by downregulation and redistribution of claudin-5 and

-8 (summarized in Figure 5). This might be clinically relevant as toxin-negative *K. oxytoca* strains are commonly also found in patients with diarrhea or colitis other than AAHC [7,37].

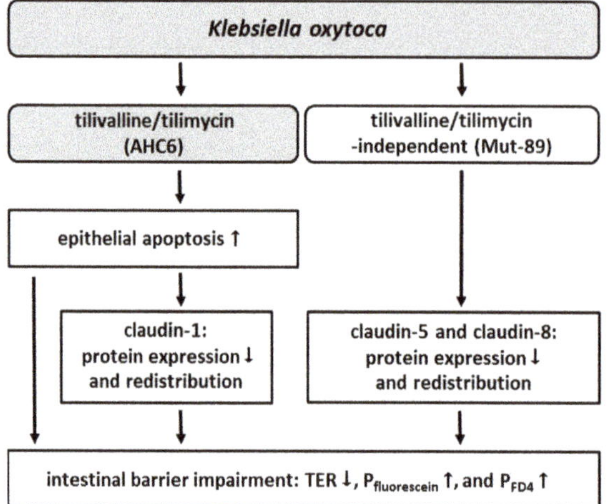

Figure 5. Overview of tilivalline/tilimycin-dependent and -independent effects in *K. oxytoca*-induced barrier impairment. *K. oxytoca* causes barrier impairment via two different pathomechanisms. On the one hand, the cytotoxins tilivalline and tilimycin (AHC6) enhance epithelial apoptosis, which is additionally linked to a reduction and redistribution of the tight junction protein claudin-1. On the other hand, the sealing claudins 5 and 8 are reduced and redistributed independently from tilivalline/tilimycin (Mut-89). In consequence, barrier function is impaired, indicated by reduced TER and increased permeability to fluorescein (322 Da) and FD4 (4 kDa).

4. Material and Methods

4.1. Cell Culture and Challenge

The epithelial colon carcinoma cell line T84 was cultured as published before [38]. Cells were grown on permeable Millicell PCF filters (0.6 cm^2 effective area; 3 µm pores, Millicell PCF, Millipore, Schwalbach, Germany) to form polarized and tight monolayers (range 1150–2020 $\Omega \cdot cm^2$). For experiments, confluent monolayers were shifted to serum-free medium, containing 100 µg/mL gentamicin in case of challenge with bacterial culture supernatants.

Klebsiella oxytoca strains #204 (AAHC wild type, tox^+) [37], AHC6 (AAHC wild type, tox^+), and Mut89 (Kan^R; *npsB*::*Tn5*, tox^-) [8] were cultured in Tryptic Soy Broth (TSB, Merck, KGaA, Darmstadt, Germany) overnight at 37 °C. For infection, overnight cultures were diluted (1:8) and grown for an additional 2 h. T84 monolayers were infected with vital bacteria (multiplicity of infection 10) from the apical side. To avoid bacterial overgrowth, 100 µg/mL gentamicin was added 2.5 h after infection. Bacterial culture supernatant was prepared as described earlier by Joaining and colleagues [7]. In short, 30 mL of TSB was inoculated with a single bacterial colony, and cultures were grown under gentle agitation (180 rpm) for 14–16 h at 37 °C. At an optical density at 600 nm of 4 to 6, the cultures were centrifuged (5000 rpm, 4 °C, 20 min) and supernatants were filtered through 0.2 µm cellulose acetate filters (Millipore). Monolayers were challenged with 150 µL bacterial culture filtrate or TSB for control from the apical side. To inhibit epithelial apoptosis, 10 µM of Q-VD-OPh, non-O-methylated (Merck) was added to both sides 3 h before challenge [39].

4.2. TER Measurement

Transepithelial electrical resistance was measured with chopstick electrodes at 37 °C, as published earlier [40]. Initial TER values were assessed before addition of bacterial supernatant or TSB. TER is given in percent of the initial value.

4.3. Permeability Measurements

Permeability of 0.1 mM fluorescein (332 Da, Sigma-Aldrich, Schnelldorf, Germany) or 0.4 mM dialyzed fluorescein-iso-thio-cyanate-dextran-4000 (FD4; TdB Consultancy, Uppsala, Sweden) was assessed by flux measurements in Ussing chambers under voltage-clamp conditions. The bathing solution contained in mM: 140 Na^+, 123.8 Cl^-, 5.4 K^+, 1.2 Ca^{2+}, 1.2 Mg^{2+}, 2.4 $HPO_4{}^{2-}$, 0.6 $H_2PO_4{}^-$, 21 $HCO_3{}^-$, 10 D(+)-glucose. The marker molecule was added to the apical side of the monolayer, and samples were collected from the basolateral side at defined time points. Fluorescence was measured in a spectrofluorimeter (Infinite M200, Tecan, Männedorf, Austria) at 525 nm, and permeability p ($cm \cdot s^{-1}$) was calculated from the ratio of flux J ($mol \cdot h^{-1} \cdot cm^{-2}$) over concentration Δc (mol/L) of the marker molecule in the Ussing chamber: $p = J/\Delta c$.

4.4. Measurement of Para- and Transcellular Resistance

Differentiation of paracellular (R^{para}) and transcellular (R^{trans}) contributions to the overall epithelial resistance (R^{epi}, TER) was performed as described earlier [39,41]. Briefly, cellular T84 monolayers were mounted in Ussing-type chambers, and transepithelial conductance (G^{epi}) and fluxes of the paracellular marker fluorescein were measured under voltage-clamp conditions. Changes in R^{para} were forced by chelating extracellular Ca^{2+} using ethylene glycol tetra acetic acid. Calculation was done from conductivity and fluxes before and after Ca^{2+} removal.

4.5. Western Blot

Proteins were extracted from epithelial T84 monolayers with ice-cold lysis buffer (in mM: 100 imidazole, 100 KCl, 300 sucrose, 2 $MgCl_2$, 10 EGTA, 1 NaF, 1 $NaVO_3$, 1 Na_2MO_4, 0.2% Triton X-100, and complete protease inhibitor cocktail (Roche, Mannheim, Germany)). For analysis of caspase-3 activation, cell lysis was performed as described earlier [39]. Twenty–thirty micrograms of protein was separated by SDS-polyacrylamid gel electrophoreses and blotted on polyvinylidene difluoride (PVDF)-membranes. For immunodetection, the following antibodies were used: anti-claudin-1, -2, -4, -5, and -8, anti-tricellulin (Thermo Fisher Scientific, Bremen, Germany), anti-cleaved Caspase-3 (Cell Signaling Technology, Denvers; USA), and anti-β-actin (Sigma-Aldrich). Bound antibodies were detected with peroxidase-conjugated goat anti-rabbit IgG or goat anti-mouse IgG antibodies using the chemiluminescence substrate Lumi-Light^{PLUS} (Roche) and the chemiluminescence detection system FX7 (Vilber Lourmat, Eberhardzell, Germany). Densitometry was performed with AIDA quantification software (Raytest, Straubenhardt, Germany). All values were normalized to β-actin.

4.6. Immunofluorescence Staining and Confocal Laser scanning Microscopy

T84 monolayers were washed with PBS and fixed with 1% paraformaldehyde. Paraffin-embedded mouse tissues were derived from experiments performed by Schneditz and colleagues. In short, antibiotic-induced dysbiosis was simulated in C57BL/6 mice by amoxicillin/clavulanate and indometacin treatment and intragastrical infection with AHC6 or Mut-89, leading to *K. oxytoca* overgrowth in the colon of infected mice [8]. Caecum sections were cut, deparaffinized, and heated in sodium citrate buffer solutions at pH 6.0 for epitope retrieval. After permeabilization with 0.05% Triton X-100, immunostaining of TJ proteins was carried out using anti-ZO-1 or anti-occludin, anti-claudin-1, -5, or -8 (1:100), followed by labelled Alexa Fluor 488 goat anti-mouse or rabbit IgG and Alexa Fluor 594 goat anti-mouse or rabbit IgG (1:1000; Thermo Fisher Scientific). Staining of ZO-1 and claudin-1 on mouse tissue sections was limited to high background staining and weak antibody binding. Therefore, occludin was used in this case. Nuclei were DAPI stained (1:5000). Fluorescence staining was visualized by confocal laser scanning microscopy (LSM 780, Zeiss, Jena, Germany), and tight junction proteins were localized by z-stack imaging. Distribution profiles of claudin signals were generated using Zen software (ZEN 2.3 lite, Zeiss, Oberkochen, Germany).

4.7. Statistical Analysis

All values are given as means ± SEM. Data were compared with Student's t-test and Bonferroni–Holm adjustment in cases of multiple testing. $p < 0.05$ was considered significant (* $p < 0.05$; ** $p < 0.01$; *** $p < 0.001$).

Author Contributions: Conceptualization, H.T.; Data curation, N.A.H.; Formal analysis, N.A.H. and A.F.; Funding acquisition, E.Z. and J.-D.S.; Investigation, N.A.H., A.F. and G.G.; Methodology, N.A.H. and A.F.; Project administration, H.T.; Resources, E.Z., C.H. and J.-D.S.; Validation, R.B., E.Z., C.H. and J.-D.S.; Writing—original draft, N.A.H.; Writing—review & editing, A.F., R.B., E.Z., C.H., M.F. and J.-D.S.

Funding: This work was funded by Deutsche Forschungsgemeinschaft (Schu 559/11); recipient JDS. Work in the Zechner Laboratory was funded by the FWF DK Molecular Enzymology (W901) and BioTechMed-Graz Flagship Secretome.

Acknowledgments: We thank In-Fah M. Lee for her excellent technical assistance.

Conflicts of Interest: All authors declare no competing interests. The sponsors had no role in the design, execution, interpretation, or writing of the study.

Abbreviations

AAHC	antibiotic-associated hemorrhagic colitis
FD4	fluorescein-iso-thio-cyanate-dextran-4000
LSM	laser scanning microscopy
K. oxytoca	*Klebsiella oxytoca*
QV	Q-VD-Oph
R^{epi}	epithelial resistance
R^{para}	paracellular resistance
R^{trans}	transcellular resistance
TER	transepithelial resistance
TSB	tryptic soy broth
ZO-1	zonula occludens-1

References

1. Bartlett, J.G. Clostridium difficile-associated Enteric Disease. *Curr. Infect. Dis. Rep.* **2002**, *4*, 477–483. [CrossRef] [PubMed]
2. Hgenauer, C.; Langner, C.; Beubler, E.; Lippe, I.T.; Schicho, R.; Gorkiewicz, G.; Krause, R.; Gerstgrasser, N.; Krejs, G.J.; Hinterleitner, T.A. Klebsiella oxytoca as a causative organism of antibiotic-associated hemorrhagic colitis. *N. Engl. J. Med.* **2006**, *355*, 2418–2426. [CrossRef] [PubMed]
3. Beaugerie, L.; Metz, M.; Barbut, F.; Bellaiche, G.; Bouhnik, Y.; Raskine, L.; Nicolas, J.C.; Chatelet, F.P.; Lehn, N.; Petit, J.C.; et al. Klebsiella oxytoca as an agent of antibiotic-associated hemorrhagic colitis. *Clin. Gastroenterol. Hepatol.* **2003**, *1*, 370–376. [CrossRef]
4. Zollner-Schwetz, I.; Hogenauer, C.; Joainig, M.; Weberhofer, P.; Gorkiewicz, G.; Valentin, T.; Hinterleitner, T.A.; Krause, R. Role of Klebsiella oxytoca in antibiotic-associated diarrhea. *Clin. Infect. Dis.* **2008**, *47*, e74–e78. [CrossRef] [PubMed]
5. Podschun, R.; Ullmann, U. Klebsiella spp. as nosocomial pathogens: Epidemiology, taxonomy, typing methods, and pathogenicity factors. *Clin. Microbiol. Rev.* **1998**, *11*, 589–603. [CrossRef] [PubMed]
6. Livermore, D.M. beta-Lactamases in laboratory and clinical resistance. *Clin. Microbiol. Rev.* **1995**, *8*, 557–584. [CrossRef] [PubMed]

7. Joainig, M.M.; Gorkiewicz, G.; Leitner, E.; Weberhofer, P.; Zollner-Schwetz, I.; Lippe, I.; Feierl, G.; Krause, R.; Hinterleitner, T.; Zechner, E.L.; et al. Cytotoxic effects of Klebsiella oxytoca strains isolated from patients with antibiotic-associated hemorrhagic colitis or other diseases caused by infections and from healthy subjects. *J. Clin. Microbiol.* **2010**, *48*, 817–824. [CrossRef] [PubMed]
8. Schneditz, G.; Rentner, J.; Roier, S.; Pletz, J.; Herzog, K.A.; Bucker, R.; Troeger, H.; Schild, S.; Weber, H.; Breinbauer, R.; et al. Enterotoxicity of a nonribosomal peptide causes antibiotic-associated colitis. *Proc. Natl. Acad. Sci. USA* **2014**, *111*, 13181–13186. [CrossRef] [PubMed]
9. Tse, H.; Gu, Q.; Sze, K.H.; Chu, I.K.; Kao, R.Y.; Lee, K.C.; Lam, C.W.; Yang, D.; Tai, S.S.; Ke, Y.; et al. A tricyclic pyrrolobenzodiazepine produced by Klebsiella oxytoca is associated with cytotoxicity in antibiotic-associated hemorrhagic colitis. *J. Biol. Chem.* **2017**, *292*, 19503–19520. [CrossRef] [PubMed]
10. Unterhauser, K.; Poltl, L.; Schneditz, G.; Kienesberger, S.; Glabonjat, R.A.; Kitsera, M.; Pletz, J.; Josa-Prado, F.; Dornisch, E.; Lembacher-Fadum, C.; et al. Klebsiella oxytoca enterotoxins tilimycin and tilivalline have distinct host DNA-damaging and microtubule-stabilizing activities. *Proc. Natl. Acad. Sci. USA* **2019**, *116*, 3774–3783. [CrossRef] [PubMed]
11. Dornisch, E.; Pletz, J.; Glabonjat, R.A.; Martin, F.; Lembacher-Fadum, C.; Neger, M.; Hogenauer, C.; Francesconi, K.; Kroutil, W.; Zangger, K.; et al. Biosynthesis of the Enterotoxic Pyrrolobenzodiazepine Natural Product Tilivalline. *Angew. Chem. Int. Ed. Engl.* **2017**, *56*, 14753–14757. [CrossRef] [PubMed]
12. von Tesmar, A.; Hoffmann, M.; Abou Fayad, A.; Huttel, S.; Schmitt, V.; Herrmann, J.; Muller, R. Biosynthesis of the Klebsiella oxytoca Pathogenicity Factor Tilivalline: Heterologous Expression, in Vitro Biosynthesis, and Inhibitor Development. *ACS Chem. Biol.* **2018**, *13*, 812–819. [CrossRef] [PubMed]
13. Quigley, E.M. Leaky gut-concept or clinical entity? *Curr. Opin. Gastroenterol.* **2016**, *32*, 74–79. [CrossRef] [PubMed]
14. Gitter, A.H.; Bendfeldt, K.; Schulzke, J.D.; Fromm, M. Leaks in the epithelial barrier caused by spontaneous and TNF-alpha-induced single-cell apoptosis. *Faseb. J.* **2000**, *14*, 1749–1753. [CrossRef] [PubMed]
15. Nielsen, H.L.; Nielsen, H.; Ejlertsen, T.; Engberg, J.; Gunzel, D.; Zeitz, M.; Hering, N.A.; Fromm, M.; Schulzke, J.D.; Bucker, R. Oral and fecal Campylobacter concisus strains perturb barrier function by apoptosis induction in HT-29/B6 intestinal epithelial cells. *PLoS ONE* **2011**, *6*, e23858. [CrossRef] [PubMed]
16. Bucker, R.; Krug, S.M.; Moos, V.; Bojarski, C.; Schweiger, M.R.; Kerick, M.; Fromm, A.; Janssen, S.; Fromm, M.; Hering, N.A.; et al. Campylobacter jejuni impairs sodium transport and epithelial barrier function via cytokine release in human colon. *Mucosal. Immunol.* **2018**, *11*, 75–577.
17. Bucker, R.; Troeger, H.; Kleer, J.; Fromm, M.; Schulzke, J.D. Arcobacter butzleri induces barrier dysfunction in intestinal HT-29/B6 cells. *J. Infect. Dis.* **2009**, *200*, 756–764. [CrossRef] [PubMed]
18. Steinhusen, U.; Badock, V.; Bauer, A.; Behrens, J.; Wittman-Liebold, B.; Dorken, B.; Bommert, K. Apoptosis-induced cleavage of beta-catenin by caspase-3 results in proteolytic fragments with reduced transactivation potential. *J. Biol. Chem.* **2000**, *275*, 16345–16353. [CrossRef] [PubMed]
19. Steinhusen, U.; Weiske, J.; Badock, V.; Tauber, R.; Bommert, K.; Huber, O. Cleavage and shedding of E-cadherin after induction of apoptosis. *J. Biol. Chem.* **2001**, *276*, 4972–4980. [CrossRef] [PubMed]
20. Bojarski, C.; Weiske, J.; Schoneberg, T.; Schroder, W.; Mankertz, J.; Schulzke, J.D.; Florian, P.; Fromm, M.; Tauber, R.; Huber, O. The specific fates of tight junction proteins in apoptotic epithelial cells. *J. Cell. Sci.* **2004**, *117*, 2097–2107. [CrossRef] [PubMed]
21. Mineta, K.; Yamamoto, Y.; Yamazaki, Y.; Tanaka, H.; Tada, Y.; Saito, K.; Tamura, A.; Igarashi, M.; Endo, T.; Takeuchi, K.; et al. Predicted expansion of the claudin multigene family. *FEBS Lett.* **2011**, *585*, 606–612. [CrossRef] [PubMed]
22. Raleigh, D.R.; Marchiando, A.M.; Zhang, Y.; Shen, L.; Sasaki, H.; Wang, Y.; Long, M.; Turner, J.R. Tight junction-associated MARVEL proteins marveld3, tricellulin, and occludin have distinct but overlapping functions. *Mol. Biol. Cell.* **2010**, *21*, 1200–1213. [CrossRef] [PubMed]
23. Krug, S.M.; Schulzke, J.D.; Fromm, M. Tight junction, selective permeability, and related diseases. *Semin. Cell. Dev. Biol.* **2014**, *36*, 166–176. [CrossRef] [PubMed]
24. De Ryck, T.; Vanlancker, E.; Grootaert, C.; Roman, B.I.; De Coen, L.M.; Vandenberghe, I.; Stevens, C.V.; Bracke, M.; Van de Wiele, T.; Vanhoecke, B. Microbial inhibition of oral epithelial wound recovery: Potential role for quorum sensing molecules? *AMB Express.* **2015**, *5*, 27. [CrossRef] [PubMed]
25. Higaki, M.; Chida, T.; Takano, H.; Nakaya, R. Cytotoxic component(s) of Klebsiella oxytoca on HEp-2 cells. *Microbiol. Immunol.* **1990**, *34*, 147–151. [CrossRef] [PubMed]

26. Jang, H.M.; Lee, H.J.; Jang, S.E.; Han, M.J.; Kim, D.H. Evidence for interplay among antibacterial-induced gut microbiota disturbance, neuro-inflammation, and anxiety in mice. *Mucosal. Immunol.* **2018**, *11*, 1386–1397. [CrossRef] [PubMed]
27. Gunzel, D.; Fromm, M. Claudins and other tight junction proteins. *Compr. Physiol.* **2012**, *2*, 1819–1852. [PubMed]
28. Hering, N.A.; Fromm, A.; Kikhney, J.; Lee, I.F.; Moter, A.; Schulzke, J.D.; Bucker, R. Yersinia enterocolitica Affects Intestinal Barrier Function in the Colon. *J. Infect. Dis.* **2016**, *213*, 1157–1162. [CrossRef] [PubMed]
29. Zeissig, S.; Burgel, N.; Gunzel, D.; Richter, J.; Mankertz, J.; Wahnschaffe, U.; Kroesen, A.J.; Zeitz, M.; Fromm, M.; Schulzke, J.D. Changes in expression and distribution of claudin 2, 5 and 8 lead to discontinuous tight junctions and barrier dysfunction in active Crohn's disease. *Gut* **2007**, *56*, 61–72. [CrossRef] [PubMed]
30. Bucker, R.; Krug, S.M.; Rosenthal, R.; Gunzel, D.; Fromm, A.; Zeitz, M.; Chakraborty, T.; Fromm, M.; Epple, H.J.; Schulzke, J.D. Aerolysin from Aeromonas hydrophila perturbs tight junction integrity and cell lesion repair in intestinal epithelial HT-29/B6 cells. *J. Infect. Dis.* **2011**, *204*, 1283–1292. [CrossRef] [PubMed]
31. Veshnyakova, A.; Protze, J.; Rossa, J.; Blasig, I.E.; Krause, G.; Piontek, J. On the interaction of Clostridium perfringens enterotoxin with claudins. *Toxins (Basel)* **2010**, *2*, 1336–1356. [CrossRef] [PubMed]
32. Yu, A.S.; McCarthy, K.M.; Francis, S.A.; McCormack, J.M.; Lai, J.; Rogers, R.A.; Lynch, R.D.; Schneeberger, E.E. Knockdown of occludin expression leads to diverse phenotypic alterations in epithelial cells. *Am. J. Physiol. Cell. Physiol.* **2005**, *288*, C1231–C1241. [CrossRef] [PubMed]
33. Yu, D.; Marchiando, A.M.; Weber, C.R.; Raleigh, D.R.; Wang, Y.; Shen, L.; Turner, J.R. MLCK-dependent exchange and actin binding region-dependent anchoring of ZO-1 regulate tight junction barrier function. *Proc. Natl. Acad. Sci. USA* **2010**, *107*, 8237–8241. [CrossRef] [PubMed]
34. Odenwald, M.A.; Choi, W.; Buckley, A.; Shashikanth, N.; Joseph, N.E.; Wang, Y.; Warren, M.H.; Buschmann, M.M.; Pavlyuk, R.; Hildebrand, J.; et al. ZO-1 interactions with F-actin and occludin direct epithelial polarization and single lumen specification in 3D culture. *J. Cell. Sci.* **2017**, *130*, 243–259. [CrossRef] [PubMed]
35. Zyrek, A.A.; Cichon, C.; Helms, S.; Enders, C.; Sonnenborn, U.; Schmidt, M.A. Molecular mechanisms underlying the probiotic effects of Escherichia coli Nissle 1917 involve ZO-2 and PKCzeta redistribution resulting in tight junction and epithelial barrier repair. *Cell Microbiol.* **2007**, *9*, 804–816. [CrossRef] [PubMed]
36. Potgens, S.A.; Brossel, H.; Sboarina, M.; Catry, E.; Cani, P.D.; Neyrinck, A.M.; Delzenne, N.M.; Bindels, L.B. Klebsiella oxytoca expands in cancer cachexia and acts as a gut pathobiont contributing to intestinal dysfunction. *Sci. Rep.* **2018**, *8*, 12321. [CrossRef] [PubMed]
37. Herzog, K.A.; Schneditz, G.; Leitner, E.; Feierl, G.; Hoffmann, K.M.; Zollner-Schwetz, I.; Krause, R.; Gorkiewicz, G.; Zechner, E.L.; Hogenauer, C. Genotypes of Klebsiella oxytoca isolates from patients with nosocomial pneumonia are distinct from those of isolates from patients with antibiotic-associated hemorrhagic colitis. *J. Clin. Microbiol.* **2014**, *52*, 1607–1616. [CrossRef] [PubMed]
38. Kreusel, K.M.; Fromm, M.; Schulzke, J.D.; Hegel, U. Cl- secretion in epithelial monolayers of mucus-forming human colon cells (HT-29/B6). *Am. J. Physiol.* **1991**, *261*. [CrossRef] [PubMed]
39. Hering, N.A.; Luettig, J.; Krug, S.M.; Wiegand, S.; Gross, G.; van Tol, E.A.; Schulzke, J.D.; Rosenthal, R. Lactoferrin protects against intestinal inflammation and bacteria-induced barrier dysfunction in vitro. *Ann. N. Y. Acad. Sci.* **2017**, *1405*, 177–188. [CrossRef] [PubMed]
40. Heller, F.; Florian, P.; Bojarski, C.; Richter, J.; Christ, M.; Hillenbrand, B.; Mankertz, J.; Gitter, A.H.; Burgel, N.; Fromm, M.; et al. Interleukin-13 is the key effector Th2 cytokine in ulcerative colitis that affects epithelial tight junctions, apoptosis, and cell restitution. *Gastroenterology* **2005**, *129*, 550–564. [CrossRef] [PubMed]
41. Krug, S.M.; Fromm, M.; Gunzel, D. Two-path impedance spectroscopy for measuring paracellular and transcellular epithelial resistance. *Biophys. J.* **2009**, *97*, 2202–2211. [CrossRef] [PubMed]

Article

Use of Modified *Clostridium perfringens* Enterotoxin Fragments for Claudin Targeting in Liver and Skin Cells

Laura-Sophie Beier [1], Jan Rossa [2], Stephen Woodhouse [3], Sophia Bergmann [4], Holger B. Kramer [5,†] Jonas Protze [2], Miriam Eichner [1], Anna Piontek [2], Sabine Vidal-y-Sy [4], Johanna M. Brandner [4], Gerd Krause [2], Nicole Zitzmann [3] and Jörg Piontek [1,*]

1. Charité-Universitätsmedizin Berlin, Institute of Clinical Physiology, Hindenburgdamm 30, 12203 Berlin, Germany; Laura-Sophie.Beier@charite.de (L.-S.B.); miriam.eichner@googlemail.com (M.E.)
2. Leibniz-Forschungsinstitut für Molekulare Pharmakologie (FMP), Robert-Rössle-Straße 10, 13125 Berlin, Germany; JanRossa@gmx.de (J.R.); Protze@fmp-berlin.de (J.P.); dr.anna.piontek.ext@bayer.com (A.P.); GKrause@fmp-berlin.de (G.K.)
3. Oxford Glycobiology Institute, Department of Biochemistry, University of Oxford, South Parks Road, Oxford OX1 3QU, UK; Stephen.woodhouse@cardiov.ox.ac.uk (S.W.); nicole.zitzmann@bioch.ox.ac.uk (N.Z.)
4. Department of Dermatology and Venerology, University Hospital Hamburg Eppendorf, 20246 Hamburg, Germany; s.bergmann@t-online.de (S.B.); s.vidal-y-sy@uke.de (S.V.-y.-S.); brandner@uke.de (J.M.B.)
5. Oxford Centre for Gene Function, Department of Physiology, Anatomy and Genetics, University of Oxford, Oxford OX3 7DQ, UK; h.kramer@lms.mrc.ac.uk

* Correspondence: joerg.piontek@charité.de

† Current Address: London Institute of Medical Sciences, Biological Mass Spectrometry and Proteomics Facility, London W12 0NN, UK.

Received: 17 August 2019; Accepted: 22 September 2019; Published: 26 September 2019

Abstract: Claudins regulate paracellular permeability in different tissues. The claudin-binding domain of *Clostridium perfringens* enterotoxin (cCPE) is a known modulator of a claudin subset. However, it does not efficiently bind to claudin-1 (Cldn1). Cldn1 is a pharmacological target since it is (i) an essential co-receptor for hepatitis C virus (HCV) infections and (ii) a key element of the epidermal barrier limiting drug delivery. In this study, we investigated the potential of a Cldn1-binding cCPE mutant (i) to inhibit HCV entry into hepatocytes and (ii) to open the epidermal barrier. Inhibition of HCV infection by blocking of Cldn1 with cCPE variants was analyzed in the Huh7.5 hepatoma cell line. A model of reconstructed human epidermis was used to investigate modulation of the epidermal barrier by cCPE variants. In contrast to cCPEwt, the Cldn1-binding cCPE-S305P/S307R/S313H inhibited infection of Huh7.5 cells with HCV in a dose-dependent manner. In addition, TJ modulation by cCPE variant-mediated targeting of Cldn1 and Cldn4 opened the epidermal barrier in reconstructed human epidermis. cCPE variants are potent claudin modulators. They can be applied for mechanistic *in vitro* studies and might also be used as biologics for therapeutic claudin targeting including HCV treatment (host-targeting antivirals) and improvement of drug delivery.

Keywords: *Clostridium perfringens* enterotoxin; claudin-1; Hepatitis C Virus; viral entry; epidermal barrier; reconstructed human epidermis; claudin targeting

---

1. Introduction

Claudins, tetraspan transmembrane proteins, are key structural and sealing components of tight junctions (TJ). In endo- and epithelia, they seal the paracellular space and regulate the paracellular ion permselectivity of the tissue and contribute to membrane polarity [1]. Besides their role in the apical TJ, claudins are also localized outside of the TJ complex, indicating that claudins might function apart from their classical role regulating paracellular permeability. Claudins have been shown to be important factors in cell signaling, cell adhesion, epithelial to mesenchymal transition, cell migration, mucosal homeostasis as well as disease pathogenesis (for review see [2]).

Various claudins serve as cell-surface receptors for pathogens, e.g., the *Clostridium perfringens* enterotoxin. The C-terminal domain of this toxin (cCPE) interacts with a subgroup of the claudin protein family (CPE-receptor claudins, e.g., claudin-3 (Cldn3), -4, -6, -9) [3]. cCPE was used in several studies to modulate the claudin-based barrier function of TJs [4]. To overcome the limitation of targeting CPE-receptor claudins only, claudin-subtype specificity of cCPE has been adapted by structure-guided mutagenesis and baculoviral display system. This enabled targeting of Cldn1-9 with the broad specificity binder cCPE-S305P/S307R/S313H (cCPE-SSS) [5,6], preferential binding to Cldn4 with cCPE-L254A/S256A/I258A/D284A [7] or preferential binding to Cldn5 with cCPE-Y306W/S313H and cCPE-N218Q/Y306W/S313H [6,8]. These variants bind with nanomolar affinities to their respective receptors [4,7,9,10]. The high affinity for their receptors as well as the claudin-specific binding renders cCPE variants appropriate tools for claudin targeting. The comparatively low production cost of the small recombinant protein and the concomitant targeting of a defined set of claudins with one molecule are convenient advantages over monoclonal antibodies against claudins.

Similar to other junctional proteins [11,12], claudins are employed as virus receptors. Cldn1 is an essential co-factor in hepatitis C virus (HCV) entry into hepatocytes [13]. Blocking of viral entry is a potent target for anti-HCV therapies with pan-genomic activity against multiple HCV subtypes; however, appropriate agents are not yet clinically available [14]. The multi-claudin binder cCPE-SSS might be a suitable agent to inhibit HCV receptor formation by binding Cldn1 and thereby blocking HCV entry into hepatocytes. HCV is a hepatotropic virus that has a high propensity to establish a persistent infection in human liver. HCV primarily infects human hepatocytes, which over time leads to chronic inflammation, progressive fibrosis, and development of hepatocellular carcinoma [15].

According to the World Health Organization, 3% of the world's population is infected with HCV [16]. About 15 to 45% of infected people spontaneously clear the virus within six months. The remaining 60 to 80%, currently ~71 million people worldwide, develop a chronic HCV infection, which, in a significant number of cases, will lead to cirrhosis or liver cancer. Each year, around 400,000 people die from hepatitis C-related cirrhosis or hepatocellular carcinoma [17]. Thus, even with the high success-rates of new direct-acting antivirals, HCV remains a major burden on global health.

HCV entry into hepatocytes is the first step of the viral life cycle resulting in productive viral infection [13,18]. The entry pathway consists of three steps: (1) viral attachment to hepatocytes; (2) receptor-mediated endocytosis of viral particle; and (3) endosomal fusion [19]. Viral entry is mediated by the viral envelope glycoproteins E1 and E2 and several host entry factors. These include heparan sulfate, tetraspanin cluster of differentiation 81 (CD81), scavenger receptor class B type I (SR-BI), and the TJ proteins Cldn1 [20] and occludin [21,22]. Subsequent studies demonstrated that Cldn6 and -9, displaying high levels of homology with Cldn1, are also able to mediate HCV entry into non-permissive cell lines [23,24]. However, in hepatoma cells, Cldn6 and Cldn9 appeared to function weaker than Cldn1 [24]. In addition, low expression levels of Cldn6 and Cldn9 in human hepatocytes seem to impede their efficient use as HCV entry factors [25].

It was suggested that Cldn1 plays a dual role in HCV entry [25]: (i) formation of a HCV co-receptor complex with CD81 [26] and (ii) direct interaction with the virus envelope E1E2 proteins [27]. While predominantly localized at the TJ, Cldn1 also localizes on the basolateral membrane of hepatocytes. Notably, this pool of Cldn1 outside of the TJ co-localizes with CD81 and allows the formation of the CD81-Cldn1 co-receptor complex for HCV entry [26,28].

HCV entry is a major target of the host neutralizing response [15,19,29] and a target for antiviral immunopreventive and therapeutic strategies (for review see [13,19,30,31]). In particular, use of blocking anti-Cldn1 antibodies showed promising results in blocking viral cell entry and cell-to-cell transmission in vitro and in vivo [14,30,32]. Treatment with an anti-Cldn1 monoclonal antibody cured chronically infected human liver chimeric mice [33]. As an alternative HCV treatment, in this study, we investigated the potential of claudin-binding cCPE variants to block HCV infection.

Cldn1 is also one of the main sealing claudins in human epidermis [34,35], a stratified epithelium, where multiple layers of cells protect the body against water loss, ultraviolet radiation and penetration of harmful xenobiotics [36]. There are two physiologic barriers serving this function: dead keratinocytes containing lipids and proteins, i.e., the *stratum corneum* (SC), and the TJ between the viable cells of the *stratum granulosum* (SG) [36,37]. Claudin targeting by cCPE variants might modify the latter barrier [37–39]. Thus, we used a skin model in addition to the HCV-hepatoma model to test cCPE variants in two different Cdn1-dependent cellular processes.

We utilized models of reconstructed human epidermis (RHE), which recreate all epidermal layers including the TJ-containing SG and the biologically dead yet highly dynamic SC. In order to develop a functional SC, the apical side of the keratinocytes is exposed to air, creating an air-liquid interface (ALI) [40]. RHE enables the analysis of keratinocyte-specific responses within the epidermis, making it an ideal model for this study, in which the capability of the cCPE variants to open the epidermal barrier was tested.

In this study, we investigated the potential of cCPE variants to target processes that depend on extra-junctional or junctional claudins. Targeting of extra-junctional Cldn1 in hepatocytes, which serves as a co-receptor for HCV infection, by cCPE-SSS efficiently inhibited HCV infection. In RHE, on the other hand, we were able to modify the junctional epidermal barrier by targeting claudins in the SG. Thus, we showed that cCPE variants are suitable for directed claudin targeting.

2. Results

2.1. Structure Homology Modeling and FRET Analyses Suggest that cCPE-Cldn1 Interaction Interferes with Hepatitis C Virus Entry into Host Cells

HCV entry into hepatocytes depends on Cldn1 as a HCV co-receptor. We hypothesized that HCV infection might be inhibited by a Cldn1-binding cCPE variant since its interaction with Cldn1 on the cell surface might block the virus entry. The interaction mechanism of cCPE with defined claudin subtypes was well characterized by binding, mutagenesis and modeling studies. The motif (N/D(P-1) P(P) L/M/V(P+1) V/T(P+2) P/A/D(P+3)) in the extracellular segment 2 (ECS2) of claudins binds to the claudin-binding pocket of cCPE consisting of a triple tyrosine and a triple leucine pit [6,7,10,41]. This mechanism was confirmed by complex structures of cCPE bound to Cldn19 or Cldn4 [42,43]. Using these structures, a homology model of a complex between the Cldn1-binding cCPE-SSS and Cldn1 was generated (Figure 1). The model clearly indicates that binding of cCPE-SSS to the extracellular domains of Cldn1 buries most of the residues, which were shown to be involved in Cldn1-CD81 HCV receptor complex formation [20,26]. In particular, this is the case for D38, E48, T42, N72 in ECS1 and M 152 and V155 in ECS2. Other residues involved in Cldn1-CD81 association include W30, G49 and W51, which are part of the claudin consensus sequence and presumably involved in the general fold of claudins [44]. In summary, the modeling suggests that cCPE binding to Cldn1 interferes with formation of the HCV receptor/entry factor complex which includes Cldn1 and CD81. Thus, cCPE-SSS might block HCV entry.

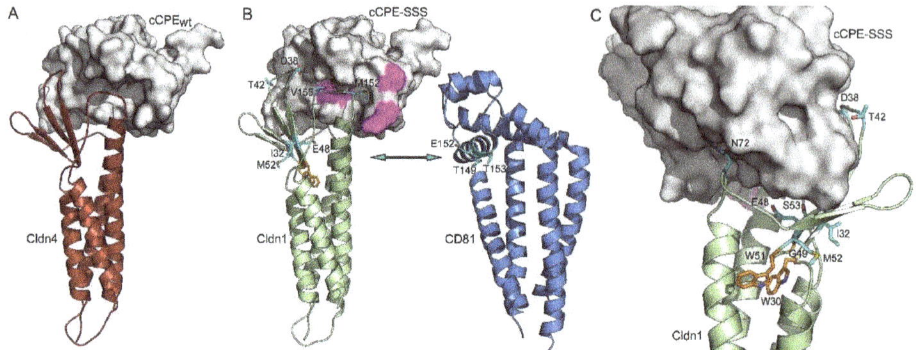

Figure 1. Homology modeling indicates that cCPE-Cldn1 interaction interferes with formation of HCV receptor complex. (**A**) X-ray structure (PDB ID: 5B2G, [43]) of Cldn4 (cartoon, red) interacting with cCPEwt (surface, gray) (**B**) Binding of cCPE-SSS to Cldn1 likely buries large parts of the extracellular domain of Cldn1 preventing Cldn1-CD81 interaction (arrow). Homology model of Cldn1 in complex with cCPE-SSS (template PDB ID: 5B2G) and CD81 (PDB ID: 5TCX). CD81 (blue) and Cldn1 (green) shown as cartoon; cCPE-SSS shown as the surface in gray and the substitutions (S305P, S307R, S313H) enhancing Cldn1 binding in magenta. Cldn1-CD81 interaction-relevant residues shown as cyan sticks (**C**) Other perspective highlighting Cldn1 residues involved in HCV entry [20,26] as cyan sticks. D38, T42, E48, S53, N72, M152, V155, reported to be involved in HCV infection are masked by cCPE. Involvement of W30, G49, W51 (sticks, orange) in HCV infection is likely to be due to a role in claudin folding [44]. I32 and M52 reported to be HCV relevant are not covered by cCPE.

Cldn1 and CD81 interact with each other in the plasma membrane outside of TJ. This interaction can be detected by fluorescence resonance energy transfer (FRET) [28]. Using human embryonic kidney cells (HEK-293), that were co-transfected with cyan fluorescent protein-fused Cldn1 and yellow fluorescent protein-fused CD81, we could show that FRET efficiency was significantly decreased after treatment with 10 μg/mL cCPE-SSS compared to the untreated control (Figure S1), indicating a reduced Cldn1-CD81 interaction. These results suggested that cCPE-SSS indeed interferes with Cldn1-CD81 co-receptor formation.

2.2. S305P/S307R/S313H Substitution in cCPE Improves Its Binding to Cldn1 of Huh7 Hepatoma Cells

The cCPE-variant cCPE-S305P/S307R/S313H (cCPE-SSS) binds more strongly to Cldn1 than cCPEwt [5,6]. To confirm this interaction in a hepatocyte-derived cell culture model, we first used human hepatocarcinoma Huh7 cells [23]. Binding of cCPE variants to Cldn1 endogenously expressed in Huh7 cells was analyzed by means of a pull down assay using GST-cCPE and cell lysates [7]. cCPEwt bound weakly yet specifically to Cldn1 whereas negative control cCPE-Y306A/L315A (cCPE-YALA), which does not interact with claudins [6,45], showed no binding (Figure 2). In contrast, cCPE-SSS showed strong binding to Cldn1 (Figure 2). These data demonstrate that cCPE-SSS binds efficiently to Cldn1 endogenously expressed in hepatocyte model cells.

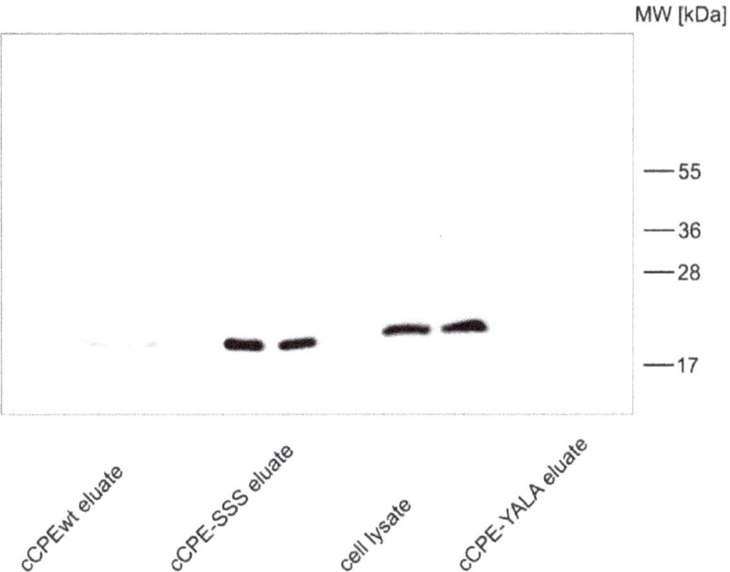

Figure 2. Compared to cCPEwt, cCPE-SSS interacts more strongly with Cldn1 expressed in Huh7 cells. Pull down assays using Huh7 lysates and GST-cCPEwt, -cCPE-SSS or -cCPE-YALA as negative control. Eluates and cell lysate were analyzed in duplicate by SDS-PAGE and Western blot using anti-Cldn1 antibodies. Strong immunoreactive bands in the expected range of ~ 20 kDa were detected for pull down with cCPE-SSS, whereas for cCPEwt, only weak bands and for cCPE-YALA, no bands were detected. Representative results are shown.

2.3. cCPE-S305P/S307R/S313H Specifically Inhibits HCV infection Of Huh7.5 Cells in a Dose-dependent Manner

Next, we tested whether cCPE-SSS is able to block Cldn1 as a HCV co-receptor and thus to inhibit HCV infection. To this end, infection of human hepatocarcinoma Huh7.5 cells with cell culture-derived HCV particles (HCVcc) was used as an established in vitro HCV infection model [46]. Compared to the parental cell line Huh7, the clone Huh7.5 is more permissive to HCV replication [47]. The infection assay was performed in the absence (control) or presence of cCPE variants. Strikingly, cCPE-SSS inhibited HCV infection of Huh7.5 cells in a dose-dependent manner (0.5–50 µg/mL), whereas the non-binding negative control cCPE-YALA had no inhibitory effect (Figure 3A). Surprisingly, CPEwt significantly enhanced HCV infection at low concentrations (0.5 and 2.5 µg/mL). Increased concentrations of cCPEwt (10 and 50 µg/mL) abolished this enhancing effect but never had an inhibiting effect (Figure 3A).

The inhibitory effect of cCPE-SSS on HCV infection was even more pronounced after extension of HCV incubation of the cells from 2 h to 26 h. Under these conditions, 10 µg/mL cCPE-SSS reduced the amount of HCV-infected cells to 0.095 ± 0.016-fold of the cells which were infected in the non-treated control. In contrast, at this concentration, neither cCPEwt nor cCPE-YALA had a significant effect (Figure 3B). Furthermore, 50 µg/mL cCPE-SSS inhibited the infection of HCV almost completely (0.013 ± 0.002-fold compared to the untreated control).

To exclude that these effects were caused by a reduced cell viability, Huh7.5 cells were treated with different concentrations of cCPEwt or –SSS and cell viability was analyzed with a tetrazolium salt-based colorimetric assay (MTS assay). Neither cCPEwt nor –SSS had adverse effects on Huh7.5 cell viability (Figure S2). This is in accordance with similar results previously obtained with claudin-transfected HEK-293 cells and brain endothelial cells, which were treated with different cCPE variants [8].

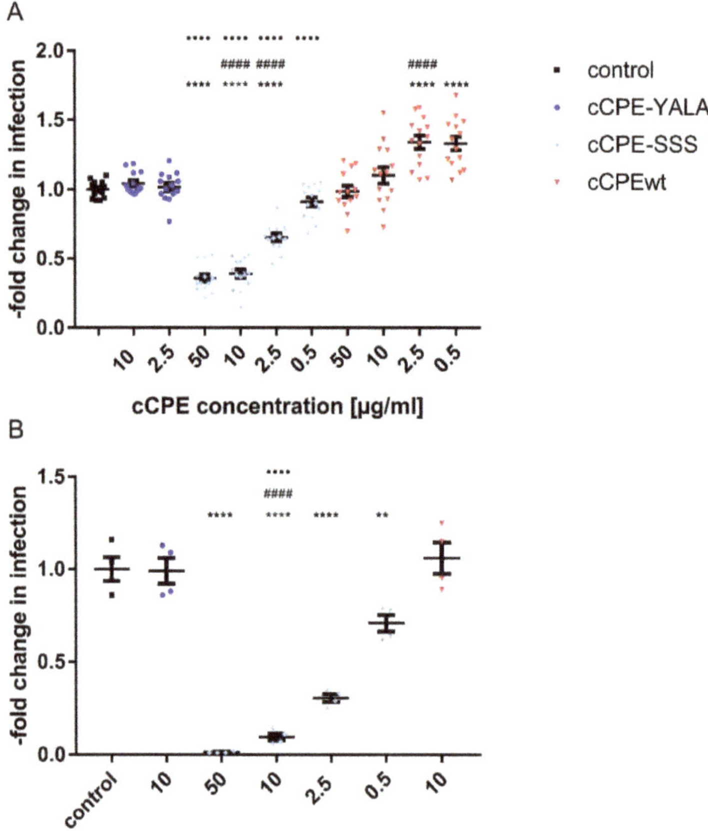

Figure 3. HCVcc infection of Huh7.5 cells in the presence of cCPE variants. Huh7.5 cells were pre-incubated for 2 h at 37 °C with cCPEwt, cCPE-SSS, cCPE–YALA (negative binding control) or not treated prior to infection (control). (**A**) HCVcc was added for 2 h, the medium was changed and 24 h later, HCV infection was analyzed by quantification of infection-positive cells after staining with an anti HCV NS5A antibody. Results are expressed as a percentage of infected cells compared to the control. Results of four experiments with total $n = 14$, for wt 50 $n = 13$. Results shown as means ± SEM. (**B**) HCVcc was added for 26 h. After this, HCV infection was analyzed by quantification of infection-positive cells after staining with an anti HCV NS5A antibody. Results are expressed as a percentage of infected cells compared to control. $n = 4$ Results shown as means ± SEM. One-way ANOVA with post-hoc Sidak's test was performed to analyze differences between treatment groups. ** $p \leq 0.01$ vs. control; **** $p \leq 0.0001$ vs. control; #### $p \leq 0.0001$ vs. cCPE-YALA; ●●●● $p < 0{,}0001$ vs. cCPEwt. cCPE-SSS strongly and specifically inhibits Cldn1-mediated HCV infection of Huh7.5 cells in a dose-dependent manner.

Together, the data demonstrate that cCPE-SSS specifically inhibits infection of Huh7.5 cells with HCV.

2.4. Both cCPE-SSS and cCPEwt Bind to Huh7.5 Cells

To further investigate the differential effects on HCV infection caused by cCPEwt and cCPE-SSS, the capability of cCPE variants to bind to Huh7.5 cells was analyzed. Confocal microscopy after immunostaining revealed binding of cCPEwt and cCPE-SSS but not of the cCPE-YALA control on the surface of Huh7.5 cells (Figure 4).

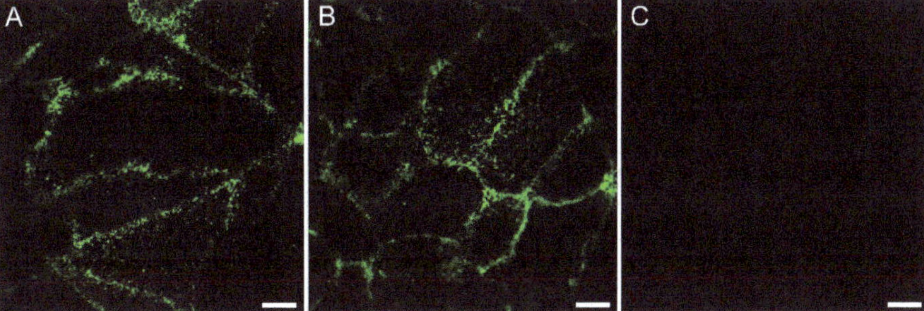

Figure 4. cCPEwt and cCPE-SSS but not cCPE-YALA bind to the surface of Huh7.5 cells. Immunostaining of Huh7.5 cells after 30 min of incubation with cCPE-constructs (10 µg/mL). Detection of GST-cCPEwt (**A**), GST-cCPE-SSS (**B**) and GST-cCPE-YALA (**C**) with anti-GST antibodies. Scale bar 10 µm.

cCPEwt and cCPE–SSS differ in their affinity towards Cldn1. Their similar binding to Huh7.5 cells indicated that, in addition to Cldn1, other CPE-receptor claudins are expressed in Huh7.5 cells. Expression of Cldn1, -2, and -4, but not of Cldn3, -6, -7, or -9 detected by Western blotting was reported [24]. Since Cldn4 but not Cldn2 is a high affinity CPE-receptor, Cldn4 was a candidate to contribute strongly to the binding of the cCPE variants to Huh7.5 cells. To confirm the presence of cCPE-binding claudins in these cells, pull down assay was performed using the GST-cCPE variants and lysates of Huh7.5 cells and analyzed by Western blot (Figure 5). Cldn1 was clearly detectable in the Huh7.5 lysates. cCPE-SSS interacted more strongly than cCPEwt with Cldn1 whereas no binding was detected with the cCPE-YALA negative control (Figure 5). Hence, concerning the cCPE-Cldn1 interaction with Huh7.5 cells similar results were obtained as for Huh7 cells (Figure 2). However, for Cldn4 no signal was obtained by Western blot neither in the cell lysate nor in the cCPE-bound fraction; for other claudins, also no clear and consistent signals were obtained by Western blot (data not shown).

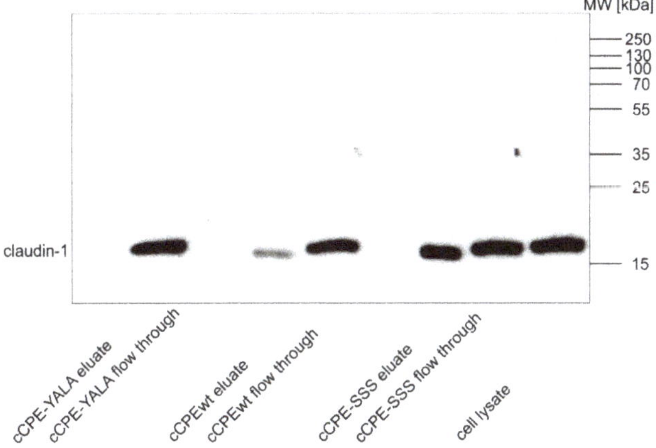

Figure 5. cCPE variants bind to Cldn1 of Huh7.5 cells. Pull down assays using GST-cCPE variants and lysates of Huh7.5 cells were performed. Eluate and flow through were separated by SDS-PAGE and analyzed by Western blot. A strong anti-Cldn1 immunoreactive band in the expected range of ~20 kDa was detected in the pull-down eluates for cCPE-SSS, whereas for cCPEwt, only a weak band and for cCPE-YALA, no band was detected. Representative results are shown.

2.5. Huh7.5 Cells Express Cldn1 and Cldn6 as Receptors for cCPE Variants

To overcome the limitations connected to sensitivity, specificity and cross-reactivity of anti-claudin antibodies, mass spectrometry was used to identify claudins which are expressed in Huh7.5 cells and able to bind to cCPEwt. A pull-down assay was performed with GST-cCPEwt and lysates of Huh7.5 cells, the cCPEwt-bound fraction as well as the non-bound fraction separated by SDS-PAGE, gel slices corresponding to areas between 15–25 kDa cut and processed for mass spectrometry. Consistent with the Western blots, Cldn1 was detected in the cCPEwt-bound and in the non-bound fraction (Table 1). In contrast, Cldn4 was not found in either fraction. However, Cldn6 was detected in the cCPE-bound fraction (Table 1). Subsequently, literature search-based purchase of another anti-Cldn6 antibody enabled confirmation of Cldn6 expression also by Western blot (Figure 6A). Cldn6 is known to be a CPE-receptor [6,10,48] and a weak entry factor for HCV [23,24]. Expression of Cldn6 mRNA in Huh7.5 cells was previously reported [29]. However, reports on Cldn6 protein expression were not completely consistent. Here, we clearly verify the expression of Cldn6 protein in Huh7.5 cells. Hence, cCPEwt can use Cldn6 and Cldn1 as receptors in Huh7.5 cells. We cannot entirely exclude that other claudins were not detected by mass spectrometry, due to methodological reasons, e.g., unsuitable peptide fragments. However, using mass spectrometry we were previously able to detect claudins such as Cldn1,-3,-4,-7 in liver cells [49].

Table 1. Mass spectrometry confirmed Cldn1 and Cldn6 protein expression in Huh7.5 cells. Pull down assay with GST-cCPEwt and lysates of Huh7.5 cells. cCPEwt-bound and non-bound fractions were separated by SDS-PAGE; gel slices corresponding to areas between 15–25 kDa were cut and processed for mass spectrometry.

Fraction	Cldn	Mascot Score	Sequence Cover-Age	Peptides	Position	Ions Score	Mass Determined	Mass Expected
E1: ~15kD	6	96	12%	K.VYDSLLALPQDLQAAR.A	66–81	46	1771.8914	1771.9468
				R.DFYNPLVAEAQK.R	146–157	90	1393.6274	1393.6878
E2: ~17KD	1	144	14%	K.VFDSLLNLSSTLQATR.A	66–81	91	1763.9194	1763.9418
				R.IVQEFYDPMTPVNAR.Y	144–158	56	1778.7754	1778.8662
				R.IVQEFYDPMTPVNAR.Y Oxidation M	144–158	80	1794.7974	1794.8611
E2: ~17KD	6	111	17%	K.VYDSLLALPQDLQAAR.A	66–81	77	1771.8914	1771.9468
				R.DFYNPLVAEAQK.R	146–157	71	1393.6614	1393.6878
				R.YSTSAPAISR.G	200–209	43	1051.4934	1051.5298
E3: ~18KD	6	88	17%	K.VYDSLLALPQDLQAAR.A	66–81	57	1771.8774	1771.9468
				R.DFYNPLVAEAQK.R	146–157	66	1393.6734	1393.6878
				R.YSTSAPAISR.G	200–209	32	1051.5194	1051.5298
FT9: ~17KD	1	97	14%	K.VFDSLLNLSSTLQATR.A	66–81	82	1763.9034	1763.9418
				R.IVQEFYDPMTPVNAR.Y	144–158	55	1778.7994	1778.8662

cCPEwt binds much more strongly to Cldn6 than to Cldn1 [6,9,10]. cCPE-SSS has a much higher affinity for Cldn1 than cCPEwt (Figure 6C) [5,6]. In contrast to cCPEwt, cCPE-SSS binds with high affinity (K_d~10 nM) both Cldn1 and to Cldn6 (Figure 6B,C). Taken together, the data indicate that high affinity cCPE-SSS binding to Cldn1 is efficiently inhibiting HCV infection of Huh7.5 cells whereas high affinity cCPEwt binding to Cldn6 is not sufficient to inhibit the infection efficiently.

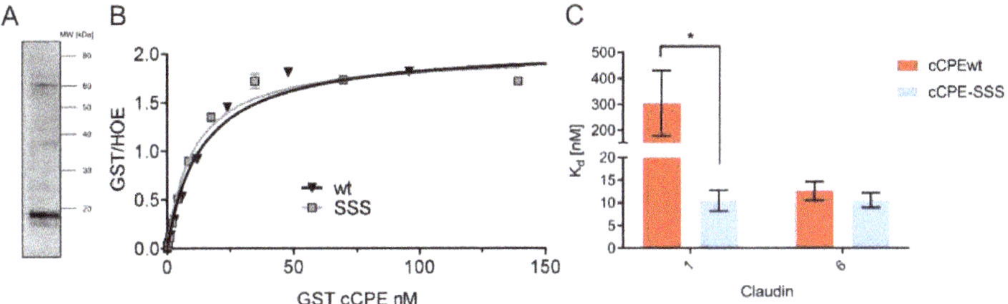

Figure 6. (**A**) Detection of Cldn6 expression in Huh7.5 cells by Western Blot. Huh7.5 cell pellets were lysed, the membrane protein fraction extracted, separated by SDS-PAGE and analyzed by Western blot using anti-Cldn6 antibodies. Cldn6 immunoreactive bands were observed in the expected range (~18 kDa). (**B**) cCPEwt and cCPE–SSS bind with similar affinities to Cldn6 on the cell surface. Binding of GST-cCPEwt and –cCPE-SSS to Cldn6. HEK-293 cells were transfected with Cldn6. Binding of GST-cCPE variants was analyzed using a cellular binding assay with varying cCPE concentrations, fixation, staining with phycoerythrin-coupled anti-GST antibody and Hoechst 33342 (nuclei) for normalization. (**C**) cCPE-SSS binds with similar high affinities to Cldn1 and -6, whereas cCPEwt has a very low affinity for Cldn1. The derived respective affinities of cCPEwt and cCPE–SSS for Cldn6 are not significantly different, but cCPE-SSS binds with a significantly stronger affinity to Cldn1 than cCPEwt does. Dissociation constants (K_d) were calculated from the binding assays using a non-linear regression analysis in GraphPad Prism 6 (model: Saturation binding, equation: one site—specific binding) [6]. Results are shown as means ± SEM. Student's t-Test, data from 3 (Cldn1) and 5 to 6 (Cldn6) independent K_d calculation experiments, *, $p < 0.05$.

2.6. Treatment of Reconstructed Human Epidermis with cCPE Variants Leads to Barrier Opening

To test claudin modulation by cCPE variants in another Cldn1-expressing tissue we analyzed TJ-modulation by cCPE variants in a human in vitro skin model. RHE expressing Cldn1 and Cldn4 [50] was treated with 50 µg/mL of the cCPE variants for 48 h at two different time points (day 4 and day 8). Subsequently, the barrier to ions (trans-epithelial resistance, TER) and the permeability for the tracer molecule Lucifer Yellow (LY) were measured to analyze the barrier function of the RHE.

The treatment with cCPEwt (binding strongly to Cldn4) led to a significant decrease in TER to 0.55 ± 0.07-fold ($p < 0.0001$ compared to untreated RHE and cCPE-YALA, Figure 7A) and a significant increase in LY permeability to 3.42 ± 0.24-fold ($p < 0.0001$ compared to the negative control and $p < 0.001$ compared to cCPE-YALA) on day four. However, on day eight, only a slight reduction in TER (0.753 ± 0.09-fold, Figure 7A, $p < 0.01$ compared to to cCPE-YALA) and a slight increase in LY permeability (1.97 ± 0.27-fold), could be observed (Figure 7B).

On day four, compared to cCPEwt, cCPE-SSS (binding strongly to Cldn1 and Cldn4) had a similar effect on TER (0.49 ± 0.04-fold, $p < 0.0001$ compared to untreated control and cCPE-YALA, Figure 7A) and a significantly greater effect on LY permeability (5.23 ± 0.44-fold, $p < 0.0001$ compared to untreated control and cCPE-YALA, $p < 0.001$ compared to cCPEwt, Figure 7B). On day eight, TER was significantly decreased after treatment with cCPE-SSS (0.66 ± 0.08-fold, $p = 0.001$ compared to untreated control, Figure 7A) and there was a slight tendency for an increased outside-to-inside passage of LY (1.78 ± 0.32%, Figure 7B) for cCPE-SSS.

The different cCPE variants had no effect on the expression of the relevant claudins in the RHE, (Cldn1, -3, -4 and -7), on day 4 or on day 8 (data not shown).

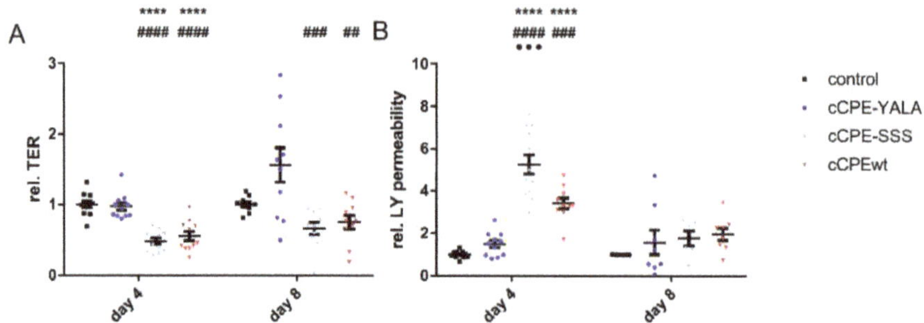

Figure 7. Effect of cCPE variants on (**A**) TER and (**B**) Lucifer Yellow permeability in reconstructed human epidermis. Average relative trans-epithelial resistance (**A**) and relative LY permeability (**B**) after 48 h incubation with 50 µg/mL of either cCPEwt, cCPE-SSS, cCPE-YALA or untreated control on days 4 and 8 after a change to ALI. Results were standardized to respective untreated control. Results are shown as means ± SEM. Statistical significances of one-way ANOVA: * compared to control, # compared to cCPE-YALA, ● compared to cCPEwt. ## $p < 0.01$; ###, ●●● $p < 0.001$; ****, #### $p < 0.0001$. $n = 5–12$.

In summary, the data showed that cCPE variants can be used as TJ modulators in the skin and that removal of Cldn1 and Cldn4 from the TJ impairs the barrier properties of the human epidermis.

3. Discussion

In this study, we showed that Cldn1 expressed in liver and skin can be efficiently targeted with the broad-specificity claudin binder cCPE-SSS. In contrast to cCPEwt, cCPE-SSS strongly inhibited Cldn1-dependent HCV infection in a hepatocyte in vitro model. In addition, cCPE-SSS improved paracellular permeability for small molecules in an in vitro human skin model to a larger extend than cCPEwt.

Claudins have been suggested as targets for diverse therapeutic application throughout numerous studies, including targeting of claudin-overexpressing cancer entities [51,52], visualization of upregulated claudin expression in colon polyps [53], opening of tissue barriers, e.g., the blood-brain-barrier [8] or intestinal barriers [54,55].

Use of monoclonal antibodies for claudin targeting led to very promising in vivo results; in particular with anti-Cldn18 or -Cldn1 antibodies for treatment of gastric cancer or HCV infections, respectively [14,18,32,33,56,57]. As an alternative for claudin targeting, application of CPE or cCPE variants has been suggested [8,58–60]. Not only is the bacterial production of these proteins very cost-effective compared to antibody production, but cCPE can also be adjusted to target several claudin subtypes. The latter is advantageous e.g., in viral infections like hepatitis C, where a pan-genomic treatment and prevention of resistance toward treatment (e.g., due to viral use of multiple claudins as receptors) is desirable. However, as claudins are expressed throughout the entire body, a thorough evaluation of possible adverse effects of claudin-targeting agents is necessary. Full-length CPE is known to have a high systemic toxicity. In contrast, more than ten weekly intravenous administration of 5 mg/kg cCPE did not increase biochemical markers of kidney and liver injury in mice [61]. These findings suggest that cCPE variants have limited and controllable adverse side effects even when applied systemically. Nevertheless, in vivo effects on tissue barriers have to be investigated in more detail. At least, blocking of extra-junctional HCV co-receptors might be achieved at much milder cCPE doses than opening of preformed junctional barrier, as indicated by the data (Figure 3, Figure 7).

Host-targeting agents are promising therapeutic approaches for novel anti-viral treatments for patients that do not respond to conventional therapies of direct-acting antivirals, or combination of ribavirin and pegylated interferon. So far, antibodies recognizing the extracellular domains of various cellular HCV receptors, for example Cldn1 [14,32,33], CD81 [62] and SR-BI [63] have been successfully used in vitro and in vivo to prevent HCV infection.

We showed that the broad-specificity claudin binder cCPE-SSS can effectively inhibit HCV infection in the hepatoma cell line Huh7.5. The substitution S305P/S307R/S313H improved cCPE binding to Cldn1 of Huh7 cells significantly compared to cCPEwt, providing a tool to target hepatoma cells potently. In contrast to cCPEwt, cCPE-SSS inhibited HCVcc infection of hepatoma Huh7.5 cells in a dose-dependent manner (Figure 3). Surprisingly, low concentrations of cCPEwt even aggravated HCV infection. Due to its much lower affinity to Cldn1 than to Cldn6, cCPEwt should bind mainly to Cldn6 under these conditions. We assume that Cldn1 and Cldn6 compete with each other for CD81 association, that cCPEwt binding to Cldn6 interferes with Cldn6-CD81 interaction (similar as for cCPE-SSS – Cldn1, indicated by homology modeling) and that in turn formation of the more potent Cldn1-CD81 HCV co-receptor complex is favored. Such a competition mechanism might explain aggravation of infection by cCPEwt.

Cldn1-specific neutralizing antibodies only partially inhibited infection of Huh7.5 cells by HCV strains with broad claudin tropisms, suggesting a possible escape from Cldn1-specific targeting molecules through Cldn6 or Cldn9 in cell culture [64]. In addition, a previously Cldn1-dependent HCV strain was shown to evolve the capacity to use Cldn6 as a host entry factor via a single amino acid mutation in the HCV E1 protein [65]. cCPE-SSS concomitantly targets Cldn1 and Cldn6 on the surface of Huh7.5 with nanomolar affinities (Figures 5 and 6). Hence, use of cCPE-SSS as a broad spectrum claudin binder might avoid the escape from Cldn1-specific antibodies, making it a more powerful agent in HCV treatment.

In summary, we showed that cCPE can be modified and used to prevent infection of hepatoma cells with HCV by binding to claudins including Cldn1, thus inhibiting formation of the viral entry complex.

Additionally, we analyzed the potential of cCPE variants to modulate the epidermal barrier by targeting claudins of the SG in RHE models. A purposeful and directed opening of the epidermal barrier can be useful to improve transdermal drug delivery [61,66].

Transdermal drug delivery is a desirable method of application. Its numerous benefits include a non-invasive technique, avoidance of first-pass metabolism by digestive enzymes and thus a better bioavailability of poorly orally bioavailable drugs, a constant plasma drug concentration as well as enhancement of patient compliance [61,66]. Treatment of RHE with cCPE variants prevents consecutive incorporation of claudins in TJs and allowed us to study the effects of TJ rearrangements on the epidermal barrier.

cCPEwt removes Cldn3, -4, -6, -7, -8 and -14 from the TJ. However, its affinity for Cldn1 is very low. We show here that cCPEwt resulted in an impaired epidermal barrier to small molecule tracers (LY) in the RHE at day 4, i.e., during formation of the SC. Impairment of the epidermal barrier to molecular tracers after cCPEwt treatment during formation of SC has been described earlier [39], and is in accordance with our results. In the study by Yuki et al., 10 or 50 μg/mL cCPEwt was added to the culture medium at the beginning of SC formation. After a four-day incubation period, the human skin equivalent models showed a significant increase in LY permeability, indicating a weakened epidermal barrier. [39]. In addition, we demonstrate that the ion barrier was weakened by cCPEwt on day 4. On day 8, when SC was more mature, cCPEwt did not have a significant effect even though there was a trend.

cCPE-SSS has a higher affinity for Cldn1-9 [5,6], which should result in a greater barrier impairment compared to cCPEwt because Cldn-1 plays a major role in epidermal barrier function [35]. Indeed, in the four-day model, the cCPE-SSS treatment led to a significant decrease in TER and a significant increase in LY permeability. This effect was greater than the effect of cCPEwt.

In the eight-day model, the TER was significantly lower after cCPE-SSS treatment compared to the other groups. This indicates that loss of Cldn1 from the TJ weakens the ionic barrier even in a well-developed SC.

All observed effects were weaker in the eight-day compared to the four-day model although the cCPE variants were applied from the basolateral side and thus were not held up by the SC. This finding might indicate a reduced accessibility of claudins in the more mature epithelium. If the extracellular domains of claudins are involved in trans-interactions, cCPE cannot bind anymore [67]. Likewise, the contribution of the TJ to epidermal barrier sealing might be less strong in the presence of a mature and intact SC.

The most effective approach to improve transdermal drug delivery should modify both epidermal barriers, the SC as well as the TJ. Hence, cCPE-based biologics might be used in combination with a chemical absorption enhancer, like surfactants (surface-active agents), fatty-acid esters, terpenes and solvents. Chemical enhancers increase skin permeability by various means, including solubilization of lipids and denaturing of keratin as well as fluidizing the crystalline structure of the SC [68]. Combining these methods with the directed claudin-targeting by cCPE could have a therapeutically significant effect on the epidermal barrier.

In conclusion, cCPE variants are promising tools for claudin targeting in diverse tissues including liver and skin. Their claudin binding properties can be adjusted to fit a certain demand, their production is very cost-effective compared to monoclonal antibodies and their systemic administration so far has been without fatal adverse side effects.

4. Materials and Methods

4.1. Structural Bioinformatics and Molecular Modeling

Homology model of the Cldn1-cCPE-SSS complex was built based on the Cldn4-cCPE complex (PDB ID: 5B2G; [43]). All manipulations, optimizations of the models, as well as calculations of hydrophobic and electrostatic potentials on the molecular surfaces, were performed with Sybyl X2.1.1 (Certara USA Inc., St. Louis, MO, USA). Models were energetically minimized using the AMBER7 FF99 force field. Structure images were generated with PyMOL Molecular Graphics System 1.8.4.1 (Schrödinger LLC, 2017).

4.2. Chemicals and Solutions

Antibodies: Phycolink® anti-GST R-phycoerythrin-conjugated antibodies (Europa Bioproducts Ltd., Cambridge, UK), rabbit anti-Cldn1 and anti-Cldn4, (Invitrogen, Carlsbad, USA), rabbit polyclonal anti-Cldn6 (Immuno-Biological Laboratories, Fujioka, JP), mouse anti-GST (Sigma-Aldrich, Steinheim, DE), mouse anti-NS5a (ABCAM, Cambridge, UK).

Chemicals: Lucifer Yellow: Sigma-Aldrich Chemie GmbH (München, DE).

4.3. Cells and Tissues

4.3.1. Cultivation of Cells

Primary human keratinocytes were isolated from juvenile foreskin as described previously [50]. The use of these samples was in accordance with approval of the ethics committee of the Ärztekammer Hamburg (WF 61/12). Samples were used anonymously and all investigations were conducted in accordance to the principles of the Declaration of Helsinki.

Human Huh7 cells [23] kindly provided by Dr. Nora Gehne (FMP), were cultured in Dulbecco's modified Eagle's medium supplemented with 10% (v/v) FBS and 1% (v/v) penicillin/streptomycin solution (10,000 U/mL penicillin, 10,000 µg/mL streptomycin, Gibco™, Thermo Fisher Scientific, Waltham, MA, USA). Huh7.5. cells (Apath, LLC, St. Louis, MO, USA) were cultured as described previously [69].

HEK-293 cells (CRL-1573, A.T.C.C., Manassas, VA, USA), [70] were kindly provided by Professor Otmar Huber (Friedrich Schiller University of Jena), cultivated in Minimal Essentail Medium with Earle's salts and L-glutamine (Merck KGaA, Darmstadt, DE) supplemented with 10% (v/v) FBS and 1% (v/v) penicillin/streptomycin solution (10000 U/mL penicillin, 10,000 µg/mL streptomycin, Gibco™, Thermo Fisher Scientific, Waltham, MA, USA). For transfection, cells were seeded onto coverslips with a diameter of 32 mm (Menzel Gläser, Thermo Fisher Scientific, Waltham, MA, USA). At 80% confluency, cells were transfected with 2.9 ng plasmid encoding N-terminally YFP-fused human CD81 and 1.1 ng plasmid encoding N-terminally CFP-fused Cldn1 using polyethlenimine (PEI MAX 40K, Polysciences, Inc., Warrington, PA, USA). Cells were further incubated at 37 °C for 2 days before measuring FRET efficiency. In the case of cCPE-SSS treatment, cells were incubated with 10 µg/mL of the protein for 30 min prior to starting the measurement.

4.3.2. HCVcc Infection Assay

Two to three days after plating on 48 well plates, Huh7.5 cells were incubated for 2 h with GST-cCPEwt, GST-cCPE-SSS or GST-cCPE-YALA at concentrations between 0.5–50 µg/mL at 37 °C or not further treated (control). After incubation the cells were infected with Gt2a HCV J6CF-JFH1 (Jc1) at a multiplicity of infection (moi) of 1.0 at 37 °C. The medium was changed 2 h after infection and Huh7.5 cells were incubated at 37 °C for further 24 h or Huh7.5 cells were incubated at 37 °C for 26 h without medium exchange.

Huh7.5 cells were fixed with acetone/methanol (1:1 ratio), incubated for 1 h at room temperature with anti NS5a antibodies, and incubated for one hour at room temp with Alexa Fluor 488-conjugated goat anti-mouse secondary antibody (washed with PBS between steps) and counterstained with 4′,6-diamidino-2-phenylindole (DAPI). Infection levels were determined using a Nikon Eclipse TE2000-U microscope (Nikon Corporation, Tokyo, Japan), and expressed as a fold change of the cCPE untreated control [46,69].

4.3.3. MTS Cell Proliferation Assay

Cellular toxicity was measured using a Cell Titer 96 aqueous nonradioactive cell proliferation assay kit, according to the manufacturer's instructions (Promega, Madison, WI, USA) as described previously [71]. Briefly, 2×10^4 Huh7.5 cells were seeded in a 96 well plate. After cultivation for 72 h, cells were incubated with cCPEwt and cCPE-SSS at concentrations ranging from 0 to 100 ng/µL for 2 h. 3-(4,5-dimethylthiazol-2-yl)-5-(3-carboxymethoxyphenyl)-2-(4-sulfophenyl)-2*H*-tetrazolium) (MTS)-phenazine methosulfate solution (40 µL) was added to each well and the samples were incubated at 37 °C in a humidified 5% CO_2 atmosphere for 3 h. Absorbance was measured at 490 nm with a UVmax plate reader (Molecular Devices, San José, CA, USA). Each sample was analyzed in triplicate.

4.3.4. Live Cell Imaging, Acceptor Photobleaching and FRET Data Analysis

Live cell imaging and photo bleaching were performed using the laser scanning microscope LSM 780 (Carl Zeiss Microscopy GmbH, Jena, DE). Experiments were performed as described previously [72].

FRET efficiency as calculated using the following formula: $E_{FRET} = (CFP_A - CFP_B)/CFP_A$; CFP_A and CFP_B referring to CFP fluorescence intensity after and before photobleaching, respectively.

4.3.5. Cultivation of Keratinocytes and Creation of Reconstructed Human Epidermis

Reconstructed human epidermis was generated from primary human keratinocytes as already described [50]. Briefly, the 3×10^5 cells were seeded in 500 µL EpiLife medium with 1.5 mM $CaCl_2$ onto Millicell cell culture inserts (Merck Millipore, Tullagreen, Ireland) and 2.5 mL cell culture medium was applied to the basal compartment. The cells were cultivated at 37 °C in humidified atmosphere for 30 h before lifting to the air-liquid interface (ALI). Concurrently, the basal medium was exchanged with 1.5 mL of EpiLife medium containing 1.5 mM $CaCl_2$, 92 µg/mL ascorbic acid (Sigma, Darmstadt, DE) and 10 ng/mL recombinant human keratinocyte growth factor (R&D Systems, Wiesbaden-Nordenstadt, DE). Cells derived from an individual donor were cultured as duplicates.

4.3.6. Treatment of Reconstructed Human Epidermis with cCPE and Barrier Assays

For cCPE treatment, 50 µg/mL of the respective cCPE mutants were added to the basal compartment of the RHE models at day 2 or 6 after transitioning to ALI. After 48 h of incubation at 37 °C in a humidified atmosphere, the transepidermal resistance as a measure for the epidermal ion barrier was measured using chopstick electrodes. Relative TER values were used due to intrinsic variation in starting absolute TER values between the different RHE cultures since they were derived from different human donors. The absolute TER values of the control RHE cultures were between 0.53 and 2.26 kOhm × cm^2 at day 4 and 0.71 and 2.92 kOhm × cm^2 at day 8.

To examine the outside-to-inside epidermal barrier for small molecules, the RHE models were transferred to 6-well plates with 1.5 mL of medium. Then, 200 µL of a 1 mM Lucifer Yellow solution was added to the apical side. After 6 h incubation at 37 °C, 100 µL of the basal medium was transferred into a microtiter plate. The LY fluorescence was measured with the Infinite M2000 fluorescence reader (Tecan Group Ltd., Männedorf, Switzerland) using the following settings: excitation wavelength 425 nm, emissions wavelength 550 nm, 10 flashes, 50-fold gain. By measuring the fluorescence of LY standard samples of known concentration, the concentration of the basal media samples could be calculated. With this, the LY permeability was calculated using the following formula: $P_{APP}\left[\frac{cm}{s}\right] = \frac{flux\left[\frac{\mu g}{s \cdot cm^2}\right]}{LY\ concentration\left[\frac{\mu g}{ml}\right]}$.

4.4. Plasmids

Plasmid encoding GST-CPE194-319wt (GST-cCPE), GST-cCPE-S305P/S307R/S313H and GST-cCPE–Y306A/L315A fusion protein were reported previously [6], as well as plasmid encoding N-terminally CFP-tagged Cldn1 [72]. pEZ-M15-NYFP/huCD81 was obtained from GeneCopoeia™ (Rockville, MD, USA).

4.5. Expression and Purification of cCPE-Constructs

CPE194-319 (cCPE) with N-terminal GST-fusions as well as GST (control) were expressed in *E. coli* BL21 as reported previously [6]. Briefly, bacteria were grown to A_{600} = 0.6–0.8, expression induced by addition of 1 mM isopropyl-β-D-thiogalactopyranoside. 3 h after induction, bacteria were harvested (10 min, 20,000× *g*, 4 °C) and lysed in PBS (phosphate buffered saline) with 1% (v/v) Triton X-100, 0.1 mM PMSF, 1 mM EDTA, protease inhibitor cocktail (Sigma-Aldrich, Steinheim, DE) and sonicated with Vibra Cell™ Model 72434 (BioBlock Scientific, Strasbourg, France) by 10 × 1 s pulses. Insoluble cell debris was removed by centrifugation (20,000× *g* for 1 h at 4 °C). Proteins were purified from supernatants using glutathione-agarose (Sigma-Aldrich, Steinheim, DE) and dialyzed against PBS. Protein concentration was determined with BCA Protein Kit (Thermo-Scientific, Rockford, USA).

4.6. Pull-down Assay

Pull-down assay was performed as described previously [10]. Briefly, Huh7 and Huh7.5 cells were lysed with 1% Triton X-100, EDTA-free protease inhibitor cocktail (Roche, Mannheim, DE) in PBS. The 10,000× *g* supernatant was incubated with GST-cCPE constructs bound to Glutathione Sepharose beads (GE Healthcare, Uppsala, Sweden) for 2 h on a shaker at 4 °C. Beads were washed 3 times with PBS containing 0.5% Triton X-100, bound proteins eluted with Laemmli buffer and analyzed by SDS-PAGE and Western blot with anti-Cldn1, or anti-Cldn4 antibodies. After stripping, membranes were incubated with anti-GST antibodies to verify that similar amounts of GST-cCPE were bound to the beads.

4.7. Immunostaining

Immunocytochemistry was performed as described [10]. Briefly, 2–3 days after plating, Huh7.5 cells were incubated with 10 µg/mL GST-cCPEwt, GST-cCPE-SSS or GST-cCPE-YALA for 30 min at 37 °C and washed with PBS. Cells were fixed with methanol/acetone 1:1 ratio, blocked with 1% BSA in PBS, incubated with anti-GST antibodies, washed, and incubated with Alexa Fluor 488-conjugated goat anti-mouse secondary antibodies. Cells were analyzed with the LSM 510 META confocal microscope (Zeiss, Jena, DE).

4.8. Mass Spectrometry

4.8.1. In-gel Trypsin Digestion

Gel bands of interest were excised after Coomassie blue staining and cut into 1–2 mm^3 gel pieces, which were placed into 1.5 mL sample tubes. Gel pieces were rinsed twice with wash solution for 18 h in total (200 μL, 50% methanol, 5% acetic acid). The solutions were removed and gel pieces were dehydrated in acetonitrile (200 μL, 5 min). Supernatants were removed and gel pieces were dried in a vacuum centrifuge for 3 min. Disulfide reduction was performed with 10 mM DTT (30 μL) for 0.5 h, followed by alkylation with 100 mM iodoacetamide (30 μL) for 0.5 h. Supernatants were removed from the gel samples and dehydration with acetonitrile and evaporation performed as described above. Gel pieces were washed with 100 mM ammonium bicarbonate (200 μL, 10 min). Supernatants were removed and dehydration performed with acetonitrile and evaporation as above. The gel samples were then rehydrated on ice with freshly prepared trypsin solution (30 μL, 20 ng/μL sequencing grade trypsin (Promega, Madison, WI, USA) in 50 mM ammonium bicarbonate). After rehydration, excess trypsin solution was removed and 50 mM ammonium bicarbonate (10 μL) was added to prevent dehydration of gel pieces. Gel samples were digested at 37 °C for 18 h. The gel pieces were then extracted sequentially with 50 mM ammonium bicarbonate (60 μL), 50% acetonitrile, 5% formic acid (60 μL) and 85% acetonitrile, 5% formic acid (60 μL). The combined extracts were evaporated in a vacuum centrifuge and were re-dissolved in 5% acetonitrile, 0.1% formic acid (20 μL) on an ultrasonic bath and transferred into LC-MS sample vials.

4.8.2. LC-MS/MS Analysis

For the analysis of in-gel digested protein material, liquid chromatography was performed using an Ultimate 3000 nano-HPLC system (Dionex, Sunnyvale, CA, USA) comprising a WPS-3000 micro auto sampler, a FLM-3000 flow manager and column compartment, a UVD-3000 UV detector, an LPG-3600 dual-gradient micro-pump, and an SRD-3600 solvent rack controlled by Hystar (Bruker Daltonics, Billerica, MA, USA) and DCMSLink 2.0 software (Dionex, Sunnyvale, CA, USA). Samples were concentrated on a trapping column Dionex (Sunnyvale, CA, USA), 300 μm i.d., 0.1 cm) at a flow rate of 20 μL/min. For the separation with a C18 Pepmap column (75 μm i.d., 15 cm, Dionex), a flow rate of 250 nL/min was used as generated by a cap-flow splitter cartridge (1/1000). Peptides were eluted by the application of a 30 min multi-step gradient using solvents A (98% H_2O, 2% acetonitrile, 0.1% formic acid) and B (80% acetonitrile, 20% H_2O, 0.1% formic acid) (Table 2):

Table 2. Elution protocol of peptides. A C18 Pepmap column (75 μm i.d., 15 cm) was used for separation of peptide fragments. Solvent A (98% H_2O, 2% acetonitrile, 0.1% formic acid) and solvent B (80% acetonitrile, 20% H_2O, 0.1% formic acid) were used in the ratio indicated.

Composition (% Solvent B)	Run Time (min)
2–10	0–3
10–25	3–18
25–50	18–30
50–90	30–30.2

The liquid chromatography was interfaced directly with a 3D high capacity ion trap mass spectrometer (amaZon; Bruker Daltonics, Billerica, MA, USA) utilizing 10 μm i.d. distal coated SilicaTips (New Objective, Woburn, MA, USA) and nano-ESI mode. SPS parameter settings on the ion trap were tuned for a target mass of 850 *m/z*, compound stability 100% and a smart ICC target of 250,000. MS/MS analysis was initiated on a contact closure signal triggered by HyStar software (version 3.2).

Up to five precursor ions were selected per cycle with active exclusion (0.5 min) in collision-induced dissociation (CID) mode. CID fragmentation was achieved using helium gas and a 30–200% collision energy sweep with amplitude 1.0 (ions are ejected from the trap as soon as they fragment).

4.8.3. Data Processing and Database Searching

Raw LC-MS/MS data were processed and Mascot compatible files were created using DataAnalysis 4.0 software (Bruker Daltonics, Billerica, MA, USA). Database searches were performed using the Mascot algorithm (version 2.4) and the UniProt_SwissProt database with mammalian taxonomy restriction (v2012.09.17, number of entries 537,505, after taxonomy filter: 66,032). The following parameters were applied: 2+, 3+ and 4+ ions, peptide mass tolerance 0.3 Da, $^{13}C = 2$, fragment mass tolerance 0.6 Da, number of missed cleavages: two, instrument type: ESI-TRAP, fixed modifications: Carbamidomethylation (Cys), variable modifications: Oxidation (Met).

4.9. Statistical Analysis

Results are expressed as means ± standard error of the mean (SEM). Gaussian distribution was tested using the D´Agostino-Pearson omnibus test. Statistical analyses were performed using one-way ANOVA with post-hoc Sidak's test or Student´s t-test (in case of failed normality test: Mann-Whitney test), as indicated. $p < 0.05$ was considered statistically significant.

Supplementary Materials: Supplementary materials can be found at http://www.mdpi.com/1422-0067/20/19/4774/s1.

Author Contributions: Conceptualization, J.R., N.Z. and J.P.; Data curation, L.-S.B. and J.R.; Formal analysis, N.Z. and J.P.; Funding acquisition, J.M.B., G.K., N.Z. and J.P.; Investigation, L.-S.B., J.R., S.W., S.B., H.B.K., J.P., M.E., A.P. and S.V.-y.-S.; Methodology, L.-S.B., J.R., S.W., S.B., H.B.K., J.P., M.E., A.P., N.Z. and J.P.; Project administration, N.Z. and J.P.; Resources, H.B.K., M.E., A.P., J.M.B., G.K., N.Z. and J.P.; Software, H.B.K., J.P., N.Z. and J.P.; Supervision, A.P., J.M.B., G.K., N.Z. and J.P.; Validation, L.-S.B., J.R., A.P., J.M.B., N.Z. and J.P.; Visualization, H.B.K., J.P., N.Z. and J.P.; Writing—original draft, L.-S.B., J.R., S.W., S.B., H.B.K., J.P., A.P., J.M.B., G.K., N.Z. and J.P.; Writing—review & editing, L.-S.B., J.R., S.W., S.B., H.B.K., J.P., A.P., J.M.B., G.K., N.Z. and J.P.

Funding: This research was funded by the Deutsche Forschungsgemeinschaft (DFG), Research Training Group GRK 2318 (A2), DFG PI 837/2-1and the Open Access Publication Fund of the Charité–Universitätsmedizin Berlin.

Conflicts of Interest: The authors declare no conflict of interest.

Abbreviations

ALI	air-liquid interface
ANOVA	analysis of variance
CD81	cluster of differentiation 81
CFP	cyan fluorescent protein
Cldn	claudin
cCPE	amino acids 194–319 of *Clostridium perfringens* enterotoxin
cCPE-SSS	cCPE-S305P/S307R/S313H
cCPEwt	wildtype cCPE
cCPE-YALA	cCPE-Y306A/L315A
CPE	*Clostridium perfringens* enterotoxin
ECS	extracellular segment
FBS	fetal bovine serum
FRET	fluorescence resonance energy transfer
GST	Glutathione S-transferase
HCV	hepatitis C virus
HCVcc	cell culture-derived hepatitis C virus particles
LC/MS-MS	liquid chromatography - tandem mass spectrometry
LY	Lucifer yellow
MTS	3-[4,5,dimethylthiazol-2-yl]-5-[3-carboxymethoxy-phenyl]-2-[4-sulfophenyl]-2H -tetrazolium
RHE	reconstructed human epidermis
SC	*stratum corneum*
SEM	standard error of the mean
SG	*stratum granulosum*

SR-BI	scavenger receptor class B type I
TER	trans-epithelial resistance
TJ	tight junction
YFP	yellow fluorescent protein

References

1. Krause, G.; Winkler, L.; Mueller, S.L.; Haseloff, R.F.; Piontek, J.; Blasig, I.E. Structure and function of claudins. *Biochim. Biophys. Acta.* **2008**, *1778*, 631–645. [CrossRef] [PubMed]
2. Hagen, S.J. Non-canonical functions of claudin proteins: Beyond the regulation of cell-cell adhesions. *Tissue Barriers* **2017**, *5*, e1327839. [CrossRef] [PubMed]
3. Freedman, J.C.; Shrestha, A.; McClane, B.A. Clostridium perfringens Enterotoxin: Action, Genetics, and Translational Applications. *Toxins* **2016**, *8*, 73. [CrossRef] [PubMed]
4. Veshnyakova, A.; Protze, J.; Rossa, J.; Blasig, I.E.; Krause, G.; Piontek, J. On the interaction of Clostridium perfringens enterotoxin with claudins. *Toxins* **2010**, *2*, 1336–1356. [CrossRef] [PubMed]
5. Takahashi, A.; Saito, Y.; Kondoh, M.; Matsushita, K.; Krug, S.M.; Suzuki, H.; Tsujino, H.; Li, X.; Aoyama, H.; Matsuhisa, K.; et al. Creation and biochemical analysis of a broad-specific claudin binder. *Biomaterials* **2012**, *33*, 3464–3474. [CrossRef] [PubMed]
6. Protze, J.; Eichner, M.; Piontek, A.; Dinter, S.; Rossa, J.; Blecharz, K.G.; Vajkoczy, P.; Piontek, J.; Krause, G. Directed structural modification of Clostridium perfringens enterotoxin to enhance binding to claudin-5. *Cell Mol. Life Sci.* **2015**, *72*, 1417–1432. [CrossRef] [PubMed]
7. Veshnyakova, A.; Piontek, J.; Protze, J.; Waziri, N.; Heise, I.; Krause, G. Mechanism of Clostridium perfringens enterotoxin interaction with claudin-3/-4 protein suggests structural modifications of the toxin to target specific claudins. *J. Biol. Chem.* **2012**, *287*, 1698–1708. [CrossRef] [PubMed]
8. Neuhaus, W.; Piontek, A.; Protze, J.; Eichner, M.; Mahringer, A.; Subileau, E.A.; Lee, I.M.; Schulzke, J.D.; Krause, G.; Piontek, J. Reversible opening of the blood-brain barrier by claudin-5-binding variants of Clostridium perfringens enterotoxin's claudin-binding domain. *Biomaterials* **2018**, *161*, 129–143. [CrossRef] [PubMed]
9. Fujita, K.; Katahira, J.; Horiguchi, Y.; Sonoda, N.; Furuse, M.; Tsukita, S. Clostridium perfringens enterotoxin binds to the second extracellular loop of claudin-3, a tight junction integral membrane protein. *FEBS Lett.* **2000**, *476*, 258–261. [CrossRef]
10. Winkler, L.; Gehring, C.; Wenzel, A.; Muller, S.L.; Piehl, C.; Krause, G.; Blasig, I.E.; Piontek, J. Molecular determinants of the interaction between Clostridium perfringens enterotoxin fragments and claudin-3. *J. Biol. Chem.* **2009**, *284*, 18863–18872. [CrossRef] [PubMed]
11. Petermann, P.; Rahn, E.; Thier, K.; Hsu, M.J.; Rixon, F.J.; Kopp, S.J.; Knebel-Morsdorf, D. Role of Nectin-1 and Herpesvirus Entry Mediator as Cellular Receptors for Herpes Simplex Virus 1 on Primary Murine Dermal Fibroblasts. *J. Virol.* **2015**, *89*, 9407–9416. [CrossRef]
12. Rahn, E.; Thier, K.; Petermann, P.; Rubsam, M.; Staeheli, P.; Iden, S.; Niessen, C.M.; Knebel-Morsdorf, D. Epithelial Barriers in Murine Skin during Herpes Simplex Virus 1 Infection: The Role of Tight Junction Formation. *J. Invest Dermatol.* **2017**, *137*, 884–893. [CrossRef] [PubMed]
13. Evans, M.J.; von Hahn, T.; Tscherne, D.M.; Syder, A.J.; Panis, M.; Wolk, B.; Hatziioannou, T.; McKeating, J.A.; Bieniasz, P.D.; Rice, C.M. Claudin-1 is a hepatitis C virus co-receptor required for a late step in entry. *Nature* **2007**, *446*, 801–805. [CrossRef] [PubMed]
14. Fukasawa, M.; Nagase, S.; Shirasago, Y.; Iida, M.; Yamashita, M.; Endo, K.; Yagi, K.; Suzuki, T.; Wakita, T.; Hanada, K.; et al. Monoclonal antibodies against extracellular domains of claudin-1 block hepatitis C virus infection in a mouse model. *J. Virol.* **2015**, *89*, 4866–4879. [CrossRef] [PubMed]
15. Wahid, A.; Helle, F.; Descamps, V.; Duverlie, G.; Penin, F.; Dubuisson, J. Disulfide bonds in hepatitis C virus glycoprotein E1 control the assembly and entry functions of E2 glycoprotein. *J. Virol.* **2013**, *87*, 1605–1617. [CrossRef] [PubMed]
16. Horsley-Silva, J.L.; Vargas, H.E. New Therapies for Hepatitis C Virus Infection. *Gastroenterol Hepatol. (N Y)* **2017**, *13*, 22–31.
17. Organization, W.H.; Hepatits, C. Available online: https://www.who.int/news-room/fact-sheets/detail/hepatitis-c (accessed on 1 May 2019).

18. Krieger, S.E.; Zeisel, M.B.; Davis, C.; Thumann, C.; Harris, H.J.; Schnober, E.K.; Mee, C.; Soulier, E.; Royer, C.; Lambotin, M.; et al. Inhibition of hepatitis C virus infection by anti-claudin-1 antibodies is mediated by neutralization of E2-CD81-claudin-1 associations. *Hepatology* **2010**, *51*, 1144–1157. [CrossRef] [PubMed]
19. Zhu, Y.Z.; Qian, X.J.; Zhao, P.; Qi, Z.T. How hepatitis C virus invades hepatocytes: The mystery of viral entry. *World J. Gastroenterol* **2014**, *20*, 3457–3467. [CrossRef] [PubMed]
20. Cukierman, L.; Meertens, L.; Bertaux, C.; Kajumo, F.; Dragic, T. Residues in a highly conserved claudin-1 motif are required for hepatitis C virus entry and mediate the formation of cell-cell contacts. *J. Virol.* **2009**, *83*, 5477–5484. [CrossRef]
21. Sarhan, M.A.; Pham, T.N.; Chen, A.Y.; Michalak, T.I. Hepatitis C virus infection of human T lymphocytes is mediated by CD5. *J. Virol.* **2012**, *86*, 3723–3735. [CrossRef]
22. Zona, L.; Tawar, R.G.; Zeisel, M.B.; Xiao, F.; Schuster, C.; Lupberger, J.; Baumert, T.F. CD81-receptor associations–impact for hepatitis C virus entry and antiviral therapies. *Viruses* **2014**, *6*, 875–892. [CrossRef] [PubMed]
23. Zheng, A.; Yuan, F.; Li, Y.; Zhu, F.; Hou, P.; Li, J.; Song, X.; Ding, M.; Deng, H. Claudin-6 and claudin-9 function as additional coreceptors for hepatitis C virus. *J. Virol.* **2007**, *81*, 12465–12471. [CrossRef] [PubMed]
24. Meertens, L.; Bertaux, C.; Cukierman, L.; Cormier, E.; Lavillette, D.; Cosset, F.L.; Dragic, T. The tight junction proteins claudin-1, -6, and -9 are entry cofactors for hepatitis C virus. *J. Virol.* **2008**, *82*, 3555–3560. [CrossRef] [PubMed]
25. Tawar, R.G.; Colpitts, C.C.; Lupberger, J.; El-Saghire, H.; Zeisel, M.B.; Baumert, T.F. Claudins and pathogenesis of viral infection. *Semin. Cell Dev. Biol.* **2015**, *42*, 39–46. [CrossRef] [PubMed]
26. Harris, H.J.; Davis, C.; Mullins, J.G.; Hu, K.; Goodall, M.; Farquhar, M.J.; Mee, C.J.; McCaffrey, K.; Young, S.; Drummer, H.; et al. Claudin association with CD81 defines hepatitis C virus entry. *J. Biol. Chem.* **2010**, *285*, 21092–21102. [CrossRef] [PubMed]
27. Douam, F.; Dao Thi, V.L.; Maurin, G.; Fresquet, J.; Mompelat, D.; Zeisel, M.B.; Baumert, T.F.; Cosset, F.L.; Lavillette, D. Critical interaction between E1 and E2 glycoproteins determines binding and fusion properties of hepatitis C virus during cell entry. *Hepatology* **2014**, *59*, 776–788. [CrossRef]
28. Harris, H.J.; Farquhar, M.J.; Mee, C.J.; Davis, C.; Reynolds, G.M.; Jennings, A.; Hu, K.; Yuan, F.; Deng, H.; Hubscher, S.G.; et al. CD81 and claudin 1 coreceptor association: Role in hepatitis C virus entry. *J. Virol.* **2008**, *82*, 5007–5020. [CrossRef] [PubMed]
29. Fofana, I.; Zona, L.; Thumann, C.; Heydmann, L.; Durand, S.C.; Lupberger, J.; Blum, H.E.; Pessaux, P.; Gondeau, C.; Reynolds, G.M.; et al. Functional analysis of claudin-6 and claudin-9 as entry factors for hepatitis C virus infection of human hepatocytes by using monoclonal antibodies. *J. Virol.* **2013**, *87*, 10405–10410. [CrossRef]
30. Timpe, J.M.; Stamataki, Z.; Jennings, A.; Hu, K.; Farquhar, M.J.; Harris, H.J.; Schwarz, A.; Desombere, I.; Roels, G.L.; Balfe, P.; et al. Hepatitis C virus cell-cell transmission in hepatoma cells in the presence of neutralizing antibodies. *Hepatology* **2008**, *47*, 17–24. [CrossRef]
31. Zeisel, M.B.; Fofana, I.; Fafi-Kremer, S.; Baumert, T.F. Hepatitis C virus entry into hepatocytes: Molecular mechanisms and targets for antiviral therapies. *J. Hepatol.* **2011**, *54*, 566–576. [CrossRef]
32. Fofana, I.; Krieger, S.E.; Grunert, F.; Glauben, S.; Xiao, F.; Fafi-Kremer, S.; Soulier, E.; Royer, C.; Thumann, C.; Mee, C.J.; et al. Monoclonal anti-claudin 1 antibodies prevent hepatitis C virus infection of primary human hepatocytes. *Gastroenterology* **2010**, *139*, 953–964. [CrossRef]
33. Mailly, L.; Xiao, F.; Lupberger, J.; Wilson, G.K.; Aubert, P.; Duong, F.H.T.; Calabrese, D.; Leboeuf, C.; Fofana, I.; Thumann, C.; et al. Clearance of persistent hepatitis C virus infection in humanized mice using a claudin-1-targeting monoclonal antibody. *Nat. Biotechnol.* **2015**, *33*, 549–554. [CrossRef] [PubMed]
34. Brandner, J.M.; Kief, S.; Grund, C.; Rendl, M.; Houdek, P.; Kuhn, C.; Tschachler, E.; Franke, W.W.; Moll, I. Organization and formation of the tight junction system in human epidermis and cultured keratinocytes. *Eur. J. Cell Biol.* **2002**, *81*, 253–263. [CrossRef] [PubMed]
35. Furuse, M.; Hata, M.; Furuse, K.; Yoshida, Y.; Haratake, A.; Sugitani, Y.; Noda, T.; Kubo, A.; Tsukita, S. Claudin-based tight junctions are crucial for the mammalian epidermal barrier: A lesson from claudin-1-deficient mice. *J. Cell Biol.* **2002**, *156*, 1099–1111. [CrossRef] [PubMed]
36. Basler, K.; Bergmann, S.; Heisig, M.; Naegel, A.; Zorn-Kruppa, M.; Brandner, J.M. The role of tight junctions in skin barrier function and dermal absorption. *J. Control Release* **2016**, *242*, 105–118. [CrossRef] [PubMed]

37. Nakajima, M.; Nagase, S.; Iida, M.; Takeda, S.; Yamashita, M.; Watari, A.; Shirasago, Y.; Fukasawa, M.; Takeda, H.; Sawasaki, T.; et al. Claudin-1 Binder Enhances Epidermal Permeability in a Human Keratinocyte Model. *J. Pharmacol. Exp. Ther.* **2015**, *354*, 440–447. [CrossRef]
38. Zhang, J.; Ni, C.; Yang, Z.; Piontek, A.; Chen, H.; Wang, S.; Fan, Y.; Qin, Z.; Piontek, J. Specific binding of Clostridium perfringens enterotoxin fragment to Claudin-b and modulation of zebrafish epidermal barrier. *Exp. Dermatol.* **2015**, *24*, 605–610. [CrossRef]
39. Yuki, T.; Komiya, A.; Kusaka, A.; Kuze, T.; Sugiyama, Y.; Inoue, S. Impaired tight junctions obstruct stratum corneum formation by altering polar lipid and profilaggrin processing. *J. Dermatol. Sci.* **2013**, *69*, 148–158. [CrossRef]
40. Poumay, Y.; Dupont, F.; Marcoux, S.; Leclercq-Smekens, M.; Herin, M.; Coquette, A. A simple reconstructed human epidermis: Preparation of the culture model and utilization in in vitro studies. *Arch. Dermatol. Res.* **2004**, *296*, 203–211. [CrossRef]
41. Robertson, S.L.; Smedley, J.G., 3rd; McClane, B.A. Identification of a claudin-4 residue important for mediating the host cell binding and action of Clostridium perfringens enterotoxin. *Infect. Immun.* **2010**, *78*, 505–517. [CrossRef]
42. Saitoh, Y.; Suzuki, H.; Tani, K.; Nishikawa, K.; Irie, K.; Ogura, Y.; Tamura, A.; Tsukita, S.; Fujiyoshi, Y. Tight junctions. Structural insight into tight junction disassembly by Clostridium perfringens enterotoxin. *Science* **2015**, *347*, 775–778. [CrossRef] [PubMed]
43. Shinoda, T.; Shinya, N.; Ito, K.; Ohsawa, N.; Terada, T.; Hirata, K.; Kawano, Y.; Yamamoto, M.; Kimura-Someya, T.; Yokoyama, S.; et al. Structural basis for disruption of claudin assembly in tight junctions by an enterotoxin. *Sci. Rep.* **2016**, *6*, 33632. [CrossRef] [PubMed]
44. Suzuki, H.; Nishizawa, T.; Tani, K.; Yamazaki, Y.; Tamura, A.; Ishitani, R.; Dohmae, N.; Tsukita, S.; Nureki, O.; Fujiyoshi, Y. Crystal structure of a claudin provides insight into the architecture of tight junctions. *Science* **2014**, *344*, 304–307. [CrossRef] [PubMed]
45. Takahashi, A.; Komiya, E.; Kakutani, H.; Yoshida, T.; Fujii, M.; Horiguchi, Y.; Mizuguchi, H.; Tsutsumi, Y.; Tsunoda, S.; Koizumi, N.; et al. Domain mapping of a claudin-4 modulator, the C-terminal region of C-terminal fragment of Clostridium perfringens enterotoxin, by site-directed mutagenesis. *Biochem. Pharmacol.* **2008**, *75*, 1639–1648. [CrossRef] [PubMed]
46. Pollock, S.; Nichita, N.B.; Bohmer, A.; Radulescu, C.; Dwek, R.A.; Zitzmann, N. Polyunsaturated liposomes are antiviral against hepatitis B and C viruses and HIV by decreasing cholesterol levels in infected cells. *Proc. Natl. Acad. Sci. USA* **2010**, *107*, 17176–17181. [CrossRef] [PubMed]
47. Feigelstock, D.A.; Mihalik, K.B.; Kaplan, G.; Feinstone, S.M. Increased susceptibility of Huh7 cells to HCV replication does not require mutations in RIG-I. *Virol. J.* **2010**, *7*, 44. [CrossRef] [PubMed]
48. Lal-Nag, M.; Battis, M.; Santin, A.D.; Morin, P.J. Claudin-6: A novel receptor for CPE-mediated cytotoxicity in ovarian cancer. *Oncogenesis* **2012**, *1*, e33. [CrossRef]
49. Lohrberg, D.; Krause, E.; Schumann, M.; Piontek, J.; Winkler, L.; Blasig, I.E.; Haseloff, R.F. A strategy for enrichment of claudins based on their affinity to Clostridium perfringens enterotoxin. *BMC Mol. Biol.* **2009**, *10*, 61. [CrossRef]
50. Basler, K.; Galliano, M.F.; Bergmann, S.; Rohde, H.; Wladykowski, E.; Vidal, Y.S.S.; Guiraud, B.; Houdek, P.; Schuring, G.; Volksdorf, T.; et al. Biphasic influence of Staphylococcus aureus on human epidermal tight junctions. *Ann. N Y Acad. Sci.* **2017**, *1405*, 53–70. [CrossRef]
51. English, D.P.; Santin, A.D. Claudins overexpression in ovarian cancer: Potential targets for Clostridium Perfringens Enterotoxin (CPE) based diagnosis and therapy. *Int. J. Mol. Sci.* **2013**, *14*, 10412–10437. [CrossRef]
52. Walther, W.; Petkov, S.; Kuvardina, O.N.; Aumann, J.; Kobelt, D.; Fichtner, I.; Lemm, M.; Piontek, J.; Blasig, I.E.; Stein, U.; et al. Novel Clostridium perfringens enterotoxin suicide gene therapy for selective treatment of claudin-3- and -4-overexpressing tumors. *Gene Ther.* **2012**, *19*, 494–503. [CrossRef] [PubMed]
53. Rabinsky, E.F.; Joshi, B.P.; Pant, A.; Zhou, J.; Duan, X.; Smith, A.; Kuick, R.; Fan, S.; Nusrat, A.; Owens, S.R.; et al. Overexpressed Claudin-1 Can Be Visualized Endoscopically in Colonic Adenomas In Vivo. *Cell Mol. Gastroenterol Hepatol.* **2016**, *2*, 222–237. [CrossRef] [PubMed]
54. Krug, S.M.; Amasheh, M.; Dittmann, I.; Christoffel, I.; Fromm, M.; Amasheh, S. Sodium caprate as an enhancer of macromolecule permeation across tricellular tight junctions of intestinal cells. *Biomaterials* **2013**, *34*, 275–282. [CrossRef] [PubMed]

55. Radloff, J.; Cornelius, V.; Markov, A.G.; Amasheh, S. Caprate Modulates Intestinal Barrier Function in Porcine Peyer's Patch Follicle-Associated Epithelium. *Int. J. Mol. Sci.* **2019**, *20*, 1418. [CrossRef] [PubMed]
56. Singh, P.; Toom, S.; Huang, Y. Anti-claudin 18.2 antibody as new targeted therapy for advanced gastric cancer. *J. Hematol. Oncol.* **2017**, *10*, 105. [CrossRef] [PubMed]
57. Colpitts, C.C.; Tawar, R.G.; Mailly, L.; Thumann, C.; Heydmann, L.; Durand, S.C.; Xiao, F.; Robinet, E.; Pessaux, P.; Zeisel, M.B.; et al. Humanisation of a claudin-1-specific monoclonal antibody for clinical prevention and cure of HCV infection without escape. *Gut* **2018**, *67*, 736–745. [CrossRef] [PubMed]
58. Hashimoto, Y.; Yagi, K.; Kondoh, M. Roles of the first-generation claudin binder, Clostridium perfringens enterotoxin, in the diagnosis and claudin-targeted treatment of epithelium-derived cancers. *Pflugers. Arch.* **2017**, *469*, 45–53. [CrossRef] [PubMed]
59. Gao, Z.; McClane, B.A. Use of Clostridium perfringens Enterotoxin and the Enterotoxin Receptor-Binding Domain (C-CPE) for Cancer Treatment: Opportunities and Challenges. *J. Toxicol.* **2012**, *2012*, 981626. [CrossRef]
60. Black, J.D.; Lopez, S.; Cocco, E.; Schwab, C.L.; English, D.P.; Santin, A.D. Clostridium perfringens enterotoxin (CPE) and CPE-binding domain (c-CPE) for the detection and treatment of gynecologic cancers. *Toxins* **2015**, *7*, 1116–1125. [CrossRef]
61. Suzuki, H.; Kondoh, M.; Li, X.; Takahashi, A.; Matsuhisa, K.; Matsushita, K.; Kakamu, Y.; Yamane, S.; Kodaka, M.; Isoda, K.; et al. A toxicological evaluation of a claudin modulator, the C-terminal fragment of Clostridium perfringens enterotoxin, in mice. *Pharmazie* **2011**, *66*, 543–546.
62. Meuleman, P.; Hesselgesser, J.; Paulson, M.; Vanwolleghem, T.; Desombere, I.; Reiser, H.; Leroux-Roels, G. Anti-CD81 antibodies can prevent a hepatitis C virus infection in vivo. *Hepatology* **2008**, *48*, 1761–1768. [CrossRef] [PubMed]
63. Meuleman, P.; Catanese, M.T.; Verhoye, L.; Desombere, I.; Farhoudi, A.; Jones, C.T.; Sheahan, T.; Grzyb, K.; Cortese, R.; Rice, C.M.; et al. A human monoclonal antibody targeting scavenger receptor class B type I precludes hepatitis C virus infection and viral spread in vitro and in vivo. *Hepatology* **2012**, *55*, 364–372. [CrossRef] [PubMed]
64. Haid, S.; Grethe, C.; Dill, M.T.; Heim, M.; Kaderali, L.; Pietschmann, T. Isolate-dependent use of claudins for cell entry by hepatitis C virus. *Hepatology* **2014**, *59*, 24–34. [CrossRef] [PubMed]
65. Hopcraft, S.E.; Evans, M.J. Selection of a hepatitis C virus with altered entry factor requirements reveals a genetic interaction between the E1 glycoprotein and claudins. *Hepatology* **2015**, *62*, 1059–1069. [CrossRef] [PubMed]
66. Iqbal, B.; Ali, J.; Baboota, S. Recent advances and development in epidermal and dermal drug deposition enhancement technology. *Int. J. Dermatol.* **2018**, *57*, 646–660. [CrossRef] [PubMed]
67. Krause, G.; Protze, J.; Piontek, J. Assembly and function of claudins: Structure-function relationships based on homology models and crystal structures. *Semin. Cell Dev. Biol.* **2015**, *42*, 3–12. [CrossRef] [PubMed]
68. Prausnitz, M.R.; Mitragotri, S.; Langer, R. Current status and future potential of transdermal drug delivery. *Nat. Rev. Drug Discov.* **2004**, *3*, 115–124. [CrossRef] [PubMed]
69. Pollock, S.; Antrobus, R.; Newton, L.; Kampa, B.; Rossa, J.; Latham, S.; Nichita, N.B.; Dwek, R.A.; Zitzmann, N. Uptake and trafficking of liposomes to the endoplasmic reticulum. *FASEB J.* **2010**, *24*, 1866–1878. [CrossRef]
70. Piontek, J.; Winkler, L.; Wolburg, H.; Muller, S.L.; Zuleger, N.; Piehl, C.; Wiesner, B.; Krause, G.; Blasig, I.E. Formation of tight junction: Determinants of homophilic interaction between classic claudins. *FASEB J.* **2008**, *22*, 146–158. [CrossRef] [PubMed]
71. Woodhouse, S.D.; Smith, C.; Michelet, M.; Branza-Nichita, N.; Hussey, M.; Dwek, R.A.; Zitzmann, N. Iminosugars in combination with interferon and ribavirin permanently eradicate noncytopathic bovine viral diarrhea virus from persistently infected cells. *Antimicrob. Agents Chemother.* **2008**, *52*, 1820–1828. [CrossRef]
72. Milatz, S.; Piontek, J.; Schulzke, J.D.; Blasig, I.E.; Fromm, M.; Gunzel, D. Probing the cis-arrangement of prototype tight junction proteins claudin-1 and claudin-3. *Biochem. J.* **2015**, *468*, 449–458. [CrossRef]

International Journal of
Molecular Sciences

Review

Potential for Tight Junction Protein–Directed Drug Development Using Claudin Binders and Angubindin-1

Yosuke Hashimoto [1], Keisuke Tachibana [1], Susanne M. Krug [2], Jun Kunisawa [1,3,4,5,6], Michael Fromm [2] and Masuo Kondoh [1,*]

1 Graduate School of Pharmaceutical Sciences, Osaka University, Osaka 565-0871, Japan
2 Institute of Clinical Physiology, Charité–Universitätsmedizin Berlin, 12203 Berlin, Germany
3 Laboratory of Vaccine Materials, Center for Vaccine and Adjuvant Research and Laboratory of Gut Environmental System, National Institutes of Biomedical Innovation, Health and Nutrition (NIBIOHN), Osaka 567-0085, Japan
4 International Research and Development Center for Mucosal Vaccines, The Institute of Medical Sciences, The University of Tokyo, Tokyo 108-8639, Japan
5 Department of Microbiology and Immunology, Kobe University Graduate School of Medicine, Kobe 650-0017, Japan
6 Graduate School of Medicine and Graduate School of Dentistry, Osaka University, Osaka 565-0871, Japan
* Correspondence: masuo@phs.osaka-u.ac.jp; Tel./Fax: +81-6-6879-8195

Received: 24 July 2019; Accepted: 14 August 2019; Published: 17 August 2019

Abstract: The tight junction (TJ) is an intercellular sealing component found in epithelial and endothelial tissues that regulates the passage of solutes across the paracellular space. Research examining the biology of TJs has revealed that they are complex biochemical structures constructed from a range of proteins including claudins, occludin, tricellulin, angulins and junctional adhesion molecules. The transient disruption of the barrier function of TJs to open the paracellular space is one means of enhancing mucosal and transdermal drug absorption and to deliver drugs across the blood–brain barrier. However, the disruption of TJs can also open the paracellular space to harmful xenobiotics and pathogens. To address this issue, the strategies targeting TJ proteins have been developed to loosen TJs in a size- or tissue-dependent manner rather than to disrupt them. As several TJ proteins are overexpressed in malignant tumors and in the inflamed intestinal tract, and are present in cells and epithelia conjoined with the mucosa-associated lymphoid immune tissue, these TJ-protein-targeted strategies may also provide platforms for the development of novel therapies and vaccines. Here, this paper reviews two TJ-protein-targeted technologies, claudin binders and an angulin binder, and their applications in drug development.

Keywords: tight junction; claudin; angulin; drug development; angubindin-1; *Clostridium perfringens* enterotoxin; *Clostridium perfringens* iota-toxin; antibody

1. Introduction

The boundaries between the inside of the body and the outside environment in the airway and gastrointestinal tract, and between the systemic circulation and tissues in the brain, eye, testis, and placenta, are separated by epithelial and endothelial cell sheets, respectively. The paracellular spaces between the adjacent cells in these sheets are sealed by a structural and functional component called the tight junction (TJ) [1]. TJs control the diffusion of ions, solutes, and water across the paracellular space to maintain homeostasis, and the loss of TJ integrity appears to be associated with the development of intestinal diseases [2,3], atopic dermatitis [4], and psychiatric disorders [5,6]. TJs also prevent mucosal and epidermal absorption of drugs and the delivery of drugs from the systemic circulation to the brain, eye, testis, and placenta.

A freeze-fracture replica electron microscopy analysis has shown that TJs consist of a meshwork of proteins called TJ strands [7]. In epithelial cells, these TJ strands are located at the apical side of the lateral membrane. TJs include various membrane proteins—including claudins, TJ-associated MARVEL proteins (occludin, tricellulin, and marvelD3), junctional adhesion molecules, and angulins—and these membrane proteins are anchored to intracellular scaffold proteins, e.g., of the zonula occludens protein family [8,9]. The physiological characteristics of TJs are determined by the specific combinations and mixing ratios of these TJ proteins [10–12]. TJ strands are dynamic structures that are repeatedly breaking and annealing, which transiently loosens the TJ seal and allows the stepwise diffusion of solutes across the meshwork and through the paracellular space [13].

There are two types of TJs: Bicellular, where two cells meet, and tricellular, where three cells meet. Bicellular TJ strands extend horizontally along the apical membrane but extend vertically when they reach a tricellular contact. Tricellular TJs seal the tubular structure created at tricellular contacts by the three vertically extending TJ strands and the three adjoining cell membranes [14].

A modulation of the structure of TJs to loosen the paracellular space can be used to increase mucosal and epidermal drug absorption, as well as drug delivery to the brain. Currently, sodium caprate and mannitol are used clinically to enhance paracellular drug absorption and drug delivery to the brain, respectively [15,16]. However, sodium caprate causes mucosal damage and lacks tissue-specificity [15,17]. The mannitol widened the interendothelial TJs to a radius of approximately 20 nm, followed by deliver chemicals, peptides, antibodies, and viral vectors to the brain [18]. Research into understanding the biochemical structure of TJs and the physiological roles of the various TJ proteins has provided insights that have been applied to the development of TJ protein–targeted drugs. Here, the efficacy and safety of claudin and angulin binders for the development of TJ-directed drugs is reviewed.

2. Claudins and Angulins

2.1. Claudins

Claudins were identified in 1998 as components of TJ strands that are crucial for the sealing of the intercellular space [19]. Currently, the mammalian claudin family comprises 27 proteins [20]. Since 2014, the crystal structures of these claudins have been gradually elucidated [21–24]. Claudins are tetra-transmembrane proteins containing two loops: The first contains four β-strands and an α-helix (extracellular helix), and the second contains a β-strand and the cell-surface-exposed transmembrane 3 domain (Figure 1a). Almost all claudins have the zonula occludens-1 binding motif at their C-terminal end. Claudins have *cis*-interactions within a cell membrane as well as *trans*-interactions with claudins in adjacent cell membranes. Both the *cis*- and *trans*-interactions are required for the building of TJ strands. The crystal structures of claudins have revealed that the *cis*-interaction between the extracellular helix and the transmembrane 3 domain leads to the formation of claudin strands within the cell membrane [21–24]. Claudins have two flexible loops, defined as variable regions: The first is between the β1- and β2-strands in the first extracellular domain, and the second is between the transmembrane 3 domain and the β5-strand in the second extracellular domain, with these flexible loops being involved in the formation of *trans*-interactions between claudin strands [25,26]. These flexible regions also appear to determine the characteristics of claudin-based TJs, because their homology is low among claudin members [21].

2.2. Angulins

The tricellular TJ is a specialized structure at the point of contact among three cells. Tricellulin, a member of the tight junction-associated MARVEL protein (TAMP) family, is an essential component of tricellular TJs [27]. Tricellulin does not form homophilic *trans*-interactions nor does it localize at tricellular contacts by itself [10,28]. Instead, it has to be recruited to tricellular contacts by angulins. Angulins are type I trans-membrane proteins with an extracellular immunoglobulin-like domain and a cytosolic tail (Figure 1b) [29]. There are three angulins: Angulin-1 (also known as lipolysis-stimulated lipoprotein receptor, LSR); angulin-2 (also known as immunoglobulin-like domain containing receptor 1, ILDR-1); and angulin-3 (also known as immunoglobulin-like domain containing receptor 2, ILDR-2). The interaction between phosphorylated Ser288 in the C-terminal tail of angulin-1 and the C-terminal cytosolic tail of tricellulin is responsible for the recruitment of tricellulin to the tricellular TJ [28,30]. Angulin-1 and -2 have a much greater ability to recruit tricellulin than angulin-3 [29]. The extracellular domain of angulin-2 may form a trimeric structure at the tricellular contact. However, the underlying mechanism remains unclear yet [31].

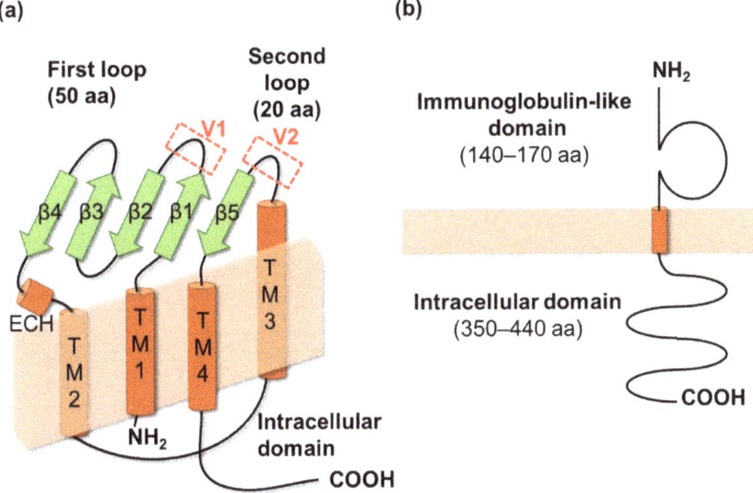

Figure 1. The representative structures of a claudin and an angulin. (**a**) Claudins are 20–27 kDa proteins containing four transmembrane (TM) domains and two extracellular loops. Extracellular loop 1 and 2 contains approximately 50 and 25 amino acids, respectively. TM domain 3 is much longer than the other three TM domains. Claudins also contain an extracellular helix (ECH) and two variable regions (V1 and V2). (**b**) Angulins are 60–70 kDa type I TM proteins containing an extracellular immunoglobulin-like domain and an intracellular tail. There is currently no structural information available for angulins. The secondary structural elements are shown as cylinders (α-helices) and arrows (β-strands). aa, amino acid.

3. Claudin and Angulin Binders

Currently, most binders that target TJ proteins are either fragments of bacterial toxins (first-generation binders) or monoclonal antibodies (mAbs; second-generation binders) [32–34].

3.1. First-Generation Binders: Fragments of Bacterial Toxins

Clostridium perfringens enterotoxin (CPE) has two domains: The N-terminal cytotoxic domain, which is involved in oligomerization and pore formation, and the C-terminal receptor binding domain (C-CPE) [35] (Figure 2a). The CPE receptor (CPE-R) was identified, and CPE-R has significant similarity to the rat androgen withdrawal apoptosis protein (RVP1) in 1997 [36]. Two years after the identification of CPE-R and RVP1, claudin-3 and -4 have been identified to be RVP1 and CPE-R, respectively [37]. C-CPE binds to claudin-3 and -4 [38]. However, C-CPE also binds to claudin-6, -7, -8, -9, -14, and -19 [22,39,40]. The affinity of C-CPE to claudin-4 is approximately 0.5 nM [41]. The treatment of MDCK cells with C-CPE decreases TJ integrity [38]. The research into the generation of claudin binders using C-CPE as a template for site-directed mutagenesis has produced several C-CPE mutants: One broad-spectrum binder to claudin-1 to -5 and four relatively specific binders for claudin-3, -4, and -5 (Table 1).

Table 1. List of claudin-binding mutants of C-terminal receptor binding domain of *Clostridium perfringens* enterotoxin.

Binder Type	Mutated Regions	Ref.
Negative-binding mutant	Y306A/L315A	[42]
Broad-spectrum binder (m19)	S304A/S305P/S307R/N309H/S313H	[43]
Enhancing specificity to claudin-3	L223A/D225A/R227A	[44]
Enhancing specificity to claudin-4	L254A/S256A/I258A/D284A	[44]
Enhancing specificity to claudin-5	Y306W/S313H	[45]
Improved specificity to claudin-5	N218Q/Y306W/S313H	[46]

Clostridium perfringens iota-toxin is a binary toxin composed of a cytotoxic domain (Ia) and a receptor-binding domain (Ib). Ib domain can be further classified into an Ia-binding domain (domain 1), oligomerization domain (domain 2), pore-formation domain (domain 3), and receptor-binding domain (domain 4). Domains 1 to 3 are involved in cytotoxicity [47]. Domain 4, which comprises amino acids 421–664, binds to the lipolysis-stimulated lipoprotein receptor without inducing cytotoxicity [48,49]. In 2013, the lipolysis-stimulated lipoprotein receptor was re-identified as angulin-1, the determining factor for the localization of tricellulin at tricellular TJs [28,29], and domain 4, the first tricellular TJ–specific modulator reported, was named angubindin-1 [50].

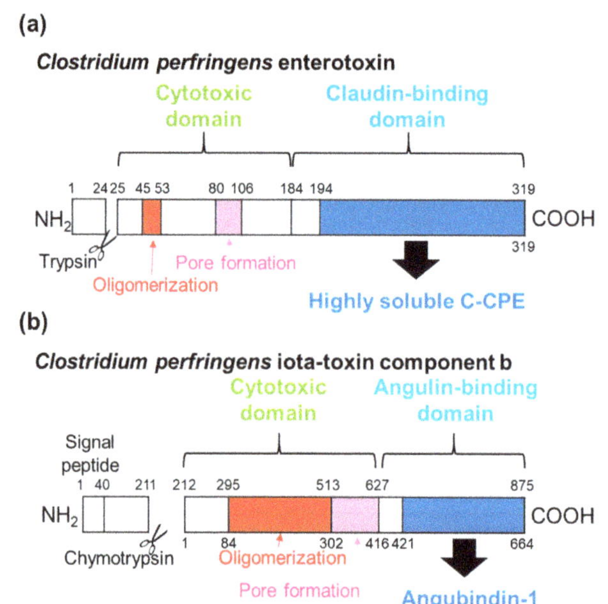

Figure 2. Schematic diagram showing the domains of first-generation claudin binders made from bacterial toxins. (**a**) *Clostridium perfringens* enterotoxin and (**b**) *C. perfringens* iota-toxin. Regions involved in oligomerization (red) and pore formation (pink) are indicated. Regions used as first-generation claudin binders are indicated in blue. The accession numbers for *C. perfringens* enterotoxin and iota-toxin are AOD41705 and CAA51960, respectively.

3.2. Second-Generation Binders: Monoclonal Antibodies

The TJ components are promising targets for the development of drugs to treat cancers and inflammatory bowel disease, for preventing infection by hepatitis C virus, and for the development of regenerative medicines. Antibodies are promising therapeutics for TJ-directed drug development because they bind to target proteins with high affinity and high specificity [51,52]. Thus, mAbs against the extracellular domains of the TJ components have been generated and their pharmaceutical activities are being investigated (Table 2).

Table 2. List of current monoclonal antibodies against tight junction components.

Indicated Application or Disease	Target	Monoclonal Antibody Name	Ref. or ClinicalTrials.gov Identifier
Modulation of epidermal barrier	Claudin-1	7A5	[53]
Modulation of blood–brain barrier	Claudin-5	R9, R2, 2B12	[54,55]
Inflammatory bowel disease	Claudin-2	1A2	[56]
Hepatitis C virus infection	Claudin-1 Occludin	OM-7D3-B3, 3A2 1-3, 67-2	[57,58] [59,60]
Gastric cancer (phase III study)	Claudin-18.2	IMAB362	NCT03504397
Pancreatic cancer (phase II study)	Claudin-18.2	IMAB362	NCT03816163
Germ cell tumor (phase II study)	Claudin-6	IMAB027	NCT03760081
Cancers (phase I or pre-clinical study)	Claudin-1 Claudin-2 Claudin-3 Claudin-4 Angulin-1 Angulin-3	3A2, 6F6 1A2 KMK3953, IgGH6 KM3934, 5D12 #1-25 BAY1905254	[61,62] [63] [64,65] [66,67] [68] NCT03666273
Regenerative medicine	Claudin-6	clone 342927	[69]

4. Drug Delivery Using Claudin and Angulin Binders

The authors have recently completed a series of studies examining TJ binders that has provided new insights for TJ-directed drug development. Here, the proofs-of-concept for TJ-directed drug development using first- and second-generation claudin and angulin binders are introduced.

4.1. Mucosal Absorption

Drug absorption across epithelia is either transcellular or paracellular [70]. One strategy for paracellular drug absorption is to loosen the TJs between adjacent epithelial cells. C-CPE increased jejunal absorption of dextran (4 kDa) 400-fold compared with sodium caprate, an absorption enhancer in current clinical use [17]. C-CPE also enhanced jejunal, nasal, and pulmonary absorption of a biologically active peptide [41]. The treatment of cells with angubindin-1 enhances the permeability of tricellular TJs to solutes up to 10 kDa (Figure 3). Angubindin-1 also enhanced jejunal absorption of dextran (4 kDa) [50]. These results demonstrate that modulation of bicellular and tricellular TJs could be useful strategy for the development of non-invasive drug-delivery systems.

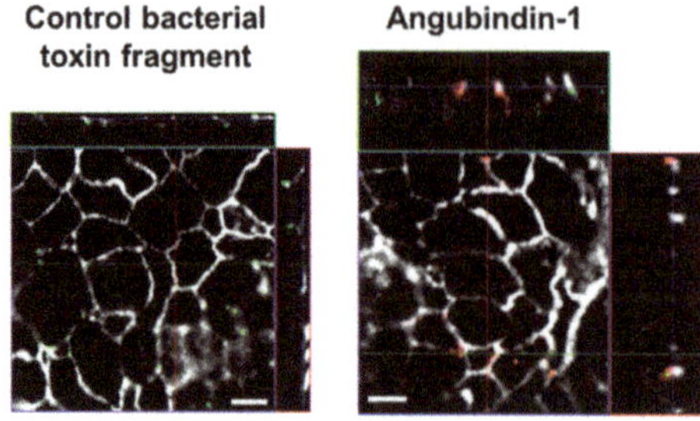

Figure 3. Visualization of the permeation of a macromolecule through tricellular TJs. HT-29/B6 cells were treated with control bacterial toxin fragment or angubindin-1 for 48 h. The cells were incubated with avidin and then with biotin-labeled tetramethylrhodamine-dextran 10 kDa (red) on the apical side of the insert for 1 h. The cells were then fixed and subjected to immunofluorescence analysis. The green signal represents tricellulin and the gray signal represents zonula occludens-1. Bars = 5 μm. The figure is reproduced from reference [50] with slight modifications and permission from the copyright holder.

4.2. Epidermal Absorption

The epidermis covers the outer body, preventing the passage of solutes and the absorption of drugs. However, epidermal administration is a potentially useful route of administration because it is noninvasive, can easily be stopped, and avoids first-pass metabolism [71]. The epidermal barrier comprises the stratum corneum and TJs in the stratum granulosum [72]. The analyses using knockout mice have revealed that claudin-1 is critical for TJ integrity in the stratum granulosum [73]. The treatment with an anti-claudin-1 mAb (7A5) weakened TJ integrity and enhanced the permeation of dextran (4 kDa) in an in vitro human epidermal model [53]. Thus, claudin-1-mediated modulation of the permeability of TJs in the stratum granulosum is a promising means of increasing epidermal drug absorption.

4.3. Cancer Targeting

Claudins are aberrantly expressed in many malignant tumors [74]. For example, in epithelium-derived tumors, TJ strands are disorganized and TJ proteins are distributed throughout the cell surface [75]. Claudin-3 and -4 are the most frequently overexpressed claudins in malignant tumors of prostate [76], breast [77], pancreatic [78], and ovarian cancers [79]. Claudin-1 and -2 are frequently overexpressed in colorectal cancers [80,81] and are associated with the promotion of tumor proliferation and invasiveness via the activation of intracellular signaling cascades [82–84].

The various claudin-targeting molecules, including toxins, toxin fragments, and antibodies, have been generated for claudin-targeted cancer therapy [32,33,85]. For example, CPE has been used as an anti-cancer agent against pancreatic cancer overexpressing claudin-4 [86]. One issue with claudin-targeted therapies is that claudins are expressed not only in malignant tissues, but also in non-malignant tissues. However, most claudins in non-malignant tissues are localized within TJ complexes, whereas their localization is often dysregulated from TJ complexes to the cell surface in malignant tissues [87,88]. This study found that C-CPE fused with protein synthesis inhibitory factor (C-CPE-PSIF) may recognize claudins with aberrant localization, resulting in less binding and therefore, less toxicity to the normal cells [89]. Caco-2 is a human colon carcinoma cell line. Caco-2 cells form a polarized cell monolayer with well-developed TJs when confluent, and they are frequently used as a model of polarized normal epithelial cells. The claudin-4 protein level in the confluent culture (normal epithelial-like cells) was higher than in the subconfluent culture (carcinoma cells), and C-CPE-PSIF was cytotoxic to preconfluent (immature TJs) but not to postconfluent Caco-2 cells (mature TJs) (Figure 4) [89,90]. Similarly, anti-claudin-4 mAbs systemically administered to mice preferentially accumulated in tumor tissue rather than in normal tissue [67,91]. Together, these results suggest that claudins are potential targets for the development of cancer-targeted therapies.

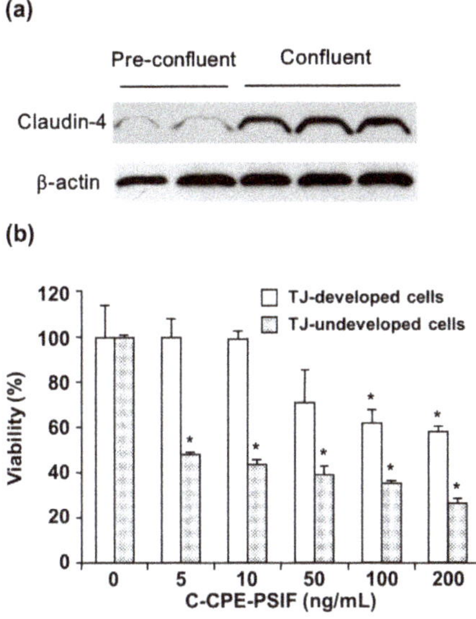

Figure 4. The effect of tight junction (TJ) maturity on the cytotoxicity of C-CPE fused with protein synthesis inhibitory factor (C-CPE-PSIF). Caco-2 cells were cultured to confluence or to preconfluence for 3 days to obtain cells with mature or immature TJs, respectively. (**a**) The cell lysates were subjected to western blotting. (**b**) The cells were treated with the indicated concentrations of C-CPE-PSIF for 48 h, and then cell viability was measured by WST-8 assay. The data are representative of at least three independent experiments. The data are shown as the mean ± S.D. ($n = 3$). * $p < 0.05$. The data are reproduced from reference [89] with slight modifications and permission from the copyright holder.

4.4. Targeting Tissues Involved in Immunological Processes

Mucosal vaccination may be a useful immunization strategy because it is non-invasive and it activates both the mucosal and systemic immune responses. Epithelial cells associated with the mucosa-associated lymphoid tissues (MALT) include Peyer's patches and nasopharynx and play pivotal roles in preventing the invasion of pathological microorganisms into the body by inducing the secretion of IgA [92]. MALT comprises various immune cells, including T cells, B cells, and dendritic cells, and is covered by follicle-associated epithelium. M cells are specialized epithelial cells in the follicle-associated epithelium that transport luminal antigens to immune cells in MALT by transcytosis [93]. In general, when antigen alone is orally or nasally administered, it fails to reach the MALT and so immune responses are not induced. Thus, the efficient delivery of antigen to MALT may provide effective mucosal vaccines.

Follicle-associated epithelium contains claudin-4-expressing cells, some of which are highly capable of capturing luminal antigen [94,95]. Claudin-4 is also expressed on the luminal surface of M cells [96]. Thus, claudin-4 targeting may be a promising strategy for delivering antigens to MALT. A nasally administered ovalbumin fused with C-CPE induced mucosal IgA production, systemic IgG production, and antigen-specific immune responses for preventing tumor growth (Figure 5) [97]. Of note, a simple mixture of C-CPE and antigen did not induce IgA production, indicating that the vaccination efficacy may be depending on the binding affinity of the C-CPE to claudins [97]. Nasal immunization with chimeric C-CPE-antigen did not induce mucosal injury [98]. The augmentation of the antigenicity of the first-generation binder C-CPE has been used to develop an adjuvant-free bivalent food poisoning vaccine [99]

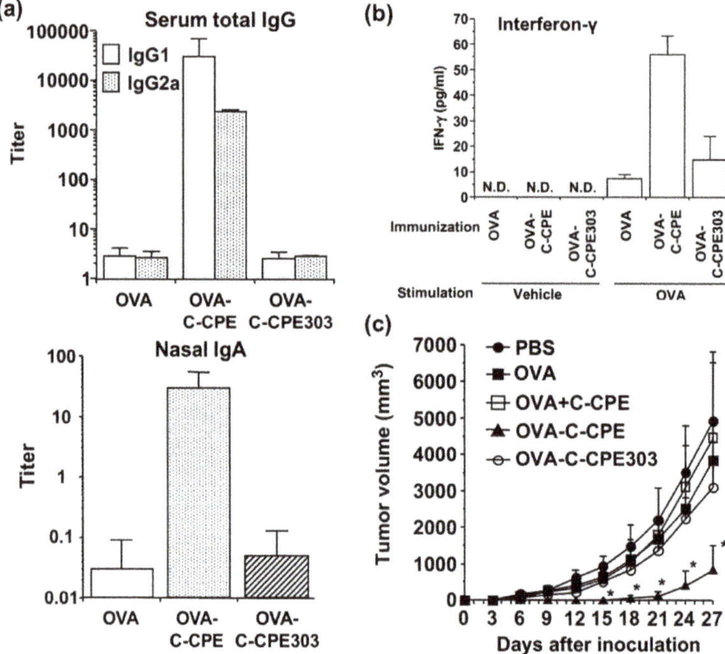

Figure 5. The effect of claudin-4-targeted mucosal immunization on systemic and mucosal immunity. (a and b) Mice were nasally immunized with ovalbumin (OVA), OVA-C-CPE, or OVA-C-CPE303 (which lacks the minimal claudin-binding region) (5 μg OVA in each formulation) once a week for 3 weeks. Seven days after the last immunization, serum and splenocytes were harvested. (**a**) The levels of serum IgG1 and IgG2a (upper panel) and nasal IgA (lower panel) were measured. (**b**) The splenocytes were stimulated with vehicle or OVA (1 mg/mL) for 24 h, and interferon-γ in the supernatant was measured. (c) Mice were nasally immunized with vehicle, OVA, a mixture of OVA and C-CPE, OVA-C-CPE, or OVA-C-CPE303 (5 μg OVA in each vaccine) once a week for 3 weeks. Seven days after the final immunization, the mice were injected subcutaneously with 1×10^6 OVA-expressing EL4 (H-2b) cells. The tumor volumes were measured over time. The data are shown as the mean ± S.D. (n = 4). * $p < 0.05$. The data are reproduced from reference [97] with slight modifications and permission from the copyright holder.

4.5. Targeting Inflamed Tissues

Ulcerative colitis is a chronic, relapsing inflammatory bowel disease characterized by severe diarrhea and mucosal inflammation in the colon. The disruption of the colonic mucosal barrier leads to the activation of immune responses against bacterial and food fragments in the colon, followed by the development of ulcerative colitis [2,3]. Although claudin-2 is rarely expressed in normal colonic epithelial cells, its expression in the colon is upregulated in ulcerative colitis patients [80]. Inflammatory cytokines, including tumor necrosis factor-α (TNF-α), decrease the epithelial barrier integrity and upregulate the expression of claudin-2 [100]. Claudin-2 decreases the integrity of TJs by facilitating the formation of discontinuous TJ strands [101]. This suggests that the inhibition of claudin-2 may restore the disrupted mucosal barrier. Indeed, an anti-claudin-2 mAb (1A2) ameliorated TNF-α-induced reduction of TJ integrity in Caco-2 cells. Moreover, the co-treatment of the cells with anti-claudin-2 mAb and an anti-TNF-α mAb showed an additive effect on the restoration of the barrier [56].

4.6. Drug Deliavery to the Brain

Unlike peripheral capillaries, those in the brain lack fenestrations and have well-developed TJs that form the blood–brain barrier (BBB). More than 98% of small-molecular-weight drugs cannot pass the BBB [102]. Claudin-5 and angulin-1 are abundantly expressed by brain endothelial cells in mice [103]. Claudin-5- or angulin-1-knockout mice have a size-selectively loosened BBB [104,105]. These data suggest that claudin-5 and angulin-1 are candidate targets for opening the BBB. Indeed, a C-CPE mutant that can bind to claudin-5, angubindin-1, and anti-claudin-5 mAb was able to reduce the transepithelial/transendothelial electrical resistance (TER) in an in vitro model of the BBB [54,106]. Furthermore, in the mice, an angubindin-1-, but not claudin-5-binding C-CPE mutant increased the permeability of the BBB to allow passage of a 16-mer gapmer antisense oligonucleotide (5.3 kDa) [106]. No obvious adverse effects were observed in the mice in these experiments.

5. Safety of Claudin- and Angulin-Targeted Therapies

A series of proof-of-concept studies examining claudin and angulin targeting has provided insights into enhancing drug absorption, treating cancer and inflammatory diseases, improving vaccines, and obtaining drug delivery to the brain. No apparent adverse effects were observed in these studies. However, claudins and angulins play roles in the formation of the intercellular seal between and among epithelial cells and endothelial cells in many tissues. Therefore, ensuring the safety of claudin- and angulin-targeted drugs is critical for future drug development.

The knockout and knockdown analyses of the genes encoding claudins and angulins have shown that there are risks associated with claudin- and angulin-targeted therapeutics (Table 3). For instance, the inhibitors of claudin-1 and -5 may induce atopic dermatitis and schizophrenia-like symptoms via the inhibition of the epidermal barrier and the BBB, respectively [4,6]. Claudin-2, -4, and angulin-2-targeted drugs may induce renal impairment with respect to the reabsorption of ions and water [107–109]. A deletion in exon 1 of claudin-1 results in neonatal ichthyosis and sclerosing cholangitis syndrome in humans [110]. A deletion of 1.5 to 3.0 Mb of human chromosome 22q11.2 that includes the *claudin-5* gene is associated with the development of schizophrenia [111]. A single nucleotide polymorphism in *claudin-5* is also associated with the development of schizophrenia [112,113].

Table 3. Phenotypes of representative claudin- or angulin-knockout or -knockdown mice.

	Phenotype of Knockout (KO) or Knockdown (KD) Mice	Ref.
Claudin-1	Atopic dermatitis (KD)	[4]
Claudin-2	Impaired renal Na^+, Cl^-, and water reabsorption (KO)	[108]
Claudin-3	Increased hepatocyte permeability to phosphate ion (KO)	[114]
Claudin-4	Impaired renal Ca^{2+} and Cl^- reabsorption (KO)	[107]
Claudin-5	Schizophrenia-like symptoms (KD)	[6]
Angulin-2	Impaired renal water reabsorption and colonic water absorption (KO)	[109]

The expression profiles of claudins and angulins differ among tissues (Table 4). The specific claudin ratio is critical for the functions of TJs [115]. Thus, the toxicity of claudin-directed drugs should be carefully investigated, especially if the target claudins are also expressed in non-target tissues. Claudins and angulins in TJs are embedded in the lateral cell membranes and extend into the intercellular space. The TJ cavity is estimated to be 0.5 nm under physiological conditions [116,117]. Large binders, such as antibodies, which are unable to access proteins embedded in TJs, are a promising modality for treating cancers and for improving the effectiveness of vaccines because in these conditions the target claudins are exposed on the cell surfaces [34,87,88,96]. TJ components are potent targets for the development of many novel therapies, but targets and drug modalities must be optimized to afford an acceptable risk–benefit balance.

Table 4. Claudin and angulin expression in representative tissues in a mouse or rat.

	Claudin					Angulin			Ref.
	1	2	3	4	5	1	2	3	
Epidermal cells (stratum granulosum)	+	-	-	+	-	+	-	-	[29,73]
Nasal epithelial cells	+	-	+	+	+	+	+	-	[29,118]
Lung (alveoli)	+	-	+	+	+	+	-	-	[29,119]
Small intestine (jejunum)	+	+	+	-	+	+	-	-	[29,109,120]
Colon (surface)	+	-	+	+	+	-	+	-	[29,109,120]
Liver	+	+	+	-	-	+	-	-	[29,121,122]
Kidney (glomerulus)	+	+	-	-	-	-	+	-	[29,123]
Kidney (proximal tube)	-	+	-	-	-	+	-	-	[29,123]
Kidney (thin ascending limb of the loop of Henle)	-	-	+	+	-	-	+	-	[29,123]
Kidney (collecting duct)	-	-	+	+	-	-	+	-	[29,109,123]
Brain endothelial cells	-	-	-	-	+	+	-	+	[103,105]
Brain ependymal cells	+	+	+	-	-	-	-	+	[29,124]
Lung endothelial cells	-	-	-	-	+	-	-	-	[125]

+, expressed; -, not detected.

6. Conclusions

TJ binders are classified as first-generation binders (toxins and their fragments), and second-generation binders (antibodies) [32,33,85]. The augmentation of the antigenicity of the first-generation binder, C-CPE, has been used to develop an adjuvant-free vaccine [97,99]. Second-generation binders are being used to develop cancer therapies (Table 2). For example, an anti-claudin-18.2 mAb is undergoing clinical study for the use in the treatment of gastric (phase III study) and pancreatic cancer (phase II study) [NCT03504397; NCT03816163].

The other application of TJ binders is to enhance the mucosal and epidermal absorption of drugs and to deliver drugs to the brain by modulating the permeability of TJs. The currently available TJ binders are toxin fragments and antibodies. However, the generation of novel TJ binders, such as peptides and chemicals, is now needed because of the potential antigenicity of toxins and the costs associated with antibody preparation. A high-throughput screening system for claudin-4 binders based on the time resolved fluorescence resonance energy transfer in a chemical library was developed [126]. In the future, the generations of peptide- and chemical-type of binders are expected to accelerate.

Author Contributions: Wrote or contributed to the writing of the manuscript: Y.H., K.T., S.M.K., J.K., M.F., and M.K.

Funding: This study was supported in part by Grants-in-Aid for Scientific Research from the Ministry of Education, Culture, Sports, Science and Technology of Japan (19H04468, 18K19400, 18H03190, 16K13044, and 24390042); a grant for Research on Development of New Drugs from the Japan Agency for Medical Research and Development (AMED); Co-Create Knowledge for Pharma Innovation with Takeda (COCKPI-T); Project MEET, Osaka University Graduate School of Medicine and Mitsubishi Tanabe Pharma Corporation; a grant of Research Support from Astellas; and grants from the Platform Project for Supporting Drug Discovery and Life Science Research (Basis for Supporting Innovative Drug Discovery and Life Science Research [BINDS]) of AMED (JP19am0101077, JP19am0101084, JP19am0101090).

Acknowledgments: We thank all members of Kondoh Lab and Kunisawa Lab for their useful comments and discussions.

Conflicts of Interest: The authors declare no conflicts of interest.

Abbreviations

BBB	Blood–brain barrier
CPE	Clostridium perfringens enterotoxin
CPE-R	CPE receptor
C-CPE	C-terminal domain of CPE
C-CPE-PSIF	C-CPE fused with protein synthesis inhibitory factor
ECH	Extracellular helix
KD	Knockdown
KO	Knockout
mAb	Monoclonal antibody
MALT	Mucosa-associated lymphoid tissues
RVP1	Rat androgen withdrawal apoptosis protein
TER	Transepithelial/transendothelial electrical resistance
TJ	Tight junction
TM	Transmembrane
TNF-α	Tumor necrosis factor-α

References

1. Farquhar, M.G.; Palade, G.E. Junctional complexes in various epithelia. *J. Cell Biol.* **1963**, *17*, 375–412. [CrossRef] [PubMed]
2. de Souza, H.S.P.; Fiocchi, C. Immunopathogenesis of IBD: Current state of the art. *Nat. Rev. Gastro. Hepat.* **2016**, *13*, 13–27. [CrossRef] [PubMed]
3. Madsen, K.L.; Malfair, D.; Gray, D.; Doyle, J.S.; Jewell, L.D.; Fedorak, R.N. Interleukin-10 gene-deficient mice develop a primary intestinal permeability defect in response to enteric microflora. *Inflamm. Bowel Dis.* **1999**, *5*, 262–270. [CrossRef] [PubMed]
4. Tokumasu, R.; Yamaga, K.; Yamazaki, Y.; Murota, H.; Suzuki, K.; Tamura, A.; Bando, K.; Furuta, Y.; Katayama, I.; Tsukita, S. Dose-dependent role of claudin-1 in vivo in orchestrating features of atopic dermatitis. *Proc. Natl. Acad. Sci. USA* **2016**, *113*, E4061–E4068. [CrossRef] [PubMed]
5. Menard, C.; Pfau, M.L.; Hodes, G.E.; Kana, V.; Wang, V.X.; Bouchard, S.; Takahashi, A.; Flanigan, M.E.; Aleyasin, H.; LeClair, K.B.; et al. Social stress induces neurovascular pathology promoting depression. *Nat. Neurosci.* **2017**, *20*, 1752–1760. [CrossRef] [PubMed]
6. Greene, C.; Kealy, J.; Humphries, M.M.; Gong, Y.; Hou, J.; Hudson, N.; Cassidy, L.M.; Martiniano, R.; Shashi, V.; Hooper, S.R.; et al. Dose-dependent expression of claudin-5 is a modifying factor in schizophrenia. *Mol. Psychiatry* **2018**, *23*, 2156–2166. [CrossRef] [PubMed]
7. Staehelin, L.A.; Mukherjee, T.M.; Williams, A.W. Freeze-etch appearance of tight junctions in epithelium of small and large intestine mice. *Protoplasma* **1969**, *67*, 165–184. [CrossRef] [PubMed]
8. Van Itallie, C.M.; Tietgens, A.J.; Anderson, J.M. Visualizing the dynamic coupling of claudin strands to the actin cytoskeleton through ZO-1. *Mol. Biol. Cell* **2017**, *28*, 524–534. [CrossRef]
9. Umeda, K.; Ikenouchi, J.; Katahira-Tayama, S.; Furuse, K.; Sasaki, H.; Nakayama, M.; Matsui, T.; Tsukita, S.; Furuse, M.; Tsukita, S. ZO-1 and ZO-2 independently determine where claudins are polymerized in tight-junction strand formation. *Cell* **2006**, *126*, 741–754. [CrossRef]
10. Cording, J.; Berg, J.; Kading, N.; Bellmann, C.; Tscheik, C.; Westphal, J.K.; Milatz, S.; Günzel, D.; Wolburg, H.; Piontek, J.; et al. In tight junctions, claudins regulate the interactions between occludin, tricellulin and marvelD3, which, inversely, modulate claudin oligomerization. *J. Cell Sci.* **2013**, *126*, 554–564. [CrossRef]
11. Yamazaki, Y.; Tokumasu, R.; Kimura, H.; Tsukita, S. Role of claudin species-specific dynamics in reconstitution and remodeling of the zonula occludens. *Mol. Biol. Cell* **2011**, *22*, 1495–1504. [CrossRef] [PubMed]
12. Rossa, J.; Ploeger, C.; Vorreiter, F.; Saleh, T.; Protze, J.; Günzel, D.; Wolburg, H.; Krause, G.; Piontek, J. Claudin-3 and claudin-5 protein folding and assembly into the tight junction are controlled by non-conserved residues in the transmembrane 3 (TM3) and extracellular loop 2 (ECL2) segments. *J. Biol. Chem.* **2014**, *289*, 7641–7653. [CrossRef] [PubMed]
13. Tervonen, A.; Ihalainen, T.O.; Nymark, S.; Hyttinen, J. Structural dynamics of tight junctions modulate the properties of the epithelial barrier. *PLoS ONE* **2019**, *14*, e0214876. [CrossRef] [PubMed]

14. Staehelin, L.A. Further observations on fine-structure of freeze-cleaved tight junctions. *J. Cell Sci.* **1973**, *13*, 763–786. [PubMed]
15. Lindmark, T.; Söderholm, J.D.; Olaison, G.; Alvan, G.; Ocklind, G.; Artursson, P. Mechanism of absorption enhancement in humans after rectal administration of ampicillin in suppositories containing sodium caprate. *Pharm. Res.* **1997**, *14*, 930–935. [CrossRef] [PubMed]
16. Doolittle, N.D.; Miner, M.E.; Hall, W.A.; Siegal, T.; Hanson, E.J.; Osztie, E.; McAllister, L.D.; Bubalo, J.S.; Kraemer, D.F.; Fortin, D.; et al. Safety and efficacy of a multicenter study using intraarterial chemotherapy in conjunction with osmotic opening of the blood-brain barrier for the treatment of patients with malignant brain tumors. *Cancer* **2000**, *88*, 637–647. [CrossRef]
17. Kondoh, M.; Masuyama, A.; Takahashi, A.; Asano, N.; Mizuguchi, H.; Koizumi, N.; Fujii, M.; Hayakawa, T.; Horiguchi, Y.; Watanbe, Y. A novel strategy for the enhancement of drug absorption using a claudin modulator. *Mol. Pharmacol.* **2005**, *67*, 749–756. [CrossRef]
18. Rapoport, S. Osmotic opening of the blood-brain barrier: Principles, mechanism, and therapeutic applications. *Cell. Mol. Neurobiol.* **2000**, *20*, 217–230. [CrossRef]
19. Furuse, M.; Fujita, K.; Hiiragi, T.; Fujimoto, K.; Tsukita, S. Claudin-1 and -2: Novel integral membrane proteins localizing at tight junctions with no sequence similarity to occludin. *J. Cell Biol.* **1998**, *141*, 1539–1550. [CrossRef]
20. Mineta, K.; Yamamoto, Y.; Yamazaki, Y.; Tanaka, H.; Tada, Y.; Saito, K.; Tamura, A.; Igarashi, M.; Endo, T.; Takeuchi, K.; et al. Predicted expansion of the claudin multigene family. *FEBS Lett.* **2011**, *585*, 606–612. [CrossRef]
21. Suzuki, H.; Nishizawa, T.; Tani, K.; Yamazaki, Y.; Tamura, A.; Ishitani, R.; Dohmae, N.; Tsukita, S.; Nureki, O.; Fujiyoshi, Y. Crystal structure of a claudin provides insight into the architecture of tight junctions. *Science* **2014**, *344*, 304–307. [CrossRef] [PubMed]
22. Saitoh, Y.; Suzuki, H.; Tani, K.; Nishikawa, K.; Irie, K.; Ogura, Y.; Tamura, A.; Tsukita, S.; Fujiyoshi, Y. Structural insight into tight junction disassembly by *Clostridium perfringens* enterotoxin. *Science* **2015**, *347*, 775–778. [CrossRef] [PubMed]
23. Shinoda, T.; Shinya, N.; Ito, K.; Ohsawa, N.; Terada, T.; Hirata, K.; Kawano, Y.; Yamamoto, M.; Kimura-Someya, T.; Yokoyama, S.; et al. Structural basis for disruption of claudin assembly in tight junctions by an enterotoxin. *Sci. Rep.* **2016**, *6*, 33632. [CrossRef] [PubMed]
24. Nakamura, S.; Irie, K.; Tanaka, H.; Nishikawa, K.; Suzuki, H.; Saitoh, Y.; Tamura, A.; Tsukita, S.; Fujiyoshi, Y. Morphologic determinant of tight junctions revealed by claudin-3 structures. *Nat. Commun.* **2019**, *10*, 816. [CrossRef] [PubMed]
25. Piontek, J.; Winkler, L.; Wolburg, H.; Müller, S.L.; Zuleger, N.; Piehl, C.; Wiesner, B.; Krause, G.; Blasig, I.E. Formation of tight junction: Determinants of homophilic interaction between classic claudins. *FASEB J.* **2008**, *22*, 146–158. [CrossRef] [PubMed]
26. Suzuki, H.; Tani, K.; Tamura, A.; Tsukita, S.; Fujiyoshi, Y. Model for the architecture of claudin-based paracellular ion channels through tight junctions. *J. Mol. Biol.* **2015**, *427*, 291–297. [CrossRef] [PubMed]
27. Ikenouchi, J.; Furuse, M.; Furuse, K.; Sasaki, H.; Tsukita, S.; Tsukita, S. Tricellulin constitutes a novel barrier at tricellular contacts of epithelial cells. *J. Cell Biol.* **2005**, *171*, 939–945. [CrossRef] [PubMed]
28. Masuda, S.; Oda, Y.; Sasaki, H.; Ikenouchi, J.; Higashi, T.; Akashi, M.; Nishi, E.; Furuse, M. LSR defines cell corners for tricellular tight junction formation in epithelial cells. *J. Cell Sci.* **2011**, *124*, 548–555. [CrossRef]
29. Higashi, T.; Tokuda, S.; Kitajiri, S.; Masuda, S.; Nakamura, H.; Oda, Y.; Furuse, M. Analysis of the 'angulin' proteins LSR, ILDR1 and ILDR2-tricellulin recruitment, epithelial barrier function and implication in deafness pathogenesis. *J. Cell Sci.* **2013**, *126*, 966–977. [CrossRef]
30. Nakatsu, D.; Kano, F.; Taguchi, Y.; Sugawara, T.; Nishizono, T.; Nishikawa, K.; Oda, Y.; Furuse, M.; Murata, M. JNK1/2-dependent phosphorylation of angulin-1/LSR is required for the exclusive localization of angulin-1/LSR and tricellulin at tricellular contacts in EpH4 epithelial sheet. *Genes Cells* **2014**, *19*, 565–581. [CrossRef]
31. Kim, N.K.D.; Higashi, T.; Lee, K.Y.; Kim, A.R.; Kitajiri, S.; Kim, M.Y.; Chang, M.Y.; Kim, V.; Oh, S.H.; Kim, D.; et al. Downsloping high-frequency hearing loss due to inner ear tricellular tight junction disruption by a novel ILDR1 mutation in the Ig-like domain. *PLoS ONE* **2015**, *10*, e0116931. [CrossRef] [PubMed]
32. Hashimoto, Y.; Yagi, K.; Kondoh, M. Current progress in a second-generation claudin binder, anti-claudin antibody, for clinical applications. *Drug Discov. Today* **2016**, *21*, 1711–1718. [CrossRef] [PubMed]

33. Hashimoto, Y.; Yagi, K.; Kondoh, M. Roles of the first-generation claudin binder, *Clostridium perfringens* enterotoxin, in the diagnosis and claudin-targeted treatment of epithelium-derived cancers. *Pflugers Arch.* **2017**, *469*, 45–53. [CrossRef] [PubMed]
34. Hashimoto, Y.; Okada, Y.; Shirakura, K.; Tachibana, K.; Sawada, M.; Yagi, K.; Doi, T.; Kondoh, M. Anti-claudin antibodies as a concept for development of claudin-directed drugs. *J. Pharmacol. Exp. Ther.* **2019**, *368*, 179–186. [CrossRef] [PubMed]
35. Kitadokoro, K.; Nishimura, K.; Kamitani, S.; Fukui-Miyazaki, A.; Toshima, H.; Abe, H.; Kamata, Y.; Sugita-Konishi, Y.; Yamamoto, S.; Karatani, H.; et al. Crystal structure of *Clostridium perfringens* enterotoxin displays features of beta-pore-forming toxins. *J. Biol. Chem.* **2011**, *286*, 19549–19555. [CrossRef] [PubMed]
36. Katahira, J.; Inoue, N.; Horiguchi, Y.; Matsuda, M.; Sugimoto, N. Molecular cloning and functional characterization of the receptor for Clostridium perfringensenterotoxin. *J. Cell Biol.* **1997**, *136*, 1239–1247. [CrossRef] [PubMed]
37. Morita, K.; Furuse, M.; Fujimoto, K.; Tsukita, S. Claudin multigene family encoding four-transmembrane domain protein components of tight junction strands. *Proc. Natl. Acad. Sci. USA* **1999**, *96*, 511–516. [CrossRef]
38. Sonoda, N.; Furuse, M.; Sasaki, H.; Yonemura, S.; Katahira, J.; Horiguchi, Y.; Tsukita, S. *Clostridium perfringens* enterotoxin fragment removes specific claudins from tight junction strands: Evidence for direct involvement of claudins in tight junction barrier. *J. Cell Biol.* **1999**, *147*, 195–204. [CrossRef]
39. Winkler, L.; Gehring, C.; Wenzel, A.; Müller, S.L.; Piehl, C.; Krause, G.; Blasig, I.E.; Piontek, J. Molecular determinants of the interaction between *Clostridium perfringens* enterotoxin fragments and claudin-3. *J. Biol. Chem.* **2009**, *284*, 18863–18872. [CrossRef]
40. Fujita, K.; Katahira, J.; Horiguchi, Y.; Sonoda, N.; Furuse, M.; Tsukita, S. *Clostridium perfringens* enterotoxin binds to the second extracellular loop of claudin-3, a tight junction integral membrane protein. *FEBS Lett.* **2000**, *476*, 258–261. [CrossRef]
41. Uchida, H.; Kondoh, M.; Hanada, T.; Takahashi, A.; Hamakubo, T.; Yagi, K. A claudin-4 modulator enhances the mucosal absorption of a biologically active peptide. *Biochem. Pharmacol.* **2010**, *79*, 1437–1444. [CrossRef] [PubMed]
42. Takahashi, A.; Komiya, E.; Kakutani, H.; Yoshida, T.; Fujii, M.; Horiguchi, Y.; Mizuqchi, H.; Tsutsumi, Y.; Tsunoda, S.I.; Koizumi, N.; et al. Domain mapping of a claudin-4 modulator, the C-terminal region of C-terminal fragment of *Clostridium perfringens* enterotoxin, by site-directed mutagenesis. *Biochem. Pharmacol.* **2008**, *75*, 1639–1648. [CrossRef] [PubMed]
43. Takahashi, A.; Saito, Y.; Kondoh, M.; Matsushita, K.; Krug, S.M.; Suzuki, H.; Tsujino, H.; Li, X.R.; Aoyama, H.; Matsuhisa, K.; et al. Creation and biochemical analysis of a broad-specific claudin binder. *Biomaterials* **2012**, *33*, 3464–3474. [CrossRef] [PubMed]
44. Veshnyakova, A.; Piontek, J.; Protze, J.; Waziri, N.; Heise, I.; Krause, G. Mechanism of *Clostridium perfringens* enterotoxin interaction with claudin-3/-4 protein suggests structural modifications of the toxin to target specific claudins. *J. Biol. Chem.* **2012**, *287*, 1698–1708. [CrossRef] [PubMed]
45. Protze, J.; Eichner, M.; Piontek, A.; Dinter, S.; Rossa, J.; Blecharz, K.C.Z.; Vajkoczy, P.; Piontek, J.; Krause, G. Directed structural modification of *Clostridium perfringens* enterotoxin to enhance binding to claudin-5. *Cell. Mol. Life Sci.* **2015**, *72*, 1417–1432. [CrossRef] [PubMed]
46. Neuhaus, W.; Piontek, A.; Protze, J.; Eichner, M.; Mahringer, A.; Subileau, E.A.; Lee, I.F.M.; Schulzke, J.D.; Krause, G.; Piontek, J. Reversible opening of the blood-brain barrier by claudin-5-binding variants of *Clostridium perfringens* enterotoxin's claudin-binding domain. *Biomaterials* **2018**, *161*, 129–143. [CrossRef]
47. Nagahama, M.; Umezaki, M.; Oda, M.; Kobayashi, K.; Tone, S.; Suda, T.; Ishidoh, K.; Sakurai, J. *Clostridium perfringens* iota-toxin b induces rapid cell necrosis. *Infect. Immun.* **2011**, *79*, 4353–4360. [CrossRef]
48. Nagahama, M.; Yamaguchi, A.; Hagiyama, T.; Ohkubo, N.; Kobayashi, K.; Sakurai, J. Binding and internalization of *Clostridium perfringens* iota-toxin in lipid rafts. *Infect. Immun.* **2004**, *72*, 3267–3275. [CrossRef]
49. Papatheodorou, P.; Carette, J.E.; Bell, G.W.; Schwan, C.; Guttenberg, G.; Brummelkamp, T.R.; Aktories, K. Lipolysis-stimulated lipoprotein receptor (LSR) is the host receptor for the binary toxin *Clostridium difficile* transferase (CDT). *Proc. Natl. Acad. Sci. USA* **2011**, *108*, 16422–16427. [CrossRef]
50. Krug, S.M.; Hayaishi, T.; Iguchi, D.; Watari, A.; Takahashi, A.; Fromm, M.; Nagahama, M.; Takeda, H.; Okada, Y.; Sawasaki, T.; et al. Angubindin-1, a novel paracellular absorption enhancer acting at the tricellular tight junction. *J. Control. Release* **2017**, *260*, 1–11. [CrossRef]

51. Souriau, C.; Hudson, P.J. Recombinant antibodies for cancer diagnosis and therapy. *Expert Opin. Biol. Ther.* **2003**, *3*, 305–318. [CrossRef] [PubMed]
52. Espiritu, M.J.; Collier, A.C.; Bingham, J.P. A 21st-century approach to age-old problems: The ascension of biologics in clinical therapeutics. *Drug Discov. Today* **2014**, *19*, 1109–1113. [CrossRef] [PubMed]
53. Nakajima, M.; Nagase, S.; Iida, M.; Takeda, S.; Yamashita, M.; Watari, A.; Shirasago, Y.; Fukasawa, M.; Takeda, H.; Sawasaki, T.; et al. Claudin-1 binder enhances epidermal permeability in a human keratinocyte model. *J. Pharmacol. Exp. Ther.* **2015**, *354*, 440–447. [CrossRef] [PubMed]
54. Hashimoto, Y.; Shirakura, K.; Okada, Y.; Takeda, H.; Endo, K.; Tamura, M.; Watari, A.; Sadamura, Y.; Sawasaki, T.; Doi, T.; et al. Claudin-5-binders enhance permeation of solutes across the blood-brain barrier in a mammalian model. *J. Pharmacol. Exp. Ther.* **2017**, *363*, 275–283. [CrossRef] [PubMed]
55. Hashimoto, Y.; Zhou, W.; Hamauchi, K.; Shirakura, K.; Doi, T.; Yagi, K.; Sawasaki, T.; Okada, Y.; Kondoh, M.; Takeda, H. Engineered membrane protein antigens successfully induce antibodies against extracellular regions of claudin-5. *Sci. Rep.* **2018**, *8*, 8383. [CrossRef]
56. Takigawa, M.; Iida, M.; Nagase, S.; Suzuki, H.; Watari, A.; Tada, M.; Okada, Y.; Doi, T.; Fukasawa, M.; Yagi, K.; et al. Creation of a claudin-2 binder and its tight junction-modulating activity in a human intestinal models. *J. Pharmacol. Exp. Ther.* **2017**, *363*, 444–451. [CrossRef] [PubMed]
57. Fofana, I.; Krieger, S.E.; Grunert, F.; Glauben, S.; Xiao, F.; Fafi-Kremer, S.; Soulier, E.; Royer, C.; Thumann, C.; Mee, C.J.; et al. Monoclonal anti-claudin 1 antibodies prevent hepatitis C virus infection of primary human hepatocytes. *Gastroenterology* **2010**, *139*, 953–964. [CrossRef]
58. Fukasawa, M.; Nagase, S.; Shirasago, Y.; Iida, M.; Yamashita, M.; Endo, K.; Yagi, K.; Suzuki, T.; Wakita, T.; Hanada, K.; et al. Monoclonal antibodies against extracellular domains of claudin-1 block hepatitis C virus infection in a mouse model. *J. Virol.* **2015**, *89*, 4866–4879. [CrossRef]
59. Shimizu, Y.; Shirasago, Y.; Kondoh, M.; Suzuki, T.; Wakita, T.; Hanada, K.; Yagi, K.; Fukasawa, M. Monoclonal antibodies against occludin completely prevented hepatitis C virus infection in a mouse model. *J. Virol.* **2018**, *92*. [CrossRef]
60. Okai, K.; Ichikawa-Tomikawa, N.; Saito, A.C.; Watabe, T.; Sugimoto, K.; Fujita, D.; Ono, C.; Fukuhara, T.; Matsuura, Y.; Ohira, H.; et al. A novel occludin-targeting monoclonal antibody prevents hepatitis C virus infection in vitro. *Oncotarget* **2018**, *9*, 16588–16598. [CrossRef]
61. Hashimoto, Y.; Tada, M.; Iida, M.; Nagase, S.; Hata, T.; Watari, A.; Okada, Y.; Doi, T.; Fukasawa, M.; Yagi, K.; et al. Generation and characterization of a human-mouse chimeric antibody against the extracellular domain of claudin-1 for cancer therapy using a mouse model. *Biochem. Biophys. Res. Commun.* **2016**, *477*, 91–95. [CrossRef] [PubMed]
62. Cherradi, S.; Ayrolles-Torro, A.; Vezzo-Vie, N.; Gueguinou, N.; Denis, V.; Combes, E.; Boissiere, F.; Busson, M.; Canterel-Thouennon, L.; Mollevi, C.; et al. Antibody targeting of claudin-1 as a potential colorectal cancer therapy. *J. Exp. Clin. Cancer Res.* **2017**, *36*, 89. [CrossRef] [PubMed]
63. Hashimoto, Y.; Hata, T.; Tada, M.; Iida, M.; Watari, A.; Okada, Y.; Doi, T.; Kuniyasu, H.; Yagi, K.; Kondoh, M. Safety evaluation of a human chimeric monoclonal antibody that recognizes the extracellular loop domain of claudin-2. *Eur. J. Pharm. Sci.* **2018**, *117*, 161–167. [CrossRef] [PubMed]
64. Ando, H.; Suzuki, M.; Kato-Nakano, M.; Kawamoto, S.; Misaka, H.; Kimoto, N.; Furuya, A.; Nakamura, K. Generation of specific monoclonal antibodies against the extracellular loops of human claudin-3 by immunizing mice with target-expressing cells. *Biosci. Biotechnol. Biochem.* **2015**, *79*, 1272–1279. [CrossRef] [PubMed]
65. Romani, C.; Cocco, E.; Bignotti, E.; Moratto, D.; Bugatti, A.; Todeschini, P.; Bandiera, E.; Tassi, R.; Zanotti, L.; Pecorelli, S.; et al. Evaluation of a novel human IgG1 anti-claudin3 antibody that specifically recognizes its aberrantly localized antigen in ovarian cancer cells and that is suitable for selective drug delivery. *Oncotarget* **2015**, *6*, 34617–34628. [CrossRef]
66. Suzuki, M.; Kato-Nakano, M.; Kawamoto, S.; Furuya, A.; Abe, Y.; Misaka, H.; Kimoto, N.; Nakamura, K.; Ohta, S.; Ando, H. Therapeutic antitumor efficacy of monoclonal antibody against Claudin-4 for pancreatic and ovarian cancers. *Cancer Sci.* **2009**, *100*, 1623–1630. [CrossRef]
67. Hashimoto, Y.; Kawahigashi, Y.; Hata, T.; Li, X.; Watari, A.; Tada, M.; Ishii-Watabe, A.; Okada, Y.; Doi, T.; Fukasawa, M.; et al. Efficacy and safety evaluation of claudin-4-targeted antitumor therapy using a human and mouse cross-reactive monoclonal antibody. *Pharmacol. Res. Perspect.* **2016**, *4*, e00266. [CrossRef]

68. Hiramatsu, K.; Serada, S.; Enomoto, T.; Takahashi, Y.; Nakagawa, S.; Nojima, S.; Morimoto, A.; Matsuzaki, S.; Yokoyama, T.; Takahashi, T.; et al. LSR antibody therapy inhibits ovarian epithelial tumor growth by inhibiting lipid uptake. *Cancer Res.* **2018**, *78*, 516–527. [CrossRef]
69. Ben-David, U.; Nudel, N.; Benvenisty, N. Immunologic and chemical targeting of the tight-junction protein Claudin-6 eliminates tumorigenic human pluripotent stem cells. *Nat. Commun.* **2013**, *4*, 1992. [CrossRef]
70. Powell, D.W. Barrier Function of Epithelia. *Am. J. Physiol.* **1981**, *241*, G275–G288. [CrossRef]
71. Tran, T.N. Cutaneous drug delivery: An update. *J. Investig. Dermatol. Symp. Proc.* **2013**, *16*, S67–S69. [CrossRef] [PubMed]
72. Tsukita, S.; Furuse, M. Claudin-based barrier in simple and stratified cellular sheets. *Curr. Opin. Cell Biol.* **2002**, *14*, 531–536. [CrossRef]
73. Furuse, M.; Hata, M.; Furuse, K.; Yoshida, Y.; Haratake, A.; Sugitani, Y.; Noda, T.; Kubo, A.; Tsukita, S. Claudin-based tight junctions are crucial for the mammalian epidermal barrier: A lesson from claudin-1-deficient mice. *J. Cell Biol.* **2002**, *156*, 1099–1111. [CrossRef] [PubMed]
74. Osanai, M.; Takasawa, A.; Murata, M.; Sawada, N. Claudins in cancer: Bench to bedside. *Pflugers Arch.* **2017**, *469*, 55–67. [CrossRef] [PubMed]
75. Oliveira, S.S.; Morgado-Diaz, J.A. Claudins: Multifunctional players in epithelial tight junctions and their role in cancer. *Cell. Mol. Life Sci.* **2007**, *64*, 17–28. [CrossRef] [PubMed]
76. Long, H.Y.; Crean, C.D.; Lee, W.H.; Cummings, O.W.; Gabig, T.G. Expression of *Clostridium perfringens* enterotoxin receptors claudin-3 and claudin-4 in prostate cancer epithelium. *Cancer Res.* **2001**, *61*, 7878–7881. [PubMed]
77. Kominsky, S.L.; Vali, M.; Korz, D.; Gabig, T.G.; Weitzman, S.A.; Argani, P.; Sukumar, S. *Clostridium perfringens* enterotoxin elicits rapid and specific cytolysis of breast carcinoma cells mediated through tight junction proteins claudin 3 and 4. *Am. J. Pathol.* **2004**, *164*, 1627–1633. [CrossRef]
78. Michl, P.; Barth, C.; Buchholz, M.; Lerch, M.M.; Rolke, M.; Holzmann, K.H.; Menke, A.; Fensterer, H.; Giehl, K.; Lohr, M.; et al. Claudin-4 expression decreases invasiveness and metastatic potential of pancreatic cancer. *Cancer Res.* **2003**, *63*, 6265–6271.
79. Zhu, Y.H.; Brannstrom, M.; Janson, P.O.; Sundfeldt, K. Differences in expression patterns of the tight junction proteins, claudin 1, 3, 4 and 5, in human ovarian surface epithelium as compared to epithelia in inclusion cysts and epithelial ovarian tumours. *Int. J. Cancer* **2006**, *118*, 1884–1891. [CrossRef]
80. Weber, C.R.; Nalle, S.C.; Tretiakova, M.; Rubin, D.T.; Turner, J.R. Claudin-1 and claudin-2 expression is elevated in inflammatory bowel disease and may contribute to early neoplastic transformation. *Lab. Investig.* **2008**, *88*, 1110–1120. [CrossRef]
81. Kinugasa, T.; Huo, Q.; Higash, D.; Shibaguchi, H.; Kuroki, M.; Tanaka, T.; Futami, K.; Yamashita, Y.; Hachimine, K.; Maekawa, S.; et al. Selective up-regulation of claudin-1 and claudin-2 in colorectal cancer. *Anticancer Res.* **2007**, *27*, 3729–3734. [CrossRef]
82. Dhawan, P.; Singh, A.B.; Deane, N.G.; No, Y.; Shiou, S.R.; Schmidt, C.; Neff, J.; Washington, M.K.; Beauchamp, R.D. Claudin-1 regulates cellular transformation and metastatic behavior in colon cancer. *J. Clin. Investig.* **2005**, *115*, 1765–1776. [CrossRef] [PubMed]
83. Dhawan, P.; Ahmad, R.; Chaturvedi, R.; Smith, J.J.; Midha, R.; Mittal, M.K.; Krishnan, M.; Chen, X.; Eschrich, S.; Yeatman, T.J.; et al. Claudin-2 expression increases tumorigenicity of colon cancer cells: Role of epidermal growth factor receptor activation. *Oncogene* **2011**, *30*, 3234–3247. [CrossRef] [PubMed]
84. Tabaries, S.; Dong, Z.; Annis, M.G.; Omeroglu, A.; Pepin, F.; Ouellet, V.; Russo, C.; Hassanain, M.; Metrakos, P.; Diaz, Z.; et al. Claudin-2 is selectively enriched in and promotes the formation of breast cancer liver metastases through engagement of integrin complexes. *Oncogene* **2011**, *30*, 1318–1328. [CrossRef] [PubMed]
85. Hashimoto, Y.; Fukasawa, M.; Kuniyasu, H.; Yagi, K.; Kondoh, M. Claudin-targeted drug development using anti-claudin monoclonal antibodies to treat hepatitis and cancer. *Ann. N. Y. Acad. Sci.* **2017**, *1397*, 5–16. [CrossRef] [PubMed]
86. Michl, P.; Buchholz, M.; Rolke, M.; Kunsch, S.; Lohr, M.; McClane, B.; Tsukita, S.; Leder, G.; Adler, G.; Gress, T.M. Claudin-4: A new target for pancreatic cancer treatment using *Clostridium perfringens* enterotoxin. *Gastroenterology* **2001**, *121*, 678–684. [CrossRef] [PubMed]
87. Morin, P.J. Claudin proteins in human cancer: Promising new targets for diagnosis and therapy. *Cancer Res.* **2005**, *65*, 9603–9606. [CrossRef] [PubMed]
88. Kominsky, S.L. Claudins: Emerging targets for cancer therapy. *Expert Rev. Mol. Med.* **2006**, *8*, 1–11. [CrossRef]

89. Saeki, R.; Kondoh, M.; Kakutani, H.; Tsunoda, S.; Mochizuki, Y.; Hamakubo, T.; Tsutsumi, Y.; Horiguchi, Y.; Yagi, K. A novel tumor-targeted therapy using a claudin-4-targeting molecule. *Mol. Pharmacol.* **2009**, *76*, 918–926. [CrossRef]
90. Meunier, V.; Bourrie, M.; Berger, Y.; Fabre, G. The human intestinal epithelial cell line Caco-2; pharmacological and pharmacokinetic applications. *Cell Biol. Toxicol.* **1995**, *11*, 187–194. [CrossRef]
91. Torres, J.B.; Knight, J.C.; Mosley, M.J.; Kersemans, V.; Koustoulidou, S.; Allen, D.; Kinchesh, P.; Smart, S.; Cornelissen, B. Imaging of claudin-4 in pancreatic ductal adenocarcinoma using a radiolabelled anti-claudin-4 monoclonal antibody. *Mol. Imaging Biol.* **2018**, *20*, 292–299. [CrossRef] [PubMed]
92. Kunisawa, J.; Fukuyama, S.; Kiyono, H. Mucosa-associated lymphoid tissues in the aerodigestive tract: Their shared and divergent traits and their importance to the orchestration of the mucosal immune system. *Curr. Mol. Med.* **2005**, *5*, 557–572. [CrossRef] [PubMed]
93. Kraehenbuhl, J.P.; Neutra, M.R. Epithelial M cells: Differentiation and function. *Annu. Rev. Cell Dev. Biol.* **2000**, *16*, 301–332. [CrossRef] [PubMed]
94. Nagatake, T.; Fujita, H.; Minato, N.; Hamazaki, Y. Enteroendocrine cells are specifically marked by cell surface expression of claudin-4 in mouse small intestine. *PLoS ONE* **2014**, *9*, e90638. [CrossRef] [PubMed]
95. Tamagawa, H.; Takahashi, I.; Furuse, M.; Yoshitake-Kitano, Y.; Tsukita, S.; Ito, T.; Matsuda, H.; Kiyono, H. Characteristics of claudin expression in follicle-associated epithelium of Peyer's patches: Preferential localization of claudin-4 at the apex of the dome region. *Lab. Investig.* **2003**, *83*, 1045–1053. [CrossRef]
96. Ye, T.; Yue, Y.; Fan, X.M.; Dong, C.S.; Xu, W.; Xiong, S.D. M cell-targeting strategy facilitates mucosal immune response and enhances protection against CVB3-induced viral myocarditis elicited by chitosan-DNA vaccine. *Vaccine* **2014**, *32*, 4457–4465. [CrossRef]
97. Kakutani, H.; Kondoh, M.; Fukasaka, M.; Suzuki, H.; Hamakubo, T.; Yagi, K. Mucosal vaccination using claudin-4-targeting. *Biomaterials* **2010**, *31*, 5463–5471. [CrossRef]
98. Suzuki, H.; Kakutani, H.; Kondoh, M.; Watari, A.; Yagi, K. The safety of a mucosal vaccine using the C-terminal fragment of *Clostridium perfringens* enterotoxin. *Pharmazie* **2010**, *65*, 766–769. [CrossRef]
99. Suzuki, H.; Hosomi, K.; Nasu, A.; Kondoh, M.; Kunisawa, J. Development of adjuvant-free bivalent food poisoning vaccine by augmenting the antigenicity of *Clostridium perfringens* enterotoxin. *Front. Immunol.* **2018**, *9*, 2320. [CrossRef]
100. Mankertz, J.; Amasheh, M.; Krug, S.M.; Fromm, A.; Amasheh, S.; Hillenbrand, B.; Tavalali, S.; Fromm, M.; Schulzke, J.D. TNF alpha up-regulates claudin-2 expression in epithelial HT-29/B6 cells via phosphatidylinositol-3-kinase signaling. *Cell Tissue Res.* **2009**, *336*, 67–77. [CrossRef]
101. Furuse, M.; Furuse, K.; Sasaki, H.; Tsukita, S. Conversion of zonulae occludentes from tight to leaky strand type by introducing claudin-2 into Madin-Darby canine kidney I cells. *J. Cell Biol.* **2001**, *153*, 263–272. [CrossRef]
102. Pardridge, W.M. The blood-brain barrier: Bottleneck in brain drug development. *NeuroRx* **2005**, *2*, 3–14. [CrossRef]
103. Vanlandewijck, M.; He, L.; Mae, M.A.; Andrae, J.; Ando, K.; Del Gaudio, F.; Nahar, K.; Lebouvier, T.; Lavina, B.; Gouveia, L.; et al. A molecular atlas of cell types and zonation in the brain vasculature. *Nature* **2018**, *554*, 475–480. [CrossRef]
104. Nitta, T.; Hata, M.; Gotoh, S.; Seo, Y.; Sasaki, H.; Hashimoto, N.; Furuse, M.; Tsukita, S. Size-selective loosening of the blood-brain barrier in claudin-5-deficient mice. *J. Cell Biol.* **2003**, *161*, 653–660. [CrossRef]
105. Sohet, F.; Lin, C.; Munji, R.N.; Lee, S.Y.; Ruderisch, N.; Soung, A.; Arnold, T.D.; Derugin, N.; Vexler, Z.S.; Yen, F.T.; et al. LSR/angulin-1 is a tricellular tight junction protein involved in blood-brain barrier formation. *J. Cell Biol.* **2015**, *208*, 703–711. [CrossRef]
106. Zeniya, S.; Kuwahara, H.; Daizo, K.; Watari, A.; Kondoh, M.; Yoshida-Tanaka, K.; Kaburagi, H.; Asada, K.; Nagata, T.; Nagahama, M.; et al. Angubindin-1 opens the blood-brain barrier in vivo for delivery of antisense oligonucleotide to the central nervous system. *J. Control. Release* **2018**, *283*, 126–134. [CrossRef]
107. Fujita, H.; Hamazaki, Y.; Noda, Y.; Oshima, M.; Minato, N. Claudin-4 deficiency results in urothelial hyperplasia and lethal hydronephrosis. *PLoS ONE* **2012**, *7*, e52272. [CrossRef]
108. Muto, S.; Hata, M.; Taniguchi, J.; Tsuruoka, S.; Moriwaki, K.; Saitou, M.; Furuse, K.; Sasaki, H.; Fujimura, A.; Imai, M.; et al. Claudin-2-deficient mice are defective in the leaky and cation-selective paracellular permeability properties of renal proximal tubules. *Proc. Natl. Acad. Sci. USA* **2010**, *107*, 8011–8016. [CrossRef]

109. Gong, Y.F.; Himmerkus, N.; Sunq, A.; Milatz, S.; Merkel, C.; Bleich, M.; Hou, J.H. ILDR1 is important for paracellular water transport and urine concentration mechanism. *Proc. Natl. Acad. Sci. USA* **2017**, *114*, 5271–5276. [CrossRef]
110. Hadj-Rabia, S.; Baala, L.; Vabres, P.; Hamel-Teillac, D.; Jacquemin, E.; Fabre, M.; Lyonnet, S.; De Prost, Y.; Munnich, A.; Hadchouel, M.; et al. Claudin-1 gene mutations in neonatal sclerosing cholangitis associated with lchthyosis: A tight junction disease. *Gastroenterology* **2004**, *127*, 1386–1390. [CrossRef]
111. Arinami, T. Analyses of the associations between the genes of 22q11 deletion syndrome and schizophrenia. *J. Hum. Genet.* **2006**, *51*, 1037–1045. [CrossRef]
112. Wei, J.; Hemmings, G.P. A study of the combined effect of the CLDN5 locus and the genes for the phospholipid metabolism pathway in schizophrenia. *Prostaglandins Leukot. Essent. Fatty Acids* **2005**, *73*, 441–445. [CrossRef]
113. Sun, Z.Y.; Wei, J.; Xie, L.; Shen, Y.; Liu, S.Z.; Ju, G.Z.; Shi, J.P.; Yu, Y.Q.; Zhang, X.; Xu, Q.; et al. The CLDN5 locus may be involved in the vulnerability to schizophrenia. *Eur. Psychiatry* **2004**, *19*, 354–357. [CrossRef]
114. Tanaka, H.; Imasato, M.; Yamazaki, Y.; Matsumoto, K.; Kunimoto, K.; Delpierre, J.; Meyer, K.; Zerial, M.; Kitamura, N.; Watanabe, M.; et al. Claudin-3 regulates bile canalicular paracellular barrier and cholesterol gallstone cores formation in mice. *J. Hepatol.* **2018**, *69*, 1308–1316. [CrossRef]
115. Furuse, M.; Sasaki, H.; Tsukita, S. Manner of interaction of heterogeneous claudin species within and between tight junction strands. *J. Cell Biol.* **1999**, *147*, 891–903. [CrossRef]
116. Knipp, G.T.; Ho, N.F.H.; Barsuhn, C.L.; Borchardt, R.T. Paracellular diffusion in Caco-2 cell monolayers: Effect of perturbation on the transport of hydrophilic compounds that vary in charge and size. *J. Pharm. Sci.* **1997**, *86*, 1105–1110. [CrossRef]
117. Watson, C.J.; Rowland, M.; Warhurst, G. Functional modeling of tight junctions in intestinal cell monolayers using polyethylene glycol oligomers. *Am. J. Physiol. Cell Physiol.* **2001**, *281*, C388–C397. [CrossRef]
118. Steinke, A.; Meier-Stiegen, S.; Drenckhahn, D.; Asan, E. Molecular composition of tight and adherens junctions in the rat olfactory epithelium and fila. *Histochem. Cell Biol.* **2008**, *130*, 339–361. [CrossRef]
119. Schlingmann, B.; Overgaard, C.E.; Molina, S.A.; Lynn, K.S.; Mitchell, L.A.; Dorsainvil White, S.; Mattheyses, A.L.; Guidot, D.M.; Capaldo, C.T.; Koval, M. Regulation of claudin/zonula occludens-1 complexes by hetero-claudin interactions. *Nat. Commun.* **2016**, *7*, 12276. [CrossRef]
120. Markov, A.G.; Veshnyakova, A.; Fromm, M.; Amasheh, M.; Amasheh, S. Segmental expression of claudin proteins correlates with tight junction barrier properties in rat intestine. *J. Comp. Physiol. B* **2010**, *180*, 591–598. [CrossRef]
121. Inai, T.; Sengoku, A.; Guan, X.; Hirose, E.; Iida, H.; Shibata, Y. Heterogeneity in expression and subcellular localization of tight junction proteins, claudin-10 and -15, examined by RT-PCR and immunofluorescence microscopy. *Arch. Histol. Cytol.* **2005**, *68*, 349–360. [CrossRef]
122. Rahner, C.; Mitic, L.L.; Anderson, J.M. Heterogeneity in expression and subcellular localization of claudins 2, 3, 4, and 5 in the rat liver, pancreas, and gut. *Gastroenterology* **2001**, *120*, 411–422. [CrossRef]
123. Kiuchi-Saishin, Y.; Gotoh, S.; Furuse, M.; Takasuga, A.; Tano, Y.; Tsukita, S. Differential expression patterns of claudins, tight junction membrane proteins, in mouse nephron segments. *J. Am. Soc. Nephrol.* **2002**, *13*, 875–886.
124. Steinemann, A.; Galm, I.; Chip, S.; Nitsch, C.; Maly, L.P. Claudin-1, -2 and -3 are selectively expressed in the epithelia of the choroid plexus of the mouse from early development and into adulthood while claudin-5 is restricted to endothelial cells. *Front. Neuroanat.* **2016**, *10*, 16. [CrossRef]
125. He, L.Q.; Vanlandewijck, M.; Mae, M.A.; Andrae, J.; Ando, K.; Del Gaudio, F.; Nahar, K.; Lebouvier, T.; Lavina, B.; Gouveia, L.; et al. Single-cell RNA sequencing of mouse brain and lung vascular and vessel-associated cell types. *Sci. Data* **2018**, *5*, 180160. [CrossRef]
126. Watari, A.; Kodaka, M.; Matsuhisa, K.; Sakamoto, Y.; Hisaie, K.; Kawashita, N.; Takagi, T.; Yamagishi, Y.; Suzuki, H.; Tsujino, H.; et al. Identification of claudin-4 binder that attenuates tight junction barrier function by TR-FRET-based screening assay. *Sci. Rep.* **2017**, *7*, 14514. [CrossRef]

International Journal of
Molecular Sciences

Article

IL-13 Impairs Tight Junctions in Airway Epithelia

Hanna Schmidt [1,†], Peter Braubach [2,3,†], Carolin Schilpp [1], Robin Lochbaum [1], Kathrin Neuland [1], Kristin Thompson [1], Danny Jonigk [2,3], Manfred Frick [1], Paul Dietl [1] and Oliver H. Wittekindt [1,*]

1 Institute of General Physiology, Ulm University, Albert-Einstein-Allee 11, 89081 Ulm, Germany
2 Institute of Pathology, Hannover Medical School, Carl-Neuberg-Str. 130625 Hannover, Germany
3 German Center of Lung Research (DZL), Partnersite BREATH, 306245 Hannover, Germany
* Correspondence: oliver.wittekindt@uni-ulm.de; Tel.: +49-731-5002-3247; Fax: +49-731-5002-3242
† These authors contributed equally to this work.

Received: 11 June 2019; Accepted: 27 June 2019; Published: 30 June 2019

Abstract: Interleukin-13 (IL-13) drives symptoms in asthma with high levels of T-helper type 2 cells (T_h2-cells). Since tight junctions (TJ) constitute the epithelial diffusion barrier, we investigated the effect of IL-13 on TJ in human tracheal epithelial cells. We observed that IL-13 increases paracellular permeability, changes claudin expression pattern and induces intracellular aggregation of the TJ proteins zonlua occludens protein 1, as well as claudins. Furthermore, IL-13 treatment increases expression of ubiquitin conjugating E2 enzyme UBE2Z. Co-localization and proximity ligation assays further showed that ubiquitin and the proteasomal marker PSMA5 co-localize with TJ proteins in IL-13 treated cells, showing that TJ proteins are ubiquitinated following IL-13 exposure. UBE2Z upregulation occurs within the first day after IL-13 exposure. Proteasomal aggregation of ubiquitinated TJ proteins starts three days after IL-13 exposure and transepithelial electrical resistance (TEER) decrease follows the time course of TJ-protein aggregation. Inhibition of JAK/STAT signaling abolishes IL-13 induced effects. Our data suggest that that IL-13 induces ubiquitination and proteasomal aggregation of TJ proteins via JAK/STAT dependent expression of UBE2Z, resulting in opening of TJs. This may contribute to barrier disturbances in pulmonary epithelia and lung damage of patients with inflammatory lung diseases.

Keywords: lung; epithelia; interleukin 13; tight junction; UBE2Z; ubiquitin

1. Introduction

Asthma is a complex disease that involves environmental interactions, genetic risk factors and chronic airway inflammation. According to the underlying pathophysiological mechanisms, asthma syndrome was proposed to be subdivided into several endotypes [1]. Depending on the dominating inflammatory response, the endotypes can be grouped in those with low levels of T-helper type 2 cells (T_h2-cells), the T_h2 low endotype, and those with high T_h2-cell levels, described as T_h2 high endotype [2]. The latter subsumes allergic asthma phenotypes with mainly T_h2 cell driven symptoms. T_h2 cells were referred to as T-helper cells that produce and release high amounts of interleukins (IL) -3, -4, -5, -6, -10 and -13 [3]. In particular, IL-13 plays a pivotal role in allergic asthma [4,5]. Novel treatment strategies follow approaches, where IL-13 signaling is inhibited either by immuno-inhibition of IL-13 dependent receptor signaling [6] or by immuno-neutralization of IL-13 itself [6–10]. Both strategies can improve asthma symptoms and reduce risk of exacerbations.

Cigarette smoke is another trigger of pulmonary inflammation, which causes damage and barrier dysfunction [11]. Moreover, cigarette smoke induced inflammatory and metabolic conditions are associated with lung cancer development [12,13]. In this process, IL-13 signaling is considered to be involved via the IL-13 receptor α2 that promotes malignant transformation [14].

IL-13 has multiple effects on pulmonary epithelia. It induces airway hyperresponsiveness [15,16], mucus hypersecretion and airway remodeling [4,17–20], all of which are the hallmarks of allergic asthma. Additionally, lateral intercellular contacts and tight junctions (TJ) of the pulmonary epithelium are disturbed in patients with allergic asthma [21]. T_h2 cytokines IL-4 and IL-13 have been found to compromise barrier function of the pulmonary epithelium [22–26].

The barrier function of an epithelium can be described as its ability to limit the diffusion-dependent exchange of solvent and solute between apical and basolateral compartments. Airway epithelial cells are linked by several intercellular protein complexes that connect lateral cell membranes of neighboring cells. TJ form the most apically localized junction complex. They function as the major limiting diffusion barrier as well as a fence that separates the apical and lateral membranes of epithelial cells [27]. TJs are comprised of different claudin (cldn) isoforms and the isoform composition determines the specific functional properties of TJs, specifically their permeability, reflection coefficient and permselectivity [27]. The cldn composition of TJs in the pulmonary epithelium changes along the respiratory tree and adjusts epithelial function to the segment specific requirements [28].

T_h2 cytokines, particularly IL-4 and IL-13, perturb the protein composition of TJ in pulmonary epithelia via different mechanisms. First, they alter expression levels of several TJ proteins. This has been reported for patients with allergic rhinosinusitis where expression levels of TJ proteins occludin (ocldn) and junctional adhesion molecule A (JAM-A) were reduced and expression of cldn2, a cldn that increases TJ's permeability in sinonasal epithelial cells, was increased [29]. Similarly, Cldn18 expression levels were significantly reduced in brush samples of human epithelial cells from patients with T_h2-high asthma endotype that was linked to IL-13 signaling [24]. Second, T_h2 cytokines decrease protein density at the TJs without affecting expression levels of TJ proteins [24,25].

Many studies revealed that IL-13 is the major effector for perturbation of TJ permeability in T_h2-high asthma patients [21–26]. In addition, breakdown of the pulmonary barrier function is considered to be a hall mark in developing respiratory distress [30]. However, the IL-13 mediated perturbations of TJ are poorly understood. Our study aims at elucidating the effect of prolonged IL-13 exposure on primary human bronchial epithelial cells cultivated at air-liquid interface (ALI) conditions. We demonstrate that IL-13 alters expression levels of cldn8, cldn9 and cldn16, upregulates the E2 ubiquitin conjugating enzyme UBE2Z and induces proteasomal aggregation of TJ proteins together with ubiquitin (UBQ) and UBE2Z. This reduces protein density at the TJ and increases TJ permeability for ions and small molecules and might be an important contributor to patients' predisposition to asthma exacerbations and subsequent respiratory distress.

2. Results

2.1. *Effect of IL-13 on hTEpC Epithelia*

Human primary tracheal epithelial cells (hTEpC) were cultivated at air-liquid interface (ALI) for 21 days. Exposure of hTEpC epithelia to 10 ng/mL IL-13 or IL-4 for the entire ALI cultivation time resulted in a transepithelial electrical resistance (TEER) decrease (Figure 1A). Both cytokines are characteristic for T_H2 high asthma endotype [31]. IL-13 reduced TEER by approximately 85%. The TEER decrease caused by IL-4 was lower and only the effect of IL-13 was statistically significant. Hence, we focused on the epithelial response to IL-13 in all subsequent experiments. The reduction of TEER by IL-13 was in line with an increased apparent permeability coefficient (P_{app}) for sodium fluorescein (Figure 1B). This result indicated that IL-13 affects the paracellular diffusion barrier. Increased mucin 5ac (Muc5ac) expression levels (Figure 1C) confirm the biological activity and efficiency of IL-13. Inhibition of JAK/STAT signaling by Tofacitinib [32,33] or S-Ruxolitinib [34,35] (both added at concentration of 2.5 μM) attenuated IL-13 effects on TEER and paracellular permeability (Figure 1D,E). This indicates that IL-13 acts via JAK/STAT signaling on paracellular permeability.

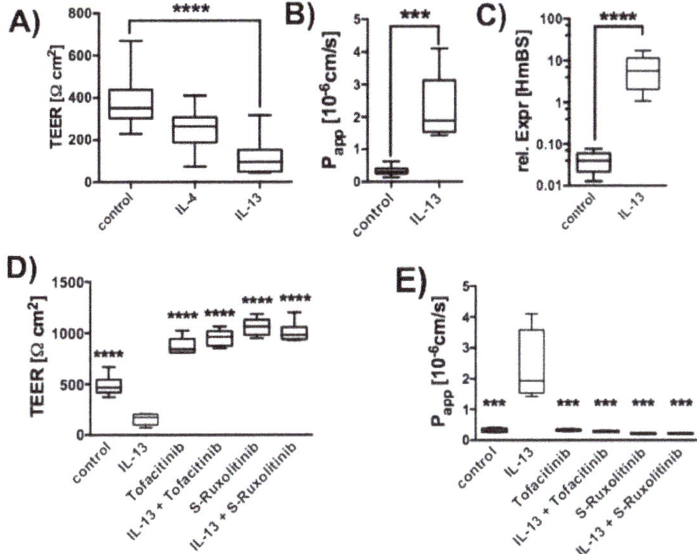

Figure 1. IL-13 increased paracellular permeability. Human tracheal epithelial cells were cultivated at air-liquid interface conditions in the absence (control) and presence of 10 ng/mL IL-13 (IL-13). Box plots summarize data as median values, the boxes represent percentiles, the whiskers indicate the minimum/maximum. (**A**) Response of human tracheal epithelial cells to T_H2 cytokines IL-4 and IL-13 (10 ng/mL each). Both cytokines reduced TEER. However, TEER reduction by IL-4 did not reach significance (Kruska-Wallis test with Dunn's correction for multiple testing, $p = 0.086$, IL-4 ($N = 8$) versus control cells (control), $N = 20$). IL-13 reduced TEER significantly (Kruska-Wallis test with Dunn's correction for multiple testing, $p < 0.0001$, IL-13, N = 17 versus control, $N = 20$). (**B**) IL-13 increases apparent permeability coefficient (P_{app}) for the small molecular marker sodium fluorescein (Mann-Whitney test, $p = 0.0002$, $N = 7$). Inhibition of JAK/STAT pathway by Tofacitinib and S-Ruxolitinib abolishes IL-13 induced effects on (**C**) Relative Expression of Muc5ac in human tracheal epithelial cells. Upregulation of mucus production was confirmed by detecting Muc5ac upregulation in IL-13 treated epithelia. (**D**) TEER of control epithelia (control), epithelia cultivated in the presence of IL-13 (IL-13), 2.5 μM Tofacitinib (Tofacitinib) or S- Ruxolitinib (S-Ruxolitinib) and in the presence of IL-13 + Tofactinib or IL-13 + S- Ruxolitinib. (all versus IL-13, ANOVA with Holm-Sidak correction for multiple testing, $p < 0.0001$, $N = 6$) and on (**E**) P_{app} (all versus IL-13, ANOVA with Holm-Sidak correction for multiple testing, $N = 4$, control $p = 0.0003$, Tofacitinib $p = 0.0003$, Tofacitinib + IL-13 $p = 0.0002$, S-Ruxolitinib $p = 0.0002$ and S-Ruxolitinib + IL-13 $p = 0.0002$). *** and **** indicate $p < 0.001$ and $p < 0.0001$, respectively.

Paracellular permeability depends on cldn composition of the tight junctions (TJ) [28]. Hence, we investigated the effect of IL-13 on expression levels of claudins and also on occludin (ocldn) as a TJ protein that seals the paracellular pathway. Semi quantitative reverse transcriptase PCR (RT-PCR) experiments revealed a significant reduction of expression levels in IL-13 treated versus control epithelia for cldn8, cldn9 and cldn16 by a factor of 7.0, 3.5 and 6.3, respectively (Figure 2A). Hence, we performed additional experiments to further investigate the effect of IL-13 on cldn8, cldn9, cldn16 expression. Cldn4, although not affected by IL-13, was included in subsequent experiments, since it interacts with cldn8 to form paracellular Cl^- pores in TJs [36] (Figure 2B–E). In line with the initial screening, IL-13 treatment reduces cldn8, cldn9 and cldn16 expression levels. Inhibition of JAK/STAT signaling by Tofacitinib [32,33] and S-Ruxolitinib [34,35] abolished the down-regulation of these cldns. Cldn4 expression was not affected neither by IL-13 nor by JAK/STAT inhibition. These experiments confirm that IL-13 specifically changes expression of cldn8, 9 and 16 via JAK/STAT signaling.

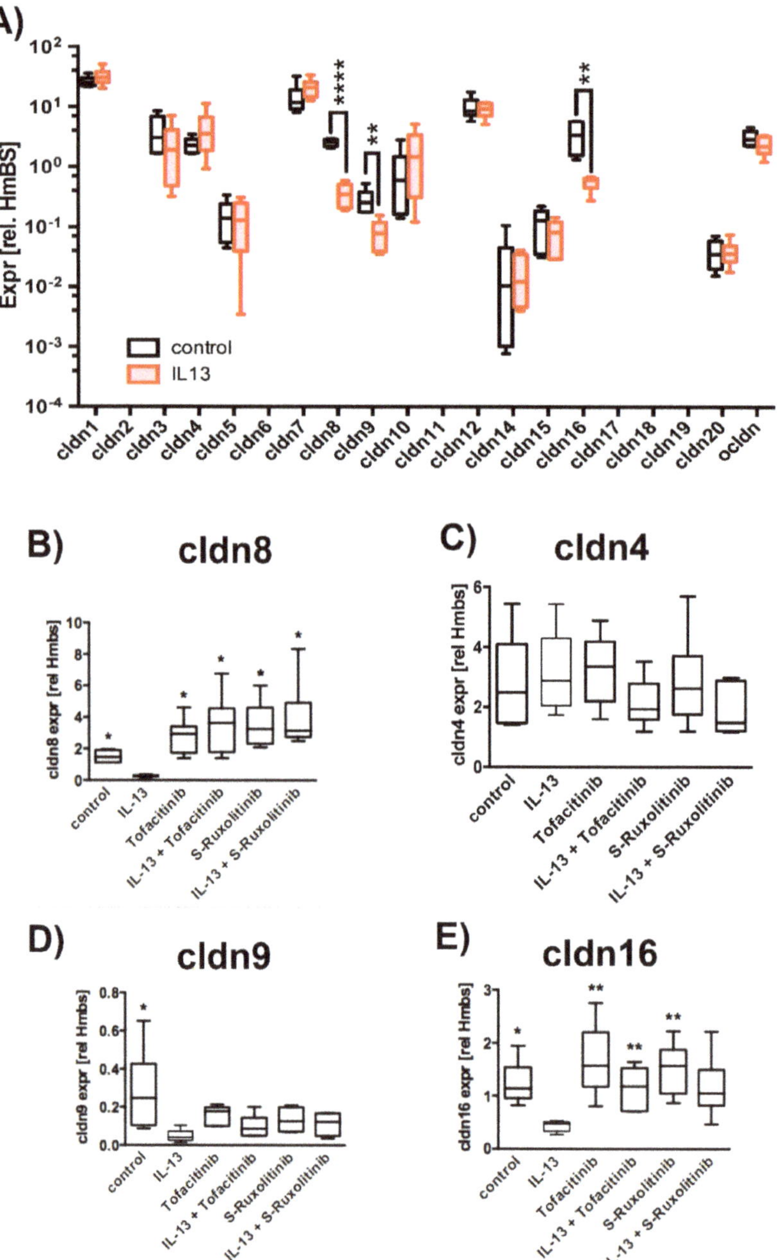

Figure 2. Effect of IL-13 on claudin and occludin expression levels. Human tracheal epithelial cells were cultivated at air-liquid interface conditions. Expression levels were normalized to HmBS. The box plot summarizes data as median values, the boxes represent percentiles, the whiskers indicate minimum/maximum. (**A**) Claudin expression levels in epithelia cultivated in the absence (control) or

presence of 10 ng/mL IL-13 (IL-13). Multiple *T*-test, all IL-13 versus control, cldn8 $p < 0.0001$, cldn9 $p = 0.005$ and cldn16 $p = 0.005$, $N = 6$). Transcripts encoding claudins 2, 6, 11, 17, 18 and 19 were not detected. Changes in cldn expression levels were confirmed and the effect of inhibition of JAK/STAT pathway was tested in subsequent independent experiments (control epithelia = control, epithelia cultivated in the presence of IL-13 = IL-13, 2.5 μM Tofacitinib = Tofacitinib or S-Ruxolitinib = S-Ruxolitinib and in the presence of IL-13 = IL-13 + Tofactinib or IL-13 + S- Ruxolitinib, respectively) (**B**) cldn8 expression (Mann Whitney test with Bonferroni correction all versus IL-13 and $N = 6$, control $p = 0.02$, Tofacitinib $p = 0.02$, IL-13 + Tofacitinib $p = 0.02$, Ruxolitinib $p = 0.02$ and IL-13 + Ruxolitinib $p = 0.02$) (**C**) cldn4 expression was not affected neither by IL-13 nor by Tofacitinib or Ruxolitinib. (**D**) cldn9 expression (Mann-Whitney test with Bonferroni correction, $N = 6$, control $p = 0.04$, Tofacitinib $p = 0.08$, IL-13 + Tofacitinib $p = 0.13$, Ruxolitinib $p = 0.07$ and IL-13 + Ruxolitinib $p = 0.2$) and (**E**) cldn16 expression (Mann-Whitney test with Bonferroni correction, $N = 6$, control $p = 0.02$, Tofacitinib $p = 0.01$, IL-13 + Tofacitinib $p = 0.01$, Ruxolitinib $p = 0.01$ and IL-13 + Ruxolitinib $p = 0.07$). *, ** and **** indicate $p < 0.05$, $p < 0.01$ and $p < 0.0001$, respectively.

To further elucidate the mechanisms underlying the IL-13-induced modulation of paracellular permeability, we investigated the cellular localization of TJ proteins (Figure 3A–C). We focused on cldn8, since we recently identified cldn8 as a claudin that regulates protein assembly at the TJs [37]. Cldn4 was chosen as the control, as its expression level is not modulated by neither IL-13 treatment nor inhibitors of JAK/STAT signaling (see above). Finally we investigated ocldn localization, since there is evidence that its localization at TJ in pulmonary epithelia is modulated by cldn8 [37]. In control epithelia, all the investigated TJ proteins colocalized with tight junction marker zonula occludens 1 (ZO1) (Figure 3A–C). Staining for TJ proteins revealed an almost continuous contouring of the cells, indicating their lateral assembly at the TJs. IL-13 treatment resulted in a highly diminished lateral localization of all TJ proteins, including ZO1. In these epithelia, the staining of TJ proteins resulted in a patchy cell contouring that indicates massively impaired protein assembly at the TJ. Moreover, TJ proteins were observed to form intracellular aggregates. This effect was not restricted to IL-13 regulated cldn8 but to all investigated TJ proteins, including ZO1. Inhibition of JAK/STAT signaling using 2.5 μM Tofacitinib [32,33] or 2.5 μM S-Ruxolitinib [34,35] abolished IL-13 induced intracellular localization of TJ proteins. Whereas TJ protein localization was affected by IL-13 treatment, the lateral localization of the adherens junction protein E-cadherin (E-cad) [38] remained almost unaffected by IL-13 (Figure 3D,E). This indicates that the IL-13 effect on cell–cell contacts is restricted to the TJ. Furthermore, IL-13 affects all TJ proteins but not exclusively those proteins, which are regulated at transcriptional levels by IL-13 treatment.

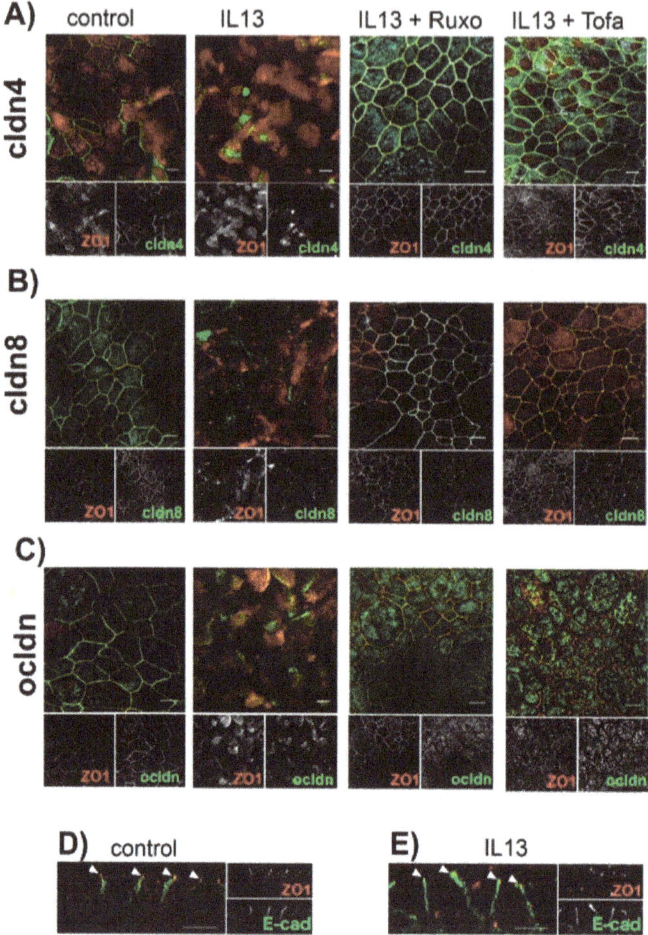

Figure 3. Effect of IL-13 on protein localization at TJ. Human tracheal epithelial cells were cultivated at air-liquid interface conditions. Epithelia were cultivated in the absence of compounds (control), in the presence of IL-13 (IL-13), in the presence of IL-13 + 2.5 µM S-Ruxolitinib (IL-13 + Ruxo) or in the presence of 2.5 µM Tofacitinib (IL-13 + Tofa). Scale bars represent 10 µm. (**A**) Intracellular localization of claudin 4 (cldn4, green channel). Zonula occludens 1 (ZO1, red channel) was used for counterstaining for tight junctions. (**B**) Intracellular localization of claudin 8 (cldn8, green channel). Counterstaining for tight junctions by ZO1 (red channel). (**C**) Intracellular localization of occludin (ocldn, green channel). Counterstaining for tight junctions by ZO1 (red channel). (**D**,**E**) Intracellular localization of E-cadherin (E-cad, green channel) as a marker for lateral adherens junction. ZO1 (red channel) was used for counterstaining of tight junctions.

2.2. *IL-13 Activates the Ubiquitin-Proteasome Pathway*

There is emerging evidence that the ubiquitin-proteasome pathway [39] controls TJ protein turnover and TJ permeability [40–45]. The observed localization of TJ proteins at intracellular granules upon IL-13 treatment may hint at the possibility that internalized proteins accumulate at proteasomal compartments. To address the question of whether IL-13 activates the ubiquitin-proteasome pathway, we initially screened for modulation of ubiquitin E1 activating and ubiquitin E2 conjugating enzymes. Semi-quantitative RT-PCR experiments revealed that IL-13 treatment specifically up-regulates expression of the ubiquitin E2 conjugating enzyme, UBE2Z (Figure 4A) 1.6 ± 0.16 fold (mean ± SD, $N = 3$; T-test $p = 0.01$). The detected upregulation of Muc5AC expression by IL-13 confirms its efficiency on the tested epithelia (Figure 4B). Subsequent experiments confirmed the upregulation of UBE2Z by IL-13, which was attenuated by inhibition of the JAK/STAT pathway (Figure 4C). The number of cells was limited due to the ALI cultivation on filters and hence limits the applicability of methods like pull down assays. Therefore, to further evaluate ubiquitination of TJ proteins, we performed proximity ligation assays (PLA, Figure 4D–G). In these experiments we investigated if cldn4 (Figure 4C), cldn8 (Figure 4D) and ZO1 (Figure 4E) form protein complexes together with ubiquitin (UBQ) in IL-13 treated cells. Indeed, compared to control cells, segmental analysis (Figure 4F) of cells revealed that IL-13 treatment results in increased number of fluorescent granules that reflect positive PLA signals. This indicates that ubiquitinated proteins and TJ-proteins localize within the same protein complex and suggests that IL-13 induces the ubiquitination of TJ-proteins. Staining for protein complexes that were positive for E-cadherin and UBQ revealed positive PLA signals in control as well as in IL-13 treated cells (Figure 4H). Also, abundance of PLA positive structures did not differ in control versus IL-13 treated cells (Figure 4I). Evidently, E-cad ubiquitination occurs even at basal conditions, presumably as part of protein turnover to maintain proteostasis. This basal E-cad ubiquitination is not affected by IL-13 and agrees with the observation that subcellular E-cad localization remained unaffected by IL-13 treatment shown above, given the results of the immunostaining experiments.

We investigated the intracellular localization of TJ proteins and markers of the ubiquitin-proteasome pathway in subsequent immunocytochemical experiments (Figure 5) and studied their colocalization using van Steensel's cross-correlation analysis [46]. In control cells, the investigated TJ proteins cldn8 and ZO1 colocalized (Figure 5A,C), resulting in a sharp peak when cross-correlation coefficients were plotted against lateral translocation. Hardly any colocalization could be observed in control cells for the TJ proteins with UBE2Z (Figure 5A) or UBQ (Figure 5C). The exposure of epithelia to IL-13 resulted in an overlapping localization of TJ proteins with UBE2Z (Figure 5B) as well as with UBQ (Figure 5D). In these cases, the van Steensel's plots revealed increases in the cross-correlation coefficients for all invested proteins. However, the broadening of peak when cross-correlation coefficients are plotted against the lateral translocation reflects the increased and inhomogeneous size of structures where TJ proteins, UBE2Z and UBQ colocalize. No colocalization of ZO1 with the proteasomal marker PSMA5 nor with UBQ or PSMA5 with UBQ could be observed in control epithelia (Figure 5E). In IL-13 treated epithelia (Figure 5F) ZO1 colocalized with UBQ and PSMA5. In these epithelia UBQ colocalizes with PSMA5 as well. Again van Steensel's plot revealed enlarged non-homogenous size of structures at which colocalization is detected.

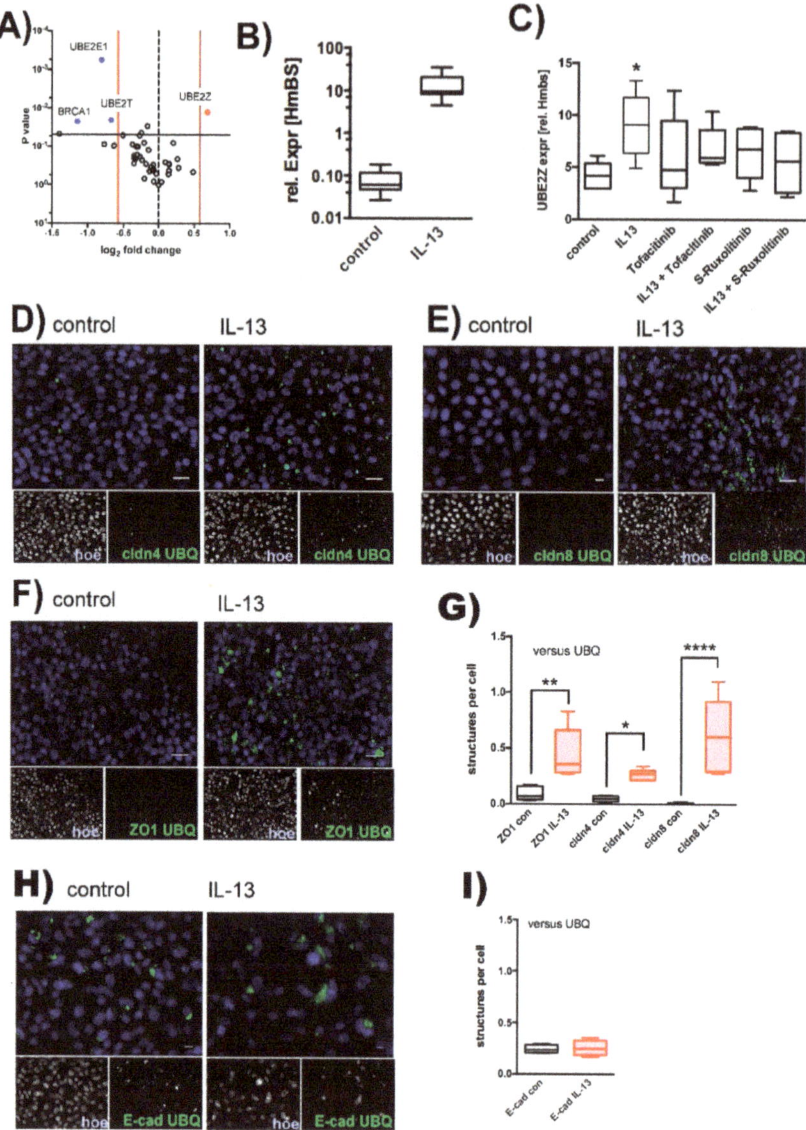

Figure 4. Ubiquitination of TJ proteins. Human tracheal epithelial cells were cultivated at air-liquid interface conditions in the absence (control) and presence of IL-13 (IL-13). (**A**) Screening for ubiquitin activating E1 and ubiquitin conjugating E2 enzyme by semi-quantitative RT-PCR. Volcano plot summarizes changes in expression levels (control vs. IL-13, each $N = 3$). Log2 fold changes are given as mean values. Vertical red lines give threshold of ±1.5 fold change. Horizontal black line gives threshold of $p = 0.05$. Genes that were considered to be up regulated are depicted in red (UBE2Z) and those to be down regulated are given in blue (BRCA1, UBE2T, UBE2E1). (**B**) Upregulation of UBE2Z by IL-13 was confirmed and effect of JAK/STAT pathway inhibition was tested in subsequent independent experiments. Relative expression of UBE2Z to HmBS of: control epithelia (control, $p = 0.021$), epithelia cultivated in the presence of IL-13 (IL-13), 2.5 μM Tofacitinib (Tofacitinib, $p = 0.035$) or S- Ruxolitinib (S-Ruxolitinib, $p = 0.16$) and in the presence of IL-13 + Tofactinib ($p = 0.16$) or IL-13 + S- Ruxolitinib ($p = 0.09$). All versus IL-13 and $N = 6$, ANOVA with Holm-Sidak correction for multiple testing. (**C**) IL-13

stimulation of epithelia used in "B" was confirmed by detecting Muc5ac upregulation (Mann-Whitney test $N = 6$, $p = 0.004$). Ubiquitination of TJ proteins were investigated in proximity ligation assays for (**D**) ubiquitination of claudin 4 (cldn4 UBQ, green channel, (**E**) ubiquitination of claudin 8 (cldn8 UBQ, green channel) and (**F**) ubiquitination of ZO1 (ZO1 UBQ, green channel). Nuclei were counterstained by Hoechst 33342 (blue channels). (**G**) PLA positive structures were identified as green fluorescent particles with an area of 75 μm^2 and above. Box plot summarizes quantification of PLA positive structures per cell as median, boxes represent percentile and whiskers give minimum/maximum (IL-13 versus control, Anova test with Holm-Sidaks correction for multiple testing, all $N = 4$, ZO1 $p = 0.0014$, cldn4 $p = 0.03$ and cldn8 $p < 0.0001$). (**H**) Ubiquitination of E-cadherin (E-cad UBQ, green channel). (**I**) PLA positive structures per cell summarized as median values, boxes represent percentiles and whiskers provide minimum/maximum values (E-cadherin = E-cad, ubiquitin = UBQ). No difference in PLA signal abundance could be detected in control versus IL-13 treated cells ($N = 4$, Mann-Whitney test $p > 0.9999$). *, ** and **** indicate $p < 0.05$, $p < 0.01$ and $p < 0.0001$, respectively. Scale bars in (**D**–**F**) and (**H**) represent 10 μM.

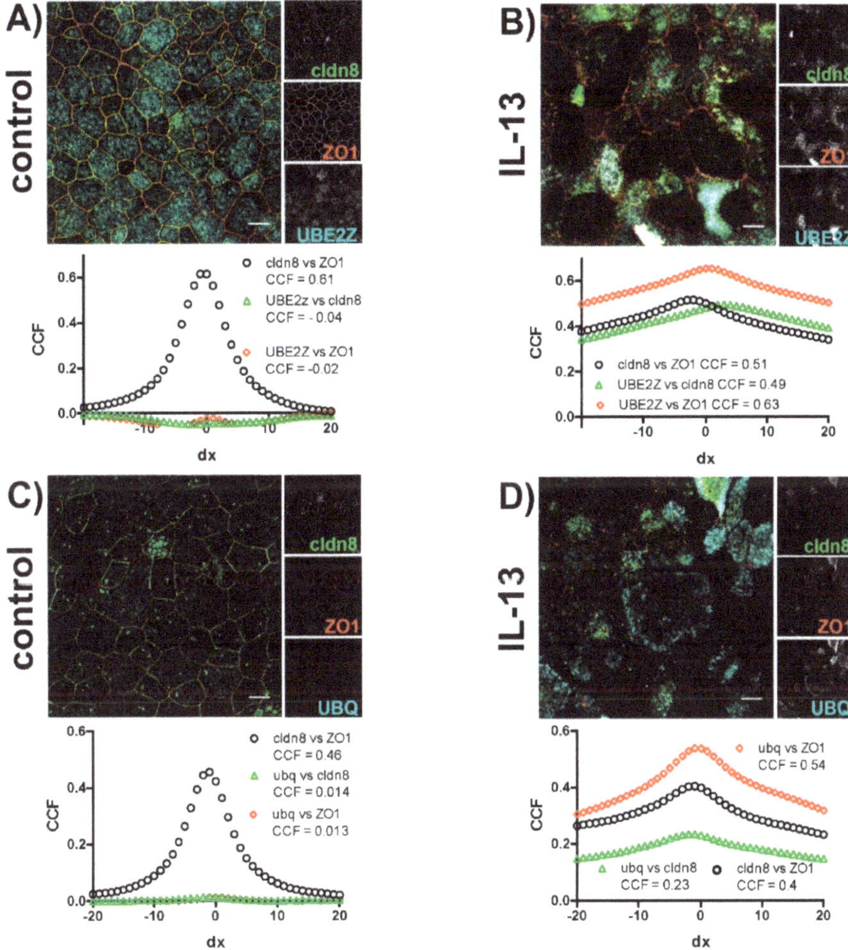

Figure 5. *Cont.*

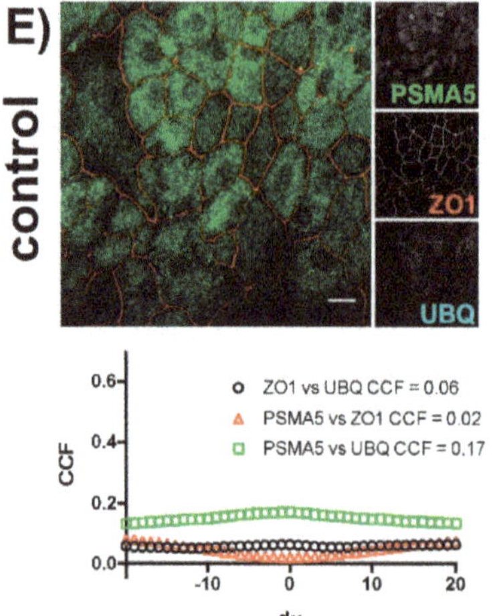

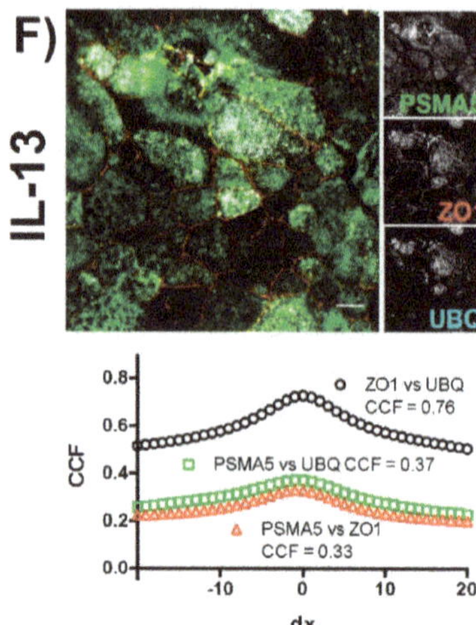

Figure 5. Colocalization of TJ proteins and proteins of the ubiquitin-proteasome machinery. Human tracheal epithelial cells were cultivated at air-liquid interface in the absence (control) and in the presence of IL-13 (IL-13). Images show merged fluorescence channels. Insets at the right give gray scale image for each channel. Van Steensel's plot below each image gives a cross-correlation coefficient (CCF) plotted against lateral delocalization (dx = 0 to ±20 pixel). Maxima of cross-correlation coefficient plot are given in the legend for each plot. (**A**) Control epithelia and (**B**) IL-13 treated epithelia immune-stained for Claudin 8 (cldn8, green channel), zonula occludens 1 (ZO1, red channel) and UBE2Z (cyan channel). (**C**) Control epithelia and (**D**) IL-13 treated epithelia immune-stained for Claudin 8 (cldn8, green channel), zonula occludens 1 (ZO1, red channel) and Ubiquitin (UBQ, cyan channel). (**E**) Control epithelia and (**F**) IL-13 treated epithelia immune-stained for proteasome subunit α Type 5 (PSMA5, green channel), zonula occludens 1 (ZO1, red channel) and ubiquitin (UBQ, cyan channel). Scale bars in (**A**–**F**) represent 10 μM.

2.3. *Effect of IL-13 on Mature hTEpC Epithelia*

In order to test if IL-13 also acts on mature epithelia, we added IL-13 at a final concentration of 10 ng/mL to hTEpC epithelia on day 18 after ALI establishment. Measurements were performed on the following days as indicated (Figure 6). In mature hTEpC epithelia, IL-13 reduced TEER from day 21 onwards (Figure 6A). The drop in TEER reached statistical significance on day 28 (control vs. IL-13 treated cells, $N = 12$, Mann-Whitney test $p = 0.0011$). The time course of TEER decrease correlated with time course of cldn8 expression levels, which decreased from day 21 onwards, three days after IL-13 exposure (Figure 6B). The decrease in cldn8 expression levels reached statistical significance from day 24 onwards. The increase in UBE2Z expression levels preceded TEER decrease and occurred already within the first day after IL-13 exposure (Figure 6C). It reached significant levels from day 19 onwards.

Time course of ubiquitination of TJ proteins was investigated in immunocytochemical experiments (Figure 6D–I). In these experiments, we used van Steensel's cross-correlation analysis [46] to quantify colocalization of ZO1 with cldn8 as well as of UBQ with ZO1 and cldn8. Time course of colocalization was visualized by plotting the maximum Pearson cross correlation coefficient detected from the van Steensel's plot against cultivation time.

In control cells cross-correlation coefficient for ZO1 and cldn8 colocalization increased from day 19 onwards (Figure 6H). This may reflect gradual tightening of TJ in mature epithelia during ALI cultivation, as it correlates with the slightly increasing TEER (compare Figure 6A). Furthermore, in these control cells no change in cross-correlation over time was observed for ZO1 and cldn8 with UBQ (Figure 6H).

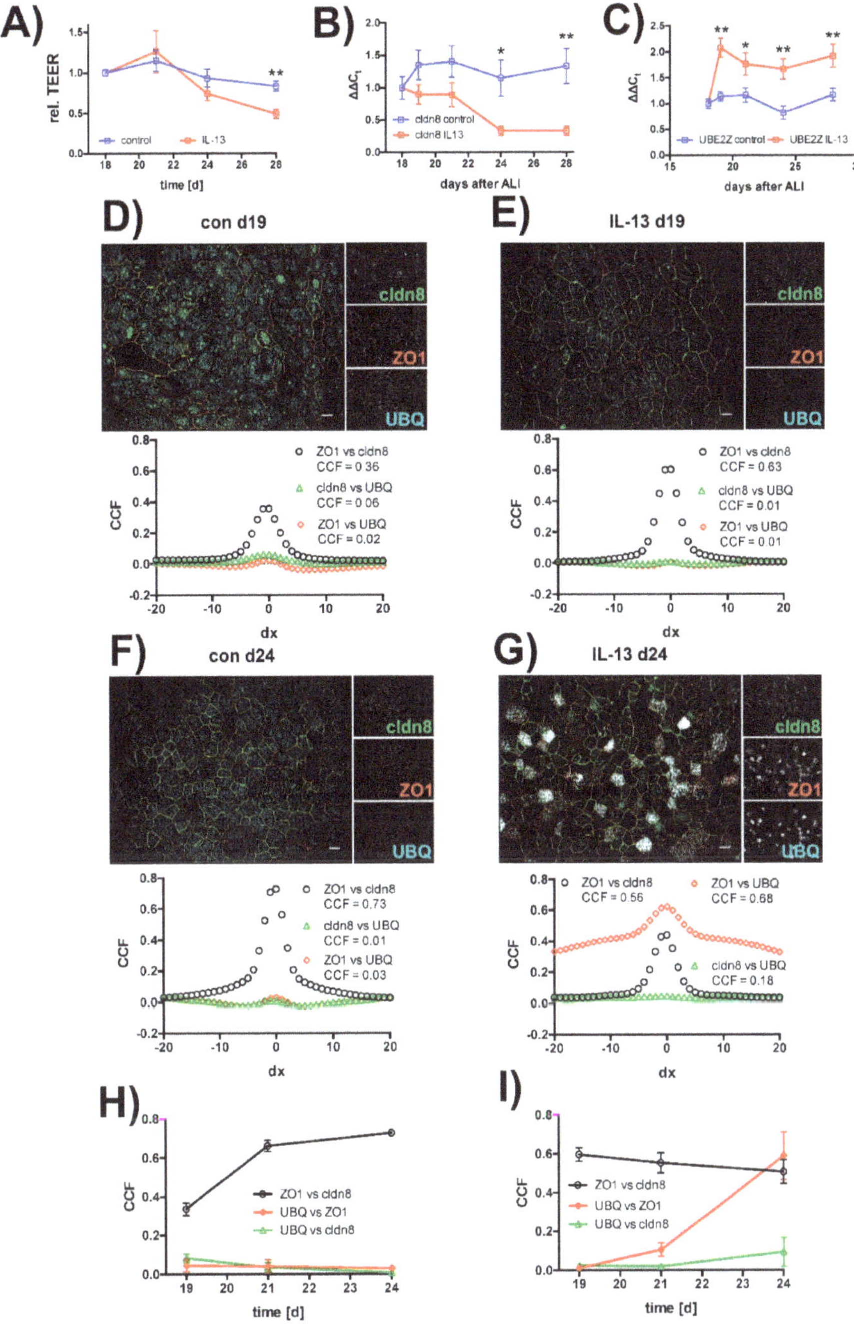

Figure 6. Time course of IL-13 induced effects on human tracheal epithelia. Human tracheal epithelial cells were seeded on Transwell filters and proliferated for 4 days. Afterwards, apical volume was removed and cells were cultivated for additional 18 days at air-liquid interface conditions. On day 18, IL-13 was added to the medium (IL-13). Control cells were not exposed to IL-13 (control). (**A**) Changes

of transepithelial electrical resistance (TEER) over time. TEER was normalized to mean TEER measured on day 18 directly before IL-13 exposure and relative TEER was plotted against cultivation time. Rel. TEER decreases from day 21 onwards. Decrease reached statistical significance on day 28. Data points summarize measurements as mean ±SEM (all N = 10 to 12, Mann-Whitney test $p = 0.001$). (**B**) Changes of claudin 8 (cldn8) expression levels over time. Changes in cldn8 expression are given as ΔΔCt values relative to the mean expression level on day 18. Data points give mean ΔΔCt ± SEM. Cldn8 expression drops from day 21 onwards. It reached significance on days 24 and 28 (all N = 9, Mann-Whitney test $p = 0.008$ and $p = 0.0002$, day 24 and 28, respectively). (**C**) Time course of UBE2Z expression. Changes in UBE2Z expression are given as ΔΔCt values relative to mean expression level on day 18 as mean ΔΔCt ± SEM. UBE2Z expression levels rose directly upon IL-13 exposure an reached significance from day 19 onwards (all N = 9, Mann-Whitney test $p = 0.0012$ day 19, $p = 0.039$ day 21, $p = 0.0027$ day 24 and $p = 0.0019$ day 28). Colocalization of tight junction proteins claudin 8 (cldn8, green channel) and zonula occludens 1 (ZO1, red channel) with ubiquitin (UBQ, cyan channel) was analyzed in van Steensel's plot (below each image). Examples are given for (**D**) control cells on day 19 (con d19) and (**E**) for cells exposed to IL-13 on day 19 (IL-13 d19) as well as (**F**) for control cells on day 24 (con d24) and (**G**) for cells exposed to IL-13 on day 24 (IL-13 d24). Changes of colocalization over time are analyzed by plotting the maximum of cross-correlation coefficients (CCF) from van Steensel's plot against the cultivation time. Changes of maximum CCF over time (**H**) in control cells and (**I**) in IL-13 treated cells. Black curves give CCF for zonula occludens 1 and claudin 8 (ZO1 vs. cldn8), red curves show CCF for ubiquitin and ZO1 (UBQ vs. ZO1) and green curves show CCF for UBQ and cldn8 (UBQ vs. CLDN8). Data points summarize CCF as mean ± SD all N = 3). * and ** indicate $p < 0.05$ and $p < 0.01$, respectively. Scale bars in (**D–G**) represent 10 μM.

In case of IL-13 treated epithelia, the cross-correlation coefficient for the colocalization of cldn8 with ZO1 did not change over time. In contrast to that and to control cells, cross-correlation coefficients for UBQ colocalization with ZO1 and cldn8 both increase, starting three days after IL-13 exposure, from day 21 onwards. However, the change in colocalization for cldn8 with UBQ was smaller compared to ZO1 with UBQ (Figure 6I) and most likely reflects a weaker staining efficiency and/or protein abundance of cldn8.

The onset of UBQ colocalization with TJ proteins correlates with onset of TEER decrease. The formation of protein aggregates does not start in all cells simultaneously. Instead, we observed an increasing amount of cells over time that were positive for granules containing tight junction proteins colocalizing with ubiquitin. On day 19 almost no cells were positive for protein aggregates. On day 21, when TEER starts to drop and the cross correlation coefficient starts to rise, 9.3 ± 5.1% of IL-13 treated cells formed aggregates (mean ± SD, N = 3). The amount of aggregate containing cells further increased to 23.6 ± 4.6% (mean ± SD, N = 3) on day 24. Thus, the time course of protein aggregate formation correlated with time course of epithelial permeability increase. This gives evidence that protein aggregate formation is the TJ damaging step in IL-13 exposed epithelia. These experiments indicate that IL-13 activates the ubiquitin-proteasome pathway that also involves UBE2Z in mature hTEpC epithelia.

2.4. Ubiquitination of TJ Proteins in Sinunasal Epithelia

To test whether the detected IL-13 induced ubiquitination and proteasomal aggregation also occurs in human airway epithelia, we investigated sinunasal biopsies from patients suffering from chronic inflammation with and without eosinophilia using PLA (Figure 7). These experiments revealed the formation of protein complexes that contain ZO1, cldn4 and cldn8 together with UBQ and UBE2Z. Especially in the absence of bacterial infection, eosinophilic inflammation is a fairly good and specific surrogate for high IL-13 levels [47]. Since measuring IL-13 levels in paraffin embedded tissue biopsies is erroneous, we used eosinophilia as a marker for elevated IL-13 tissue concentrations instead. There was no difference in number of PLA-positive particles detectable in eosinophilic inflammation in comparison to non-eosinophilic inflammation (all as mean ±SD in counts/μm^2 and N = 3 patients.

Eosinophilic rhinosinositis: UBQ/cldn4: 0.64 ± 0.18, UBQ/cldn8: 0.78 ± 0.41, UBQ/ZO1: 0.29 ± 0.14 and UBE2Z/cldn4: 0.38 ± 0.02, UBE2Z/cldn8: 0.4 ± 0.19, UBE2Z/ZO1: 0.51 ± 0.26 and non-eosinophilic rhinosinostis: UBQ/cldn4: 0.83 ± 0.28, UBQ/cldn8: 0.56 ± 0.14, UBQ/ZO1: 0.48 ± 0.33 and UBE2Z/cldn4: 0.41 ± 0.06, UBE2Z/cldn8: 1.15 ± 0.89, UBE2Z/ZO1: 0.3 ± 0.05). However, these experiments confirm that UBE2Z dependent ubiquitination is not restricted to in vitro models of airway epithelia but also occurs in vivo.

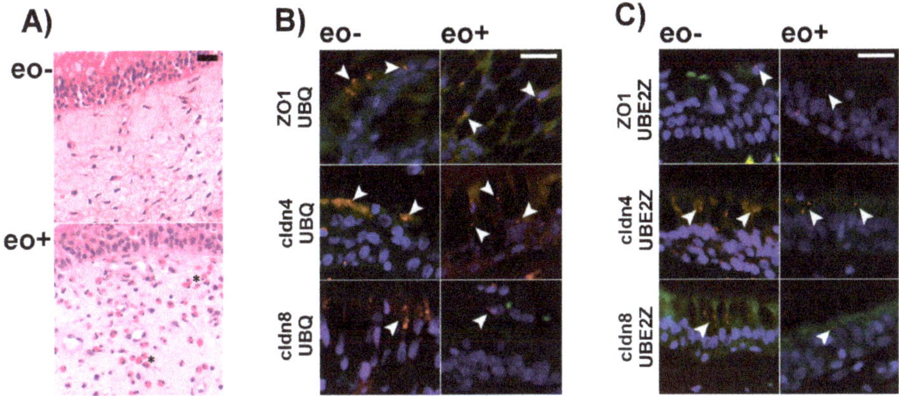

Figure 7. Proximity ligation assays (PLA) were performed on formalin fixed and paraffin embedded tissue from the sinunasal tract containing respiratory type epithelium of patients with and without eosinophilic inflammation. (**A**) shows representative Hematoxilin and Eosin stained images from nasal mucosa with (eo+) and without (eo−) eosinophilic inflammation. Eosinophilic granulocytes can be detected as reddish stained cells in eo+ mucosa (examples are highlighted by asterix) but were not detected in eo- mucosa. For the PLA merge images of nuclear counterstain (blue), autofluorescence (green) and PLA signals (red) are shown. Antibodies specific for ZO1, Claudin 4 and Claudin 8 were combined with antibodies detecting Ubiquitin (**A**) and Ube2Z (**B**). Specific PLA reaction signals (depicted by arrow heads) can be detected in the epithelium under all conditions, suggesting association between ZO1, Claudin 4 and Claudin 8 with Ubiquitin and UBE2Z respectively. Scale bars are 20 µm in panels (**A–C**).

3. Discussion

Early studies on transgenic mice revealed the dominating role of IL-13 in allergic asthma [4]. A more recent study identified IL-13 to modulate TJ permeability via altering claudin expression levels [48]. In accordance with these studies, we demonstrated that long-term exposure of epithelial cells to IL-13 results in perturbed claudin expression pattern and reduced protein densities at the TJs itself. A reduced protein density at the apico-lateral junctions of airway epithelia as a result of T_h2 cytokine exposure was already described [25,48] and was recently demonstrated to be linked to IL-13/IL-4 induced activation of the JAK/STAT signaling pathway [25], as we also demonstrated herein. Inhibition of either cytokines or their receptors have beneficial effects in patients with allergic asthma (summarized in reference [49]). However, most studies measured the reduction of inflammatory markers or parameters of lung function to quantify the therapeutic effect of IL-4/IL-13 inhibition but not changes in tightness of TJ. Hence these studies cannot be directly compared with the herein reported observations. In inflammatory bowel diseases, IL-13 acts predominantly on transcriptional levels. It seems to increase intestinal epithelial permeability by increasing cldn2 expression levels via activating STAT6, PI3K and MAPK pathways and in addition via reduction of tricellulin expression as a result of JNK, AP1 and ERK1/2 signaling [50].

Herein we demonstrate that IL-13 activates the ubiquitin-proteasome system, which involves UBE2Z and finally results in proteasomal aggregation of TJ proteins.

The ubiquitin-proteasome system was formerly described as a system that regulates non-lysosomal protein degradation [39]. It involves the marking of proteins that are targeted for degradation by E1-activating enzymes, E2 conjugases and finally E3 ubiquitin ligases. This sequential process results in a highly regulated and substrate specific degradation of proteins [39]. Later on, investigations revealed that UBQ dependent mechanisms also modulate the abundance of membrane spanning proteins at the plasma membrane and controls their lysosomal as well as non-lysosomal degradation [51–53]. Nowadays, it is commonly accepted that the ubiquitin-proteasome system is pivotal in maintaining cellular protein homeostasis via controlling protein turnover, in quality control of protein folding and in guiding of proteins to different subcellular destinations [54]. The ubiquitin-proteasome system has been shown to regulate protein abundance at TJ of cldn1, cldn2 and cldn4 in Madin-Darby Kidney cells (MDCK) cells [55], of cldn5 in endothelial cells [43], of cldn8 in kidney epithelia [40], of cldn4 in rat salivary gland epithelial cells [56], of ocldn in porcine epithelial cells [45], in vascular endothelial cells [57] and in Sertoli cells [42] and of zonula occludens protein 2 in MDCK cells [58]. These studies unequivocally demonstrate that ubiquitin dependent mechanisms are involved in regulating protein composition of TJ and thereby adjust their functional properties. However, the function of ubiquitination of TJ proteins is complex. On one hand, ubiquitination has been shown to label cldn1, cldn2 and ocldn for removal from the TJ via endocytosis and their transport along the endosomal and lysosomal pathway [55,57]. On the other hand, ubiquitination targets cldn5, cldn4 and also ocldn for proteasomal degradation [42,43,45,56]. Thereby, ubiquitination does not only act as a control mechanism to ensure appropriate protein folding or to adjust the protein half-life. It also becomes activated in response to environmental factors to adjust functional properties of TJs, as has been demonstrated for the cAMP dependent cldn4 regulation in Sertoli cells [42] and for ocldn modulation upon muscarinic acetylcholine receptor stimulation [56]. We herein identified IL-13 as another factor that activates the ubiquitin-proteasome system to increase the permeability of TJs. PLA revealed that upon IL-13 exposure, ubiquitinated TJ proteins localized intracellularly rather than at the TJ or anywhere else at the plasma membrane. Sorting and translocation of internalized TJ proteins from endosomal compartments towards the proteasome upon ubiquitination was recently reported for cldn5 [43]. In this study, the authors proposed a derlin1 dependent pathway [43] that presumably involves retrograde transport along endosomal compartments to the endoplasmic reticulum and employs endoplasmic reticulum quality control machinery to translocate the proteins to the proteasome [59,60]. Similar mechanisms could be involved in IL-13 induced proteasomal accumulation of TJ proteins in lung epithelia.

We observed an instantaneous upregulation of the E2 conjugase UBE2Z. The cDNA of UBE2Z was recently cloned and identified to encode an ubiquitin conjugating enzyme [61]. Later reports suggested using UBE2Z to regulate ubiquitin charging of ubiquitin E3 ligases and provide substrate specificity of protein ubiquitination [62,63]. Our experiments strongly support the hypothesis that IL-13 induces ubiquitination and proteasomal accumulation of TJ proteins that are involved in modulation of the paracellular diffusion barrier. Subcellular localization of E-cadherin that forms adherens junctions via homotypic trans-interaction and mediates the dynamic connection of adjacent lateral membranes [64] remained almost unaffected by IL-13. This indicates that lateral inter-cellular adhesion and integrity of epithelial cell layer is not preliminarily disturbed by IL-13. This specificity to TJ is possibly mediated by UBE2Z.

Time course of ubiquitin-mediated regulation of protein abundance at the TJs is reported as a rather fast process, operating within a few hours after stimulation [43,45,56–58]. In line with these studies, we observed an instantaneous upregulation of UBE2Z expression, indicating that lung epithelial cells rapidly respond to IL-13 exposure. However, measurable effects on TJ permeability were delayed and correlated with the reduction of protein abundance at the TJ as well as their accumulation at the proteasome. As recently reviewed, aggregation of poly-ubiquitinated proteins at the proteasome occurs in post-ischemic neurons and results from down-regulation of proteasome activity [65]. Indeed, proteasomal hyperactivity reduces aggregation of poly-ubiqutinated proteins [66]. Proteasomal activity depends on intra-cellular ATP [67]. A drop in intra-cellular ATP levels as a result of inflammatory stress conditions (as was introduced herein by IL-13 exposure) in combination with elevated protein transport to the proteasome and proteasomal degradation could exhaust proteasome activity to such an extent that proteins accumulate and aggregate. Impaired proteasomal activity is discussed to affect cellular proteostasis and therefore cause cellular stress and/or damage and also interfere with the location of proteins and the composition of polymeric protein complexes [68]. Such an assumption is in line with a recent study providing evidence that proteasome inhibition in mouse models of inflammatory bowel disease interrupts epithelial barrier function in colonic mucosa [69] and would also explain our observation of the delayed TJ damage as a consequence of impaired intracellular proteostasis.

Because of its distinct function in controlling intra-cellular proteostasis, the ubiquitin proteasome system is believed to play a pivotal role in inflammatory lung diseases and lung injury [70]. Genes involved in protein ubiquitination are up-regulated in the lung tissue of patients with chronic obstructive pulmonary disease (COPD) [71]. This goes in line with the observations that protein ubiquitination is increased in the lung tissue of COPD patients with emphysema [72]. Furthermore, cigarette smoke extract also induces aggregation of poly-ubiquitinated proteins in bronchial epithelial cells [73] and increases protein ubiquitination in lung endothelial cells [74]. These findings underscore the involvement of dysregulation of the ubiquitin proteasome system in the pathogenesis of chronic inflammatory lung diseases like COPD. Our findings are in line with these former observations. They demonstrate that IL-13 results in an incomplete proteasomal degradation of ubiquitinated proteins that coincide with TJ disturbance. They further point towards a yet unknown function of UBE2Z as an effector in IL-13 driven lung inflammation acting on TJs and thereby on paracellular barrier function in pulmonary epithelia. Opening of TJs of the pulmonary epithelium is a hallmark in the pathogenesis of alveolar edema and exudate formation and subsequent respiratory distress [30]. Disturbances of TJs are a predisposing condition for the development of acute respiratory distress syndrome [75] and lung injury for instance following mechanical ventilation [76,77].

Our results gives evidence that UBE2Z dependent ubiquitination of TJ-proteins also occurs in vivo. We used chronic inflammation with eosinophilic granulocytes as surrogate marker for an IL-13 high phenotype [47]. However, there was no significant difference in levels TJ-protein ubiquitination between samples with and without eosinophilic inflammation. Detection of eosinophilia does not measure cytokine concentrations. IL-13 concentrations for eosinophilic non-bacterial lung inflammation was recently measured to reach concentrations of 10.4 pg/mL in the sputum of humans [47] and around 100 ng/mL in bronchoalveolar lavage of IL-13 overexpressing transgenic mice [78]. These measurements may give an estimation of the range of IL-13 tissue levels during eosinophilic lung inflammation. However, it must be considered that eosinophilia reflects chronic inflammation, while a commonly used nasal steroid therapy in patients with chronic rhinosinusitis potentially reduces inflammatory response and cytokine levels. Under these conditions, IL-13 concentrations are probably below the levels that are needed to initiate the observed TJ protein ubiqutination in the herein investigated in vitro model. However, IL-13 tissue levels might increase and become effective to induce the observed TJ protein ubiquitination during exacerbation episodes. Nonetheless, our results demonstrate that TJ protein ubiquitination that involves UBE2Z also occurs in native tissue.

4. Materials and Methods

4.1. Compounds

IL-4 and IL-13 were obtained from Merck Millipore, Darmstadt, Germany. S-.Ruxolitinib and Tofacitinib, both were obtained from Selleckchem, distributed by Absource Diagnostics, Munich, Germany. If not mentioned otherwise, IL-13 was added at a concentration of 10 ng/mL. S-Ruxolitinib and Tofacitinib was used at concentrations of 2.5 μM each.

4.2. Cell Culture

Transwell permeable supports (Polyester membrane diameter 6.5 mm, 0.4 µm pore size, Costar distributed by Omnilab, Bremen, Germany) were overlaid with 100 µL Collagen I and II solution (StemCell Technologies, Cologne, Germany) diluted 1:100 in phosphate buffered Saline (PBS, Gibco distributed by Thermo Fisher, Darmstadt, Germany) and dried over night at room temperature. Dried filters were UV irradiated and stored until use at 4 °C.

Cryo-preserved passage 1 primary human tracheal epithelial cells (Promocell, Heidelberg Germany and Epithelix, Genève Switzerland) were proliferated in T75 flasks (Sarstedt, Nümbrecht, Germany) using Airway Epithelial Growth Medium (Promocell, Heidelberg, Germany) at 37 °C, 5% CO_2 and 95% humidity. Upon reaching 80% confluence, cells were detached using DetachKit (Promocell, Heidelberg, Germany) and suspended in Airway Epithelial Growth medium. 3.5×10^4 cells in 200 µL Airway Epithelial Growth Medium were seeded onto collagen-coated filters. Filters were placed in 24 well plates filled with 500 µL Airway Epithelial Growth Medium and incubated at 37 °C, 5% CO_2, 95% humidity for two days. Afterwards, air-liquid interface (ALI) conditions were established by removing the medium from the apical surface and replacing Airway Epithelial Growth Medium by 500 µL of fresh ALI-Medium. ALI-medium contained DMEM-H and LHC basal medium (both from Gibco via Thermo Fisher, Darmstadt, Germany) at a 1:1 ratio and was supplemented with: insulin 0.87 µM, hydrocortisone 0.21 µM, epidermal growth factor 0.5 ng/mL, triiodothyronine 0.01 µM, transferrin 0.125 µM, epinephrine 2.5 µM, bovine pituitary extract 10 µg/mL, bovine serum albumin 0.5 mg/mL (all from Promocell, Heidelberg, Germany), phosphorylethanolamine 0.5 µM, ethanolamine 0.5 µM, zinc sulfate 3 µM, retinoic acid 0.05 µM, ferrous sulfate 1.5 nM, calcium chloride 0.6 µM, magnesium chloride 0.11 µM, sodium selenite 30 µM, manganese chloride 1 µM, sodium silicate 0.5 nM, ammonium molybdate tetrahydrate 1 µM, ammonium metavanadate 5 µM, nickel sulfate 1 µM, tin chloride 0,5 µM (all from Sigma-Aldrich GmbH, Steinheim, Germany), 100 U/mL Penicillin and 100 µg/mL Streptomycin (both from Gibco via Thermo Fisher, Darmstadt, Germany). Medium was replaced every second day. Cells were incubated at 37 °C, 5% CO_2 and 95% humidity for the time given in the text.

4.3. TEER Measurement

Transepithelial electrical resistance (TEER) was measured by impedance spectroscopy using the cellZscope (NanoAnalytics, Münster, Germany). For measurements, the basal electrode was overlaid by 500 µL equilibrated ALI medium. After inserting filters, 100 µL of ALI medium were added to the apical surface and the apical electrode was placed into the apical liquid. The measurements were performed immediately after positioning of apical electrodes., The software package provided with the instrument (NanoAnalytics, Münster, Germany) was used for data acquisition and analysis.

4.4. Paracellular Permeability

The paracellular permeability was measured as the apparent permeability coefficient (P_{app}) of Na^+-fluorescein (Sigma-Aldrich GmbH, Steinheim, Germany). Na^+-fluorescein was added to the basolateral medium at a concentration of 150 µM. The Apical surface was overlaid with 100 µL of isotonic NaCl-solution. After incubation for 30 min at 37 °C, 95% humidity and 5% CO_2, the Na^+-fluorescein concentration was measured in the apical solution using an Infinite 200 plate reader with iControl software package (Tecan, Männedorf, Switzerland). P_{app} was calculated according to the equation: $P_{app} = (c_{ap} \times V_{ap})/(c_0 \times t \times A)$ with c_{ap} = Na^+-fluorescein concentration measured in the apical compartment after incubation, V_{ap} = Volume of isotonic NaCl-solution added to the apical side, C_0 = Na^+-fluorescein concentration in the basolateral compartment at time point t = 0, t = incubation time and A = epithelial surface area.

4.5. qRT-PCR

Total RNA was isolated from cells grown at ALI conditions in the absence or presence of IL-13, as indicated. Cell lysis was performed using my-Budget RNA Mini Kit (Biobudget, Krefeld, Germany) without prior cell suspension using the lysis buffer provided with the kit. RNA was isolated according to manufacturers' protocol. First strand synthesis was carried out on 0.6 to 0.8 µg total RNA using the VILO SuperScript cDNA Synthesis Kit and qRT-PCR was performed using SYBR GreenER qPCR Supermix (all from Thermo Fisher, Darmstadt, Germany) according to manufactures' protocols with Quantitect RT-PCR primer assays (Qiagen, Hilden, Germany). Quantitect primer assays: Cldn1 QT00225764, Cldn2 QT00089481, Cldn3 QT00201376, Cldn4 QT00241073, Cldn5 QT01681232, Cldn6 QT00235193, Cldn7 QT00236061, Cldn8 QT00212268, Cldn9 QT00209482, Cldn10 QT00031101, Cldn11 QT00008085, Cldn12 QT01012186, Cldn14 QT00234731, Cldn15 QT01017723, Cldn16 QT00039655, Cldn17 QT01018507, Cldn18 QT00039550, Cldn19 QT00083475, Cldn20 QT00218057, Occludin QT00081844, Hmbs QT00014462, UBE2Z QT0082516 (all from Qiagen, Hilden, Germany). To screen for transcripts encoding ubiquitin activating enzymes E1 and ubiquitin conjugating enzymes E2 ubiquitination RT^2 Profiler PCR array and RT^2 SYBR Green qPCR Mastermix were used (all from Qiagen, Hilden, Germany). RealPlex2 thermocycler with latest version of Realplex data acquisition and analysis software were used (Eppendorf, Hamburg, Germany). PCR cycling protocol: Initial denaturation and Taq activation 95 °C for 2 min, 40 cycles with denaturing 95 °C for 15 s, primer annealing and amplification 60 °C 75 s. PCR products were verified by determining melting curves, final denaturation 95 °C for 15 s, annealing at 60 °C for 15 sec and heat ramping from 60 to 95 °C with a ramp duration of 20 min. Crossing points were determined using the CalPlex method provided with the data acquisition and analysis software. Each target gene expression was determined as a triplet. PCR efficiency, correction for efficiency and relative expression were calculated as previously described [79].

4.6. Immunocytochemistry

Epithelial cells cultivated as ALI with and without exposure to IL-13 were washed with ice cold PBS and immediately fixed with ice-cold methanol for 1 min. After methanol removal and washing with PBS, filters and cells were incubated in blocking buffer (PBS with 1% Triton X-100 and 5% BSA) for 5 min at room temperature (RT). After removal of blocking buffer, primary antibody incubation was performed (1 h, RT). Antibodies were diluted in Diluent buffer (PBS with 0.2% Triton X-100 and 0.5% BSA). The following antibodies were used: anti-claudin 4 (ab53156), anti-ZO1 (ab99462), anti-ubiquitin (ab7254), anti-proteasome α-subunit type 5 (PSMA5, ab189855), anti-E-cadherin (ab1416) (all from abcam, Cambridge, UK), anti-claudin 8 (LS-B5505) (Life Span Bioscience, Seattle, WA, USA), anti UBE2Z (H00065264-M01) (Abnova, Taipei, Taiwan) and anti-occludin (MAB7074) (R&D Systems, Minneapolis, MN, USA). Antibodies were removed by three-times washing with PBS buffer and filters were incubated for 30 min at RT with secondary antibodies diluted in Diluent buffer. Secondary antibodies were: Alexa Fluor 488 donkey anti-rabbit IgG, Alexa Fluor 568 donkey anti-goat IgG and Alexa Fluor 647 donkey anti-mouse IgG (all from Invitrogen, Karlsruhe, Germany). Hoechst 33342 (Invitrogen, Karlsruhe, Germany) was used for staining of the nuclei. Afterwards, filters were washed with PBS and mounted on coverslips using ibidi mounting medium (ibidi, Martinsried, Germany). Images of individually cultivated epithelia were taken from a randomly selected area close to the filter center on an inverted confocal microscope (Leica TCS SP5, Leica, Germany) using a 40× lens (Leica HCX PL APO CS 40× 1.25 oil). Images for the blue (DAPI), green (AlexaFluor 488), red (AlexaFluor 568) and far-red (AlexaFluor 647) channels were taken in sequential mode using appropriate excitation and emission settings. Images represent sections with 90 to 150 cells each. Image analysis was performed using latest version of ImageJ software (NIH, Bethesda, MD, USA) [80]. Colocalization analysis was performed according to the method established by van Steensel [46]. In brief: Cross-correlation of a dual labeling image was calculated by shifting one channel image pixelwise over a distance of ±20 pixel with respect to the other channel. Pearson's correlation coefficient was calculated and the CCF-value was obtained by plotting Pearson's correlation coefficient against the translocation of channels, Δx. A maximum of this function at $\Delta x = 0$ indicates overlapping fluorescence signals, a minimum of this function at $\Delta x = 0$ excludes overlapping fluorescence signals. A randomly distribution of fluorescence signals was indicated by a function without any minima/maxima.

4.7. Proximity Ligation Assay

Adjacent proteins were identified by proximity ligation assays (PLA) using the Duolink kit (Sigma Aldrich, Steinheim, Germany) on methanol fixed and permeabilized epithelia (primary antibodies see above). Secondary antibodies: Anti-Mouse MINUS DUO92004, Anti-Rabbit PLUS DUO92002, Goat Plus 82023 (supplied with the Duolink kit). PLA was performed according to the manufacturer's protocol. IMIC digital microscope (Till Photonics, Muenchen, Germany) equipped with UApo/340 40× /1.35 oil objective (Olympus, Hamburg, Germany) and latest version of Live Acquisition were used for microscopy, (Till Photonics, Muenchen, Germany). Images were taken for each individually cultivated filter from a single randomly selected region close to the filter center. PLA signals were quantified using latest version of ImageJ software [80] after binarization of particles with an area of 75 μm^2 and above.

4.8. Immunohistochemistry

Formalin-fixed paraffin embedded tissue of sinunasal biopsies with chronic inflammation with and without eosinophilia were selected randomly from the archives of the Institute for Pathology (Hannover Medical School). For evaluation 4 µm thick sections were taken and stained with the proximity ligation assay Duolink (Sigma-Aldrich GmbH, Steinheim, Germany). Primary and secondary antibodies are given above. PLA was performed according to manufacturer's protocol. Afterwards, the reaction to specimens was embedded with a medium containing DAPI. Representative microscopy images of respiratory epithelia were taken on a Keyence fluorescence microscope (BZ9000) with a 20× magnification lens using the appropriate filter sets to detect DAPI (Excitation 377/50 nm, Beamsplitter 409 nm, Emission 447/60 nm), green (autofluorescence) (Excitation 472/50 nm, Beamsplitter 495 nm, Emission 520/35 nm) and far red (Excitation 628/40 nm, Beamsplitter 660 nm, Emission 692/40 nm) signals. Microscopy images were fused using ImageJ. After linear intensity adjustment, a region of interest containing in-focus and intact respiratory epithelium was marked. PLA signals within this area were manually counted. Recorded values are given as counts per area. All experiments were performed with approval of the ethics committee of the Hannover Medical School (Project no. 2699-2015, approved April 2015).

4.9. Statistical Analysis

GraphPad Prims6 software (GraphPad, La Jolla, CA, USA) was used for statistical analysis (tests are given with the text). For each experiment cells were obtained from 2 to 4 individual donors. Number of experiments (N) gives number of individual cultivated filters. Differences were considered to be significant if calculated p-values were $p \leq 0.05$. Significance levels are given within the diagrams as: $0.05 \geq p > 0.01$ = *, $0.01 \geq p > 0.001$ = **, $0.001 \geq .001$ = *, p-levels are given as 0.00001 = ****. Used test methods and exact p-values are given within the figure legends.

Author Contributions: Conceptualization, H.S. and O.H.W.; Data curation, H.S., P.B. and O.H.W.; Formal analysis, P.B., C.S., R.L. and O.H.W.; Funding acquisition, M.F., P.D. and O.H.W.; Investigation, H.S., P.B., C.S., R.L., K.N., K.T. and O.H.W.; Methodology, H.S.; Resources, D.J., M.F., P.D. and O.H.W.; Supervision, P.D.; Writing—original draft, H.S. and P.B.; Writing—review & editing, K.T., D.J., M.F., P.D. and O.H.W.

Funding: This work was supported by the Ministry of Science and Arts of Baden-Würthemberg (32-7533-6-10/15/5), German Research Foundation (DFG) (Project Nr. 175083951) and Pulmosens GRK 2203.

Conflicts of Interest: The authors declare no conflict of interest.

Abbreviations

ALI	air-liquid interface
CCF	cross-correlation coefficient
cldn	claudin
E-cad	E-cadherin
hTEpC	human tracheal epithelial cells
IL	interleukin
JAK	janus kinase
JAM-A	junctional adhesion molecule A
MDCK	Madin-Darby kidney cells
muc5ac	mucin 5AC
ocldn	ocludin
P_{app}	apparent permeability coefficient
PLA	proximity ligation assay
PSMA5	proteasomal subunit alpha type 5
RT-PCR	reverse transcriptase PCR
STAT	signal transducer and activator of transcription
TEER	transepithelial electrical resistance
T_h2	T-helper cell type 2
TJ	tight junction
UBE2Z	ubiquitin conjugating enzyme 2Z
UBQ	ubiquitin
ZO1	zonula occludens 1 protein

References

1. Lötvall, J.; Akdis, C.A.; Bacharier, L.B.; Bjermer, L.; Casale, T.B.; Custovic, A.; Lemanske, R.F., Jr.; Wardlaw, A.J.; Wenzel, S.E.; Greenberger, P.A. Asthma endotypes: A new approach to classification of disease entities within the asthma syndrome. *J. Allergy Clin. Immunol.* **2011**, *127*, 355–360. [CrossRef] [PubMed]
2. Robinson, D.; Humbert, M.; Buhl, R.; Cruz, A.A.; Inoue, H.; Korom, S.; Hanania, N.A.; Nair, P. Revisiting Type 2-high and Type 2-low airway inflammation in asthma: Current knowledge and therapeutic implications. *Clin. Exp. Allergy* **2017**, *47*, 161–175. [CrossRef] [PubMed]
3. Del Prete, G. The concept of type-1 and type-2 helper T cells and their cytokines in humans. *Int. Rev. Immunol.* **1998**, *16*, 427–455. [CrossRef] [PubMed]
4. Zhu, Z.; Homer, R.J.; Wang, Z.; Chen, Q.; Geba, G.P.; Wang, J.; Zhang, Y.; Elias, J.A. Pulmonary expression of interleukin-13 causes inflammation, mucus hypersecretion, subepithelial fibrosis, physiologic abnormalities, and eotaxin production. *J. Clin. Investig.* **1999**, *103*, 779–788. [CrossRef] [PubMed]
5. Wills-Karp, M.; Luyimbazi, J.; Xu, X.; Schofield, B.; Neben, T.Y.; Karp, C.L.; Donaldson, D.D. Interleukin-13: Central mediator of allergic asthma. *Science* **1998**, *282*, 2258–2261. [CrossRef] [PubMed]
6. Hanania, N.A.; Noonan, M.; Corren, J.; Korenblat, P.; Zheng, Y.; Fischer, S.K.; Cheu, M.; Putnam, W.S.; Murray, E.; Scheerens, H.; et al. Lebrikizumab in moderate-to-severe asthma: Pooled data from two randomised placebo-controlled studies. *Thorax* **2015**, *70*, 748–756. [CrossRef] [PubMed]
7. Brightling, C.E.; Chanez, P.; Leigh, R.; O'Byrne, P.M.; Korn, S.; She, D.; May, R.D.; Streicher, K.; Ranade, K.; Piper, E. Efficacy and safety of tralokinumab in patients with severe uncontrolled asthma: A randomised, double-blind, placebo-controlled, phase 2b trial. *Lancet Respir. Med.* **2015**, *3*, 692–701. [CrossRef]
8. Baverel, P.G.; White, N.; Vicini, P.; Karlsson, M.O.; Agoram, B. Dose-exposure-response relationship of the investigational anti-interleukin-13 monoclonal antibody tralokinumab in patients with severe, uncontrolled asthma. *Clin. Pharmacol. Ther.* **2018**, *103*, 826–835. [CrossRef]
9. Blease, K.; Jakubzick, C.; Westwick, J.; Lukacs, N.; Kunkel, S.L.; Hogaboam, C.M. Therapeutic effect of IL-13 immunoneutralization during chronic experimental fungal asthma. *J. Immunol.* **2001**, *166*, 5219–5224. [CrossRef]

10. Tripp, C.S.; Cuff, C.; Campbell, A.L.; Hendrickson, B.A.; Voss, J.; Melim, T.; Wu, C.; Cherniack, A.D.; Kim, K. RPC4046, a novel anti-interleukin-13 antibody, blocks IL-13 binding to IL-13 α1 and α2 receptors: A randomized, double-blind, placebo-controlled, dose-escalation first-in-human study. *Adv. Ther.* **2017**, *34*, 1364–1381. [CrossRef]
11. Aghapour, M.; Raee, P.; Moghaddam, S.J.; Hiemstra, P.S.; Heijink, I.H. Airway epithelial barrier dysfunction in chronic obstructive pulmonary disease: Role of cigarette smoke exposure. *Am. J. Respir. Cell Mol. Biol.* **2018**, *58*, 157–169. [CrossRef] [PubMed]
12. Pezzuto, A.; Carico, E. Role of HIF-1 in cancer progression: Novel insights. A review. *Curr. Mol. Med.* **2018**, *18*, 343–351. [CrossRef] [PubMed]
13. Tonini, G.; D'Onofrio, L.; Dell'Aquila, E.; Pezzuto, A. New molecular insights in tobacco-induced lung cancer. *Future Oncol.* **2013**, *9*, 649–655. [CrossRef] [PubMed]
14. Meng, M.; Liao, H.; Zhang, B.; Pan, Y.; Kong, Y.; Liu, W.; Yang, P.; Huo, Z.; Cao, Z.; Zhou, Q. Cigarette smoke extracts induce overexpression of the proto-oncogenic gene interleukin-13 receptor α2 through activation of the PKA-CREB signaling pathway to trigger malignant transformation of lung vascular endothelial cells and angiogenesis. *Cell. Signal.* **2017**, *31*, 15–25. [CrossRef] [PubMed]
15. Venkayya, R.; Lam, M.; Willkom, M.; Grünig, G.; Corry, D.B.; Erle, D.J. The Th2 lymphocyte products IL-4 and IL-13 rapidly induce airway hyperresponsiveness through direct effects on resident airway cells. *Am. J. Respir. Cell Mol. Biol.* **2002**, *26*, 202–208. [CrossRef] [PubMed]
16. Morse, B.; Sypek, J.P.; Donaldson, D.D.; Haley, K.J.; Lilly, C.M. Effects of IL-13 on airway responses in the guinea pig. *Am. J. Physiol. Lung Cell. Mol. Physiol.* **2002**, *282*, L44–L49. [CrossRef] [PubMed]
17. Winkelmann, V.E.; Thompson, K.E.; Neuland, K.; Jaramillo, A.-M.; Fois, G.; Schmidt, H.; Wittekindt, O.H.; Han, W.; Tuvim, M.J.; Dickey, B.F.; et al. Inflammation-induced upregulation of P2X4 expression augments mucin secretion in airway epithelia. *Am. J. Physiol. Lung Cell. Mol. Physiol.* **2019**, *316*, L58–L70. [CrossRef]
18. Kanoh, S.; Tanabe, T.; Rubin, B.K. IL-13-induced MUC5AC production and goblet cell differentiation is steroid resistant in human airway cells. *Clin. Exp. Allergy* **2011**, *41*, 1747–1756. [CrossRef]
19. Kistemaker, L.E.M.; Hiemstra, P.S.; Bos, I.S.T.; Bouwman, S.; van den Berge, M.; Hylkema, M.N.; Meurs, H.; Kerstjens, H.A.M.; Gosens, R. Tiotropium attenuates IL-13-induced goblet cell metaplasia of human airway epithelial cells. *Thorax* **2015**, *70*, 668–676. [CrossRef]
20. Zhang, F.-Q.; Han, X.-P.; Zhang, F.; Ma, X.; Xiang, D.; Yang, X.-M.; Ou-Yang, H.-F.; Li, Z. Therapeutic efficacy of a co-blockade of IL-13 and IL-25 on airway inflammation and remodeling in a mouse model of asthma. *Int. Immunopharmacol.* **2017**, *46*, 133–140. [CrossRef]
21. Ohashi, Y.; Motojima, S.; Fukuda, T.; Makino, S. Airway hyperresponsiveness, increased intracellular spaces of bronchial epithelium, and increased infiltration of eosinophils and lymphocytes in bronchial mucosa in asthma. *Am. Rev. Respir. Dis.* **1992**, *145*, 1469–1476. [CrossRef] [PubMed]
22. Wise, S.K.; Laury, A.M.; Katz, E.H.; Den Beste, K.A.; Parkos, C.A.; Nusrat, A. Interleukin-4 and interleukin-13 compromise the sinonasal epithelial barrier and perturb intercellular junction protein expression. *Int. Forum Allergy Rhinol.* **2014**, *4*, 361–370. [CrossRef] [PubMed]
23. Wawrzyniak, P.; Wawrzyniak, M.; Wanke, K.; Sokolowska, M.; Bendelja, K.; Rückert, B.; Globinska, A.; Jakiela, B.; Kast, J.I.; Idzko, M.; et al. Regulation of bronchial epithelial barrier integrity by type 2 cytokines and histone deacetylases in asthmatic patients. *J. Allergy Clin. Immunol.* **2017**, *139*, 93–103. [CrossRef] [PubMed]
24. Sweerus, K.; Lachowicz-Scroggins, M.; Gordon, E.; LaFemina, M.; Huang, X.; Parikh, M.; Kanegai, C.; Fahy, J.V.; Frank, J.A. Claudin-18 deficiency is associated with airway epithelial barrier dysfunction and asthma. *J. Allergy Clin. Immunol.* **2016**, *139*, 1–10. [CrossRef] [PubMed]
25. Saatian, B.; Rezaee, F.; Desando, S.; Emo, J.; Chapman, T.; Knowlden, S.; Georas, S.N. Interleukin-4 and interleukin-13 cause barrier dysfunction in human airway epithelial cells. *Tissue Barriers* **2013**, *1*, e24333. [CrossRef] [PubMed]
26. Ahdieh, M.; Vandenbos, T.; Youakim, A. Lung epithelial barrier function and wound healing are decreased by IL-4 and IL-13 and enhanced by IFN-gamma. *Am. J. Physiol. Cell Physiol.* **2001**, *281*, C2029–C2038. [CrossRef] [PubMed]
27. Günzel, D. Claudins: Vital partners in transcellular and paracellular transport coupling. *Pflug. Arch. Eur. J. Physiol.* **2017**, *469*, 35–44. [CrossRef]

28. Schlingmann, B.; Molina, S.A.; Koval, M. Claudins: Gatekeepers of lung epithelial function. *Semin. Cell Dev. Biol.* **2015**, *42*, 47–57. [CrossRef]
29. Den Beste, K.A.; Hoddeson, E.K.; Parkos, C.A.; Nusrat, A.; Wise, S.K. Epithelial permeability alterations in an in vitro air-liquid interface model of allergic fungal rhinosinusitis. *Int. Forum Allergy Rhinol.* **2013**, *3*, 19–25. [CrossRef]
30. Holter, J.F.; Weiland, J.E.; Pacht, E.R.; Gadek, J.E.; Davis, W.B. Protein permeability in the adult respiratory distress syndrome. Loss of size selectivity of the alveolar epithelium. *J. Clin. Invest.* **1986**, *78*, 1513–1522. [CrossRef]
31. Ritchie, A.I.; Jackson, D.J.; Edwards, M.R.; Johnston, S.L. Airway epithelial orchestration of innate immune function in response to virus infection. A focus on asthma. *Ann. Am. Thorac. Soc.* **2016**, *13* (Suppl. 1), S55–S63.
32. Meyer, D.M.; Jesson, M.I.; Li, X.; Elrick, M.M.; Funckes-Shippy, C.L.; Warner, J.D.; Gross, C.J.; Dowty, M.E.; Ramaiah, S.K.; Hirsch, J.L.; et al. Anti-inflammatory activity and neutrophil reductions mediated by the JAK1/JAK3 inhibitor, CP-690,550, in rat adjuvant-induced arthritis. *J. Inflamm.* **2010**, *7*, 41. [CrossRef] [PubMed]
33. Gehringer, M.; Pfaffenrot, E.; Bauer, S.; Laufer, S.A. Design and synthesis of tricyclic jak3 inhibitors with picomolar affinities as novel molecular probes. *ChemMedChem* **2014**, *9*, 277–281. [CrossRef] [PubMed]
34. Fridman, J.S.; Scherle, P.A.; Collins, R.; Burn, T.C.; Li, Y.; Li, J.; Covington, M.B.; Thomas, B.; Collier, P.; Favata, M.F.; et al. Selective inhibition of JAK1 and JAK2 is efficacious in rodent models of arthritis: Preclinical characterization of INCB028050. *J. Immunol.* **2010**, *184*, 5298–5307. [CrossRef] [PubMed]
35. Quintás-Cardama, A.; Vaddi, K.; Liu, P.; Manshouri, T.; Li, J.; Scherle, P.A.; Caulder, E.; Wen, X.; Li, Y.; Waeltz, P.; et al. Preclinical characterization of the selective JAK1/2 inhibitor INCB018424: Therapeutic implications for the treatment of myeloproliferative neoplasms. *Blood* **2010**, *115*, 3109–3117. [CrossRef] [PubMed]
36. Hou, J.; Renigunta, A.; Yang, J.; Waldegger, S. Claudin-4 forms paracellular chloride channel in the kidney and requires claudin-8 for tight junction localization. *Proc. Natl. Acad. Sci. USA* **2010**, *107*, 18010–18015. [CrossRef]
37. Kielgast, F.; Schmidt, H.; Braubach, P.; Winkelmann, V.E.; Thompson, K.E.; Frick, M.; Dietl, P.; Wittekindt, O.H. Glucocorticoids regulate tight junction permeability of lung epithelia by modulating claudin 8. *Am. J. Respir. Cell Mol. Biol.* **2016**, *54*, 707–717. [CrossRef]
38. Nelson, W.J.; Shore, E.M.; Wang, A.Z.; Hammerton, R.W. Identification of a membrane-cytoskeleton complex containing the cell adhesion molecule uvomorulin (E-Cadherin), ankyrin, and fodrin in Madin-Darby canine kidney epithelial cells. *J. Cell Biol.* **1990**, *110*, 349–357. [CrossRef]
39. Ciechanover, A. The ubiquitin-proteasome proteolytic pathway. *Cell* **1994**, *79*, 13–21. [CrossRef]
40. Gong, Y.; Wang, J.; Yang, J.; Gonzales, E.; Perez, R.; Hou, J. KLHL3 regulates paracellular chloride transport in the kidney by ubiquitination of claudin-8. *Proc. Natl. Acad. Sci. USA* **2015**, *112*, 4340–4345. [CrossRef]
41. Lui, W.Y.; Lee, W.M. Regulation of junction dynamics in the testis—Transcriptional and post-translational regulations of cell junction proteins. *Mol. Cell. Endocrinol.* **2006**, *250*, 25–35. [CrossRef] [PubMed]
42. Wing, Y.L.; Lee, W.M. cAMP perturbs inter-Sertoli tight junction permeability barrier in vitro via its effect on proteasome-sensitive ubiquitination of occludin. *J. Cell. Physiol.* **2005**, *203*, 564–572.
43. Mandel, I.; Paperna, T.; Volkowich, A.; Merhav, M.; Glass-Marmor, L.; Miller, A. The ubiquitin-proteasome pathway regulates claudin 5 degradation. *J. Cell. Biochem.* **2012**, *113*, 2415–2423. [CrossRef]
44. Van Campenhout, C.A.; Eitelhuber, A.; Gloeckner, C.J.; Giallonardo, P.; Gegg, M.; Oller, H.; Grant, S.G.N.; Krappmann, D.; Ueffing, M.; Lickert, H. Dlg3 Trafficking and apical tight junction formation is regulated by Nedd4 and Nedd4-2 E3 Ubiquitin ligases. *Dev. Cell* **2011**, *21*, 479–491. [CrossRef] [PubMed]
45. Traweger, A.; Fang, D.; Liu, Y.C.; Stelzhammer, W.; Krizbai, I.A.; Fresser, F.; Bauer, H.C.; Bauer, H. The tight junction-specific protein occludin is a functional target of the E3 ubiquitin-protein ligase itch. *J. Biol. Chem.* **2002**, *277*, 10201–10208. [CrossRef] [PubMed]
46. Van Steensel, B.; van Binnendijk, E.P.; Hornsby, C.D.; van der Voort, H.T.; Krozowski, Z.S.; de Kloet, E.R.; van Driel, R. Partial colocalization of glucocorticoid and mineralocorticoid receptors in discrete compartments in nuclei of rat hippocampus neurons. *J. Cell Sci.* **1996**, *109*, 787–792. [PubMed]
47. Ghebre, M.A.; Bafadhel, M.; Desai, D.; Cohen, S.E.; Newbold, P.; Rapley, L.; Woods, J.; Rugman, P.; Pavord, I.D.; Newby, C.; et al. Biological clustering supports both "dutch" and "british" hypotheses of asthma and chronic obstructive pulmonary disease. *J. Allergy Clin. Immunol.* **2015**, *135*, 63–72. [CrossRef] [PubMed]

48. Sugita, K.; Steer, C.A.; Martinez-Gonzalez, I.; Altunbulakli, C.; Morita, H.; Castro-Giner, F.; Kubo, T.; Wawrzyniak, P.; Rückert, B.; Sudo, K.; et al. Type 2 innate lymphoid cells disrupt bronchial epithelial barrier integrity by targeting tight junctions through IL-13 in asthmatic patients. *J. Allergy Clin. Immunol.* **2018**, *141*, 300–310. [CrossRef] [PubMed]
49. May, R.D.; Fung, M. Strategies targeting the IL-4/IL-13 axes in disease. *Cytokine* **2015**, *75*, 89–116. [CrossRef]
50. Krug, S.M.; Bojarski, C.; Fromm, A.; Lee, I.M.; Dames, P.; Richter, J.F.; Turner, J.R.; Fromm, M.; Schulzke, J.-D. Tricellulin is regulated via interleukin-13-receptor α2, affects macromolecule uptake, and is decreased in ulcerative colitis. *Mucosal Immunol.* **2018**, *11*, 345–356. [CrossRef]
51. Skieterska, K.; Rondou, P.; Van Craenenbroeck, K. Regulation of G protein-coupled receptors by ubiquitination. *Int. J. Mol. Sci.* **2017**, *18*, 923. [CrossRef] [PubMed]
52. Hicke, L. Ubiquitin-dependent internalization regulation of plasma membrane proteins. *FASEB J.* **1997**, *11*, 1215–1226. [CrossRef] [PubMed]
53. Widagdo, J.; Guntupalli, S.; Jang, S.E.; Anggono, V. Regulation of AMPA receptor trafficking by protein ubiquitination. *Front. Mol. Neurosci.* **2017**, *10*, 1–10. [CrossRef] [PubMed]
54. Ciechanover, A. Intracellular protein degradation: From a vague idea thru the lysosome and the ubiquitin-proteasome system and onto human diseases and drug targeting. *Best Pract. Res. Clin. Haematol.* **2017**, *30*, 341–355. [CrossRef] [PubMed]
55. Takahashi, S.; Iwamoto, N.; Sasaki, H.; Ohashi, M.; Oda, Y.; Tsukita, S.; Furuse, M. The E3 ubiquitin ligase LNX1p80 promotes the removal of claudins from tight junctions in MDCK cells. *J. Cell Sci.* **2009**, *122*, 985–994. [CrossRef]
56. Cong, X.; Zhang, Y.; Li, J.; Mei, M.; Ding, C.; Xiang, R.-L.; Zhang, L.-W.; Wang, Y.; Wu, L.-L.; Yu, G.-Y. Claudin-4 is required for modulation of paracellular permeability by muscarinic acetylcholine receptor in epithelial cells. *J. Cell Sci.* **2015**, *128*, 2271–2286. [CrossRef]
57. Murakami, T.; Felinski, E.A.; Antonetti, D.A. Occludin phosphorylation and ubiquitination regulate tight junction trafficking and vascular endothelial growth factor-induced permeability. *J. Biol. Chem.* **2009**, *284*, 21036–21046. [CrossRef]
58. Wetzel, F.; Mittag, S.; Cano-Cortina, M.; Wagner, T.; Krämer, O.H.; Niedenthal, R.; Gonzalez-Mariscal, L.; Huber, O. SUMOylation regulates the intracellular fate of ZO-2. *Cell. Mol. Life Sci.* **2016**, *74*, 373–392. [CrossRef]
59. Dang, H.; Klokk, T.I.; Schaheen, B.; McLaughlin, B.M.; Thomas, A.J.; Durns, T.A.; Bitler, B.G.; Sandvig, K.; Fares, H. Derlin-dependent retrograde transport from endosomes to the golgi apparatus. *Traffic* **2011**, *12*, 1417–1431. [CrossRef]
60. Schaheen, B.; Dang, H.; Fares, H. Derlin-dependent accumulation of integral membrane proteins at cell surfaces. *J. Cell Sci.* **2009**, *122*, 2228–2239. [CrossRef]
61. Gu, X.; Zhao, F.; Zheng, M.; Fei, X.; Chen, X.; Huang, S.; Xie, Y.; Mao, Y. Cloning and characterization of a gene encoding the human putative ubiquitin conjugating enzyme E2Z (UBE2Z). *Mol. Biol. Rep.* **2007**, *34*, 183–188. [CrossRef] [PubMed]
62. Liu, X.; Zhao, B.; Sun, L.; Bhuripanyo, K.; Wang, Y.; Bi, Y.; Davuluri, R.V.; Duong, D.M.; Nanavati, D.; Yin, J.; et al. Orthogonal ubiquitin transfer identifies ubiquitination substrates under differential control by the two ubiquitin activating enzymes. *Nat. Commun.* **2017**, *8*, 1–12. [CrossRef] [PubMed]
63. Jin, J.; Li, X.; Gygi, S.P.; Harper, J.W. Dual E1 activation systems for ubiquitin differentially regulate E2 enzyme charging. *Nature* **2007**, *447*, 1135–1138. [CrossRef] [PubMed]
64. Mehta, S.; Nijhuis, A.; Kumagai, T.; Lindsay, J.; Silver, A. Defects in the adherens junction complex (E-cadherin/β-catenin) in inflammatory bowel disease. *Cell Tissue Res.* **2015**, *360*, 749–760. [CrossRef] [PubMed]
65. Caldeira, M.V.; Salazar, I.L.; Curcio, M.; Canzoniero, L.M.T.; Duarte, C.B. Role of the ubiquitin-proteasome system in brain ischemia: Friend or foe? *Prog. Neurobiol.* **2014**, *112*, 50–69. [CrossRef] [PubMed]
66. Choi, W.H.; De Poot, S.A.H.; Lee, J.H.; Kim, J.H.; Han, D.H.; Kim, Y.K.; Finley, D.; Lee, M.J. Open-gate mutants of the mammalian proteasome show enhanced ubiquitin-conjugate degradation. *Nat. Commun.* **2016**, *7*, 1–12. [CrossRef] [PubMed]
67. Kanayama, H.O.; Tamura, T.; Ugai, S.; Kagawa, S.; Tanahashi, N.; Yoshimura, T.; Tanaka, K.; Ichihara, A. Demonstration that a human 26S proteolytic complex consists of a proteasome and multiple associated protein components and hydrolyzes ATP and ubiquitin-ligated proteins by closely linked mechanisms. *Eur. J. Biochem.* **1992**, *206*, 567–578. [CrossRef]

68. Balch, W.E.; Morimoto, R.I. Adapting proteostatis for disease intervention. *Science* **2008**, *319*, 916–919. [CrossRef]
69. Inoue, S.; Nakase, H.; Matsuura, M.; Mikami, S.; Ueno, S.; Uza, N.; Chiba, T. The effect of proteasome inhibitor MG132 on experimental inflammatory bowel disease. *Clin. Exp. Immunol.* **2009**, *156*, 172–182. [CrossRef]
70. Vadász, I.; Weiss, C.H.; Sznajder, J.I. Ubiquitination and proteolysis in acute lung injury. *Chest* **2012**, *141*, 763–771. [CrossRef]
71. Stepaniants, S.; Wang, I.M.; Boie, Y.; Mortimer, J.; Kennedy, B.; Elliott, M.; Hayashi, S.; Luo, H.; Wong, J.; Loy, L.; et al. Genes related to emphysema are enriched for ubiquitination pathways. *BMC Pulm. Med.* **2014**, *14*, 1–11. [CrossRef] [PubMed]
72. Min, T.; Bodas, M.; Mazur, S.; Vij, N. Critical role of proteostasis-imbalance in pathogenesis of COPD and severe emphysema. *J. Mol. Med.* **2011**, *89*, 577–593. [CrossRef] [PubMed]
73. Tran, I.; Ji, C.; Ni, I.; Min, T.; Tang, D.; Vij, N. Role of cigarette smoke-induced aggresome formation in chronic obstructive pulmonary disease-emphysema pathogenesis. *Am. J. Respir. Cell Mol. Biol.* **2015**, *53*, 159–173. [CrossRef] [PubMed]
74. Kim, S.Y.; Kim, H.J.; Park, M.K.; Huh, J.W.; Park, H.Y.; Ha, S.Y.; Shin, J.H.; Lee, Y.S. Mitochondrial E3 ubiquitin protein ligase 1 mediates cigarette smoke-induced endothelial cell death and dysfunction. *Am. J. Respir. Cell Mol. Biol.* **2016**, *54*, 284–296. [CrossRef] [PubMed]
75. Ware, L.B.; Matthay, M.A. The acute respiratory distress syndrome. *N. Engl. J. Med.* **2000**, *342*, 1334–1349. [CrossRef]
76. Li, G.; Flodby, P.; Luo, J.; Kage, H.; Sipos, A.; Gao, D.; Ji, Y.; Beard, L.M.L.; Marconett, C.N.; DeMaio, L.; et al. Knockout mice reveal key roles for Claudin 18 in alveolar barrier properties and fluid homeostasis. *Am. J. Respir. Cell Mol. Biol.* **2014**, *51*, 210–222. [CrossRef] [PubMed]
77. Kage, H.; Flodby, P.; Gao, D.; Kim, Y.H.; Marconett, C.N.; DeMaio, L.; Kim, K.-J.; Crandall, E.D.; Borok, Z. Claudin 4 knockout mice: Normal physiological phenotype with increased susceptibility to lung injury. *Am. J. Physiol. Lung Cell. Mol. Physiol.* **2014**, *307*, L524–L536. [CrossRef]
78. Fulkerson, P.C.; Fischetti, C.A.; Hassman, L.M.; Nikolaidis, N.M.; Rothenberg, M.E. Persistent effects induced by IL-13 in the lung. *Am. J. Respir. Cell Mol. Biol.* **2006**, *35*, 337–346. [CrossRef]
79. Pfaffl, M.W. A new mathematical model for relative quantification in real-time RT-PCR. *Nucleic Acids Res.* **2001**, *29*, e45. [CrossRef]
80. Schindelin, J.; Rueden, C.T.; Hiner, M.C.; Eliceiri, K.W. The ImageJ ecosystem: An open platform for biomedical image analysis. *Mol. Reprod. Dev.* **2015**, *82*, 518–529. [CrossRef]

International Journal of
Molecular Sciences

Article

Antagonistic Effects of IL-4 on IL-17A-Mediated Enhancement of Epidermal Tight Junction Function

Matthew G. Brewer [1], Takeshi Yoshida [1], Fiona I. Kuo [1], Sade Fridy [1], Lisa A. Beck [1] and Anna De Benedetto [1,2,*]

1 Department of Dermatology, University of Rochester Medical Center, Rochester, NY 14642, USA
2 Department of Dermatology, College of Medicine University of Florida, Gainesville, FL 32610, USA
* Correspondence: adebenedetto@ufl.edu

Received: 2 August 2019; Accepted: 13 August 2019; Published: 21 August 2019

Abstract: Atopic dermatitis (AD) is the most common chronic and relapsing inflammatory skin disease. AD is typically characterized by skewed T helper (Th) 2 inflammation, yet other inflammatory profiles (Th1, Th17, Th22) have been observed in human patients. How cytokines from these different Th subsets impact barrier function in this disease is not well understood. As such, we investigated the impact of the canonical Th17 cytokine, IL-17A, on barrier function and protein composition in primary human keratinocytes and human skin explants. These studies demonstrated that IL-17A enhanced tight junction formation and function in both systems, with a dependence on STAT3 signaling. Importantly, the Th2 cytokine, IL-4 inhibited the barrier-enhancing effect of IL-17A treatment. These observations propose that IL-17A helps to restore skin barrier function, but this action is antagonized by Th2 cytokines. This suggests that restoration of IL-17/IL-4 ratio in the skin of AD patients may improve barrier function and in so doing improve disease severity.

Keywords: atopic dermatitis; tight junctions; cytokines; claudin; STAT3

1. Introduction

Atopic dermatitis (AD) is a chronic, inflammatory skin disorder characterized by an overactive immune response to a host of environmental allergens and dry/itchy skin. Over the last decade, important discoveries have demonstrated that AD develops in part from genetic, and/or acquired defects in the skin barrier. The overall hypothesis is that skin barrier defects allow resident immune cells to gain access to microbes, allergens, and irritants, resulting in a robust type 2 immune response that can be characterized by the tissue recruitment of T helper (Th) 2 cells. Recently, other Th subtypes have been identified in acute (e.g., Th22/Th17) and chronic (e.g., Th22 and Th1) AD lesions, although the importance of these subsets in AD pathogenesis is still unclear [1–3].

Epidermal barrier is mediated by two structural components, the stratum corneum (SC) and tight junctions (TJs). TJs, found in the stratum granulosum, seal intercellular spaces and the 'tightness' of this structure constitutes a key physiological and dynamic function of the skin. The essential role that epidermal TJs play in maintaining the skin barrier was demonstrated by the claudin (CLDN) 1 deficient mouse, which suffers from extensive transepidermal water loss and succumbs shortly after birth to severe dehydration [4]. We and others have demonstrated functional defects in epidermal TJs from skin samples of AD subjects, which may in part be due to the reduced expression of two transmembrane TJ components, CLDN1 and CLDN23 [5–7]. Kast et al. reported that CLDN1 is the most abundant CLDN family member in the skin and in primary human keratinocytes (PHK), with lower levels of the CLDN family members 4 and 25 occurring, while CLDN 3, 5, 7, 8, 10, and 23 were only present in the skin (but not detected in primary culture models) [8].

Altered TJ protein expression has been observed in several human skin disorders including psoriasis, lichen planus, ichthyosis vulgaris, skin wounds, skin infections and in response to ultraviolet exposure; however, the functional consequences of these changes have not been fully evaluated yet [9]. Additionally, very little is known about what regulates epidermal TJ function and claudin expression in immune-mediated skin disorders. Several T-cell derived cytokines were found in AD skin (e.g., IL-4, IL-13, IL-17A, IL-22, IL-25, or TNF-α) and these cytokines were shown to reduce the expression of several SC barrier proteins such as filaggrin, loricrin, S100A11, and involucrin [10–12]. In one study, TNF-α was shown to alter TJ protein expression and localization in a skin explant model, as well as TJ function in cultured keratinocytes [9]. Another study showed contradictory results, with no effect observed with either TNF-α or IL-4 treatment, while IL-17 diminished and IL-22 enhanced TJ function in cultured keratinocytes, as measured by transepithelial electric resistance (TEER) [6]. Importantly, in vivo studies have strongly suggested that IL-17A has barrier promoting properties. This was highlighted by increased intestinal permeability in mice lacking IL-17A as well as spontaneous development of allergic skin inflammation in mice lacking the IL-17RA [13,14]. To begin to clarify the effects that AD related cytokines play in barrier function of the skin, we analyzed how the Th17 cytokine, IL-17A, influences TJ barrier properties in cultured keratinocytes and human skin explants and how Th2 cytokines may affect that response.

2. Results

2.1. IL-17A Enhances Epidermal TJ Barrier Integrity

PHK were exposed to cytokines throughout the differentiation process, thus modeling the effect they would have on TJ assembly during wound repair. In our submerged culture model, we observed that TJ barrier function, quantified by TEER, increases to a peak of approximately 2500 ohms × cm^2 around 120 ± 24 h after addition of differentiation media [15]. Notably, in our model we observed that as TEER increases, the immunoreactivity of TJ components changes from a dotted, disrupted pattern to a continuous "chicken wire" membranous staining typically observed in mature TJ of the human epidermis (Figure S1).

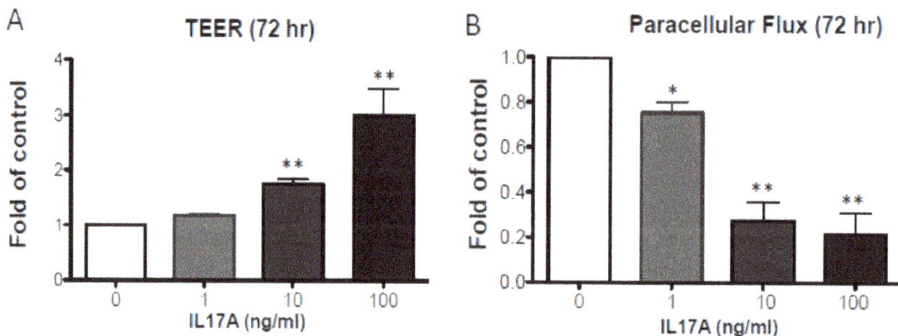

Figure 1. IL-17A enhances epidermal TJ barrier integrity. In PHK, IL-17A dose-dependently (**A**) enhanced TEER (n = 3–6) and (**B**) reduced paracellular flux of fluorescein (n = 3) 72 h after cytokine treatment. Data are shown as mean ± SEM fold of control. Significance was calculated compared to untreated controls. * $p \leq 0.05$, ** $p \leq 0.05$.

When PHK were differentiated in the presence of IL-17A, we observed a dose-dependent enhancement of the TJ barrier function by 72 h (Figure 1). This was demonstrated by a significant enhancement of TEER in IL-17A treated PHK (72 h, $p \leq 0.01$, n = 3–6; Figure 1A) and a reduction in the paracellular fluorescein flux (72 h, $p \leq 0.05$, n = 3; Figure 1B). To further validate our findings, we tested the effect of IL-17A on human epidermal samples in an ex vivo model. We tape stripped (15x) discarded healthy human skin, as a model of mechanical disruption and then isolated the epidermis and monitored epidermal TJ recovery with and without IL-17A stimulation. To do this we used a modified micro-Snapwell™ system, previously developed to study TJ function in intestinal epithelium [16], and adapted in our laboratory for epidermal sheets [17]. We observed a significant reduction in the transepidermal fluorescein flux ($p \leq 0.02$, $n = 3$; Figure 2C) after IL-17A treatment (24 h), thus confirming our barrier-enhancing findings in submerged PHK. Altogether, these observations demonstrate that IL-17A enhances the development of TJ epidermal barrier function.

2.2. IL-4 Inhibits IL-17A-Mediated TJ Barrier Enhancement

In contrast to the IL-17A effects on TEER in our model, IL-4 did not significantly alter TJ integrity of cultured PHK monolayers. Previous studies have demonstrated that Th2 cytokines antagonize IL-17A-induced production of antimicrobial peptides and S100A8 in human keratinocytes [18–20]. Therefore, we tested whether Th2 cytokines also inhibit IL-17A barrier-enhancing effects. When PHK were treated with both IL-17A and IL-4 we observed that the enhanced TEER and reduced paracellular fluorescein flux observed in response to IL-17A were completely inhibited by co-treatment with IL-4 (50 ng/mL; respectively $p = 0.024$ and $p = 0.002$, Figure 2A,B). Of note, we observed a slight, but not significant ($p = 0.08$) increase in paracellular fluorescein flux with IL-4 treatment at 72 h (Figure 2B), suggesting this cytokine might have an effect on TJ pore size. We again validated our finding in an ex vivo model, co-treatment with IL-4 was able to block the IL-17A induced increase in permeability flux of fluorescein (Figure 2C). To determine whether IL-4 blocks other known IL-17A mediated effects in our in vitro model we measured the expression of S100A7, a well-known downstream product of IL-17A signaling [21]. S100A7, also known as psoriasin, is a calcium-binding protein with chemotactic and antimicrobial properties that is expressed in AD lesions and even more so in psoriasis skin, a Th1/Th17 driven disease [22]. IL-17A-mediated S100A7 expression in PHK was blocked by co-stimulation with IL-4 (Figure S2).

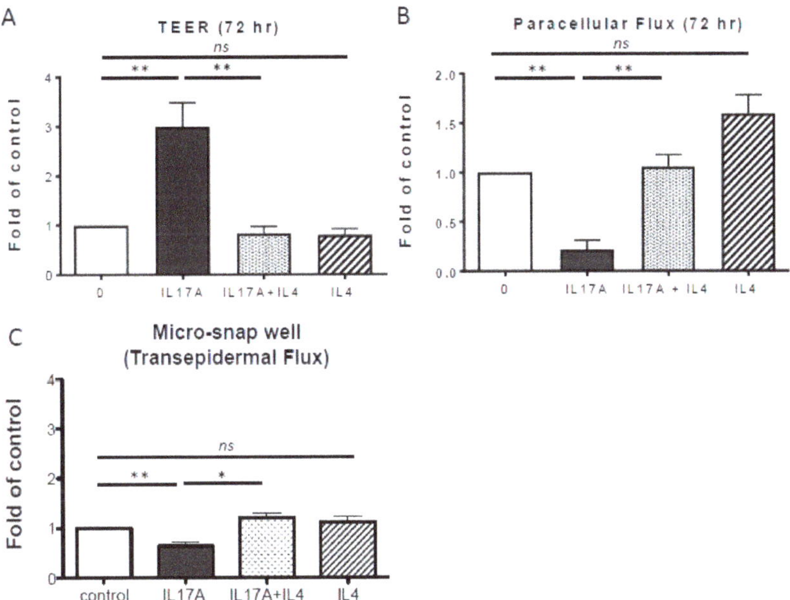

Figure 2. IL-4 inhibits IL-17A mediated barrier enhancement in PHK. Co-treatment with IL-4 (50 ng/mL) (**A**) inhibited IL-17A (100 ng/mL) increased TEER and (**B**) enhanced paracellular flux ($n = 3$–5). (**C**) IL-17A treatment reduced transepidermal flux of fluorescein (50 ng/mL 0.02% fluorescein; $n = 3$) in tape-stripped human skin samples. Data are shown as mean ± SEM fold of control. * $p \leq 0.05$, ** $p \leq 0.01$, ns: not significant.

2.3. Inhibition of STAT3 Activation Reduces IL-17A-Induced TEER

It has been suggested that Janus kinases (JAK) and mitogen-activated protein kinases (MAPK) are downstream signaling pathways activated by IL-17A [23]. Therefore, we examined if JAK and MAPK activation are involved in IL-17A mediated actions on epidermal TJ function. A major cytosolic target of JAK signaling is the phosphorylation and nuclear translocation of signal transducer and activator of transcription 3 (STAT3). In our model, we confirmed that IL-17A enhanced STAT3 activation in PHK, as demonstrated by increased STAT3 phosphorylation at amino acid Y705 (Figure 3A), suggesting the importance of the JAK-STAT3 pathway in the barrier-enhancing effect of IL-17A. Using a pan-JAK inhibitor (JAK_{Tot}) that blocks all four JAK isoforms (JAK inhibitor I; Calbiochem, 10 μM) a significant decrease in the IL-17A-dependent increase of TEER was observed in our PHK model (IL-17A vs JAK_{Tot} + IL-17A, $p = 0.025$; $n = 4$; Figure 3B). By contrast, no inhibition of IL-17A-enhanced TEER was observed from PHK treated with an upstream MAPK inhibitor (PD98056; data not shown). Treatment with JAK_{Tot} completely blocked both baseline and IL-17A-dependent STAT3 activation (Figure 3A). Notably, we observed that IL-4 reduced STAT3 phosphorylation both at baseline and in response to IL-17A treatment (Figure 3C), suggesting that IL-4's attenuation of the IL-17A barrier-enhancing effect may be mediated by the JAK-STAT3 pathway.

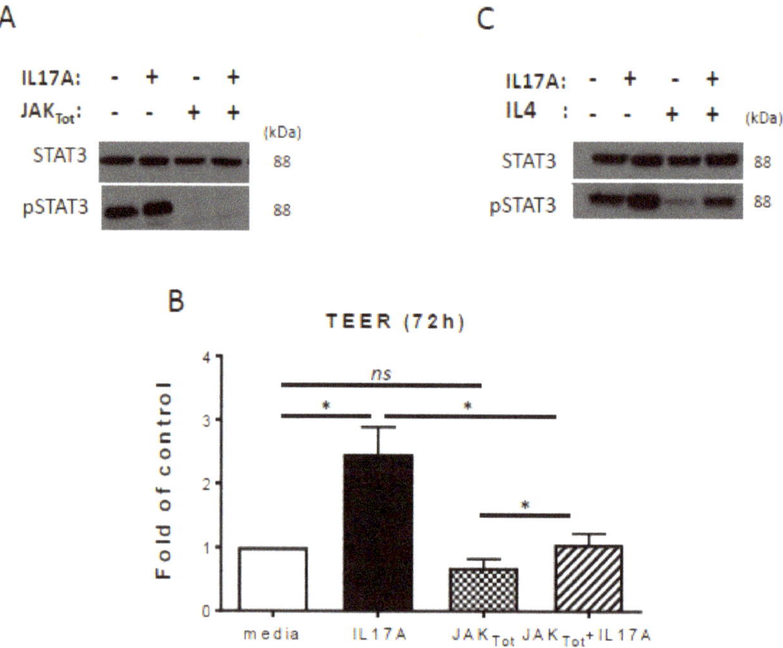

Figure 3. A pan-JAK inhibitor blocks IL-17A-induced TEER enhancement. (**A**) A representative Western blot demonstrates that pan-JAK inhibition blocks STAT3 activation (pSTAT3; EP2147Y), but has no effect on total STAT3 expression (STAAD22A) (**B**) Pretreatment with a pan-JAK inhibitor (JAKtot: 10 μM; $n = 4$) inhibited IL-17A (100 ng/mL) induced TEER in primary human keratinocytes. (**C**) IL-4 blocked STAT3 phosphorylation in PHK treated with IL-17A or media alone. Representative Western blot bands of $n = 3$ experiments are shown. All data are shown as mean ± SEM fold of control. * $p \leq 0.05$, ns: not significant.

2.4. IL-17A Enhances Claudin-4 Expression

To further investigate the mechanisms responsible for IL-17A mediated barrier enhancement, we quantified the expression of two highly expressed epidermal claudins, CLDN1 and CLDN4, in both submerged PHK and in the epidermal organotypic model. This latter model is three-dimensional and more closely mimics mature human epidermis with a well-differentiated, multilayered structure that includes both a stratum granulosum (where TJs reside) and SC layers. Using this model, we detected increased CLDN4 immunoreactivity in the presence of IL-17A (100 ng/mL; p = 0.04) (Figure 4A,B). IL-4 in contrast slightly, but significantly reduced CLDN4 expression compared to the media alone (50 ng/mL; p = 0.04). Additionally, IL-17A-induced CLDN4 expression was blocked by IL-4 co-treatment (p = 0.009; Figure 4A,B). The expression of CLDN1 was not affected by either cytokine alone, or in combination (Figure 4C). The TJ findings in the submerged PHK culture were in line with those observed in the organotypic model. IL-4 stimulation alone and in combination with IL-17A reduced CLDN4 expression as detected by Western blot analysis while CLDN1 expression was unaffected by cytokine treatment (Figure S3).

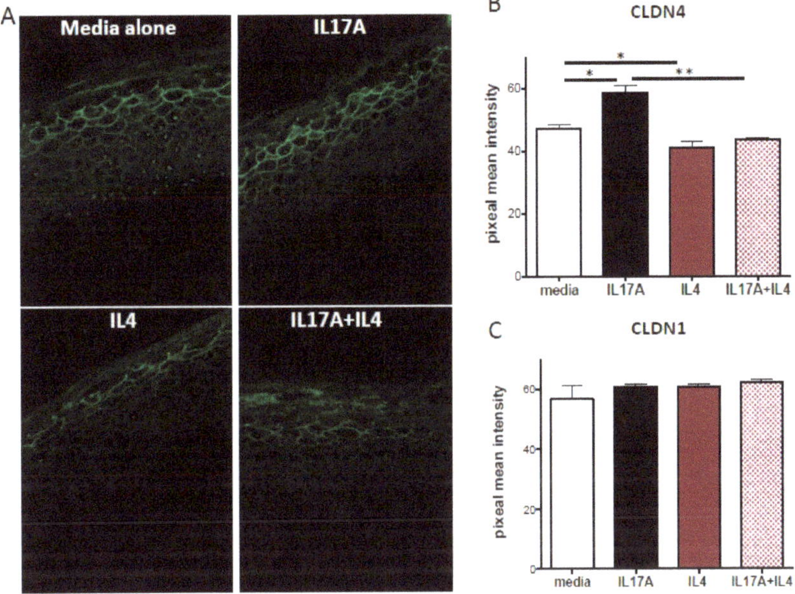

Figure 4. IL-17A increases CLDN4, but not CLDN1 in an organotypic skin model. IL-17A (100 ng/mL), IL-4 (50 ng/mL) or a combination of the two were added to the media of organotypic cultures for 72 h after 9 days of growth. (**A**) Paraffin-fixed sections were stained with either a CLDN4 (pAb) or CLDN1 (JAY.8; not shown) antibody and detected with an Alexa Fluor 488-conjugated secondary (green). (**B**) CLDN4 and (**C**) CLDN1 immunostaining intensity analysis. Images were acquired at the same instrumental setting and at 40× magnification. Mean pixel intensity (5 random fields) was measured using ImageJ. All data are shown as mean ± SEM. * $p \leq 0.05$, ** $p \leq 0.01$.

3. Discussion

There is no doubt that type 2 immunity plays a central role in atopic diseases, including AD. Less clear is the contribution of other cytokines, such as IL-17A, in AD pathogenesis. We have now unveiled a novel role for IL-17A in repairing and/or enhancing TJ epidermal barrier function. In this work, we demonstrated that IL-17A enhances TJ integrity at an early time point in differentiating PHK (Figure 1) and this was associated with enhanced expression of CLDN4 (Figure 4).

Notably, IL-17A effects may be inhibited in AD because the antagonistic effect of the Th2 cytokines in the tissue. Previous studies showed Th2 cytokines can reduce IL-17A expression by Th17 cells, as well as IL-17A-induced production of antimicrobial peptides, S100A7 and S100A8 by human keratinocytes [18–20,24]. We have added another example to strengthen the paradigm that it is the relative expression of IL-4 and IL-17A that determines the biological consequences of either of these cytokines (Figure 2). IL-17A's ability to induce antimicrobial peptides as well as recruit neutrophils strongly suggests that this cytokine plays a crucial protective role at epithelial sites. Clinical trials testing the effect of biological therapeutics targeting the Th1/Th17 pathways have been conducted in AD. Anecdotally, good responses in some AD patients were reported with Ustekinumab, an antibody that blocks Th1/Th23 signaling and thus reduces Th17 activation. However, in two independent, randomized, placebo-controlled, phase II studies Ustekinumab did not demonstrate meaningful efficacy in AD. Another trial with Secukinumab that targets the cytokine, IL-17A, was conducted in AD; however, results are not available yet. This trial will be extremely informative to better understand the role of specific cytokines in AD.

We demonstrated in this study that blocking STAT3 activation via a pan-JAK inhibitor was associated with loss of IL-17A-mediated TJ function enhancement (Figure 3). Remarkably, we observed that the IL-4 inhibitory effect on IL-17A's ability to enhance barrier was partly mediated by reduced STAT3 activation. Since another IL-17A downstream protein S100A7 (also STAT3 dependent) was inhibited by IL-4 co-stimulation, this observation further supports that IL-4 blocks IL-17A signaling at the level of transcriptional control. Activation of STAT3 has been shown to play a key role in many cellular processes such as cell growth and apoptosis. Indeed, STAT3 knockout mice are not viable [25]. K5-Cre/STAT3 transgenic mice, whose epidermal and follicular keratinocytes lack STAT3, were viable and the development of the epidermis and hair follicles appeared normal. However, hair cycle and wound healing processes were severely compromised [26]. When STAT3 is inactivated in intestinal epithelial cells through lentiviral transduction of a dominant negative STAT3 protein, barrier is severely impacted [27]. STAT3 activation was also shown to increase permeability in human retinal endothelial cells [28]. Contrasting with our data, another group showed that baseline STAT3 activation was not reduced when PHK were treated with IL-4 and -13 together [29]. This could be explained because IL-13 can also induce phosphorylation of STAT3 in PHK [30]. In humans, STAT3 mutations cause a form of the Hyper IgE syndrome (AD-HIES; OMIM: 147060). Severe AD-like phenotype, increased susceptibility to S. aureus infections and eosinophilia are some of the clinical features of this disease, in addition to reduced peripheral Th17 cells (STAT3 is involved in Th17 differentiation cells) [31]. However, involvement of the STAT3 pathway in TJ function and composition in the skin has not been directly evaluated in these patients.

An alternative and complementary, approach to investigate the effect of cytokines on TJ function is to directly look at the expression and localization of key TJ components. In this study, we investigated the effect of the IL-17A with or without IL-4 on the expression of CLDN1 and 4, the two most abundant CLDNs in human keratinocytes [8]. Our studies demonstrated that CLDN4, but not CLDN1 expression is enhanced by IL-17A, and this enhancement is prevented by co-treatment with IL-4 (Figure 4). Previous work by Kirschner et al. showed that silencing CLDN4 in PHK reduced TEER and enhanced permeability to small macromolecules (4 kDa dextran-FITC conjugated). CLDN23 is another TJ-associated molecule we hypothesized to be important in barrier function since it is down-regulated (mRNA) in the skin of AD patients [5]. Currently, the reagents for this protein are poorly validated in human skin, which is why it was not investigated. Future studies will have to clarify the specific effects of reduced expression of each of the claudins in AD subjects

Our findings in an epidermal in vitro model are in contrast with previous data showing IL-17A decreased TEER in ocular epithelial cells, suggesting impaired barrier function capabilities [32]. Gutowska-Owsiak D. et al. reported findings from a microarray study in HaCaT cells showing that IL-17A treatment (20 or 50 ng/mL for 24 h) reduced expression of several intercellular junction components, implying IL-17A could reduce keratinocyte barrier function at the level of the SC, but a link to TJs was not suggested [21]. It is important to point out that studying the mechanisms regulating epithelial TJ function and composition is made difficult by pleiotropic cytokine actions, in addition to temporal and dose-dependent variation in cell culture systems [33]. Additionally, the study of epidermal TJs is further complicated by different culture conditions employed, which also typically results in variable levels of differentiation. To minimize many of these sources of variability we chose to focus our studies on submerged keratinocyte cultures because they form robust barrier at the level of the TJ (Figure S1), which we believe more faithfully mimics the epithelium. Air liquid cultures cornify which is ideal for studying the SC, but importantly they do not form robust TJ, making this an unreliable model to study the effects of cytokines on barrier formation. This biological complexity might explain the discrepancy of data published so far in this field and as such requires further validation in vivo.

Dupilumab (Regeneron, Tarrytown, NY, USA & Sanofi, Paris, France) a fully human monoclonal antibody against the Th2 cytokine receptor IL-4Rα, is the first biologic medication approved worldwide for the treatment of AD. This has been a revolutionary treatment option for patients with AD, with a good safety profile [34]. This fully humanized monoclonal antibody targets the common receptor subunit (IL-4Rα) used by both IL-4 and IL-13 and thereby blocks signaling. We hypothesize that some of the clinical improvement and normalization of the epithelial transcriptome may be due to diminished antagonism of IL-17A. A better understanding of the complex interactions between relevant Th cytokines found in AD lesions and how they affect epidermal barrier function will be critical for a more thorough understanding of AD pathogenesis. The development and study of highly targeted AD therapies will likely provide an opportunity to look at this issue and allow us to tailor therapies to individual patient's tissue inflammation.

4. Methods

4.1. Isolation and Culture of Primary Human Foreskin Keratinocytes

Human keratinocytes were isolated from neonatal foreskin as previously described [35]. PHK were cultured in Keratinocyte-SFM (Thermo Fisher Scientific, Waltham, MA, USA) with 1% Pen/Strep, 0.2% Amphotericin B (Thermo Fisher Scientific, Waltham, MA, USA). To differentiate PHK, cells were grown in DMEM (11965092, Thermo Fisher Scientific, Waltham, MA, USA) with 1% Pen/Strep, 0.2% Amphotericin B. For cytokine experiments, the following reagents were added alone or in combination to the culture media from the time of differentiation and replaced with every 48-h media change: human IL-4 (5–50 ng/mL; R&D system, Minneapolis, MN, USA), human IL-13 (5–50 ng/mL; R&D system, Minneapolis, MN, USA), human IL-17A (1–100 ng/mL; R&D system), JAK inhibitor I (10 μM; Calbiochem, San Diego, CA, USA), and PD98056 (10 μM; Calbiochem, San Diego, CA, USA). For the epidermal organotypic model experiment, keratinocytes were grown as previously described [36]. For all cytokine treatment studies, keratinocytes were starved of growth factors for 24 h before stimulation with the indicated cytokines.

4.2. Transepithelial Electric Resistance (TEER) and Paracellular Flux

TJ integrity in cultured PHK was evaluated using two complementary assays, TEER and paracellular flux, as previously described [5]. Briefly, TEER was measured using an EVOMX voltohmmeter (World Precision Instruments, Sarasota, FL, USA). To evaluate the paracellular flux across PHK cultures, 0.02% Fluorescein (FITC) Sodium (Sigma, St. Louis, MO, USA) in HBSS was added to the upper chamber. Samples were collected from the lower chamber at different times (0.5–3 h). Paracellular permeability was presented as "Relative Fluorescence Flux" = experimental condition/media alone × 100.

We used the Micro-Snapwell™ barrier function assay (Corning Incorporated, Corning, NY, USA) to determine the permeability of ex vivo human skin. The protocol was adapted from previously described work [16]. Full thickness epidermis was isolated from discarded, deidentified skin samples. To disrupt skin barrier, the epidermis was taped stripped 15 times and then removed using a Weck blade (Goulian Skin Graft Knife Set, George Tiemann & Co. Hauppauge, NY, USA) and washed in PBS. Skin samples were mounted on filter supports (Whatman Nuclepore Track-Etch Membrane; GE Healthcare Biosciences, Pittsburgh, PA, USA). The epidermal side was faced upward and sandwiched between two custom made Plexiglas discs containing an opening of 3 mm. This sandwich was then placed in the modified Snapwell™ chambers. Samples were immersed in DMEM media and kept at

37 °C, 5% CO_2 for 30 min. Media was added to samples and to test the effect of IL-17A, 50 ng/mL of the cytokine was incorporated into the media, which was then added to both sides of the transwell. Paracellular flux of fluorescein into the lower levels of human epidermis was measured at 24 h later as described above. The Pathology Department at the University of Rochester Medical Center provided the discarded human skin, which was approved by the institutional Research Subject Review Board (RSRB 00042616; approved on 31 May 2012).

4.3. Immunoblotting

PHK were placed on ice in RIPA lysis buffer (20 mM Tris, 50mM NaCl, 2mM EDTA, 2mM EGTA, 1% sodium deoxycholate, 1% TX-100, and 2% SDS, pH 7.4) with a 100-fold dilution of both Protease Inhibitor Cocktail (Sigma, St. Louis, MO, USA) and Phosphatase Inhibitor Cocktail (Sigma, St. Louis, MO, USA) for 30 min. The samples were boiled for 10 min and then centrifuged for 15 min. Forty μg of protein, as determined by the BCA (Thermo Fisher Scientific, Waltham, MA, USA) protein quantification assay, were mixed in NuPage LDS Sample Buffer (Thermo Fisher Scientific, Waltham, MA, USA) and applied to 4-12% NuPage Bis-Tris gels (Thermo Fisher Scientific, Waltham, MA, USA). Electrophoresis was performed under reducing conditions with MES SDS Buffer (Thermo Fisher Scientific, Waltham, MA, USA). Membranes were incubated in blocking solution, which contained 5% non-fat dry milk + 0.05% Tween-20 in PBS, for 1 h at RT and then incubated with primary antibodies: claudin-4 (ab53156; Abcam, Cambridge, UK), claudin-1 (JAY.8; Thermo Fisher Scientific, Waltham, MA, USA) STAT3 (STAAD22A; Abcam, Cambridge, UK), pSTAT3 (EP2147Y, phospho-Y705; Abcam, Cambridge, UK), S100A7 (47c1068; Thermo Fisher Scientific, Waltham, MA, USA) and β-Actin (C4; Santa Cruz, Dallas, TX, USA). HRP-linked secondary antibody (GE Healthcare Biosciences, Pittsburgh, PA, USA) was used to visualize protein staining in tandem with ECL reagent (GE Healthcare Biosciences, Pittsburgh, PA, USA) by autoradiography with Kodak BioMax MR film (Kodak, Rochester, NY, USA).

4.4. Immunostaining

PHK were grown on a cover glass were fixed in methanol at −20 °C for 15 min, followed by blocking with 1% BSA in PBS and immunolabeled with the above pSTAT3 antibody for 2.5 h at 4 °C. This was followed by a 1-h incubation with secondary antibodies. A 1:1000 dilution of Alexa Fluor 568 donkey-anti-mouse IgG H+L (Thermo Fisher Scientific, Waltham, MA, USA) and 1:10,000 4′,6-diamidino-2-phenyl-indole, dihydrochloride (DAPI) (Molecular Probe, Eugene, OR, USA) was used. Fluorescent images were taken using an Olympus FV1000 laser scanning confocal microscope with the FV10-ASW 1.7 software (Olympus, Tokyo, Japan). The saturation level of fluorescence intensity was determined for the controls using the Hi-Lo feature of the Fluoview software. In each experiment, the images captured used identical settings during acquisition and no further manipulation of the images occurred before into importing into the FV1000 software. The organotypic samples were fixed in formalin then embedded in paraffin. Five μm sections were rehydrated after deparaffinization. Slides were incubated in EDTA containing solution, pH 8.0 at 95 °C for 10 min. Samples were incubated overnight at with the same primary Abs used for immunoblotting at 4 °C. After primary antibody binding a 1-h incubation with the secondary antibody Alexa Fluor 488 donkey-anti-rabbit IgG H+L (Thermo Fisher Scientific, Waltham, MA, USA) diluted 1000-fold was done. Fluorescent images were obtained with an Olympus Bx 60 microscope

4.5. Statistical Analysis

Statistical significance was tested using either a pairwise analysis T test (Mann-Whitney) or the one-way analysis of variance (Kruskal-Wallis) to determine significance. Statistical analysis was accomplished using GraphPad Prism (GraphPad Software, Inc., San Diego, CA, USA). A value of $p < 0.05$ qualified as significant.

Supplementary Materials: Supplementary materials can be found at http://www.mdpi.com/1422-0067/20/17/4070/s1.

Author Contributions: Conceptualization, A.D.B, L.B. and M.B. Methodology, A.D.B, L.B., T.Y., F.I.K, S.F. Formal Analysis, L.B., A.D.B., T.Y., F.I.K, M.B. Writing—Original Draft Preparation, A.D.B, M.B. and L.B. Writing—Review & Editing, T.Y., F.I.K, M.B. and A.D.B.; Visualization, T.Y. Funding Acquisition, A.D.B and L.B.

Funding: The work herein was funded by a National Eczema Association research grant (A.D.B. and L.A.B.) and a Dermatology Foundation research grant (A.D.B.).

Acknowledgments: The authors are thankful to S. Getsios at Northwestern University – Skin Disease Research Center Keratinocyte Core who assisted in the keratinocyte organotypic cultures with support from the National Institutes of Health (NIH) National Institute of Arthritis and Musculoskeletal and Skin (NIAMS) Diseases grant AR057216.

Conflicts of Interest: A.D.B. has served on advisory board for Regeneron & Sanofi and Allakos. She has been an investigator for Novartis, Kiniksa, Alchem, Regeneron and Sanofi. L.B. has served on advisory board for Abbvie, Allakos, Arena Pharma, Astra-Zeneca, Connect Biopharma, Incyte, Leo Pharma, Lilly, Novan, Novartis, Pfizer, Regeneron, Sanofi, and UCB. She has been an investigator for Abbvie, Leo Pharma, Pfizer and Regeneron and she owns stock in Pfizer. F.I.K. is employee and shareholder of Incyte Corporation.

References

1. Chan, T.C.; Sanyal, R.D.; Pavel, A.B.; Glickman, J.; Zheng, X.; Xu, H.; Cho, Y.T.; Tsai, T.F.; Wen, H.C.; Peng, X.; et al. Atopic dermatitis in Chinese patients shows TH2/TH17 skewing with psoriasiform features. *J. Allergy Clin. Immunol.* **2018**, *142*, 1013–1017. [CrossRef] [PubMed]
2. Sanyal, R.D.; Pavel, A.B.; Glickman, J.; Chan, T.C.; Zheng, X.; Zhang, N.; Cueto, I.; Peng, X.; Estrada, Y.; Fuentes-Duculan, J.; et al. Atopic dermatitis in African American patients is TH2/TH22-skewed with TH1/TH17 attenuation. *Ann. Allergy Asthma Immunol.* **2019**, *122*, 99–110.
3. Gittler, J.K.; Shemer, A.; Suárez-Fariñas, M.; Fuentes-Duculan, J.; Gulewicz, K.J.; Wang, C.Q.; Mitsui, H.; Cardinale, I.; de Guzman Strong, C.; Krueger, J.G.; et al. Progressive activation of T(H)2/T(H)22 cytokines and selective epidermal proteins characterizes acute and chronic atopic dermatitis. *J. Allergy Clin. Immunol.* **2012**, *130*, 1344–1354. [CrossRef]
4. Furuse, M.; Hata, M.; Furuse, K.; Yoshida, Y.; Haratake, A.; Sugitani, Y.; Noda, T.; Kubo, A.; Tsukita, S. Claudin-based tight junctions are crucial for the mammalian epidermal barrier: A lesson from claudin-1-deficient mice. *J. Cell Biol.* **2002**, *156*, 1099–1111. [CrossRef]
5. De Benedetto, A.; Rafaels, N.M.; McGirt, L.Y.; Ivanov, A.I.; Georas, S.N.; Cheadle, C.; Berger, A.E.; Zhang, K.; Vidyasagar, S.; Yoshida, T.; et al. Tight junction defects in patients with atopic dermatitis. *J. Allergy Clin. Immunol.* **2011**, *127*, 773–786.
6. Yuki, T.; Tobiishi, M.; Kusaka-Kikushima, A.; Ota, Y.; Tokura, Y. Impaired Tight Junctions in Atopic Dermatitis Skin and in a Skin-Equivalent Model Treated with Interleukin-17. *PLoS ONE* **2016**, *11*, e0161759. [CrossRef] [PubMed]
7. Batista, D.I.S.; Perez, L.; Orfali, R.L.; Zaniboni, M.C.; Samorano, L.P.; Pereira, N.V., Sotto, M.N., Ishizaki, A.S.; Oliveira, L.M.S.; Sato, M.N.; et al. Profile of skin barrier proteins (filaggrin, claudins 1 and 4) and Th1/Th2/Th17 cytokines in adults with atopic dermatitis. *J. Eur. Acad. Dermatol. Venereol.* **2015**, *29*, 1091–1095. [PubMed]
8. Kast, J.I.; Wanke, K.; Soyka, M.B.; Wawrzyniak, P.; Akdis, D.; Kingo, K.; Rebane, A.; Akdis, C.A. The broad spectrum of interepithelial junctions in skin and lung. *J. Allergy Clin. Immunol.* **2012**, *130*, 544–547. [CrossRef]
9. Kirschner, N.; Poetzl, C.; von den Driesch, P.; Wladykowski, E.; Moll, I.; Behne, M.J.; Brandner, J.M. Alteration of tight junction proteins is an early event in psoriasis: Putative involvement of proinflammatory cytokines. *Am. J. Pathol.* **2009**, *175*, 1095–1106. [CrossRef] [PubMed]
10. Gutowska-Owsiak, D.; Schaupp, A.L.; Salimi, M.; Taylor, S.; Ogg, G.S. Interleukin-22 downregulates filaggrin expression and affects expression of profilaggrin processing enzymes. *Br. J. Dermatol.* **2011**, *165*, 492–498.
11. Hvid, M.; Vestergaard, C.; Kemp, K.; Christensen, G.B.; Deleuran, B.; Deleuran, M. IL-25 in atopic dermatitis: A possible link between inflammation and skin barrier dysfunction? *J. Investig. Dermatol.* **2011**, *131*, 150–157. [CrossRef] [PubMed]
12. Howell, M.D.; Fairchild, H.R.; Kim, B.E.; Bin, L.; Boguniewicz, M.; Redzic, J.S.; Hansen, K.C.; Leung, D.Y. Th2 cytokines act on S100/A11 to downregulate keratinocyte differentiation. *J. Investig. Dermatol.* **2008**, *128*, 2248–2258. [CrossRef] [PubMed]

13. Floudas, A.; Saunders, S.P.; Moran, T.; Schwartz, C.; Hams, E.; Fitzgerald, D.C.; Johnston, J.A.; Ogg, G.S.; McKenzie, A.N.; Walsh, P.T.; et al. IL-17 Receptor A Maintains and Protects the Skin Barrier to Prevent Allergic Skin Inflammation. *J. Immunol.* **2017**, *199*, 707–717. [CrossRef] [PubMed]
14. Lee, J.S.; Tato, C.M.; Joyce-Shaikh, B.; Gulen, M.F.; Cayatte, C.; Chen, Y.; Blumenschein, W.M.; Judo, M.; Ayanoglu, G.; McClanahan, T.K.; et al. Interleukin-23-Independent IL-17 Production Regulates Intestinal Epithelial Permeability. *Immunity* **2015**, *43*, 727–738. [CrossRef] [PubMed]
15. Nadeau, P.; Henehan, M.; De Benedetto, A. Activation of protease-activated receptor 2 leads to impairment of keratinocyte tight junction integrity. *J. Allergy Clin. Immunol.* **2018**, *142*, 281–284. [CrossRef] [PubMed]
16. El Asmar, R.; Panigrahi, P.; Bamford, P.; Berti, I.; Not, T.; Coppa, G.V.; Catassi, C.; Fasano, A. Host-dependent zonulin secretion causes the impairment of the small intestine barrier function after bacterial exposure. *Gastroenterology* **2002**, *123*, 1607–1615. [CrossRef]
17. Kuo, I.H.; Carpenter-Mendini, A.; Yoshida, T.; McGirt, L.Y.; Ivanov, A.I.; Barnes, K.C.; Gallo, R.L.; Borkowski, A.W.; Yamasaki, K.; Leung, D.Y.; et al. Activation of epidermal toll-like receptor 2 enhances tight junction function: Implications for atopic dermatitis and skin barrier repair. *J. Investig. Dermatol.* **2013**, *133*, 988–998. [PubMed]
18. Eyerich, K.; Pennino, D.; Scarponi, C.; Foerster, S.; Nasorri, F.; Behrendt, H.; Ring, J.; Traidl-Hoffmann, C.; Albanesi, C.; Cavani, A. IL-17 in atopic eczema: Linking allergen-specific adaptive and microbial-triggered innate immune response. *J. Allergy Clin. Immunol.* **2009**, *123*, 59–66. [CrossRef]
19. Nograles, K.E.; Suárez-Fariñas, M.; Shemer, A.; Fuentes-Duculan, J.; Chiricozzi, A.; Cardinale, I.; Zaba, L.C.; Kikuchi, T.; Ramon, M.; Bergman, R.; et al. Atopic dermatitis (AD) keratinocytes exhibit normal Th17 cytokine responses. *J. Investig. Dermatol.* **2010**, *125*, 744–746.
20. Bai, B.; Yamamoto, K.; Sato, H.; Sugiura, H.; Tanaka, T. Complex regulation of S100A8 by IL-17, dexamethasone, IL-4 and IL-13 in HaCat cells (human keratinocyte cell line). *J. Dermatol. Sci.* **2007**, *47*, 259–262. [CrossRef]
21. Gutowska-Owsiak, D.; Schaupp, A.L.; Salimi, M.; Selvakumar, T.A.; McPherson, T.; Taylor, S.; Ogg, G.S. IL-17 downregulates filaggrin and affects keratinocyte expression of genes associated with cellular adhesion. *Exp. Dermatol.* **2012**, *21*, 104–110. [CrossRef] [PubMed]
22. Nograles, K.E.; Zaba, L.C.; Guttman-Yassky, E.; Fuentes-Duculan, J.; Suárez-Fariñas, M.; Cardinale, I.; Khatcherian, A.; Gonzalez, J.; Pierson, K.C.; White, T.R.; et al. Th17 cytokines interleukin (IL)-17 and IL-22 modulate distinct inflammatory and keratinocyte-response pathways. *Br. J. Dermatol.* **2008**, *159*, 1092–1102. [CrossRef] [PubMed]
23. Huang, F.; Kao, C.Y.; Wachi, S.; Thai, P.; Ryu, J.; Wu, R. Requirement for both JAK-mediated PI3K signaling and ACT1/TRAF6/TAK1-dependent NF-kappaB activation by IL-17A in enhancing cytokine expression in human airway epithelial cells. *J. Immunol.* **2007**, *179*, 6504–6513. [CrossRef] [PubMed]
24. Newcomb, D.C.; Boswell, M.G.; Zhou, W.; Huckabee, M.M.; Goleniewska, K.; Sevin, C.M.; Hershey, G.K.K.; Kolls, J.K.; Peebles, R.S., Jr. Human TH17 cells express a functional IL-13 receptor and IL-13 attenuates IL-17A production. *J. Allergy Clin. Immunol.* **2011**, *127*, 1006–1013. [CrossRef] [PubMed]
25. Takeda, K.; Noguchi, K.; Shi, W.; Tanaka, T.; Matsumoto, M.; Yoshida, N.; Kishimoto, T.; Akira, S. Targeted disruption of the mouse Stat3 gene leads to early embryonic lethality. *Proc. Natl. Acad. Sci. USA* **1997**, *94*, 3801–3804. [CrossRef] [PubMed]
26. Sano, S.; Itami, S.; Takeda, K.; Tarutani, M.; Yamaguchi, Y.; Miura, H.; Yoshikawa, K.; Akira, S.; Takeda, J. Keratinocyte-specific ablation of Stat3 exhibits impaired skin remodeling, but does not affect skin morphogenesis. *EMBO J.* **1999**, *18*, 4657–4668. [CrossRef]
27. Wang, Z.; Li, R.; Tan, J.; Peng, L.; Wang, P.; Liu, J.; Xiong, H.; Jiang, B.; Chen, Y. Syndecan-1 Acts in Synergy with Tight Junction Through Stat3 Signaling to Maintain Intestinal Mucosal Barrier and Prevent Bacterial Translocation. *Inflamm. Bowel Dis.* **2015**, *21*, 1894–1907. [CrossRef]
28. Yun, J.H.; Park, S.W.; Kim, K.J.; Bae, J.S.; Lee, E.H.; Paek, S.H.; Kim, S.U.; Ye, S.; Kim, J.H.; Cho, C.H. Endothelial STAT3 Activation Increases Vascular Leakage Through Downregulating Tight Junction Proteins: Implications for Diabetic Retinopathy. *J. Cell. Physiol.* **2017**, *232*, 1123–1134. [CrossRef]
29. Amano, W.; Nakajima, S.; Kunugi, H.; Numata, Y.; Kitoh, A.; Egawa, G.; Dainichi, T.; Honda, T.; Otsuka, A.; Kimoto, Y.; et al. The Janus kinase inhibitor JTE-052 improves skin barrier function through suppressing signal transducer and activator of transcription 3 signaling. *J. Allergy Clin. Immunol.* **2015**, *136*, 667–677. [CrossRef]

30. Mitamura, Y.; Nunomura, S.; Nanri, Y.; Ogawa, M.; Yoshihara, T.; Masuoka, M.; Tsuji, G.; Nakahara, T.; Hashimoto-Hachiya, A.; Conway, S.J.; et al. The IL-13/periostin/IL-24 pathway causes epidermal barrier dysfunction in allergic skin inflammation. *Allergy* **2018**, *73*, 1881–1891. [CrossRef]
31. Ma, C.S.; Chew, G.Y.; Simpson, N.; Priyadarshi, A.; Wong, M.; Grimbacher, B.; Fulcher, D.A.; Tangye, S.G.; Cook, M.C. Deficiency of Th17 cells in hyper IgE syndrome due to mutations in STAT3. *J. Exp. Med.* **2008**, *205*, 1551–1557. [CrossRef]
32. Chen, Y.; Yang, P.; Li, F.; Kijlstra, A. The effects of Th17 cytokines on the inflammatory mediator production and barrier function of ARPE-19 cells. *PLoS ONE* **2011**, *6*, e18139. [CrossRef] [PubMed]
33. Capaldo, C.T.; Nusrat, A. Cytokine regulation of tight junctions. *Biochim. Biophys. Acta* **2009**, *1788*, 864–871. [CrossRef] [PubMed]
34. Beck, L.A.; Thaçi, D.; Hamilton, J.D.; Graham, N.M.; Bieber, T.; Rocklin, R.; Ming, J.E.; Ren, H.; Kao, R.; Simpson, E.; et al. Dupilumab treatment in adults with moderate-to-severe atopic dermatitis. *N. Engl. J. Med.* **2014**, *371*, 130–139. [CrossRef] [PubMed]
35. Poumay, Y.; Roland, I.H.; Leclercq-Smekens, M.; Leloup, R. Basal detachment of the epidermis using dispase: Tissue spatial organization and fate of integrin alpha 6 beta 4 and hemidesmosomes. *J. Investig. Dermatol.* **1994**, *102*, 111–117. [CrossRef] [PubMed]
36. Gordon, K.; Kochkodan, J.J.; Blatt, H.; Lin, S.Y.; Kaplan, N.; Johnston, A.; Swindell, W.R.; Hoover, P.; Schlosser, B.J.; Elder, J.T.; et al. Alteration of the EphA2/Ephrin-A signaling axis in psoriatic epidermis. *J. Investig. Dermatol.* **2013**, *133*, 712–722. [CrossRef] [PubMed]

International Journal of
Molecular Sciences

Article

Celiac Disease Monocytes Induce a Barrier Defect in Intestinal Epithelial Cells

Deborah Delbue, Danielle Cardoso-Silva, Federica Branchi, Alice Itzlinger, Marilena Letizia, Britta Siegmund and Michael Schumann *

Department of Gastroenterology, Infectious Diseases and Rheumatology, Campus Benjamin Franklin, Charité – Universitätsmedizin Berlin, 12203 Berlin, Germany; deborah.delbue-da-silva@charite.de (D.D.); danielle.cardoso-da-silva@charite.de (D.C.-S.); federica.branchi@charite.de (F.B.); alice.itzlinger@charite.de (A.I.); marilena.letizia@charite.de (M.L.); britta.siegmund@charite.de (B.S.)

* Correspondence: michael.schumann@charite.de

Received: 16 October 2019; Accepted: 5 November 2019; Published: 9 November 2019

Abstract: Intestinal epithelial barrier function in celiac disease (CeD) patients is altered. However, the mechanism underlying this effect is not fully understood. The aim of the current study was to evaluate the role of monocytes in eliciting the epithelial barrier defect in CeD. For this purpose, human monocytes were isolated from peripheral blood mononuclear cells (PBMCs) from active and inactive CeD patients and healthy controls. PBMCs were sorted for expression of CD14 and co-cultured with intestinal epithelial cells (IECs, Caco2BBe). Barrier function, as well as tight junctional alterations, were determined. Monocytes were characterized by profiling of cytokines and surface marker expression. Transepithelial resistance was found to be decreased only in IECs that had been exposed to celiac monocytes. In line with this, tight junctional alterations were found by confocal laser scanning microscopy and Western blotting of ZO-1, occludin, and claudin-5. Analysis of cytokine concentrations in monocyte supernatants revealed higher expression of interleukin-6 and MCP-1 in celiac monocytes. However, surface marker expression, as analyzed by FACS analysis after immunostaining, did not reveal significant alterations in celiac monocytes. In conclusion, CeD peripheral monocytes reveal an intrinsically elevated pro-inflammatory cytokine pattern that is associated with the potential of peripheral monocytes to affect barrier function by altering TJ composition.

Keywords: barrier function; tight junction assembly; monocytes; celiac disease

1. Introduction

Celiac disease (CeD) is an autoimmune enteropathy triggered by the ingestion of gluten, affecting approximately 1% of the population in Western countries [1]. Current understanding of CeD immune pathology focusses on activation of gluten-specific TH1-cells secondary to presentation of DQ2- or DQ8-restricted gliadin peptides as the cause of the small intestinal inflammatory response. As a consequence, inflammation leads to villous atrophy and crypt hyperplasia, thereby causing the typical clinical features of intestinal malabsorption of nutrients [2,3].

A so-far unresolved issue of CeD is the nature of the associated intestinal epithelial barrier defect. It not only occurs secondary to the inflammatory process located in the lamina propria in active disease, but appears to be primary, since it is verifiable in inactive CeD patients on a gluten-free diet (GFD) and in relatives of CeD patients who do not suffer from CeD [4,5]. Moreover, Kumar et al. identified barrier-defining genes associated with CeD, thereby providing genetic proof for the relevance of barrier function in CeD pathogenesis. Importantly, the barrier function is maintained by a protein-protein complex interaction, where the main structure is the tight junction (TJ) proteins. TJs are the most apical contact between enterocytes formed by integral membrane proteins, including occludin, claudins, and scaffolding proteins as ZO-1 [6]. Although structural changes in celiac barrier function can be allocated to enterocyte TJ composition and epithelial transcytosis of gliadin peptides, research aiming to clarify how these changes arise is scarce [7–10].

Interestingly, monocytes are strongly involved in the regulation of intestinal barrier function, either by secretion of cytokines or by direct interaction with intestinal epithelial cells [11–13]. After extravasation, monocytes infiltrate the lamina propria, differentiating into macrophages and producing inflammatory mediators to combat pathogens [14]. For CeD, it has been previously described that monocytes isolated from CeD patients produced substantial amounts of TNF-α and interleukin-8 (IL-8) in a gluten-dependent manner [15]. Moreover, a gliadin-stimulated monocytic cell line was shown to have the potential to modulate intestinal epithelial barrier function [16]. Furthermore, monocytes isolated from healthy individuals that were stimulated with IL-15 as a celiac-mimicking cytokine milieu were capable of secreting pro-inflammatory cytokines that are known to induce barrier defects [17].

In the current work, we aimed to analyze the potency of celiac monocytes to perturb intestinal barrier function. For this purpose, we isolated monocytes from CeD patients and co-cultured these cells with intestinal epithelial cells to analyze epithelial barrier function.

2. Results

2.1. Monocytes Derived from Celiac Patients Induce a Barrier Defect in Intestinal Epithelial Cells

To evaluate whether monocytes exert an effect on epithelial barrier function, intestinal epithelial cells (IECs) were co-cultured with human primary monocytes. Peripheral blood mononuclear cells (PBMCs) were isolated from peripheral blood of CeD patients with different disease status and of healthy donors and then sorted for CD14 expression. IECs were grown on transwell filters, where they reached confluence and built up a stable barrier function, whereas the monocytes were placed underneath the filters. Thus, interaction between monocytes and IECs was possible mostly through soluble factors that pass through the filter membranes. Interestingly, transepithelial resistance (TER) of IECs measured after 48 h of co-culture was reduced when they were co-cultured with celiac monocytes, compared to the co-culture with healthy controls. This effect added up to approx. 70% of the control level and was irrespective of CeD activity (Figure 1).

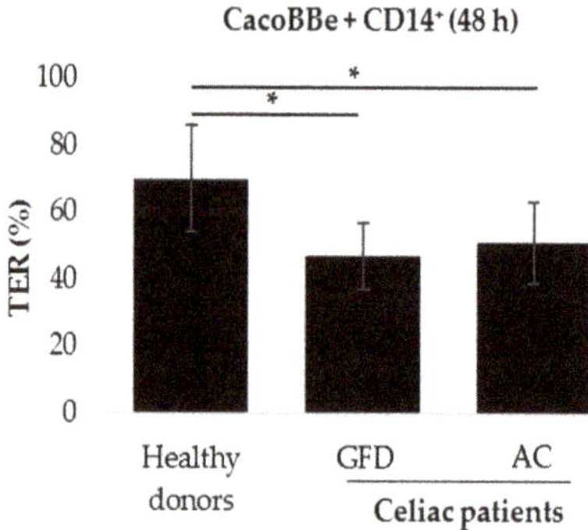

Figure 1. Epithelial barrier function after co-culture of intestinal epithelial cells (IECs) with monocytes derived from celiac disease patients. After peripheral blood mononuclear cell (PBMC) isolation and CD14+ cell-sorting, epithelial cells were co-cultured with monocytes from healthy donors. celiac patients on a gluten-free diet (GFD), or AC (active celiac disease (CeD)) patients. Subsequently, the TER was measured after 48 h of co-culture (% of TER prior to addition of monocytes). Mean of $n = 36$ (healthy donors), $n = 15$ (GFD), and $n = 20$ (AC) individual filters measurements. Monocytes used for these experiments were isolated from $n = 8$ (healthy donors), $n = 4$ (GFD) and $n = 5$ (active CeD). Mann-Whitney U * $p < 0.05$, comparison between co-cultures with monocytes from healthy donors and CeD patients.

Co-culturing IECs with unsorted (i.e., total) PBMCs caused a similar decrease in TER (Figure S1). To exclude possible direct effects of gliadin or IL-15 on the epithelium, CacoBBe cells were exposed to IL-15/Tglia alone, with monocytes or PBMCs. In IECs alone, we did not observe a decrease in TER with only IL-15/Tglia addition. Nevertheless, in the cells exposed to monocytes, TER decreased at the same levels as with monocytes plus IL-15/Tglia. (Figure S2). These results showed that effects observed in TER are independent of IL-15/Tglia stimulation and that this is rather directly associated with monocytes. In summary, this experiment uncovered the potential of celiac monocytes to alter epithelial barrier function.

2.2. Celiac Monocytes Alter IEC-TJ Structure

As a next step, we aimed to evaluate whether $CD14^+$ cells alter IEC barrier function through changes in tight junction (TJ) integrity. First, IECs that had completed 48 h of co-culture with $CD14^+$ monocytes were immunostained for TJ proteins ZO-1 and occludin. As shown in Figure 2, no effect was found regarding TJ localization or expression in IEC layers that had been co-cultured with $CD14^+$ monocytes isolated from healthy donors. However, for IECs co-cultured with monocytes derived from CeD patients, lower levels of occludin and a mosaic expression pattern of ZO-1 was found, with ZO-1 being significantly reduced in expression in some, but not all, regions of the filter. Moreover, TJs appeared to be irregular regarding loop-like linings, which was not observed when IECs were exposed to monocytes of healthy donors (Figure 2A). Additionally to the aforementioned reduction in expression level of ZO-1, XZ-projections revealed an uneven structure of the apical membrane in IECs that were co-cultured with CeD monocytes (Figure 2B). Furthermore, protein levels of occludin and the TJ-sealing claudin-5, which has previously been implicated in the CeD barrier defect, were analyzed (Figure 2C) [9]. IEC protein levels of occludin and claudin-5, after exposure to celiac monocytes, were reduced compared to protein levels of the respective healthy control monocytes. These data show evidence that CeD monocytes exert effects on the TJ structure of co-cultured IECs.

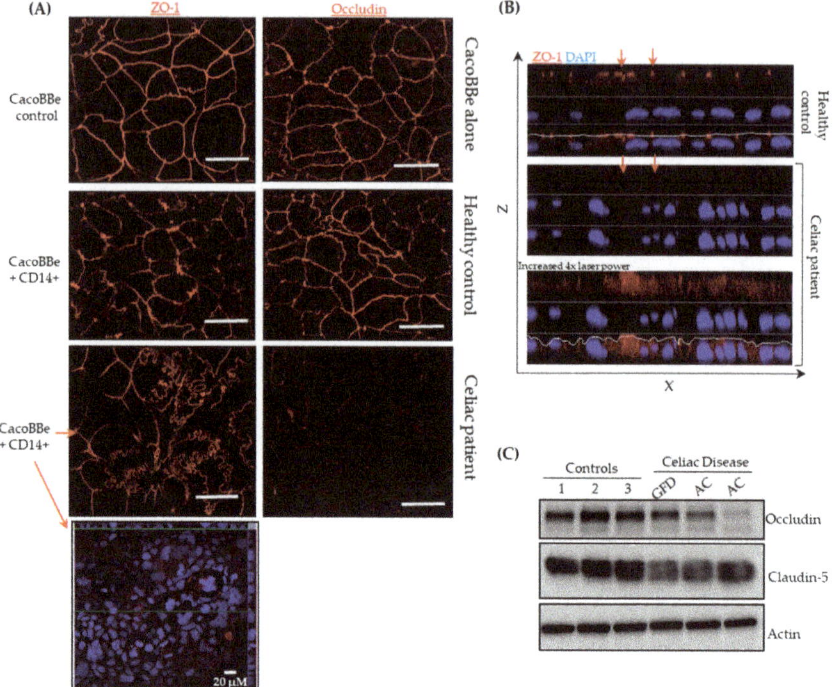

Figure 2. Tight junction (TJ) structure and protein composition after co-culture with monocytes derived from celiac disease patients. (**A**) Cellular localization of occludin and ZO-1 were investigated using confocal laser scanning microscopy after immunostaining. Representative images from $n = 5$ (healthy donors), $n = 3$ (CeD on GFD) and $n = 3$ (Active CeD patients). Scale bar: 50 µM. The two effects of CD14 co-culture in ZO-1 expression is pointed out by red arrows (**B**) Collapsed XZ-projections. ZO-1 staining reveals apical junctional complexes at approx. identical Z heights, as illustrated by the white lining in the merged image. When CacoBBe cells co-cultured with celiac monocytes were immunostained, lining was comparably irregular, and ZO-1 level was reduced. (**C**) IEC protein levels by Western blotting of occludin and claudin-5 after co-culture with monocytes.

2.3. Monocytes Derived from Celiac Disease Patients Present Higher Levels of Proinflammatory Cytokine Production

Next, we characterized isolated human monocytes that had previously been sorted for CD14 to uncover potential differences regarding cytokine and surface marker expression between celiac and healthy control monocytes. First, we analyzed the expression of surface markers that are characteristic of classically and non-classically activated macrophages. Surface marker expression was analyzed after CD14-sorting (Figure S3) and 24 h of culture—media including Granulocyte macrophage colony stimulating factor (GM-CSF)—by immunostaining. Using a gating strategy revealed in Figure 3A, monocyte populations were detected, doublets were excluded, and the viable population (DAPI-negative cells) was analyzed. Then, frequency of positivity for CD11b, CD80, HLA-DR, CD163, and CD16 was evaluated. No significant differences in the frequency of any of the examined surface markers were found (Figure 3B-G; Figure S3). The frequency of cells revealing a double positivity for CD80 and HLA-DR (i.e., expression of both inflammatory markers within the same cell) was also analyzed but turned out to be not significantly different. Although CD16 frequency was not significantly increased, a tendency toward higher frequencies of CD16-positive cells was observed. This points to an increased fraction of intermediate/non-classical monocytes in the peripheral blood of CeD patients.

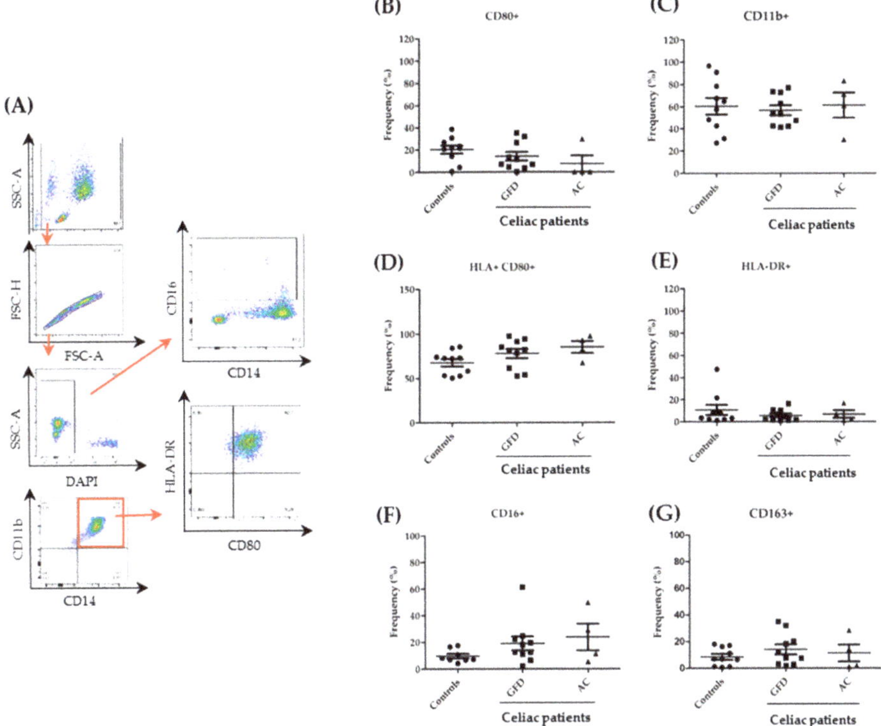

Figure 3. Expression of surface markers in peripheral monocytes from celiac patients. PBMCs were isolated from 10 controls, 10 GFD and 4 active CeD (AC) patients and sorted for CD14. Subsequently, cells were cultured in the presence of granulocyte macrophage colony stimulating factor (GM-CSF; 10 ng/ml) for 24 h and evaluated by flow cytometry. (**A**) Gating strategy used to determine surface marker expression. The stepwise gating approach is highlighted by various steps of analysis that are interconnected by red arrows. Representative plots from a healthy control are shown. (**B–G**) Results for the expression of single surface markers are shown. Each dot represents the surface marker expression result of a single patient. Mean values ± standard error of the mean (SEM) are shown. The Mann–Whitney U test revealed no significant differences.

Subsequently, we determined the concentration of a set of cytokines within the supernatant of the monocytes at the end of the 24 h time period following isolation and CD14-sorting. To guarantee survival, they were cultured at this time in the presence of GM-CSF (10 mg/ml). As illustrated in Figure 4A, concentrations of IFN-α2, IFN-λ, IL-19, IL-12p70, IL-17A, IL-18, IL-23, and IL-33 did not reach the detection level of the assay, neither in the control nor the CeD group. More interestingly, levels of IL-1β, TNF-α, IL-8, IL-10, IL-6, and MCP-1 were found at detectable levels. Although concentrations of IL-1β, TNF-α, IL-8, and IL-10 were not significantly different between the groups of patients and healthy controls, a tendency for higher levels of the pro-inflammatory cytokines was observed in CeD supernatants (Figure 4B–E). Interestingly, IL-6 and MCP-1 (monocyte chemotactic protein-1, synonymous: CC-chemokine ligand-2, CCL2) were found at significantly higher levels in the supernatant of monocytes from CeD patients who had received a GFD, compared to monocytes from healthy controls (Figure 4E,F). In summary, these data reveal that $CD14^+$ monocytes isolated from CeD patients carry a more pro-inflammatory phenotype, an effect that appears to be independent of disease activity.

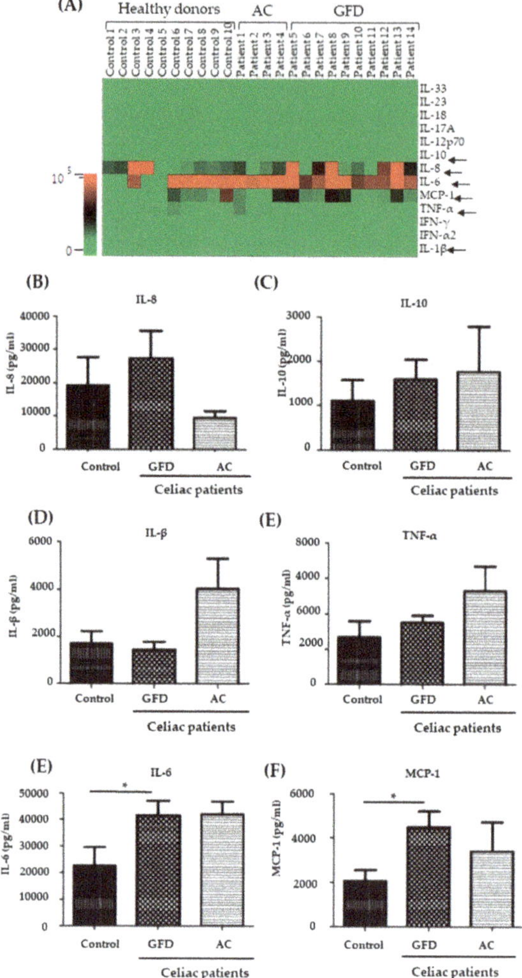

Figure 4. Increased levels of pro-inflammatory cytokines in the supernatant of monocytes derived from celiac disease patients. (**A**) PBMCs were isolated from 10 controls, 10 GFD, and 4 active CeD patients, and monocytes were sorted using CD14 MACS MicroBeads. Subsequently, cells were kept in the incubator for 24 h in the presence of GM-CSF (10 ng/mL). Supernatants of monocyte cultures were collected and cytokine levels were determined. Data are illustrated as a heat map revealing color-coded concentrations of cytokines (green: low concentration; red: high concentration). (**B–F**) Results for individual cytokine measurements are shown: Interleukin- (IL-)8, IL-10, IL-1β, TNF-α, IL-6, and MCP-1. Mean values ± SEM are shown. Mann–Whitney test: *, $p < 0.05$.

3. Discussion

CeD is an autoimmune enteropathy triggered by the ingestion of dietary gluten [3]. Although the central role for on the one hand gluten-specific, and on the other hand tissue-resident, cytotoxic T-cells is undisputed, results from genome-wide association studies and from functional data collected in non-affected relatives of CeD patients point to a primary defect of the epithelial barrier as a discrete pathophysiologic entity within the overall immune pathology of CeD [2,5,18]. Nevertheless, it is mostly unclear how the barrier defect in CeD is triggered. Since monocytes are major mediators in mucosal barrier defects, we aimed to elucidate the potential of peripheral human monocytes to alter intestinal epithelial barrier function. The rationale for using peripheral blood monocytes comes from early data that convincingly showed that in intestinal inflammation, monocytes infiltrate the mucosa and then further differentiate to macrophages secreting pro-inflammatory cytokines, rather than showing that resident tissue-macrophages re-differentiate into pro-inflammatory macrophages [19].

$CD14^+$ monocytes derived from peripheral blood of CeD patients revealed an intrinsically higher secretion of IL-6 and MCP-1. Moreover, a non-significant tendency for increased expression of TNF-α and IL-1β was observed. A similar pattern of cytokine expression was previously described for intestinal (i.e., not peripheral) macrophages in IBD [20,21]. Specifically, Kamada et al. described the CD14+ macrophage population secondary to their cytokine expression (e.g., TNF-α, IL-6, IL-12/23p40, and IL-23p19) as pro-inflammatory and non-resident, compared to $CD14^-$ resident macrophages that are found primarily in healthy gut and do not express pro-inflammatory cytokines. Phenotypically similar macrophages were shown to elicit a barrier defect on IECs in a co-culture model, similar to that used in the current work [11]. However, in the current work, expression analysis of IL-1β and TNF-α, which were previously shown by Lissner et al. to be mostly responsible for the IEC barrier defect in the M1- and M0-polarized macrophage model, only revealed a non-significant tendency towards higher levels of these cytokines. On the other hand, IL-6 and MCP-1 were significantly increased. MCP-1 (synonymous CCL-2) is secreted by various cells, including monocytes, as a chemoattractant for monocytes, T-cells, and dendritic cells, and enables via the CCL2-CCR2 axis the extravasation of monocytes into the lamina propria [14,22]. Our data on MCP-1 are in line with data from Italy that revealed a higher expression of CCL-2 by PBMCs after stimulation with the p31-43 gliadin peptide [23]. IL-6, a pro-inflammatory cytokine also secreted by intestinal macrophages, has previously been described to induce an intestinal barrier defect by increasing expression of pore-forming claudins, including claudin-2 [24]. Interestingly, IL-6 secreted by macrophages as the cause for a reduced intestinal barrier function was also discussed as a mechanism for the epithelial barrier defect found in liver cirrhosis [25].

Although a difference in cytokine production was found between celiac monocytes and monocytes isolated from healthy controls, which was in line with the functional barrier data revealing a defect only inducible by peripheral monocytes from CeD patients, we did not observe significant differences in monocyte surface marker expression. CD16 as a marker for alternatively activated macrophages showed a non-significant tendency towards higher expression in celiac monocytes. However, this aspect of our work remains somewhat inconclusive.

Monocytes isolated from CeD patients induced a functional barrier defect on IECs. The reduced barrier function was measured by TER and was associated with an altered TJ morphology on confocal laser scanning microscopy (LSM) as well as TJ protein expression in Western blotting. Celiac barrier defects that are localized to the TJ have been described previously [9,26]. In those studies, the alteration of the celiac TJ was complex, involving reduced expression of occludin, claudin-3, -5, and -7, and altered phosphorylation of ZO-1. This is mostly in line with our work, since celiac monocytes induced a reduction of occludin and claudin-5 expression, and confocal LSM revealed alterations in cellular ZO-1 distribution. Nevertheless, we should mention that we and others have shown that apoptosis of IECs, which was not analyzed in the current study, might also contribute to a defective epithelial barrier and that induction of (apoptotic or non-apoptotic) cell death might be a conceivable fate of an IEC that is exposed to monocytes [9,27].

Taken together, our results suggest that in CeD patients, peripheral blood monocytes have the potential to induce an epithelial defect of the intestinal mucosa. This IEC reaction to monocyte exposure is presumably related to action of monocytic IL-6 on IECs. Tight junctional alterations in the intestinal epithelial cell layer are at least partially responsible for the functional barrier defect. These effects might be primary in nature. However, work herein is not sufficient to prove the primary nature. In the future, this could be approached by analyzing monocytes of first-degree relatives to CeD patients to determine if these cells also show an impact on barrier function.

4. Materials and Methods

4.1. Human Material

The study was approved by the Ethics Committees of the Charité – Universitätsmedizin Berlin, Germany (protocol number EA4/116/18, accepted on Jan 22nd, 2019). Heparinized whole blood samples were collected from healthy individuals and CeD patients. Inactive (GFD) patients received a GFD for >1 year. All patients declared their informed consent (signed consent form). Healthy controls were individuals without a history of enteropathy and without clinical signals of CeD or other autoimmune diseases. For further characteristics of CeD patients, refer to Table 1.

Table 1. Characteristics of CeD patients.

Number of Subjects	17
Female/Male	14/3
Age at enrolment, median (range)	46 (23-83)
Age at CeD diagnosis, median (range)	32 (6-73)
Marsh Grade at enrolment, *n* (%)	
0	5 (29)
1	2 (12)
2	0 (0)
3a	1 (6)
3b	3 (18)
3c	0 (0)
not available	6 (35)
tTG at enrolment, *n* (%)	
positive	6 (35)
negative	2 (12)
not available	9 (53)
HLA-DQ status, *n* (%)	
DQ2+	11 (65)
DQ8+	0 (0)
not available	6 (35)
GFD status, *n* (%)	
Active CeD	6 (35)
New CeD diagnosis	4
CeD, non-compliant to GFD	2
CeD on GFD	11 (65)

CeD: celiac disease; tTG: transglutaminase antibodies; GFD: gluten-free diet.

4.2. Cell Line

The human colorectal cell line, CacoBBe (C2BBe1 [clone of Caco-2] ATCC®CRL-2102™), was maintained in Dulbecco's Modified Eagle's Medium (DMEM) + GlutaMAX (Gibco), with 10% fetal bovine serum (Gibco), 1% penicillin and streptomycin (Corning), 10μM HEPES-buffer and 1M non-essential amino acids (Merck Millipore). Cells were kept at 37 °C in a 5% CO_2 environment. Culture medium was changed three times per week.

4.3. PBMCs Isolation and CD14+ Sorting

Peripheral blood mononuclear cells (PBMCs) were isolated from peripheral blood of healthy donors and celiac disease patients by Biocoll (Merk Millipore) separating solution and centrifugation, as previously described [28]. Subsequently, PBMCs were sorted using CD14 MACS MicroBeads (Miltenyi Biotech, Bergisch Gladbach, Germany). As determined by flow cytometry, preparations contained >90% CD14+ cells. Monocytes were plated in 24-well dishes with RPMI-1640 as media (Gibco), supplemented with 10% of fetal bovine serum (Gibco) and 1% of penicillin and streptomycin (Corning).

CD14+ cells received human granulocyte macrophage colony stimulating factor (10ng/mL; GM-CSF) for 24 h prior to co-culture. Cell culture supernatants were collected for flow cytometry analysis.

4.4. Co-Culture and TER Measurement

Intestinal epithelial cells (IECs) were plated on permeable transwell polycarbonate filter supports (0.4 μm; 0.6 cm^2, Merck Millipore) and kept at 37 °C in a 5% CO_2 environment. Culture medium was changed three times per week. On days 10 to 16 after plating, filters were transferred to 24-well dishes containing $CD14^+$ cells (5×10^5 cells per well). In addition, IL-15/Tglia (10 mg/mL) was added (gift by W. Dieterich; LPS-free) to filters with IECs. Subsequently, the transepithelial resistance (TER) was measured, as described previously [11], for 48 h of co-culture with monocytes from CeD patients or healthy donors (controls). For immunofluorescence experiments, filters were fixed using 1% paraformaldehyde for 15 minutes at room temperature.

4.5. Immunofluorescence

Epithelial cell layers were stained using the following primary antibodies: ZO-1 (1:100; BD Biosciences, NJ, USA). The secondary antibodies used were Alexa Fluor 488 goat anti-mouse or rabbit IgG, and Alexa Fluor 594 goat anti-mouse or rabbit IgG (1:500; Thermo Fisher Scientific, MA, USA). To determine occludin expression and cellular distribution, an occludin mouse monoclonal antibody (OC-3F10) was used as an Alexa Fluor®594 Conjugate (Thermofischer). Nuclei were stained using DAPI (4′,6-Diamidin-2-phenylindol, conc. 1:2000). Immunofluorescence staining was analyzed by confocal laser scanning microscopy (LSM 780, Carl Zeiss, Jena, Germany) as previously described [9,11].

4.6. Western Blotting

For Western blotting analysis, the protocol was followed as previously described [9]. The primary antibodies used were occludin (1:1000; Sigma Aldrich, St. Louis, MO, USA), claudin-5 (1:1000; Invitrogen, Carlsbad, CA, USA) and actin (1:1000; Sigma Aldrich, St. Louis, MO, USA). The peroxidase-conjugated secondary antibodies used were goat anti-rabbit IgG or goat anti-mouse IgG (Jackson ImmunoResearch, Ely, UK). SuperSignal West Pico PLUS Stable Peroxide Solution (Thermo Scientific, Waltham, MA, USA) was used for protein detection, and Fusion FX7 imaging system (Vilber Lourmat Deutschland GmbH, Eberhardzell, Germany) was used to detect protein signal levels.

4.7. Flow Cytometric Assessment—Surface Markers and Cytokine Expression

$CD14^+$ cells were washed twice with PBS, and the surface markers were checked. The following fluorochrome-coupled antibodies were applied: anti-CD14 (61D3) from BD Biosciences, anti- anti-CD16 (3G8), anti-CD11b (ICRF44), anti-CD163 (GHI/61) from Biolegend, CD80 (2D10.4), and anti-HLA-DR (LN3) from eBioscence. Dead cells were excluded by DAPI staining. Samples were assessed by flow cytometry using a FACSCanto II and the FACS Diva software (version 6; BD Biosciences). Supernatants of the cultures after 24 h of culture with GM-CSF (10 ng/ml) were tested for cytokine expression (IL-1β, IFN-α2, IFN-λ, TNF-α, MCP-1, IL-6, IL-8, IL-19, IL-12p70, IL-17A, IL-18, IL-23, and IL-33) using the LEGENDplex Multi-Analyte Flow Assay kit–Human Inflammation Panel (13-plex) (Biolegend) according to the manufacturer's protocol. FACS data were analyzed using FlowJo (v10.6.1) and LEGENDplex v8.0 software (BioLegend, San Diego, CA, USA).

4.8. Statistical Analysis

Statistical analysis was performed using GraphPad Prism software (GraphPad Software, La Jolla, CA) by a non-parametric Mann–Whitney test to analyze differences between the control and the CeD patients. One-way ANOVA was used to compare differences in all groups analyzed. All data are expressed as mean values ± standard error of the mean (SEM). A $p < 0.05$ was considered significant.

Supplementary Materials: Supplementary materials can be found at http://www.mdpi.com/1422-0067/20/22/5597/s1.

Author Contributions: D.D. planned and carried out the experiments and wrote the manuscript. D.C.-S. performed cell culture and immunostainings. F.B. identified patients and characterized them further. M.L. performed the FACS analysis. A.I. partially performed the co-culture and the immunostainings. B.S. planned the experiments and wrote the manuscript. M.S. planned experiments, identified patients, and wrote the manuscript.

Funding: This research was funded by the Deutsche Forschungsgemeinschaft TRR241 ("IEC in IBD") and Deutsche Forschungsgemeinschaft Graduiertenkolleg "TJ-train" GRK 2318, as well as a grant for "Polarity in Celiac Disease" by the company Dr. Schär and the research prize for celiac disease by the Deutsche Zöliakie Gesellschaft.

Acknowledgments: We acknowledge the very skilled support given by Claudia Heldt and the helpful support with Tglia given by Walburga Dieterich (Erlangen, Germany).

Conflicts of Interest: The authors declare no conflict of interest.

Abbreviations

AC	Active celiac
CeD	Celiac disease
CCL2	CC-chemokine ligand-2
GFD	Gluten-free diet
GM-CSF	Granulocyte macrophage colony stimulating factor
IL-1β	Interleukin-1β
IL-6	Interleukin-6
IL-8	Interleukin-8
IL-10	Interleukin-10
IL-12p70	Interleukin-12p70
IL-15	Interleukin-15
IL-17A	Interleukin-17A
IL-18	Interleukin-18
IL-23	Interleukin-23
IL-33	Interleukin-33
IFN-γ	Interferon-γ
IFNα2	Interferon-α2
MACS	Magnetic cell sorting
MCP-1	Monocyte chemotactic protein-1
PBMCs	Peripheral blood mononuclear colony stimulating factor
TER	Transepithelial resistance
Tglia	Trypsinized gliadin
TJ	Tight junctions

References

1. Mustalahti, K.; Catassi, C.; Reunanen, A.; Fabiani, E.; Heier, M.; McMillan, S.; Murray, L.; Metzger, M.H.; Gasparin, M.; Bravi, E.; et al. The prevalence of celiac disease in Europe: results of a centralized, international mass screening project. *Ann. Med.* **2010**, *42*, 587–595. [CrossRef] [PubMed]
2. Jabri, B.; Sollid, L.M. T Cells in Celiac Disease. *J. Immunol.* **2017**, *198*, 3005–3014. [CrossRef] [PubMed]
3. Ludvigsson, J.F.; Leffler, D.A.; Bai, J.C.; Biagi, F.; Fasano, A.; Green, P.H.; Hadjivassiliou, M.; Kaukinen, K.; Kelly, C.P.; Leonard, J.N.; et al. The Oslo definitions for coeliac disease and related terms. *Gut* **2013**, *62*, 43–52. [CrossRef]
4. Schumann, M.; Siegmund, B.; Schulzke, J.D.; Fromm, M. Celiac Disease: Role of the Epithelial Barrier. *Cell Mol. Gastroenterol. Hepatol.* **2017**, *3*, 150–162. [CrossRef]
5. van Elburg, R.M.; Uil, J.J.; Mulder, C.J.; Heymans, H.S. Intestinal permeability in patients with coeliac disease and relatives of patients with coeliac disease. *Gut* **1993**, *34*, 354–357. [CrossRef]
6. Cardoso-Silva, D.; Delbue, D.; Itzlinger, A.; Moerkens, R.; Withoff, S.; Branchi, F.; Schumann, M. Intestinal Barrier Function in Gluten-Related Disorders. *Nutrients* **2019**, *11*, 2325. [CrossRef] [PubMed]

7. Matysiak-Budnik, T.; Moura, I.C.; Arcos-Fajardo, M.; Lebreton, C.; Menard, S.; Candalh, C.; Ben-Khalifa, K.; Dugave, C.; Tamouza, H.; van Niel, G.; et al. Secretory IgA mediates retrotranscytosis of intact gliadin peptides via the transferrin receptor in celiac disease. *J. Exp. Med.* **2008**, *205*, 143–154. [CrossRef]
8. Menard, S.; Lebreton, C.; Schumann, M.; Matysiak-Budnik, T.; Dugave, C.; Bouhnik, Y.; Malamut, G.; Cellier, C.; Allez, M.; Crenn, P.; et al. Paracellular versus transcellular intestinal permeability to gliadin peptides in active celiac disease. *Am. J. Pathol.* **2012**, *180*, 608–615. [CrossRef]
9. Schumann, M.; Gunzel, D.; Buergel, N.; Richter, J.F.; Troeger, H.; May, C.; Fromm, A.; Sorgenfrei, D.; Daum, S.; Bojarski, C.; et al. Cell polarity-determining proteins Par-3 and PP-1 are involved in epithelial tight junction defects in coeliac disease. *Gut* **2012**, *61*, 220–228. [CrossRef]
10. Schumann, M.; Richter, J.F.; Wedell, I.; Moos, V.; Zimmermann-Kordmann, M.; Schneider, T.; Daum, S.; Zeitz, M.; Fromm, M.; Schulzke, J.D. Mechanisms of epithelial translocation of the alpha(2)-gliadin-33mer in coeliac sprue. *Gut* **2008**, *57*, 747–754. [CrossRef]
11. Lissner, D.; Schumann, M.; Batra, A.; Kredel, L.I.; Kuhl, A.A.; Erben, U.; May, C.; Schulzke, J.D.; Siegmund, B. Monocyte and M1 Macrophage-induced Barrier Defect Contributes to Chronic Intestinal Inflammation in IBD. *Inflamm. Bowel. Dis.* **2015**, *21*, 1297–1305. [CrossRef] [PubMed]
12. Managlia, E.; Liu, S.X.L.; Yan, X.; Tan, X.D.; Chou, P.M.; Barrett, T.A.; De Plaen, I.G. Blocking NF-kappaB Activation in Ly6c(+) Monocytes Attenuates Necrotizing Enterocolitis. *Am. J. Pathol.* **2019**, *189*, 604–618. [CrossRef]
13. Morhardt, T.L.; Hayashi, A.; Ochi, T.; Quiros, M.; Kitamoto, S.; Nagao-Kitamoto, H.; Kuffa, P.; Atarashi, K.; Honda, K.; Kao, J.Y.; et al. IL-10 produced by macrophages regulates epithelial integrity in the small intestine. *Sci. Rep.* **2019**, *9*, 1223. [CrossRef] [PubMed]
14. Bain, C.C.; Schridde, A. Origin, Differentiation, and Function of Intestinal Macrophages. *Front Immunol.* **2018**, *9*, 2733. [CrossRef] [PubMed]
15. Cinova, J.; Palova-Jelinkova, L.; Smythies, L.E.; Cerna, M.; Pecharova, B.; Dvorak, M.; Fruhauf, P.; Tlaskalova-Hogenova, H.; Smith, P.D.; Tuckova, L. Gliadin peptides activate blood monocytes from patients with celiac disease. *J. Clin Immunol.* **2007**, *27*, 201–209. [CrossRef] [PubMed]
16. Barilli, A.; Rotoli, B.M.; Visigalli, R.; Ingoglia, F.; Cirlini, M.; Prandi, B.; Dall'Asta, V. Gliadin-mediated production of polyamines by RAW264.7 macrophages modulates intestinal epithelial permeability in vitro. *Biochim. Biophys. Acta* **2015**, *1852*, 1779–1786. [CrossRef]
17. Harris, K.M.; Fasano, A.; Mann, D.L. Monocytes differentiated with IL-15 support Th17 and Th1 responses to wheat gliadin: implications for celiac disease. *Clin Immunol.* **2010**, *135*, 430–439. [CrossRef]
18. Kumar, V.; Gutierrez-Achury, J.; Kanduri, K.; Almeida, R.; Hrdlickova, B.; Zhernakova, D.V.; Westra, H.J.; Karjalainen, J.; Ricano-Ponce, I.; Li, Y.; et al. Systematic annotation of celiac disease loci refines pathological pathways and suggests a genetic explanation for increased interferon-gamma levels. *Hum Mol. Genet* **2015**, *24*, 397–409. [CrossRef]
19. Grimm, M.C.; Pullman, W.E.; Bennett, G.M.; Sullivan, P.J.; Pavli, P.; Doe, W.F. Direct evidence of monocyte recruitment to inflammatory bowel disease mucosa. *J. Gastroenterol. Hepatol.* **1995**, *10*, 387–395. [CrossRef]
20. Kamada, N.; Hisamatsu, T.; Okamoto, S.; Chinen, H.; Kobayashi, T.; Sato, T.; Sakuraba, A.; Kitazume, M.T.; Sugita, A.; Koganei, K.; et al. Unique CD14 intestinal macrophages contribute to the pathogenesis of Crohn disease via IL-23/IFN-gamma axis. *J. Clin Invest.* **2008**, *118*, 2269–2280.
21. Lampinen, M.; Waddell, A.; Ahrens, R.; Carlson, M.; Hogan, S.P. CD14+CD33+ myeloid cell-CCL11-eosinophil signature in ulcerative colitis. *J. Leukoc Biol.* **2013**, *94*, 1061–1070. [CrossRef]
22. Smith, P.D.; Smythies, L.E.; Shen, R.; Greenwell-Wild, T.; Gliozzi, M.; Wahl, S.M. Intestinal macrophages and response to microbial encroachment. *Mucosal Immunol.* **2011**, *4*, 31–42. [CrossRef]
23. Vincentini, O.; Maialetti, F.; Gonnelli, E.; Silano, M. Gliadin-dependent cytokine production in a bidimensional cellular model of celiac intestinal mucosa. *Clin Exp. Med.* **2015**, *15*, 447–454. [CrossRef]
24. Suzuki, T.; Yoshinaga, N.; Tanabe, S. Interleukin-6 (IL-6) regulates claudin-2 expression and tight junction permeability in intestinal epithelium. *J. Biol. Chem.* **2011**, *286*, 31263–31271. [CrossRef]
25. Du Plessis, J.; Vanheel, H.; Janssen, C.E.; Roos, L.; Slavik, T.; Stivaktas, P.I.; Nieuwoudt, M.; van Wyk, S.G.; Vieira, W.; Pretorius, E.; et al. Activated intestinal macrophages in patients with cirrhosis release NO and IL-6 that may disrupt intestinal barrier function. *J. Hepatol.* **2013**, *58*, 1125–1132. [CrossRef]

26. Ciccocioppo, R.; Finamore, A.; Ara, C.; Di Sabatino, A.; Mengheri, E.; Corazza, G.R. Altered expression, localization, and phosphorylation of epithelial junctional proteins in celiac disease. *Am. J. Clin. Pathol.* **2006**, *125*, 502–511. [CrossRef]
27. Bojarski, C.; Gitter, A.H.; Bendfeldt, K.; Mankertz, J.; Schmitz, H.; Wagner, S.; Fromm, M.; Schulzke, J.D. Permeability of human HT-29/B6 colonic epithelium as a function of apoptosis. *J. Physiol.* **2001**, *535 Pt 2*, 541–552. [CrossRef]
28. Kredel, L.I.; Batra, A.; Stroh, T.; Kuhl, A.A.; Zeitz, M.; Erben, U.; Siegmund, B. Adipokines from local fat cells shape the macrophage compartment of the creeping fat in Crohn's disease. *Gut* **2013**, *62*, 852–862. [CrossRef]

International Journal of
Molecular Sciences

Review

Contributions of Myosin Light Chain Kinase to Regulation of Epithelial Paracellular Permeability and Mucosal Homeostasis

Wei-Qi He [1,*], Jing Wang [1], Jian-Ying Sheng [1], Juan-Min Zha [1], W. Vallen Graham [2,3] and Jerrold R. Turner [4,*]

1 Jiangsu Key Laboratory of Neuropsychiatric Diseases and Cambridge-Suda (CAM-SU) Genomic Resource Center, Medical College of Soochow University, Suzhou 215123, China; 20184250017@stu.suda.edu.cn (J.W.); 20184250012@stu.suda.edu.cn (J.-Y.S.); zhajuanmin@suda.edu.cn (J.-M.Z.)
2 Thelium Therapeutics, New York, NY 10014, USA; vallen@theliumtx.com
3 Laboratory of Chemical Biology & Signal Transduction, The Rockefeller University, New York, NY 10065, USA
4 Laboratory of Mucosal Barrier Pathobiology, Department of Pathology, Brigham and Women's Hospital and Harvard Medical School, Boston, MA 02115, USA
* Correspondence: whe@suda.edu.cn (W.-Q.H.); jrturner@bwh.harvard.edu (J.R.T.); Tel.: +86-512-6588-3545 (W.-Q.H.); +1-6175258165 (J.R.T.)

Received: 21 January 2020; Accepted: 30 January 2020; Published: 3 February 2020

Abstract: Intestinal barrier function is required for the maintenance of mucosal homeostasis. Barrier dysfunction is thought to promote progression of both intestinal and systemic diseases. In many cases, this barrier loss reflects increased permeability of the paracellular tight junction as a consequence of myosin light chain kinase (MLCK) activation and myosin II regulatory light chain (MLC) phosphorylation. Although some details about MLCK activation remain to be defined, it is clear that this triggers perijunctional actomyosin ring (PAMR) contraction that leads to molecular reorganization of tight junction structure and composition, including occludin endocytosis. In disease states, this process can be triggered by pro-inflammatory cytokines including tumor necrosis factor-α (TNF), interleukin-1β (IL-1β), and several related molecules. Of these, TNF has been studied in the greatest detail and is known to activate long MLCK transcription, expression, enzymatic activity, and recruitment to the PAMR. Unfortunately, toxicities associated with inhibition of MLCK expression or enzymatic activity make these unsuitable as therapeutic targets. Recent work has, however, identified a small molecule that prevents MLCK1 recruitment to the PAMR without inhibiting enzymatic function. This small molecule, termed Divertin, restores barrier function after TNF-induced barrier loss and prevents disease progression in experimental chronic inflammatory bowel disease.

Keywords: tight junction; barrier function; inflammatory bowel disease; drug development; mucosal immunology; cytokines; ZO-1; occludin; claudin; actomyosin

---

1. Structure of Epithelial Intercellular Junctions

Intestinal mucosal surfaces are covered by a single layer of columnar epithelial cells required for absorptive and defensive functions. This requires cellular polarization in order to ensure appropriately-oriented membrane specializations, protein and lipid trafficking, and vectorial transport of ions and larger solutes [1–3]. For example, the dense forests of microvilli that increase apical surface area and facilitate nutrition absorption [4] are not needed on the basolateral surface, which interacts with adjacent epithelial cells, immune cells, and the basement membrane.

In order to maintain polarized function and regulate the stimuli to which apical and basolateral surfaces are exposed, intestinal epithelial cells must establish a physical barrier that prevents free diffusion across the paracellular shunt pathway [5–14]. This barrier is formed by a series of junctions that provide different types of intercellular connections. This apical junctional complex cannot be resolved by light microscopy but can be seen as the terminal bar, a dense spot or bar where apical and lateral membranes meet [5,15–17]. Transmission electron microscopy allows visualization of distinct regions such as, from apical to basal, tight junctions (zonulae occludens), adherens junctions (zonulae adherens), and desmosomes [5] (Figure 1A,B). Tight junctions are sites of close apposition of adjacent cell membranes, termed kiss sites, where the adjacent plasma membrane outer leaflets appear to fuse into a single layer. This corresponds to the site at which paracellular flux of macromolecular probes is blocked. Although some microfilaments are associated with tight junctions, they are most concentrated at adherens junctions [18–20], which do not form barriers to paracellular flux but are critical to intercellular adhesion and maintenance of the apical junctional complex. Finally, desmosomes can be recognized as electron dense membrane structures associated with intermediate, i.e., cytokeratin, filaments [21–24].

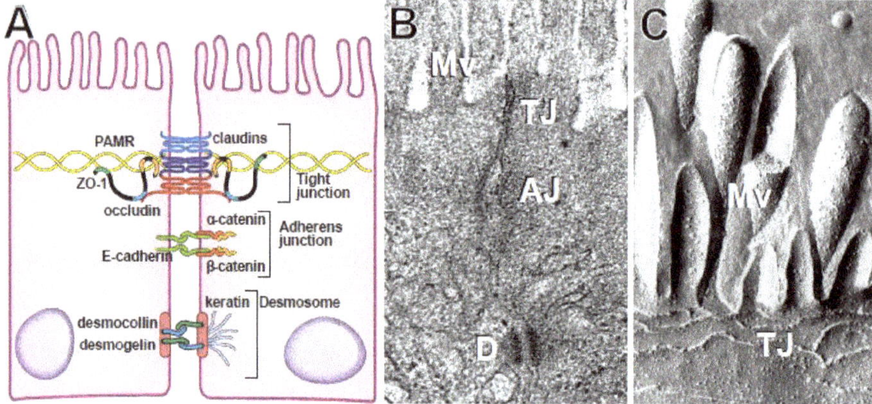

Figure 1. The structure of epithelial intercellular junctions. (**A**) Schematic showing interactions between the perijunctional actomyosin ring (PAMR), zonula occludens-1 (ZO-1), occludin, claudins at the tight junction; E-cadherin, α-catenin, and β-catenin at the adherens junction; and desmogelin, desmocollin, and intermediate filaments at the desmosome. (**B**) Transmission electron micrograph showing the tight junction (TJ), adherens junction (AJ), desmosome (D), and microvilli (Mv). From Turner. Nat Rev Immunol 2009. (**C**) Freeze-fracture electron micrograph of intramembranous tight junction strands. From Shen et al. *Annu. Rev. Physiol.* 2011.

Tight junction structure is far more interesting when viewed by freeze-fracture electron microscopy [25–27]. This reveals an anastomosing, mesh-like network of intramembranous strands (Figure 1C). Closer examination shows that the strands are composed of individual particles, causing some observers to compare the appearance to a string of pearls. The particles are thought to represent tight junction protein complexes that include polymers of claudin family proteins [7,28–30]. Consistent with this, alterations in the ensemble of claudin proteins expressed can modify the architecture of the strand network [31]. Although lipids must also be associated with tight junction structures, these are less well-characterized. It is, however, known that tight junctions are cholesterol- and sphingolipid-rich microdomains and that cholesterol depletion reduces both strand network complexity and paracellular barrier function [32–34].

2. The Paracellular "Shunt" Pathway

The intestinal mucosa confines potentially injurious contents within the lumen. The paracellular barrier, however, cannot be absolute; it must be selectively permeable to water, ions, small nutrients, and selected macromolecules in order to facilitate passive transport that is essential for nutrition and metabolism. Permeability of tight junction flux pathways must, therefore, be precisely regulated. For example, tight junction permeability is increased during nutrient absorption. This is triggered by Na^+–nutrient cotransport, which increases paracellular permeability by activating myosin light chain kinase (MLCK) to cause perijunctional actomyosin ring (PAMR) remodeling [35–39] (Figure 2). In the context of nutrient absorption, these permeability increases are limited to small, nutrient-sized molecules [35,40]. This couples with the transepithelial gradients established by active, transcellular transport, i.e., Na^+ and nutrient release into the basal extracellular milieu, to drive passive paracellular fluid absorption [37,41,42]. The absorbed fluid, from the unstirred layer, which contains high concentrations of nutrient monomers as a consequence of brush border hydrolase, e.g., disaccharidase and peptidase, activity [43,44]. Fluid absorption therefore carries nutrients, against their concentration gradient, by the mechanisms of solvent drag [42,44,45]. Increased tight junction permeability amplifies this process and allows total transepithelial nutrient absorption to exceed the maximum capacity of transcellular transport pathways [37,38,41,45–48]. A similar process allows claudin-2-mediated paracellular Na^+ transport to complement transcellular Na^+ transport and enhance the efficiency of Na^+ reabsorption in the renal proximal tubule [49].

In contrast to Na^+–nutrient cotransport [35,40], MLCK activation by inflammatory stimuli, e.g., tumor necrosis factor α (TNF), increases paracellular permeability to larger macromolecules, up to ~125 Å in diameter, thereby activating the low capacity leak pathway [50–55] (Figure 2). The differences between these two forms of MLCK-dependent barrier regulation are incompletely understood, but it is notable that occludin endocytosis occurs in response to TNF but not Na^+–nutrient cotransport (Figure 2).

Some claudin proteins, e.g., claudin-2, form actively-gated paracellular channels that define the pore pathway [52,53,56] In contrast to the leak pathway, the high capacity pore pathway channels are exquisitely size- and charge-selective, with a cutoff of ~8 Å diameter [57,58]. This limits the pore pathway to small ions and water and is too small to accommodate even small nutrients, e.g., glucose and amino acids. The pore pathway is, however, essential for nutrient transport as it allows Na^+ ions within the lamina propria, i.e., beneath the epithelial cells, to leak back into the gut lumen [59,60]. This provides the lumenal Na^+ that is required for Na^+–nutrient cotransport, the dominant route of intestinal nutrient absorption. Thus, mice lacking the two principal claudins that form paracellular cation channels within the intestinal epithelium die in the first few weeks of life as a result of nutrient malabsorption [59]. The remainder of this review will focus on the leak pathway. Claudin channels and the pore pathway are discussed elsewhere [61–64].

Na^+–nutrient cotransport at the apical brush border activates MLCK. Nutrients and Na^+ exit across the basolateral membrane via diffusive exchangers and the Na^+/K^+ ATPase, respectively. Although not indicated here, activation of other transporters, e.g., apical NHE3-mediated Na^+ absorption, further increases basolateral Na^+ [65–69]. Together, these events increase lamina propria osmolarity [70] to draw fluid across the tight junction. The modest, size-selective, increases in leak pathway permeability elicited by MLCK allow small nutrient-sized molecules to be carried with this fluid via solvent drag [44]. The end result is amplification of transcellular nutrient absorption by paracellular water and nutrient absorption. MLCK is also activated by tumor necrosis factor (TNF), but, in this case, myosin light chain phosphorylation triggers caveolar endocytosis of occludin that causes much greater increases in macromolecular permeability. Although not shown here, TNF also inhibits apical NHE3-mediated Na^+ absorption, in part explaining why fluid secretion accompanies TNF-induced increases in paracellular permeability [71].

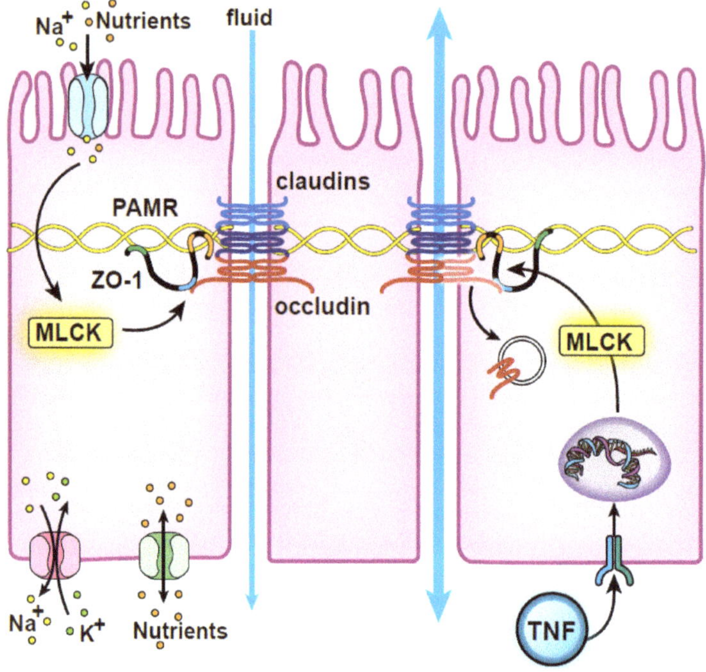

Figure 2. Roles of myosin light chain kinase (MLCK) in physiological and pathophysiological tight junction regulation. PAMR: perijunctional actomyosin ring; TNF: tumor necrosis factor.

3. MLCK, ZO-1, and Occludin Regulate the Leak Pathway

3.1. MLCK Regulates Leak Pathway Permeability

Morphological analyses of rodent mucosae identified the appearance of intrajunctional dilatations and PAMR condensation as the morphological correlates of increased paracellular permeability induced by Na^+–nutrient cotransport [37–39]. In vitro models were then used to show that MLCK-mediated phosphorylation of myosin II regulatory light chain (MLC) was required for paracellular permeability increases that follow activation of Na^+–nutrient cotransport [35], enteropathogenic E. coli infection [72], or TNF stimulation [50]. MLCK inhibition also prevented Na^+–nutrient cotransport-induced paracellular permeability increases in rodent mucosae [35] and was associated with Na^+–nutrient cotransport-induced permeability increases in human intestine [36]. Finally, transgenic expression of constitutive-active MLCK increased paracellular permeability in vitro [73,74] and in vivo [75]. MLCK is, therefore, a key signaling node in physiological and pathophysiological regulation of epithelial tight junctions.

3.2. MLCK Regulates Tight Junction Protein Interactions and Structure

The importance of MLCK to tight junction regulation in vivo was initially demonstrated in the context of acute, systemic T cell activation [51]. Administration of T cell-activating anti-CD3 antibodies to mice and humans induces a cytokine storm that includes massive systemic release of TNF [76–78]. This results in an acute, self-limited, TNF-dependent diarrhea [79,80]. Ultrastructural examination revealed PAMR condensation similar to that induced by Na^+–nutrient cotransport; intrajunctional dilatations were not detected and are likely a consequence of the massive paracellular water absorption driven by Na^+–nutrient cotransport [51]. Further study confirmed increased intestinal epithelial MLC phosphorylation following T cell activation. Notably, the peak of epithelial MLC phosphorylation coincided with maximal intestinal fluid accumulation, and both MLC phosphorylation and fluid accumulation resolved with similar time courses [51]. Analysis of tight junction protein distributions by immunofluorescence microscopy demonstrated subtle changes in ZO-1 localization that included reduced ZO-1 staining and increased waviness of bicellular tight junction profiles within intestinal epithelia of anti-CD3-treated mice [51]. Genetic MLCK activation in vitro also induced undulations within ZO-1-labeled tight junctions [74]. Enzymatic MLCK inhibition reversed these changes, both in vivo and in vitro [51,74]. It may be that these morphological alterations reflect changes in ZO-1 and ZO-2 phase separation [81].

In vitro ZO-1 reorganization induced by MLCK activation was accompanied by similar changes in occludin and F-actin profiles, and these proteins continued to be closely associated [74]. In contrast, claudins 1 and 2 appeared to remain at tight junctions at sites of ZO-1, occludin, and F-actin invaginations [74]. Consistent with this, a previous ultrastructural study of rodent mucosae demonstrated ZO-1 displacement from junctional fibrils, which we now understand to be comprised of claudin polymers, in the context of Na^+–nutrient cotransport-induced increases in paracellular permeability [82]. Thus, structural changes induced by MLCK-dependent MLC phosphorylation include reorganization of tight junction protein complexes.

Recognition that tight junction protein complexes undergo continuous remodeling, even at steady-state, i.e., in the absence of exogenous stimuli [83], led to the hypothesis that MLCK-mediated leak pathway regulation might be a consequence of altered remodeling dynamics. In vitro MLCK inhibition had no effect on anchoring and exchange of claudin-1, occludin, or F-actin and epithelial tight junctions [84]. In contrast, ZO-1 exchange was markedly reduced following MLCK inhibition [84]. Remarkably, enzymatic MLCK inhibition also reduced ZO-1 exchange in vivo [84]. This increase in ZO-1 anchoring at the tight junction following MLCK inhibition was mapped to the actin binding region (ABR) of ZO-1 [84]. Mutation of ZO-1 to delete the ABR caused a modest increase in basal mobile fraction but rendered ZO-1 insensitive to MLCK inhibition. Moreover, the free ABR was able to act as a dominant negative inhibitor of MLCK-mediated barrier regulation [84]. These data suggest that one mechanism by which MLCK regulates tight junction structure and function involves interactions mediated by the ZO-1 ABR.

3.3. MLCK Activation Triggers Tight-Junction Protein Endocytosis

In addition to subtle ZO-1 reorganization, TNF induced endocytosis of the tight junction protein occludin in vitro and in vivo [51,54,85] (Figure 2). Although TNF also triggered endocytosis of other tight junction proteins in vitro, only occludin was internalized in vivo. This was, initially, difficult to understand, as intestinal barrier function in occludin knockout mice has been reported to be similar to that of wildtype mice [86,87]. MLCK-dependent occludin endocytosis was, however, the first change that accompanied actin depolymerization-induced barrier loss in vitro [88]. This internalization occurred via caveolae, and inhibition of caveolar endocytosis prevented actin depolymerization-induced barrier loss in vitro [88]. MLCK inhibition blocked the caveolin-1 dependent endocytosis of occludin.

Further investigation showed that occludin endocytosis triggered by TNF and related cytokines was mediated by caveolae, both in vitro and in vivo [85,89,90]. Moreover, inhibition of caveolar endocytosis blocked this inflammation-induced barrier loss. Most strikingly, neither occludin endocytosis nor barrier loss occurred after TNF treatment of caveolin-1 knockout mice [85]. This does not, however, demonstrate that occludin endocytosis is essential for TNF-induced barrier loss; it only shows that occludin is a reliable marker of the caveolar endocytosis that drives such barrier loss. To determine the specific contribution(s) of occludin to TNF-induced barrier loss, responses of transgenic mice expressing enhanced green fluorescent protein (EGFP)-occludin within intestinal epithelial cells were analyzed. TNF did not induce diarrhea, i.e., net fluid secretion, in these occludin overexpressing mice [85]. Barrier loss was also markedly attenuated in EGFP-occludin transgenic mice [85]. Thus, removal of occludin from the tight junction is required for TNF-induced increases in tight junction permeability.

3.4. Interactions Mediated by the Occludin OCEL Domain Regulate Leak Pathway Barrier Function

Despite the reported absence of barrier dysfunction in occludin knockout mice, several studies have shown that occludin knockdown in vitro increases paracellular permeability to macromolecules [89,91]. Similarly, acute occludin downregulation by miR-122a transfection in vivo increased paracellular permeability to 10 kDa dextran [92]. It may, therefore, be that compensation by other members of the tight junction associated Marvel protein (TAMP) family [93–98] supports normal macromolecular barrier function in occludin knockout mice.

Detailed analyses of Caco-2 intestinal epithelial cell monolayers demonstrated that occludin knockdown specifically increased paracellular permeability by a pathway with a theoretical diameter of 125 Å, presumably the leak pathway [89]. Similar studies of occludin-knockdown MDCK cell monolayers demonstrated increased permeability to molecules with diameters greater than ~4 Å, although no upper limit was defined [91]. These MDCK cells were protected from cytokine-induced increases in 3 kDa dextran flux [99]. Conversely, occludin overexpression augmented such cytokine-induced macromolecular flux [99]. However, the studies found that cytokine treatment paradoxically increased transepithelial electrical resistance (TER) in MDCK monolayers and that these TER increases were either attenuated or exaggerated by occludin knockdown or overexpression, respectively [99]. Subsequent studies of Caco-2 monolayers demonstrated the opposite, that occludin knockdown prevented TNF-induced TER loss [89]. TNF was also unable to increase paracellular macromolecular permeability of occludin knockdown Caco-2 monolayers. Thus, macromolecular permeability of TNF-treated, occludin-sufficient Caco-2 monolayers was similar to that of occludin-deficient monolayers regardless of TNF treatment [89]. The impact of occludin on barrier function required the C-terminal coiled-coil occludin/ELL domain (OCEL) domain, as barrier function of occludin-deficient Caco-2 monolayers was enhanced by EGFP-occludin, but not EGFP-occludin$^{\Delta OCEL}$, expression [89]. EGFP-occludin expression also restored TNF-sensitivity to occludin-knockdown Caco-2 monolayers, but monolayers expressing EGFP-occludin$^{\Delta OCEL}$ remained insensitive to TNF [89]. These data indicate that the same barrier defects are induced by genetic occludin knockdown or TNF-induced occludin removal from the tight junction [89]. Thus, while not essential for tight junction assembly, occludin is a critical regulator of the macromolecular, leak pathway barrier.

3.5. MLCK-Induced Occludin Endocytosis Requires ZO-1 Interactions with the Occludin OCEL Domain

Despite a close functional relationship, molecular sites of interactions between occludin and microfilaments have not been defined. It is, therefore, not clear how MLCK-mediated actomyosin contraction induces occludin endocytosis. One possibility is that ZO-1 acts as an intermediate that links occludin to F-actin. Although this has not been explored in detail, the free OCEL domain, which mediates occludin binding to ZO-1, does act as a dominant negative to prevent TNF-induced occludin endocytosis [89]. Moreover, occludin K433, located within the ZO-1 binding site, is critical to this dominant negative OCEL activity [89]. Thus, although further study is needed, ZO-1 may link occludin endocytosis to MLCK-dependent actomyosin contraction. This may also be related to tight junction-dependent mechanosensation, which involves ZO-1, ZO-2, the transcription factor DbpA/ZONAB, and their interactions with the cytoskeleton [100,101].

4. Regulation of MLCK Expression and Localization

4.1. Regulation of MLCK Transcription

The critical role of MLCK in acute, TNF-induced barrier loss was initially demonstrated in vitro using pharmacological inhibitors [50,54,55] and in vivo using a combination of pharmacological inhibitors and mice lacking epithelial long MLCK [51]. Further study demonstrated that, beyond MLCK enzymatic activity, transcriptional upregulation of long MLCK expression was essential to TNF-induced barrier loss [54]. TNF activated long MLCK transcription via the high-affinity TNF receptor TNFR2, thereby explaining how extremely low, nanogram concentrations of TNF were sufficient to trigger MLCK upregulation and barrier dysfunction in vitro [102]. Moreover, these data demonstrated that requirement for interferon-γ (IFN-γ) pretreatment before some cell monolayers were able to respond to TNF reflected IFN-γ-induced TNFR2 transcription [54,102]. Consistent with this, in vivo studies demonstrated that TNFR2 signaling was required for epithelial long MLCK upregulation during disease progression in T cell transfer-induced, experimental, chronic inflammatory bowel disease [103].

In vitro studies from two groups conflict as to whether TNF-induced MLCK upregulation depended on NFκB or AP-1 signaling [54,104]. One group found that the NFκB inhibitors curcumin, triptolide, and pyrrolidine dithiocarbamate prevented TNF-induced MLCK upregulation and barrier loss [104,105]. A second group found that a series of NFκB inhibitors, including MG132, capsaicin, curcumin, and triptolide were unable to prevent TNF-induced barrier loss [54]; MG132 and triptolide actually enhanced TNF-induced barrier loss. This group found that sulfasalazine was able to block TNF-induced barrier loss but did so in a dose dependent manner; barrier loss was prevented at 0.5 mM sulfasalazine but exaggerated at 2 mM sulfasalazine. Biochemical studies of NFκB RelA translocation to the nucleus and luciferase expression from an NFκB-responsive promoter showed that MG132 and 2 mM sulfasalazine, but not 0.5 mM sulfasalazine, inhibited TNF-induced NFκB activation. In contrast, 0.5 mM sulfasalazine, but not 2 mM sulfasalazine or MG132, blocked TNF-induced MLCK upregulation and MLC phosphorylation [54]. These data clearly separated NFκB from MLCK upregulation and suggested that another signaling pathway activated by TNF was responsible for barrier loss. Although both of these groups used Caco-2 cells in their studies, one possible explanation could be that the first group used a relatively undifferentiated Caco-2 clone while the second group used the differentiated, enterocyte-like Caco-2_{BBe} subclone [106].

Both groups went on to clone the human long MLCK promoter, which contained functional binding sites for both NFκB and AP-1 [107,108]. However, results again differed between the groups. The first group used a human genome database search to identify the 2 kb upstream of the human long MLCK transcriptional start site and eight NFκB binding sites within that region [104,107]. They found that two of these were active, with one repressing and the other increasing transcription ~2-fold [104]. In contrast, the second group used 5′ Rapid Amplification of cDNA Ends (5′-RACE) to identify two different long MLCK transcriptional start sites [108]. The upstream 4 kb sequence contained three functional AP-1 sites and two functional NFκB sites, all of which could regulate long MLCK transcription [108]. However, the intact 4 kb long MLCK promoter was not responsive to TNF prior to 14 days after confluence, suggesting that Caco-2 differentiation, which occurs progressively in the 2 weeks following growth to confluence, was required for full promoter activation [108]. From 3 to 14 days after confluence, activity of the MLCK promoter in response to TNF increased ~7-fold and transcription from an AP-1-responsive promoter increased ~2-fold, but transcription from an NFκB -responsive promoter decreased ~2-fold. Taken as a whole, these data suggest a unified model that explains the data from both groups. In this model, undifferentiated intestinal epithelial cells modestly upregulate long MLCK via NFκB signaling, while TNF induces more extensive MLCK upregulation via AP-1 signaling within differentiated cells. Consistent with this, the first group has gone on to show that AP-1-activating elements, such as mitogen activated protein (MAP) kinases, are involved in cytokine-induced MLCK upregulation [109,110].

4.2. MLCK Expression in Chronic Intestinal Disease

Quantitative morphometry of biopsies from human patients showed that MLCK expression and activity were increased in ulcerative colitis and Crohn's disease [111]. Consistent with a role for TNF signaling, the magnitude of MLCK activation correlated directly with the degree of inflammatory activity in these biopsies.

Further analyses of the specific contributions of MLCK activation to disease were performed using transgenic mice expressing constitutively-active MLCK within the intestinal epithelium or global long MLCK knockout mice. Mice expressing constitutively-active MLCK displayed increased intestinal paracellular leak pathway permeability and increased MLC phosphorylation relative to non-transgenic littermates [75]. Both of these could be corrected by enzymatic MLCK inhibition [75]. Although these barrier defects were insufficient to induce spontaneous disease, compensatory mucosal immune activation including increased lamina propria T cell numbers, mild Th1 polarization, and heightened acute responses to infectious pathogens were detected [75,112]. However, when studied using the T cell transfer model of chronic inflammatory bowel disease, the transgenic mice developed disease more rapidly than non-transgenic littermates [75]. Moreover, overall disease severity was greater in transgenic mice and survival was markedly reduced. Thus, MLCK upregulation can impact progression of experimental inflammatory bowel disease.

Conversely, global long MLCK knockout mice were markedly protected from T cell transfer-induced colitis [103]. Although reduced disease could, potentially, reflect deletion of long MLCK in cells other than intestinal epithelium, the protection afforded by long MLCK knockout could be overcome by intestinal epithelial-specific constitutively-active MLCK expression [75]. Thus, MLCK inhibition is a potential target in chronic inflammatory bowel disease. It should, however, be noted that MLCK has many functions beyond tight junction regulation. For example, MLCK activation is critical to epithelial migration and wound repair [113–115]. Consistent with this, long MLCK knockout mice fared worse than their wild type counterparts when subjected to dextran sulfate sodium (DSS)-induced epithelial damage [103]. Similarly, in vivo knockout of nonmuscle myosin II markedly disrupts intestinal homeostasis, as demonstrated by goblet cell and barrier loss; as expected, these severely compromised mice are hypersensitive to DSS-induced injury [116]. Finally, MLCK loss may also have other consequences, either directly or as a secondary effect following disruption of cytoskeletal organization, protein trafficking, and signal transduction [117]. Thus, while intestinal epithelial MLCK inhibition and tight junction barrier preservation may be helpful in immune-mediated disease, they may also be associated with toxicities when epithelial damage predominates.

4.3. Enzymatic MLCK Inhibition is Not Feasible as a Therapeutic Intervention

Three distinct MLCK genes are present in mammals. These are *MYLK*, the smooth muscle/non-muscle MLCK located on human chromosome 3; skeletal muscle *MYLK2* on human chromosome 20; and cardiac *MYLK3* on human chromosome 16. *MYLK* encodes a long MLCK expressed in intestinal epithelium as well as short MLCK and telokin [118,119]. Telokin has regulatory functions but cannot phosphorylate MLC [120–122]. In contrast, both short MLCK, an ~110-130 kDa protein expressed in smooth muscle, and long MLCK, an ~215 kDa protein expressed in non-muscle cells, including epithelium, share identical catalytic and calmodulin-binding regulatory domains (Figure 3A). Thus, neither chemical nor genetic inhibitors can distinguish between short and long MLCK.

Enzymatic MLCK inhibition could not be used therapeutically since, as noted above, long and short MLCK have identical catalytic domains. In mice, short MLCK knockout leads to death in the early perinatal period [123]. Development of inducible knockout mice with loxP sites flanking the *MYLK* sequence encoding the catalytic domain overcame this limitation [124]. However, tissue-specific knockout in smooth muscle resulted in hypotension, bladder dysfunction, and severe intestinal dysmotility, and death [124]. Thus, while targeted inhibition of MLCK-mediated tight junction barrier regulation might have therapeutic benefit, loss of MLCK enzymatic activity in smooth muscle [124] and non-muscle [103,113] cells would have unacceptable toxicities.

4.4. TNF Induces IgCAM3-Mediated Long MLCK1 Recruitment to the Perijunctional Actomyosin Ring

The difference between long and short MLCK is the presence of a long stretch of N-terminal sequence in long MLCK [125–127]. The exons that encode that part of the protein undergo extensive alternative splicing, resulting in a number of different long MLCK splice variants [127]. Only isoforms MLCK1 and MLCK2, which differ by a single 207 nucleotide exon, are expressed in intestinal epithelia [125]. The differential functions of these and other long MLCK splice variants are not well-defined. However, it has been reported that the 69 amino acids encoded by the 207 nucleotide exon that distinguishes MLCK1 from MLCK2 include a src kinase target site whose phosphorylation can regulate MLCK activity [126].

Tissue analysis using an antibody specific for long MLCK1 demonstrated that, relative to total epithelial long MLCK, long MLCK1 was concentrated within the perijunctional actomyosin ring [125] (Figure 3B). Moreover, targeted knockdown of long MLCK1 within Caco-2 monolayers reduced paracellular permeability [125]. Further study showed that, in addition to increasing long MLCK expression, TNF specifically induced long MLCK1 recruitment to the perijunctional actomyosin ring [128]. This selective effect on MLCK1 was only post-translational, as intestinal epithelial MLCK1 and MLCK2 transcripts were comparably increased by TNF (unpublished data, Graham and Turner).

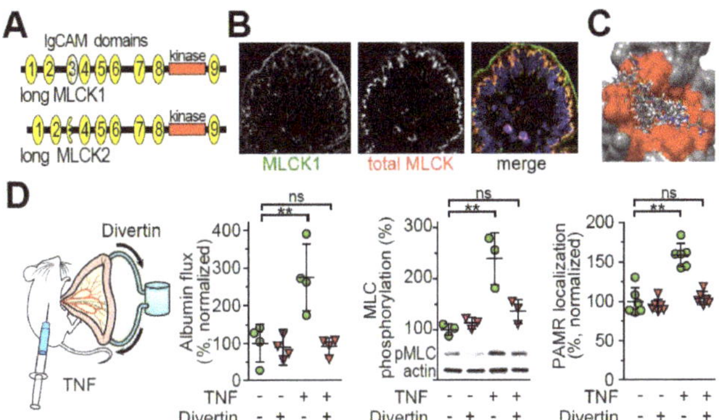

Figure 3. Specific targeting of long MLCK isoform 1 (MLCK1) prevents TNF-induced barrier loss in vivo. (**A**) Protein domain structure of long MLCK isoforms 1 and 2. Immunoglobulin-cell adhesion molecule (IgCAM) domains are numbered from the amino terminus. (**B**) Long MLCK1 (green), total MLCK (red), and nuclei (blue) in normal human jejunum. MLCK1 is preferentially-localized to the perijunctional actomyosin ring. (**C**) Virtually screened compounds docked to a binding pocket within the unique IgCAM3 domain of MLCK1. (**D**) Mice were injected with vehicle or TNF, and jejunal loops were perfused with either saline-containing vehicle or Divertin. TNF-induced increases in albumin flux (from bloodstream into the gut lumen) were blocked by Divertin. Divertin also blocked TNF-induced myosin II regulatory light chain (MLC) phosphorylation and MLCK1 recruitment to the PAMR. Notably, Divertin does not inhibit MLCK enzymatic activity. ** $p < 0.01$ by ANOVA with Dunn's multiple comparison test. From Graham et al. *Nat. Med.* 2019.

Structural analysis indicated that the 69 amino acids unique to long MLCK1 completed the immunoglobulin-cell adhesion molecule (IgCAM) domain 3, one of nine IgCAM domains within long MLCK1 [128]. Because this is the only difference between MLCK1 and MLCK2, it stands to reason that the key features responsible for TNF-induced MLCK 1 recruitment to the perijunctional actomyosin ring reside within IgCAM3. Consistent with this, expression of the IgCAM3 domain alone attenuated TNF-induced MLCK1 recruitment to the perijunctional actomyosin ring and barrier regulation, suggesting that it disrupted interactions between MLCK1 and other intracellular molecules (unpublished data, He and Turner).

4.5. TNF Induces IgCAM3-Mediated Long MLCK1 Recruitment to the Perijunctional Actomyosin Ring

The observation that an enzymatically inactive region, i.e., IgCAM3, was responsible for long MLCK1 recruitment to the perijunctional actomyosin ring presented an opportunity for blocking MLCK-dependent tight junction regulation in disease. After solving the IgCAM3 crystal structure [128,129], a hydrophobic drug-binding pocket within IgCAM3, but not the other long MLCK1 IgCAM domains, was identified. Importantly, this drug binding pocket was conserved between human and mouse long MLCK1 [128]. An in silico screen using molecular docking software identified a small number of candidate molecules with putative binding to the targeted hydrophobic pocket (Figure 3C). In vitro screening demonstrated that one of these was able to increase intestinal epithelial barrier function without inhibiting MLCK enzymatic activity in a cell-free assay, disrupting smooth muscle contraction, or interfering with epithelial wound repair, i.e., migration. This molecule, termed Divertin, was, however, able to displace long MLCK1 from the perijunctional actomyosin ring, reverse TNF-induced barrier loss, and reduce MLC phosphorylation in Caco-2 monolayers [128].

Consistent with its targeted effect on MLCK1-dependent tight junction regulation, mice could receive daily Divertin administration for 30 days without any apparent toxicity. Divertin was, however, able to prevent MLCK1 recruitment to the perijunctional actomyosin ring, MLC phosphorylation, occludin internalization, and leak pathway barrier loss in response to acute TNF challenge in mice (Figure 3D). Divertin also prevented perijunctional MLCK1 recruitment, MLC phosphorylation, and occludin internalization in human intestinal biopsies treated with TNF in vitro. Thus, Divertin was able to reverse and prevent acute, TNF-induced MLC phosphorylation and barrier loss in vitro and in vivo [128] (Figure 3D).

The efficacy of Divertin and restoring intestinal barrier function in chronic disease was initially assessed using IL-10 knockout mice, which develop in intestinal barrier defect early in the course of disease [128,130]. Although Divertin did not affect intestinal permeability in wild type mice, it was able to restore the intestinal barrier in IL-10 knockout mice [128]. Moreover, Divertin was markedly effective in limiting disease during T cell transfer colitis regardless of whether it was administered just prior to or after clinical disease presentation [128] (Figure 4). Remarkably, all measures assessed showed that Divertin was superior or equivalent to anti-TNF in limiting T cell transfer colitis severity [128] (Figure 4). Some data did, however, indicate that the beneficial effects of Divertin might be additive to, or even synergistic with, those of anti-TNF. Thus, preventing MLCK1 recruitment to the perijunctional actomyosin ring may be a non-toxic approach to limiting or preventing immune-mediated colitis either alone or in combination with immunomodulatory agents, e.g., anti-TNF [128] (Figure 4).

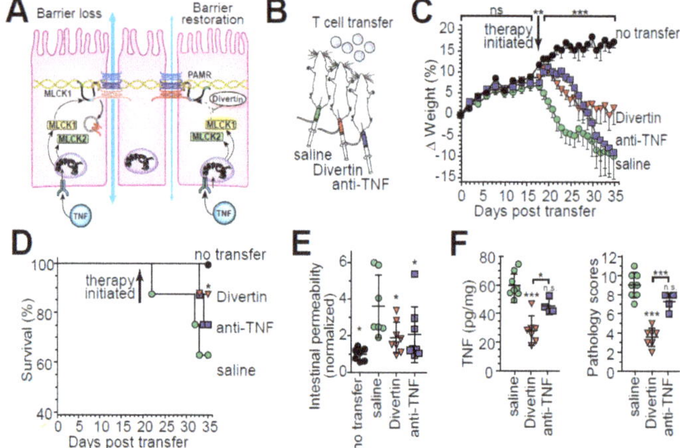

Figure 4. Inhibition of MLCK1 recruitment to the perijunctional actomyosin ring attenuates immune-mediated colitis. (**A**) Proposed mechanism of Divertin action. (**B**) Immunodeficient mice received naïve CD4+ effector T cells. Therapy with saline (vehicle), divertin, anti-TNF, or combined Divertin and anti-TNF was initiated after definitive features of disease developed (day 19). (**C**) Divertin limited weight loss after T cell transfer and was superior to anti-TNF antibody treatment. ** $p < 0.01$ by two-tailed t test for no transfer vs. all other mice at day 18. *** $p < 0.001$ by ANOVA with Tukey's multiple comparison test over the interval from 19–35 days. (**D**) Divertin enhanced survival after T cell transfer. * $p < 0.05$, versus saline-treated mice, by Gehan–Breslow–Wilcoxon test. (**E**) Divertin limited intestinal barrier loss. * $p < 0.05$, versus saline-treated mice, by ANOVA with Newman–Keuls multiple comparison test. (**F**) Divertin was superior to anti-TNF antibody treatment in limiting mucosal cytokine production and histopathology after T cell transfer. * $p < 0.05$; *** $p < 0.001$ by ANOVA with Bonferroni correction. From Graham et al. *Nat. Med.* 2019.

5. Perspective and Future Directions

The past 25 years have seen tremendous growth in our understanding of tight junction structure, cell biology, and pathobiology. Most recently, this has resulted in the develop of a proof-of-concept molecule that may provide a foundation for creation of actual therapeutic agents. Despite the enthusiasm this has engendered, there remains much to be learned. Specific topics include elucidation of mechanisms that regulate MLCK1 recruitment to the perijunctional actomyosin ring, further definition of pore pathway function in health and disease, and characterization of the structural and functional properties of defined and to-be-discovered tight junction components.

Author Contributions: Manuscript conceptualization, W.-Q.H., and J.R.T.; initial draft, W.-Q.H., J.W., J.-Y.S., and J.R.T.; figure preparation, W.-Q.H., J.-M.Z., W.V.G., and J.R.T.; manuscript revision, W.-Q.H., J.-M.Z., W.V.G., and J.R.T.; All authors have read and agreed to the published version of the manuscript.

Funding: Supported by The National Natural Science Foundation of China (31971062), the Natural Science Foundation of Jiangsu Province (BK20190043), the Natural Science Foundation of the Jiangsu Higher Education Institutions of China (19KJB320003), the Tang Scholar of Soochow University, the Research Fund of State Key Laboratory of Pharmaceutical Biotechnology, Nanjing University (KF-GN-202004), the National Institute of Diabetes Digestive and Kidney Disease (R01DK61931, R01DK068271, R24DK099803, R43DK123904), and the Harvard Digestive Disease Center (P30DK034854).

Conflicts of Interest: W.V.G. is an employee of Thelium Therapeutics, Inc. W.V.G. and J.R.T. are co-founders of Thelium Therapeutics, Inc.

References

1. Blasky, A.J.; Mangan, A.; Prekeris, R. Polarized protein transport and lumen formation during epithelial tissue morphogenesis. *Annu. Rev. Cell Dev. Biol.* **2015**, *31*, 575–591. [CrossRef] [PubMed]
2. Venkatasubramanian, J.; Ao, M.; Rao, M.C. Ion transport in the small intestine. *Curr. Opin. Gastroenterol.* **2010**, *26*, 123–128. [CrossRef] [PubMed]
3. Cereijido, M.; Contreras, R.G.; Shoshani, L.; Flores-Benitez, D.; Larre, I. Tight junction and polarity interaction in the transporting epithelial phenotype. *Biochim. Biophys. Acta* **2008**, *1778*, 770–793. [CrossRef] [PubMed]
4. Crawley, S.W.; Mooseker, M.S.; Tyska, M.J. Shaping the intestinal brush border. *J. Cell Biol.* **2014**, *207*, 441–451. [CrossRef] [PubMed]
5. Farquhar, M.; Palade, G. Junctional complexes in various epithelia. *J. Cell Biol.* **1963**, *17*, 375–412. [CrossRef] [PubMed]
6. Cereijido, M.; Robbins, E.S.; Dolan, W.J.; Rotunno, C.A.; Sabatini, D.D. Polarized monolayers formed by epithelial cells on a permeable and translucent support. *J. Cell Biol.* **1978**, *77*, 853–880. [CrossRef]
7. Martinez-Palomo, A.; Meza, I.; Beaty, G.; Cereijido, M. Experimental modulation of occluding junctions in a cultured transporting epithelium. *J. Cell Biol.* **1980**, *87*, 736–745. [CrossRef]
8. Otani, T.; Nguyen, T.P.; Tokuda, S.; Sugihara, K.; Sugawara, T.; Furuse, K.; Miura, T.; Ebnet, K.; Furuse, M. Claudins and JAM-A coordinately regulate tight junction formation and epithelial polarity. *J. Cell Biol.* **2019**, *218*, 3372–3396. [CrossRef]
9. Nalle, S.C.; Zuo, L.; Ong, M.; Singh, G.; Worthylake, A.M.; Choi, W.; Manresa, M.C.; Southworth, A.P.; Edelblum, K.L.; Baker, G.J.; et al. Graft-versus-host disease propagation depends on increased intestinal epithelial tight junction permeability. *J. Clin. Investig.* **2019**, *129*, 902–914. [CrossRef]
10. Kuo, W.T.; Shen, L.; Zuo, L.; Shashikanth, N.; Ong, M.; Wu, L.; Zha, J.; Edelblum, K.L.; Wang, Y.; Wang, Y.; et al. Inflammation-induced Occludin Downregulation Limits Epithelial Apoptosis by Suppressing Caspase-3 Expression. *Gastroenterology* **2019**, *157*, 1323–1337. [CrossRef]
11. Kohno, T.; Konno, T.; Kojima, T. Role of Tricellular Tight Junction Protein Lipolysis-Stimulated Lipoprotein Receptor (LSR) in Cancer Cells. *Int. J. Mol. Sci.* **2019**, *20*, 3555. [CrossRef] [PubMed]
12. Camilleri, M. Leaky gut: Mechanisms, measurement and clinical implications in humans. *Gut* **2019**, *68*, 1516–1526. [CrossRef] [PubMed]
13. Brazil, J.C.; Quiros, M.; Nusrat, A.; Parkos, C.A. Innate immune cell-epithelial crosstalk during wound repair. *J. Clin. Investig.* **2019**, *129*, 2983–2993. [CrossRef] [PubMed]
14. Powell, D.W. Barrier function of epithelia. *Am. J. Physiol.* **1981**, *241*, G275–G288. [CrossRef] [PubMed]

15. Kottra, G.; Fromter, E. Functional properties of the paracellular pathway in some leaky epithelia. *J. Exp. Biol.* **1983**, *106*, 217–229. [PubMed]
16. Overton, J.; Shoup, J. Fine Structure of Cell Surface Specializations in the Maturing Duodenal Mucosa of the Chick. *J. Cell Biol.* **1964**, *21*, 75–85. [CrossRef]
17. Tsubouchi, S.; Leblond, C.P. Migration and turnover of entero-endocrine and caveolated cells in the epithelium of the descending colon, as shown by radioautography after continuous infusion of 3H-thymidine into mice. *Am. J. Anat.* **1979**, *156*, 431–451. [CrossRef]
18. Harris, T.J.; Tepass, U. Adherens junctions: From molecules to morphogenesis. *Nat. Rev. Mol. Cell Biol.* **2010**, *11*, 502–514. [CrossRef]
19. Pinheiro, D.; Bellaiche, Y. Mechanical Force-Driven Adherens Junction Remodeling and Epithelial Dynamics. *Dev. Cell* **2018**, *47*, 3–19. [CrossRef]
20. Tsukita, S.; Tsukita, S.; Nagafuchi, A.; Yonemura, S. Molecular linkage between cadherins and actin filaments in cell-cell adherens junctions. *Curr. Opin. Cell Biol.* **1992**, *4*, 834–839. [CrossRef]
21. Brooke, M.A.; Nitoiu, D.; Kelsell, D.P. Cell-cell connectivity: Desmosomes and disease. *J. Pathol.* **2012**, *226*, 158–171. [CrossRef] [PubMed]
22. Capaldo, C.T.; Farkas, A.E.; Nusrat, A. Epithelial adhesive junctions. *F1000prime Rep.* **2014**, *6*, 1. [CrossRef] [PubMed]
23. Nava, P.; Laukoetter, M.G.; Hopkins, A.M.; Laur, O.; Gerner-Smidt, K.; Green, K.J.; Parkos, C.A.; Nusrat, A. Desmoglein-2: A novel regulator of apoptosis in the intestinal epithelium. *Mol. Biol. Cell* **2007**, *18*, 4565–4578. [CrossRef] [PubMed]
24. Zen, K.; Babbin, B.A.; Liu, Y.; Whelan, J.B.; Nusrat, A.; Parkos, C.A. JAM-C is a component of desmosomes and a ligand for CD11b/CD18-mediated neutrophil transepithelial migration. *Mol. Biol. Cell* **2004**, *15*, 3926–3937. [CrossRef] [PubMed]
25. Chalcroft, J.P.; Bullivant, S. An interpretation of liver cell membrane and junction structure based on observation of freeze-fracture replicas of both sides of the fracture. *J. Cell Biol.* **1970**, *47*, 49–60. [CrossRef] [PubMed]
26. Claude, P.; Goodenough, D.A. Fracture faces of zonulae occludentes from "tight" and "leaky" epithelia. *J. Cell Biol.* **1973**, *58*, 390–400. [CrossRef]
27. Wade, J.B.; Karnovsky, M.J. The structure of the zonula occludens. A single fibril model based on freeze-fracture. *J. Cell Biol.* **1974**, *60*, 168–180. [CrossRef]
28. Furuse, M.; Fujita, K.; Hiiragi, T.; Fujimoto, K.; Tsukita, S. Claudin-1 and -2: Novel integral membrane proteins localizing at tight junctions with no sequence similarity to occludin. *J. Cell Biol.* **1998**, *141*, 1539–1550. [CrossRef]
29. Furuse, M.; Sasaki, H.; Fujimoto, K.; Tsukita, S. A single gene product, claudin-1 or -2, reconstitutes tight junction strands and recruits occludin in fibroblasts. *J. Cell Biol.* **1998**, *143*, 391–401. [CrossRef]
30. Furuse, M.; Sasaki, H.; Tsukita, S. Manner of interaction of heterogeneous claudin species within and between tight junction strands. *J. Cell Biol.* **1999**, *147*, 891–903. [CrossRef]
31. Yamazaki, Y.; Tokumasu, R.; Kimura, H.; Tsukita, S. Role of claudin species-specific dynamics in reconstitution and remodeling of the zonula occludens. *Mol. Biol. Cell* **2011**, *22*, 1495–1504. [CrossRef] [PubMed]
32. Francis, S.A.; Kelly, J.M.; McCormack, J.; Rogers, R.A.; Lai, J.; Schneeberger, E.E.; Lynch, R.D. Rapid reduction of MDCK cell cholesterol by methyl-beta-cyclodextrin alters steady state transepithelial electrical resistance. *Eur. J. Cell Biol.* **1999**, *78*, 473–484. [CrossRef]
33. Nusrat, A.; Parkos, C.A.; Verkade, P.; Foley, C.S.; Liang, T.W.; Innis-Whitehouse, W.; Eastburn, K.K.; Madara, J.L. Tight junctions are membrane microdomains. *J. Cell Sci.* **2000**, *113*, 1771–1781.
34. Nusrat, A.; von Eichel-Streiber, C.; Turner, J.R.; Verkade, P.; Madara, J.L.; Parkos, C.A. Clostridium difficile toxins disrupt epithelial barrier function by altering membrane microdomain localization of tight junction proteins. *Infect. Immun.* **2001**, *69*, 1329–1336. [CrossRef]
35. Turner, J.R.; Rill, B.K.; Carlson, S.L.; Carnes, D.; Kerner, R.; Mrsny, R.J.; Madara, J.L. Physiological regulation of epithelial tight junctions is associated with myosin light-chain phosphorylation. *Am. J. Physiol.* **1997**, *273*, C1378–C1385. [CrossRef]
36. Berglund, J.J.; Riegler, M.; Zolotarevsky, Y.; Wenzl, E.; Turner, J.R. Regulation of human jejunal transmucosal resistance and MLC phosphorylation by Na(+)-glucose cotransport. *Am. J. Physiol.- Gastrointest. Liver Physiol.* **2001**, *281*, G1487–G1493. [CrossRef]

37. Madara, J.L.; Pappenheimer, J.R. Structural basis for physiological regulation of paracellular pathways in intestinal epithelia. *J. Membr. Biol.* **1987**, *100*, 149–164. [CrossRef]
38. Atisook, K.; Carlson, S.; Madara, J.L. Effects of phlorizin and sodium on glucose-elicited alterations of cell junctions in intestinal epithelia. *Am. J. Physiol.* **1990**, *258*, C77–C85. [CrossRef]
39. Atisook, K.; Madara, J.L. An oligopeptide permeates intestinal tight junctions at glucose-elicited dilatations. *Gastroenterology* **1991**, *100*, 719–724. [CrossRef]
40. Turner, J.R.; Cohen, D.E.; Mrsny, R.J.; Madara, J.L. Noninvasive in vivo analysis of human small intestinal paracellular absorption: Regulation by Na+-glucose cotransport. *Dig. Dis. Sci.* **2000**, *45*, 2122–2126. [CrossRef]
41. Pappenheimer, J.R. Physiological regulation of transepithelial impedance in the intestinal mucosa of rats and hamsters. *J. Membr. Biol.* **1987**, *100*, 137–148. [CrossRef] [PubMed]
42. Pappenheimer, J.R.; Reiss, K.Z. Contribution of solvent drag through intercellular junctions to absorption of nutrients by the small intestine of the rat. *J. Membr. Biol.* **1987**, *100*, 123–136. [CrossRef]
43. Pappenheimer, J.R. Role of pre-epithelial "unstirred" layers in absorption of nutrients from the human jejunum. *J. Membr. Biol.* **2001**, *179*, 185–204. [CrossRef] [PubMed]
44. Pappenheimer, J.R. On the coupling of membrane digestion with intestinal absorption of sugars and amino acids. *Am. J. Physiol.* **1993**, *265*, G409–G417. [CrossRef] [PubMed]
45. Pappenheimer, J.R.; Dahl, C.E.; Karnovsky, M.L.; Maggio, J.E. Intestinal absorption and excretion of octapeptides composed of D amino acids. *Proc. Natl. Acad. Sci. USA* **1994**, *91*, 1942–1945. [CrossRef]
46. Pappenheimer, J.R. Physiological regulation of epithelial junctions in intestinal epithelia. *Acta Physiol. Scand. Suppl.* **1988**, *571*, 43–51.
47. Pappenheimer, J.R. Paracellular intestinal absorption of glucose, creatinine, and mannitol in normal animals: Relation to body size. *Am. J. Physiol.* **1990**, *259*, G290–G299. [CrossRef]
48. Meddings, J.B.; Westergaard, H. Intestinal glucose transport using perfused rat jejunum in vivo: Model analysis and derivation of corrected kinetic constants. *Clin. Sci.* **1989**, *76*, 403–413. [CrossRef]
49. Pei, L.; Solis, G.; Nguyen, M.T.; Kamat, N.; Magenheimer, L.; Zhuo, M.; Li, J.; Curry, J.; McDonough, A.A.; Fields, T.A.; et al. Paracellular epithelial sodium transport maximizes energy efficiency in the kidney. *J. Clin. Investig.* **2016**, *126*, 2509–2518. [CrossRef]
50. Zolotarevsky, Y.; Hecht, G.; Koutsouris, A.; Gonzalez, D.E.; Quan, C.; Tom, J.; Mrsny, R.J.; Turner, J.R. A membrane-permeant peptide that inhibits MLC kinase restores barrier function in in vitro models of intestinal disease. *Gastroenterology* **2002**, *123*, 163–172. [CrossRef]
51. Clayburgh, D.R.; Barrett, T.A.; Tang, Y.; Meddings, J.B.; Van Eldik, L.J.; Watterson, D.M.; Clarke, L.L.; Mrsny, R.J.; Turner, J.R. Epithelial myosin light chain kinase-dependent barrier dysfunction mediates T cell activation-induced diarrhea in vivo. *J. Clin. Investig.* **2005**, *115*, 2702–2715. [CrossRef] [PubMed]
52. Turner, J.R. Intestinal mucosal barrier function in health and disease. *Nat. Rev. Immunol.* **2009**, *9*, 799–809. [CrossRef]
53. Anderson, J.M.; Van Itallie, C.M. Physiology and function of the tight junction. *Cold Spring Harb. Perspect Biol.* **2009**, *1*, a002584. [CrossRef] [PubMed]
54. Wang, F.; Graham, W.V.; Wang, Y.; Witkowski, E.D.; Schwarz, B.T.; Turner, J.R. Interferon-gamma and tumor necrosis factor-alpha synergize to induce intestinal epithelial barrier dysfunction by up-regulating myosin light chain kinase expression. *Am. J. Pathol.* **2005**, *166*, 409–419. [CrossRef]
55. Ma, T.Y.; Boivin, M.A.; Ye, D.; Pedram, A.; Said, H.M. Mechanism of TNF-{alpha} modulation of Caco-2 intestinal epithelial tight junction barrier: Role of myosin light-chain kinase protein expression. *Am. J. Physiol. Gastrointest. Liver Physiol.* **2005**, *288*, G422–G430. [CrossRef] [PubMed]
56. Weber, C.R.; Liang, G.H.; Wang, Y.; Das, S.; Shen, L.; Yu, A.S.; Nelson, D.J.; Turner, J.R. Claudin-2-dependent paracellular channels are dynamically gated. *eLife* **2015**, *4*, e09906. [CrossRef]
57. Yu, A.S.; Cheng, M.H.; Angelow, S.; Gunzel, D.; Kanzawa, S.A.; Schneeberger, E.E.; Fromm, M.; Coalson, R.D. Molecular basis for cation selectivity in claudin-2-based paracellular pores: Identification of an electrostatic interaction site. *J. Gen. Physiol.* **2009**, *133*, 111–127. [CrossRef]
58. Li, J.; Zhuo, M.; Pei, L.; Rajagopal, M.; Yu, A.S. Comprehensive cysteine-scanning mutagenesis reveals Claudin-2 pore-lining residues with different intrapore locations. *J. Biol. Chem.* **2014**, *289*, 6475–6484. [CrossRef]

59. Wada, M.; Tamura, A.; Takahashi, N.; Tsukita, S. Loss of claudins 2 and 15 from mice causes defects in paracellular Na+ flow and nutrient transport in gut and leads to death from malnutrition. *Gastroenterology* **2013**, *144*, 369–380. [CrossRef]
60. Tamura, A.; Hayashi, H.; Imasato, M.; Yamazaki, Y.; Hagiwara, A.; Wada, M.; Noda, T.; Watanabe, M.; Suzuki, Y.; Tsukita, S. Loss of claudin-15, but not claudin-2, causes Na+ deficiency and glucose malabsorption in mouse small intestine. *Gastroenterology* **2011**, *140*, 913–923. [CrossRef]
61. Suzuki, H.; Tani, K.; Fujiyoshi, Y. Crystal structures of claudins: Insights into their intermolecular interactions. *Ann. N. Y. Acad. Sci.* **2017**, *1397*, 25–34. [CrossRef] [PubMed]
62. Rosenthal, R.; Gunzel, D.; Theune, D.; Czichos, C.; Schulzke, J.D.; Fromm, M. Water channels and barriers formed by claudins. *Ann. N. Y. Acad. Sci.* **2017**, *1397*, 100–109. [CrossRef] [PubMed]
63. Garcia-Hernandez, V.; Quiros, M.; Nusrat, A. Intestinal epithelial claudins: Expression and regulation in homeostasis and inflammation. *Ann. N. Y. Acad. Sci.* **2017**, *1397*, 66–79. [CrossRef]
64. Muto, S. Physiological roles of claudins in kidney tubule paracellular transport. *Am. J. Physiol. Renal Physiol.* **2016**. [CrossRef] [PubMed]
65. Turner, J.R.; Black, E.D. NHE3-dependent cytoplasmic alkalinization is triggered by Na(+)-glucose cotransport in intestinal epithelia. *Am. J. Physiol. Cell Physiol.* **2001**, *281*, C1533–C1541. [CrossRef]
66. Zhao, H.; Shiue, H.; Palkon, S.; Wang, Y.; Cullinan, P.; Burkhardt, J.K.; Musch, M.W.; Chang, E.B.; Turner, J.R. Ezrin regulates NHE3 translocation and activation after Na+-glucose cotransport. *Proc. Natl. Acad. Sci. USA* **2004**, *101*, 9485–9490. [CrossRef]
67. Shiue, H.; Musch, M.W.; Wang, Y.; Chang, E.B.; Turner, J.R. Akt2 phosphorylates ezrin to trigger NHE3 translocation and activation. *J. Biol. Chem.* **2005**, *280*, 1688–1695. [CrossRef]
68. Hu, Z.; Wang, Y.; Graham, W.V.; Su, L.; Musch, M.W.; Turner, J.R. MAPKAPK-2 is a critical signaling intermediate in NHE3 activation following Na^+-glucose cotransport. *J. Biol. Chem.* **2006**, *281*, 24247–24253. [CrossRef]
69. Lin, R.; Murtazina, R.; Cha, B.; Chakraborty, M.; Sarker, R.; Chen, T.E.; Lin, Z.; Hogema, B.M.; de Jonge, H.R.; Seidler, U.; et al. D-glucose acts via sodium/glucose cotransporter 1 to increase NHE3 in mouse jejunal brush border by a Na+/H+ exchange regulatory factor 2-dependent process. *Gastroenterology* **2011**, *140*, 560–571. [CrossRef]
70. Bohlen, H.G. Na+-induced intestinal interstitial hyperosmolality and vascular responses during absorptive hyperemia. *Am. J. Physiol.* **1982**, *242*, H785–H789. [CrossRef]
71. Clayburgh, D.R.; Musch, M.W.; Leitges, M.; Fu, Y.X.; Turner, J.R. Coordinated epithelial NHE3 inhibition and barrier dysfunction are required for TNF-mediated diarrhea in vivo. *J. Clin. Investig.* **2006**, *116*, 2682–2694. [CrossRef] [PubMed]
72. Yuhan, R.; Koutsouris, A.; Savkovic, S.D.; Hecht, G. Enteropathogenic Escherichia coli-induced myosin light chain phosphorylation alters intestinal epithelial permeability. *Gastroenterology* **1997**, *113*, 1873–1882. [CrossRef]
73. Hecht, G.; Pestic, L.; Nikcevic, G.; Koutsouris, A.; Tripuraneni, J.; Lorimer, D.D.; Nowak, G.; Guerriero, V., Jr.; Elson, E.L.; Lanerolle, P.D. Expression of the catalytic domain of myosin light chain kinase increases paracellular permeability. *Am. J. Physiol.* **1996**, *271*, C1678–C1684. [CrossRef] [PubMed]
74. Shen, L.; Black, E.D.; Witkowski, E.D.; Lencer, W.I.; Guerriero, V.; Schneeberger, E.E.; Turner, J.R. Myosin light chain phosphorylation regulates barrier function by remodeling tight junction structure. *J. Cell Sci.* **2006**, *119*, 2095–2106. [CrossRef] [PubMed]
75. Su, L.; Shen, L.; Clayburgh, D.R.; Nalle, S.C.; Sullivan, E.A.; Meddings, J.B.; Abraham, C.; Turner, J.R. Targeted epithelial tight junction dysfunction causes immune activation and contributes to development of experimental colitis. *Gastroenterology* **2009**, *136*, 551–563. [CrossRef]
76. Charpentier, B.; Hiesse, C.; Lantz, O.; Ferran, C.; Stephens, S.; O'Shaugnessy, D.; Bodmer, M.; Benoit, G.; Bach, J.F.; Chatenoud, L. Evidence that antihuman tumor necrosis factor monoclonal antibody prevents OKT3-induced acute syndrome. *Transplantation* **1992**, *54*, 997–1002. [CrossRef]
77. Ferran, C.; Dy, M.; Merite, S.; Sheehan, K.; Schreiber, R.; Leboulenger, F.; Landais, P.; Bluestone, J.; Bach, J.F.; Chatenoud, L. Reduction of morbidity and cytokine release in anti-CD3 MoAb-treated mice by corticosteroids. *Transplantation* **1990**, *50*, 642–648. [CrossRef]

78. Ferran, C.; Sheehan, K.; Dy, M.; Schreiber, R.; Merite, S.; Landais, P.; Noel, L.H.; Grau, G.; Bluestone, J.; Bach, J.F.; et al. Cytokine-related syndrome following injection of anti-CD3 monoclonal antibody: Further evidence for transient in vivo T cell activation. *Eur. J. Immunol.* **1990**, *20*, 509–515. [CrossRef]
79. Musch, M.W.; Clarke, L.L.; Mamah, D.; Gawenis, L.R.; Zhang, Z.; Ellsworth, W.; Shalowitz, D.; Mittal, N.; Efthimiou, P.; Alnadjim, Z.; et al. T cell activation causes diarrhea by increasing intestinal permeability and inhibiting epithelial Na+/K+-ATPase. *J. Clin. Investig.* **2002**, *110*, 1739–1747. [CrossRef]
80. Tang, Y.; Clayburgh, D.R.; Mittal, N.; Goretsky, T.; Dirisina, R.; Zhang, Z.; Kron, M.; Ivancic, D.; Katzman, R.B.; Grimm, G.; et al. Epithelial NF-kappaB enhances transmucosal fluid movement by altering tight junction protein composition after T cell activation. *Am. J. Pathol.* **2010**, *176*, 158–167. [CrossRef]
81. Beutel, O.; Maraspini, R.; Pombo-Garcia, K.; Martin-Lemaitre, C.; Honigmann, A. Phase Separation of Zonula Occludens Proteins Drives Formation of Tight Junctions. *Cell* **2019**, *179*, 923–936.e911. [CrossRef]
82. Madara, J.L.; Carlson, S.; Anderson, J.M. ZO-1 maintains its spatial distribution but dissociates from junctional fibrils during tight junction regulation. *Am. J. Physiol.* **1993**, *264*, C1096–C1101. [CrossRef] [PubMed]
83. Shen, L.; Weber, C.R.; Turner, J.R. The tight junction protein complex undergoes rapid and continuous molecular remodeling at steady state. *J. Cell Biol.* **2008**, *181*, 683–695. [CrossRef] [PubMed]
84. Yu, D.; Marchiando, A.M.; Weber, C.R.; Raleigh, D.R.; Wang, Y.; Shen, L.; Turner, J.R. MLCK-dependent exchange and actin binding region-dependent anchoring of ZO-1 regulate tight junction barrier function. *Proc. Natl. Acad. Sci. USA* **2010**, *107*, 8237–8241. [CrossRef] [PubMed]
85. Marchiando, A.M.; Shen, L.; Graham, W.V.; Weber, C.R.; Schwarz, B.T.; Austin, J.R., 2nd; Raleigh, D.R.; Guan, Y.; Watson, A.J.; Montrose, M.H.; et al. Caveolin-1-dependent occludin endocytosis is required for TNF-induced tight junction regulation in vivo. *J. Cell Biol.* **2010**, *189*, 111–126. [CrossRef] [PubMed]
86. Saitou, M.; Furuse, M.; Sasaki, H.; Schulzke, J.D.; Fromm, M.; Takano, H.; Noda, T.; Tsukita, S. Complex phenotype of mice lacking occludin, a component of tight junction strands. *Mol. Biol. Cell* **2000**, *11*, 4131–4142. [CrossRef] [PubMed]
87. Schulzke, J.D.; Gitter, A.H.; Mankertz, J.; Spiegel, S.; Seidler, U.; Amasheh, S.; Saitou, M.; Tsukita, S.; Fromm, M. Epithelial transport and barrier function in occludin-deficient mice. *Biochim. Biophys. Acta* **2005**, *1669*, 34–42. [CrossRef]
88. Shen, L.; Turner, J.R. Actin depolymerization disrupts tight junctions via caveolae-mediated endocytosis. *Mol. Biol. Cell* **2005**, *16*, 3919–3936. [CrossRef]
89. Buschmann, M.M.; Shen, L.; Rajapakse, H.; Raleigh, D.R.; Wang, Y.; Wang, Y.; Lingaraju, A.; Zha, J.; Abbott, E.; McAuley, E.M.; et al. Occludin OCEL-domain interactions are required for maintenance and regulation of the tight junction barrier to macromolecular flux. *Mol. Biol. Cell* **2013**, *24*, 3056–3068. [CrossRef]
90. Schwarz, B.T.; Wang, F.; Shen, L.; Clayburgh, D.R.; Su, L.; Wang, Y.; Fu, Y.X.; Turner, J.R. LIGHT signals directly to intestinal epithelia to cause barrier dysfunction via cytoskeletal and endocytic mechanisms. *Gastroenterology* **2007**, *132*, 2383–2394. [CrossRef]
91. Yu, A.S.; McCarthy, K.M.; Francis, S.A.; McCormack, J.M.; Lai, J.; Rogers, R.A.; Lynch, R.D.; Schneeberger, E.E. Knockdown of occludin expression leads to diverse phenotypic alterations in epithelial cells. *Am. J. Physiol. Cell Physiol.* **2005**, *288*, C1231–C1241. [CrossRef] [PubMed]
92. Ye, D.; Guo, S.; Al-Sadi, R.; Ma, T.Y. MicroRNA regulation of intestinal epithelial tight junction permeability. *Gastroenterology* **2011**, *141*, 1323–1333. [CrossRef]
93. Cording, J.; Berg, J.; Kading, N.; Bellmann, C.; Tscheik, C.; Westphal, J.K.; Milatz, S.; Gunzel, D.; Wolburg, H.; Piontek, J.; et al. In tight junctions, claudins regulate the interactions between occludin, tricellulin and marvelD3, which, inversely, modulate claudin oligomerization. *J. Cell Sci.* **2013**, *126*, 554–564. [CrossRef] [PubMed]
94. Raleigh, D.R.; Marchiando, A.M.; Zhang, Y.; Shen, L.; Sasaki, H.; Wang, Y.; Long, M.; Turner, J.R. Tight junction-associated MARVEL proteins marveld3, tricellulin, and occludin have distinct but overlapping functions. *Mol. Biol. Cell* **2010**, *21*, 1200–1213. [CrossRef] [PubMed]
95. Steed, E.; Elbediwy, A.; Vacca, B.; Dupasquier, S.; Hemkemeyer, S.A.; Suddason, T.; Costa, A.C.; Beaudry, J.B.; Zihni, C.; Gallagher, E.; et al. MarvelD3 couples tight junctions to the MEKK1-JNK pathway to regulate cell behavior and survival. *J. Cell Biol.* **2014**, *204*, 821–838. [CrossRef] [PubMed]
96. Steed, E.; Rodrigues, N.T.; Balda, M.S.; Matter, K. Identification of MarvelD3 as a tight junction-associated transmembrane protein of the occludin family. *BMC Cell Biol.* **2009**, *10*, 95. [CrossRef] [PubMed]

97. Krug, S.M.; Amasheh, S.; Richter, J.F.; Milatz, S.; Gunzel, D.; Westphal, J.K.; Huber, O.; Schulzke, J.D.; Fromm, M. Tricellulin forms a barrier to macromolecules in tricellular tight junctions without affecting ion permeability. *Mol. Biol. Cell* **2009**, *20*, 3713–3724. [CrossRef] [PubMed]
98. Krug, S.M.; Bojarski, C.; Fromm, A.; Lee, I.M.; Dames, P.; Richter, J.F.; Turner, J.R.; Fromm, M.; Schulzke, J.D. Tricellulin is regulated via interleukin-13-receptor alpha2, affects macromolecule uptake, and is decreased in ulcerative colitis. *Mucosal. Immunol.* **2018**, *11*, 345–356. [CrossRef]
99. Van Itallie, C.M.; Fanning, A.S.; Holmes, J.; Anderson, J.M. Occludin is required for cytokine-induced regulation of tight junction barriers. *J. Cell Sci.* **2010**, *123*, 2844–2852. [CrossRef]
100. Schwayer, C.; Shamipour, S.; Pranjic-Ferscha, K.; Schauer, A.; Balda, M.; Tada, M.; Matter, K.; Heisenberg, C.P. Mechanosensation of Tight Junctions Depends on ZO-1 Phase Separation and Flow. *Cell* **2019**, *179*, 937–952 e918. [CrossRef]
101. Spadaro, D.; Le, S.; Laroche, T.; Mean, I.; Jond, L.; Yan, J.; Citi, S. Tension-Dependent Stretching Activates ZO-1 to Control the Junctional Localization of Its Interactors. *Curr. Biol.* **2017**, *27*, 3783–3795 e3788. [CrossRef] [PubMed]
102. Wang, F.; Schwarz, B.T.; Graham, W.V.; Wang, Y.; Su, L.; Clayburgh, D.R.; Abraham, C.; Turner, J.R. IFN-gamma-induced TNFR2 expression is required for TNF-dependent intestinal epithelial barrier dysfunction. *Gastroenterology* **2006**, *131*, 1153–1163. [CrossRef] [PubMed]
103. Su, L.; Nalle, S.C.; Shen, L.; Turner, E.S.; Singh, G.; Breskin, L.A.; Khramtsova, E.A.; Khramtsova, G.; Tsai, P.Y.; Fu, Y.X.; et al. TNFR2 activates MLCK-dependent tight junction dysregulation to cause apoptosis-mediated barrier loss and experimental colitis. *Gastroenterology* **2013**, *145*, 407–415. [CrossRef] [PubMed]
104. Ye, D.; Ma, I.; Ma, T.Y. Molecular mechanism of tumor necrosis factor-alpha modulation of intestinal epithelial tight junction barrier. *Am. J. Physiol. Gastrointest. Liver Physiol.* **2006**, *290*, G496–G504. [CrossRef] [PubMed]
105. Ma, T.Y.; Iwamoto, G.K.; Hoa, N.T.; Akotia, V.; Pedram, A.; Boivin, M.A.; Said, H.M. TNF-alpha-induced increase in intestinal epithelial tight junction permeability requires NF-kappa B activation. *Am. J. Physiol. Gastrointest. Liver Physiol.* **2004**, *286*, G367–G376. [CrossRef] [PubMed]
106. Peterson, M.D.; Mooseker, M.S. Characterization of the enterocyte-like brush border cytoskeleton of the C2BBe clones of the human intestinal cell line, Caco-2. *J. Cell Sci.* **1992**, *102*, 581–600. [PubMed]
107. Ye, D.; Ma, T.Y. Cellular and molecular mechanisms that mediate basal and tumour necrosis factor-alpha-induced regulation of myosin light chain kinase gene activity. *J. Cell Mol. Med.* **2008**, *12*, 1331–1346. [CrossRef] [PubMed]
108. Graham, W.V.; Wang, F.; Clayburgh, D.R.; Cheng, J.X.; Yoon, B.; Wang, Y.; Lin, A.; Turner, J.R. Tumor necrosis factor-induced long myosin light chain kinase transcription is regulated by differentiation-dependent signaling events. Characterization of the human long myosin light chain kinase promoter. *J. Biol. Chem.* **2006**, *281*, 26205–26215. [CrossRef]
109. Al-Sadi, R.; Guo, S.; Ye, D.; Dokladny, K.; Alhmoud, T.; Ereifej, L.; Said, H.M.; Ma, T.Y. Mechanism of IL-1beta modulation of intestinal epithelial barrier involves p38 kinase and activating transcription factor-2 activation. *J. Immunol.* **2013**, *190*, 6596–6606. [CrossRef]
110. Al-Sadi, R.; Guo, S.; Ye, D.; Ma, T.Y. TNF-alpha Modulation of Intestinal Epithelial Tight Junction Barrier Is Regulated by ERK1/2 Activation of Elk-1. *Am. J. Pathol.* **2013**. [CrossRef]
111. Blair, S.A.; Kane, S.V.; Clayburgh, D.R.; Turner, J.R. Epithelial myosin light chain kinase expression and activity are upregulated in inflammatory bowel disease. *Lab. Investig.* **2006**, *86*, 191–201. [CrossRef] [PubMed]
112. Edelblum, K.L.; Sharon, G.; Singh, G.; Odenwald, M.A.; Sailer, A.; Cao, S.; Ravens, S.; Thomsen, I.; El Bissati, K.; McLeod, R.; et al. The microbiome activates CD4 T-cell-mediated immunity to compensate for increased intestinal permeability. *Cell Mol. Gastroenterol. Hepatol.* **2017**, *4*, 285–297. [CrossRef] [PubMed]
113. Russo, J.M.; Florian, P.; Shen, L.; Graham, W.V.; Tretiakova, M.S.; Gitter, A.H.; Mrsny, R.J.; Turner, J.R. Distinct temporal-spatial roles for rho kinase and myosin light chain kinase in epithelial purse-string wound closure. *Gastroenterology* **2005**, *128*, 987–1001. [CrossRef] [PubMed]
114. Tamada, M.; Perez, T.D.; Nelson, W.J.; Sheetz, M.P. Two distinct modes of myosin assembly and dynamics during epithelial wound closure. *J. Cell Biol.* **2007**, *176*, 27–33. [CrossRef]
115. Bement, W.M.; Forscher, P.; Mooseker, M.S. A novel cytoskeletal structure involved in purse string wound closure and cell polarity maintenance. *J. Cell Biol.* **1993**, *121*, 565–578. [CrossRef]

116. Lechuga, S.; Ivanov, A.I. Disruption of the epithelial barrier during intestinal inflammation: Quest for new molecules and mechanisms. *Biochim. Biophys. Acta* **2017**, *1864*, 1183–1194. [CrossRef]
117. Isobe, K.; Raghuram, V.; Krishnan, L.; Chou, C.L.; Yang, C.R.; Knepper, M.A. CRISPR-Cas9/ Phosphoproteomics Identifies Multiple Non-canonical Targets of Myosin Light Chain Kinase. *Am. J. Physiol. Renal Physiol.* **2020**. [CrossRef]
118. Kamm, K.E.; Stull, J.T. Dedicated myosin light chain kinases with diverse cellular functions. *J. Biol. Chem.* **2001**, *276*, 4527–4530. [CrossRef]
119. Garcia, J.G.; Lazar, V.; Gilbert-McClain, L.I.; Gallagher, P.J.; Verin, A.D. Myosin light chain kinase in endothelium: Molecular cloning and regulation. *Am. J. Respir. Cell Mol. Biol.* **1997**, *16*, 489–494. [CrossRef]
120. Khromov, A.S.; Wang, H.; Choudhury, N.; McDuffie, M.; Herring, B.P.; Nakamoto, R.; Owens, G.K.; Somlyo, A.P.; Somlyo, A.V. Smooth muscle of telokin-deficient mice exhibits increased sensitivity to Ca2+ and decreased cGMP-induced relaxation. *Proc. Natl. Acad. Sci. USA* **2006**, *103*, 2440–2445. [CrossRef]
121. Choudhury, N.; Khromov, A.S.; Somlyo, A.P.; Somlyo, A.V. Telokin mediates Ca2+-desensitization through activation of myosin phosphatase in phasic and tonic smooth muscle. *J. Muscle Res. Cell Motil.* **2004**, *25*, 657–665. [CrossRef] [PubMed]
122. Nieznanski, K.; Sobieszek, A. Telokin (kinase-related protein) modulates the oligomeric state of smooth-muscle myosin light-chain kinase and its interaction with myosin filaments. *Biochem. J.* **1997**, *322*, 65–71. [CrossRef] [PubMed]
123. Somlyo, A.V.; Wang, H.; Choudhury, N.; Khromov, A.S.; Majesky, M.; Owens, G.K.; Somlyo, A.P. Myosin light chain kinase knockout. *J. Muscle Res. Cell Motil.* **2004**, *25*, 241–242. [CrossRef] [PubMed]
124. He, W.Q.; Peng, Y.J.; Zhang, W.C.; Lv, N.; Tang, J.; Chen, C.; Zhang, C.H.; Gao, S.; Chen, H.Q.; Zhi, G.; et al. Myosin light chain kinase is central to smooth muscle contraction and required for gastrointestinal motility in mice. *Gastroenterology* **2008**, *135*, 610–620. [CrossRef] [PubMed]
125. Clayburgh, D.R.; Rosen, S.; Witkowski, E.D.; Wang, F.; Blair, S.; Dudek, S.; Garcia, J.G.; Alverdy, J.C.; Turner, J.R. A differentiation-dependent splice variant of myosin light chain kinase, MLCK1, regulates epithelial tight junction permeability. *J. Biol. Chem.* **2004**, *279*, 55506–55513. [CrossRef] [PubMed]
126. Birukov, K.G.; Csortos, C.; Marzilli, L.; Dudek, S.; Ma, S.F.; Bresnick, A.R.; Verin, A.D.; Cotter, R.J.; Garcia, J.G. Differential regulation of alternatively spliced endothelial cell myosin light chain kinase isoforms by p60(Src). *J. Biol. Chem.* **2001**, *276*, 8567–8573. [CrossRef] [PubMed]
127. Lazar, V.; Garcia, J.G. A single human myosin light chain kinase gene (MLCK.; MYLK). *Genomics* **1999**, *57*, 256–267. [CrossRef]
128. Graham, W.V.; He, W.; Marchiando, A.M.; Zha, J.; Singh, G.; Li, H.S.; Biswas, A.; Ong, M.; Jiang, Z.H.; Choi, W.; et al. Intracellular MLCK1 diversion reverses barrier loss to restore mucosal homeostasis. *Nat. Med.* **2019**, *25*, 690–700. [CrossRef]
129. Graham, W.V.; Magis, A.T.; Bailey, K.M.; Turner, J.R.; Ostrov, D.A. Crystallization and preliminary X-ray analysis of the human long myosin light-chain kinase 1-specific domain IgCAM3. *Acta Crystallogr. Sect. F Struct Biol. Cryst Commun.* **2011**, *67*, 221–223. [CrossRef]
130. Madsen, K.L.; Malfair, D.; Gray, D.; Doyle, J.S.; Jewell, L.D.; Fedorak, R.N. Interleukin-10 gene-deficient mice develop a primary intestinal permeability defect in response to enteric microflora. *Inflamm. Bowel Dis.* **1999**, *5*, 262–270. [CrossRef]

International Journal of
Molecular Sciences

Article

Temporal Effects of Quercetin on Tight Junction Barrier Properties and Claudin Expression and Localization in MDCK II Cells

Enrique Gamero-Estevez [1,2], Sero Andonian [3], Bertrand Jean-Claude [2,4], Indra Gupta [1,2,5] and Aimee K. Ryan [1,2,5,*]

1. Department of Human Genetics, McGill University, Montreal, QC H4A 3J1, Canada; egameroestevez@gmail.com (E.G.-E.); guptalab@gmail.com (I.G.)
2. Research Institute of the McGill University Health Centre, Glen Site, Montreal, QC H4A 3J1, Canada; bertrandj.jean-claude@mcgill.ca
3. Division of Urology, McGill University, Montreal, QC H4A 3J1, Canada; sero.andonian@mcgill.ca
4. Department of Medicine, McGill University, Montreal, QC H4A 3J1, Canada
5. Departments of Pediatrics, McGill University, Montreal, QC H4A 3J1, Canada

* Correspondence: aimee.ryan@mcgill.ca

Received: 2 August 2019; Accepted: 29 September 2019; Published: 2 October 2019

Abstract: Kidney stones affect 10% of the population. Yet, there is relatively little known about how they form or how to prevent and treat them. The claudin family of tight junction proteins has been linked to the formation of kidney stones. The flavonoid quercetin has been shown to prevent kidney stone formation and to modify claudin expression in different models. Here we investigate the effect of quercetin on claudin expression and localization in MDCK II cells, a cation-selective cell line, derived from the proximal tubule. For this study, we focused our analyses on claudin family members that confer different tight junction properties: barrier-sealing (Cldn1, -3, and -7), cation-selective (Cldn2) or anion-selective (Cldn4). Our data revealed that quercetin's effects on the expression and localization of different claudins over time corresponded with changes in transepithelial resistance, which was measured continuously throughout the treatment. In addition, these effects appear to be independent of PI3K/AKT signaling, one of the pathways that is known to act downstream of quercetin. In conclusion, our data suggest that quercetin's effects on claudins result in a tighter epithelial barrier, which may reduce the reabsorption of sodium, calcium and water, thereby preventing the formation of a kidney stone.

Keywords: tight junctions; claudins; kidney stones; paracellular transport; ion reabsorption; quercetin

1. Introduction

Kidney stones affect 10% of the population and recur in 50% of adults. They are associated with renal failure in children and adults. They cause extreme pain and have a significant financial burden to society [1–3]. Given these morbidities, surprisingly little is known about why stones recur and how they can be prevented. Most stones are calcium-based with calcium oxalate more frequently observed than calcium phosphate stones [1]. Kidney stones result from the accumulation of salts along the kidney nephron, where an important factor in salt reabsorption is the epithelial tight junction barrier. Tight junctions are the most apical junction between apposing cells, where they compartmentalize the apical and basolateral intercellular space and regulate the passive paracellular movement of water and solutes [4]. The ion specificity of the tight junction barrier is determined by the claudin family of tetraspanin proteins, which contains close to 30 members [5–7]. Claudins can bind hetero- and homotypically with claudins in the same cell through their transmembrane domains or to claudins in the apposing cell through their extracellular loops [6]. The combination of claudins expressed within an epithelium determines the tightness and selectivity of the tight junction barriers [8].

The claudin composition of tight junctions varies along the different segments of the nephron and, consequently, ions and salts are differentially reabsorbed in the different nephron segments [8]. Reabsorption of calcium from the urine filtrate predominantly occurs in two segments: the proximal tubule reabsorbs 70% and the thick ascending limb reabsorbs 25% of the calcium. Several claudin family members are implicated in calcium reabsorption. A common sequence variant in CLDN14 and rare mutations in CLDN16 and -19, all of which are expressed in the thick ascending limb, are risk factors for the formation of calcium-based kidney stones in humans [9–11]. Both Cldn2 and -10 null mice develop nephrocalcinosis [12,13], while Cldn16 and -19 knockdown mice develop hypercalciuria [14,15]. Cldn7 is also essential for salt reabsorption: Cldn7 null mice die shortly after birth due to salt loss and dehydration [16]. Although not known to participate in ion exchange, Cldn3 has been shown to form a complex with Cldn16 and -19 in the thick ascending limb, which is essential for calcium and magnesium reabsorption in this segment [17]. Cldn4 has a critical role in chloride reabsorption in the kidney collecting duct where it forms a pore with Cldn8 [18]. The role of Cldn1 in ion reabsorption is less clear; however, it is present in different kidney segments and may be important in diabetic nephropathy [19]. Because claudins are essential for sodium, calcium, and water reabsorption in the nephron, targeting this claudin-based epithelial barrier may be a successful approach to decrease salt reabsorption and prevent kidney stone formation.

The flavonoid quercetin prevents kidney stone formation in a rat model [20]. Quercetin alters claudin expression in the Caco-2 intestine-derived cell line and in LLC-PK1 cells, an anion-selective cell line derived from the renal proximal tubule [21–24]. To date, no one has studied the effect of quercetin in a cation-selective cell line that models an epithelial barrier which is permeable to calcium and sodium. We hypothesize that quercetin may prevent kidney stones through its effects on claudins at tight junctions in the epithelial barrier of the nephron. For our study, we used Madin-Darby Canine Kidney cells (MDCK II), which are cation-selective and derived from the proximal tubule where the majority of sodium and calcium reabsorption occurs. MDCK II cells express Cldn1, -2, -3, -4, and -7 [25,26], which allows us to investigate the effect of quercetin on claudins with different properties: barrier-sealing (Cldn1, -3 and -7), cation-selective (Cldn2), or anion-selective (Cldn4) within a cation-selective cell line [27–30]. The MDCK II cell line has been widely studied and its TER and cation permeability properties are well characterized. Cldn2 confers the leaky barrier properties of MDCK II cells. If Cldn2 is depleted, TER increases and sodium and chloride potentials are significantly reduced [31].

We found that quercetin differentially affected both the expression and the localization of some claudins over time. Quercetin also significantly increased transepithelial resistance. Our data suggest that after treatment with quercetin, the barrier becomes tighter, which could lead to a decrease in cation and water reabsorption. This may result in a more favorable urinary filtrate that is less prone to crystal supersaturation and the formation of a stone.

2. Results

2.1. Quercetin Increased Transepithelial Resistance of MDCK II Cells

Transepithelial resistance (TER) is a measure of the electrical resistance of a cell monolayer and is modulated by cell confluence, barrier permeability, and tight junction composition. For these studies, we used the cellZscope (NanoAnalytics) to continuously measure TER in MDCK II cells from the time of plating until several days after confluence when stable, mature tight junctions are formed. In cation-selective cell lines, TER increases as cells become closer or when paracellular movement of cations is reduced [30]. As predicted, there was an increase in TER immediately after seeding, when cells are dividing and confluence is increasing (Figure S1A). Under control conditions, MDCK II cells reached a peak resistance of 130 $\Omega \cdot cm^2$ ~40 h after seeding. Once maximum confluence is achieved, contact inhibition and tight junction remodeling takes place, which leads to a decrease in the TER that eventually stabilizes and becomes constant as the cells form stable mature tight junctions [32,33]. In MDCK II cells, this translated into a decrease in TER that then remained at 50–80 Ω·cm^2, which is characteristic for mature tight junctions in this model [33]. Therefore, this was the time point selected for treatment with quercetin since it best represents the mature renal proximal tubule epithelium.

We confirmed that 400 µM quercetin, the concentration used in previous studies [21,22], was also the most optimal for our experiments through a dose response curve (Figure S1). We monitored the effects of quercetin on TER for ~96 h after treatment. Figure 1A shows the TER profile for one of the three biological replicates, which were each done in triplicate. Analysis of all data showed a small expected increase in TER immediately after treatment, due to the removal of the cells from the incubator and change of media. In control cells, the TER stabilized within 4–5 h and then remained at a resistance of ~50 Ω·cm^2 (Figure 1A). Two-way ANOVA analysis showed that TER changed significantly over time and in a treatment-dependent manner (P_{int} = 0.0405; P_{time} < 0.0001; P_{treat} < 0.0001). Cells treated with 400 µM quercetin exhibited a progressive increase in TER, reaching significance at 3 h when the TER was ~15 Ω·cm^2 higher than the control cells (p = 0.049; quercetin-treated = 90.04 ± 4.01 Ω·cm^2 versus control cells = 70.7 ± 1.62 Ω·cm^2). The TER remained significantly increased until 5 h post-treatment (p = 0.046; quercetin-treated = 86.33 ± 2.94 Ω·cm^2 versus control cells = 66.86 ± 3.59 Ω·cm^2) and then progressively decreased to ~5 Ω·cm^2 below control levels 15 h after treatment, which was not significant (15 h: p > 0.99; control 60.08 ± 3.61, quercetin 62.21 ± 2.37). Following the decrease, TER increased again, reaching a steady state level of 10 Ω·cm^2 above control, 36 h after treatment, which was statistically significant and remained significantly increased for the duration of the experiment (36 h: p = 0.0071; control 54.7 ± 2.31, quercetin 78.05 ± 5.19) (48 h: p < 0.0001; control 53.35 ± 1.8, quercetin 85.68 ± 2.55) (Figure 1).

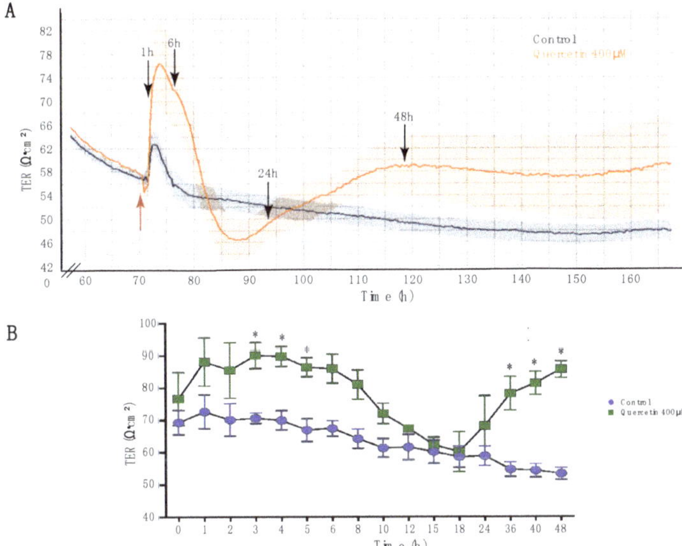

Figure 1. Quercetin caused oscillations in transepithelial resistance (TER) of MDCK II cells. (**A**) Representative plot of TER in control cells (black) and cells treated with 400 µM quercetin (orange) from one biological replicate performed in triplicate. Red arrow indicates when quercetin was added to the culture medium. Black arrows indicate the time points taken for western blot and immunofluorescence analysis. (**B**) TER of control and quercetin-treated cells at different time points after treatment from three independent experiments performed in triplicate. Two-way ANOVA was performed (P_{int} = 0.04; P_{time} < 0.0001; P_{treat} < 0.0001). Mean and SEM are plotted. * Denotes significance, p < 0.05.

2.2. Quercetin Treatment Caused Claudin-Specific Changes in Expression and Membrane Localization

To determine if the changes in TER caused by quercetin treatment corresponded with different claudin profiles, cells were collected 1, 6, 24, and 48 h after treatment. Claudin expression was assessed by western blot analysis and localization to the tight junction barrier was assessed by immunofluorescence. Immunofluorescence also provided a qualitative assessment of claudin expression. Five claudins expressed in MDCK II cells were studied: Cldn1, -3, and -7 that have barrier-sealing functions, Cldn2 that is involved in cation pore formation, and Cldn4 that is involved in anion pore formation. For all experiments, cells were cultured for 72 h before treatment with 400 μM quercetin to ensure that the cells had established mature tight junctions.

2.2.1. Cldn1

Western blot analysis revealed a significant decrease in Cldn1 expression over time in both controls and quercetin-treated cells ($P_{time} = 0.021$). Quercetin treatment significantly lowered Cldn1 levels at 48 h compared to controls ($p = 0.038$; control 1.47 ± 0.55; quercetin 0.44 ± 0.18). A change in the relative abundance in the two migratory bands was observed at 24 h, although the total amount of Cldn1 was not affected (Figure 2A,B). Immunofluorescence analysis revealed decreased levels of Cldn1 at 1, 6, and 48 h (Figure 2C). A reduction in Cldn1 co-localization with ZO1 can be seen at 1 and 48 h, although it was not significant ($p = 0.3$ and $p = 0.2$, respectively) (Figure 2D). These data suggest than even though general levels of Cldn1 were decreased, the remaining Cldn1 still co-localized with ZO1.

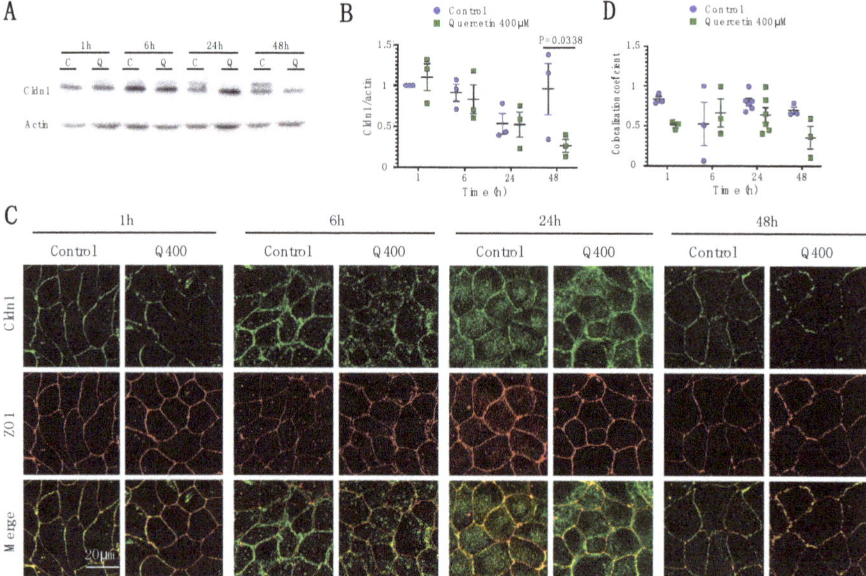

Figure 2. Analysis of Cldn1 expression and localization in MDCK II cells following quercetin treatment. (**A**) Western blot analysis of Cldn1 expression in cell lysates from control and 400 μM quercetin-treated MDCK II cells. Actin was used as a loading control. (**B**) Densitometry measurements of Cldn1 and actin intensity on western blot were normalized to expression at 1h. Normalized Cldn1 expression relative to normalized actin expression is plotted ($P_{int} = 0.11$; $P_{time} = 0.022$; $P_{treat} = 0.16$). (**C**) Immunofluorescence of control and 400 μM of quercetin-treated MDCK II cells at 1, 6, 24, and 48 h. Cldn1 is shown in green and ZO1 is shown in red. (**D**) Localization of claudin at the tight junction was assessed by determining the amount of ZO1 that was co-localized with claudin expression ($P_{int} = 0.24$; $P_{time} = 0.35$; $P_{treat} = 0.055$). For all graphs, each point corresponds to an independent experiment; mean and SEM are shown. C = Control and Q = Quercetin. Scale bar, 20 μm.

2.2.2. Cldn2

Two-way ANOVA analysis on Cldn2 expression showed that Cldn2 decreased significantly with quercetin treatment (P_{treat} = 0.0021), while changes in time and interaction only suggested significance (P_{int} = 0.0525; P_{time} = 0.096). Sidak's multiple comparison test showed that the decrease observed at 48 h after treatment with quercetin was significant (p = 0.0011; control 2.0 ± 0.79; quercetin 0.06 ± 0.057). At 24 h, no Cldn2 was observed by western blot analysis in quercetin-treated cells (Figure 3A,B). As previously reported, Cldn2 detection by immunofluorescence was patchy within the epithelial monolayer of control MDCK II cells [34], with clusters of cells showing high levels of Cldn2 expression and adjacent groups of cells showing virtually no expression (Figure 3C). Localization of Cldn2 was significantly changed after treatment with quercetin (P_{treat} = 0.0055). At 1 h and 6 h after quercetin treatment, no changes in localization of Cldn2 were observed. However, at 24 h and 48 h after treatment, quercetin-treated cells showed a significant reduction both in the amount of Cldn2 and in the portion of Cldn2 that co-localized with ZO1 (24h: p = 0.036; control 0.54 ± 0.11, quercetin 0.14 ± 0.08; 48 h: p = 0.034; control 0.52 ± 0.12, quercetin 0.12 ± 0.05) (Figure 3C,D).

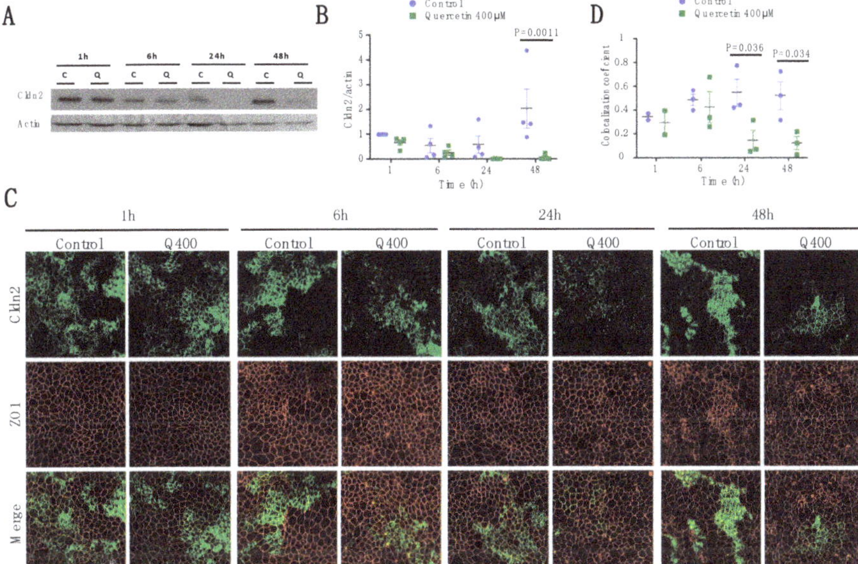

Figure 3. Analysis of Cldn2 expression and localization in MDCK II cells following quercetin treatment. (**A**) Western blot analysis of Cldn2 expression in cell lysates from control and 400 μM quercetin-treated MDCK II cells. Actin was used as a loading control. (**B**) Densitometry measurements of Cldn2 and actin intensity on western blot were normalized to control at 1 h. Normalized Cldn2 expression relative to normalized actin is plotted (P_{int} = 0.052; P_{time} = 0.096; P_{treat} = 0.0021). (**C**) Immunofluorescence of control and 400 μM of quercetin-treated MDCK II cells at 1, 6, 24, and 48 h. Cldn2 is shown in green and ZO1 is shown in red. (**D**) Localization of claudin at the tight junction was assessed by determining the amount of ZO1 that co-localized with claudin expression (P_{int} = 0.15; P_{time} = 0.46; P_{treat} = 0.0055). For all graphs, each point corresponds to an independent experiment; mean and SEM are shown. C = Control and Q = Quercetin. Scale bar, 20 μm.

2.2.3. Cldn3

Western blot analysis showed that quercetin caused a significant treatment-dependent increase in Cldn3 expression (P_{treat} = 0.0395). In marked contrast to the effects on Cldn1 and -2, Cldn3 expression was increased at 24 h and 48 h after treatment, although it was only statistically significant at 48 h

($p = 0.04$; control 0.29 ± 0.1; quercetin 2.4 ± 0.98) (Figure 4A,B). Although there appeared to be a general decrease in Cldn3 detection by immunofluorescence after 1 h and 6 h of treatment (Figure 4C), co-localization analysis showed that there was no effect on Cldn3 co-localization with ZO1 at 1 h and 6 h ($p = 0.79$ and $p = 0.99$, respectively) (Figure 4D). Cldn3 also co-localized with ZO1 at later time points.

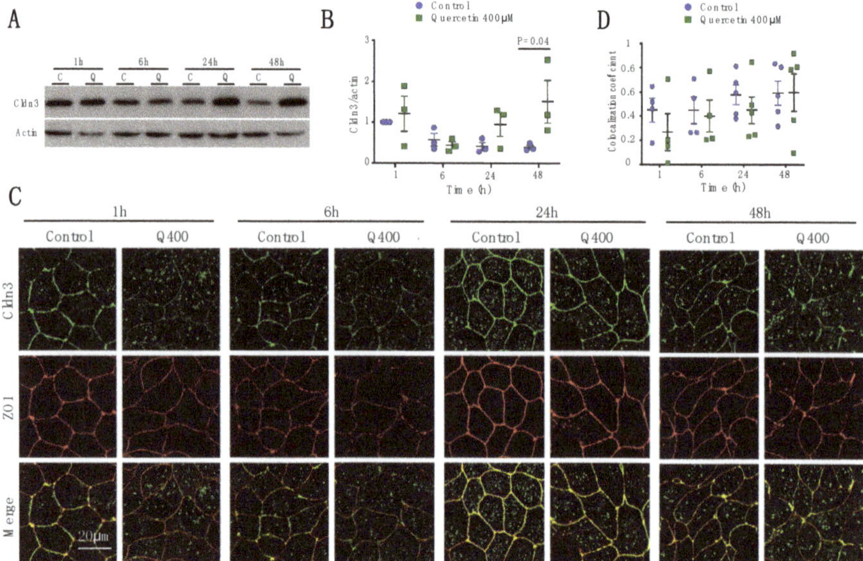

Figure 4. Analysis of Cldn3 expression and localization in MDCK II cells following quercetin treatment. (**A**) Western blot analysis of Cldn3 expression in cell lysates from control and 400 μM quercetin-treated MDCK II cells. Actin was used as a loading control. (**B**) Densitometry measurements of Cldn3 and actin were normalized to control at 1 h. Normalized Cldn3 expression relative to normalized actin is plotted ($P_{int} = 0.16$; $P_{time} = 0.16$; $P_{treat} = 0.039$). (**C**) Immunofluorescence of control and 400 μM of quercetin-treated MDCK II cells at 1, 6, 24, and 48 h. Cldn3 is shown in green and ZO1 is shown in red. (**D**) Localization of claudin at the tight junction was assessed by determining the amount of ZO1 that co-localized with claudin expression ($P_{int} = 0.87$; $P_{time} = 0.24$; $P_{treat} = 0.31$). For all graphs each point corresponds to an independent experiment; mean and SEM are shown. C = Control and Q = Quercetin. Scale bar, 20 μm.

2.2.4. Cldn4

Cldn4 expression changed over time following quercetin treatment ($P_{int} < 0.0001$; $P_{time} = 0.0017$; $P_{treat} < 0.0001$). At 24 h and 48 h, Cldn4 expression was significantly increased by western blot analysis (control = 0.67 ± 0.37; quercetin = 2.74 ± 0.36, $p < 0.0001$ and control = 0.79 ± 0.36; quercetin = 2.93 ± 0.37, $p < 0.0001$, respectively) (Figure 5A,B). In contrast, by immunofluorescence, qualitatively, Cldn4 was downregulated at 48 h (Figure 5C), but co-localized with ZO1 (Figure 5D). This could be a consequence of possible claudin modifications that may avoid correct recognition of the epitope by the antibody. At this point, we cannot explain the discrepancy in Cldn4 expression as assessed by western blot and immunofluorescence.

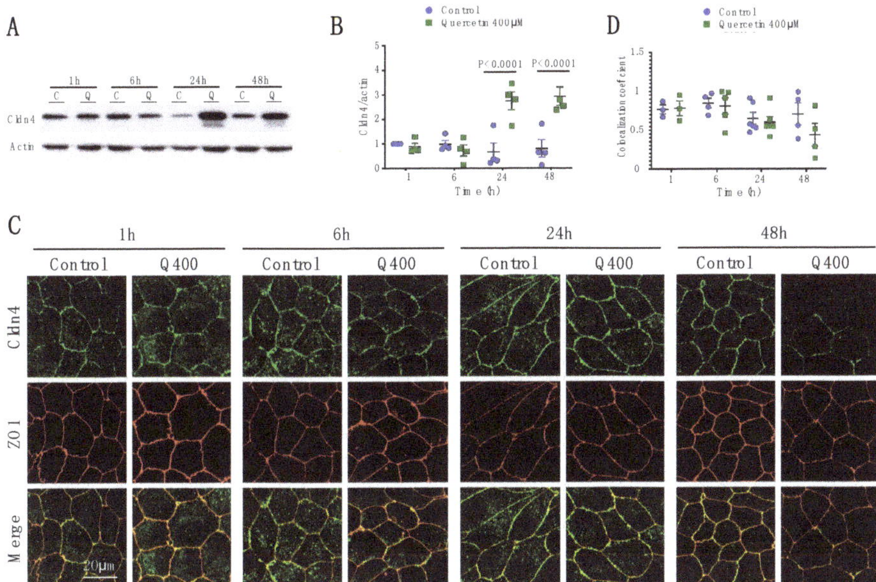

Figure 5. Analysis of Cldn4 expression and localization in MDCK II cells following quercetin treatment. (**A**) Western blot analysis of Cldn4 expression in cell lysates from control and 400 μM quercetin-treated MDCK II cells. Actin was used as a loading control. (**B**) Densitometry measurements of Cldn4 and actin intensity on western blot were normalized to control levels at 1 h. Normalized Cldn4 expression relative to normalized actin is plotted ($P_{int} < 0.0001$; $P_{time} = 0.0017$; $P_{treat} < 0.0001$). (**C**) Immunofluorescence of control and 400 μM of quercetin-treated MDCK II cells at 1, 6, 24, and 48 h. Cldn4 is shown in green and ZO1 is shown in red. (**D**) Localization of claudin at the tight junction was assessed by determining the amount of ZO1 that co-localized with claudin expression ($P_{int} = 0.54$; $P_{time} = 0.051$; $P_{treat} = 0.24$). For all graphs, each point corresponds to an independent experiment; mean and SEM are shown. C = Control and Q = Quercetin. Scale bar represent 20 μm.

2.2.5. Cldn7

In contrast to the other claudin family members examined, Cldn7 expression and localization were not affected by quercetin treatment (Figure 6). At 24h, there was an increased level of Cldn7 immunofluorescence in the cytoplasm (Figure 6C) but this was not associated with increased co-localization with ZO1 at the tight junction (Figure 6D). This may suggest that Cldn7 incorporation into the tight junction complex is saturated at this time point.

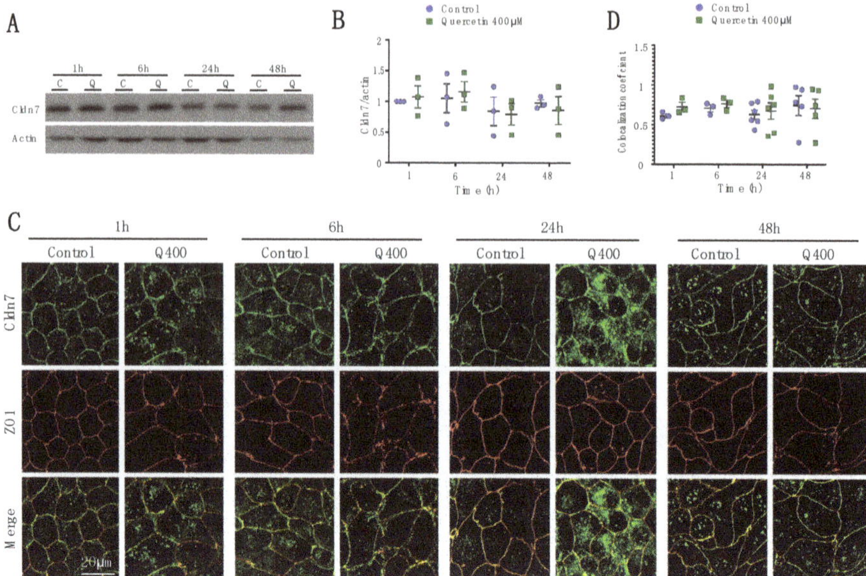

Figure 6. Analysis of Cldn7 expression and localization in MDCK II cells following quercetin treatment. (**A**) Western blot analysis of Cldn7 expression in cell lysates from control and 400 μM quercetin-treated MDCK II cells. Actin was used as a loading control. (**B**) Densitometry measurements of Cldn7 and actin intensity on western blot were normalized to control at 1h. Normalized Cldn7 expression relative to normalized actin is plotted (P_{int} = 0.916; P_{time} = 0.41; P_{treat} = 0.98). (**C**) Immunofluorescence of control and 400 μM of quercetin-treated MDCK II cells at 1, 6, 24, and 48 h. Cldn7 is shown in green and ZO1 is shown in red. (**D**) Localization of claudin at the tight junction was assessed by determining the amount of ZO1 that co-localized with claudin expression (P_{int} = 0.89; P_{time} = 0.8; P_{treat} = 0.55). For all graphs each point corresponds to an independent experiment; mean and SEM are shown. C = Control and Q = Quercetin. Scale bar, 20 μm.

2.3. *Quercetin's Effects on Claudin Expression and Localization is Independent of the PI3K/AKT Pathway*

Quercetin has been identified as a strong inhibitor of the PI3K/AKT pathway [35,36]. Given that receptor tyrosine kinases and pAKT phosphorylate and regulate claudins both directly and indirectly [37,38], we hypothesized that quercetin affects claudin expression through the PI3K/AKT pathway. Western blot analysis of pPI3K, AKT, and pAKT was performed following quercetin treatment. The levels of pPI3K and AKT in MDCK II cells did not change significantly after quercetin treatment. pAKT was reduced in quercetin-treated cells at 1 h and 6 h after treatment. pAKT levels could not be detected at 1 h and 6 h after quercetin treatment limiting the statistical analysis at these time points. At 24 h and 48 h, pAKT was decreased, but this was not statistically significant (p = 0.99 and p = 0.7, respectively). These data suggest that quercetin inhibits the AKT/pAKT pathway in MDCK type II cells (Figure 7A,B).

To query if quercetin's effects on claudin expression were due to impairment of the PI3K/AKT pathway, we inhibited or activated the PI3K/AKT pathways using wortmannin or SC79, respectively, alone or in combination with quercetin. As predicted, treatment with wortmannin caused a transient decrease in pAKT in MDCK II cells at 1 h and 6 h and SC79 increased pAKT at 1 h and 6 h (Figure 7C–F). In both cases, effects were greatly attenuated by 24 h. However, neither activation nor inhibition of the pAKT pathway recapitulated the effects on claudins observed following quercetin treatment (Figure 8), and neither wortmannin nor SC79 rescued the effects of quercetin. Therefore, our data suggest that quercetin is not acting through the PI3K/AKT pathway to effect changes in claudin protein expression.

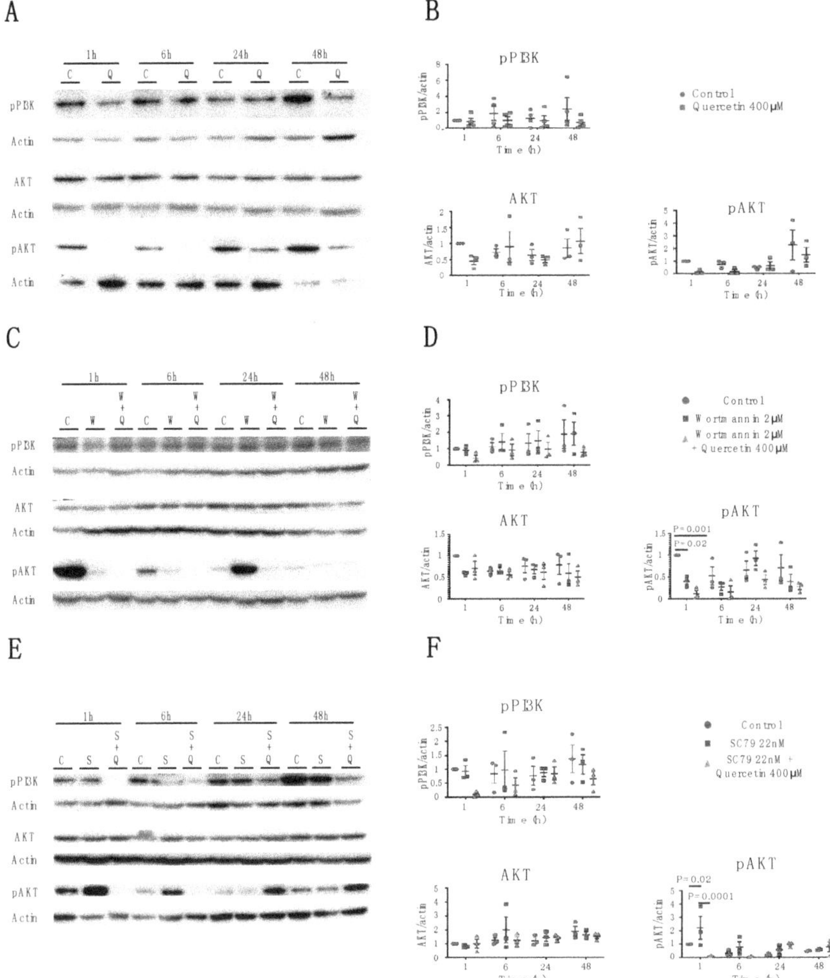

Figure 7. Analysis of pPI3K, pAKT, and AKT expression in MDCK II cells following quercetin treatment. (**A**) Western blot analysis of claudin expression in cell lysates from control and 400 μM quercetin-treated MDCK II cells. Actin was used as a loading control. (**B**) Densitometry measurements of band intensity on western blot were normalized to control at 1 h and protein expression was plotted relative to normalized actin expression. pPI3K (P_{int} = 0.67; P_{time} = 0.79; P_{treat} = 0.14), AKT (P_{int} = 0.44; P_{time} = 0.49; P_{treat} = 0.68), and pAKT (P_{int} = 0.68; P_{time} = 0.027; P_{treat} = 0.142) measurements were plotted. (**C**) Western blot analysis of pPI3K, pAKT, and AKT expression in cell lysates from control, 2 μM wortmannin-treated and 2 μM wortmannin plus 400 μM quercetin-treated MDCK II cells. Actin was used as a loading control. (**D**) Densitometry measurements of band intensity on western blot were normalized to control at 1 h and protein expression was plotted relative to normalized actin expression. pPI3K (P_{int} = 0.97; P_{time} = 0.03; P_{treat} = 0.13), AKT (P_{int} = 0.79; P_{time} = 0.53; P_{treat} = 0.099), and pAKT (P_{int} = 0.15; P_{time} = 0.064; P_{treat} = 0.0008) measurements were plotted. (**E**) Western blot analysis of pPI3K, pAKT, and AKT expression in cell lysates from control 22 nM SC79-treated and 22 nM SC79 plus 400 μM quercetin-treated MDCK II cells. Actin was used as a loading control. (**F**) Densitometry measurements of band intensity on western blot were normalized to control at 1 h and protein expression was plotted relative to normalized actin expression. pPI3K (P_{int} = 0.81; P_{time} = 0.51; P_{treat} = 0.078), AKT (P_{int} = 0.79; P_{time} = 0.52; P_{treat} = 0.099), and pAKT (P_{int} = 0.16; P_{time} = 0.063; P_{treat} = 0.0008) measurements were plotted. Each point corresponds to an independent experiment; mean and SEM are shown. W = Wortmannin; S = SC79; C = Control; and Q = Quercetin.

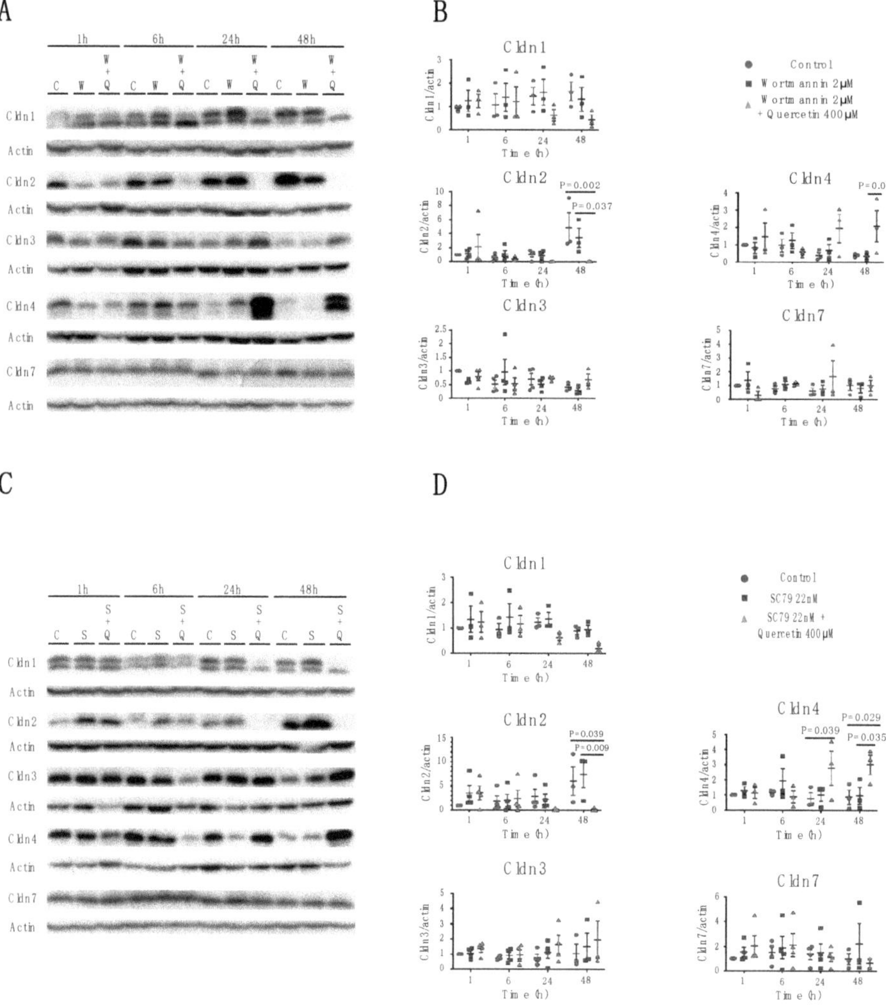

Figure 8. Analysis of Cldn1, -2, -3, -4, and -7 expression in MDCK II cells following wortmannin or SC79 treatment. (**A**) Western blot analysis of claudin expression in cell lysates from control, 2 μM wortmannin-treated and 2 μM wortmannin plus 400 μM quercetin-treated MDCK II cells. Actin was used as a loading control. (**B**) Densitometry measurements of band intensity on western blot were normalized to control at 1 h and protein expression was plotted relative to normalized actin expression. Cldn1 (P_{int} = 0.6; P_{time} = 0.98; P_{treat} = 0.24), Cldn2 (P_{int} = 0.059; P_{time} = 0.016; P_{treat} = 0.097), Cldn3 (P_{int} = 0.29; P_{time} = 0.26; P_{treat} = 0.78), Cldn4 (P_{int} = 0.21; P_{time} = 0.97; P_{treat} = 0.041), and Cldn7 (P_{int} = 0.41; P_{time} = 0.98; P_{treat} = 0.83) measurements were plotted. (**C**) Western blot analysis of claudin expression in cell lysates from control 22 nM SC79-treated and 22 nM SC79 plus 400 μM quercetin-treated MDCK II cells. Actin was used as a loading control. (**D**) Densitometry measurements of band intensity on western blot were normalized to control at 1 h and protein expression was plotted relative to normalized actin expression. Cldn1 (P_{int} = 0.6; P_{time} = 0.15; P_{treat} = 0.11), Cldn2 (P_{int} = 0.09; P_{time} = 0.14; P_{treat} = 0.11), Cldn3 (P_{int} = 0.99; P_{time} = 0.46; P_{treat} = 0.21), Cldn4 (P_{int} = 0.082; P_{time} = 0.73; P_{treat} = 0.047), and Cldn7 (P_{int} = 0.87; P_{time} = 0.78; P_{treat} = 0.5) measurements were plotted. Each point corresponds to an independent experiment; mean and SEM are shown. W = Wortmannin; S = SC79; C = Control; and Q = Quercetin.

3. Discussion

The purpose of this study was to determine the effect of quercetin on the cation-selective tight junction barrier in MDCK II cells, which models the proximal tubule of the kidney. We showed that the effects of quercetin on the MDCK barrier were stabilized 48 h after treatment, resulting in an increased TER of at least 20% over the TER observed in untreated cells. The quercetin-dependent increase in TER correlates with decreased expression of Cldn1 and decreased expression and localization of Cldn2 to the tight junction. Although a significant increase in Cldn3 and -4 was observed by western blot analysis, tight junction localization of Cldn3, -4, and -7 were not affected at this timepoint.

In the mature kidney, transepithelial resistance in the proximal tubule is <10 $\Omega \cdot cm^2$, indicating that this nephron segment has a leaky barrier, which correlates with its important role in salt and water reabsorption [30]. Therefore, the increase of 10–20 $\Omega \cdot cm^2$ observed following quercetin treatment of MDCK II cells would be very relevant in the context of the proximal tubule, where similar oscillations could completely change its barrier properties.

The TER oscillation observed during the first 48 h after quercetin treatment reflects dynamic tight junction remodeling and correlates with the changes in claudin expression that we observed. The decrease in TER observed at 6h coincided with a decrease in the immunofluorescent levels of barrier-sealing claudins, Cldn1 and -3. This is predicted to translate into a 'leakier' barrier, and consequently, a reduced TER. The increased TER at 24 h and 48 h coincided primarily with a decrease in the expression and tight junction localization of Cldn2, a cation-selective pore-forming claudin. Decreased Cldn2 at the tight junction is predicted to reduce paracellular movement of cations [31]—including sodium and calcium, and water—and may contribute to the increase in TER, as previously described for Cldn4 in MDCK [30].

Our data and those obtained by other groups, suggest that the effects of quercetin are cell-line dependent [21–24,31,39]. When Tokuda and Furuse depleted Cldn2 from MDCK II cells, they observed increased TER and decreased sodium and chloride potential [31]. The increase in TER was significantly higher than what we observed (>1000 Ω cm^2), perhaps due to compensation by other claudins in response to removal of Cldn2. In their experiment, depletion of Cldn2 led to increased tight junction localization of Cldn1, -3, -4, and -7, which would effectively tighten the barrier and increase TER. In our quercetin-treated cells, the 90% reduction of Cldn2, was not accompanied by an increased localization of other claudins to the tight junction: Cldn1 expression was decreased, Cldn3 and -4 expression was increased but localization to the tight junction was unchanged, and Cldn7 was unchanged. Thus, quercetin causes a different net effect on claudin expression/localization compared to depletion of Cldn2 alone and, as a result, there is a more modest increase in TER.

Quercetin also increased TER of the anion-selective barrier in LLC-PK1 cells, here, expression of Cldn2 and -3 were downregulated while Cldn4, -5, and -7 were upregulated [21]. However, claudin localization was not assessed. An oscillation and final increase in TER following quercetin treatment, has been observed in Caco-2 cells [23,24], where TER oscillated similarly to what we observed in MDCK II cells during the first 48 h after treatment. In this case Cldn1 was displaced to the cytoskeletal fraction, Cldn4 expression was increased, while Cldn3 was unperturbed. In contrast, Valenzano et al., did not report any effect on TER after 17 h of quercetin treatment on Caco-2 cells, although Cldn2, -4, and -5 were increased [22]. These data together with our findings indicate that quercetin is able to tighten tight junction barriers, but does so by differentially affecting different claudin family members, in a cell-type specific manner.

We observed two migratory bands for Cldn1 and -4 in our western blot analyses. Previous studies have linked these different migratory bands to either claudin degradation [40] or claudin phosphorylation [41,42]. Claudins are known to undergo post-translational modifications, including phosphorylation, glycosylation, palmitoylation, and ubiquitination [43]. These modifications are thought to rapidly change claudin stability at the tight junction. For instance, dephosphorylation of Cldn1 and -2 is a signal for degradation and reduces their presence at tight junctions [44,45]. In the case of Cldn4, studies have shown that the effects depend on the site of phosphorylation. In some cases, phosphorylation leads to the disruption of the tight junction [46,47], while in others, phosphorylation increases the stability of Cldn4 at the tight junction [41]. Further experiments are required to elucidate if the two migratory bands are a consequence of different posttranslational modifications, or if they correspond to degradation products [40].

Other studies that have looked at quercetin's effects on tight junction barrier properties and claudin expression have been based only on western blot analyses. The lack of congruence between our western blot and immunofluorescence expression data for some of the claudins may reflect the availability of claudin epitopes for detection using these two methods. It could also reflect disparities in how posttranslational modifications or protein degradation can be assessed by western blot versus immunofluorescence. For instance, in the case of Cldn1, at 24 h and 48 h, the western blot data shows a loss of the slower migrating band and decreased Cldn1 expression in the quercetin-treated cells. In contrast, the levels by immunofluorescence are equivalent to the control cells. This could suggest that the immunofluorescence signal is primarily due to recognition of the faster migrating band.

Cldn2 immunofluorescence exhibited quite a different pattern compared to other claudin family members. In contrast to most claudins, Cldn2 was highly expressed in some regions and almost absent in other regions within confluent cell layer. This has been previously described and high Cldn2 levels appear to correlate with increased cell confluence [34]. The increased Cldn2 during tight junction maturation, can be clearly seen in the control cells by both western blot analysis and by colocalization with ZO1 at tight junctions.

The mechanism(s) by which quercetin impacts the tight junction barrier through its effects on claudin expression and localization remains unresolved. Quercetin is known to bind to several membrane-bound nutraceutical receptors that signal through different pathways to reduce oxidative stress. Quercetin also interacts directly with several kinases [48–52]. Some studies show that it is a potent inhibitor of PI3K/AKT pathway [36,53]; although there is some controversy [49,54,55], while others show that it inhibits the ERK and NF-κβ pathways [56], and activates the AMPK pathway [49,57]. However, these studies have been done on different cell lines, using different conditions and measuring different outcomes. Therefore, it remains unclear if quercetin acts upstream of all of these pathways, or if its effects are cell-type specific. We showed that quercetin inhibits PI3K/AKT pathway in MCDK II cells. However, inhibition the PI3K/AKT pathway was not sufficient to recapitulate quercetin's effects on claudin expression.

So, how does this translate to the potential therapeutic effects of quercetin to prevent kidney stone formation in the context of the proximal tubule? The site of kidney stone formation is not well understood. Some studies suggest that kidney stones are formed in the interstitium, while others suggest that they form in the lumen of the nephron [2,58,59]. Here, we showed that quercetin led to a relative increase in barrier claudins at the tight junction relative to the pore-forming claudins in MDCK II cells. Although direct measurements of cation potential and water permeability were not performed, our data suggest that quercetin may be beneficial by preventing sodium, calcium, and water within the urine from crossing the epithelial barrier to enter the interstitium. We believe that the effects seen for Cldn2 are essential to explain what may be happening in the nephron. Cldn2 is a cation-selective pore forming claudin that promotes the paracellular movement of sodium, calcium, and water [29]. A decrease in Cldn2 is predicted to lead to a reduction in reabsorption of these substances. This would then prevent the transport of sodium and calcium in the interstitium which can drive stone formation. Alternatively, quercetin may effectively tighten the proximal tubular epithelium to maintain the water content within the urinary lumen of this segment to keep calcium and sodium solubilized [60]. This mechanism could correlate with the beneficial effects that high-water diets or diuretic drugs have in the prevention of kidney stones. In vivo experiments are necessary to discern which specific scenario is taking place and whether this effect could be beneficial for the prevention of kidney stones.

4. Material and Methods

4.1. Cell Culture

Madin–Darby canine kidney cells II (MDCK II) cells were obtained from ATCC. They were incubated at 37 °C and 5% CO2 in Dulbecco's modified Eagle's medium (DMEM) supplemented with 10% FBS, 51 IU penicillin, 50 µg/mL streptomycin and 16 µg/mL of gentamicin (Wisent BioProducts, Quebec, QC, Canada).

4.2. Chemicals and Antibodies

Quercetin (Sigma-Aldrich, Darmstadt, Germany) was added to the culture media, and heated at 37 °C with continuous stirring for at least 30 min in order to ensure that it was dissolved. It was used at the working concentration of 400 µM. PI3K/AKT pathway modulators Wortmannin (Abcam, 120148, Cambridge, UK) and SC79 (Abcam, 146428, Cambridge, UK) were used at working concentrations of 2 µM and 22 nM respectively.

Primary antibodies used for immunofluorescence and western blotting were: Cldn1 (Invitrogen, 374900, Carlsbad, CA, USA), Cldn2 (Invitrogen, 516100, Carlsbad, CA, USA), Cldn3 (Abcam, 15102, Cambridge, UK), Cldn4 (Invitrogen, 364800, Carlsbad, CA, USA), Cldn7 (Spring Bioscience, E10594, Pleasanton, CA, USA), ZO1 (Invitrogen, 339100, Carlsbad, CA, USA), pPI3K (Cell signaling, 4228, Danvers, MA, USA), pAKT (Cell signaling, 9271, Danvers, MA, USA), AKT (Cell signaling, 9272, Danvers, MA, USA), and pan-actin (Cell signaling, 4968, Danvers, MA, USA,). In addition, secondary goat anti-rabbit (Alexa Fluor 595 and 488, Invitrogen, Carlsbad, CA, USA), goat anti mouse (Alexa Fluor 595 and 488, Invitrogen, Carlsbad, CA, USA), goat anti-rabbit-HRT conjugated (Cell Signaling, 70748, Danvers, MA, USA), and goat anti mouse peroxidase conjugated (Jackson ImmunoResearch, 115-035-146, West Grove, PA, USA) were used.

4.3. Transepithelial Resistance (TER)

Cells were seeded on sterile 0.4 µm pore size, 12-well transparent polyethylene terephthalate inserts (Corning, NY, USA) at a cell density of 0.15×10^6 cells/insert and placed in the automated cell monitoring CellZscope® system (NanoAnalytics, Münster, Germany). The CellZscope® system provides noninvasive, continuous monitoring of cell monolayers impedance, capacitance, and resistance.

The tissue culture inserts were placed into a 12-well cell module, and the system was incubated at 37 °C and 5% CO2. TER measurements, expressed in ohm square centimeters, were recorded in real time every 20 min. Insert background resistance was automatically subtracted by the CellZscope® system. DMEM in both the apical and basal compartments was replaced with fresh DMEM (control) or DMEM plus 400 µM quercetin at 72 h, at which point mature tight junctions were established and stabilized (Resistance ≅ 70 $\Omega{\cdot}cm^2$). Measurements were recorded in real time every 20 min for at least 90 h following treatment. Each time the machine was removed from the incubator or the media was changed, a small increase in TER is expected which then stabilizes.

4.4. Immunofluorescence Staining

Cells were seeded on coverslips in 12-well plates at a cell density of 0.1×10^6 cells/well. Once confluence and TJ maturity were achieved (≈ 72 h), cells were treated with DMEM (control) or DMEM plus 400 µM quercetin and collected at 1, 6, 24, or 48 h. Cell layers were rinsed with phosphate buffer saline (PBS) and fixed in 10% trichloroacetic acid for 15 min at 4 °C. Cells were then blocked with 10% normal goat serum in phosphate buffer saline and 0.3% Triton-100 (PBST) and incubated overnight at 4 °C with 5% normal goat serum in PBST and primary antibodies: Cldn1, -2, -3, and -4 (1:100), Cldn7 (1:50), and ZO1 (1:150). Cells were washed with PBS. Alexa Fluor-conjugated secondary antibodies (1:500) in PBST were added for 1 h at RT. Coverslips were washed with PBS and then placed on slides with Slowfade Gold with DAPI (Invitrogen, Carlsbad, CA, USA). Z-stacks were imaged using a Zeiss LSM780 laser scanning confocal microscope. To determine the amount of claudin localized in the membrane, maximum intensity projections were obtained for each stack of images and then ZEN software colocalization analysis was performed. Only M2 results are presented; M2 makes reference to how many of the red pixels (ZO1) coincide with green pixels (claudin). A high M2 indicates high claudin at the membrane while a low M2 indicates high ZO1 alone, which relates to a low claudin at the membrane. M2 is not a reading of abundance or intensity levels and therefore it only provides information about localization.

4.5. Western Blot Analysis

Cells were seeded in 10 cm plates at a cell density of 2×10^6 cells/plate. Once confluence and TJ maturity was reached (≈ 72 h), cells were treated with DMEM (control) or DMEM plus 400 µM quercetin for 1, 6, 24, or 48 h, and then collected by physical scraping into lysis buffer (25 mM Tris-HCl Ph7.4; 10 mM sodium pyrophosphate; 25 mM NaCl; 10 mM sodium fluoride; 2 mM EGTA; 2 mM EDTA; 1% NP40; 0.1% SDS; 1.45 nM Pepstatin A; 2.1 nM Leupeptin; 0.15 nM Aprotinin, and 0.57 µM PMSF). Protein concentration was determined by Bradford assay and 50 µg per sample were separated by 12.5% SDS-PAGE and transferred onto PVDF membranes. Membranes were blocked for 1 h with 5% non-fat milk in PBST or with 5% BSA in Tris-NaCl buffer and 0.3% of Tween-20 (TBST). Membranes were incubated overnight at 4 °C with primary antibodies in PBST or TBST: Cldn1, -2, -3, -4, -7, pPI3K, pAKT, and AKT (1:1000) or pan-actin (1:2000). Membranes were then washed with PBST or TBST and then incubated with goat-anti-rabbit-HRT conjugated or goat-anti-mouse-peroxidase conjugated (1:5000) secondary antibodies for 1h at RT. Membranes were revealed using Clarity Western ECL substrate (Biorad laboratories, 1705061, Hercules, CA, USA), and imaged using Amersham imager 600 (GE Healthcare, Little Chalfont, United Kingdom). Band densities were quantified by densitometry using ImageJ, normalized against the 1 h control sample on each blot and then to normalized actin. Because some variability was observed in actin, we performed a statistical analysis of actin expression from all experiments and determined that there was no significant change over time or in response to quercetin treatment.

4.6. Statistics

All plotted results are mean ± SEM. Statistical analyses were performed using Graph-Pad PRISM (version 8.02, GraphPad Software, Inc., California, USA). Two-way ANOVA analysis was used for effects of treatment, time, and interaction. Sidak's multiple comparison test was used to compare the effect of treatment at the different time points. For TER Figure S1B, unpaired T-test were performed at the specific time points. Significance is considered if $p < 0.05$.

Supplementary Materials: Supplementary materials can be found at http://www.mdpi.com/1422-0067/20/19/4889/s1.

Author Contributions: Conceptualization, E.G.-E., S.A., B.J.-C., I.G. and A.K.R.; Formal analysis, E.G.-E.; Funding acquisition, S.A., I.G. and A.K.R.; Investigation, E.G.-E., I.G. and A.K.R.; Methodology, E.G.-E., I.G. and A.K.R.; Writing—original draft, E.G.-E.; Writing—review and editing, I.G. and A.K.R.

Funding: This work was supported by a New Directions Grant (Montreal Children's Hospital Foundation; A.K.R., I.G.), SGBKids and Montreal Children's Hospital Foundation (I.G., A.K.R.) and the Kidney Foundation of Canada (I.G., A.K.R., S.A.). A.K.R., I.G. and B.J.-C. are members of the RI-MUHC, which is supported in part by the FRQS.

Acknowledgments: We thank L. Jerome-Majewska, S. Kumar, L. McCaffrey and Y. Yamanaka for helpful discussions, A. Naumova for help with the statistical analysis, and M. Fu and Shi-Bo Feng (RI-MUHC Imaging Platform) for technical assistance. EGE is the recipient of a doctoral studentship from Fonds de recherche du Québec – Santé (FRQS). EGE has also been supported by CRRD, RI-MUHC and McGill University Faculty of Medicine studentships.

Conflicts of Interest: The authors declare no conflict of interest.

References

1. Walker, V. Phosphaturia in kidney stone formers: Still an enigma. *Adv. Clin. Chem.* **2019**, *90*, 133–196. [CrossRef] [PubMed]
2. Khan, S.R.; Pearle, M.S.; Robertson, W.G.; Gambaro, G.; Canales, B.K.; Doizi, S.; Traxer, O.; Tiselius, H.G. Kidney stones. *Nat. Rev. Dis. Primers* **2016**, *2*, 16008. [CrossRef] [PubMed]
3. Evan, A.P. Physiopathology and etiology of stone formation in the kidney and the urinary tract. *Pediatr. Nephrol.* **2010**, *25*, 831–841. [CrossRef] [PubMed]
4. Anderson, J.M.; Van Itallie, C.M. Physiology and function of the tight junction. *Cold Spring Harb. Perspect. Biol.* **2009**, *1*, a002584. [CrossRef] [PubMed]
5. Van Itallie, C.M.; Anderson, J.M. Claudins and epithelial paracellular transport. *Annu. Rev. Physiol.* **2006**, *68*, 403–429. [CrossRef] [PubMed]
6. Gupta, I.R.; Ryan, A.K. Claudins: Unlocking the code to tight junction function during embryogenesis and in disease. *Clin. Genet.* **2010**, *77*, 314–325. [CrossRef] [PubMed]
7. Tsukita, S.; Tanaka, H.; Tamura, A. The Claudins: From Tight Junctions to Biological Systems. *Trends Biochem. Sci.* **2019**, *44*, 141–152. [CrossRef]
8. Yu, A.S. Claudins and the kidney. *J. Am. Soc. Nephrol.* **2015**, *26*, 11–19. [CrossRef] [PubMed]
9. Simon, D.B.; Lu, Y.; Choate, K.A.; Velazquez, H.; Al-Sabban, E.; Praga, M.; Casari, G.; Bettinelli, A.; Colussi, G.; Rodriguez-Soriano, J.; et al. Paracellin-1, a renal tight junction protein required for paracellular Mg^{2+} resorption. *Science* **1999**, *285*, 103–106. [CrossRef]
10. Konrad, M.; Schaller, A.; Seelow, D.; Pandey, A.V.; Waldegger, S.; Lesslauer, A.; Vitzthum, H.; Suzuki, Y.; Luk, J.M.; Becker, C.; et al. Mutations in the tight-junction gene claudin 19 (CLDN19) are associated with renal magnesium wasting, renal failure, and severe ocular involvement. *Am. J. Hum. Genet.* **2006**, *79*, 949–957. [CrossRef]
11. Thorleifsson, G.; Holm, H.; Edvardsson, V.; Walters, G.B.; Styrkarsdottir, U.; Gudbjartsson, D.F.; Sulem, P.; Halldorsson, B.V.; de Vegt, F.; d'Ancona, F.C.; et al. Sequence variants in the CLDN14 gene associate with kidney stones and bone mineral density. *Nat. Genet.* **2009**, *41*, 926–930. [CrossRef] [PubMed]
12. Breiderhoff, T.; Himmerkus, N.; Stuiver, M.; Mutig, K.; Will, C.; Meij, I.C.; Bachmann, S.; Bleich, M.; Willnow, T.E.; Muller, D. Deletion of claudin-10 (Cldn10) in the thick ascending limb impairs paracellular sodium permeability and leads to hypermagnesemia and nephrocalcinosis. *Proc. Natl. Acad. Sci. USA* **2012**, *109*, 14241–14246. [CrossRef] [PubMed]
13. Muto, S.; Hata, M.; Taniguchi, J.; Tsuruoka, S.; Moriwaki, K.; Saitou, M.; Furuse, K.; Sasaki, H.; Fujimura, A.; Imai, M.; et al. Claudin-2-deficient mice are defective in the leaky and cation-selective paracellular permeability properties of renal proximal tubules. *Proc. Natl. Acad. Sci. USA* **2010**, *107*, 8011–8016. [CrossRef] [PubMed]
14. Himmerkus, N.; Shan, Q.; Goerke, B.; Hou, J.; Goodenough, D.A.; Bleich, M. Salt and acid-base metabolism in claudin-16 knockdown mice: Impact for the pathophysiology of FHHNC patients. *Am. J. Physiol. Renal Physiol.* **2008**, *295*, F1641–F1647. [CrossRef] [PubMed]
15. Hou, J.; Renigunta, A.; Gomes, A.S.; Hou, M.; Paul, D.L.; Waldegger, S.; Goodenough, D.A. Claudin-16 and claudin-19 interaction is required for their assembly into tight junctions and for renal reabsorption of magnesium. *Proc. Natl. Acad. Sci. USA* **2009**, *106*, 15350–15355. [CrossRef] [PubMed]
16. Tatum, R.; Zhang, Y.; Salleng, K.; Lu, Z.; Lin, J.J.; Lu, Q.; Jeansonne, B.G.; Ding, L.; Chen, Y.H. Renal salt wasting and chronic dehydration in claudin-7-deficient mice. *Am. J. Physiol. Renal Physiol.* **2010**, *298*, F24–F34. [CrossRef]
17. Plain, A.; Alexander, R.T. Claudins and nephrolithiasis. *Curr. Opin. Nephrol. Hypertens.* **2018**, *27*, 268–276. [CrossRef]
18. Gong, Y.; Hou, J. Claudins in barrier and transport function-the kidney. *Pflugers Arch.* **2017**, *469*, 105–113. [CrossRef]
19. Wakino, S.; Hasegawa, K.; Itoh, H. Sirtuin and metabolic kidney disease. *Kidney Int.* **2015**, *88*, 691–698. [CrossRef]
20. Zhu, W.; Xu, Y.F.; Feng, Y.; Peng, B.; Che, J.P.; Liu, M.; Zheng, J.H. Prophylactic effects of quercetin and hyperoside in a calcium oxalate stone forming rat model. *Urolithiasis* **2014**, *42*, 519–526. [CrossRef]

21. Mercado, J.; Valenzano, M.C.; Jeffers, C.; Sedlak, J.; Cugliari, M.K.; Papanikolaou, E.; Clouse, J.; Miao, J.; Wertan, N.E.; Mullin, J.M. Enhancement of tight junctional barrier function by micronutrients: Compound-specific effects on permeability and claudin composition. *PLoS ONE* **2013**, *8*, e78775. [CrossRef] [PubMed]
22. Valenzano, M.C.; DiGuilio, K.; Mercado, J.; Teter, M.; To, J.; Ferraro, B.; Mixson, B.; Manley, I.; Baker, V.; Moore, B.A.; et al. Remodeling of Tight Junctions and Enhancement of Barrier Integrity of the CACO-2 Intestinal Epithelial Cell Layer by Micronutrients. *PLoS ONE* **2015**, *10*, e0133926. [CrossRef] [PubMed]
23. Suzuki, T.; Hara, H. Quercetin enhances intestinal barrier function through the assembly of zonula [corrected] occludens-2, occludin, and claudin-1 and the expression of claudin-4 in Caco-2 cells. *J. Nutr.* **2009**, *139*, 965–974. [CrossRef] [PubMed]
24. Suzuki, T.; Tanabe, S.; Hara, H. Kaempferol enhances intestinal barrier function through the cytoskeletal association and expression of tight junction proteins in Caco-2 cells. *J. Nutr.* **2011**, *141*, 87–94. [CrossRef]
25. Dukes, J.D.; Whitley, P.; Chalmers, A.D. The MDCK variety pack: Choosing the right strain. *BMC Cell Biol.* **2011**, *12*, 43. [CrossRef] [PubMed]
26. Hou, J.; Gomes, A.S.; Paul, D.L.; Goodenough, D.A. Study of claudin function by RNA interference. *J. Biol. Chem.* **2006**, *281*, 36117–36123. [CrossRef]
27. Gunzel, D.; Fromm, M. Claudins and other tight junction proteins. *Compr. Physiol.* **2012**, *2*, 1819–1852. [CrossRef] [PubMed]
28. Borovac, J.; Barker, R.S.; Rievaj, J.; Rasmussen, A.; Pan, W.; Wevrick, R.; Alexander, R.T. Claudin-4 forms a paracellular barrier, revealing the interdependence of claudin expression in the loose epithelial cell culture model opossum kidney cells. *Am. J. Physiol. Cell Physiol.* **2012**, *303*, C1278–C1291. [CrossRef]
29. Rosenthal, R.; Gunzel, D.; Krug, S.M.; Schulzke, J.D.; Fromm, M.; Yu, A.S. Claudin-2-mediated cation and water transport share a common pore. *Acta Physiol.* **2017**, *219*, 521–536. [CrossRef]
30. Van Itallie, C.M.; Fanning, A.S.; Anderson, J.M. Reversal of charge selectivity in cation or anion-selective epithelial lines by expression of different claudins. *Am. J. Physiol. Renal Physiol.* **2003**, *285*, F1078–F1084. [CrossRef]
31. Tokuda, S.; Furuse, M. Claudin-2 knockout by TALEN-mediated gene targeting in MDCK cells: Claudin-2 independently determines the leaky property of tight junctions in MDCK cells. *PLoS ONE* **2015**, *10*, e0119869. [CrossRef] [PubMed]
32. Srinivasan, B.; Kolli, A.R.; Esch, M.B.; Abaci, H.E.; Shuler, M.L.; Hickman, J.J. TEER measurement techniques for in vitro barrier model systems. *J. Lab. Autom.* **2015**, *20*, 107–126. [CrossRef] [PubMed]
33. Dembla, S.; Hasan, N.; Becker, A.; Beck, A.; Philipp, S.E. Transient receptor potential A1 channels regulate epithelial cell barriers formed by MDCK cells. *FEBS Lett.* **2016**, *590*, 1509–1520. [CrossRef] [PubMed]
34. Amoozadeh, Y.; Anwer, S.; Dan, Q.; Venugopal, S.; Shi, Y.; Branchard, E.; Liedtke, E.; Ailenberg, M.; Rotstein, O.D.; Kapus, A.; et al. Cell confluence regulates claudin-2 expression: Possible role for ZO-1 and Rac. *Am. J. Physiol. Cell Physiol.* **2018**, *314*, C366–C378. [CrossRef] [PubMed]
35. Welker, M.E.; Kulik, G. Recent syntheses of PI3K/Akt/mTOR signaling pathway inhibitors. *Bioorg. Med. Chem.* **2013**, *21*, 4063–4091. [CrossRef]
36. Jiang, W.; Luo, T.; Li, S.; Zhou, Y.; Shen, X.Y.; He, F.; Xu, J.; Wang, H.Q. Quercetin Protects against Okadaic Acid-Induced Injury via MAPK and PI3K/Akt/GSK3beta Signaling Pathways in HT22 Hippocampal Neurons. *PLoS ONE* **2016**, *11*, e0152371. [CrossRef]
37. Khan, N.; Binder, L.; Pantakani, D.V.K.; Asif, A.R. MPA Modulates Tight Junctions' Permeability via Midkine/PI3K Pathway in Caco-2 Cells: A Possible Mechanism of Leak-Flux Diarrhea in Organ Transplanted Patients. *Front. Physiol.* **2017**, *8*, 438. [CrossRef]
38. Lin, X.; Shang, X.; Manorek, G.; Howell, S.B. Regulation of the Epithelial-Mesenchymal Transition by Claudin-3 and Claudin-4. *PLoS ONE* **2013**, *8*, e67496. [CrossRef]
39. Piegholdt, S.; Pallauf, K.; Esatbeyoglu, T.; Speck, N.; Reiss, K.; Ruddigkeit, L.; Stocker, A.; Huebbe, P.; Rimbach, G. Biochanin A and prunetin improve epithelial barrier function in intestinal CaCo-2 cells via downregulation of ERK, NF-kappaB, and tyrosine phosphorylation. *Free Radic. Biol. Med.* **2014**, *70*, 255–264. [CrossRef]
40. Horng, S.; Therattil, A.; Moyon, S.; Gordon, A.; Kim, K.; Argaw, A.T.; Hara, Y.; Mariani, J.N.; Sawai, S.; Flodby, P.; et al. Astrocytic tight junctions control inflammatory CNS lesion pathogenesis. *J. Clin. Investig.* **2017**, *127*, 3136–3151. [CrossRef]

41. Aono, S.; Hirai, Y. Phosphorylation of claudin-4 is required for tight junction formation in a human keratinocyte cell line. *Exp. Cell Res.* **2008**, *314*, 3326–3339. [CrossRef] [PubMed]
42. Fujibe, M.; Chiba, H.; Kojima, T.; Soma, T.; Wada, T.; Yamashita, T.; Sawada, N. Thr203 of claudin-1, a putative phosphorylation site for MAP kinase, is required to promote the barrier function of tight junctions. *Exp. Cell Res.* **2004**, *295*, 36–47. [CrossRef] [PubMed]
43. Shigetomi, K.; Ikenouchi, J. Regulation of the epithelial barrier by post-translational modifications of tight junction membrane proteins. *J. Biochem.* **2018**, *163*, 265–272. [CrossRef] [PubMed]
44. Fujii, N.; Matsuo, Y.; Matsunaga, T.; Endo, S.; Sakai, H.; Yamaguchi, M.; Yamazaki, Y.; Sugatani, J.; Ikari, A. Hypotonic Stress-induced Down-regulation of Claudin-1 and -2 Mediated by Dephosphorylation and Clathrin-dependent Endocytosis in Renal Tubular Epithelial Cells. *J. Biol. Chem.* **2016**, *291*, 24787–24799. [CrossRef] [PubMed]
45. Van Itallie, C.M.; Tietgens, A.J.; LoGrande, K.; Aponte, A.; Gucek, M.; Anderson, J.M. Phosphorylation of claudin-2 on serine 208 promotes membrane retention and reduces trafficking to lysosomes. *J. Cell Sci.* **2012**, *125*, 4902–4912. [CrossRef] [PubMed]
46. Tanaka, M.; Kamata, R.; Sakai, R. EphA2 phosphorylates the cytoplasmic tail of Claudin-4 and mediates paracellular permeability. *J. Biol. Chem.* **2005**, *280*, 42375–42382. [CrossRef] [PubMed]
47. D'Souza, T.; Indig, F.E.; Morin, P.J. Phosphorylation of claudin-4 by PKCepsilon regulates tight junction barrier function in ovarian cancer cells. *Exp. Cell Res.* **2007**, *313*, 3364–3375. [CrossRef] [PubMed]
48. Zhu, Y.; Teng, T.; Wang, H.; Guo, H.; Du, L.; Yang, B.; Yin, X.; Sun, Y. Quercetin inhibits renal cyst growth in vitro and via parenteral injection in a polycystic kidney disease mouse model. *Food Funct.* **2018**, *9*, 389–396. [CrossRef]
49. Jiang, H.; Yamashita, Y.; Nakamura, A.; Croft, K.; Ashida, H. Quercetin and its metabolite isorhamnetin promote glucose uptake through different signalling pathways in myotubes. *Sci. Rep.* **2019**, *9*, 2690. [CrossRef]
50. Gamero-Estevez, E.; Baumholtz, A.I.; Ryan, A.K. Developing a link between toxicants, claudins and neural tube defects. *Reprod. Toxicol.* **2018**, *81*, 155–167. [CrossRef]
51. Chuenkitiyanon, S.; Pengsuparp, T.; Jianmongkol, S. Protective effect of quercetin on hydrogen peroxide-induced tight junction disruption. *Int. J. Toxicol.* **2010**, *29*, 418–424. [CrossRef] [PubMed]
52. Tan, C.; Meng, F.; Reece, E.A.; Zhao, Z. Modulation of nuclear factor-kappaB signaling and reduction of neural tube defects by quercetin-3-glucoside in embryos of diabetic mice. *Am. J. Obstet. Gynecol.* **2018**, *219*, 197.e1–197.e8. [CrossRef] [PubMed]
53. Maurya, A.K.; Vinayak, M. PI-103 and Quercetin Attenuate PI3K-AKT Signaling Pathway in T- Cell Lymphoma Exposed to Hydrogen Peroxide. *PLoS ONE* **2016**, *11*, e0160686. [CrossRef] [PubMed]
54. Wang, X.Q.; Yao, R.Q.; Liu, X.; Huang, J.J.; Qi, D.S.; Yang, L.H. Quercetin protects oligodendrocyte precursor cells from oxygen/glucose deprivation injury in vitro via the activation of the PI3K/Akt signaling pathway. *Brain Res. Bull.* **2011**, *86*, 277–284. [CrossRef] [PubMed]
55. Du, G.; Zhao, Z.; Chen, Y.; Li, Z.; Tian, Y.; Liu, Z.; Liu, B.; Song, J. Quercetin attenuates neuronal autophagy and apoptosis in rat traumatic brain injury model via activation of PI3K/Akt signaling pathway. *Neurol. Res.* **2016**, *38*, 1012–1019. [CrossRef] [PubMed]
56. Wu, L.; Zhang, Q.; Dai, W.; Li, S.; Feng, J.; Li, J.; Liu, T.; Xu, S.; Wang, W.; Lu, X.; et al. Quercetin Pretreatment Attenuates Hepatic Ischemia Reperfusion-Induced Apoptosis and Autophagy by Inhibiting ERK/NF-kappaB Pathway. *Gastroenterol. Res. Pract.* **2017**, *2017*, 9724217. [CrossRef] [PubMed]
57. Qiu, L.; Luo, Y.; Chen, X. Quercetin attenuates mitochondrial dysfunction and biogenesis via upregulated AMPK/SIRT1 signaling pathway in OA rats. *Biomed. Pharmacother.* **2018**, *103*, 1585–1591. [CrossRef]
58. Bird, V.Y.; Khan, S.R. How do stones form? Is unification of theories on stone formation possible? *Arch. Esp. Urol.* **2017**, *70*, 12–27. [PubMed]
59. Coe, F.L.; Evan, A.P.; Worcester, E.M.; Lingeman, J.E. Three pathways for human kidney stone formation. *Urol. Res.* **2010**, *38*, 147–160. [CrossRef] [PubMed]
60. Tokuda, S.; Hirai, T.; Furuse, M. Effects of Osmolality on Paracellular Transport in MDCK II Cells. *PLoS ONE* **2016**, *11*, e0166904. [CrossRef]

International Journal of
Molecular Sciences

Article

Curcumin Mitigates Immune-Induced Epithelial Barrier Dysfunction by *Campylobacter jejuni*

Fábia Daniela Lobo de Sá [1], Eduard Butkevych [1], Praveen Kumar Nattramilarasu [1], Anja Fromm [1], Soraya Mousavi [2], Verena Moos [3], Julia C. Golz [4], Kerstin Stingl [4], Sophie Kittler [5], Diana Seinige [5], Corinna Kehrenberg [6], Markus M. Heimesaat [2], Stefan Bereswill [2], Jörg-Dieter Schulzke [1,*] and Roland Bücker [1]

1. Institute of Clinical Physiology/Nutritional Medicine, Medical Department, Division of Gastroenterology, Infectiology, Rheumatology, Charité–Universitätsmedizin Berlin, 12203 Berlin, Germany
2. Institute of Microbiology, Infectious Diseases and Immunology, Charité–Universitätsmedizin Berlin, Campus Benjamin Franklin, 14195 Berlin, Germany
3. Medical Department, Division of Gastroenterology, Infectiology and Rheumatology, Charité–Universitätsmedizin Berlin, 12203 Berlin, Germany
4. German Federal Institute for Risk Assessment (BfR), Department of Biological Safety, National Reference Laboratory for Campylobacter, 12277 Berlin, Germany
5. University of Veterinary Medicine Hannover, Research Center for Emerging Infections and Zoonoses, 30559 Hannover, Germany
6. Institute for Veterinary Food Science, Justus-Liebig-University, 35392 Giessen, Germany

* Correspondence: joerg.schulzke@charite.de

Received: 5 August 2019; Accepted: 26 September 2019; Published: 28 September 2019

Abstract: *Campylobacter jejuni* (*C. jejuni*) is the most common cause of foodborne gastroenteritis worldwide. The bacteria induce diarrhea and inflammation by invading the intestinal epithelium. Curcumin is a natural polyphenol from turmeric rhizome of *Curcuma longa*, a medical plant, and is commonly used in curry powder. The aim of this study was the investigation of the protective effects of curcumin against immune-induced epithelial barrier dysfunction in *C. jejuni* infection. The indirect *C. jejuni*-induced barrier defects and its protection by curcumin were analyzed in co-cultures with HT-29/B6-GR/MR epithelial cells together with differentiated THP-1 immune cells. Electrophysiological measurements revealed a reduction in transepithelial electrical resistance (TER) in infected co-cultures. An increase in fluorescein (332 Da) permeability in co-cultures as well as in the germ-free IL-$10^{-/-}$ mouse model after *C. jejuni* infection was shown. Curcumin treatment attenuated the *C. jejuni*-induced increase in fluorescein permeability in both models. Moreover, apoptosis induction, tight junction redistribution, and an increased inflammatory response—represented by TNF-α, IL-1β, and IL-6 secretion—was observed in co-cultures after infection and reversed by curcumin. In conclusion, curcumin protects against indirect *C. jejuni*-triggered immune-induced barrier defects and might be a therapeutic and protective agent in patients.

Keywords: *Campylobacter jejuni*; curcumin; tight junction; claudin; apoptosis; co-culture; mouse colon; cytokines; TNF; NFκB

1. Introduction

Campylobacter jejuni is the most prevalent pathogenic bacterium of zoonotic gastroenteritis [1]. Typical symptoms provoked by *C. jejuni* are watery to bloody diarrhea, abdominal pain, fever, and nausea [2]. This human pathogen is present in the intestinal microbiota of farm animals, especially poultry, which is the main source of infection for humans by ingestion of contaminated or undercooked food [1]. The bacteria adhere to the mucus and the surface of intestinal epithelial cells, invade the intestinal epithelium, while they pass the cells via the transcellular or paracellular route [3,4]. Consequently, direct epithelial barrier defects such as dysregulation of tight junction (TJ) proteins, induction of epithelial lesions or also indirect effects by inflammatory responses of epithelial or immune cells occur [5,6].

Epithelial lesions can go along with single-cell lesions by increased apoptosis, mid-sized leaks like focal leaks, as well as erosions or even ulcerations. These pathological findings could explain the type of diarrhea for the *Campylobacter* infection as leak flux pathomechanism, which is characterized by a loss of water and solutes from the organism into the intestinal lumen through a leaky epithelium [6,7]. For a physiological intestinal integrity, an intact epithelium is essential. The main components of intestinal epithelial barrier are the TJs. They are the apical components of the intercellular seal and divide the epithelial cell membrane into an apical and basolateral compartment (fence function) [8]. Tight junctions form a barrier for water, ions, and solutes (barrier function) [9] and consist of various molecular transmembrane proteins like claudins and occludin, junction adhesion molecules (JAM), tricellulin, and cytoplasmatic scaffolding proteins as zonula occludens protein-1 (ZO-1) [8,9].

Recently, the number of *C. jejuni* infections surpasses *Salmonella* infections as the leading cause of zoonotic bacterial gastroenteritis in the US and Europe. Moreover, high levels of fluoroquinolone resistance have been reported [10] and levels of multi drug resistances (MDR) are increasing [11]. In consequence, new therapeutic targets as well as innovative strategies to reduce the amount of *C. jejuni* in farm animals and environment are urgently required. A potent natural substance thought to interfere with *Campylobacter* infection is curcumin.

Curcumin is a polyphenolic compound found in the turmeric root of the *Curcuma longa* plant (commonly known as turmeric), a member of the ginger family [12–14]. It is used as coloring agent, cosmetics, and it is the principal ingredient in oriental food spice like curry powder [12,13]. For centuries, curcumin has been used in Asian traditional medicine as a common nontoxic and active agent against different gastrointestinal and digestive disorders [12,14,15]. In addition, curcumin is known to possess anti-inflammatory, antioxidative, antibacterial, anticarcinogenic [16], antiviral [15,17], antiapoptotic, and antiproliferative properties [12]. The antibacterial effect was shown against *Salmonella* spp. [18], *Helicobacter pylori* [16,19,20], methicillin-resistant *Staphylococcus aureus*, *Bacillus subtilis*, and *Escherichia coli* [21]. Various studies have shown that curcumin modulates inflammatory cytokine pathways like IL-1β, TNF-α, IL-6, IL-8, IL-10 [12,13,22,23], and inhibit signaling pathways like NFκB [12,17,19].

The aim of the presented study was to evaluate the barrier-protective and anti-inflammatory properties of curcumin on intestinal epithelial barrier function, which is compromised by the *C. jejuni*-mediated immune response. Therefore, a novel in vitro co-culture model consisting of intestinal epithelial HT-29/B6-GR/MR and immune THP-1 cells was established. This co-culture model enables the screening of potential barrier-protective and anti-inflammatory compounds in *C. jejuni* infection, as a result of which curcumin was identified as being effective.

2. Results

2.1. Establishment of a Co-Culture with Colon Epithelial Cells HT-29/B6-GR/MR and Immune THP-1 Cells

To analyze the impact of the immune response during a *Campylobacter* infection on epithelial barrier function, human colon epithelial cells and immune cells were co-cultured. Colonic epithelial cells (HT-29/B6-GR/MR) were seeded on cell culture filter supports (0.4 µm pore size) and represented the epithelial compartment. Under the membrane of the epithelial cell filter, THP-1 immune cells were seeded on the bottom of the cell culture well and represented the subepithelial compartment. Basal bacterial infection with *C. jejuni* decreased transepithelial electrical resistance (TER) when colonic epithelial and immune cells (differentiated THP-1 cells) were co-cultured (Figure 1). From the apical supernatant, no bacteria could be cultured after the experiment, indicating that the bacteria could not pass the filter membrane with 0.4 µm pore size. As a result of this, and in contrast to the results in the co-culture model, *C. jejuni* infection at the basal membrane had no effect on TER in HT-29/B6-GR/MR mono-culture (Figure 1).

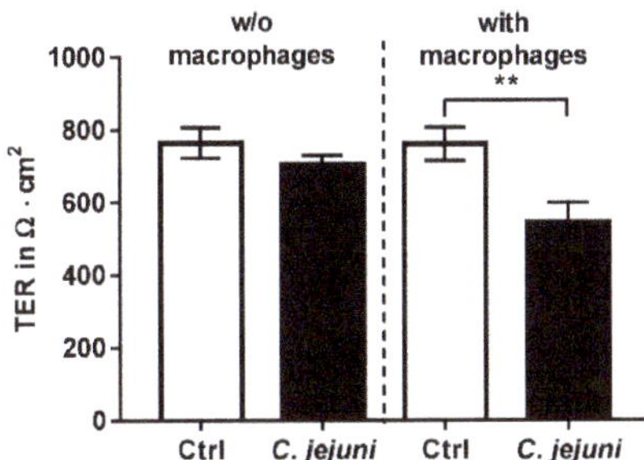

Figure 1. *C. jejuni* infection in human intestinal epithelial cells (HT-29/B6-GR/MR) or in co-culture together with human immune cells (differentiated THP-1 cells). Transepithelial electrical resistance (TER) in co-culture after *C. jejuni* infection from the basal side decreased 48 h post infection ($n = 7–9$, ** $p < 0.01$, unpaired Student's *t*-test).

2.2. Curcumin Improves Disturbed Intestinal Barrier Function in the Co-Culture System

Using mRNA sequencing and subsequent ingenuity pathway analysis (IPA) of human mucosa biopsies from *C. jejuni*-infected patients, inhibition of curcumin-dependent pathways indicated that the presence of curcumin might prevent the *C. jejuni*-induced changes in gene expression. Based on this bioinformatic prediction of curcumin's effect on host gene expression, the addition of curcumin should activate downstream pathways and might reduce the effects of *Campylobacter* infections (Supplementary Table S1). In order to test the barrier protective and anti-inflammatory properties of curcumin in *Campylobacter*-induced barrier dysfunction, a curcumin solution was added to the infected co-culture model. As shown in Figure 2, *C. jejuni* reduced TER significantly 48 h after infection, whereas this change was inhibited by the presence of curcumin.

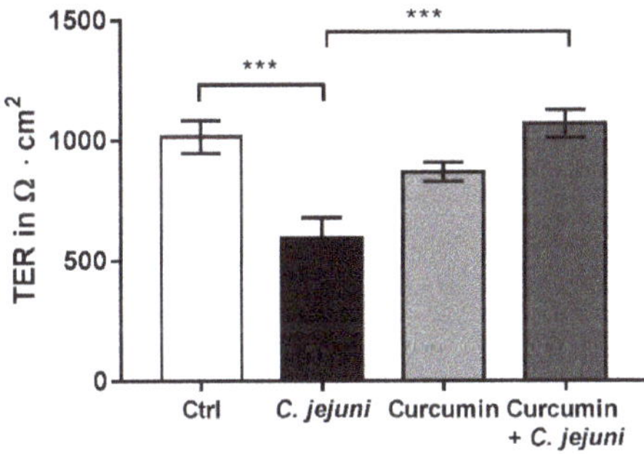

Figure 2. Curcumin protects against *C. jejuni*-induced barrier disruption in co-culture. Cells were treated with 50 μM curcumin and infected with *C. jejuni* for 48 h ($n = 13$, *** $p < 0.001$, one-way ANOVA with Bonferroni correction for multiple comparisons).

Permeability studies were performed with the paracellular flux marker fluorescein (332 Da). Compared to controls, *C. jejuni* infection caused an increase in permeability to fluorescein. Application of curcumin not only ameliorated the permeability increase for fluorescein, but even resulted in an improved barrier function for paracellular passage of macromolecules when exceeded the control level (Figure 3).

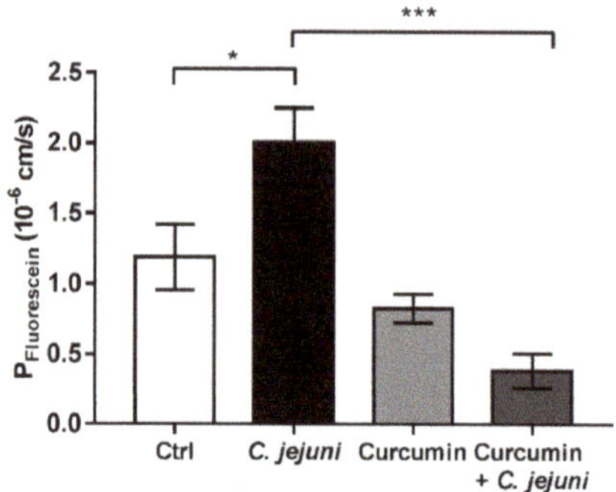

Figure 3. Curcumin improves the *C. jejuni*-induced increase in permeability to fluorescein (332 Da) in co-cultures with colon epithelial HT-29/B6-GR/MR and immune THP-1 cells. The co-cultures were infected with *C. jejuni* after incubation with 50 µM curcumin. Fluorescein permeabilities were measured 48 h post infection (n = 7–8, * $p < 0.05$, *** $p < 0.001$, one-way ANOVA with Bonferroni correction for multiple comparisons).

2.3. *Effect of Curcumin In Vivo on Intestinal Barrier Regulation in IL-10$^{-/-}$ Mice*

Secondary abiotic IL-10$^{-/-}$ mice were infected with *C. jejuni* and treated with 0.5 mg/mL of curcumin, starting 4 days prior infection until the end of the observation period (i.e., day 6 post infection). Epithelial barrier function of distal colon specimens from IL-10$^{-/-}$ mice was analyzed in Ussing chambers. Conductance (G, Figure 4A) and permeability to fluorescein (Figure 4B) was significant higher in *C. jejuni*-infected mice than in untreated infected mice. Mice receiving curcumin showed a significantly lower conductance and permeability to fluorescein than untreated infected mice. The electrophysiological results obtained in vivo (Figure 4) resemble the results from the in vitro co-culture system (Figure 3).

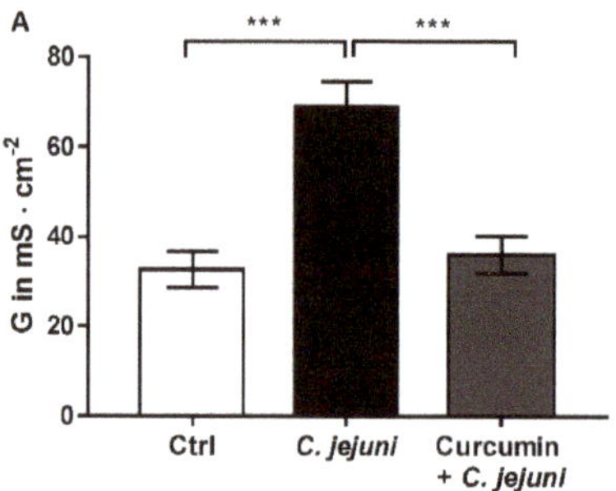

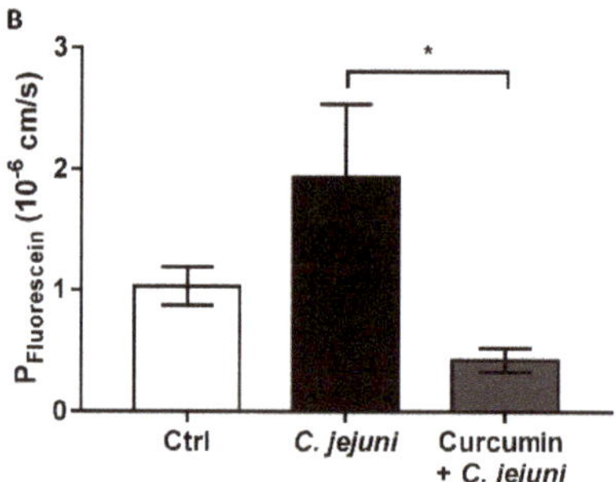

Figure 4. Curcumin ameliorates the *C. jejuni*-induced barrier dysfunction in colon of IL-10$^{-/-}$ mice. Colon specimens of IL-10$^{-/-}$ mice were mounted in Ussing chambers designed for impedance spectroscopy. (**A**) Epithelial conductance (G) (n = 5–18, *** $p < 0{,}001$, Mann–Whitney *U*-test) and (**B**) fluorescein (332 Da) permeability (n = 5–14, * $p < 0.05$, Mann–Whitney *U*-test) of IL-10$^{-/-}$ mouse colon were measured in Ussing chambers. IL-10$^{-/-}$ mice were either infected with *C. jejuni* (per oral gavage with 10 ^{8} CFU in 0.3 mL) or infected and treated for 6 days with Curcumin (0.5 mg/mL).

2.4. *Effect of Curcumin on Bacterial Integrity*

To determine whether curcumin had a direct effect on the viability of *C. jejuni* in the concentration used in our experiments, the minimal inhibitory concentration (MIC) for curcumin was determined. The MIC of curcumin in *C. jejuni* 81–176 amounted to 87 µM under physiological conditions at a pH of 7.4, and is thus above our experimental curcumin concentrations in the co-culture. Moreover, if curcumin is used as treatment, it is necessary to know whether the substance exerts any stress to the bacterium, leading to enhanced adaption by, e.g., stimulation of natural transformation capacity. Therefore, competence development and DNA uptake of *C. jejuni* was investigated under the treatment of curcumin using a single cell assay. After curcumin treatment, no change in the ratio of competent bacteria was registered (control 26% ± 2% versus treated 25.5% ± 0.5% with 50 μM curcumin, $n = 3$, n.s., unpaired Student's *t*-test). In addition, also the DNA uptake processes in competent *C. jejuni* was unaffected after curcumin treatment (control 100% versus treated 86.9% ± 6.7%, $n = 3$, n.s., unpaired Student´s *t*-test).

2.5. Cytokine Secretion in C. jejuni-Infected Co-Cultures

To identify which cytokines were released by the immune cells and affected epithelial cells function, cytometric bead arrays (CBA) from supernatants of treated co-cultures were performed. TNF-α, IL-1β, and IL-6 were increased in *C. jejuni*-infected co-cultures, while this changes were prevented by curcumin treatment (Figure 5).

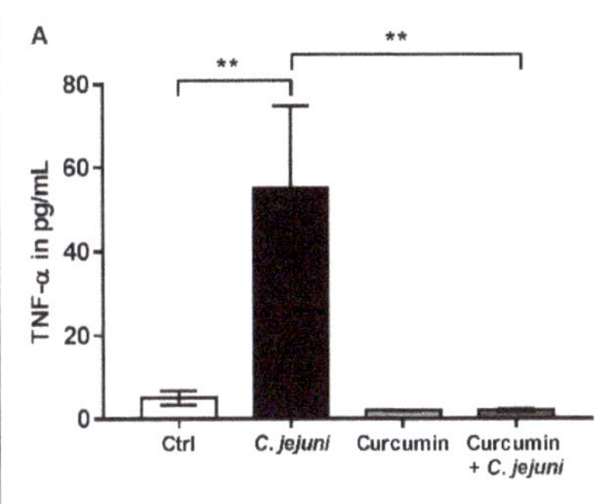

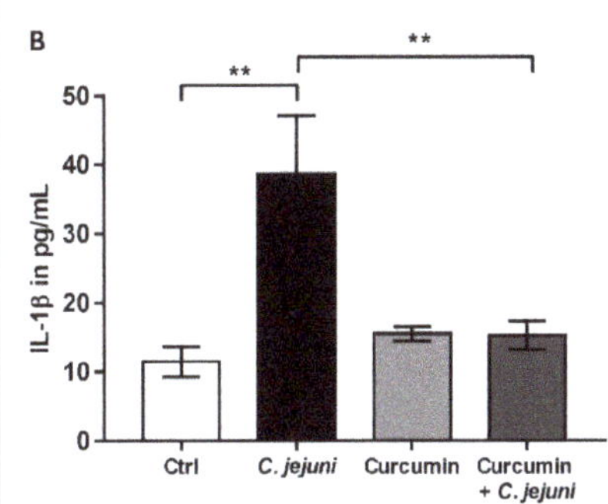

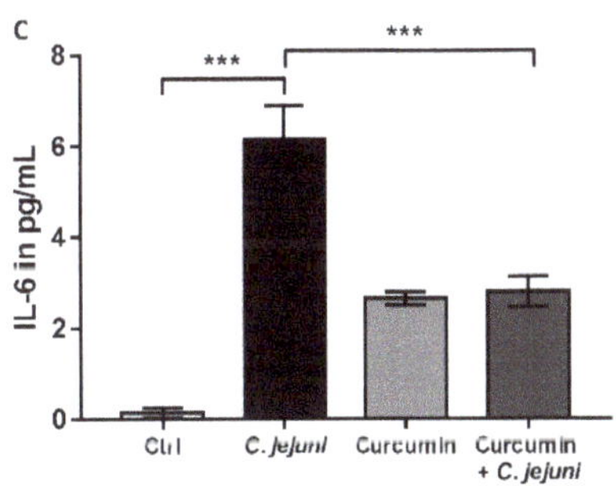

Figure 5. Cytokine release in co-cultures after *C. jejuni* infection. (**A**) Tumor necrosis factor-α (TNF-α), (**B**) Interleukin-1β (IL-1β), and (**C**) Interleukin-6 (IL-6) secretion after treatment with curcumin in *C. jejuni*-infected co-cultures ($n = 5$–6, * $p < 0.05$, ** $p < 0.01$, *** $p < 0.001$, one-way ANOVA with Bonferroni correction for multiple comparisons).

2.6. NFκB Signaling Pathway is Involved in C. jejuni-Induced Barrier Dysfunction

Furthermore, we evaluated the possible relevance of the NFκB pathway for barrier function in our cell model using the NFκB inhibitor BAY 11-7082 (BAY). BAY inhibited the *C. jejuni*-induced reduction in TER, similar to the effect of curcumin (Figure 6A). In addition, the increase in permeability to fluorescein could be reversed after BAY inhibition and almost reached control values (Figure 6B).

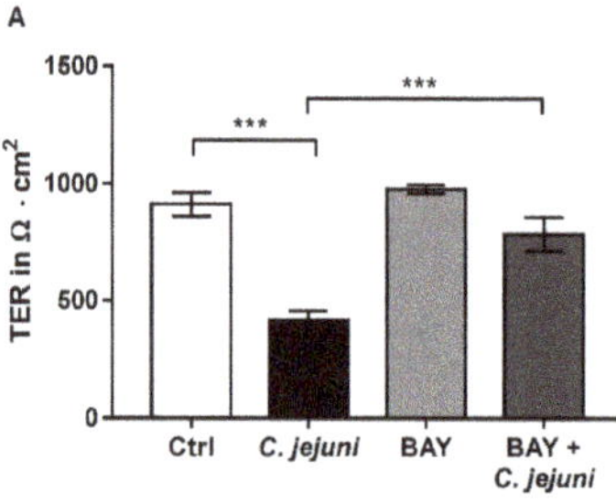

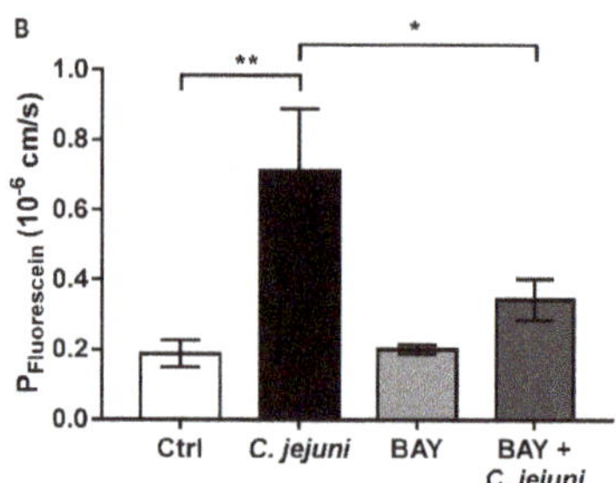

Figure 6. The NFκB pathway is barrier-relevant. Inhibitory effect of BAY 11-7082 in *C. jejuni*-infected co-cultures. (**A**) Transepithelial electrical resistance (TER) and (**B**) permeability to fluorescein (332 Da) after incubation with 10 μM BAY 11-7082 and infection with *C. jejuni* for 48 h ($n = 5$, * $p < 0.05$, ** $p < 0.01$, *** $p < 0.001$, one-way ANOVA with Bonferroni correction for multiple comparisons).

2.7. Epithelial Apoptosis is Blocked by Curcumin in the Co-Culture Model

Given that the induction of epithelial apoptosis constitutes a barrier-relevant pathomechanism, and *Campylobacter* infections have been shown to induce epithelial apoptosis, the impact of curcumin on apoptosis was analyzed using cleaved caspase-3 staining and visualized by confocal laser-scanning microscopy (CLSM) (Figure 7).

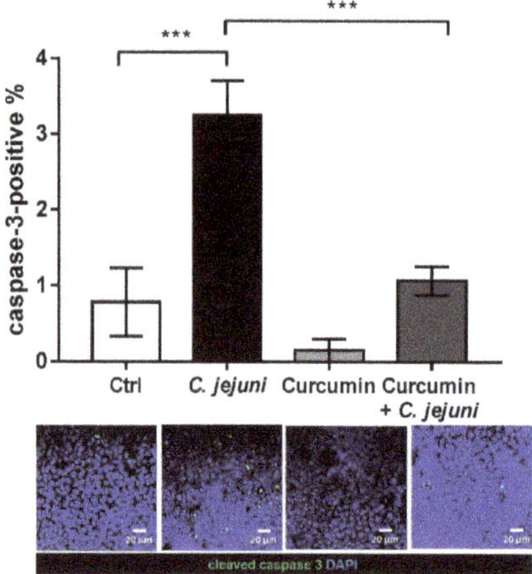

Figure 7. Apoptosis induction in *C. jejuni*-infected co-cultures. Cleaved caspase-3 staining indicates the number of apoptotic cells and is visualized by confocal microscopy. Monolayers were stained with antibodies against cleaved caspase-3 (green) and nuclei were colored with 4′-d-diamidino-2-phenylidole dihydrochloride (DAPI, blue). Signals of cleaved caspase-3-positive cells in CLSM were counted in the top view pictures and related to the number of DAPI stained nuclei (~300 nuclei/frame) to calculate the percentage of apoptotic cells in the monolayers ($n = 6$, *** $p < 0.001$, one-way ANOVA with Bonferroni correction for multiple comparisons). Representative pictures are shown for HT-29/B6-GR/MR cells, obtained from co-culture after treatment with 50 μM curcumin and infection with *C. jejuni*.

After infection with *C. jejuni*, the apoptotic ratio was increased 3-fold in comparison to control. Incubation with curcumin during the infection with *C. jejuni* reduced the apoptotic ratio to that of controls (Figure 7) (control 0.7% ± 0.4% versus infected 3.3% ± 0.4% versus treated 1.1% ± 0.2%, $n = 6$, ** $p < 0.01$, *** $p < 0.001$, one-way ANOVA with Bonferroni correction for multiple comparisons).

2.8. Influence of C. jejuni and Curcumin on Tight Junction Protein Expression in the Co-Culture Model

To define the *C. jejuni*-induced tight junction (TJ) expression changes, protein lysates were generated from colon epithelial cells of co-cultures after treatment with curcumin and infection with *C. jejuni* (Figure 8). The expression change was quantified by densitometric analysis with β-actin as loading control, which showed an increase in protein expression of claudin-1 protein expression after *C. jejuni* infection. Claudin-1 expression seemed to decrease again after curcumin treatment in the *C. jejuni* infected group. However, this did not reach statistical significance. No further expression changes of the tested TJ proteins could be observed. A reduction by trend after *C. jejuni* infection in occludin and claudin-8 was detectable, but without reaching statistical significance after multiple test correction (claudin-8 infected versus curcumin-treated $p = 0.046$, uncorrected Student´s *t*-test).

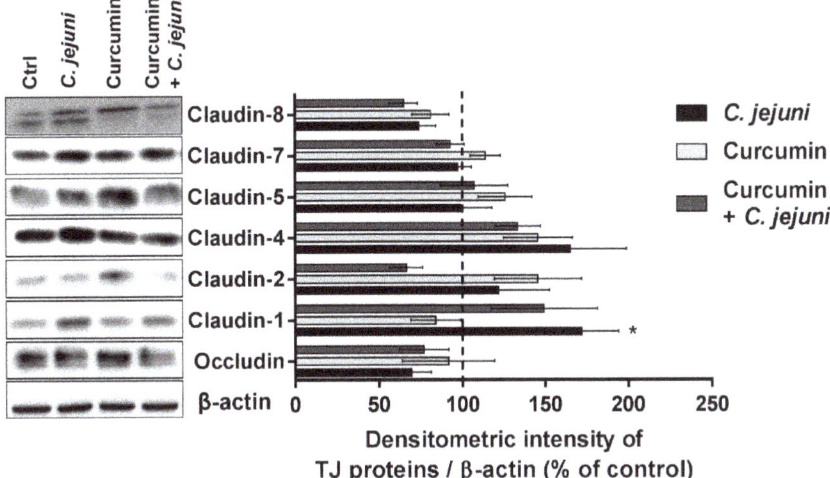

Figure 8. Tight junction (TJ) protein expression in western blot. Representative western blots and densitometry of TJ protein expression from HT-29/B6-GR/MR cells obtained from co-culture after incubation with 50 µM curcumin and infection with *C. jejuni*. TJ proteins were normalized to the level of β-actin ($n = 6$–9, * $p < 0.05$, two-way ANOVA with Bonferroni correction for multiple comparisons. Control value is marked with a dashed line.

2.9. Influence of C. jejuni and Curcumin on Subcellular TJ Protein Distribution in Co-Cultures in Confocal Laser-Scanning Microscopy

Since the composition and presence of TJ proteins in the TJ strands influence barrier function, the subcellular distribution of the TJs was analyzed with confocal laser-scanning microscopy (CLSM). Z-stacks were generated from different TJ proteins of the immunostained epithelial cell monolayers, and intensity-distance plots were generated. The TJ strands after *C. jejuni* infection indicated a redistribution of claudin-4 (Figure 9A) and claudin-8 (Figure 9B). Claudin-4 and claudin-8 signals were retracted from the TJ strands. Moreover, claudin-4 and claudin-8 were no longer co-localized with zonula occludens protein-1 (ZO-1) after *C. jejuni* infection. Intensity–distance plots with high background intensities indicate re-localization into the intracellular regions. After treatment with curcumin, the appearance of claudin-4 and claudin-8 improved. As a consequence, the infected and curcumin treated group showed co-localization (merging), with high peak intensities in the areas of TJ contact, comparable to the control. Curcumin control groups showed no difference compared to the untreated controls (Supplementary Figure S2).

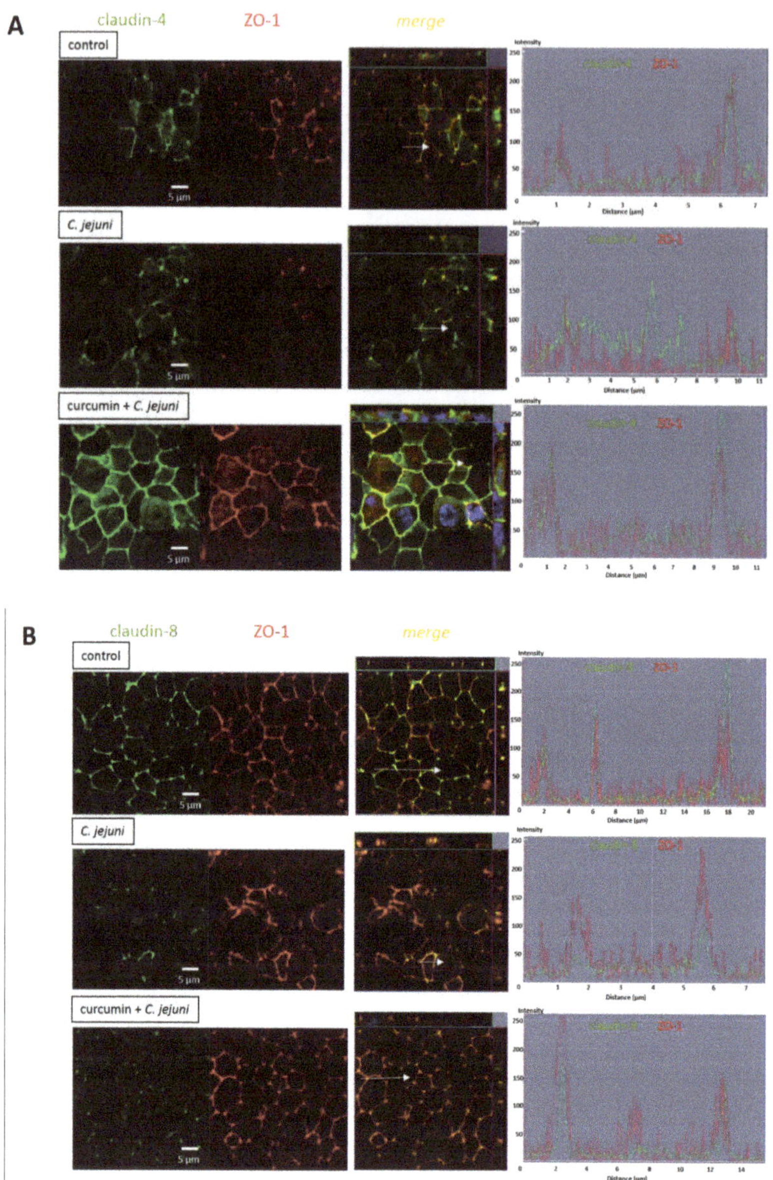

Figure 9. Tight junction distribution in *C. jejuni* infection after treatment with curcumin. Representative confocal laser-scanning microscopy pictures of HT-29/B6-GR/MR after co-culturing together with immune cells. (**A**) Claudin-4 (green) and zonula occludens protein-1 (ZO-1, red), and (**B**) claudin-8 (green) and ZO-1 (red). Nuclei are stained in blue with 4′-6-diamidino-2-phenylindole dihydrochloride (DAPI).

3. Discussion

It is important to contain the *Campylobacter jejuni* infection because of its impact on gastroenteritis worldwide and resulting sequelae, such as irritable bowel syndrome, reactive arthritis, or Guillian–Barré syndrome. Therefore, we aimed to study the underlying pathomechanisms for barrier dysfunction and the influence of the immune system in more detail. For this purpose, we used a new in vitro co-culture model and an in vivo mouse model to study the protective properties of the natural polyphenol curcumin.

Curcumin is a vitamin D receptor (VDR) ligand [24], and we found the downstream signaling pathway of curcumin to be inhibited in the mucosa of acutely infected *C. jejuni* patients by means of the bioinformatic prediction in Qiagen ingenuity pathway analysis (IPA) from RNA-sequencing (Supplementary Table S1). This prediction indicated that curcumin could be a potential therapeutic or preventive target against *C. jejuni* infections. Based on this bioinformatics prediction, we tested the effects of curcumin on *C. jejuni* infection in vitro and in vivo. The inhibitory effect of another top hit upstream regulator in IPA analysis on the mucosa of *C. jejuni*-infected patients was calcitriol (active form of vitamin D). Calcitriol exhibited a high significance value comparable to curcumin (Appendix A, Table S1). Annulment of *C. jejuni*-induced cytotoxicity after calcitriol treatment has been demonstrated in HT-29/B6 cells previously, supporting the view that the bacterium affects intestinal barrier function via a VDR-dependent pathway [6].

As predicted by the pathway analysis in acute human campylobacteriosis, curcumin was effective in our experiments. It could prevent the decrease in TER as well as the increase in fluorescein permeability in our co-culture model, indicating improved barrier function by curcumin in immune-mediated epithelial barrier dysfunction by *C. jejuni*.

Because we expected an essential contribution of the immune system, we established a co-culture model with basal infection. In this model, the *C. jejuni*-triggered immune response reduced the TER after 48 h. The apical bathing medium did not contain bacteria even after 48 h, which suggests that the effect on resistance was caused indirectly by the *C. jejuni*-induced immune response. The decrease in TER may therefore reflected an increase in paracellular permeability secondary to epithelial barrier defects, which would increase permeability to small molecules and toxins. Furthermore, this change in barrier function may increase the access of bacteria to the underlying tissue and potentiate the immune response, causing additional barrier disruption, with loss of cellular integrity [6,25]. Concomitantly, a leak-flux type of diarrhea is caused that represents a main diarrheal mechanism.

Depending on the cell type and the experimental infection conditions, *C. jejuni* was more or less able to modulate TER in previous studies. No effects on TER have been found in MKN-28 cells [26] or Caco-2 monolayers [4]. In contrast, exposure to *C. jejuni* has been reported to reduce TER in T84 and MDCK-I cells [27]. In our HT-29/B6-GR/MR cells, TER was also not different from control even after 48 h basal infection, although a decrease in TER after basolateral infection has been reported to be more pronounced than after apical infection in T84 cells [28]. That this was not seen in our HT-29/B6-GR/MR cells may reflect differences in cell type and/or the 0.4 μm filter pore size of our filters, limiting the bacterial access to the epithelial cells.

Within 6 days following peroral *C. jejuni* infection, secondary abiotic IL-$10^{-/-}$ mice develop acute ulcerative enterocolitis mimicking key features of severe campylobacteriosis in humans, with a specific cytokine response [29]. We were able to show for the first time, using impedance spectroscopy in Ussing chambers, compromised barrier integrity with an increased epithelial conductance and permeability to macromolecules in this mouse model. These data reflect the results of our co-culture model, in which the barrier defects could be similarly prevented by curcumin. Bücker et al. [6] incubated colonic mucosal biopsies from *C. jejuni*-infected patients in culture medium and quantified cytokine release into the supernatant. Secretion of TNF-α, IFN-γ, IL-1β, and IL-13 was enhanced in comparison to controls [6]. It was concluded that this mucosal cytokine storm led to disruption of intestinal TJs and the induction of epithelial apoptosis. In our infected co-cultures, release of cytokines was also mainly responsible for the barrier defects, even though THP-1 cells produce and secrete a lower amount and spectrum of cytokines in comparison to fresh isolated peripheral blood mononuclear cells (PBMC). After *C. jejuni* infection, levels of the pro-inflammatory cytokine TNF-α, IL-1β, and IL-6 increased remarkably, and these changes were largely prevented by curcumin. Curcumin not only reduced local inflammation in the intestine, it also reduced systemic inflammation triggered by LPS [23]. In another study, curcumin nanoparticles suppressed the expression of mRNAs encoding pro-inflammatory mediators, including TNF-α, IL-1β, and IL-6 [14].

C. jejuni was found to activate the NFκB signaling pathway [28], which was suppressed by curcumin [12,17,19]. To determine the role of the NFκB pathway in maintaining barrier function, inhibition studies with the NFκB-inhibitor BAY 11-7082 were performed in *C. jejuni*-infected co-cultures. BAY 11-7082 treatment inhibited both the *C. jejuni*-induced decrease in TER and the increase in fluorescein permeability. This points to the involvement of NFκB in the regulation of epithelial barrier function. Similar results were obtained by our group previously with the bioactive ginger ingredient 6-shogaol after TNF-α stimulation in Caco-2 and HT-29/B6 cells [30].

In our co-culture, *C. jejuni* provoked paradoxical upregulation of claudin-1 (Claudin-1 paradox); that is, upregulation of the barrier-forming TJ protein claudin-1 while epithelial resistance was reduced, which we explained by the re-distribution of claudin-1 off the tight junction domain [6]. Claudin-1 induction has also been observed before in human colon biopsies after *C. jejuni* infection [6], *C. fetus* or *C. coli* infection in HT-29/B6 cells [31], and pro-inflammatory cytokine stimulation in HT-29/B6 cells [32,33]. A correlation between increased claudin-1 protein expression and apoptosis induction also occurs in HT-29/B6 cells [34]. In addition, curcumin induced claudin-4 mRNA expression in Caco-2 cells [8]. However, in our co-culture model, curcumin did not induce claudin-4 protein expression. Cell type-dependent differences in TJ expression were also found for quercetin, another plant-derived polyphenol with barrier-improving properties, which increased claudin-4 protein expression in Caco-2, but not in HT-29/B6 cells [35,36].

In order to exclude antibacterial effects of curcumin in our study, the MIC was determined with 87 μM (at pH 7.4). This suggests that the protective effects of 50 μM curcumin were not based on its antibacterial properties. Thus, we conclude that curcumin had direct barrier-protective and anti-inflammatory properties in the host cells.

To further exclude interfering properties of curcumin with bacterial fitness and adaptive response, we investigated if curcumin changes natural transformation capacity. Therefore, we used a single cell DNA uptake assay and challenged the bacteria with 50 μM curcumin. *C. jejuni* is naturally competent for DNA uptake from the environment. Generally, this leads to genetic recombinations between bacterial strains, and as a consequence, to more diversity [37]. Curcumin did not influence the competence development and horizontal DNA uptake of *C. jejuni* in our experimental concentrations.

One of the major challenges to the development of therapeutic approaches is the remarkable diversity between different *Campylobacter* strains in animals, the environment, and food [2]. Consequently, the efficacy of curcumin should also be tested using other *C. jejuni* field strains and, subsequently, in other *Campylobacter* species such as *C. coli* or *C. concisus*. Curcumin is a natural agent, and should be more appropriate than synthetic drugs [12]. Curcumin has been shown to be beneficial in maintaining remission in ulcerative colitis patients when used as a complementary therapeutic substance together with the antiphlogistic compound mesalazine in several randomized clinical studies [38–40]. Since the immune response plays an important role in pathogenesis in IBD as well as in *C. jejuni* infection, curcumin may be equally effective in immune-induced barrier dysfunctions by *C. jejuni* as an add-on to other anti-inflammatory drugs. Indeed, multimodal therapies may have an even greater protective effect, and synergism has been reported between curcumin and quercetin [41], genistein [42], epigallocatechin-3-gallate [15,43], and antioxidants [12,44]. Thus, curcumin alone or in combination with other agents may be useful in preventing, treating, and combating *Campylobacter* infection in humans, *C. jejuni*-associated sequelae as well as colonization of farm animals and especially poultry.

Taken together, our results suggest that a significant part of the *C. jejuni*-induced barrier defect is mediated indirectly by the immune cells. Curcumin abolishes functional epithelial disorders as well as pro-inflammatory cytokine secretion. Further barrier-protective or anti-inflammatory compounds, as well as curcumin combinations with other therapies, should be tested in the co-culture model.

4. Materials and Methods

4.1. Epithelial Cell Culture and Differentiation of THP-1 Cells

The human colon carcinoma cell line HT-29/B6-GR/MR [45], a subclone of the HT-29/B6 line [46] was cultured in 25 cm^2 culture flasks in RPMI 1640 culture medium (Sigma Aldrich, St. Louis, MO, USA). HT-29/B6-GR/MR cells expressing the human glucocorticoid receptor α and the human mineralocorticoid receptor, were established earlier by stable transfection [45,47]. The culture medium was supplemented with 10% fetal calf serum (FCS; Gibco, Carlsbad, CA, USA), 1% penicillin/streptomycin (Corning, Wiesbaden, Germany), G418-BC (300 μg/mL; Invitrogen, Carlsbad, CA, USA) and hygromycin B (200 μg/mL; Biochrom GmbH, Berlin, Germany), and cells were passaged every week. Cells were seeded on Millicell PCF filters membranes (Merck Millipore, Billerica, MA, USA; 0.4 μm pore size and an effective growth area of 0.6 cm^2). The cell medium in the culture flask, as well as in the filters, were changed every other day. Experiments were performed between 7 and 9 days after seeding, when polarized cells formed confluent monolayers and the transepithelial electrical resistance was 600–900 $\Omega \cdot cm^2$. These cells were chosen because of their stable growth behavior and the high resistance. The human monocyte leukemia cell line THP-1 (ATCC TIB-202) was provided by Verena Moos. THP-1 cells were cultured in RPMI 1640 supplemented with 10% heat-inactivated FCS and 1% penicillin/streptomycin. Cultures were maintained by the addition of fresh medium and were subcultured before reaching a density of 8×10^5 cells/mL once a week. THP-1 cells were re-suspended in antibiotic-free medium supplemented with phorbol 12-myristate 13-acetate (PMA; Sigma Aldrich, St. Louis, MO, USA; solved in DMSO) at a final concentration of 100 nM, seeded with a density of 1.8×10^5 in 12-well plates, and incubated for 24 h for differentiation to macrophages [48]. After incubation, PMA-containing medium was removed and the differentiated and adherent macrophages-like cells were cultured together with epithelial monolayers grown on filters in co-culture (Figure 10). All cells were cultured at 37 °C in a humidified 5% CO_2 atmosphere.

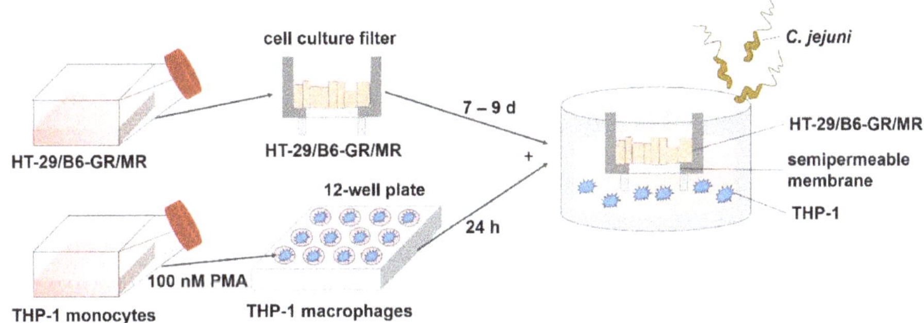

Figure 10. Experimental setting for co-culture of colon epithelial HT-29/B6-GR/MR and THP-1 immune cells. For the continuous culture of epithelial and immune cells, cells were cultured in culture flask. Epithelial cells were then seeded on filter membranes and differentiated over 7–9 days to a polarized monolayer with tight junctions. 24 h before co-culture started, immune THP-1 cells were stimulated with phorbol 12-myristate 13-acetate (PMA) to allow a differentiation to adherent macrophage-like immune cells. After the differentiation of epithelial and immune cells, filters were placed into the 12-well plate with epithelial cells in the apical and immune cells in the basal compartment. Infection with *C. jejuni* was performed in the basal compartment.

4.2. Growth Conditions of *C. jejuni*, Treatment and Infection Procedure In Vitro

Campylobater jejuni wildtype (wt) 81-176 reference strain was pre-cultured on blood agar plates (Oxoid, Thermo Scientific, Waltham, MA, USA) under microaerobic conditions (5–10% O_2, 10% CO_2, 85% N_2) at 37 °C and re-cultured a second time before infection experiments. Microaerobic conditions were generated in a plastic jar with CampyGen gaspacks from Oxoid (Oxoid, Thermo Scientific, Waltham, MA, USA). Bacteria were then cultured in Mueller–Hinton broth for at least 2.5 h at 37 °C under microaerobic conditions, centrifuged (2 min, 5000 *g*, 10 °C), and re-suspended in the cell culture medium. The optical density at OD_{600} was adjusted to 1 and the cells where infected from basal side with a multiplicity of infection (MOI) of 100. At least 1.5 h before the infection, cells were washed three times with antibiotic-free culture medium supplemented with 10% heat-inactivated FCS. To analyze the barrier-protective and anti-inflammatory properties of curcumin (Sigma Aldrich, St. Louis, MO, USA; final concentration 50 μM), different conditions were tested: control, *C. jejuni* infected, curcumin control (50 μM), and curcumin in combination with *C. jejuni*. Curcumin control showed no adverse effects on the cell viability of HT-29/B6-GR/MR monolayers at the concentration tested. Cells were pre-incubated with curcumin on both sides for 2 h, and infected with *C. jejuni* from the basal side for a direct immune cell infection and cytokine-induced barrier effect. Curcumin was added to the apical and basal side, since functional measurements depend on identical solutions on both the sides of the monolayer to avoid, e.g., potential osmotic effects. Moreover, curcumin readily accessed the basolateral compartment via *C. jejuni*-induced leaks. After infection, cells were incubated at 37 °C under microaerobic conditions favorable to the bacteria for 48 h.

4.3. Generation of Secondary Abiotic IL10$^{-/-}$ Mice, Treatment Infection

Permeability to fluorescein (332 Da; Sigma Aldrich, St. Louis, MO, USA) measurements were performed in Ussing chambers with colon of secondary abiotic IL-10$^{-/-}$ mice suffering from acute enterocolitis within 6 days following peroral *C. jejuni* infection [49]. IL-10$^{-/-}$ mice (in C57BL/6j background) were held under specific pathogen free (SPF) conditions in the animal facilities of the Forschungseinrichtung für Experimentelle Medizin (Charité–Universitätsmedizin Berlin). To remove the commensal gut microbiota, mice were transferred to sterile cages and treated for 8 weeks with an antibiotic cocktail in the drinking water ad libitum containing ampicillin/sulbactam (1.5 g/L), ciprofloxacin (200 mg/L), imipenem/cilastatin (250 mg/L), metronidazole (1 g/L), and vancomycin (500 mg/L) as described previously [50]. Four days before infection, the antibiotic cocktail was replaced by curcumin (Sigma Aldrich, St. Louis, MO, USA; 0.5 mg/mL; solved in 2% Carboxymethyl cellulose (Sigma Aldrich, St. Louis, MO, USA) in phosphate-buffered saline (PBS; pH7.4; Sigma Aldrich, St. Louis, MO, USA)) in autoclaved drinking water. Mice were then infected via oral gavage with 10^8 colony forming units (CFU) of *C. jejuni* strain 81-176 in a volume of 0.3 mL PBS. Six days after infection mice were sacrificed by isoflurane inhalation and colon samples were removed for tracer flux measurements analysis.

4.4. Ethics Statement

The animal experiments were carried out in our animal facility according to the German animal protection law (LaGeSo Berlin; approval number G0172/16, 13th Oct. 2016).

4.5. Electrophysiological Studies

Transepithelial electrical resistance (TER) was measured in vitro with a chopstick electrode pair under sterile conditions at 37 °C. The TER-values were corrected with the resistance of an empty cell filter and the bath solution. Colon samples from mice were mounted in modified miniaturized Ussing chambers (Institute of Clinical Physiology, Charité, Berlin, with an effective area of 0.049 cm^2) in modified Ringer´s solution. Ringer´s solution for the Ussing experiments contained (in mM): NaCl (113.6), $NaHCO_3$ (21), KCL (5.4), Na_2HPO_4 (2.4), $MgCl_2$ (1.2), $CaCl_2$ (1.2), NaH_2PO_4 (0.6), D(+)-glucose (10), D(+)-mannose (10) beta-hydroxybutyric acid (0.5), L-glutamine (2.5), and the antibiotics piperacillin (50 mg/L) and imipenem (4 mg/L), and was equilibrated for 15 min with carbogen gas to a pH of 7.4. One-path impedance spectroscopy measurements were performed as described previously [51] to delineate between epithelial and subepithelial conductance (G).

4.6. Epithelial Permeability

For permeability measurements, unidirectional flux studies were conducted with the paracellular marker fluorescein (332 Da; 100 μM) from the mucosal to the serosal compartment, either directly in 12-well plates or under short circuit current (I_{SC}) conditions in Ussing chambers, as described earlier [7]. I_{SC} was recorded under voltage clamp conditions by an automatic clamp device (CVC6, Fiebig Hard & Software, Berlin, Germany) at 37 °C over 1.5 h. Fluorescein was dissolved either in Ringer´s solution or in media. Samples were taken every 15 min for one hour from the basolateral side, and fluorescence was measured in a spectrophotometer (Tecan GmbH, Maennedorf, Switzerland). Permeability was calculated from flux over concentration difference.

4.7. Cytometric Bead Array

Supernatants of co-cultures were collected 48 h after infection and analyzed using the Cytometric Bead Assay (CBA; BD Biosciences, Franklin Lakes, NJ, USA) according to manufacturer´s instructions (human Th1, Th2, Th17 Kit, Flex Set IL-1β) to determine the secretion of cytokines TNF-α, IL-1β, IL-6. Flow cytometric measurement were performed with FACS CantoII (BD Biosciences; Franklin Lakes, NJ, USA) and analyzed with FACP Array^{TM} software v3.0 (BD Biosciences, Franklin Lakes, NJ, USA).

4.8. Western Blot Analysis

For protein quantification, epithelial cells were washed twice with ice-cold PBS. Whole cell lysates were extracted with ice-cold lysis buffer. Whole cell lysis buffer was prepared with 150 mM NaCl, 10 mM Tris buffer pH of 7.5, 0.5% Triton X-100, and 1% SDS. A volume of 10 mL lysis buffer was supplemented with one Complete Protease Inhibitor Cocktail tablet (Roche AG, Basel, Switzerland). After lysis, cells were scraped from the filters, incubated for 60 min on ice, and vortexed every 10 min. After centrifugation (30 min, 15,000× *g* at 4 °C), the supernatant was collected. Sonification of the lysate followed by further centrifugation. Total protein quantification was performed by Pierce BCA assay (Thermo Scientific, Waltham, MA, USA) according to the product instructions using a Tecan plate reader (Tecan GmbH, Maennedorf, Switzerland) at an absorbance of 562 nm. Protein samples (10–20 μg) were mixed with 5xLaemmli buffer and loaded on a SDS polyacrylamide gel (for claudins 12.5%, for occludin 10% polyacrylamide). After electrophoretic separation of the proteins and transfer to a nitrocellulose membrane, membranes were blocked for 2 h at room temperature with 1% PVP-40 (Polyvinylpyrrolidone; Sigma Aldrich, St. Louis, MO, USA) in TBST/0.05%Tween-20 buffer. Primary antibodies anti-occludin (1:100; Sigma Aldrich, St. Louis, MO, USA), anti-claudin-1, -2, -4, -5, -7, -8 (1:100; Invitrogen, Carlsbad, CA, USA), and anti-β-actin (1:5000; Sigma Aldrich, St. Louis, MO, USA) as internal loading control were incubated overnight at 4 °C. Peroxidase conjugated secondary antibodies goat anti-rabbit IgG or goat anti-mouse IgG (Jackson ImmunoResearch, Ely, UK) were incubated for 2 h at room temperature. For protein detection, SuperSignal West Pico PLUS Stable Peroxide Solution (Thermo Scientific, Waltham, MA, USA) was used and signals were detected with Fusion FX7 imaging system (Vilber Lourmat Deutschland GmbH, Eberhardzell, Germany). Densitometric quantification was performed using ImageJ software 1.48v/Java 1.6.0_20 (Rasband, W. S., ImageJ, NIH, Bethesda, MD, USA), and the values were normalized to β-actin.

4.9. Immunofluorescence Staining

Cells grown on filters were rinsed twice with PBS and fixed for 30 min in 2% paraformaldehyde (PFA; Electron Microscopy Sciences, Hatfield, PA, USA) for immunostaining and microscopic analysis. Afterwards, the cells were washed twice with PBS, permeabilized with 0.5% Triton X-100 (Sigma Aldrich, St. Louis, MO, USA) for 7 min, and blocked for 10 min with 1% goat serum (Gibco, Carlsbad, CA, USA). The cells were incubated with the primary antibodies anti-claudin-4 and -8 (1:100; Invitrogen, Carlsbad, CA, USA), anti-ZO-1 (1:100; BD Biosciences, Franklin Lakes, NJ, USA) or cleaved caspase-3 (1:100; Cell Signaling Technology, Cambridge, UK) for 1 h at room temperature, followed by the secondary anti-rabbit or anti-mouse antibody for 1 h (1:500; Invitrogen, Carlsbad, CA, USA). Secondary antibodies were conjugated to Alexa-Fluor 488 or 594. The nuclei were stained with 4′-6-diamidino-2-phenylindole dihydrochloride (DAPI, 1:1000, Roche AG, Basel, Switzerland). Subsequently, the cells on the filters were washed with water and ethanol, then embedded in ProTaq Mount Fluor (Biocyc, Luckenwalde, Germany). The subcellular distribution of the tight junctions was analyzed by confocal laser-scanning microscopy (CLSM, Zeiss LSM780, Jena, Germany). Stained apoptoses with cleaved caspase-3 were counted microscopically.

4.10. Determination of Minimal Inhibitory Concentration Values

Determinations of minimal inhibitory concentration values (MICs) of *Campylobacter* strain 81-176 were performed in a broth microdilution assay. The wells of microtiter plates contained 2-fold serial dilutions of curcumin with a test range of 1–1024 μg/mL. For MIC determinations, inoculum level, growth medium, incubation time, and conditions were performed in accordance with the recommendations given in the Clinical and Laboratory Standards Institute (CLSI) document VET01-A4. In brief, *Campylobacter* colonies were taken from a blood agar plate and transferred into a tube containing 5 mL of 0.85% saline. The suspension was adjusted to the turbidity equivalent of 0.5 McFarland standard, and diluted 1:100 with growth medium. Subsequently, 50 μL of the diluted inoculum was added to each test well (50 μL) in the dilution series (containing substances at double the desired final concentration) and mixed. The microtiter plates were incubated for 24 to 48 h at 42 °C under microaerobic conditions to obtain sufficient growth. They were analyzed visually, and the lowest concentration preventing visible growth of bacteria was defined as the MIC. All susceptibility tests were performed in growth medium adjusted to pH of 7.4. *Campylobacter jejuni* reference strain DSM 4688 was used for quality control purposes. The MIC values of the quality control strain were determined in advance in three independent experiments using the broth microdilution and macrodilution method.

4.11. DNA-Uptake Assay

C. jejuni strain 81-176 was streaked out from a −80 °C stock and subcultured on Columbia blood agar containing 5% sheep blood (Oxoid, Thermo Scientific, Waltham, MA, USA). Plates were incubated overnight under an atmosphere containing 3.5% H_2, 6% O_2, 7% CO_2, rest N_2 at 37 °C. Cells were inoculated at OD_{ini} ~0.3 in sterile-filtered brain heart infusion broth (BHI; Oxoid, Thermo Scientific, Waltham, MA, USA), and incubated under the same atmosphere at 37 °C for 6–9 h at 140 rpm. Liquid cultures were subcultured under the same conditions in 5 mL BHI with or without 50 μM curcumin. A suitable OD_{ini} was chosen to reach optical densities of 0.1–0.6 after 17 h (± 3 h). 500 μL (± 300 μL) of the cell suspension was harvested by centrifugation (~16000× *g* for 5 min). The pellet was resuspended in 100 μL BHI with or without 50 μM curcumin. DNA labeling and uptake analysis were performed as previously described for *H. pylori* [52]. In short, *C. jejuni* BfR-CA-14430 genomic DNA was extracted using the PureLink Kit (Life Technologies, Thermo Scientific, Waltham, MA, USA), and labeled with fluorescein in a 1:1 (volume:weight) ratio of Label IT reagent to nucleic acid according to the manufacturer's protocol (Mirus Label IT Fluorescein, Mirus Bio LLC, Madison, WI, USA). One μL of labeled DNA (100 ng/μL) was added to the cell suspension and incubated at 5% O_2, 10% CO_2, rest N_2 for 30 min at 37 °C. Subsequently, cells were centrifuged (~16000× *g* for 5 min) and resuspended in 15 μL BHI supplemented with at least 3 U DNaseI (Roche AG, Basel, Switzerland). DNaseI digestion was performed at 37 °C for 5–10 min. The fluorescence microscope Axio Observer ZI (Zeiss, Jena, Germany) with a plan apochromatic 63*x*/1.4 objective and differential interference contrast (DIC) was used for the analysis of *Campylobacter* cells immobilized on 1.5% agarose pads. A metal halide light source (HXP120C) and a filter set with excitation at 470 ± 20 nm and emission at 525 ± 25 nm were used to visualize fluorescein. Exposing times varied between 250 ms and 750 ms. Images were taken by the 12-bit monochromatic AxioCam MRm camera. Cells with at least one fluorescent focus were considered active for DNA uptake, i.e., competent for natural transformation. For each condition, the fraction of competent cells was calculated and data are presented from three independent experiments.

4.12. Cytotoxicity

The CCK-8 assay (Cell Counting Kit-8, Thermo Scientific, Waltham, MA, USA) was performed in 96-well plates with colon epithelial cells HT-29/B6-GR/MR 1 day after seeding in accordance with the manufactures' instructions. The cells were treated with different curcumin concentrations 2 h before infection. The absorption at 450 nm was measured by a spectrophotometer (Tecan GmbH, Maennedorf, Switzerland), and the percentage of viable cells calculated. In HT-29/B6-GR/MR cells, *C. jejuni* induced no significant loss in viability in a pre-test (control 100 ± 18% and DMSO control 81 ± 7% versus *C. jejuni*-infected 81 ± 4%, n = 3–4, *n.s.*, one-way ANOVA with Bonferroni´s multiple comparison). No effect on cell viability was also seen at curcumin concentrations 30 µM, 40 µM, and 50 µM (85 ± 8%, 64 ± 4%, and 66 ± 6%, n = 3–4, *n.s.*, one-way ANOVA with Bonferroni´s multiple comparison).

4.13. Ingenuity Pathways Analysis

Ingenuity Pathways Analysis (IPA, Qiagen Silicon Valley, Reswood, CA, USA) was used to evaluate curcumin-dependent expression data from a dataset from human colon biopsies, generated, and analyzed previously [6].

4.14. Statistical Analysis

All data are expressed as mean values ± standard error of the mean (SEM). Statistical analyses were performed with GraphPad Prism (version 7.0, GraphPad Software, Inc., San Diego, CA, USA) using one-way ANOVA with Bonferroni adjustment for multiple comparison or unpaired Student´s *t*-test. For data that were not normally distributed, the nonparametric Mann–Whitney *U*-test was used. $p < 0.05$ was considered to be statistically significant.

Supplementary Materials: Supplementary materials are available online at http://www.mdpi.com/1422-0067/20/19/4830/s1.

Author Contributions: Conceptualization, F.D.L.d.S., J.-D.S., and R.B.; Data curation, F.D.L.d.S., E.B., P.K.N., A.F., S.M., V.M., J.C.G., D.S., M.M.H., and S.B.; Formal analysis, F.D.L.d.S., E.B., P.K.N., A.F., S.M., V.M., J.C.G., K.S., S.K., D.S., C.K., M.M.H., and S.B.; Funding acquisition, J.-D.S. and R.B.; Investigation, F.D.L.d.S., J.-D.S., R.B.; Methodology, J.-D.S. and R.B.; Project administration, J.-D.S. and R.B.; Resources, V.M.; Supervision, J.-D.S. and R.B.; Writing—original draft, F.D.L.d.S.; Writing—review and editing, E.B., P.K.N., A.F., S.M., V.M., J.C.G., K.S., S.K., D.S., C.K., M.M.H., S.B., J.-D.S., and R.B.

Funding: The German Federal Ministries of Education and Research (BMBF) funded the study in connection with the zoonoses research consortium PAC-CAMPY to R.B. and J.D.S. (IP8/01KI1725D), S.B. and M.M.H. (IP7/01KI1725D), K.S. (IP3/01KI1725B), and S.K. (IP3/01KI1725C). The funders had no influence in the study design, data collection, analysis, decision to publish, or preparation of the manuscript. We acknowledge support from the German Research Foundation (DFG) and the Open Access Publication Fund of the Charité–Universitätsmedizin Berlin.

Acknowledgments: In-Fah Maria Lee is gratefully acknowledged for her excellent technical support. We thank Geoffrey I. Sandle, (University of Leeds, Leeds, UK) for his critical discussion and support of the study.

Conflicts of Interest: The authors declare no conflict of interest.

Abbreviations

BHI	Brain heart infusion broth
CBA	Cytometric bead array
CFU	Colony forming units
CLSM	Confocal laser-scanning microscopy
DAPI	4´-6-diamidino-2-phenylindole dihydrochloride
IFN	Interferon
IL	Interleukin
LPS	Lipopolysaccharide
MIC	Minimal inhibitory concentration
PBS	Phosphate buffered saline
PMA	Phorbol 12-myristate 13-acetate

TER	Transepithelial electrical resistance
TJ	Tight junction
TNF	Tumor necrosis factor
ZO	Zonula occludens

Appendix A

Acute Campylobacteriosis Contains Activated Signaling Pathways for which Curcumin Has Counter-Regulatory Properties in IPA Analysis

From acute infected *C. jejuni* patients a RNA-Sequencing (RNA-Seq) analysis with concomitant ingenuity pathway analysis (IPA, Qiagen Silicon Valley) was performed. Once gene expression modifications were revealed, a bioinformatic prediction about possible inhibitors of the changed gene regulation could be carried out [6]. The hypothesis is, that upstream regulators, which have an inhibited activation pattern, may re-activate in *Campylobacter* infection, when the substance is applied during infection. Consequently, inhibited upstream regulators could be protective or therapeutic approaches in *Campylobacter* infection by activation of the corresponding downstream pathways. Different potential candidates were screened for barrier-protective and anti-inflammatory properties in *C. jejuni* infection. One promising predicted regulator candidate that might counter-regulate the *C. jejuni*-induced downstream pathways was curcumin. Curcumin showed a significant effect on downstream target genes, with a p-value of $2.06E^{-5}$ and an activation ^{z}-score of −3.489, and might therefore be another promising barrier-protecting or potential therapeutic substance in campylobacteriosis (Table S1). The *C. jejuni*-induced target genes in the dataset that could be counter-regulated by curcumin belong mainly to pro-inflammatory pathways, such as TNF-α or IL-1β. Another promising and studied candidate against *C. jejuni* infections is calcitriol (active vitamin D). Vitamin D shows in the RNA-Seq analysis with concomitant IPA analysis from patients in contrast to curcumin an even higher significance value (overlap p-value of $8.97E^{-25}$, z-score −6.25; with negative expression direction) [6].

References

1. Kaakoush, N.O.; Castaño-Rodríguez, N.; Mitchell, H.M.; Man, S.M. Global Epidemiology of *Campylobacter* Infection. *Clin. Microbiol. Rev.* **2015**, *28*, 687–720. [CrossRef] [PubMed]
2. Burnham, P.M.; Hendrixson, D.R. *Campylobacter jejuni*: Collective components promoting a successful enteric lifestyle. *Nat. Rev. Microbiol.* **2018**, *16*, 551–565. [CrossRef] [PubMed]
3. Konkel, M.E.; Mead, D.J.; Hayes, S.F.; Cieplak, W., Jr. Translocation of *Campylobacter jejuni* across human polarized epithelial cell monolayer cultures. *J. Infect. Dis.* **1992**, *166*, 308–315. [CrossRef] [PubMed]
4. Brás, A.M.; Ketley, J.M. Transcellular translocation of *Campylobacter jejuni* across polarised epithelial monolayers. *FEMS Microbiol. Lett.* **1999**, *179*, 209–215. [CrossRef] [PubMed]
5. Ketley, J.M. Pathogenesis of enteric infection by *Campylobacter*. *Microbiology* **1997**, *143*, 5–21. [CrossRef] [PubMed]
6. Bücker, R.; Krug, S.M.; Moss, V.; Bojarski, C.; Schweiger, M.R.; Kerick, M.; Fromm, A.; Janßen, S.; Fromm, M.; Hering, N.A.; et al. *Campylobacter jejuni* impairs sodium transport and epithelial barrier function via cytokine release in human colon. *Mucosal Immunol.* **2018**, *11*, 575–577. [CrossRef] [PubMed]
7. Bücker, R.; Schulz, E.; Günzel, D.; Bojarski, C.; Lee, I.F.; John, L.J.; Wiegand, S.; Janßen, T.; Wieler, L.H.; Dobrindt, U.; et al. α-Haemolysin of *Escherichia coli* in IBD: A potentiator of inflammatory activity in the colon. *Gut* **2014**, *63*, 1893–1901. [CrossRef] [PubMed]
8. Watari, A.; Yagi, K.; Kondoh, M. A simple reporter assay for screening Claudin-4 modulators. *Biochem. Biophys. Res. Commun.* **2012**, *426*, 454–460. [CrossRef]
9. Chiba, H.; Osanai, M.; Murata, M.; Kojima, T.; Sawada, N. Transmembrane proteins of tight junctions. *Biochim. Biophys. Acta* **2008**, *1778*, 588–600. [CrossRef]
10. Hakanen, A.; Jousimies-Somer, H.; Siitonen, A.; Huovinen, P.; Kotilainen, P. Fluoroquinolone resistance in *Campylobacter jejuni* isolates in travelers returning to Finland: Association of ciprofloxacin resistance to travel destination. *Emerg. Infect. Dis.* **2003**, *9*, 267–270. [CrossRef]
11. Hakanen, A.J.; Lehtopolku, M.; Siitonen, A.; Huovinen, P.; Kotilainen, P. Multidrug resistance in *Campylobacter jejuni* strains collected from Finnish patients during 1995–2000. *J. Antimicrob. Chemother.* **2003**, *52*, 1035–1039. [CrossRef] [PubMed]
12. Goel, A.; Kunnumakkara, A.B.; Aggarwal, B.B. Curcumin a "Curecumin": From kitchen to clinic. *Biochem. Pharmacol.* **2008**, *75*, 787–809. [CrossRef] [PubMed]

13. Cho, J.A.; Park, E. Curcumin utilizes the anti-inflammatory response pathway to protect the intestine against bacterial invasion. *Nutr. Res. Pract.* **2015**, *9*, 117–122. [CrossRef] [PubMed]
14. Masahi, O.; Nishida, A.; Sugitani, Y.; Nishino, K.; Inatomi, O.; Sugimoto, M.; Kawahara, M.; Andoh, A. Nanoparticle curcumin ameliorated experimental colitis via modulation of gut microbiota and induction of regulatory T cells. *PLoS ONE* **2017**, *12*, e0185999. [CrossRef]
15. Anggakusuma; Colpitts, C.C.; Schang, L.M.; Rachmawati, H.; Frentzen, A.; Pfaender, S.; Behrendt, P.; Brown, R.J.; Bankwitz, D.; Steinmann, J.; et al. Tumeric curcumin inhibits entry of all hepatitis C virus genotypes into human liver cells. *Gut* **2014**, *63*, 1137–1149. [CrossRef]
16. Mahady, G.B.; Pendland, S.L.; Yun, G.; Lu, Z.Z. Turmeric (*Curcuma longa*) and curcumin inhibit the growth of *Helicobacter pylori*, a group 1 carcinogen. *Anticancer Res.* **2002**, *22*, 4179–4181.
17. Chen, D.-Y.; Shien, J.-H.; Tiley, L.; Chiou, S.S.; Wang, S.Y.; Chang, T.J.; Lee, Y.J.; Chan, K.W.; Hsu, W.L. Curcumin inhibits influenza virus infection and haemagglutination activity. *Food Chem.* **2010**, *119*, 1346–1351. [CrossRef]
18. Marathe, S.A.; Balakrishnan, A.; Negi, V.D.; Sakorey, D.; Chandra, N.; Chakravortty, D. Curcumin reduces the motility of *Salmonella enterica* serovar typhymurium by binding to the flagella, thereby leading to flagellar fragility and shedding. *J. Bacteriol.* **2016**, *198*, 1798–1811. [CrossRef]
19. Moghadamtousi, S.Z.; Kadir, H.A.; Hassandarvish, P.; Tajik, H.; Abubakar, S.; Zandi, K. A review on antibacterial, antiviral, and antifungal activity of curcumin. *Biomed. Res. Int.* **2014**, *2014*, 186864. [CrossRef]
20. Santos, A.M.; Lopes, T.; Oleastro, M.; Gato, I.V.; Floch, P.; Benejat, L.; Chaves, P.; Pereira, T.; Seixas, E.; Machado, J.; et al. Curcumin inhibits gastric inflammation induced by *Helicobacter pylori* infection in a mouse model. *Nutrients* **2015**, *7*, 306–320. [CrossRef]
21. Gunes, H.; Gulen, D.; Mutlu, R.; Gumus, A.; Tas, T.; Topkaya, A.E. Antibacterial effects of curcumin: An in vitro minimum inhibitory concentration study. *Toxicol. Ind. Health* **2016**, *32*, 246–250. [CrossRef] [PubMed]
22. Sharma, R.A.; Geschner, A.J.; Steward, W.P. Curcumin: The story so far. *Eur J. Cancer* **2005**, *41*, 1955–1968. [CrossRef] [PubMed]
23. Wang, J.; Ghosh, S.S.; Ghosh, S. Curcumin improves intestinal barrier function: Modulation of intracellular signaling, and organization of tight junctions. *Am. J. Physiol.-Cell. Physiol.* **2017**, *312*, C438–C445. [CrossRef] [PubMed]
24. Bartik, L.; Whitfield, G.K.; Kaczmarska, M.; Lowmiller, C.L.; Moffet, E.W.; Furmick, J.K.; Hernandez, Z.; Haussler, C.A.; Haussler, M.R.; Jurutka, P.W. Curcumin: A novel nutritionally derived ligand of the vitamin D receptor with implications for colon cancer chemoprevention. *J. Nutr. Biochem.* **2010**, *21*, 1153–1161. [CrossRef] [PubMed]
25. Backert, S.; Boehm, M.; Wessler, S.; Tegtmeyer, N. Transmigration route of *Campylobacter jejuni* across polarized intestinal epithelial cells: Paracellular, transcellular or both? *Cell Commun. Signal.* **2013**, *11*, 72. [CrossRef] [PubMed]
26. Boehm, M.; Hoy, B.; Rohde, M.; Tegtmeyer, N.; Bæk, K.T.; Oyarzabal, O.A.; Brøndsted, L.; Wessler, S.; Backert, S. Rapid paracellular transmigration of *Campylobacter jejuni* across polarized epithelial cells without affecting TER: Role of proteolytic-active HtrA cleaving E-cadherin but not fibronectin. *Gut Pathog.* **2012**, *4*, 3. [CrossRef] [PubMed]
27. Wine, E.; Chan, V.L.; Sherman, P.M. *Campylobacter jejuni* Mediated Disruption of Polarized Epithelial Monolayers is Cell-Type Specific, Time Dependent, and Correlates with Bacterial Invasion. *Pediatric Res.* **2008**, *64*, 599–604. [CrossRef] [PubMed]
28. Chen, M.L.; Ge, Z.; Fox, J.G.; Schauer, D.B. Disruption of tight junctions and induction of proinflammatory cytokine responses in colonic epithelial cells by *Campylobacter jejuni*. *Infect. Immun.* **2006**, *74*, 6581–6589. [CrossRef]
29. Heimesaat, M.M.; Grundmann, U.; Alutis, M.E.; Fischer, A.; Bereswill, S. Small Intestinal Pro-Inflammatory Immune Responses Following *Campylobacter jejuni* Infection of Secondary Abiotic IL-10−/− Mice Lacking Nucleotide-Oligomerization-Domain-2. *Eur. J. Microbiol. Immunol.* **2017**, *7*, 138–145. [CrossRef]
30. Luettig, J.; Rosenthal, R.; Lee, I.-F.M.; Krug, S.M.; Schulzke, J.D. The ginger component 6-shogaol prevents TNF-α-induced barrier loss via inhibition of PI3K/Akt and NF-κB signaling. *Mol. Nutr. Food Res.* **2016**, *60*, 2576–2586. [CrossRef]

31. Bücker, R.; Krug, S.M.; Fromm, A.; Nielsen, H.L.; Fromm, M.; Nielsen, H.; Schulzke, J.-D. *Campylobacter fetus* impairs barrier function in HT-29/B6 cells through focal tight junction alterations and leaks. *Ann. N. Y. Acad. Sci.* **2017**, *1405*, 189–201. [CrossRef] [PubMed]
32. Amasheh, M.; Fromm, A.; Krug, S.M.; Amasheh, S.; Andres, S.; Zeitz, M.; Fromm, M.; Schulzke, J.-D. TNFalpha-induced and berberine-antagonized tight junction barrier impairment via tyrosine kinase, Akt and NFkappaB signaling. *J. Cell Sci.* **2010**, *123*, 4145–4155. [CrossRef]
33. Rosenthal, R.; Luettig, J.; Hering, N.A.; Krug, S.M.; Albrecht, U.; Fromm, M.; Schulzke, J.D. Myrrh exerts barrier-stabilising and –protective effects in HT-29/B6 and Caco-2 intestinal epithelial cells. *Int. J. Colorectal Dis.* **2017**, *32*, 623–634. [CrossRef]
34. Bojarski, C.; Weiske, J.; Schöneberg, T.; Schröder, W.; Mankertz, J.; Schulzke, J.-D.; Florian, P.; Fromm, M.; Tauber, R.; Huber, O. The specific fates of tight junction proteins in apoptotic epithelial cells. *J. Cell Sci.* **2004**, *117*, 2097–2107. [CrossRef]
35. Amasheh, M.; Schlichter, S.; Amasheh, S.; Mankertz, J.; Zeitz, M.; Fromm, M.; Schulzke, J.D. Quercetin enhances epithelial barrier function and increases claudin-4 expression in Caco-2 cells. *J. Nutr.* **2008**, *138*, 1067–1073. [CrossRef] [PubMed]
36. Amasheh, M.; Luettig, J.; Amasheh, S.; Zeitz, M.; Fromm, M.; Schulzke, J.-D. Effects of quercetin studied in colonic HT-29/B6 cells and rat intestine in vitro. *Ann. N. Y. Acad. Sci.* **2012**, *1258*, 100–107. [CrossRef] [PubMed]
37. Young, K.T.; Davis, L.M.; Dirita, V.J. *Campylobacter jejuni*: Molecular biology and pathogenesis. *Nat. Rev. Microbiol.* **2007**, *5*, 665–679. [CrossRef] [PubMed]
38. Hanai, H.; Iida, T.; Takeuchi, K.; Watanabe, F.; Maruyama, Y.; Andoh, A.; Tsujikawa, T.; Fujiyama, Y.; Mitsuyama, K.; Sata, M.; et al. Curcumin maintenance therapy for ulcerative colitis: Randomized, multicenter, double-blind, placebo-controlled trial. *Clin. Gastroenterol. Hepatol.* **2006**, *4*, 1502–1506. [CrossRef] [PubMed]
39. Singla, V.; Pratap Mouli, V.; Garg, S.K.; Rai, T.; Choudhury, B.N.; Verma, P.; Deb, R.; Tiwari, V.; Rohatgi, S.; Dhingra, R.; et al. Induction with NCB-02 (curcumin) enema for mild-to-moderate distal ulcerative colitis—A randomized, placebo-controlled, pilot study. *J. Crohn's Colitis* **2014**, *8*, 208–214. [CrossRef] [PubMed]
40. Lang, A.; Salomon, N.; Wu, J.C.; Kopylov, U.; Lahat, A.; Har-Noy, O.; Ching, J.Y.; Cheong, P.K.; Avidan, B.; Gamus, D.; et al. Curcumin in Combination With Mesalamine Induces Remission in Patients With Mild-to-Moderate Ulcerative Colitis in a Randomized Controlled Trial. *Clin. Gastroenterol. Hepatol.* **2015**, *13*, 1444–1449. [CrossRef] [PubMed]
41. Cruz-Correa, M.; Shoskes, D.A.; Sanchez, P.; Zhao, R.; Hylind, L.M.; Wexner, S.D.; Giardiello, F.M. Combination treatment with curcumin and quercetin of adenomas in familial adenomatous polyposis. *Clin. Gastroenterol. Hepatol.* **2006**, *4*, 1035–1038. [CrossRef] [PubMed]
42. Verma, S.P.; Salamone, E.; Goldin, B. Curcumin and genistein, plant natural products, show synergistic inhibitory effects on the growth of human breast cancer MCF-7 cells induced by estrogenic pesticides. *Biochem. Biophys. Res. Commun.* **1997**, *233*, 692–696. [CrossRef] [PubMed]
43. Balasubramanian, S.; Eckert, R.L. Green tea polyphenol and curcumin inversely regulate human involucrin promoter activity via opposing effects on CCAAT/enhancer-binding protein function. *J. Biol. Chem.* **2004**, *279*, 24007–24014. [CrossRef] [PubMed]
44. Wang, Y.J.; Pan, M.H.; Cheng, A.-L.; Lin, L.-I.; Ho, Y.-S.; Hsieh, C.-Y.; Lin, J.-K. Stability of curcumin in buffer solutions and characterization of its degradation products. *J. Pharm. Biomed. Anal.* **1997**, *15*, 1867–1876. [CrossRef]
45. Bergann, T.; Plöger, S.; Fromm, A.; Zeissig, S.; Borden, S.A.; Fromm, M.; Schulzke, J.D. A colonic mineralocorticoid receptor cell model expressing epithelial Na+ channels. *Biochem. Biophys. Res. Comm.* **2009**, *381*, 280–285. [CrossRef] [PubMed]
46. Kreusel, K.M.; Fromm, M.; Schulzke, J.D.; Hegel, U. Cl- secretion in epithelial monolayers of mucus-forming human colon cells (HT-29/B6). *Am. J. Physiol.* **1991**, *261*, C574–C582. [CrossRef] [PubMed]
47. Zeissig, S.; Bürgel, N.; Günzel, D.; Richter, J.; Mankertz, J.; Wahnschaffe, U.; Kroesen, A.J.; Zeitz, M.; Fromm, M.; Schulzke, J.D. Changes in expression and distribution of claudin 2, 5 and 8 lead to discontinuous tight junctions and barrier dysfunction in active Crohn's disease. *Gut* **2007**, *56*, 61–72. [CrossRef] [PubMed]
48. Kämpfer, A.A.M.; Urbán, P.; Gioria, S.; Kanase, N.; Stone, V.; Kinsner-Ovaskainen, A. Development of an in vitro co-culture model to mimic the human intestine in healthy and diseased state. *Toxicol. In Vitro* **2017**, *45*, 31–43. [CrossRef]

49. Haag, L.M.; Fischer, A.; Otto, B.; Plickert, R.; Kühl, A.A.; Göbel, U.B.; Bereswill, S.; Heimesaat, M.M. Intestinal microbiota shifts towards elevated commensal *Escherichia coli* loads abrogate colonization resistance against Campylobacter jejuni in mice. *PLoS ONE* **2012**, *7*, e35988. [CrossRef]
50. Bereswill, S.; Plickert, R.; Fischer, A.; Kühl, A.A.; Loddenkemper, C.; Batra, A.; Siegmund, B.; Gobel, U.B.; Heimesaat, M.M. What you eat is what you get: Novel *Campylobacter* models in the quadrangle relationship between nutrition, obesity, microbiota and susceptibility to infection. *Eur. J. Microbiol. Immunol.* **2011**, *1*, 237–248. [CrossRef]
51. Gitter, A.H.; Schulzke, J.-D.; Sorgenfrei, D.; Fromm, M. Ussing chamber for high-frequency transmural impedance analysis of epithelial tissues. *J. Biochem. Biophys. Methods* **1997**, *35*, 81–88. [CrossRef]
52. Krüger, N.J.; Knüver, M.T.; Zawilak-Pawlik, A.; Appel, B.; Stingl, K. Genetic Diversity as Consequence of a Microaerobic and Neutrophilic Lifestyle. *PLoS Pathog.* **2016**, *12*, e1005626. [CrossRef] [PubMed]

International Journal of
Molecular Sciences

Article

Caprate Modulates Intestinal Barrier Function in Porcine Peyer's Patch Follicle-Associated Epithelium

Judith Radloff [1,2], Valeria Cornelius [1], Alexander G. Markov [3] and Salah Amasheh [1,*]

1 Institute of Veterinary Physiology, Department of Veterinary Medicine, Freie Universität Berlin, 14163 Berlin, Germany; judith.radloff@vetmeduni.ac.at (J.R.); valeria.cornelius@fu-berlin.de (V.C.)
2 Department for Biomedical Sciences, University of Veterinary Medicine, 1210 Vienna, Austria
3 Department of General Physiology, St. Petersburg State University, Universitetskaya nab. 7/9, 199034 St. Petersburg, Russia; markov_51@mail.ru
* Correspondence: salah.amasheh@fu-berlin.de; Tel.: +49-30-838-62602

Received: 7 February 2019; Accepted: 19 March 2019; Published: 20 March 2019

Abstract: Background: Many food components influence intestinal epithelial barrier properties and might therefore also affect susceptibility to the development of food allergies. Such allergies are triggered by increased antibody production initiated in Peyer's patches (PP). Usually, the presentation of antigens in the lumen of the gut to the immune cells of the PP is strongly regulated by the follicle-associated epithelium (FAE) that covers the PP. As the food component caprate has been shown to impede barrier properties in villous epithelium, we hypothesized that caprate also affects the barrier function of the PP FAE, thereby possibly contributing a risk factor for the development of food allergies. Methods: In this study, we have focused on the effects of caprate on the barrier function of PP, employing in vitro and ex vivo experimental setups to investigate functional and molecular barrier properties. Incubation with caprate induced an increase of transepithelial resistance, and a marked increase of permeability for the paracellular marker fluorescein in porcine PP to 180% of control values. These effects are in accordance with changes in the expression levels of the barrier-forming tight junction proteins tricellulin and claudin-5. Conclusions: This barrier-affecting mechanism could be involved in the initial steps of a food allergy, since it might trigger unregulated contact of the gut lumen with antigens.

Keywords: tissue barrier; tight junction; claudins; tricellulin

---

1. Introduction

The gastrointestinal mucosa is one of the largest surface areas of the body and constantly interacts with possibly harmful substances or non-hazardous food components. The intestinal epithelial cells and the associated immune cells of the gut-associated-lymphocytic tissue (GALT) need to be able to distinguish the category of a molecule.

Peyer's patches (PP) represent the dominating part of the GALT. They are structured into a germinal center that consists of macrophages, lymphocytes, and dendritic cells (DC) and a covering epithelial cell layer, namely the follicle-associated epithelium (FAE) [1]. The FAE of rat and pig has recently been characterized in detail, giving a greater insight into the barrier properties directly above the PP [2,3]. Generally, the FAE has a more porous basal membrane and contains fewer goblet cells, resulting in a thinner mucus layer, than the villous epithelium (VE) [4,5]. The FAE also includes microfold cells (M-cells), which are specialized for antigen translocation to the underlying cells of the PP [4]. The presentation of antigens can also be initiated via DC that reach through the epithelial cell layer into the gut lumen and directly sample bacteria. During this process, the barrier integrity is preserved, as DC also express tight junction (TJ) proteins [6]. Overall, the barrier of the FAE can be classified as tight when compared with the surrounding leaky VE, as sealing claudins such as claudin-1, -3, -4, -5, and -8 are predominantly expressed in PP samples of various species [2,3,7].

Numerous food components have been shown to influence barrier integrity in the VE, including the medium chain fatty acids laurate and caprate [8–10], the amino acids glutamine and glycine [11,12], and plant components quercetin and berberine [13,14], all which lead to a change of transepithelial resistance (TER) and paracellular permeability. In turn, this can enhance or limit the presentation of common food components to the immune cells of the PP, triggering or impeding the excessive formation of immunoglobulin E (IgE) [15]. The consequence is commonly known as food allergy.

Food allergies are defined as an adverse health effect arising from a specific immune response that occurs reproducibly on exposure to a given food [16]. With food allergies being more prevalent than ever, the need for an effective treatment or prevention mechanism is obvious [17]. This remains difficult for as long as the precise causes remain unidentified.

One of these causes might involve the barrier-weakening effect of the medium chain fatty acid caprate. Caprate has been shown to decrease the TER reversibly, leading to an increase of paracellular permeability for fluorescein and various fluorescein isothyocyanate (FITC)-dextrans (FD4, FD10) in the intestinal cell line HT29-B6 [8]. This can be attributed to the reversible decrease of the sealing TJ proteins claudin-5 and tricellulin, an effect opening the paracellular pathway [8]. Because of its reversible barrier-weakening properties, caprate is used as an absorption enhancer, improving membrane permeation for pharmaceutically active agents [18]. Since caprate is also contained in common foods, such as milk and coconut oil, it is widely consumed by most people, albeit in low concentrations [19].

Despite the numerous studies on the effect of caprate on intestinal barrier function [8,20,21], the effect of caprate on the FAE has not yet been investigated. Therefore, we have focused on the barrier-influencing properties of caprate on the FAE of the porcine PP and on the consequences with regard to gastrointestinal immunology.

2. Results

2.1. Cell Culture Study

Initial studies on the effect of caprate on the porcine intestinal barrier function were carried out by employing the Intestinal Porcine Epithelial Cell line-J2 (IPEC-J2) and 5 mM caprate.

Incubation of IPEC-J2 cells with 5 mM caprate led to a rapid and consistent decrease of barrier function (from 5.85 ± 0.38 kΩ to 3.68 ± 0.59 kΩ, ** $p < 0.01$, $n = 20$) when compared with the control group (not significant (n. s.), $n = 20$; Figure 1). This decrease was evident during the whole timespan of the experiment (120 min). After a recovery period of 24–48 h, TER values of the caprate group were comparable with control levels (4.78 ± 0.63 kΩ to 5.05 ± 0.67 kΩ, $n = 20$).

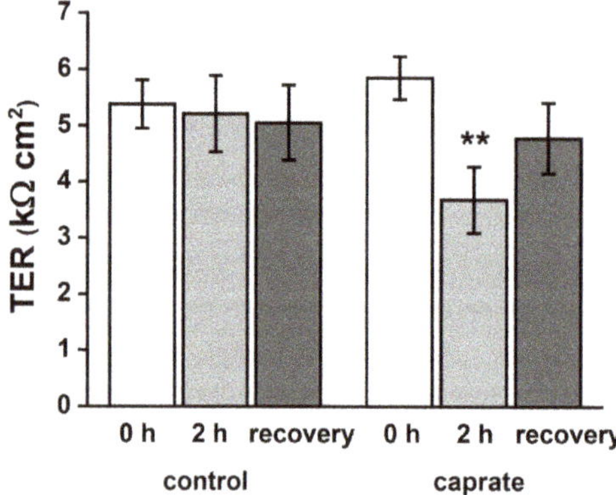

Figure 1. Transepithelial electrical resistance (TER) of IPEC-J2 cell monolayers. Incubation with 5 mM caprate leads to a reversible decrease of TER (** $p < 0.01$, $n = 20$).

2.2. Ussing Chamber Experiments

The effect of caprate on PP tissue taken from adult pigs was analyzed by the Ussing chamber technique. Tissue was incubated with 5 mM caprate while TER values and paracellular permeability measurements for sodium fluorescein were carried out.

In contrast to the IPEC-J2 experiments, the TER of PP increased significantly after incubation of the tissue with caprate (240 min; 92.97 ± 4.94% to 116.73 ± 7.55%, * $p < 0.05$, n = 11–12) (Figure 2A). However, the permeability measurements of sodium fluorescein revealed a marked increase to 180.24 ± 29.52% (* $p < 0.05$, n = 8; Figure 2B).

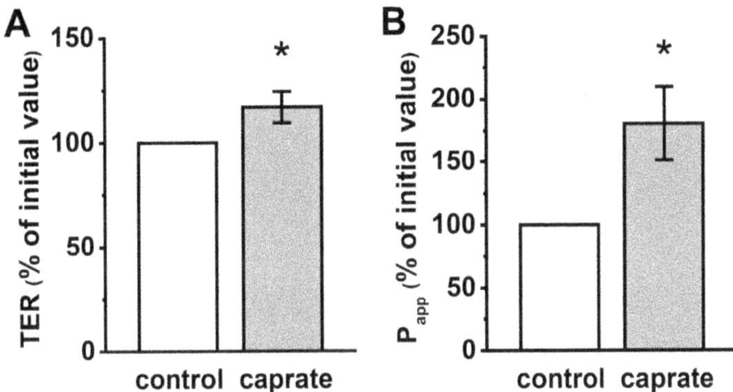

Figure 2. (**A**) Transepithelial electrical resistance (TER) and (**B**) apparent permeability (P_{app}) of the paracellular flux marker fluorescein in porcine Peyer's patches (PP). 5 mM caprate lead to markedly higher TER values (* $p < 0.05$, n = 11–12), and increased permeability for fluorescein (n = 8, * $p < 0.05$).

2.3. Immunoblotting

To account for the detected functional differences in TER, the expression levels of sealing TJ proteins were analyzed; the immunoblots revealed major changes (Figure 3). Whereas claudin-3 only showed a positive trend in expression levels (162.7 ± 31.23%, p = 0.064, n = 8), claudin-5 was significantly increased (198.82 ± 40.11%, $p < 0.05$, n = 8). Tricellulin on the other hand was significantly reduced (66.64 ± 13.69%, $p < 0.05$, n = 6). Sealing claudin-1 in PP tissue (139.32 ± 24.62%, n = 8) was not affected by caprate incubation.

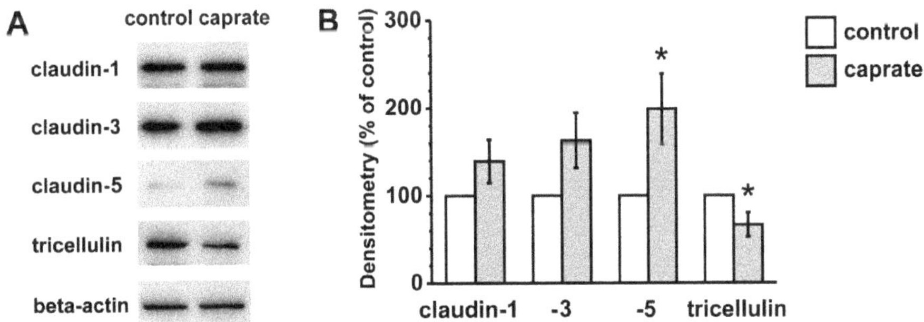

Figure 3. (**A**) Immunoblots and (**B**) densitometry of TJ proteins in PP tissue (control versus caprate). Densitometric analysis of TJ protein signals after caprate incubation reveals the lower expression of tricellulin, whereas stronger expression of claudin-5 can be detected (n = 6–8, * $p < 0.05$).

2.4. Immunohistochemistry

To analyze the location of single TJ proteins, immunoflourescent staining of tissue specimens was performed following the Ussing chamber experiments. Immunohistochemistry of these self-same tissues revealed specific signals for TJ proteins in all tested PP tissue specimens, as shown for claudin-3 and claudin-5 (Figure 4).

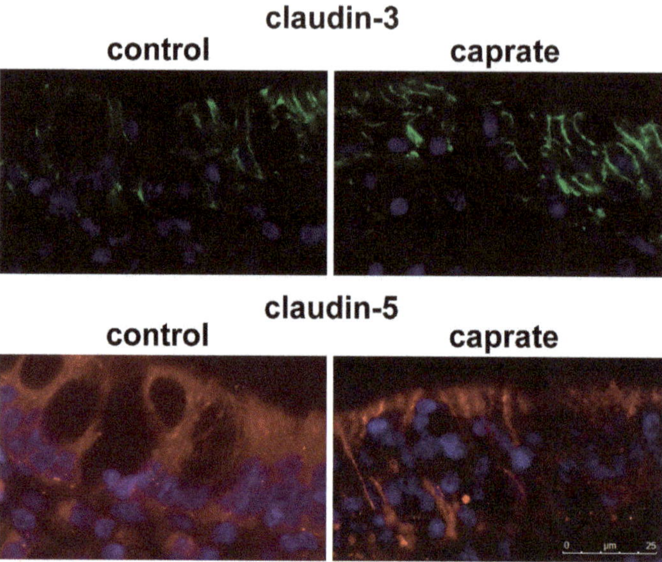

Figure 4. Immunohistochemistry. TJ proteins claudin-3 and claudin-5 are detectable as paracellular signals in the surface epithelium in PP tissue (bar: 25 μm).

3. Discussion

3.1. Suitability of a Porcine Model for Human Gastrointestinal Pathophysiology

Porcine models exhibit certain benefits when exploring molecular mechanisms of intestinal pathophysiological conditions. In 2007, Humphray et al. showed that 60% of the porcine and human genome sequence match [22]. The morphology and anatomy of the human gastrointestinal tract is comparable with that of a pig [23]. Moreover, the physiology of digestive processes and the subsequent absorption of nutrients are also similar [24,25]. Last but not least, the pig is the only commonly used laboratory animal that has comparable nutritional needs with humans, on account of it being classified as an omnivore [26].

Since its origin in 1989, the IPEC-J2 cell line has been extensively used for studies on the gastrointestinal barrier and the effects of nutrients. It represents one of the best cell lines of non-human origin for intestinal research because of its non-transformed properties [27,28]. Taking all of these aspects into account, we chose porcine intestinal epithelial tissue specimens and the IPEC-J2 cell line as promising models for studying the effects of caprate on barrier function with regard to immunologically relevant mechanisms, which might also provide a perspective as a model for porcine PP FAE.

3.2. Caprate in Modern Foods

Coconut oil is one of the main contributors of caprate in Western nutrition. Since it is currently regarded as a superfood, the intake of coconut oil, and therefore caprate, has increased tremendously. The health benefits of coconut oil include weight loss, cardioprotective effects, and improved immunity and memory [29]. The milk of various mammalian species is the other main source of caprate in current nutrition [19,30]. Existing data on the health benefits of milk include preventive effects against diabetes mellitus, cancer, hypertension, and obesity [31].

Even without the specific ingestion of caprate as an absorption enhancer, the daily intake of caprate can be high because of the recommended consumption of 2.5–3 servings of dairy products [32]. Digestion effects can be expectedly rather limited, as the medium chain fatty acid absorbed is non-modified, even devoid pancreatic enzymes [33]. Moreover, medium-chain triglycerides are degraded into fatty acids and the glycerol backbone, a mechanism adding to luminal caprate concentration. After absorption, the medium chain fatty acids are transferred into the portal vein for transport to the liver for hepatic metabolism [34]. However, prior to absorption, luminal caprate effects might even add up with effects of other bioactive compounds, e.g., laurate, which then might reach barrier-affecting concentrations earlier [9].

3.3. *Effects of Caprate*

Since foods containing caprate are consumed regularly, the effects of the fatty acid need to be brought into focus. Currently, caprate is used as an absorption enhancer to improve the bioavailability of pharmaceutically active agents [18]. A human in vitro cell culture experiment, employing the colonic HT-29/B6 cell line, has revealed the cellular mechanisms involving the selective removal of claudin-5 and tricellulin from the TJ [8]. Recent in vitro experiments have provided evidence of the antibacterial effect of caprate in porcine epithelial cells when challenged with *Salmonella enteritidis* or an enteropathogenic *Escherichia coli* strain [35]. Furthermore, caprate has also been shown to reduce body fat and therefore, body weight [36].

Data concerning the intestinal barrier function obtained in this study has shown that the effect of caprate can also occur in other species. Significantly lower TER values were recorded for the porcine IPEC-J2 cell line after caprate incubation (5 mM). This effect was reversible in both species (compare with [8] for human cells). However, transferring this functional result to the ex vivo setup using the Ussing chamber technique was not possible, as the TER markedly increased in PP tissue. This apparent discrepancy might reflect a limited suitability of the plain IPEC-J2 model for modeling the PP FAE. However, it might still provide a suitable matrix for, e.g., co-culturing experiments approaching a functional FAE model, in vitro.

Permeability measurements in PP tissue revealed a higher permeability for fluorescein, which can be compared with glucose in size (330 Da). A higher permeability for fluorescein, but also for FITC-dextran (4k, 10k), was observed after incubation with 10 mM caprate in HT-29/B6 cells [8].

Increased paracellular permeability might give rise to unregulated contact between the immune cells of the PP and food components. The most common foods that provoke allergic reactions are nuts, eggs, milk, soy, wheat, and shell fish [37]. The uptake of antigens within the PP is strictly regulated, as it is usually undertaken by specific binding to receptors, through specialized epithelial cells, the M-cells, or via the extending dendrites of DC belonging to the underlying PP [4,6,38]. An impairment of the intestinal barrier function would interrupt these controlled processes.

3.4. *Caprate Effects on Tight Junction Protein Expression*

The study has shown that claudin-5 and tricellulin have been removed from bi- and tricellular TJ through caprate incubation [8]. TJ are apicolateral epithelial cell membrane contacts determining paracellular permeability [39–42]. Claudin-5 is classically regarded as a sealing claudin, since it increases barrier properties by effects on TER [43,44].

Tricellulin was the first barrier-forming TJ protein to be found in tricellular TJ, and supression of tricellulin leads to the disruption of bicellular and tricellular TJ [45]. Overexpression of tricellulin increases barrier function and decreases the permeability of ions and larger solutes [46]. As caprate induces a reversible removal of claudin-5 and tricellulin from the TJ, functional results under caprate incubation can easily be attributed to structural changes of TJ formation.

The immunoblots after Ussing chamber experiments in our study also revealed changes in claudin-5 and tricellulin. Quantitative analysis of PP tissue samples showed an increased expression level of claudin-5 and decreased levels of tricellulin. These results suggest the differential sensitivity of the FAE towards the effects of caprate when compared with the intestinal epithelial monolayers that were used in the previously performed initial incubation experiment [8]. This is understandable when we keep the main task of the PP in mind. A structural change in TJ formation prevents the accidental passage of antigen to the underlying PP. Our immunohistological assessment has shown the paracellular localization of claudin-3 and claudin-5. Although a decrease of claudin-5 signals could not be observed in the current study, the overall increase of its expression level might have contributed to a compensation, in accordance with the observed increase of TER. This apparent inconsistency might also be explained by the different contribution of single tight junction proteins to transepithelial resistance and permeability of flux markers, as tricellulin is affecting the paracellular passage of molecules [46], whereas claudin-5 primarily affects the passage of ions [44]. However, our results indicate that it is possible to functionally uncouple resistance and solute permeability by affecting different tight junction proteins in parallel, which includes the possibility that an increased expression of a sealing protein such as claudin-5 does not necessarily provide an expected primary functional effect.

However, further studies, particularly in humans, are required to demonstrate that caprate may affect the susceptibility to the development of food allergies.

4. Materials and Methods

4.1. Cell Culturing and Experiments

The porcine jejunal cell line IPEC-J2 is derived from the jejunum of unsuckled piglets and is commonly regarded as an in vitro model for the analysis of the intestinal barrier function in both pigs and humans [27]. For routine cultivation, Dulbecco's Modified Eagle Medium (MEM)/ Nutrient F 12 Ham's (Biochrom, Berlin, Germany) was used as the medium basis, with 10% fetal calf serum, 1% penicillin-streptomycin, 5 ng /mL epithelial growth factor (Biochrom, Berlin, Germany) and insulin, transferrin and selenium (Life Technologies, Carlsbad, CA, USA) as supplements. Cells were passaged twice a week and fed every other day.

For experiments, cells (10^5) were seeded on cell culture inserts with a pore size of 0.45 μm (Millicell, Darmstadt, Germany). The seeded inserts were placed into multiwell plates and cultivated for approximately 14 days until a functional barrier had formed. This was assessed via measurements of TER employing a chopstick electrode and an Epithelial Volt/Ohm Meter (EVOM) measuring device (World Precision Instruments, Hertfordshire, United Kingdom). Only filters that showed TER values above 2 kΩ were chosen for experimentation. Passages ranging from 9 to 15 were used in this experimental setup. TER values were measured initially in plain culture medium. Subsequently, the culture media was replaced with medium A (plain culture medium) or medium B (culture medium with 5 mM caprate). During the incubation period of 120 min, TER values were measured every 20 min, after which the medium was replaced with plain culture medium. Following a recovery period of 24–48 h, TER was measured again. The cells were kept at 37 °C during the recovery period.

4.2. Tissue Samples

Samples of the distal small intestine of 12 adult pigs were taken immediately after slaughter at a commercial slaughterhouse (Salzbrunn, Germany). PP tissue was transported in cold (4–6 °C) buffer solution containing (in mmol·L-1): Na^+ (149.9), Cl^- (128.8), K^+ (5), Ca^{2+} (1.2), Mg^{2+} (1.2), HCO_3^- (25), $H_2PO_4^{2-}$ (0.6), HPO_4^- (1.2), and D-glucose (10). The pH was set to 7.4 before transport.

4.3. Ussing Chamber Experiments

Tissue samples were mounted in a common Ussing chamber setup (Mussler Scientific Instruments, Aachen, Germany), and data recording commenced after 45 min of preincubation with fluorescein, but only once electrophysiological values were stable. The incubation experiment itself was carried out under voltage clamp conditions. The experimental buffer contained (in mmol·L-1): Na^+ (149.9), Cl^- (128.8), K^+ (5), Ca^{2+} (1.2), Mg^{2+} (1.2), HCO_3^- (25), $H_2PO_4^{2-}$ (0.6), HPO_4^- (1.2), and D-glucose (10). It was constantly gassed with 95% O_2 and 5% CO_2, resulting in a pH of 7.4. After 30 min, 5 mM caprate was added to the apical side. TER was recorded continuously throughout the overall incubation time of 4 h. Electrophysiological data are expressed in relation to the initial value before the addition of caprate.

4.4. Permeability Measurements

Once the tissue had been mounted in the Ussing chamber, fluorescein (100 μmol·L^{-1}) was added to the apical side of each chamber. During the incubation experiment, permeability samples were taken every 60 min from the basolateral side. The volume that was taken was replaced with fresh buffer solution. Samples were measured photometrically at 514 nm by employing the EnSpire Multimode Plate Reader (Perkin Elmer, Waltham, MA, USA). Paracellular flux and permeability were then calculated by using the following equations.

$$c = (c_{t-1} \times V_S + c_t \times V_K)/V_K \tag{1}$$

$$J = (c_t - c_{t-1}) \times V_K/\Delta t \times A \tag{2}$$

$$P_{app} = J/c_a. \tag{3}$$

In Equation (1), the measured concentration of fluorescein in the basolateral sample has to be corrected for the dilution that takes effect because of the replacement of the experimental buffer, with ct being the measured concentration, c_{t-1} the measured concentration of the previous period, V_S the sample volume, and V_K the volume of the Ussing chamber. Equation (2) gives the paracellular flux (J) for fluorescein, with Δt being the duration of the measurement period, and A being the area within the Ussing chamber. Finally, the paracellular permeability can be calculated by using Equation (3), with c_a being the fluorescein concentration in the apical side of the Ussing chamber. Results were expressed as a fraction of the control values.

4.5. Protein Extraction and Quantification

Samples for protein extraction were taken out of the Ussing chamber and directly placed into liquid nitrogen. These samples were homogenized in a Tris-buffer containing in mmol·L -1: Tris (10), NaCl (150), Triton X-100 (0.5), SDS (0.1), and enzymatic protease inhibitors (Complete, Boehringer, Mannheim, Germany). After homogenization, samples were centrifuged for 1 min at 13,000 rpm (Eppendorf centrifuge 5418, Eppendorf AG, Hamburg, Germany). The supernatant was then cooled on ice for 30 min and retrieved after a second centrifugation step for 15 min at 15,000 g at 4 °C (sigma 3-30ks, Sigma-Aldrich, Munich, Germany).

Protein quantification of PP tissues was carried out by using Bradford reagent (Bio-Rad Laboratories GmbH, Munich, Germany) as instructed in the leaflet provided, and the EnSpire Multimode Plate Reader (Perkin Elmer, Waltham, MA, USA) was used for detection.

4.6. Immunoblotting and Densitometry

Protein (4.1 μg) and Laemmli buffer (Bio-Rad Laboratories GmbH, Munich, Germany) were mixed and loaded onto a 12.5% polyacrylamide gel. Electrophoresis was allowed to run for 5 min at 200 mV, followed by 60 min at 100 mV. Afterwards, the samples were transferred onto a PVDF membrane, which was blocked for 60 min in 5% milk (in Tris-buffered saline with 0.1% Tween 20) and later incubated with the antibodies for claudin-1, claudin-3, claudin-5, tricellulin, or beta-actin (Life Technologies, Carlsbad, CA, USA) overnight at 4 °C. Antibodies were used as recommended by the manufacturer (1 μg/mL). To enable visual detection, peroxidase-conjugated goat anti-mouse or anti-rabbit antibodies (Cell Signaling Technology, Danvers, MA, USA) were employed to bind the primary antibodies. Signals were then visualized by employing the chemiluminescence detection system of Clarity Western ECL Blotting Substrate (Bio-Rad Laboratories GmbH, Munich, Germany) and a luminescence imager (ChemiDoc MP, Munich, Germany).

The density of individual immunoblotting bands was analyzed by using the imager-associated software (ImageLab, BioRad, Hercules, CA, USA). The measured values were divided by the coherent beta-actin band signal. Finally, the signal for the caprate challenge was expressed in relation to the control values for VE and PP, which were set to 100%.

4.7. Immunohistochemistry

Tissue samples were placed into 4% formaldehyde (Roti-Histofix, Karlsruhe, Germany) for 24 h, dehydrated by using an increasing alcohol gradient, transferred to xylol, and embedded in paraffin. Paraffin blocks were cut into 3-µm-thick sections on a Leica RM 2245 microtome (Leica Microsystems Heidelberg, Germany). Before staining, the sections on the slides were warmed for 30 min at 60 °C (incubator), deparaffinized in xylol, and taken through a decreasing alcohol gradient. Depending on the protein, EDTA or citrate buffer was used for heat-induced antigen retrieval (45 min). Following the permeabilization of the samples with Triton X-100 (Carl Roth, Karlsruhe, Germany), sections were incubated with 5% goat serum in PBS and subsequently with antibodies raised against claudin-3 and -5 (1:100, Life Technologies, Carlsbad, CA, USA) for 60 min at 37 °C. After a washing step in blocking solution, samples were incubated with goat anti-mouse and -rabbit Alexa Fluor-488, and -594 (1:1000, Life Technologies, Carlsbad, CA, USA) and 4′,6-diamidino-2-phenylindole (DAPI, 1:5000) for 60 min at 37 °C. Sections were then mounted with ProTags Mount Fluor (Biocyc, Luckenwalde, Germany). Analyses and imaging of these sections was carried out by employing a Leica microscope of the DMI 6000 series (Leica Microsystems, Heidelberg, Germany).

4.8. Statistical Analysis

Results are always expressed as the mean ± standard error of the mean, with n being the number of animals used. For in vitro experiments, n represents the number of experiments carried out. When indicated, relative values were used for statistical evaluation.

Data from the cell culture experiments were analyzed by employing the paired Student's t-test, whereas the results of the Ussing chamber experiments were evaluated with the unpaired Student's t-test. Significant difference was considered at $p < 0.05$.

5. Conclusions

The results of this study show an increased permeability in the PP and a reorganization of the TJ structure after caprate incubation. Since caprate is a prevalent food component, these data imply a constant increasing effect on the paracellular permeability of the PP FAE. This might also accelerate the development of allergies, since more molecules are presented to the immune cells of the PP.

Author Contributions: J.R. and V.C. performed the experiments, J.R., A.G.M. and S.A. discussed the data and wrote the manuscript.

Funding: This work was supported by the Deutsche Forschungsgemeinschaft (grant No. AM141/11-1) and RFBR (grant No. 16-04-01661).

Acknowledgments: We thank Martin Grunau, Uwe Tietjen, and Evgeny Falchuk for excellent technical assistance.

Conflicts of Interest: The authors declare no conflict of interest.

Abbreviations

Cldn	Claudin
DC	Dendritic cells
FAE	Follicle-associated epithelium
GALT	Gut-associated lymphocytic tissue
IgE	Immunoglobulin E
M-cell	Microfold cell
PP	Peyer's patch
TER	Transepithelial electrical resistance
VE	Villous epithelium

References

1. Neutra, M.R.; Mantis, N.J.; Kraehenbuhl, J.P. Collaboration of epithelial cells with organized mucosal lymphoid tissues. *Nat. Immunol.* **2001**, *2*, 1004–1009. [CrossRef] [PubMed]
2. Markov, A.G.; Falchuk, E.F.; Kruglova, N.M.; Radloff, J.; Amasheh, S. Claudin expression in follicle-associated epithelium of rat Peyer's patches defines a major restriction of the paracellular pathway. *Acta Physiol.* **2016**, *216*, 112–119. [CrossRef] [PubMed]
3. Radloff, J.; Falchuk, E.L.; Markov, A.G.; Amasheh, S. Molecular Characterization of Barrier Properties in Follicle-Associated Epithelium of Porcine Peyer's Patches Reveals Major Sealing Function of Claudin-4. *Front. Physiol.* **2017**, *8*, 579. [CrossRef] [PubMed]
4. Owen, R.L.; Jones, A.L. Epithelial cell specialization within human Peyer's patches: An ultrastructural study of intestinal lymphoid follicles. *Gastroenterology* **1974**, *66*, 189–203. [CrossRef]
5. Takeuchi, T.; Gonda, T. Distribution of the pores of epithelial basement membrane in the rat small intestine. *J. Vet. Med. Sci.* **2004**, *66*, 695–700. [CrossRef] [PubMed]
6. Rescigno, M.; Urbano, M.; Valzasina, B.; Francolini, M.; Rotta, G.; Bonasio, R.; Granucci, F.; Kraehenbuhl, J.P.; Ricciardi-Castagnoli, P. Dendritic cells express tight junction proteins and penetrate gut epithelial monolayers to sample bacteria. *Nat. Immunol.* **2001**, *2*, 361–367. [CrossRef]
7. Tamagawa, H.; Takahashi, I.; Furuse, M.; Yoshitake-Kitano, Y.; Tsukita, S.; Ito, T.; Matsuda, H.; Kiyono, H. Characteristics of claudin expression in follicle-associated epithelium of Peyer's patches: Preferential localization of claudin-4 at the apex of the dome region. *Lab. Invest.* **2003**, *83*, 1045–1053. [CrossRef]
8. Krug, S.M.; Amasheh, M.; Dittmann, I.; Christoffel, I.; Fromm, M.; Amasheh, S. Sodium caprate as an enhancer of macromolecule permeation across tricellular tight junctions of intestinal cells. *Biomaterials* **2013**, *34*, 275–282. [CrossRef]
9. Dittmann, I.; Amasheh, M.; Krug, S.M.; Markov, A.G.; Fromm, M.; Amasheh, S. Laurate permeates the paracellular pathway for small molecules in the intestinal epithelial cell model HT-29/B6 via opening the tight junctions by reversible relocation of claudin-5. *Pharm. Res.* **2014**, *31*, 2539–2548. [CrossRef]
10. Radloff, J.; Zakrzewski, S.S.; Pieper, R.; Markov, A.G.; Amasheh, S. Porcine milk induces a strengthening of barrier function in porcine jejunal epithelium in vitro. *Ann. N. Y. Acad. Sci.* **2017**, *1397*, 110–118. [CrossRef]
11. Li, W.; Sun, K.; Ji, Y.; Wu, Z.; Wang, W.; Dai, Z.; Wu, G. Glycine regulates expression and distribution of claudin-7 and ZO-3 proteins in intestinal porcine epithelial cells. *J. Nutr.* **2016**, *146*, 964–969. [CrossRef] [PubMed]
12. Wang, B.; Wu, Z.; Ji, Y.; Sun, K.; Dai, Z.; Wu, G. L-Glutamine enhances tight junction integrity by activating CaMK kinase 2-AMP-activated protein kinase signaling in intestinal porcine epithelial cells. *J. Nutr.* **2016**, *146*, 501–508. [CrossRef]
13. Amasheh, M.; Schlichter, S.; Amasheh, S.; Mankertz, J.; Zeitz, M.; Fromm, M.; Schulzke, J.D. Quercetin enhances epithelial barrier function and increases claudin-4 expression in Caco-2 cells. *J. Nutr.* **2008**, *138*, 1067–1073. [CrossRef]
14. Amasheh, M.; Fromm, A.; Krug, S.M.; Amasheh, S.; Andres, S.; Zeitz, M.; Fromm, M.; Schulzke, J.D. TNFalpha-induced and berberine-antagonized tight junction barrier impairment via tyrosine kinase, Akt and NFkappaB signaling. *J. Cell Sci.* **2010**, *123*, 4145–4155. [CrossRef] [PubMed]
15. Berin, M.C. Pathogenesis of IgE-mediated food allergy. *Clin. Exp. Allergy* **2015**, *45*, 1483–1496. [CrossRef] [PubMed]

16. Boyce, J.A.; Assa'ad, A.; Burks, A.W.; Jones, S.M.; Sampson, H.A.; Wood, R.A.; Plaut, M.; Cooper, S.F.; Fenton, M.J.; Arshad, S.H.; et al. Guidelines for the Diagnosis and Management of Food Allergy in the United States: Summary of the NIAID-Sponsored Expert Panel Report. *J. Allergy Clin. Immunol.* **2010**, *126*, 1105–1118. [CrossRef] [PubMed]
17. Tang, M.L.; Mullins, R.J. Food allergy: Is prevalence increasing? *Intern. Med. J.* **2017**, *47*, 256–261. [CrossRef]
18. Lindmark, T.; Söderholm, J.D.; Olaison, G.; Alvan, G.; Ocklind, G.; Artursson, P. Mechanism of Absorption Enhancement in Humans After Rectal Administration of Ampicillin in Suppositories Containing Sodium Caprate. *Pharm. Res.* **1997**, *14*, 930–935. [CrossRef] [PubMed]
19. Beare-Rogers, J.L.; Dieffenbacher, A.; Holm, J.V. Lexicon of lipid nutrition (IUPAC Technical Report). *Pure. Appl. Chem.* **2001**, *73*, 685–744. [CrossRef]
20. Tomita, M.; Hayashi, M.; Awazu, S. Absorption-enhancing mechanism of sodium caprate and decanoylcarnitine in Caco-2 cells. *J. Pharmacol. Exp. Ther.* **1995**, *272*, 739–743. [CrossRef] [PubMed]
21. Lv, X.Y.; Li, J.; Zhang, M.; Wang, C.M.; Fan, Z.; Wang, C.Y.; Chen, L. Enhancement of Sodium Caprate on Intestine Absorption and Antidiabetic Action of Berberine. *AAPS PharmSciTech.* **2010**, *11*, 372–382. [CrossRef]
22. Humphray, S.J.; Scott, C.E.; Clark, R.; Marron, B.; Bender, C.; Camm, N.; Davis, J.; Jenks, A.; Noon, A.; Patel, M.; et al. A high utility integrated map of the pig genome. *Genome Biol.* **2007**, *8*, R139. [CrossRef] [PubMed]
23. Kararli, T.T. Comparison of the gastrointestinal anatomy, physiology, and biochemistry of humans and commonly used laboratory animals. *Biopharm. Drug. Dispos.* **1995**, *16*, 351–380. [CrossRef] [PubMed]
24. Rowan, A.M.; Moughan, P.J.; Wilson, M.N.; Maher, K.; Tasman-Jones, C. Comparison of the ileal and faecal digestibility of dietary amino acids in adult humans and evaluation of the pig as a model animal for digestion studies in man. *Br. J. Nutr.* **1994**, *71*, 29–42. [CrossRef] [PubMed]
25. Patterson, J.K.; Lei, X.G.; Miller, D.D. The Pig as an Experimental Model for Elucidating the Mechanisms Governing Dietary Influence on Mineral Absorption. *Exp. Biol. Med.* **2008**, *233*, 651–664. [CrossRef] [PubMed]
26. Miller, E.R.; Ullrey, D.E. The pig as a model for human nutrition. *Annu. Rev. Nutr.* **1987**, *7*, 361–382. [CrossRef] [PubMed]
27. Schierack, P.; Nordhoff, M.; Pollmann, M.; Weyrauch, K.D.; Amasheh, S.; Lodemann, U.; Jores, J.; Tachu, B.; Kleta, S.; Blikslager, A.; et al. Characterization of a porcine intestinal epithelial cell line for in vitro studies of microbial pathogenesis in swine. *Histochem. Cell. Biol.* **2006**, *125*, 293–305. [CrossRef] [PubMed]
28. Vergauwen, H. The IPEC-J2 Cell Line. In *The Impact of Food Bioactives on Health: In Vitro and Ex Vivo Models*; Springer: Cham, Switzerland, 2015. [CrossRef]
29. Sankararaman, S.; Sferra, T.J. Are We Going Nuts on Coconut Oil? *Curr. Nutr. Rep.* **2018**, *7*, 107–115. [CrossRef] [PubMed]
30. Davies, D.T.; Holt, C.; Christie, W.W. The composition of milk. In *Biochemistry of Lactation*; Mepham, T.B., Ed.; Elsevier Science Publishers: New York, NY, USA, 1983; pp. 71–117.
31. Prentice, A.M. Dairy products in global public health. *Am. J. Clin. Nutr.* **2014**, *99*, 1212–1216. [CrossRef]
32. Guillot, E.; Lemarchal, P.; Dhorne, T.; Rerat, A. Intestinal absorption of medium chain fatty acids: In vivo studies in pigs devoid of exocrine pancreatic secretion. *Br. J. Nutr.* **1994**, *72*, 545–553. [CrossRef]
33. You, Y.Q.; Ling, P.R.; Qu, J.Z.; Bistrian, B.R. Effects of medium-chain triglycerides, long-chain triglycerides, or 2-monododecanoin on fatty acid composition in the portal vein, intestinal lymph, and systemic circulation in rats. *JPEN J. Parenter. Enteral. Nutr.* **2008**, *32*, 169–175. [CrossRef]
34. Quann, E.E.; Fulgoni, V.L., III; Auestad, N. Consuming the daily recommended amounts of dairy products would reduce the prevalence of inadequate micronutrient intakes in the United States: Diet modeling study based on NHANES 2007–2010. *Nutr. J.* **2015**, *14*, 90. [CrossRef]
35. Martínez-Vallespín, B.; Vahjen, W.; Zentek, J. Effects of medium chain fatty acids on the structure and immune response of IPEC-J2 cells. *Cytotechnology* **2016**, *68*, 1925–1936. [CrossRef]
36. Rego Costa, A.; Rosado, E.L.; Soares-Mota, M. Influence of the dietary intake of medium chain triglycerides on body composition, energy expenditure and satiety: A systematic review. *Nutr. Hosp.* **2012**, *27*, 103–108. [CrossRef]
37. Husain, Z.; Schwartz, R.A. Food allergy update: More than a peanut of a problem. *Int. J. Dermatol.* **2013**, *52*, 286–294. [CrossRef]

38. Fujita, K.; Katahira, J.; Horiguchi, Y.; Sonoda, N.; Furuse, M.; Tskuita, S. Clostridium perfringens enterotoxin binds to the second extracellular loop of claudin-3, a tight junction membrane protein. *FEBS Lett.* **2000**, *476*, 258–261. [CrossRef]
39. Farquhar, M.G.; Palade, G.E. Junctional complexes in various epithelia. *J. Cell Biol.* **1963**, *17*, 375–412. [CrossRef]
40. Martinez-Palomo, A.; Erlij, D.; Bracho, H. Localization of permeability barriers in the frog skin epithelium. *J. Cell. Biol.* **1971**, *50*, 277–287. [CrossRef]
41. Günzel, D.; Fromm, M. Claudins and other tight junction proteins. *Compr. Physiol.* **2012**, *2*, 1819–1852. [CrossRef]
42. Markov, A.G.; Veshnyakova, A.; Fromm, M.; Amasheh, M.; Amasheh, S. Segmental expression of claudin proteins correlates with tight junction barrier properties in rat intestine. *J. Comp. Physiol. B* **2010**, *180*, 591–598. [CrossRef]
43. Wen, H.; Watry, D.D.; Marcondes, M.C.; Fox, H.S. Selective decrease in paracellular conductance of tight junctions: Role of the first extracellular domain of claudin-5. *Mol. Cell. Biol.* **2004**, *24*, 8408–8417. [CrossRef] [PubMed]
44. Amasheh, S.; Schmidt, T.; Mahn, M.; Florian, P.; Mankertz, J.; Tavalali, S.; Gitter, A.H.; Schulzke, J.D.; Fromm, M. Contribution of claudin-5 to barrier properties in tight junctions of epithelial cells. *Cell Tissue Res.* **2005**, *321*, 89–96. [CrossRef] [PubMed]
45. Ikenouchi, J.; Furuse, M.; Furuse, K.; Sasaki, K.; Tsukita, S.; Tsukita, S. Tricellulin constitutes a novel barrier at tricellular contacts of epithelial cells. *J. Cell Biol.* **2005**, *171*, 939–945. [CrossRef] [PubMed]
46. Krug, S.M.; Amasheh, S.; Richter, J.F.; Milatz, S.; Günzel, D.; Westphal, J.K.; Huber, O.; Schulzke, J.D.; Fromm, M. Tricellulin forms a barrier to macromolecules in tricellular tight junctions without affecting ion permeability. *Mol. Biol. Cell.* **2009**, *20*, 3713–3724. [CrossRef] [PubMed]

International Journal of
Molecular Sciences

Review

Super-Resolution Imaging of Tight and Adherens Junctions: Challenges and Open Questions

Hannes Gonschior [1], Volker Haucke [1,2] and Martin Lehmann [1,*]

1 Leibniz-Forschungsinstitut für Molekulare Pharmakologie (FMP), 13125 Berlin, Germany; gonschior@fmp-berlin.de (H.G.); haucke@fmp-berlin.de (V.H.)
2 Faculty of Biology, Chemistry, Pharmacy, Freie Universität Berlin, 14195 Berlin, Germany
* Correspondence: mlehmann@fmp-berlin.de

Received: 11 December 2019; Accepted: 16 January 2020; Published: 23 January 2020

Abstract: The tight junction (TJ) and the adherens junction (AJ) bridge the paracellular cleft of epithelial and endothelial cells. In addition to their role as protective barriers against bacteria and their toxins they maintain ion homeostasis, cell polarity, and mechano-sensing. Their functional loss leads to pathological changes such as tissue inflammation, ion imbalance, and cancer. To better understand the consequences of such malfunctions, the junctional nanoarchitecture is of great importance since it remains so far largely unresolved, mainly because of major difficulties in dynamically imaging these structures at sufficient resolution and with molecular precision. The rapid development of super-resolution imaging techniques ranging from structured illumination microscopy (SIM), stimulated emission depletion (STED) microscopy, and single molecule localization microscopy (SMLM) has now enabled molecular imaging of biological specimens from cells to tissues with nanometer resolution. Here we summarize these techniques and their application to the dissection of the nanoscale molecular architecture of TJs and AJs. We propose that super-resolution imaging together with advances in genome engineering and functional analyses approaches will create a leap in our understanding of the composition, assembly, and function of TJs and AJs at the nanoscale and, thereby, enable a mechanistic understanding of their dysfunction in disease.

Keywords: super-resolution microscopy; structured illumination microscopy; stimulated emission depletion; single molecule localization microscopy; adherens junction; tight junction

1. Introduction

1.1. Adherens Junctions and Tight Junctions

Epithelial and endothelial cells line multiple organs to segregate internal and external compartments. Cell contacts are formed on the one hand by the adherens junction (AJ) which provides mechanical adhesion, and on the other by the tight junction (TJ) which primarily acts as a barrier to solutes and water and as a fence that separates the apical and basolateral plasma membrane domains. However, some TJ proteins form paracellular channels for small cations, anions, or water [1].

Both junctions are characterized by extracellularly interacting transmembrane proteins that bind via intracellular adapter proteins to the cytoskeleton [2].

TJs were first observed by transmission electron microscopy (TEM) as apical membrane contacts extending over 200–500 nm with electron-dense areas forming a continuous belt-like attachment by fusing the outer membrane leaflets of two neighboring cells into a ≈3 nm thick structure [3]. Later, by using freeze fracture electron microscopy (FFEM) their structure was described as an interconnecting strand network with a general distance between its single strands of 30–80 nm in epithelial cells and tissues [4]. The first protein components of TJs were identified as Zonula occludens (ZO-1) [5] and later the transmembrane proteins occludin [6], junctional adhesion molecule A (JAM-A) [7], and claudins [8].

Occludin, together with tricellulin and marvel-D3 form the tetraspan family of TJ-associated Marvel Domain Proteins (TAMPs) [9]. Claudins regulate interactions between the TAMPs and vice versa [10]. The mammalian genome codes for 27 claudin proteins with specific functions and tissue-specific expression patterns. Zonula occludens 1 and 2 (ZO-1, -2) turned out to be the major adapter proteins on the cytosolic side of TJs [5,11,12]. They act via various interaction domains as scaffolds for occludin, claudins, and other integral TJ proteins, providing their linkage to the actin cytoskeleton [13–16]. For JAM-A it could be recently shown that its assembly enables epithelial cell polarization. JAM-A is a single transmembrane protein that regulates membrane apposition. In contrast, tetra spanning proteins like claudins form via their two extracellular loops paracellular barriers or channels. So far it could be shown that claudin-2 [17] and claudin-15 [18] promote increased water permeability and both claudins, as well as claudin-10b, also promote increased cation permeability [19–21]. Whereas, claudin-10a [21] and claudin-17 [22] contribute to a higher anion permeability. It is also known that some specific claudin combinations can support paracellular ion permeability, e.g., claudin-16/-19 for cations [23] and claudin-4/-8 for small anions [24]. As recently described a knockout of multiple claudins results in increased ion permeability but does not affect macromolecular diffusion or polarity in epithelial cells indicating that other TJ-related proteins must be involved [25].

The initial strand and TJ-meshwork formation is highly dependent on claudins [26] which form via *cis-* and *trans-*interactions arrays of TJ strands which are about 10 nm in diameter. TJ strands assemble to a complex TJ meshwork at bicellular junctions. A specialized site of the TJ where macromolecules can pass has been identified to be the tricellular TJ (tTJ), the location where three cells meet [27]. The proteins of the tTJ are tricellulin [28], and three singe-transmembrane proteins, angulin-1 to -3 [29]. The tTJ is virtually impermeable at high tricellulin expression for macromolecules but becomes permeable for it at low expression as observed in the inflammatory bowel disease ulcerative colitis [30].

The AJ mediates mechanical cell–cell interactions during development and in differentiated tissues. Intercellular adhesion is mediated by *cis-* and *trans-*association of extracellular domains of cadherins coded by 20 different genes in vertebrates. Tissue specific cadherins are epithelial cadherins (E-cad), neuronal cadherin (N-cad), and vascular endothelial cadherins (VE-cad). The *trans-*interactions of E-cad extracellular domains are Ca^{2+} dependent. Intracellularly E-cad are linked via catenin [31–33], vinculin [34,35], and other actin-binding proteins to the actin cytoskeleton [36]. Two forms of AJ have been observed, first linear AJ linked to the circumferential actin belt in mature epithelial sheets and second punctuate AJ at free edges with direct actin linkage. Actin bundles in linear AJ can be contracted by actomyosins to produce planar-polarized apical constrictions during tissue remodeling. Rho GTPases, formins, and myosins regulate actin polymerization and contraction. Cadherin can be removed from the surface via clathrin mediated endocytosis (CME) that counterbalances actomyosin mediated constriction in neighboring cells [37]. Therefore, AJs notably integrate both cellular adhesion and cortical tension sensing [38] important for tissue homeostasis and remodeling during development.

In order to mediate cell adhesion and barrier formation, both AJ and TJ form dense and complex multiprotein structures. Some of their characteristics like formation of the TJ meshwork have been visualized by electron microscopy (EM) [4]. However, due to the inherent limitations of EM with respect to defining molecular and structural dynamics many key questions remain unsolved yet: for example, how are the individual protein components like channel and barrier forming claudins of TJs or the different components of AJs organized within the junction and how does their interaction contribute to the overall functionality? How are junctions remodeled, e.g., by strand reformation, endocytosis, during cell division, or upon injury? To answer these questions, it is indispensable to understand the nanoscale organization of the molecular components of TJs (claudins, TAMPs, JAMs, ZOs) and AJs (cadherins, catenins, vinculin, actin-binding proteins) in fixed cells and tissue and, most importantly, in living preparations. Such an approach requires multi-color imaging at nanoscale resolution by super-resolution microscopy (SRM). SRM offers the possibility to perform quantitative analysis of nanoclusters of AJs or measurements of the strand and meshwork organization of TJs.

Recent studies using SRM have begun to uncover the nanoscale organization and dynamics of TJs [13,39,40] and AJs and their components [41–43]. Here we provide an overview on recent advances and limitations in SRM, summarize important new insight in junction research from recent studies and provide an outlook for future research.

1.2. Resolution Limit

TJs were initially identified by electron microscopy analysis of ultrathin sections of resin embedded cells and tissues as μm-sized apical membrane contacts with characteristic electron dense areas. These membrane contacts appeared to act as a barrier to organic dyes and showed specific membrane kissing points [3]. Using freeze fracture sample preparation these close membrane appositions could be resolved as fence-like intramembrane strands with characteristic mesh sizes of ≈30–80 nm and strand morphology [4]. Using immunogold labelling on resin and freeze fracture samples the early biochemically identified components of TJs, namely claudins and occludin, were shown to be localized to these meshworks and specifically copolymerize in strands [6,8]. Since FFEM involves difficult sample preparation and the immunogold labelling of chemically fixed samples even with overexpressed proteins and with high affinity antibodies is often incomplete, a dense and specific nanoscale distribution of multiple TJ and AJ proteins was not achieved so far. Fluorescence microscopy on the other hand offers high density labelling and highly sensitive detection of AJ and TJ components in intact tissues and living cells. Generally, the resolution of fluorescence microscopes is limited by Abbe's law to 200–250 nm in the lateral and 500–700 nm in axial direction [44]. Therefore, TJs in polarized epithelia often appear as diffraction-limited spots in x-z scans. Several fluorescence microscopy techniques have overcome the resolution limit by either spatially or temporally controlling fluorescence emission of which structured illumination microscopy (SIM), time-gated stimulated emission depletion (gSTED), and single molecule localization microscopy (SMLM) have turned out to be the three main techniques to visualize biological specimens at the nanoscale (Figure 1).

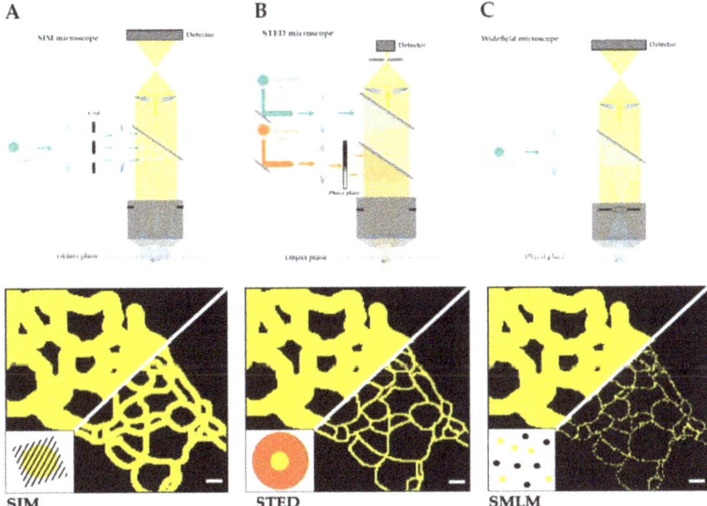

Figure 1. Schematic depiction of the main three SRM techniques. It is shown for each technique a detailed construction of the microscope and the light path as well as a schematic comparison of a reconstituted TJ meshwork resolved by structured illumination microscopy (SIM) (resolution limit: 100 nm) (**A**), time-gated stimulated emission depletion (gSTED) (resolution limit: 40 nm) (**B**), and single molecule localization microscopy (SMLM) (resolution limit: 20 nm) (**C**) in reference to confocal imaged TJ meshwork (resolution limit: 250 nm). Scale bars: 200 nm.

1.3. Structured Illumination Microscopy (SIM)

In structured illumination microscopy (SIM) spatially restricted fluorescence emissions are produced by stripe-patterned wide-field excitation (see Figure 1A) [45]. Multiple camera images are collected from an excitation pattern with different phases and orientations. The collected images contain high resolution information as moiré fringes are formed from the fluorescent sample illuminated by the stripe illumination. The high-resolution information from SIM images can be extracted from Fourier-transformed images in a process called deconvolution. Several illumination modes have been used to create SIM excitation patterns, e.g., in 3D [45], in a total internal reflection mode [46,47] at a grazing incidence angle [48] or in a lattice light sheet [49]. The final SIM images show a lateral resolution of typically 100–130 nm and an axial resolution of 100–250 nm that depends on the wavelength of emission light, the numerical aperture (NA) of the objective, the distance from the coverslip, and the illumination mode. Using an objective with the highest NA or nonlinear SIM with patterned activation resolutions of 84 and 48 nm, respectively, have been achieved [50].

SIM offers key advantages like the use of standard organic fluorophores and fluorescent proteins, low excitation power, and fast imaging speeds up to several hundred frames per second [48]. Since the image reconstruction relies heavily on mathematical processing the point-spread function (PSF) must be determined carefully, and low intensity raw images or high background can produce substantial reconstruction artefacts. Commercial wide-field SIM systems are offered by Zeiss (Elyra S1 and 7), GE Healthcare (Deltavision OMX), and Nikon (N-SIM), while lattice light sheet SIM is offered by 3i.

Point-scanning SIM or Re-Scan Microscopy uses a confocal microscope for scanning an excitation spot over the sample and collects the fluorescence emission directly on a multi-array detector in the absence of a pinhole element in the conjugate plane [51]. The signal at the central pixel and the signal distribution over multiple arrays represent accurately the PSF of the microscope and can be used to increase the resolution by a factor of 1.4 using linear deconvolution. The total signal intensity on the array is then reassigned to the scanned pixel. Since in the point-scanning SIM the pinhole of the confocal microscope is removed more light is collected, the resolution and signal-to-noise is increased by working with an accurately measured PSF and linear deconvolution. Alternatively, in a rescanning mode a camera detects the scanned emission signal with a higher magnification that also results in a 1.4-fold increase in resolution. Commercial systems are available from Zeiss (Airy Detector) and Confocal.nl (RCM) [52].

Standard SIM works best within the first 10 μm from the coverslip that would limit the observation of native TJs and AJs to certain cell types or thin tissues sections. If TJs and AJs form at flat cell contacts, e.g., punctuate AJs that are within 1 μm from the coverslip restricted excitation in total internal refection (TIRF) and GI-SIM would enable to visualize fast dynamics of TJ strands or AJ protein cluster movements with up to 250 fr/s over extended time periods at very low light doses [48]. Additionally, regarding the high detection efficiency and low light input of SIM we would see it as an ideal tool to investigate the AJ/TJ components at near endogenous levels for example by CrispR mediated knock-in of fluorescent tag or in optically cleared samples. Since SIM relies on mathematical image reconstruction with a strong influence of microscope alignment, temperature, and sample parameters, a high fluorescent signal intensity, low background, and careful calibrations are required to produce artefact-free SIM images. Overall SIM enables robust, sensitive, and fast imaging with highest contrast, but only achieves routinely resolution down to 100 nm resolution [53]. This resolution that is approximately half the diffraction limit would not enable resolving of individual TJ strands, small TJ meshes, cadherin nanoclusters, or the cortical actin organization at AJs.

1.4. Stimulated Emission Depletion (STED)

In stimulated emission depletion (STED) spatially restricted fluorescence emissions are produced in the center of a coaligned focused excitation spot surrounded by a ring-shaped depletion pattern (see Figure 1B) [54]. The emissions from the diffraction-limited excitation spot is depleted by the ring-shaped depletion focus except for the center region. By scanning the coaligned focused spots and detecting the emission in a confocal microscope, images with a lateral resolution of about 50 nm can be obtained by optimizing imaging conditions such as excitation and depletion energies. The confocal configuration eliminates out of focus light and offers both high contrast and optical sectioning. Pulsed excitation and time-gated detection have enabled resolutions below 50 nm [55]. Using a two-objective configuration or 3D STED depletion patterns a lateral and axially similar resolution of 50–70 nm has been demonstrated [56]. The overall STED resolution is mostly limited by depletion laser power, alignment quality and, most importantly, by labelling density and the availability of bright and photostable dyes with an optimal 5%–10% normalized emission intensity at the depletion laser wavelength. Commercial STED microscopes can be purchased from Leica, Picoquant, or Abberior Instruments.

Since both TJ and AJ are formed by oligomeric transmembrane proteins the lower signal to noise effect of STED can be outweighed by higher label density, e.g., by claudins forming strands by *cis*-and *trans*-interactions. Low abundant or under-labelled structures may not give enough signal for STED imaging. Multicolor STED is routinely performed with 2–3 channels. Therefore, STED imaging of TJs and AJs with optimal labelling [57,58] and optimized 2D and 3D cell culture conditions will show several TJ and AJ components as well as the actin cytoskeleton with <60 nm resolution deep inside cells [59]. Despite STED having the lowest signal to background ratio of all super-resolution techniques it produces raw images that usually do not require post-processing. During a practical comparison STED and SMLM gave similar resolutions on microtubules and vesicular structures [53]. Another advantage of STED is the possibility of imaging cellular structures in living cells and tissues with <50 nm resolution [58] and comparing the ultrastructural preservation after chemical fixation, permeabilization, and immune labelling [60]. Especially, the ultrastructure of tension-sensitive AJ, complex TJ strand morphologies, and the actin cytoskeleton could be altered by chemical fixation. Native AJs and TJs extend in the axial direction in polarized monolayers and are found several μm away from the glass surface. Here the optical sectioning of STED and the possibility to increase resolution in Z direction down to 70 nm would enable the visualization of TJ and AJ structures in a native context. For functional studies of nanoscale dynamics of AJ and TJ protein dynamics STED can be combined with fluorescence recovery after photobleaching (FRAP), fluorescence correlation spectroscopy (FCS), and fluorescence lifetime imaging microscopy (FLIM).

1.5. Single Molecule Localization Microscopy (SMLM)

Single Molecule Localization Microscopy (SMLM) increases resolution by exerting temporal control of fluorescence emission. Single molecule fluorescence emissions are detected from a sparse subset of fluorophores that show recovery from an inactive state (see Figure 1C) [61,62], photo activation [63], photo conversion, or short presence in the detection volume [64]. Depending on the brightness of the fluorescence signal the position of the single molecule can be determined with 2–50 nm precision [61,63,65]. The molecule positions are extracted from tens of thousands of images of sparse emitters, drift corrected, and plotted into a new super-resolution image. The SMLM image resolution is determined by both localization precision and localization density [66]. The limited photon output of single molecule emissions often requires strong laser excitation, sensitive detection by low noise cameras, and imaging in TIRF mode within 200 nm from the coverslip. Three-dimensional SMLM was achieved by astigmatic distortion of the PSF in z direction [67], by multi-plane detection [68] or an interferometric approach [69,70]. Both organic fluorophores and photoactivatable or blinking fluorescent proteins can be used as labels for SMLM. The movement of single biological molecules in cells could be analyzed by using tracking SMLM [71]. SMLM offers the highest lateral and axial resolution and can be performed with relatively simple setups when significant challenges of labelling density, sample drift, acquisition time, multicolor acquisition, and image analysis are overcome. A wide range of image analysis packages for 2D and 3D SMLM are available [72]. Commercial SMLM systems are offered by Leica (Leica GSD), Nikon (N-STORM), Zeiss (Zeiss Elyra P1) and Bruker (Vutara 350).

Similar to STED, the SMLM techniques offer diffraction unlimited resolution when high labelling densities can be achieved. Therefore, endogenous labelling of TJ components should be performed with polyclonal antibodies that can bind multiple epitopes. The maximal resolution of SMLM can be in the order of 10 nm that should be sufficient to resolve individual claudin strands or E-cad nanoclusters. Such high resolutions often require direct genetic labelling of proteins of interest with either photoactivatable proteins or self-labelling enzymatic tags. SMLM with photoactivatable fluorescent proteins especially in tissue is often limited by auto fluorescent and detection of rather dim fluorescence bursts and could therefore overestimate monomeric E-cad and underestimate clusters. Therefore, SMLM imaging should be performed close to the cover glass, e.g., in flat endothelial cells or in sparse cultures of epithelial cells that form flat cellular overlaps with TJ strands. Since flat portions of cellular contacts are often not within the typical depth of a TIRF, illumination of ≈200 nm SMLM should be performed with wide-field excitation and 3D SMLM. When 3D SMLM is used on AJ or TJ structures that are located several micrometers away from the glass coverslip the signal to noise ratio decreases that could result in more false-positive localization and a lower localization precision. The higher signal of SMLM in TIRF mode enables the analysis of individual AJ and TJ proteins formed in the lower plasma membrane, e.g., during early steps of polymerization, intracellular anchoring, or signal transduction outside of AJ and TJ. In order to reveal the nanoscale organization of multiple claudins in TJ meshworks or E-cad clusters together with intracellular AJ proteins multicolor SMLM is required. Major challenges for multicolor imaging are first to find spectrally different fluorophores with a similar performance as the widely used Alexa Fluor 647 dye with similar photon outputs, second harmonize buffer requirements of multiple fluorophores, and third register multicolor images after drift correction.

A general technical review [73] and comparison of SIM, STED, and SMLM can be found in [53]. In direct comparison both SMLM and STED gave the highest resolution of <50 nm compared to 110 nm for SIM, but SIM and SMLM offered the highest signal-to-noise ratio [53]. The principle and important considerations for TJ/AJ imaging are summarized in Figure 1, Figure 2 and Table 1.

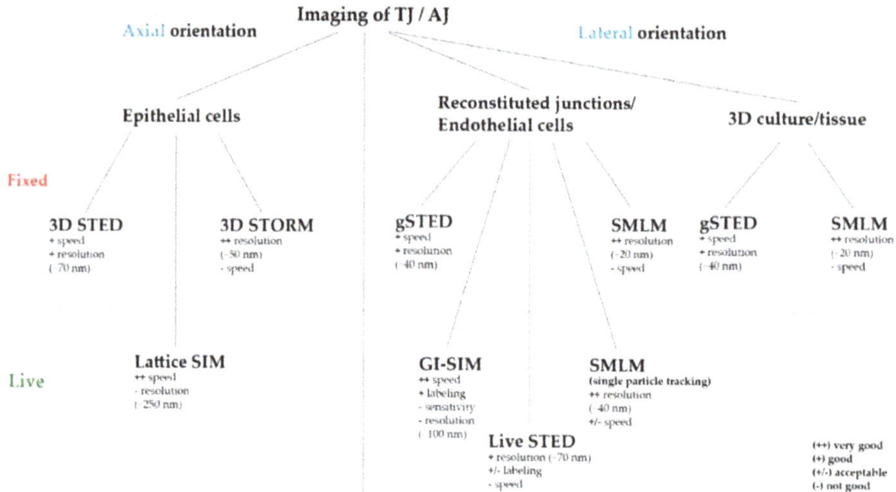

Figure 2. Practicable overview of the ideally used SRM technique for TJ and AJ imaging regarding the axial or lateral direction of the junction, the system (epi-, endothelial cells/fibroblasts/tissue), different labeling (fluorescent proteins/antibodies) and imaging (fixed/live) approaches.

Table 1. Overview and direct comparison of the key parameters of SIM, gSTED, and SMLM.

	SIM	gSTED	SMLM
Resolution limit	100 nm (lateral) 250 nm (axial)	40 nm (lateral) 70 nm (axial)	20 nm (lateral) 50 nm (axial)
Multi-color	4 colors	2–4 colors	2–4 colors
Sample preparation	Standard	Standard	Standard
Live imaging	Yes	Yes	Yes
Imaging speed	10 ms–10 s	10 s–5 min	10 min
Illumination power	10–100 W/cm^2	1–100 kW/cm^2	1–100 kW/cm^2

2. Results

2.1. *Super-Resolution Imaging of Adherens Junctions Reveals Nanoscale Clustering and Stratified Intracellular Organization*

AJs form at overlaps of endothelial cells and basolateral of TJs in epithelial cells and integrate tension sensing and adhesion remodeling [74]. In EM images AJs appear as close membrane appositions with an electron dense cytosolic plaque of actin [3] but freeze fracture images show no continuous strands of E-cad molecules. Instead, E-cad molecules were found in individual clusters using diffraction limited fluorescence microscopy [75].

At the nanoscale level endogenous E-cad in *Drosophila* embryos was found to form discrete nanometer-sized clusters by SMLM. The protein density estimations from fluorescence correlation spectroscopy calibrations and from single molecule counting were consistent with tightly packed E-cad molecules about eight nanometers apart from each other. Most E-cad was monomeric with varying cluster sizes that have been found to be regulated by dynamin-mediated endocytosis. Conversely, elimination of α-catenin or PAR3 reduced E-cad clustering, while interfering with action polymerization [41].

Similarly, when SMLM 3D-STORM was performed in epithelial cells with both bright organic dyes and photoactivatable proteins, highly homogenous cluster sizes of 50–60 nm with a fixed distance of ≈160 nm were identified [42]. E-cad clusters in epithelial cells contained six molecules similar to results in *Drosophila* embryos [41] and form independently of cadherin–cadherin interactions. Mutation in E-cad that interfered with *trans* or *cis*-interaction still produced clusters, albeit with a decreased molecular density. Coculture experiment found coexisting adhesive and non-adhesive E-cad clusters with similar size. Dual color SMLM revealed that adhesive and non-adhesive E-cad clusters are delimited by F-actin. Both the actin depolymerizing drug Latrunculin A and expression of a tail-less E-cad results in the formation of larger E-cad clusters. Therefore, SMLM revealed that E-cad forms a nanocluster surrounded by an actin fence that depends both on homophilic interactions and anchoring via the cytosolic tail. Even larger and more mature AJs are formed by groups of individual nanoclusters that can be linked to and stabilized by intracellular scaffold proteins [42]. Therefore, adhesion and tension sensing are mediated by small units of E-cad that are evenly spaced and can be finely regulated by incorporation into larger units and controlled by dynamin-dependent endocytosis.

How are the intracellular proteins of AJs organized? Using a combination of 3D PALM, surface-generated structured illumination together with biochemical perturbation on planarized biomimetic cadherin-based AJs the nanoscale architecture of AJ was analyzed [43]. Notably a plasma membrane-proximal cadherin-catenin compartment segregated from the actin cytoskeletal compartment, connected by an interface zone containing vinculin was found. The position of vinculin is determined by catenin and vinculin showed a tension and tyrosine phosphorylation dependent extended confirmation. The robust surface-generated structured illumination assay revealed the nanoscale changes of vinculin conformation in different cell types, under biochemical and pharmacological perturbations and could be verified by fluorescence resonance energy transfer

(FRET)-tension measurements. Therefore, vinculin integrates mechanical and chemical signals between cadherin and the actomyosin layer to control cell adhesion in cells and tissues [43].

Of note, a similar stratified nanoarchitecture and tension sensitive conformation of vinculin was previously found in focal adhesions (FA) [76]. Integrins form nanoclusters [77] while the tension sensing molecules paxillin and vinculin form linear and non-colocalizing areas within FA [78]. Vinculin FRET-based tension sensors revealed a high tension especially at small FA and that a general increase in cellular tension increases the size of FA [79].

2.2. *Super-Resolution Imaging Shows Claudin Meshworks, Strand Dynamics, and Molecular Composition of the TJ*

TJs are organized by claudin strands and meshworks that were revealed by FFEM of cell and tissue samples. Since EM requires strong fixation and often lacks molecular specificity SRM would be needed to confirm the nanoarchitecture of TJ in intact cells and tissues.

A first approach in this direction was done in 2003 when Sasaki et al. were able to image dynamics of claudin strands and whole TJ meshworks in claudin-1-GFP overexpressing L-fibroblasts using confocal microscopy. However, due to the previously described resolution limit a further analysis of multiple claudins and strand dynamics in denser not peripheral meshwork areas was not possible [80].

Van Itallie et al. reconstituted eGFP-claudin-2 strands in fibroblasts and imaged extended strands networks by live SIM with ≈120 nm resolution. Large meshes as well as single or double strand configurations and strand dynamics could be detected. Claudin-2 networks formed in the absence of claudin-2-ZO-1 binding and showed a higher mobility and failed to co-align with the actin cytoskeleton. Using multicolor SIM, a partial overlap between actin, ZO-1 and claudin-2 could be found, since actin and ZO-1 are also highly concentrated at AJ. Strand breaks and strand elongation occurred independently of ZO-1 binding, but preferentially at branch points. Branch points were also identified as polymerization sites of claudin and showed a weak enrichment of occludin. How other claudins copolymerize, associate with actin, ZO-1, and occludin are important questions for the nanostructure and the regulation of TJs. Since some claudins like claudin-3 form much smaller meshes in reconstituted TJ meshworks (e.g., Figure 3C) than claudin-2, super-resolution techniques with higher spatial resolution like SMLM and STED will help to resolve these questions [13].

Kaufmann et al. reconstituted TJs from C-terminally YFP tagged claudin-3 and claudin-5 in an epithelial cell line, devoid of endogenous claudin expression. SMLM with 50 nm structural resolution resolved meshworks and found two meshwork populations with ≈100/360 nm (claudin-3) and ≈60/480 nm (claudin-5) mesh sizes. Notably inhomogeneous claudin distribution along strands and in claudin proteins between strands were detected and could play a role in strand formation. Since SMLM requires >20,000 frames for image reconstruction, the dynamics of claudin meshworks were not resolved. Available multicolor SMLM could resolve mixtures of claudins or the nanoscale organization of TJ proteins in the future [39] (Figure 3).

An example for TJs resolved with STED microscopy has not been published yet. Just Cording et al. in 2017 were able to image changes of the Tricellular nanoarchitecture upon treatment with tricellulin derived peptides [81].

That STED microscopy in general is a tool to resolve TJ meshworks at nanoscale level without any further image processing is shown in Figure 4 by an example of a reconstituted meshwork of YFP-claudin-3 in Cos7 fibroblasts.

On an endogenous level, the group around Schlingmann and Koval studied the effect of alcohol on the nanostructure of alveolar epithelial tight junction using SMLM in combination with functional readouts. Alcohol increased claudin-5 levels, induced claudin-18 and actin containing spikes, and increased the paracellular flux. SMLM revealed mostly linear nanocluster arrangements of immunolabelled claudin-5, -18, and ZO1 in TJs. Under alcohol treatment, the specific nano-colocalization of claudin-18 and ZO1 decreased, while the colocalization of claudin-5 and claudin-18 increased. As molecular mechanism leading to alcohol induced paracellular flux the competition of claudin-5 and claudin-18 for interaction with ZO1 was proposed [40].

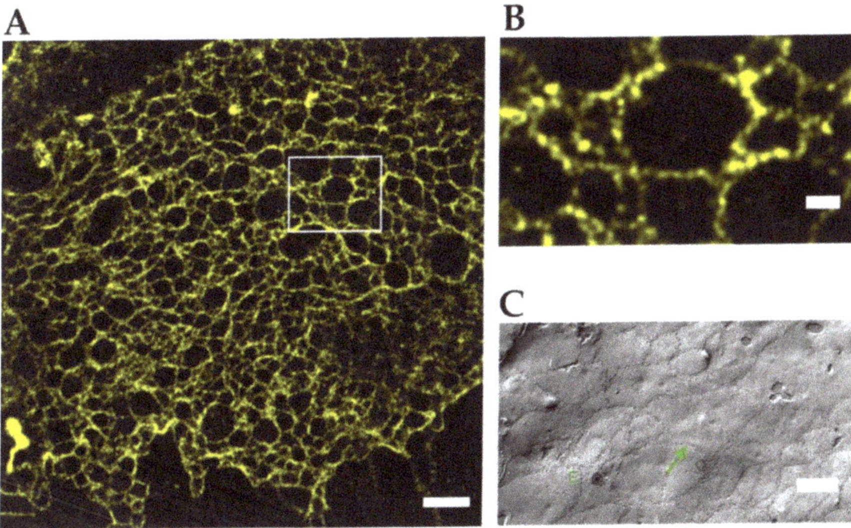

Figure 3. Reconstituted TJ networks formed by claudin3-YFP in HEK293 cells imaged by using SMLM (adapted from Kaufmann et al. 2012 [39]); (**A**) localization microscopy image of the region marked in the conventional wide-field fluorescence image with a mean effective optical resolution 48 nm. Scale bar: 1 µm; (**B**) magnification of claudin-3-YFP in HEK293 cells. Scale bar: 200 nm; (**C**) FFEM of claudin-5-YFP/claudin-3-YFP cotransfected HEK293 cells. Scale bar: 250 nm.

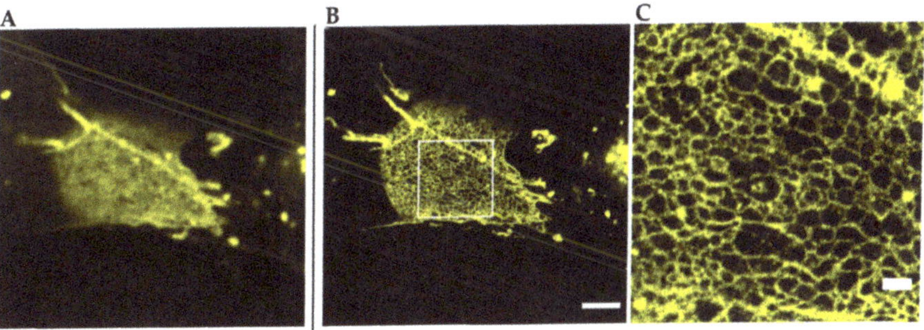

Figure 4. Reconstituted TJ networks formed by YFP-claudin-3 in Cos7 fibroblasts (unpublished data) imaged by using STED microscopy; (**A**) confocal overview image of a reconstituted TJ meshwork at the overlapping area of two transfected cells; (**B**) STED image of the same meshwork area from A. Scale bar: 2 µm; (**C**) magnification of the marked area in B. Scale bar: 500 nm.

3. Discussion and Outlook

Multicellular organisms and tissues rely on multiple cellular junction structures to mediate adherence, sense tension, and form a paracellular barrier. AJs and TJs form dense and complex multiprotein assemblies with a specific nanoscale architecture. While EM revealed claudin strands and membrane kissing points at the TJ or dense cytosolic plaques at the AJ, the intricate molecular nano-organization and dynamic remodeling of these cellular structures only begin to emerge. As shown by SMLM, E-cad forms in the AJ regularly spaced nanoclusters connected to a stratified intracellular layer of tension-sensing and actin interacting proteins at the AJ [41–43]. Revealing the dynamic nanoscale organization of AJs, especially movements of cadherins cluster, vinculin conformational changes, and actin remodeling under tension, during vascular development and endothelial tissue remodeling, should give important new insights into AJ regulation. Here, a combination of single particle tracking, and live STED could enable the study of the concerted action of multiple AJ components.

With recently developed, and now well matured, super-resolution microscopy techniques a first glimpse at claudin strand organization and dynamics at TJ was revealed. Since most TJs contain multiple claudins and other TJ associated proteins, e.g., occludin, or JAM, and various TJ-related scaffolding proteins that exchange on second timescales, fast multicolor imaging with nanoscale resolution would be required. How multiple claudins and occludin assemble into the TJ meshwork is still an open question. Since multicolor SMLM microscopy is still challenging because of slow image acquisition, low photon counts from fluorescent proteins, fluorophore incompatibilities, channel registration errors, and drift correction errors in the range of the resolution alternative multicolor techniques like STED and SIM are advantageous. STED imaging could also be applied to living cells and tissues [58]. Special care must be taken for the optimization of fixation, permeabilization, and labelling protocols as cellular nanostructures can be significantly altered during these sample preparation steps [60]. Ideally similar nanostructures should be visible in live STED of genetically tagged proteins, in fixed cells, and after endogenous labelling by antibodies.

Since real TJs in cultured epithelial cells and tissues mostly extend in the axial direction, the significantly lower z-resolution of most available super-resolution techniques sets a barrier to resolving the native TJ nano-architecture. Therefore, TJ components, namely claudins and occludin, were visualized in reconstituted flat cell culture systems that recapitulate early stages of epithelial formation. Here, claudin meshwork structures were visualized that were very similar to early FFEM studies. Importantly any observation in such a reconstituted system would have to be tested in endothelial or epithelial cells, in 3D culture models [59] and, importantly, in tissues. Nevertheless the improvement of SRM techniques which provide imaging at high resolution in thicker samples (>30 μm) and most importantly also with a sufficient z-resolution to image single strands at 50–100 nm spacing in tissue and epithelial cells will be one of the main and most important steps towards understanding the physiology and biology of TJs. One way to achieve this goal will be by using carefully aligned and refractive index matched 3D STED imaging of endogenously tagged abundant TJ proteins, like claudins, ZO-1, or occludin, providing novel insights about claudin localization within the meshwork but also its organization in the lateral membrane. Endogenous labelling still depends on the availability of specific antibodies or recently available genetic knock-in technology. Since SRM techniques resolve fluorophore positions down to a few nanometers, the sizes of antibodies become a limiting factor. Therefore, we expect that super-resolution imaging of endogenous tagged TJ, AJ, and actin will reveal the important nano-structural features of native junctions together with the associated cytoskeleton. Small, bright, and photostable fluorescent labels should be placed as close as possible to biological target structures that can be achieved via genetic tags like fluorescent proteins or self-labelling enzymes like SNAP or Halo tags [82]. Brighter fluorogenic dyes [83], more sensitive detectors, and microscopy systems like light sheet microscopes would enable faster and better resolved imaging of AJ and TJ structures. A better understanding of nano-structural details of TJs and AJs will ultimately enable discovery of new strategies to overcome or strengthen the intercellular barrier for drug delivery or during tissue damage, respectively.

Author Contributions: Conceptualization, H.G., M.L.; Funding Acquisition, V.H., M.L.; Visualization H.G. Writing—Original Draft Preparation H.G., M.L.; Writing—Review and Editing: H.G., V.H., M.L. All authors have read and agreed to the published version of the manuscript.

Funding: This research was funded by the Deutsche Forschungsgemeinschaft (DFG), Research Training Group GRK 2318 A4.

Conflicts of Interest: The authors declare no conflict of interest.

Abbreviations

AJ	Adherens junction
dSTORM	Direct stochastic optical reconstruction microscopy
E-cad	E-cad
eGFP	Enhanced green fluorescent protein
EM	Electron microscopy
FA	Focal adhesion
FCS	Fluorescence correlation spectroscopy
FFEM	Freeze fracture electron microscopy
fr	Frame
FRAP	Fluorescence recovery after photobleaching
FRET	Fluorescence resonance energy transfer
gSTED	Gated stimulated emission depletion
JAM-A	Junctional adhesion molecule A
kW	Kilowatt
min	Minute
ms	Millisecond
NA	Numerical aperture
nm	Nanometer
PALM	Photoactivated localization microscopy
PSF	Point spread function
s	Second
SIM	Structured illumination microscopy
SMLM	Single molecule localization microscopy
SRM	Super-resolution microscopy
TAMP	Tight junction-associated MARVEL protein
TEM	Transmission electron microscopy
TJ (bTJ, tTJ)	Tight junction (bicellular TJ, tricellular TJ)
YFP	Yellow fluorescent protein
ZO-1, -2, -3	Zonula occludens protein 1, 2, or 3

References

1. Günzel, D.; Fromm, M. Claudins and other tight junction proteins. *Compr. Physiol.* **2012**, 2, 1819–1852. [PubMed]
2. Zihni, C.; Mills, C.; Matter, K.; Balda, M.S. Tight junctions: From simple barriers to multifunctional molecular gates. *Nat. Rev. Mol. Cell Biol.* **2016**, 17, 564–580. [CrossRef] [PubMed]
3. Farquahr, M.G.; Palade, G.E. Junctional complexes in various epithelia. *J. Cell Biol.* **1963**, 17, 375–412. [CrossRef] [PubMed]
4. Staehelin, L.A. Further observations on the fine structure of freeze-cleaved tight junctions. *J. Cell Sci.* **1973**, 13, 763–786.
5. Stevenson, B.R.; Siliciano, J.D.; Mooseker, M.S.; Goodenough, D.A. Identification of ZO-1: A high molecular weight polypeptide associated with the tight junction (Zonula Occludens) in a variety of epithelia. *J. Cell Biol.* **1986**, 103, 755–766. [CrossRef]
6. Furuse, M.; Hirase, T.; Itoh, M.; Nagafuchi, A.; Yonemura, S.; Tsukita, S.; Tsukita, S. Occludin: A novel integral membrane protein localizing at tight junctions. *J. Cell Biol.* **1993**, 123, 1777–1788. [CrossRef]
7. Martìn-Padura, I.; Lostaglio, S.; Schneemann, M.; Williams, L.; Romano, M.; Fruscella, P.; Panzeri, C.; Stoppacciaro, A.; Ruco, L.; Villa, A.; et al. Junctional adhesion molecule, a novel member of the immunoglobulin superfamily that distributes at intercellular junctions and modulates monocyte transmigration. *J. Cell Biol.* **1998**, 142, 117–127. [CrossRef]
8. Furuse, M.; Fujita, K.; Hiiragi, T.; Fujimoto, K.; Tsukita, S. Claudin-1 and -2: Novel integral membrane proteins localizing at tight junctions with no sequence similarity to occludin. *J. Cell Biol.* **1998**, 141, 1539–1550. [CrossRef]

9. Raleigh, D.R.; Marchiando, A.M.; Zhang, Y.; Shen, L.; Sasaki, H.; Wang, Y.; Long, M.; Turner, J.R. Tight Junction–associated MARVEL Proteins MarvelD3, Tricellulin, and Occludin Have Distinct but Overlapping Functions. *Mol. Biol. Cell* **2010**, *21*, 1200–1213. [CrossRef]
10. Cording, J.; Berg, J.; Kading, N.; Bellmann, C.; Tscheik, C.; Westphal, J.K.; Milatz, S.; Gunzel, D.; Wolburg, H.; Piontek, J.; et al. In tight junctions, claudins regulate the interactions between occludin, tricellulin and marvelD3, which, inversely, modulate claudin oligomerization. *J. Cell Sci.* **2013**, *126*, 554–564. [CrossRef]
11. Jesaitis, L.A.; Goodenough, D.A. Molecular characterization and tissue distribution of ZO-2, a tight junction protein homologous to ZO-1 and the Drosophila discs-large tumor suppressor protein. *J. Cell Biol.* **1994**, *124*, 949–961. [CrossRef] [PubMed]
12. Balda, M.S.; Gonzalez-Mariscal, L.; Matter, K.; Cereijido, M.; Anderson, J.M. Assembly of the tight junction: The role of diacylglycerol. *J. Cell Biol.* **1993**, *123*, 293–302. [CrossRef] [PubMed]
13. Van Itallie, C.M.; Tietgens, A.J.; Anderson, J.M. Visualizing the dynamic coupling of claudin strands to the actin cytoskeleton through ZO-1. *Mol. Biol. Cell* **2017**, *28*, 524–534. [CrossRef] [PubMed]
14. Hirokawa, N.; Tilney, L.G. Interactions between actin filaments and between actin filaments and membranes in quick-frozen and deeply etched hair cells of the chick ear. *J. Cell Biol.* **1982**, *95*, 249–261. [CrossRef]
15. Madara, J.L.; Pappenheimer, J.R. Structural basis for physiological regulation of paracellular pathways in intestinal epithelia. *J. Membr. Biol.* **1987**, *100*, 149–164. [CrossRef]
16. Fanning, A.S.; Jameson, B.J.; Jesaitis, L.A.; Anderson, J.M. The tight junction protein ZO-1 establishes a link between the transmembrane protein occludin and the actin cytoskeleton. *J. Biol. Chem.* **1998**, *273*, 29745–29753. [CrossRef]
17. Rosenthal, R.; Milatz, S.; Krug, S.M.; Oelrich, B.; Schulzke, J.D.; Amasheh, S.; Günzel, D.; Fromm, M. Claudin-2, a component of the tight junction, forms a paracellular water channel. *J. Cell Sci.* **2010**, *123*, 1913–1921. [CrossRef]
18. Rosenthal, R.; Günzel, D.; Piontek, J.; Krug, S.M.; Ayala-Torres, C.; Hempel, C.; Theune, D.; Fromm, M. Claudin-15 forms a water channel through the tight junction with distinct function compared to claudin-2. *Acta Physiol.* **2019**, *2018*, 1–15. [CrossRef]
19. Amasheh, S.; Meiri, N.; Gitter, A.H.; Schöneberg, T.; Mankertz, J.; Schulzke, J.D.; Fromm, M. Claudin-2 expression induces cation-selective channels in tight junctions of epithelial cells. *J. Cell Sci.* **2002**, *115*, 4969–4976. [CrossRef]
20. Van Itallie, C.M.; Fanning, A.S.; Anderson, J.M. Reversal of charge selectivity in cation or anion-selective epithelial lines by expression of different claudins. *Am. J. Physiol. Ren. Physiol.* **2003**, *285*, 1078–1084. [CrossRef]
21. Van Itallie, C.M.; Rogan, S.; Yu, A.; Vidal, L.S.; Holmes, J.; Anderson, J.M. Two splice variants of claudin-10 in the kidney create paracellular pores with different ion selectivities. *Am. J. Physiol. Ren. Physiol.* **2006**, *291*, 1288–1299. [CrossRef] [PubMed]
22. Fromm, M.; Günzel, D.; Krug, S.M.; Amasheh, S.; Rosenthal, R.; Schulzke, J.D.; Fromm, A.; Conrad, M.P. Claudin-17 forms tight junction channels with distinct anion selectivity. *Cell. Mol. Life Sci.* **2012**, *69*, 2765–2778.
23. Hou, J.; Renigunta, A.; Konrad, M.; Gomes, A.S.; Schneeberger, E.E.; Paul, D.L.; Waldegger, S.; Goodenough, D.A. Claudin-16 and claudin-19 interact and form a cation-selective tight junction complex. *J. Clin. Investig.* **2008**, *118*, 619–628. [CrossRef] [PubMed]
24. Hou, J.; Renigunta, A.; Yang, J.; Waldegger, S. Claudin-4 forms paracellular chloride channel in the kidney and requires claudin-8 for tight junction localization. *Proc. Natl. Acad. Sci. USA* **2010**, *107*, 18010–18015. [CrossRef] [PubMed]
25. Otani, T.; Nguyen, T.P.; Tokuda, S.; Sugihara, K.; Sugawara, T.; Furuse, K.; Miura, T.; Ebnet, K.; Furuse, M. Claudins and JAM-A coordinately regulate tight junction formation and epithelial polarity. *J. Cell Biol.* **2019**, *218*, 3372–3396. [CrossRef] [PubMed]
26. Furuse, M.; Sasaki, H.; Fujimoto, K.; Tsukita, S. A single gene product, claudin-1 or -2, reconstitutes tight junction strands and recruits occludin in fibroblasts. *J. Cell Biol.* **1998**, *143*, 391–401. [CrossRef]
27. Krug, S.M.; Amasheh, S.; Richter, J.F.; Milatz, S.; Günzel, D.; Westphal, J.K.; Huber, O.; Schulzke, J.D.; Michael, F. Tricellulin Forms a Barrier to Macromolecules in Tricellular Tight Junctions without Affecting Ion Permeability. *Mol. Biol. Cell* **2009**, *20*, 3713–3724. [CrossRef]

28. Ikenouchi, J.; Furuse, M.; Furuse, K.; Sasaki, H.; Tsukita, S.; Tsukita, S. Tricellulin constitutes a novel barrier at tricellular contacts of epithelial cells. *J. Cell Biol.* **2005**, *171*, 939–945. [CrossRef]
29. Higashi, T.; Tokuda, S.; Kitajiri, S.I.; Masuda, S.; Nakamura, H.; Oda, Y.; Furuse, M. Analysis of the 'angulin' proteins LSR, ILDR1 and ILDR2—Tricellulin recruitment, epithelial barrier function and implication in deafness pathogenesis. *J. Cell Sci.* **2013**, *126*, 3797. [CrossRef]
30. Krug, S.M.; Bojarski, C.; Fromm, A.; Lee, I.M.; Dames, P.; Richter, J.F.; Turner, J.R.; Fromm, M.; Schulzke, J.D. Tricellulin is regulated via interleukin-13-receptor α2, affects macromolecule uptake, and is decreased in ulcerative colitis. *Mucosal Immunol.* **2018**, *11*, 345–356. [CrossRef]
31. Huber, A.H.; Weis, W.I.; West, C.D. The Structure of the ß-Catenin/E-Cadherin Complex and the Molecular Basis of Diverse Ligand Recognition by ß-Catenin. *Cell* **2001**, *105*, 391–402. [CrossRef]
32. Pokutta, S.; Weis, W.I. Structure of the Dimerization and ß-Catenin-Binding Region of a-Catenin. *Mol. Cell* **2000**, *5*, 533–543. [CrossRef]
33. Drees, F.; Pokutta, S.; Yamada, S.; Nelson, W.J.; Weis, W.I. α-Catenin Is a Molecular Switch that Binds E-Cadherin-β-Catenin and Regulates Actin-Filament Assembly. *Cell* **2005**, *123*, 903–915. [CrossRef] [PubMed]
34. Watabe-Uchida, M.; Uchida, N.; Imamura, Y.; Nagafuchi, A.; Fujimoto, K.; Uemura, T.; Vermeulen, S.; Van Roy, F.; Adamson, E.D.; Takeichi, M. α-Catenin-Vinculin Interaction Functions to Organize the Apical Junctional Complex in Epithelial Cells. *J. Cell Biol.* **1998**, *142*, 847–857. [CrossRef]
35. Weiss, E.E.; Kroemker, M.; Rüdiger, A.; Jockusch, B.M.; Rüdiger, M. Vinculin Is Part of the Cadherin–Catenin Junctional Complex: Complex Formation between α-Catenin and Vinculin. *J. Cell Biol.* **1998**, *141*, 755–764. [CrossRef]
36. Kobielak, A.; Fuchs, E.; Medical, H.H. α-Catenin: At the junction of intercellular adhesion and actin dynamics. *Nat. Rev. Mol. Cell Biol.* **2004**, *5*, 614–625. [CrossRef]
37. Nanes, B.A.; Chiasson-MacKenzie, C.; Lowery, A.M.; Ishiyama, N.; Faundez, V.; Ikura, M.; Vincent, P.A.; Kowalczyk, A.P. P120-Catenin Binding Masks an Endocytic Signal Conserved in Classical Cadherins. *J. Cell Biol.* **2012**, *199*, 365–380. [CrossRef]
38. Lecuit, T.; Yap, A.S. E-cadherin junctions as active mechanical integrators in tissue dynamics. *Nat. Cell Biol.* **2015**, *17*, 533–539. [CrossRef]
39. Kaufmann, R.; Piontek, J.; Grüll, F.; Kirchgessner, M.; Rossa, J.; Blasig, I.E.; Cremer, C. Visualization and Quantitative Analysis of Reconstituted Tight Junctions Using Localization Microscopy. *PLoS ONE* **2012**, *7*, 1–9. [CrossRef]
40. Schlingmann, B.; Overgaard, C.E.; Molina, S.A.; Lynn, K.S.; Mitchell, L.A.; Dorsainvil White, S.; Mattheyses, A.L.; Guidot, D.M.; Capaldo, C.T.; Koval, M. Regulation of claudin/zonula occludens-1 complexes by hetero-claudin interactions. *Nat. Commun.* **2016**, *7*, 12276. [CrossRef]
41. Quang, B.A.T.; Mani, M.; Markova, O.; Lecuit, T.; Lenne, P.F. Principles of E-cadherin supramolecular organization in vivo. *Curr. Biol.* **2013**, *23*, 2197–2207.
42. Wu, Y.; Kanchanawong, P.; Zaidel-Bar, R. Actin-Delimited Adhesion-Independent Clustering of E-Cadherin Forms the Nanoscale Building Blocks of Adherens Junctions. *Dev. Cell* **2015**, *32*, 139–154. [CrossRef] [PubMed]
43. Bertocchi, C.; Wang, Y.; Ravasio, A.; Hara, Y.; Wu, Y.; Sailov, T.; Baird, M.A.; Davidson, M.W.; Zaidel-bar, R.; Toyama, Y.; et al. Nanoscale architecture of cadherin-based cell adhesions. *Nat. Cell Biol.* **2017**, *19*, 28. [CrossRef] [PubMed]
44. Abbe, E. Beiträge zur Theorie des Mikroskops und der mikroskopischen Wahrnehmung. *Arch. Mikrosk. Anat.* **1873**, *9*, 413–418. [CrossRef]
45. Schermelleh, L.; Carlton, P.M.; Haase, S.; Shao, L.; Winoto, L.; Kner, P.; Burke, B.; Cardoso, M.C.; Agard, D.A.; Gustafsson, M.G.L.; et al. Subdiffraction multicolor imaging of the nuclear periphery with 3D structured illumination microscopy. *Science* **2008**, *320*, 1332–1336. [CrossRef] [PubMed]
46. Chung, E.; Kim, D.; So, P.T.C. Extended resolution wide-field optical imaging: Internal reflection fluorescence microscopy. *Opt. Lett.* **2006**, *31*, 945–947. [CrossRef]
47. Guo, M.; Chandris, P.; Giannini, J.P.; Trexler, A.J.; Fischer, R.; Chen, J.; Vishwasrao, H.D.; Rey-Suarez, I.; Wu, Y.; Wu, X.; et al. Single-shot super-resolution total internal reflection fluorescence microscopy. *Nat. Methods* **2018**, *15*, 425–428. [CrossRef]

48. Guo, Y.; Li, D.; Zhang, S.; Yang, Y.; Liu, J.J.; Wang, X.; Liu, C.; Milkie, D.E.; Moore, R.P.; Tulu, U.S.; et al. Visualizing Intracellular Organelle and Cytoskeletal Interactions at Nanoscale Resolution on Millisecond Timescales. *Cell* **2018**, *175*, 1430–1442. [CrossRef]
49. Chen, B.C.; Legant, W.R.; Wang, K.; Shao, L.; Milkie, D.E.; Davidson, M.W.; Janetopoulos, C.; Wu, X.S.; Hammer, J.A.; Liu, Z.; et al. Lattice light-sheet microscopy: Imaging molecules to embryos at high spatiotemporal resolution. *Science* **2014**, *346*, 1257998. [CrossRef]
50. Li, D.; Shao, L.; Chen, B.C.; Zhang, X.; Zhang, M.; Moses, B.; Milkie, D.E.; Beach, J.R.; Hammer, J.A.; Pasham, M.; et al. Extended-resolution structured illumination imaging of endocytic and cytoskeletal dynamics. *Science* **2015**, *349*, aab3500. [CrossRef]
51. De Luca, G.M.R.; Breedijk, R.M.P.; Brandt, R.A.J.; Zeelenberg, C.H.C.; de Jong, B.E.; Timmermans, W.; Azar, L.N.; Hoebe, R.A.; Stallinga, S.; Manders, E.M.M. Re-scan confocal microscopy: Scanning twice for better resolution. *Biomed. Opt. Express* **2013**, *4*, 2644. [CrossRef] [PubMed]
52. Huff, J. The Airyscan detector from ZEISS: Confocal imaging with improved signal-to-noise ratio and super-resolution. *Nat. Methods* **2015**, *12*, 1205. [CrossRef]
53. Wegel, E.; Göhler, A.; Lagerholm, B.C.; Wainman, A.; Uphoff, S.; Kaufmann, R.; Dobbie, I.M. Imaging cellular structures in super-resolution with SIM, STED and Localisation Microscopy: A practical comparison. *Sci. Rep.* **2015**, *6*, 1–13. [CrossRef] [PubMed]
54. Hell, S.W.; Wichmann, J. Breaking the diffraction resolution limit by stimulated emission: Stimulated-emission-depletion fluorescence microscopy. *Opt. Lett.* **1994**, *19*, 780. [CrossRef]
55. Vicidomini, G.; Moneron, G.; Han, K.Y.; Westphal, V.; Ta, H.; Reuss, M.; Engelhardt, J.; Eggeling, C.; Hell, S.W. Sharper low-power STED nanoscopy by time gating. *Nat. Methods* **2011**, *8*, 571–575. [CrossRef]
56. Curdt, F.; Herr, S.J.; Lutz, T.; Schmidt, R.; Engelhardt, J.; Sahl, S.J.; Hell, S.W. isoSTED nanoscopy with intrinsic beam alignment. *Opt. Express* **2015**, *23*, 30891–30903. [CrossRef]
57. Grimm, J.B.; Muthusamy, A.K.; Liang, Y.; Brown, T.A.; Lemon, W.C.; Patel, R.; Lu, R.; Macklin, J.J.; Keller, P.J.; Ji, N.; et al. A general method to fine-tune fluorophores for live-cell and in vivo imaging. *Nat. Methods* **2017**, *14*, 987. [CrossRef]
58. Bottanelli, F.; Kromann, E.B.; Allgeyer, E.S.; Erdmann, R.S.; Baguley, S.W.; Sirinakis, G.; Schepartz, A.; Baddeley, D.; Toomre, D.K.; Rothman, J.E.; et al. Two-colour live-cell nanoscale imaging of intracellular targets. *Nat. Commun.* **2015**, *7*, 1–5. [CrossRef]
59. Maraspini, R.; Wang, C.-H.; Honigmann, A. Optimization of 2D and 3D cell culture to study membrane organization with STED microscopy. *J. Phys. Appl. Phys.* **2020**, *53*, 014001. [CrossRef]
60. Schnell, U.; Dijk, F.; Sjollema, K.A.; Giepmans, B.N.G. Immunolabeling artifacts and the need for live-cell imaging. *Nat. Methods* **2012**, *9*, 152–158. [CrossRef]
61. Rust, M.J.; Bates, M.; Zhuang, X. Sub-diffraction-limit imaging by stochastic optical reconstruction microscopy (STORM). *Nat. Methods* **2006**, *3*, 793–795. [CrossRef] [PubMed]
62. Bates, M.; Huang, B.; Dempsey, G.T.; Zhuang, X. Multicolor super-resolution imaging with photo-switchable fluorescent probes. *Science* **2007**, *317*, 1749–1753. [CrossRef] [PubMed]
63. Betzig, E.; Patterson, G.H.; Sougrat, R.; Lindwasser, O.W.; Olenych, S.; Bonifacino, J.S.; Davidson, M.W.; Lippincott-Schwartz, J.; Hess, H.F. Imaging intracellular fluorescent proteins at nanometer resolution. *Science* **2006**, *313*, 1642–1645. [CrossRef] [PubMed]
64. Giannone, G.; Hosy, E.; Levet, F.; Constals, A.; Schulze, K.; Sobolevsky, A.I.; Rosconi, M.P.; Gouaux, E.; Tampe, R.; Choquet, D.; et al. Dynamic superresolution imaging of endogenous proteins on living cells at ultra-high density. *Biophys. J.* **2010**, *99*, 1303–1310. [CrossRef]
65. Vaughan, J.C.; Jia, S.; Zhuang, X. Ultrabright photoactivatable fluorophores created by reductive caging. *Nat. Methods* **2012**, *9*, 1181–1184. [CrossRef]
66. Jones, S.A.; Shim, S.H.; He, J.; Zhuang, X. Fast, three-dimensional super-resolution imaging of live cells. *Nat. Methods* **2011**, *8*, 499–505. [CrossRef]
67. Huang, B.; Wang, W.; Bates, M.; Zhuang, X. Three-Dimensional Super-Resolution Reconstruction Microscopy. *Science* **2008**, *319*, 810–813. [CrossRef]
68. Juette, M.F.; Gould, T.J.; Lessard, M.D.; Mlodzianoski, M.J.; Nagpure, B.S.; Bennett, B.T.; Hess, S.T.; Bewersdorf, J. Three-dimensional sub-100 nm resolution fluorescence microscopy of thick samples. *Nat. Methods* **2008**, *5*, 527–529. [CrossRef]

69. Shtengel, G.; Galbraith, J.A.; Galbraith, C.G.; Lippincott-Schwartz, J.; Gillette, J.M.; Manley, S.; Sougrat, R.; Waterman, C.M.; Kanchanawong, P.; Davidson, M.W.; et al. Interferometric fluorescent super-resolution microscopy resolves 3D cellular ultrastructure. *Proc. Natl. Acad. Sci. USA* **2009**, *106*, 3125–3130. [CrossRef]
70. Xu, K.; Babcock, H.P.; Zhuang, X. Dual-objective STORM reveals three-dimensional filament organization in the actin cytoskeleton. *Nat. Methods* **2012**, *9*, 185–188. [CrossRef]
71. Manley, S.; Gillette, J.M.; Patterson, G.H.; Shroff, H.; Hess, H.F.; Betzig, E.; Lippincott-Schwartz, J. High-density mapping of single-molecule trajectories with photoactivated localization microscopy. *Nat. Methods* **2008**, *5*, 155–157. [CrossRef] [PubMed]
72. Sage, D.; Pham, T.A.; Babcock, H.; Lukes, T.; Pengo, T.; Chao, J.; Velmurugan, R.; Herbert, A.; Agrawal, A.; Colabrese, S.; et al. Super-resolution fight club: Assessment of 2D and 3D single-molecule localization microscopy software. *Nat. Methods* **2019**, *16*, 387–395. [CrossRef] [PubMed]
73. Schermelleh, L.; Ferrand, A.; Huser, T.; Eggeling, C.; Sauer, M.; Biehlmaier, O.; Drummen, G.P.C. Super-resolution microscopy demystified. *Nat. Cell Biol.* **2019**, *21*, 72–84. [CrossRef] [PubMed]
74. Maître, J.L.; Berthoumieux, H.; Krens, S.F.G.; Salbreux, G.; Jülicher, F.; Paluch, E.; Heisenberg, C.P. Adhesion functions in cell sorting by mechanically coupling the cortices of adhering cells. *Science* **2012**, *338*, 253–256. [CrossRef] [PubMed]
75. Adams, C.L.; Nelson, W.J.; Smith, S.J. Quantitative analysis of cadherin-catenin-actin reorganization during development of cell-cell adhesion. *J. Cell Biol.* **1996**, *135*, 1899–1911. [CrossRef] [PubMed]
76. Case, L.B.; Baird, M.A.; Shtengel, G.; Campbell, S.L.; Hess, H.F.; Davidson, M.W.; Waterman, C.M. Molecular mechanism of vinculin activation and nanoscale spatial organization in focal adhesions. *Nat. Cell Biol.* **2015**, *17*, 880–892. [CrossRef] [PubMed]
77. Van Zanten, T.S.; Cambi, A.; Koopman, M.; Joosten, B.; Figdor, C.G.; Garcia-Parajo, M.F. Hotspots of GPI-anchored proteins and integrin nanoclusters function as nucleation sites for cell adhesion. *Proc. Natl. Acad. Sci. USA* **2009**, *106*, 18557–18562. [CrossRef]
78. Shroff, H.; Galbraith, C.G.; Galbraith, J.A.; White, H.; Gillette, J.; Olenych, S.; Davidson, M.W.; Betzig, E. Dual-color superresolution imaging of genetically expressed probes within individual adhesion complexes. *Proc. Natl. Acad. Sci. USA* **2007**, *104*, 20308–20313. [CrossRef]
79. Grashoff, C.; Hoffman, B.D.; Brenner, M.D.; Zhou, R.; Parsons, M.; Yang, M.T.; McLean, M.A.; Sligar, S.G.; Chen, C.S.; Ha, T.; et al. Measuring mechanical tension across vinculin reveals regulation of focal adhesion dynamics. *Nature* **2010**, *466*, 263–266. [CrossRef]
80. Sasaki, H.; Matsui, C.; Furuse, K.; Mimori-Kiyosue, Y.; Furuse, M.; Tsukita, S. Dynamic behavior of paired claudin strands within apposing plasma membranes. *Proc. Natl. Acad. Sci. USA* **2003**, *100*, 3971–3976. [CrossRef]
81. Cording, J.; Arslan, B.; Staat, C.; Dithmer, S.; Krug, S.M.; Krüger, A.; Berndt, P.; Günther, R.; Winkler, L.; Blasig, I.E.; et al. Trictide, a tricellulin-derived peptide to overcome cellular barriers. *Ann. N. Y. Acad. Sci.* **2017**, *1405*, 89–101. [CrossRef] [PubMed]
82. Liss, V.; Barlag, B.; Nietschke, M.; Hensel, M. Self-labelling enzymes as universal tags for fluorescence microscopy, super-resolution microscopy and electron microscopy. *Sci. Rep.* **2015**, *5*, 1–13. [CrossRef] [PubMed]
83. Lukinavičius, G.; Umezawa, K.; Olivier, N.; Honigmann, A.; Yang, G.; Plass, T.; Mueller, V.; Reymond, L.; Corrêa, I.R.; Luo, Z.G.; et al. A near-infrared fluorophore for live-cell super-resolution microscopy of cellular proteins. *Nat. Chem.* **2013**, *5*, 132–139. [CrossRef] [PubMed]

International Journal of
Molecular Sciences

Review

Structural Features of Tight-Junction Proteins

Udo Heinemann [1,*] and Anja Schuetz [2,*]

1 Macromolecular Structure and Interaction Laboratory, Max Delbrück Center for Molecular Medicine, 13125 Berlin, Germany
2 Protein Production & Characterization Platform, Max Delbrück Center for Molecular Medicine, 13125 Berlin, Germany
* Correspondence: heinemann@mdc-berlin.de (U.H.); anja.schuetz@mdc-berlin.de (A.S.); Tel.: +49-30-9406-3420 (U.H.); +49-30-9406-2985 (A.S.)

Received: 29 October 2019; Accepted: 26 November 2019; Published: 29 November 2019

Abstract: Tight junctions are complex supramolecular entities composed of integral membrane proteins, membrane-associated and soluble cytoplasmic proteins engaging in an intricate and dynamic system of protein–protein interactions. Three-dimensional structures of several tight-junction proteins or their isolated domains have been determined by X-ray crystallography, nuclear magnetic resonance spectroscopy, and cryo-electron microscopy. These structures provide direct insight into molecular interactions that contribute to the formation, integrity, or function of tight junctions. In addition, the known experimental structures have allowed the modeling of ligand-binding events involving tight-junction proteins. Here, we review the published structures of tight-junction proteins. We show that these proteins are composed of a limited set of structural motifs and highlight common types of interactions between tight-junction proteins and their ligands involving these motifs.

Keywords: tight junction; protein structure; protein domain; claudins; occludin; tricellulin; junctional adhesion molecule; *zonula occludens*; MAGUK proteins; PDZ domain

---

1. Introduction

A classical paper published more than half a century ago [1] clearly demonstrated that the epithelia of several glands and cavity-forming internal organs of the rat and guinea pig all share characteristic tripartite junctional complexes between adjacent cells. These junctional complexes were found in the epithelia of the stomach, intestine, gall bladder, uterus, oviduct, liver, pancreas, parotid, thyroid, salivary ducts, and kidney. Progressing from the apical to the basal side of the endothelial cell layer, the elements of the junctional complex were characterized as tight junctions (*zonulae occludens*), adherens junctions, and desmosomes. As most apical elements of the junctional complex, tight junctions (TJs) were distinct by the apparent fusion of adjacent cell membranes over variable distances and appeared as a diffuse band of dense cytoplasmic material in the electron microscope. TJs formed a continuous belt-like structure, whereas desmosomes displayed discontinuous button-like structures, and adherens junctions (AJs) were intermediate in appearance. The molecular composition of TJs was revealed in subsequent work by many laboratories, e.g., [2–5], and shown to include at least 40 different proteins.

In the pioneering work of Farquhar and Palade, TJs were proposed to function as effective diffusion barriers or seals [1]. The sealing function of TJs contributes to the formation and physiological function of the blood-brain barrier (BBB), which consists of endothelial cells sealed by apical junctional complexes including TJs. Functions of transmembrane TJ proteins at the BBB are well documented [6–8]. BBB dysfunction is linked to a number of diseases including multiple sclerosis, stroke, brain tumors, epilepsy, and Alzheimer's disease [9,10].

Here, we review the current literature regarding three-dimensional structures of TJ proteins, their domains and intermolecular interactions. We do not primarily aim at presenting each and every structure in detail, but attempt to distill common structural principles that underlie the architecture of the TJ. We apologize to those authors whose work may not have been covered in this paper for reasons of space and readability.

2. Structural Insight into Tight-Junction Proteins, Their Domains, and Interactions

In the most general, birds-eye description, the TJ consists of a set of transmembrane (TM) proteins and the cytoplasmic plaque, a complex network of scaffolding and effector proteins that connects the TM proteins to the actomyosin cytoskeleton of the cell (Figure 1). The TM proteins interact with their extracellular domains in the paracellular space, and the connection to the cytoskeleton inside the cell is structurally as yet uncharacterized [2–5,11,12]. The *zonula occludens* proteins ZO-1, ZO-2, and ZO-3 and the two mammalian polarity complexes PAR-3/PAR-6/aPKC and Crumbs/ PALS1/PATJ are central players of the cytoplasmic plaque and are described in more detail in this review together with the transmembrane TJ proteins.

2.1. *Tight-Junction Transmembrane Proteins*

TJ transmembrane proteins contain either one, three, or four TM segments. The Crumbs proteins (CRBs), the junctional adhesion molecules (JAMs), the angulin proteins, and the coxsackievirus–adenovirus receptor (CAR) are representatives of single-span TJ membrane proteins. BVES (blood-vessel epicardial substance, also known as POPDC1 for Popeye domain-containing protein-1) is a TJ-associated protein with three TM regions [13,14]. The claudins and the TAMPs (tight junction-associated MARVEL-domain proteins) occludin (MARVELD1), tricellulin (MARVELD2), and MARVELD3 are tetra-span TM proteins. MARVEL is used as a common acronym for MAL (myelin and lymphocyte) and related proteins for vesicle trafficking and membrane link [15]. Where crystal structures are available, for example the claudin family (Section 2.1.2, [16]), the TM segments were shown to be α-helical.

2.1.1. Junctional Adhesion Molecules and Other Ig-Like TJ Proteins

The JAMs are a family of adhesion molecules with immunoglobulin (Ig)-like ectodomains, localized in epithelial and endothelial cells, leukocytes, and myocardial cells [17]. The 2.5-Å crystal structure of the soluble extracellular part of mouse JAM-A provided the first structural insight into a TJ transmembrane protein [18]. In this structure, two Ig domains are connected by a short linker peptide, and a U-shaped dimer is formed by symmetrical interaction of the N-terminal Ig domains (Figure 2). This structure provided the basis for a model of homophilic interactions between the N-domains to explain the adhesive function of JAMs in the TJ. The crystal structure of the extracellular Ig domains of coxsackievirus–adenovirus receptor CAR, another component of the epithelial apical junction complex that is essential for TJ integrity [19], suggests a very similar mode of CAR homodimer formation through symmetrical interaction of its N-terminal Ig domain [20].

The extracellular portions of JAMs serve as viral attachment sites. Reoviruses attach to human cells by binding to cell–surface carbohydrates and the junctional adhesion molecule JAM-A. The crystal structure of reovirus attachment protein σ1 bound to the soluble form of JAM-A shows that σ1 disrupts the native JAM-A dimer to form a heterodimer via the same interface as used in JAM-A homodimers, but with a 1000-fold lower dissociation constant of the σ1/JAM-A heterodimer as compared to the JAM-A homodimer [21]. In cat, infection with calicivirus is initiated by binding of the minor capsid protein VP2 to feline junctional adhesion molecule A (JAM-A). High-resolution cryo-EM structures of VP2 and soluble JAM-A-decorated VP2 show formation of a large portal-like assembly, which is hypothesized to serve as a channel for the transfer of the viral genome [22].

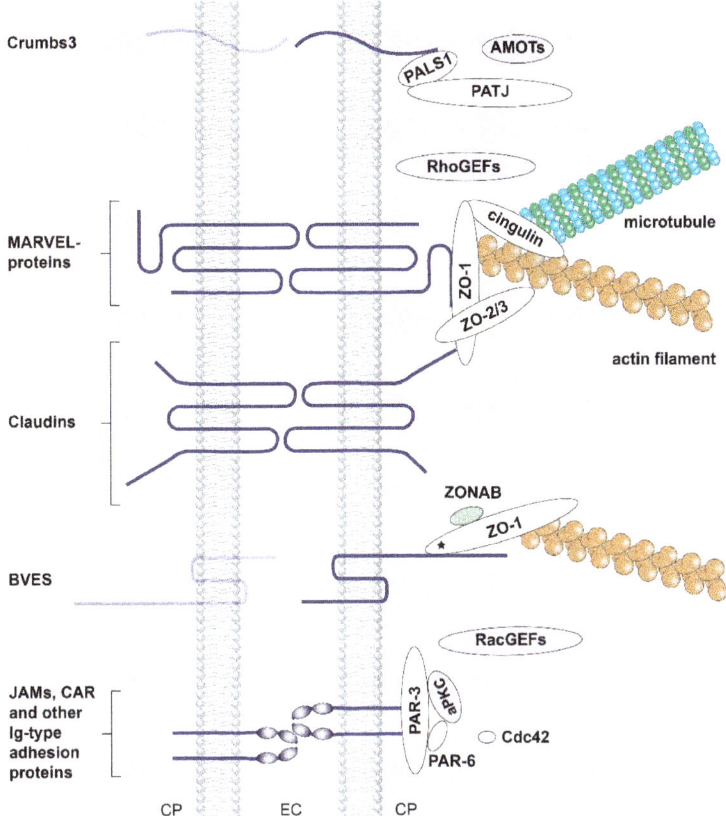

Figure 1. The tight-junction core structure. TM proteins of the TJ (dark blue) interact with a complex cytoplasmic protein network, the cytoplasmic plaque (shown on the right), providing a physical link to the cytoskeleton (microtubules, actin filaments). Cross-membrane interactions between TM proteins are indicated schematically. For some TM proteins (shown in pale blue) there is no direct evidence for a *trans* pairing interaction between TM proteins of opposing cells. The cytoplasmic plaque is composed of scaffolding proteins (yellow ovals) that associate with signaling proteins (purple ovals) and (post)transcriptional regulators (green oval), forming the zonular signalosome [23]. The three major protein complexes located in the cytoplasmic plaque are depicted. Within the ZO complex, ZO proteins are present as homodimers or ZO-1/ZO-2 and ZO-1/ZO-3 heterodimers [24] that directly associate with integral TJ membrane proteins through multiple interactions. The polarity complexes PAR-3/PAR-6/aPKC and Crumbs/PALS1/PATJ are responsible for the development of the apico-basal axis of epithelial cells and act as apical components of TJs. TM proteins: Crumbs homolog 3 (CRB3); MARVEL-domain containing proteins occludin, tricellulin, and MARVEL domain-containing protein 3 (MARVELD3); the claudins; the protein blood vessel epicardial substance (BVES); immunoglobulin (Ig) superfamily members such as junctional adhesion molecules (JAMs) and the coxsackievirus–adenovirus receptor (CAR). Cytoplasmic scaffolding proteins: *Zonula occludens* (ZO) proteins ZO-1, ZO-2, and ZO-3; partitioning defective 3/6 homologs (PAR-3, PAR-6); protein associated with Lin-7 1 (PALS1); PALS1-associated tight junction (PATJ) protein; cytoskeletal linker cingulin. Signaling proteins: Atypical protein kinase C (aPKC); proteins of the angiomotin family (AMOTs) [25]; the small Rho-GTPase Cdc42, and guanine nucleotide exchange factors for the Rho-GTPases RhoA (RhoGEFs, e.g., ARHGEF11 [26]) and Rac1 (RacGEFs, e.g., Tiam-1 [27]), respectively. Transcriptional regulator: ZO-1–associated nucleic acid-binding protein (ZONAB, YBX3 in human) [28]. Figure modified and updated after Zihni et al. [12].

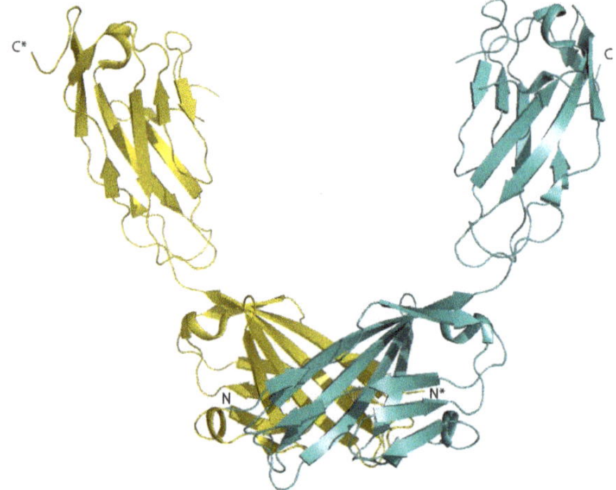

Figure 2. JAM-A dimerization via extracellular Ig domains. Crystal structure of murine JAM-A (PDB entry 1F97) [18]. The dimer is generated by crystallographic two-fold symmetry.

In addition to JAMs and CAR, other single-span Ig-like adhesion molecules such as the endothelial cell-selective adhesion molecule (ESAM), the coxsackievirus and adenovirus receptor-like membrane protein (CLMP), the brain- and testis-specific immunoglobulin superfamily protein (BT-IgSF or IgSF11) [19], and the angulin family of proteins are present at TJs. The latter comprise the proteins LSR (lipolysis-stimulated lipoprotein receptor), ILDR1, and ILDR2 (Ig-like domain-containing receptors) that complement each other at tricellular TJs and co-operate with tricellulin to mediate full barrier function in epithelial sheets [29,30]. Loss of LSR is linked to cell invasion and migration in human cancer cells [31]. To date, however, no structural data are available for any of these proteins.

2.1.2. Claudins

Members of the claudin family are the most abundant TM proteins of the TJ [32]. Claudin genes are expressed across all epithelial tissues, and in all epithelia various different claudins are expressed at the same time [12,33–36]. Tissue-specific expression of claudin genes has been documented, for example, for claudins in the kidney, inner ear, and eye [33].

At TJs, claudins are arranged to form extended strands by homophilic or heterophilic *cis* pairing within the same membrane or *trans* pairing across membranes [37,38]. In humans, 23 claudins and two claudin-like proteins are currently known (Figure 3). Although clearly homologous, the claudins share only a small number of strictly conserved residues and differ in the lengths of their N- and C-termini and the loops connecting their four TM helices. Highest sequence conservation is observed within the first extracellular loop (ECL1) where a tryptophan and two cysteine residues are strictly conserved in all sequences, suggesting formation of a disulfide bond in this region, which was experimentally verified by crystal structure analysis [16]. Various schemes for grouping claudins have been proposed. Based on sequence conservations, a grouping into classical claudins (1–10, 14, 15, 17, 19) and non-classical claudins (11–13, 16, 18, 20–24) was suggested [39], but the alignment shown in Figure 3 does not clearly support a separation into these groups. Based on function within the TJ, claudins may be grouped according to their barrier or channel forming properties with respect to different solutes [34,35,40]. Claudins 1, 3, 5, 11, 14, and 19, for example, have been characterized as predominantly sealing, whereas claudins 2, 10a, 10b, 15, and 17 were described as predominantly channel forming. For claudins 4, 7, 8, and 16 a sealing or channel-forming function has not been unequivocally determined [34]. Moreover, assignment of these functions to TM proteins of the TJ may be difficult when individual TJ proteins are functionally replaced by paralogs in certain epithelia or in the presence of post-translational modification.

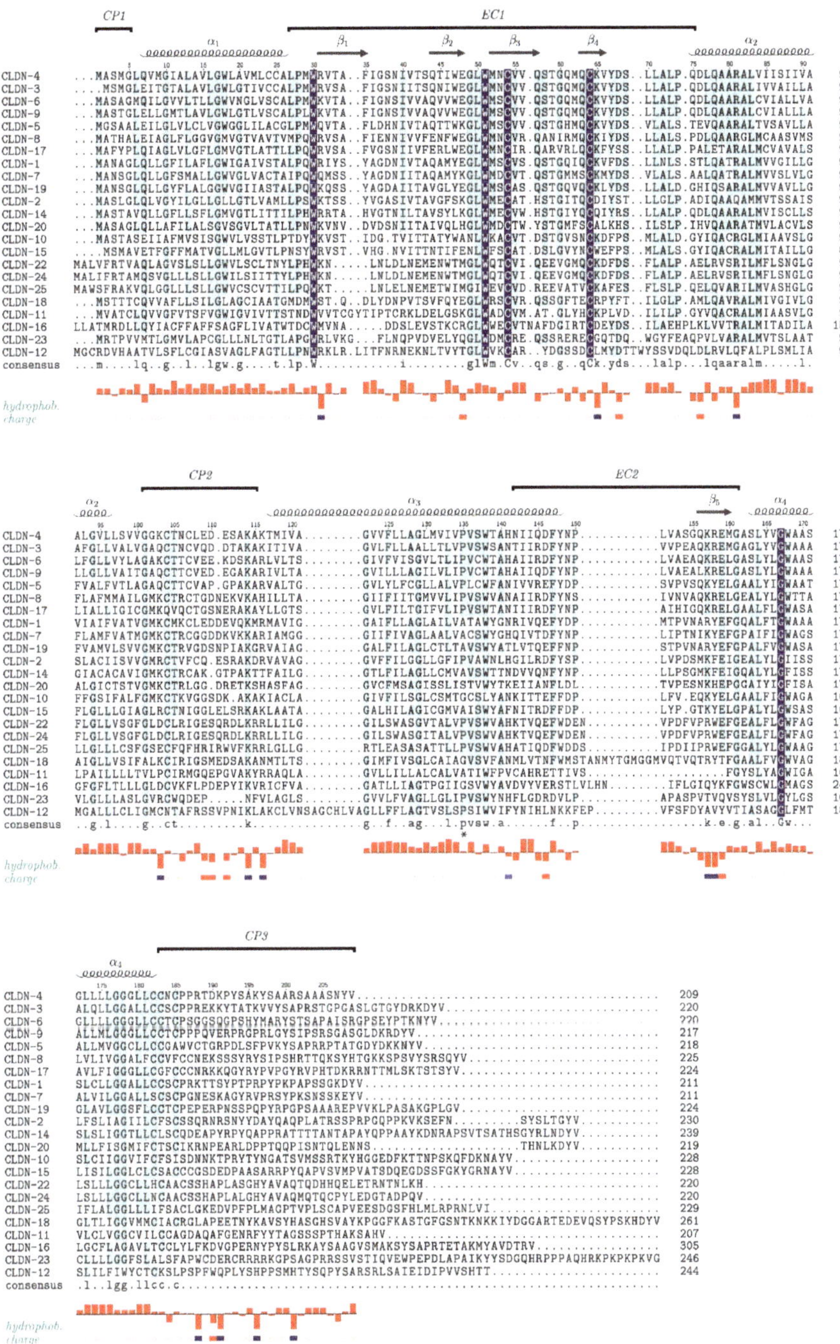

Figure 3. Sequence alignment of human claudins. The Uniprot database [41] lists 23 human claudin sequences belonging to CLDN-1 to CLDN-12 and CLDN-14 to CLDN-25; CLDN-21 is sometimes referred to as CLDN-24. In addition, there are two claudin domain-containing proteins (CLDND1 und

CLDND2), which are not included in the alignment. 66 N-terminal residues of CLDN-16 and 46 C-terminal residues of CLDN-23 are omitted from the alignment, because they have no match in any other human claudin sequence. Domain and secondary-structure annotation follows CLDN-4 for which a crystal structure is known in the presence of a bound toxin and loop EC1 is fully ordered [16,42], and the claudin sequences are listed in order of their match with the CLDN-4 sequence. Residues conserved across all human claudins are highlighted on dark blue background and residues conserved in ≥ 50% of the sequences are shown on a light blue background. EC: Extracellular, CP: Cytoplasmic. Conservation of hydrophobicity and charge (blue, positive; red, negative) is indicated at the bottom of the alignment. The asterisk marks a proline residue within α3 of claudin-3, which induces a kink in this helix and probably most other claudins. The amino-acid sequences were aligned using the Clustal Omega server [43], and TEXshade [44] was used for illustration.

A major breakthrough in TJ research was made in 2014 when the 2.4-Å resolution crystal structure of full-length mouse claudin-15 was reported [45]. As predicted from the sequence, the polypeptide chain was organized into four antiparallel TM helices with the N- and C-termini on the cytoplasmic side. On the extracellular side, a five-stranded up-and-down antiparallel β-sheet was formed by the long ECL1 (strands β1-β4) and the short ECL2 (strand β5, pairing with β1). In this crystal structure, ECL1 is partially disordered, because the loop (v1) connecting strands β1 and β2 is not represented in electron density [45], but it is ordered in the presence of bound ligand (see below). A molecular dynamics study based on the claudin-15 crystal structure [45] suggested that the protein forms a tetrameric channel in which a cage of four aspartate-15 residues acts as a selectivity filter that favors cation flux over anion flux [46].

This and other claudin structures are hoped to provide a basis for the targeted disruption of epithelial barriers in the administration of drugs [47]. The subsequently published crystal structure of mouse claudin-3 showed that proline 134 in TM helix α3 induces a bend in this helix, which is alleviated by the corresponding alanine or glycine mutations. A proline residue at this position is present in the majority of human claudin sequences; a helix bend brought about by this residue is likely to modulate the morphology and adhesiveness of TJ strands [48]. Three-dimensional structures of claudins provide the basis for *in silico* modeling of claudin based TJ self-assembly, their barrier and/or channel forming potential [49–51].

Much as the extracellular Ig domains of the JAMs are attachment sites for viruses, the extracellular loops of claudins serve as landing sites for bacterial toxins such as the *Clostridium perfringens* enterotoxin (CpE). A crystal structure of full-length claudin-19 bound to the soluble, claudin-binding C-terminal fragment of CpE (C-CPE) was determined at 3.7 Å resolution [52]. This structure showed that ligand binding leads to a stabilization of loop v1, which is now ordered, and indicated how C-CPE binding to selected claudins may lead to the disintegration of TJs and increased permeability across epithelial layers. C-CPE appears to bind different claudins with a conserved geometry and to disrupt the lateral interactions of their extracellular parts in the same way [16,52] as suggested by the crystal structure of C-CPE-bound human claudin-4 (Figure 4) [42]. Human claudin-9 (hCLDN-9) is highly expressed in the inner ear, essential for hearing and a high-affinity receptor of CpE. Two recently published 3.2-Å crystal structures of hCLDN-9 bound to C-CPE reveal structural changes in claudin epitopes involved in claudin self-assembly and suggest a mechanism for the disruption of claudin and TJ dissociation by CpE [53].

2.1.3. Occludin

Occludin and the other TAMPs of the TJ, tricellulin, and MARVELD3, share with the claudins the general architecture as tetraspan TM proteins with cytoplasmic N- and C-termini. However, the TAMPs are not homologous with claudins and differ in the length and structure of their cytoplasmic domains and extracellular loops.

The occludin cytosolic C-terminus forms a coiled-coil structure, dimerizes, and associates with all three ZO-proteins from the TJ cytoplasmic plaque [24,54,55]. Disulfide formation within the coiled-coil domain was proposed as a mechanism to influence the oligomerization of occludin [56,57]. The 1.45-Å crystal structure of the cytosolic C-terminus of occludin comprises three helices that form two separate anti-parallel coiled-coils and a loop that packs tightly against one of the coiled-coils (Figure 5a). This structure revealed a large positively charged surface that binds ZO-1 [58]. The cytoplasmic C-terminal coiled-coil region of occludin associates with mainly the GUK region of ZO-1 as shown by SAXS, NMR, and in vitro binding studies [59], which also revealed that serine phosphorylation within the acidic binding motif of the occludin coiled-coil significantly increases the binding affinity. Notably, several occludin isoforms result from alternative splicing and alternate promoter use, but neither this structural polymorphism nor the multitude of known post-translational modifications from proteolysis and serine, threonine or tyrosine phosphorylation of occludin [60] have so far been studied by X-ray or NMR methods.

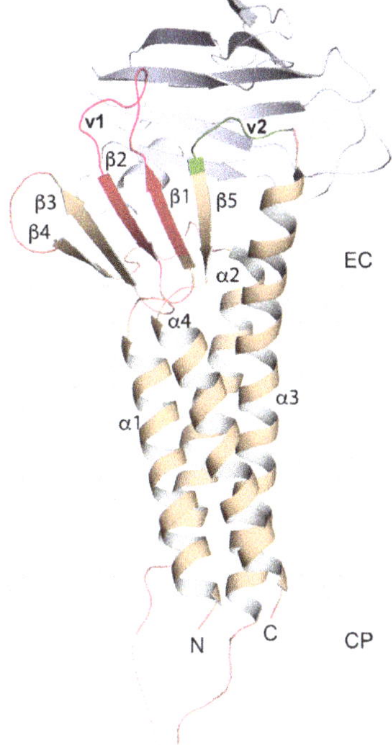

Figure 4. Crystal structure of human claudin-4. Cartoon model of the overall fold of human CLDN-4 (wheat color) in complex with the C-terminal fragment of *Clostridium perfringens* enterotoxin (C-CPE, light blue; PDB entry 5B2G) [16,42]. The extracellular variable regions of CLDN-4 that mediate hetero- and homotypic interactions are highlighted in magenta (v1, comprising β1 and β2) and green (v2, between TM-helix α3 and β5), respectively. The dotted line marks a segment of polypeptide chain not represented in electron density. The stylized lipid molecules indicate the cell membrane and are not part of the experimental structure. EC: Extracellular; CP: Cytoplasmic.

2.1.4. Tricellulin

The precise definition of TJ architecture through freeze–fracture microscopy of epithelial preparations from rat intestine revealed a modified structure at tricellular junctions [61]. Tricellular pores and bicellular strand opening contribute to allowing the passage of large molecules through the TJ in the "leak pathway" as suggested by computational structural dynamics studies [62]. TJs completely disappear during the epithelial–mesenchymal transition (EMT), where the transcriptional repressor Snail plays a central role. The protein tricellulin was identified in a screen using Snail-overexpressing epithelial cells as a protein concentrated at tricellular tight junctions (tTJs) and named for this property [63].

Tricellulin is downregulated during the EMT. The E3 ubiquitin ligase Itch binds the N-terminus of tricellulin via its WW domain (named after two signature tryptophan residues) to stimulate its ubiquitination, which is, however, not primarily involved in proteasomal breakdown of tricellulin [64]. During apoptosis, cells are extruded from epithelial cell layers. Loss of functional tricellulin contributes to dissociation of tTJs during apoptosis, when it is cleaved by caspase at aspartate residues 441 and 487 in the C-terminal coiled-coil [65]. Tricellulin is of key importance for hearing, as it was reported that mutations in the human *TRIC* gene are associated with deafness [66].

Tricellulin is localized to tTJs but also to bicellular TJs. When tricellulin is selectively overexpressed at tTJs, it decreases the permeability for large solutes up to 10 kDa, but not for ions. This seemingly paradoxical observation may be explained by the rare occurrence of tricellular junctions relative to bicellular junctions [67]. Tricellular TJs are regarded as potential weak points in the paracellular barrier. Tricellulin-dependent macromolecular passage is observed in both leaky and tight epithelia [68]. Tricellulin tightens tricellular junctions and regulates bicellular TJ proteins. The extracellular loops of tricellulin may be crucial for its sealing function, because it could be shown that a synthetic peptide (trictide) derived from the tricellulin ECL2 may increase the passage of solutes into human adenocarcinoma cells [69]. In MDCK cells, the tricellulin C-terminus is important for basolateral translocation, whereas the N-terminus directs tricellulin to tricellular contacts. There is evidence for the formation of heteromeric tricellulin–occludin contacts at elongating bicellular junctions and of homomeric tricellulin–tricellulin complexes at tricellular junctions [70].

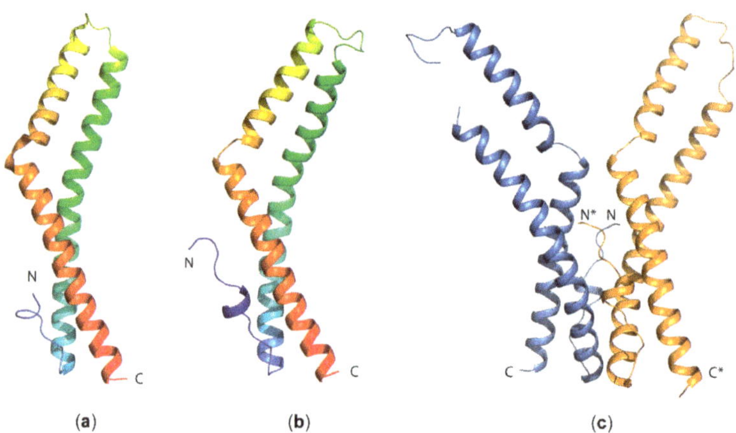

Figure 5. Structural insight into occludin and tricellulin function. Cartoon models of the overall fold of the coiled-coil domain of (**a**) human occludin (PDB entry 1XAW) [58] and (**b**) human tricellulin (PDB entry 5N7K) [71]. The molecules are colored in a gradient ranging from blue at the N-terminus (N) to red at the C-terminus. (**c**) Dimeric arrangement of the tricellulin C-terminal coiled-coil domain observed in the crystal structure [71]. The chain marked with an asterisk (*) corresponds to the second monomer within the dimer.

Tricellulin has an extended cytoplasmic N-terminus of 194 aa and a cytoplasmic C-terminal region of 195 aa, in marked contrast to occludin, where these regions include 66 aa and 256 aa, respectively. With the exception of the C-terminal coiled-coil domain, no cytoplasmic region carries a sequence signature suggesting a known domain structure in either protein. A crystal structure of the C-terminal coiled-coil domain of tricellulin was determined at 2.2-Å resolution (Figure 5b). This structure reveals a dimeric arrangement with an extended polar interface (Figure 5c), which may contribute to stabilizing tTJs [71].

2.1.5. Other Tight-Junction Transmembrane Proteins

With the exception of tricellulin, the extracellular loops and ectodomains of the abovementioned transmembrane TJ proteins are involved in *trans* pairing interactions of opposing cells (Figure 1) [5]. For the POPDC and Crumbs family of transmembrane TJ proteins, no such cross–membrane interactions are described. The POPDC family of tri-span TM proteins consists of BVES/POPDC1, POPDC2, and POPDC3. BVES protein dimers are mediated by the cytoplasmic Popeye domain, and BVES–BVES *cis* pairing interactions are necessary to maintain epithelial integrity and junctional stability. The cytoplasmic tail of BVES was shown to directly interact with ZO-1 [14], but structural information on the atomic level is still missing [14]. Crumbs was first described in *D. melanogaster* [72]; in mammals it has three homologs (CRB1, CRB2, CRB3) of which the latter is expressed in all epithelial tissues [73]. As the Crumbs protein family members are part of the cell polarity complex Crumbs/PALS1/PATJ, further information is included in Section 2.2.3.

2.2. *Proteins of the Cytoplasmic Plaque*

The proteins of the cytoplasmic plaque are characterized by recurrent protein–protein interaction (PPI) domains and frequently contain natively unfolded regions [74–76]. They are interconnected in a dynamic and multivalent PPI system, which has been partly mapped down to the domain level (Figure 6). In addition to the interactions displayed in the figure, there are multiple PPIs with regulatory and signaling proteins not covered in this review.

2.2.1. PDZ Domains

Many proteins of the cytoplasmic plaque contain one or multiple PDZ domains (Figure 6). We next discuss some key features of these ubiquitous PDZ domains. PDZ domains regulate multiple cellular processes by promoting protein–protein interactions and are abundant protein modules in TJ proteins, but also in many other proteins in all kingdoms of life. Frequently, PDZ domains are associated with WW, SH2, SH3 (Src homology 2 or 3), or PH (Pleckstrin homology) domains within one polypeptide chain [77]. The term PDZ is derived from the three founding members of the family, PSD-95 (postsynaptic density-95), the *Drosophila* tumor suppressor protein DLG-1 (discs large 1), and ZO-1. As early as 2010, > 900 PDZ domains were annotated in > 300 proteins encoded in the mouse genome, and > 200 X-ray or NMR structures of PDZ domains from various sources were known [78]. In August 2019, a PDB [79] search returned 533 entries with the keyword "PDZ domain" and 138 entries with the keyword "PDZ domain-like". Thus, extensive structural data are available for these domains. In general, PDZ domains are structured as a β-sandwich capped by two α-helices and bind ligand peptides in a shallow groove between helix α2 and strand β2 (Figure 7a). Their propensity to dimerize via domain swapping was first described for the second PDZ domain (PDZ2) of ZO-2 [80] and later also for PDZ2 of ZO-1 and ZO-3 (Figure 7b, see Section 2.2.2.).

Domain swapping is frequently observed in small β-sheet domains. Bacterial major cold-shock proteins [81,82], for example, were found to form domain-swapped dimers. A domain-swapped three-stranded segment of the *E. coli* cold-shock protein CspA is capable of recombining with a polypeptide region of ribosomal protein S1 to form a closed β-barrel recapitulating structural features of both parent proteins [73].

PDZ domains have been divided into three specificity classes according to the preferred amino acid residue at position –2 (P^{-2}) of the binding groove [72]. Typically, PDZ domains recognize sequence motifs at the extreme carboxy terminus of ligand proteins (Figure 7c), but binding of internal sequence motifs is also common (Figure 7d).

PDZ domains are regarded as promising drug targets for neurological and oncological disorders, as well as viral infections. Many structure-guided efforts are underway towards the development of small-molecule or peptidic modulators of PDZ domains [83,84], including the PDZ domains from Shank3, a central scaffolding protein of the post-synaptic density protein complex [85] and of the protein interacting with C kinase (PICK1), a regulator of AMPA receptor trafficking at neuronal synapses [86].

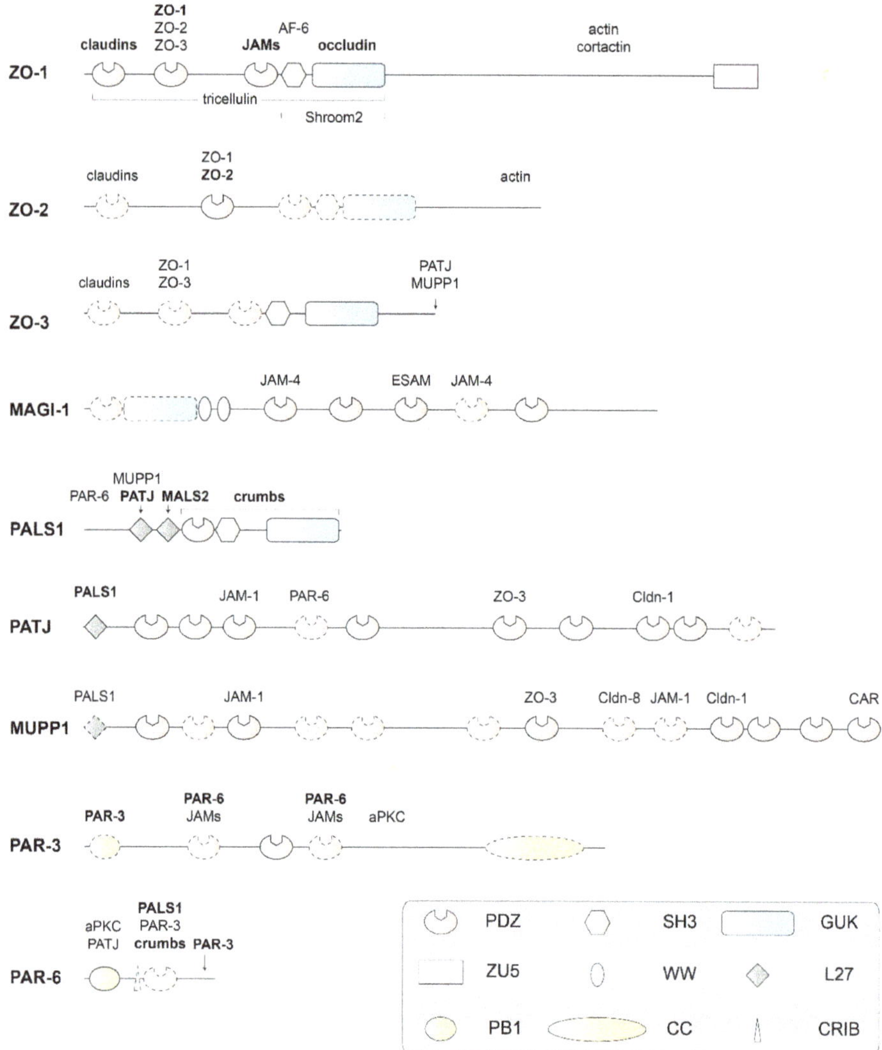

Figure 6. Domain structure of human PDZ domain-containing adapter proteins of the cytoplasmic plaque that interact physically with TM proteins of the tight junction. The proteins are scaled to the length of their amino-acid sequences. Experimental structures (usually by X-ray or NMR analysis) are available for protein domains drawn with solid contours, but not for domains drawn with dashed contours or for extended regions of polypeptide chains without domain annotation. Proteins binding to components of the human cytoplasmic plaque or their homologs are indicated above or below their interacting domains. With the exception of aPKC (as a subunit of the PAR-3/PAR-6/aPKC complex), only TM proteins or classical adapter proteins of the TJ are included as interacting proteins. Names of interacting proteins are written in bold letters, where the interaction is structurally characterized. Protein names and abbreviations are explained in the text or the legend of Figure 1. Domains are abbreviated as follows: PDZ: Initially identified in PSD-95 (postsynaptic density-95); DLG-1 (the *Drosophila* tumor suppressor protein discs large 1) and ZO-1; SH3: Src homology-3; PB1: Initially identified in PHOX and BEM1; ZU5: Present in ZO-1 and UNC5; L27: LIN-2/LIN-7; GUK: guanylate kinase homolog; WW: Named after two signature tryptophan residues; CC: coiled-coil; CRIB: CDC42/RAC interactive binding. Figure modified and updated after Guillemot et al. [3].

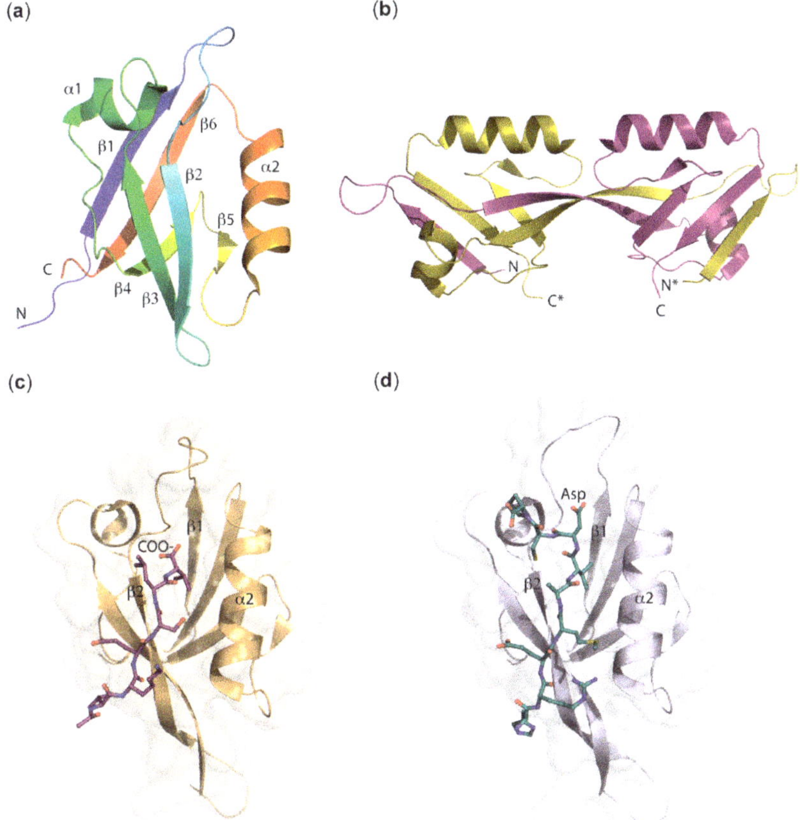

Figure 7. Structural features of PDZ domains. (**a**) Topology of a prototypical PDZ domain, here PALS 1 PDZ, PDB entry 4UU6 [87]. The polypeptide chain is drawn in rainbow colors changing from N- to C-terminus. A cleft for binding ligand peptides in an extended conformation is visible between strand β2 and helix α2. (**b**) Domain-swapped dimer formed by ZO-2 PDZ2, PDB entry 3E17 [88]. The two polypeptide chains are drawn in yellow and purple. The chain marked with an asterisk (*) corresponds to the second monomer within the dimer. Domain swapping moves the N-terminal β1 strand and half of β2 of one chain into the core structure of the other, leaving the ligand-binding geometry in both halves of the dimer intact. The PDZ2 domains in ZO-1, ZO-2, and ZO-3 are all found in the domain-swapped dimeric form [80,88–91]. (**c**) Canonical binding of a C-terminal ligand peptide to a PDZ domain, here PAR-6 PDZ bound to the hexapeptide VKRSLV, PDB entry 1RZX [92]. The terminal carboxy group is bound to the carboxylate-binding loop between strands β1 and β2 of PAR-6 PDZ and the extended ligand peptide aligns in antiparallel orientation with β2, extending the β-sheet. Note that this binding mode is energetically disfavored for PAR-6 PDZ and requires binding of CDC42 at a nearby CRIB domain (not shown). (**d**) Binding of an internal peptide to a PDZ domain, here PAR-6 PDZ bound to a dodecapeptide representing amino acids 29–40 of PALS1, PDB entry 1X8S [93]. The ligand peptide adopts an extended conformation with an aspartic acid side chain mimicking the carboxy group of the canonical C-terminal peptide ligand. Note the altered conformation of the carboxylate-binding loop.

2.2.2. MAGUK Proteins

Membrane-associated guanylate kinase homologs (MAGUKs) constitute a family of scaffolding molecules with a core MAGUK module consisting of a PDZ, SH3, and an enzymatically inactive guanylate kinase (GUK) domain [94–96]. The MAGUK protein family members ZO-1, ZO-2, and ZO-3 link the TJ membrane proteins to the cytoskeleton and provide the structural basis for the assembly of multiprotein complexes at the cytoplasmic side of TJs (Figure 1) [97]. ZO-1 is a cytoplasmic component of both TJs and AJs, and connects the TJ to the actin cytoskeleton via extended, presumably unstructured polypeptide regions near its C-terminus [98]. Direct actin binding was also reported for ZO-2 and ZO-3 [24,54,55].

ZO-1 and its paralogs ZO-2 and ZO-3 contain three N-terminal PDZ domains (Figure 6). The propensity of these PDZ domains to recognize specific C-terminal or other peptide motifs and assemble multicomponent TJ protein complexes will be highlighted below. Most of the claudins present at TJs have conserved C-terminal tails that bind to PDZ1 of ZO proteins. Compared to the single PDZ domain of the AJ protein Erbin, the ZO-1 PDZ1 domain has a broadened ligand specificity. Crystal structures of the Erbin PDZ and the ZO-1 PDZ1 revealed the structural basis for the different ligand specificities, where subtle conformational rearrangements are identified at multiple ligand-binding subsites, and support a model for ligand recognition by these domains [99].

The intracellular C-terminus of claudins binds to the N-terminal PDZ1 domain of ZO proteins with variable affinity. The affinity of claudin binding to ZO-1 PDZ1 depends on the absence or presence of a tyrosine residue at position -6 from the claudin C-terminus. Crystal structures of ZO-1 PDZ1 with empty ligand-binding groove, with a bound claudin-1 heptapeptide, which does not have a Tyr -6, or with a bound claudin-2 heptapeptide containing a Tyr -6 revealed significantly different binding geometries explaining the influence of the signature tyrosine residue on binding affinity [100]. In addition to claudin binding, the ZO-1 PDZ1 also mediates interactions with phosphoinositides. Mapping the inositol hexaphosphate binding site onto an NMR structure of ZO-1 PDZ1 revealed spatial overlap with the claudin binding surface and thus provided a structural rationale for the observed competition of both ligands for ZO-1 [101].

The second PDZ domain (PDZ2) of ZO-proteins is known to promote protein dimerization [24,54,55]. A crystal structure shows that ZO-1—PDZ2 dimerization is stabilized by extensive domain swapping of β-strands. This structural rearrangement leaves the canonical peptide-binding groove intact in both subunits of PDZ2 dimer, which are composed of elements from both monomers [89]. Domain swapping of human ZO-1 PDZ2 was subsequently confirmed by solution NMR analysis. In this study, the importance of strand β2 for the domain exchange was demonstrated by insertion mutagenesis [91]. NMR analysis clearly demonstrated that PDZ2 of ZO-2 may also dimerize by domain swapping. A 1.75 Å resolution crystal structure of the ZO-2 PDZ2 confirms formation of a domain-swapped dimer with exchange of β-strands 1 and 2 (Figure 7b) [88], and there is evidence for the formation of PDZ2-promoted domain-swapped homodimers in all three ZO proteins. Based on this observation and the high sequence similarity between the ZO-1, -2, and -3 PDZ2 domains (66% sequence identity between ZO-1 and ZO-2, 50% between ZO-1 and ZO-3, 54% between ZO-2 and ZO-3), heterodimer formation between them was proposed as a potential mechanism of forming and stabilizing the cytoplasmic plaque [80]. Structural evidence for domain-swapped heterodimers of ZO proteins is, however, still lacking.

ZO-1 PDZ2 interacts with connexins, in particular the abundant connexin43, which functions in gap junction formation and regulation. X-ray and NMR analyses showed that domain swapping of ZO-1 PDZ2 preserves the carboxylate tail-binding pockets of the PDZ domains and creates a distinct interface for connexin43 binding [90].

The third PDZ domain (PDZ3) of ZO proteins is important for the interaction with the C-terminus of transmembrane JAMs. A crystal structure of ZO-1 PDZ3 was determined at 1.45 Å resolution. This study established that ZO-1 PDZ3 preferentially binds ligands of sequence type X[D/E]XΦ_{COOH} where X may be any amino acid and Φ is a hydrophobic residue [72].

Following the three N-terminal PDZ domains, ZO proteins contain a SH3-GUK module. Crystal structure analysis of the ZO-1 SH3-GUK tandem domain confirmed independent folding of the SH3 and GUK domains, and pulldown assays identified the downstream U6 loop as an intramolecular ligand of the SH3-GUK core with a potential role in regulating TJ assembly in vivo [102]. Crystal structures of the complete MAGUK core module of ZO-1 comprising the PDZ3-SH3-GUK region and its complex with the cytoplasmic tail of adhesion molecule JAM-A revealed that residues from the adjacent SH3 domain are involved in ligand binding to the ZO-1 PDZ3 [103].

ZO-1 is distinct from its paralogs ZO-2 and ZO-3 by the presence of an extended C-terminal region harboring a ZU5 domain [104,105] and described to mediate physical interaction with the CDC42 effector kinase MRCKβ. An NMR structure showed the ZO-1 ZU5 domain to adopt a β-barrel structure, which is incomplete in comparison with homologous proteins by lacking two β-strands. Attempts to analyze the structure of a ZO-1 ZU5/MRCKβ complex remained unsuccessful, but evidence could be provided that GRINL1A (glutamate receptor, ionotropic, N-methyl-D-aspartate-like 1A combined protein) binds ZO-1 ZU5 in a very similar way as MRCKβ. NMR analysis then showed that a 22-aa GRINL1A peptide hairpin associates with the ZO-1 ZU5 domain to form a complete canonical ZU5 domain [106].

In the MAGI proteins (MAGUKs with inverted domain structure), the characteristic arrangement of PPI domains present in common MAGUK proteins is inverted. Furthermore, the MAGI proteins contain two WW domains in place of the SH3 domain found in MAGUKs (Figure 6) [107]. The family member MAGI-1 is tethered to TJs through interactions of its PDZ domains with the C-terminus of the non-classical junctional adhesion molecule JAM-4 [108]. In addition to its function in the TJ, the first PDZ domain of MAGI-1 binds peptide ligands derived from the oncoprotein E6 of human papillomavirus and the ribosomal S6 kinase 1 (RSK1) [109,110]. NMR analysis suggests the involvement of peptide regions flanking the PDZ domain in ligand binding [109]. PALS1 (protein associated with Lin seven 1, also known as MPP5—membrane-associated palmitoylated protein 5) is another member of the MAGUK family (Figure 6) and described below as part of the Crumbs/ PALS1/PATJ complex.

2.2.3. The Crumbs/PALS1/PATJ Complex

The Crumbs/PALS1/PATJ complex is involved in establishing and maintaining cell polarity and located in the cytoplasmic plaque of TJs [2,3,111]. Crumbs is a single-span TM protein, whereas the other proteins present in this complex, PALS1 and PATJ (PALS-associated tight-junction protein), are cytoplasmic scaffolding proteins (Figure 1). PALS1 functions as an adaptor protein mediating indirect interactions between Crumbs and PATJ (Figure 6). Both PALS1 and PATJ share an N-terminal L27 domain. L27 domains organize scaffold proteins into supramolecular complexes by heteromeric L27 interactions. PATJ is recruited to TJs through interactions with the C-termini of claudin-1 and ZO-3 [112].

The PATJ–PALS1 interaction is mediated by the single L27 domain of PATJ and the N-terminal L27 domain (L27N) of PALS1 [113]. A crystal structure of the PALS1-L27N/PATJ-L27 heterodimer shows that each L27 domain is composed of three α-helices and that heterodimer formation is due to formation of a four-helix bundle by the first two α-helices of the L27 domains and coiled–coil interactions between the helices α3 [114]. NMR structure analysis revealed closely similar topologies for heterotetrameric mLin-2/mLin-7 and PATJ/PALS1 complexes, suggesting a general assembly mode for L27 domains [115]. A crystal structure of a heterotrimeric complex formed by the N-terminal L27 domain of PATJ, the N-terminal tandem L27 domains of PALS1, and the N-terminal L27 domain of MALS2 (mammalian homolog-2 of Lin-7) revealed an assembly of two cognate pairs of heterodimeric L27 domains. This structure is thought to reveal a novel mechanism for tandem L27 domain-mediated supramolecular complex assembly [116].

The intracellular functions of Crumbs3 (CRB3) are mediated by its conserved 37-aa cytoplasmic tail (Crb-CT) and its interaction with PALS1 and the actin-binding protein moesin. The crystal structure of a PALS1 PDZ-SH3-GUK/Crb-CT complex shows that all three domains of PALS1 contribute to Crb-CT binding [117]. A further crystal structure of human PALS1 PDZ bound to 17-aa C-terminal CRB1 peptide shows that only the very C-terminal tetrapeptide ERLI is involved in direct binding to PALS1 PDZ. Comparison with apo-PALS1 PDZ (Figure 7a) revealed that a key phenylalanine residue in the PALS1 PDZ controls access to the ligand-binding groove [87]. To reveal the nature of the Crumbs/moesin interaction, the FERM (protein 4.1/ezrin/radixin/moesin) domain of murine moesin was co-crystallized with the soluble C-terminus of *Drosophila* Crumbs. The 1.5-Å resolution crystal structure revealed that both the FERM-binding motif, as well as the PDZ-binding motif present in the Crumbs C-terminal peptide contribute to the interaction with moesin. Phosphorylation of the Crb-CT by atypical protein kinase C (aPKC) disrupts the Crumbs/moesin association but not the Crumbs/PALS1 interaction. Crumbs may therefore act as aPKC-mediated sensor in epithelial tissues [118].

2.2.4. The PAR-3/PAR-6/aPKC Complex

Similar as the Crumbs/PALS1/PATJ complex, the evolutionarily conserved PAR-3/PAR-6/aPKC complex is associated to TJs and crucial for establishing and maintaining cell polarity. The complex formed by the PAR (partition defective)-3 and PAR-6 proteins, as well as the atypical protein kinase C (aPKC) interacts with subunits from the Crumbs/PALS1/PATJ complex and is regulated by binding to the small GTPases CDC42 and RAC1. Composition and stoichiometry of the PAR-3/PAR-6/aPKC complex are linked to cell polarity and to the cell cycle [119].

Human PAR-6 contains a single PDZ domain, which mediates binding to the C-terminus of TM receptor CRB3. Binding of C-terminal ligands to the PAR-6 PDZ depends on binding of the Rho-GTPase CDC42 to a CRIB domain adjacent to the PAR-6 PDZ. In addition, the PAR-6 PDZ also binds internal peptides, e.g., from PALS1 and its *Drosophila* homolog Stardust. The regulation of ligand binding to PAR-6 PDZ by CDC42 has been structurally characterized in a number of studies. A 2.5-Å crystal structure of a PAR-6 PDZ-bound internal dodecapeptide derived from PALS1 revealed a characteristic deformation of the carboxylate-binding loop of PAR-6 PDZ relative to the structure with bound C-terminal ligand (Figure 7d) [93]. The structural adjustments associated with regulator and ligand binding to the PAR-6 PDZ were also highlighted in a 2.1-Å crystal structure and an NMR structure of the PAR-6 PDZ domain (Figure 7c), which revealed deviations from the canonical PDZ conformation that account for low-affinity binding of C-terminal ligands. CDC42 binding to the adjacent CRIB domain triggered a structural transition to the canonical PDZ conformation and was associated with a ~13-fold increase in affinity for C-terminal ligands [92]. NMR structures of the isolated PAR-6 PDZ domain and a disulfide-stabilized CRIB-PDZ fragment identified a conformational switch in the PAR-6 PDZ domain that is linked to the increase in ligand affinity induced by CDC42 binding to PAR-6 [120]. Finally, NMR analysis of a C-terminal Crumbs peptide binding to PAR-6 and the crystal structure of the PAR-6 PDZ/peptide complex indicated why the affinity of this interaction is 6-fold higher than in previously studied PAR-6/peptide binding studies [121].

PAR-3 acts as central organizer of the PAR-3/PAR-6/aPKC complex and is thus essential for establishment and maintenance of cell polarity. In *Caenorhabditis elegans*, PAR-3 mediates TJ binding through interaction with junctional adhesion molecule (JAM) [122,123]. In cultured endothelial cells, PAR-3 associates with JAM-2 and JAM-3, but neither with the related Ig-like TM proteins ESAM nor CAR [124]. PAR-3 contains an N-terminal oligomerization domain in addition to three PDZ domains. NMR analysis showed the monomeric PAR-3 N-terminal domain (NTD) to adopt a PB1-like fold and to oligomerize into helical filaments. This interaction was proposed to facilitate the assembly of higher-order PAR-3/PAR-6/aPKC complexes [125]. The ability of the PAR-3 NTD to self-associate and form filamentous structures was further studied by crystallographic analysis of the PAR-3 NTD and analysis of the filament structure by cryo-electron microscopy (cryo-EM). Here, it was revealed that both lateral and longitudinal packing within PAR-3 NTD filaments is primarily mediated by Coulomb interactions [126].

The second PDZ domain of PAR-3 binds phosphatidylinositol (PI) lipid membranes with high affinity as shown in a biochemical and NMR study of PAR-3 PDZ2. This study also showed that the lipid phosphatase PTEN (phosphatase and tensin homolog) binds PAR-3 PDZ3 and thus cooperates with PI in regulating cell polarity through PAR-3 [127]. A three-dimensional structure of the second PDZ domain of human PAR-3 was also determined as part of an NMR structure analysis automation study [128]. A previously unknown C-terminal PDZ-binding motif, identified in PAR-6 through crystal structures and NMR binding analyses, mediates interactions with PDZ1 and PDZ3, but not with PDZ2 of PAR-3. Evidently, PAR-3 has the ability to recruit two PAR-6 molecules simultaneously, possibly facilitating the assembly of polarity protein networks through these interactions [129].

2.2.5. Other Cytoplasmic Tight-Junction Proteins

In addition to the MAGUK proteins, the subunits and regulators of the Crumbs/PALS1/PATJ and PAR-3/PAR-6/aPKC complexes, various other cytoplasmic proteins are associated with TJs. The multiple PDZ domain protein 1 (MUPP1, also known as MPDZ) contains 13 PDZ domains. MUPP1 is a paralog of PATJ, shares several TJ-binding partners (Figure 6) and a similar subcellular localization, but displays a distinct selectivity in its interactions with claudins and is dispensable for TJ formation while PATJ is not [130,131]. A crystal structure of the twelfth PDZ domain of MUPP1 was determined in a structural genomics effort to crystallize PDZ domains with self-binding C-terminal extensions [132]. Several additional MUPP1-PDZ domains were analyzed within the research program of the Center for Eukaryotic Structural Genomics [133] and submitted to the Protein Data Bank (PDB) [79], but without functional annotation. Crystal structure analysis of the mouse MUPP1-PDZ4 domain revealed a canonical PDZ fold with six β-strands and three α-helices [60]. The angiomotin (AMOT) proteins [134] were reported to interact with MUPP1, PATJ [135], ZO-1 and MAGI-1b, and ascribed a role in the assembly of endothelial cell junctions [25]. The cytoskeletal linker cingulin, a predicted dimeric coiled-coil protein of unknown three-dimensional structure, was initially characterized as a peripheral TJ component [136]. Although its amino-terminal region was reported to interact with ZO-1 in cells [137], cingulin was later shown to be dispensable for TJ integrity and epithelial barrier function [138]. The ZO-1 associated nucleic acid-binding protein ZONAB (also referred to as YBX3 or CSDA1) is a transcription factor that shuttles between TJs and the nucleus and regulates epithelial cell proliferation [139,140]. Although a crystal or NMR structure of ZONAB is not available, the conformation and nucleic acid binding of its N-terminal cold shock domain may be inferred from structures of bacterial major cold shock proteins [141] or the homologous Y-box factor YB-1 [142].

The molecular composition of TJs varies significantly between different epithelia and determines their dual functions as effective barriers for solutes or channels for particular classes of solutes [143]. Therefore, the expression patterns of TJ proteins in various tissues are kept under tight control by various transcription factors in addition to ZONAB. A large number of TJ and AJ transmembrane (TM) proteins are under transcriptional control by the Grainyhead-like proteins GRHL1 and GRHL2 or by nuclear receptors [144]. These transcription factors therefore regulate a large subset of proteins, making up the apical junction complex. Frequently, one transcription factor controls the expression of multiple genes encoding TJ or AJ proteins. For example, GRHL2 acts as transcriptional activator of both AJ and TJ components including several claudins and thus functions as regulator of epithelial differentiation [145]. Equally frequently, one TJ protein is controlled by multiple transcription factors. The *Cldn4* gene, being under transcriptional control by GRHL2, GRHL3, the androgen receptor, retinoic acid/retinoid X receptors, and p63 in different tissues [144,146,147], serves as an impressive example here. Although structural information is available for several of these factors [148–151], these proteins will not be further discussed here, where the focus is placed on proteins of the TJ core structure. Equally, the large set of signaling and effector proteins acting on the TJ [2,11,12] will not be discussed further.

3. Conclusions

Here, we have reviewed three-dimensional structures of TJ proteins, focusing on the core TJ complex. It becomes clear that our knowledge of these structures is fairly incomplete, because most TJ proteins do not lend themselves easily to structure analysis due to their size and/or the presence of TM regions, natively unfolded polypeptide segments or heterogeneous post-translational modifications [74–76]. Our knowledge of protein–protein interactions within the TJ also does not go far beyond structural features of selected binary interactions, often involving small protein fragments or peptides. At a resolution that permits construction of atomic models, very little is known about the general architecture of the TJ. Herein lies a great challenge and opportunity for future research making use of new integrative methods in structural biology, including cryo-electron microscopy [152–155], cryo-electron tomography [156,157], cross-linking mass spectrometry [158,159], small-angle X-ray and neutron scattering [160,161], and others [162]. These rapidly emerging and developing methods can be informed by the available high-resolution structures of TJ proteins, protein domains and protein–protein interactions. We may expect to see exciting results along these lines in the near future, revealing the architecture of TJs at high resolution in defined functional states.

Author Contributions: U.H. wrote the first draft of manuscript and did the final editing. A.S. designed the figures and revised the manuscript text.

Funding: This research received no external funding.

Acknowledgments: U.H. and A.S. express their gratitude to Michael Fromm, Dorothee Günzel, and Susanne Krug (Charité, Berlin) for long-standing collaboration and encouragement in our work on TJ proteins.

Conflicts of Interest: The authors declare no conflict of interest.

Abbreviations

aa	Amino acid
AJ	Adherens junction
BBB	Blood-brain barrier
CTD	C-terminal domain
ECL	Extracellular loop
EMT	Epithelial-mesenchymal transition
NTD	N-terminal domain
PDB	Protein Data Bank
PPI	Protein–protein interaction
TJ	Tight junction
tTJ	Tricellular tight junction
TM	Transmembrane

References

1. Farquhar, M.G.; Palade, G.E. Junctional complexes in various epithelia. *J. Cell Biol.* **1963**, *17*, 375–412. [CrossRef] [PubMed]
2. Balda, M.S.; Matter, K. Tight junctions at a glance. *J. Cell Sci.* **2008**, *121*, 3677–3682. [CrossRef]
3. Guillemot, L.; Paschoud, S.; Pulimeno, P.; Foglia, A.; Citi, S. The cytoplasmic plaque of tight junctions: A scaffolding and signalling center. *Biochim. Biophys. Acta* **2008**, *1778*, 601–613. [CrossRef] [PubMed]
4. Furuse, M. Molecular basis of the core structure of tight junctions. *Cold Spring Harb Perspect Biol.* **2010**, *2*, a002907. [CrossRef]
5. Van Itallie, C.M.; Anderson, J.M. Architecture of tight junctions and principles of molecular composition. *Semin. Cell Dev. Biol.* **2014**, *36*, 157–165. [CrossRef] [PubMed]
6. Haseloff, R.F.; Dithmer, S.; Winkler, L.; Wolburg, H.; Blasig, I.E. Transmembrane proteins of the tight junctions at the blood-brain barrier: Structural and functional aspects. *Semin. Cell Dev. Biol.* **2015**, *38*, 16–25. [CrossRef] [PubMed]
7. Irudayanathan, F.J.; Trasatti, J.P.; Karande, P.; Nangia, S. Molecular Architecture of the Blood Brain Barrier Tight Junction Proteins-A Synergistic Computational and In Vitro Approach. *J. Phys. Chem. B* **2016**, *120*, 77–88. [CrossRef] [PubMed]
8. Irudayanathan, F.J.; Wang, N.; Wang, X.Y.; Nangia, S. Architecture of the paracellular channels formed by claudins of the blood-brain barrier tight junctions. *Ann. New York Acad. Sci.* **2017**, *1405*, 131–146. [CrossRef] [PubMed]
9. Weiss, N.; Miller, F.; Cazaubon, S.; Couraud, P.O. The blood-brain barrier in brain homeostasis and neurological diseases. *Biochim. Biophys. Acta* **2009**, *1788*, 842–857. [CrossRef]

10. Luissint, A.C.; Artus, C.; Glacial, F.; Ganeshamoorthy, K.; Couraud, P.O. Tight junctions at the blood brain barrier: Physiological architecture and disease-associated dysregulation. *Fluids Barriers Cns.* **2012**, *9*, 23. [CrossRef]
11. Zihni, C.; Balda, M.S.; Matter, K. Signalling at tight junctions during epithelial differentiation and microbial pathogenesis. *J. Cell Sci.* **2014**, *127*, 3401–3413. [CrossRef] [PubMed]
12. Zihni, C.; Mills, C.; Matter, K.; Balda, M.S. Tight junctions: From simple barriers to multifunctional molecular gates. *Nat. Rev. Mol. Cell Biol.* **2016**, *17*, 564–580. [CrossRef] [PubMed]
13. Russ, P.K.; Pino, C.J.; Williams, C.S.; Bader, D.M.; Haselton, F.R.; Chang, M.S. Bves modulates tight junction associated signaling. *PLoS ONE* **2011**, *6*, e14563. [CrossRef] [PubMed]
14. Han, P.; Lei, Y.; Li, D.; Liu, J.; Yan, W.; Tian, D. Ten years of research on the role of BVES/ POPDC1 in human disease: A review. *Onco. Targets* **2019**, *12*, 1279–1291. [CrossRef]
15. Sanchez-Pulido, L.; Martin-Belmonte, F.; Valencia, A.; Alonso, M.A. MARVEL: A conserved domain involved in membrane apposition events. *Trends Biochem. Sci.* **2002**, *27*, 599–601. [CrossRef]
16. Suzuki, H.; Tani, K.; Fujiyoshi, Y. Crystal structures of claudins: Insights into their intermolecular interactions. *Ann. New York Acad. Sci.* **2017**, *1397*, 25–34. [CrossRef]
17. Garrido-Urbani, S.; Bradfield, P.F.; Imhof, B.A. Tight junction dynamics: The role of junctional adhesion molecules (JAMs). *Cell Tissue Res.* **2014**, *355*, 701–715. [CrossRef]
18. Kostrewa, D.; Brockhaus, M.; D'Arcy, A.; Dale, G.E.; Nelboeck, P.; Schmid, G.; Mueller, F.; Bazzoni, G.; Dejana, E.; Bartfai, T.; et al. X-ray structure of junctional adhesion molecule: Structural basis for homophilic adhesion via a novel dimerization motif. *Embo. J.* **2001**, *20*, 4391–4398. [CrossRef]
19. Matthaus, C.; Langhorst, H.; Schutz, L.; Juttner, R.; Rathjen, F.G. Cell-cell communication mediated by the CAR subgroup of immunoglobulin cell adhesion molecules in health and disease. *Mol. Cell Neurosci.* **2017**, *81*, 32–40. [CrossRef]
20. Patzke, C.; Max, K.E.; Behlke, J.; Schreiber, J.; Schmidt, H.; Dorner, A.A.; Kroger, S.; Henning, M.; Otto, A.; Heinemann, U.; et al. The coxsackievirus-adenovirus receptor reveals complex homophilic and heterophilic interactions on neural cells. *J. Neurosci.* **2010**, *30*, 2897–2910. [CrossRef]
21. Kirchner, E.; Guglielmi, K.M.; Strauss, H.M.; Dermody, T.S.; Stehle, T. Structure of reovirus sigma1 in complex with its receptor junctional adhesion molecule-A. *Plos Pathog.* **2008**, *4*, e1000235. [CrossRef] [PubMed]
22. Conley, M.J.; McElwee, M.; Azmi, L.; Gabrielsen, M.; Byron, O.; Goodfellow, I.G.; Bhella, D. Calicivirus VP2 forms a portal-like assembly following receptor engagement. *Nature* **2019**, *565*, 377–381. [CrossRef] [PubMed]
23. Citi, S.; Guerrera, D.; Spadaro, D.; Shah, J. Epithelial junctions and Rho family GTPases: The zonular signalosome. *Small Gtpases.* **2014**, *5*, 1–15. [CrossRef] [PubMed]
24. Wittchen, E.S.; Haskins, J.; Stevenson, B.R. Protein interactions at the tight junction. Actin has multiple binding partners, and ZO-1 forms independent complexes with ZO-2 and ZO-3. *J. Biol. Chem.* **1999**, *274*, 35179–35185. [CrossRef]
25. Bratt, A.; Birot, O.; Sinha, I.; Veitonmaki, N.; Aase, K.; Ernkvist, M.; Holmgren, L. Angiomotin regulates endothelial cell-cell junctions and cell motility. *J. Biol. Chem.* **2005**, *280*, 34859–34869. [CrossRef] [PubMed]
26. Itoh, M.; Tsukita, S.; Yamazaki, Y.; Sugimoto, H. Rho GTP exchange factor ARHGEF11 regulates the integrity of epithelial junctions by connecting ZO-1 and RhoA-myosin II signaling. *Proc. Natl. Acad. Sci. USA* **2012**, *109*, 9905–9910. [CrossRef] [PubMed]
27. Chen, X.; Macara, I.G. Par-3 controls tight junction assembly through the Rac exchange factor Tiam1. *Nat. Cell Biol.* **2005**, *7*, 262–269. [CrossRef]
28. Balda, M.S.; Matter, K. The tight junction protein ZO-1 and an interacting transcription factor regulate ErbB-2 expression. *Embo. J.* **2000**, *19*, 2024–2033. [CrossRef]
29. Higashi, T.; Tokuda, S.; Kitajiri, S.; Masuda, S.; Nakamura, H.; Oda, Y.; Furuse, M. Analysis of the 'angulin' proteins LSR, ILDR1 and ILDR2–tricellulin recruitment, epithelial barrier function and implication in deafness pathogenesis. *J. Cell Sci.* **2013**, *126*, 966–977. [CrossRef]
30. Shimada, H.; Satohisa, S.; Kohno, T.; Konno, T.; Takano, K.I.; Takahashi, S.; Hatakeyama, T.; Arimoto, C.; Saito, T.; Kojima, T. Downregulation of lipolysis-stimulated lipoprotein receptor promotes cell invasion via claudin-1-mediated matrix metalloproteinases in human endometrial cancer. *Oncol. Lett.* **2017**, *14*, 6776–6782. [CrossRef]

31. Shimada, H.; Abe, S.; Kohno, T.; Satohisa, S.; Konno, T.; Takahashi, S.; Hatakeyama, T.; Arimoto, C.; Kakuki, T.; Kaneko, Y.; et al. Loss of tricellular tight junction protein LSR promotes cell invasion and migration via upregulation of TEAD1/AREG in human endometrial cancer. *Sci. Rep.* **2017**, *7*, 37049. [CrossRef] [PubMed]
32. Tsukita, S.; Tanaka, H.; Tamura, A. The Claudins: From Tight Junctions to Biological Systems. *Trends Biochem. Sci.* **2019**, *44*, 141–152. [CrossRef] [PubMed]
33. Elkouby-Naor, L.; Ben-Yosef, T. Functions of claudin tight junction proteins and their complex interactions in various physiological systems. *Int. Rev. Cell Mol. Biol.* **2010**, *279*, 1–32. [PubMed]
34. Gunzel, D.; Fromm, M. Claudins and other tight junction proteins. *Compr. Physiol.* **2012**, *2*, 1819–1852. [PubMed]
35. Gunzel, D.; Yu, A.S. Claudins and the modulation of tight junction permeability. *Physiol. Rev.* **2013**, *93*, 525–569. [CrossRef] [PubMed]
36. Krause, G.; Protze, J.; Piontek, J. Assembly and function of claudins: Structure-function relationships based on homology models and crystal structures. *Semin. Cell Dev. Biol.* **2015**, *42*, 3–12. [CrossRef] [PubMed]
37. Milatz, S.; Piontek, J.; Schulzke, J.D.; Blasig, I.E.; Fromm, M.; Gunzel, D. Probing the cis-arrangement of prototype tight junction proteins claudin-1 and claudin-3. *Biochem. J.* **2015**, *468*, 449–458. [CrossRef]
38. Milatz, S.; Piontek, J.; Hempel, C.; Meoli, L.; Grohe, C.; Fromm, A.; Lee, I.F.M.; El-Athman, R.; Gunzel, D. Tight junction strand formation by claudin-10 isoforms and claudin-10a/-10b chimeras. *Ann. New York Acad. Sci.* **2017**, *1405*, 102–115. [CrossRef]
39. Krause, G.; Winkler, L.; Mueller, S.L.; Haseloff, R.F.; Piontek, J.; Blasig, I.E. Structure and function of claudins. *Biochim. Biophys. Acta* **2008**, *1778*, 631–645. [CrossRef]
40. Rosenthal, R.; Gunzel, D.; Theune, D.; Czichos, C.; Schulzke, J.D.; Fromm, M. Water channels and barriers formed by claudins. *Ann. New York Acad. Sci.* **2017**, *1397*, 100–109. [CrossRef]
41. The UniProt Consortium, UniProt: A worldwide hub of protein knowledge. *Nucleic Acids Res.* **2018**, *47*, D506–D515.
42. Shinoda, T.; Shinya, N.; Ito, K.; Ohsawa, N.; Terada, T.; Hirata, K.; Kawano, Y.; Yamamoto, M.; Kimura-Someya, T.; Yokoyama, S.; et al. Structural basis for disruption of claudin assembly in tight junctions by an enterotoxin. *Sci. Rep.* **2016**, 6. [CrossRef] [PubMed]
43. Sievers, F.; Wilm, A.; Dineen, D.; Gibson, T.J.; Karplus, K.; Li, W.; Lopez, R.; McWilliam, H.; Remmert, M.; Soding, J.; et al. Fast, scalable generation of high-quality protein multiple sequence alignments using Clustal Omega. *Mol. Syst. Biol.* **2011**, *7*, 539. [CrossRef] [PubMed]
44. Beitz, E. TEXshade: Shading and labeling of multiple sequence alignments using LATEX2 epsilon. *Bioinformatics* **2000**, *16*, 135–139. [CrossRef]
45. Suzuki, H.; Nishizawa, T.; Tani, K.; Yamazaki, Y.; Tamura, A.; Ishitani, R.; Dohmae, N.; Tsukita, S.; Nureki, O.; Fujiyoshi, Y. Crystal Structure of a Claudin Provides Insight into the Architecture of Tight Junctions. *Science* **2014**, *344*, 304–307. [CrossRef]
46. Samanta, P.; Wang, Y.T.; Fuladi, S.; Zou, J.J.; Li, Y.; Shen, L.; Weber, C.; Khalili-Araghi, F. Molecular determination of claudin-15 organization and channel selectivity. *J. Gen. Physiol.* **2018**, *150*, 949–968. [CrossRef]
47. Artursson, P.; Knight, S.D. Structural biology. Breaking the intestinal barrier to deliver drugs. *Science* **2015**, *347*, 716–717. [CrossRef]
48. Nakamura, S.; Irie, K.; Tanaka, H.; Nishikawa, K.; Suzuki, H.; Saitoh, Y.; Tamura, A.; Tsukita, S.; Fujiyoshi, Y. Morphologic determinant of tight junctions revealed by claudin-3 structures. *Nat. Commun.* **2019**, *10*. [CrossRef]
49. Weber, C.R.; Turner, J.R. Dynamic modeling of the tight junction pore pathway. *Ann. New York Acad. Sci.* **2017**, *1397*, 209–218. [CrossRef]
50. Alberini, G.; Benfenati, F.; Maragliano, L. Molecular Dynamics Simulations of Ion Selectivity in a Claudin-15 Paracellular Channel. *J. Phys. Chem. B* **2018**, *122*, 10783–10792. [CrossRef]
51. Irudayanathan, F.J.; Wang, X.Y.; Wang, N.; Willsey, S.R.; Seddon, I.A.; Nangia, S. Self-Assembly Simulations of Classic Claudins-Insights into the Pore Structure, Selectivity, and Higher Order Complexes. *J. Phys. Chem. B* **2018**, *122*, 7463–7474. [CrossRef] [PubMed]
52. Saitoh, Y.; Suzuki, H.; Tani, K.; Nishikawa, K.; Irie, K.; Ogura, Y.; Tamura, A.; Tsukita, S.; Fujiyoshi, Y. Structural insight into tight junction disassembly by Clostridium perfringens enterotoxin. *Science* **2015**, *347*, 775–778. [CrossRef] [PubMed]

53. Vecchio, A.J.; Stroud, R.M. Claudin-9 structures reveal mechanism for toxin-induced gut barrier breakdown. *Proc. Natl. Acad. Sci. USA* **2019**, 201908929. [CrossRef] [PubMed]
54. Fanning, A.S.; Jameson, B.J.; Jesaitis, L.A.; Anderson, J.M. The tight junction protein ZO-1 establishes a link between the transmembrane protein occludin and the actin cytoskeleton. *J. Biol. Chem.* **1998**, *273*, 29745–29753. [CrossRef]
55. Itoh, M.; Morita, K.; Tsukita, S. Characterization of ZO-2 as a MAGUK family member associated with tight as well as adherens junctions with a binding affinity to occludin and alpha catenin. *J. Biol. Chem.* **1999**, *274*, 5981–5986. [CrossRef]
56. Walter, J.K.; Castro, V.; Voss, M.; Gast, K.; Rueckert, C.; Piontek, J.; Blasig, I.E. Redox-sensitivity of the dimerization of occludin. *Cell Mol. Life Sci.* **2009**, *66*, 3655–3662. [CrossRef]
57. Walter, J.K.; Rueckert, C.; Voss, M.; Mueller, S.L.; Piontek, J.; Gast, K.; Blasig, I.E. The oligomerization of the coiled coil-domain of occludin is redox sensitive. *Ann. N. Y. Acad. Sci.* **2009**, *1165*, 19–27. [CrossRef]
58. Li, Y.; Fanning, A.S.; Anderson, J.M.; Lavie, A. Structure of the conserved cytoplasmic C-terminal domain of occludin: Identification of the ZO-1 binding surface. *J. Mol. Biol.* **2005**, *352*, 151–164. [CrossRef]
59. Tash, B.R.; Bewley, M.C.; Russo, M.; Keil, J.M.; Griffin, K.A.; Sundstrom, J.M.; Antonetti, D.A.; Tian, F.; Flanagan, J.M. The occludin and ZO-1 complex, defined by small angle X-ray scattering and NMR, has implications for modulating tight junction permeability. *P. Natl. Acad. Sci. USA* **2012**, *109*, 10855–10860. [CrossRef]
60. Zhu, H.; Liu, Z.; Huang, Y.; Zhang, C.; Li, G.; Liu, W. Biochemical and structural characterization of MUPP1-PDZ4 domain from Mus musculus. *Acta. Biochim. Biophys. Sin. (Shanghai)* **2015**, *47*, 199–206. [CrossRef]
61. Staehelin, L.A. Further observations on the fine structure of freeze-cleaved tight junctions. *J. Cell Sci.* **1973**, *13*, 763–786. [PubMed]
62. Tervonen, A.; Ihalainen, T.O.; Nymark, S.; Hyttinen, J. Structural dynamics of tight junctions modulate the properties of the epithelial barrier. *PLoS ONE* **2019**, *14*, e0214876. [CrossRef] [PubMed]
63. Ikenouchi, J.; Furuse, M.; Furuse, K.; Sasaki, H.; Tsukita, S.; Tsukita, S. Tricellulin constitutes a novel barrier at tricellular contacts of epithelial cells. *J. Cell Biol.* **2005**, *171*, 939–945. [CrossRef] [PubMed]
64. Jennek, S.; Mittag, S.; Reiche, J.; Westphal, J.K.; Seelk, S.; Dorfel, M.J.; Pfirrmann, T.; Friedrich, K.; Schutz, A.; Heinemann, U.; et al. Tricellulin is a target of the ubiquitin ligase Itch. *Ann. New York Acad. Sci.* **2017**, *1397*, 157–168. [CrossRef]
65. Janke, S.; Mittag, S.; Reiche, J.; Huber, O. Apoptotic Fragmentation of Tricellulin. *Int. J. Mol. Sci.* **2019**, *20*, 4882. [CrossRef]
66. Mariano, C.; Sasaki, H.; Brites, D.; Brito, M.A. A look at tricellulin and its role in tight junction formation and maintenance. *Eur. J. Cell Biol.* **2011**, *90*, 787–796. [CrossRef]
67. Krug, S.M.; Amasheh, S.; Richter, J.F.; Milatz, S.; Gunzel, D.; Westphal, J.K.; Huber, O.; Schulzke, J.D.; Fromm, M. Tricellulin forms a barrier to macromolecules in tricellular tight junctions without affecting ion permeability. *Mol. Biol. Cell.* **2009**, *20*, 3713–3724. [CrossRef]
68. Krug, S.M. Contribution of the tricellular tight junction to paracellular permeability in leaky and tight epithelia. *Ann. N. Y. Acad. Sci.* **2017**, *1397*, 219–230. [CrossRef]
69. Cording, J.; Arslan, B.; Staat, C.; Dithmer, S.; Krug, S.M.; Kruger, A.; Berndt, P.; Gunther, R.; Winkler, L.; Blasig, I.E.; et al. Trictide, a tricellulin-derived peptide to overcome cellular barriers. *Ann. N. Y. Acad. Sci.* **2017**, *1405*, 89–101. [CrossRef]
70. Westphal, J.K.; Dorfel, M.J.; Krug, S.M.; Cording, J.D.; Piontek, J.; Blasig, I.E.; Tauber, R.; Fromm, M.; Huber, O. Tricellulin forms homomeric and heteromeric tight junctional complexes. *Cell Mol. Life Sci.* **2010**, *67*, 2057–2068. [CrossRef]
71. Schuetz, A.; Radusheva, V.; Krug, S.M.; Heinemann, U. Crystal structure of the tricellulin C-terminal coiled-coil domain reveals a unique mode of dimerization. *Ann. N. Y. Acad. Sci.* **2017**, *1405*, 147–159. [CrossRef] [PubMed]
72. Ernst, A.; Appleton, B.A.; Ivarsson, Y.; Zhang, Y.; Gfeller, D.; Wiesmann, C.; Sidhu, S.S. A structural portrait of the PDZ domain family. *J. Mol. Biol.* **2014**, *426*, 3509–3519. [CrossRef] [PubMed]
73. de Bono, S.; Riechmann, L.; Girard, E.; Williams, R.L.; Winter, G. A segment of cold shock protein directs the folding of a combinatorial protein. *Proc. Natl. Acad. Sci. USA* **2005**, *102*, 1396–1401. [CrossRef] [PubMed]

74. Uversky, V.N.; Gillespie, J.R.; Fink, A.L. Why are "natively unfolded" proteins unstructured under physiologic conditions? *Proteins* **2000**, *41*, 415–427. [CrossRef]
75. Uversky, V.N. What does it mean to be natively unfolded? *Eur. J. Biochem.* **2002**, *269*, 2–12. [CrossRef]
76. Fink, A.L. Natively unfolded proteins. *Curr. Opin. Struct. Biol.* **2005**, *15*, 35–41. [CrossRef]
77. Ye, F.; Zhang, M. Structures and target recognition modes of PDZ domains: Recurring themes and emerging pictures. *Biochem. J.* **2013**, *455*, 1–14. [CrossRef]
78. Lee, H.J.; Zheng, J.J. PDZ domains and their binding partners: Structure, specificity, and modification. *Cell Commun. Signal.* **2010**, *8*, 8. [CrossRef]
79. Burley, S.K.; Berman, H.M.; Kleywegt, G.J.; Markley, J.L.; Nakamura, H.; Velankar, S. Protein Data Bank (PDB): The Single Global Macromolecular Structure Archive. *Methods Mol. Biol.* **2017**, *1607*, 627–641.
80. Wu, J.; Yang, Y.; Zhang, J.; Ji, P.; Du, W.; Jiang, P.; Xie, D.; Huang, H.; Wu, M.; Zhang, G.; et al. Domain-swapped dimerization of the second PDZ domain of ZO2 may provide a structural basis for the polymerization of claudins. *J. Biol. Chem.* **2007**, *282*, 35988–35999. [CrossRef]
81. Max, K.E.; Zeeb, M.; Bienert, R.; Balbach, J.; Heinemann, U. Common mode of DNA binding to cold shock domains. Crystal structure of hexathymidine bound to the domain-swapped form of a major cold shock protein from Bacillus caldolyticus. *Febs. J.* **2007**, *274*, 1265–1279. [CrossRef] [PubMed]
82. Mojib, N.; Andersen, D.T.; Bej, A.K. Structure and function of a cold shock domain fold protein, CspD, in Janthinobacterium sp. Ant5–2 from East Antarctica. *Fems. Microbiol. Lett.* **2011**, *319*, 106–114. [CrossRef] [PubMed]
83. Christensen, N.R.; Čalyševa, J.; Fernandes, E.F.A.; Lüchow, S.; Clemmensen, L.S.; Haugaard-Kedström, L.M.; Strømgaard, K. PDZ Domains as Drug Targets. *Adv. Ther.* **2019**, *2*, 1800143. [CrossRef]
84. Dev, K.K. Making protein interactions druggable: Targeting PDZ domains. *Nat. Rev. Drug Discov.* **2004**, *3*, 1047–1056. [CrossRef] [PubMed]
85. Saupe, J.; Roske, Y.; Schillinger, C.; Kamdem, N.; Radetzki, S.; Diehl, A.; Oschkinat, H.; Krause, G.; Heinemann, U.; Rademann, J. Discovery, structure-activity relationship studies, and crystal structure of nonpeptide inhibitors bound to the Shank3 PDZ domain. *Chem. Med. Chem.* **2011**, *6*, 1411–1422. [CrossRef] [PubMed]
86. Lin, E.Y.S.; Silvian, L.F.; Marcotte, D.J.; Banos, C.C.; Jow, F.; Chan, T.R.; Arduini, R.M.; Qian, F.; Baker, D.P.; Bergeron, C.; et al. Potent PDZ-Domain PICK1 Inhibitors that Modulate Amyloid Beta-Mediated Synaptic Dysfunction. *Sci. Rep.* **2018**, *8*, 13438. [CrossRef] [PubMed]
87. Ivanova, M.E.; Fletcher, G.C.; O'Reilly, N.; Purkiss, A.G.; Thompson, B.J.; McDonald, N.Q. Structures of the human Pals1 PDZ domain with and without ligand suggest gated access of Crb to the PDZ peptide-binding groove. *Acta. Cryst. D Biol. Cryst.* **2015**, *71*, 555–564. [CrossRef]
88. Chen, H.; Tong, S.; Li, X.; Wu, J.; Zhu, Z.; Niu, L.; Teng, M. Structure of the second PDZ domain from human zonula occludens 2. *Acta. Cryst. Sect. F Struct. Biol. Cryst. Commun.* **2009**, *65*, 327–330. [CrossRef]
89. Fanning, A.S.; Lye, M.F.; Anderson, J.M.; Lavie, A. Domain swapping within PDZ2 is responsible for dimerization of ZO proteins. *J. Biol. Chem.* **2007**, *282*, 37710–37716. [CrossRef]
90. Chen, J.; Pan, L.; Wei, Z.; Zhao, Y.; Zhang, M. Domain-swapped dimerization of ZO-1 PDZ2 generates specific and regulatory connexin43-binding sites. *Embo. J.* **2008**, *27*, 2113–2123. [CrossRef]
91. Ji, P.; Yang, G.; Zhang, J.; Wu, J.; Chen, Z.; Gong, Q.; Wu, J.; Shi, Y. Solution structure of the second PDZ domain of Zonula Occludens 1. *Proteins* **2011**, *79*, 1342–1346. [CrossRef] [PubMed]
92. Peterson, F.C.; Penkert, R.R.; Volkman, B.F.; Prehoda, K.E. Cdc42 regulates the Par-6 PDZ domain through an allosteric CRIB-PDZ transition. *Mol. Cell.* **2004**, *13*, 665–676. [CrossRef]
93. Penkert, R.R.; DiVittorio, H.M.; Prehoda, K.E. Internal recognition through PDZ domain plasticity in the Par-6-Pals1 complex. *Nat. Struct. Mol. Biol.* **2004**, *11*, 1122–1127. [CrossRef] [PubMed]
94. Woods, D.F.; Bryant, P.J. ZO-1, DlgA and PSD-95/SAP90: Homologous proteins in tight, septate and synaptic cell junctions. *Mech. Dev.* **1993**, *44*, 85–89. [CrossRef]
95. Anderson, J.M. Cell signalling: MAGUK magic. *Curr. Biol.* **1996**, *6*, 382–384. [CrossRef]
96. Gonzalez-Mariscal, L.; Betanzos, A.; Avila-Flores, A. MAGUK proteins: Structure and role in the tight junction. *Semin. Cell Dev. Biol.* **2000**, *11*, 315–324. [CrossRef]
97. Bauer, H.; Zweimueller-Mayer, J.; Steinbacher, P.; Lametschwandtner, A.; Bauer, H.C. The dual role of zonula occludens (ZO) proteins. *J. Biomed. Biotechnol.* **2010**, *2010*, 402593. [CrossRef]

98. Fanning, A.S.; Ma, T.Y.; Anderson, J.M. Isolation and functional characterization of the actin binding region in the tight junction protein ZO-1. *Faseb. J.* **2002**, *16*, 1835–1837. [CrossRef]
99. Appleton, B.A.; Zhang, Y.; Wu, P.; Yin, J.P.; Hunziker, W.; Skelton, N.J.; Sidhu, S.S.; Wiesmann, C. Comparative structural analysis of the Erbin PDZ domain and the first PDZ domain of ZO-1. Insights into determinants of PDZ domain specificity. *J. Biol. Chem.* **2006**, *281*, 22312–22320. [CrossRef]
100. Nomme, J.; Antanasijevic, A.; Caffrey, M.; Van Itallie, C.M.; Anderson, J.M.; Fanning, A.S.; Lavie, A. Structural Basis of a Key Factor Regulating the Affinity between the Zonula Occludens First PDZ Domain and Claudins. *J. Biol. Chem.* **2015**, *290*, 16595–16606. [CrossRef]
101. Hiroaki, H.; Satomura, K.; Goda, N.; Nakakura, Y.; Hiranuma, M.; Tenno, T.; Hamada, D.; Ikegami, T. Spatial Overlap of Claudin- and Phosphatidylinositol Phosphate-Binding Sites on the First PDZ Domain of Zonula Occludens 1 Studied by NMR. *Molecules* **2018**, *23*, 2465. [CrossRef] [PubMed]
102. Lye, M.F.; Fanning, A.S.; Su, Y.; Anderson, J.M.; Lavie, A. Insights into regulated ligand binding sites from the structure of ZO-1 Src homology 3-guanylate kinase module. *J. Biol. Chem.* **2010**, *285*, 13907–13917. [CrossRef] [PubMed]
103. Nomme, J.; Fanning, A.S.; Caffrey, M.; Lye, M.F.; Anderson, J.M.; Lavie, A. The Src homology 3 domain is required for junctional adhesion molecule binding to the third PDZ domain of the scaffolding protein ZO-1. *J. Biol. Chem.* **2011**, *286*, 43352–43360. [CrossRef] [PubMed]
104. Ackerman, S.L.; Kozak, L.P.; Przyborski, S.A.; Rund, L.A.; Boyer, B.B.; Knowles, B.B. The mouse rostral cerebellar malformation gene encodes an UNC-5-like protein. *Nature* **1997**, *386*, 838–842. [CrossRef]
105. Leonardo, E.D.; Hinck, L.; Masu, M.; Keino-Masu, K.; Ackerman, S.L.; Tessier-Lavigne, M. Vertebrate homologues of C. elegans UNC-5 are candidate netrin receptors. *Nature* **1997**, *386*, 833–838. [CrossRef]
106. Huo, L.; Wen, W.; Wang, R.; Kam, C.; Xia, J.; Feng, W.; Zhang, M. Cdc42-dependent formation of the ZO-1/MRCKbeta complex at the leading edge controls cell migration. *Embo. J.* **2011**, *30*, 665–678. [CrossRef]
107. Dobrosotskaya, I.; Guy, R.K.; James, G.L. MAGI-1, a membrane-associated guanylate kinase with a unique arrangement of protein-protein interaction domains. *J. Biol. Chem.* **1997**, *272*, 31589–31597. [CrossRef]
108. Hirabayashi, S.; Tajima, M.; Yao, I.; Nishimura, W.; Mori, H.; Hata, Y. JAM4, a junctional cell adhesion molecule interacting with a tight junction protein, MAGI-1. *Mol. Cell Biol.* **2003**, *23*, 4267–4282. [CrossRef]
109. Charbonnier, S.; Nomine, Y.; Ramirez, J.; Luck, K.; Chapelle, A.; Stote, R.H.; Trave, G.; Kieffer, B.; Atkinson, R.A. The structural and dynamic response of MAGI-1 PDZ1 with noncanonical domain boundaries to the binding of human papillomavirus E6. *J. Mol. Biol.* **2011**, *406*, 745–763. [CrossRef]
110. Gogl, G.; Biri-Kovacs, B.; Poti, A.L.; Vadaszi, H.; Szeder, B.; Bodor, A.; Schlosser, G.; Acs, A.; Turiak, L.; Buday, L.; et al. Dynamic control of RSK complexes by phosphoswitch-based regulation. *Febs. J.* **2018**, *285*, 46–71. [CrossRef]
111. Assemat, E.; Bazellieres, E.; Pallesi-Pocachard, E.; Le Bivic, A.; Massey-Harroche, D. Polarity complex proteins. *Biochim. Biophys. Acta* **2008**, *1778*, 614–630. [CrossRef] [PubMed]
112. Roh, M.H.; Liu, C.J.; Laurinec, S.; Margolis, B. The carboxyl terminus of zona occludens-3 binds and recruits a mammalian homologue of discs lost to tight junctions. *J. Biol. Chem.* **2002**, *277*, 27501–27509. [CrossRef] [PubMed]
113. Roh, M.H.; Makarova, O.; Liu, C.J.; Shin, K.; Lee, S.; Laurinec, S.; Goyal, M.; Wiggins, R.; Margolis, B. The Maguk protein, Pals1, functions as an adapter, linking mammalian homologues of Crumbs and Discs Lost. *J. Cell Biol.* **2002**, *157*, 161–172. [CrossRef] [PubMed]
114. Li, Y.; Karnak, D.; Demeler, B.; Margolis, B.; Lavie, A. Structural basis for L27 domain-mediated assembly of signaling and cell polarity complexes. *Embo. J.* **2004**, *23*, 2723–2733. [CrossRef] [PubMed]
115. Feng, W.; Long, J.F.; Zhang, M. A unified assembly mode revealed by the structures of tetrameric L27 domain complexes formed by mLin-2/mLin-7 and Patj/Pals1 scaffold proteins. *Proc. Natl. Acad. Sci. USA* **2005**, *102*, 6861–6866. [CrossRef] [PubMed]
116. Zhang, J.; Yang, X.; Wang, Z.; Zhou, H.; Xie, X.; Shen, Y.; Long, J. Structure of an L27 domain heterotrimer from cell polarity complex Patj/Pals1/Mals2 reveals mutually independent L27 domain assembly mode. *J. Biol. Chem.* **2012**, *287*, 11132–11140. [CrossRef] [PubMed]
117. Li, Y.; Wei, Z.; Yan, Y.; Wan, Q.; Du, Q.; Zhang, M. Structure of Crumbs tail in complex with the PALS1 PDZ-SH3-GK tandem reveals a highly specific assembly mechanism for the apical Crumbs complex. *Proc. Natl. Acad. Sci. USA* **2014**, *111*, 17444–17449. [CrossRef]

118. Wei, Z.; Li, Y.; Ye, F.; Zhang, M. Structural basis for the phosphorylation-regulated interaction between the cytoplasmic tail of cell polarity protein crumbs and the actin-binding protein moesin. *J. Biol. Chem.* **2015**, *290*, 11384–11392. [CrossRef]
119. Dickinson, D.J.; Schwager, F.; Pintard, L.; Gotta, M.; Goldstein, B. A Single-Cell Biochemistry Approach Reveals PAR Complex Dynamics during Cell Polarization. *Dev. Cell.* **2017**, *42*, 416–434 e11. [CrossRef]
120. Whitney, D.S.; Peterson, F.C.; Volkman, B.F. A conformational switch in the CRIB-PDZ module of Par-6. *Structure* **2011**, *19*, 1711–1722. [CrossRef]
121. Whitney, D.S.; Peterson, F.C.; Kittell, A.W.; Egner, J.M.; Prehoda, K.E.; Volkman, B.F. Binding of Crumbs to the Par-6 CRIB-PDZ Module Is Regulated by Cdc42. *Biochemistry* **2016**, *55*, 1455–1461. [CrossRef] [PubMed]
122. Ebnet, K.; Suzuki, A.; Horikoshi, Y.; Hirose, T.; Meyer Zu Brickwedde, M.K.; Ohno, S.; Vestweber, D. The cell polarity protein ASIP/PAR-3 directly associates with junctional adhesion molecule (JAM). *Embo. J.* **2001**, *20*, 3738–3748. [CrossRef] [PubMed]
123. Itoh, M.; Sasaki, H.; Furuse, M.; Ozaki, H.; Kita, T.; Tsukita, S. Junctional adhesion molecule (JAM) binds to PAR-3: A possible mechanism for the recruitment of PAR-3 to tight junctions. *J. Cell Biol.* **2001**, *154*, 491–497. [CrossRef] [PubMed]
124. Ebnet, K.; Aurrand-Lions, M.; Kuhn, A.; Kiefer, F.; Butz, S.; Zander, K.; Meyer zu Brickwedde, M.K.; Suzuki, A.; Imhof, B.A.; Vestweber, D. The junctional adhesion molecule (JAM) family members JAM-2 and JAM-3 associate with the cell polarity protein PAR-3: A possible role for JAMs in endothelial cell polarity. *J. Cell Sci.* **2003**, *116*, 3879–3891. [CrossRef] [PubMed]
125. Feng, W.; Wu, H.; Chan, L.N.; Zhang, M. The Par-3 NTD adopts a PB1-like structure required for Par-3 oligomerization and membrane localization. *Embo. J.* **2007**, *26*, 2786–2796. [CrossRef] [PubMed]
126. Zhang, Y.; Wang, W.; Chen, J.; Zhang, K.; Gao, F.; Gao, B.; Zhang, S.; Dong, M.; Besenbacher, F.; Gong, W.; et al. Structural insights into the intrinsic self-assembly of Par-3 N-terminal domain. *Structure* **2013**, *21*, 997–1006. [CrossRef]
127. Wu, H.; Feng, W.; Chen, J.; Chan, L.N.; Huang, S.; Zhang, M. PDZ domains of Par-3 as potential phosphoinositide signaling integrators. *Mol. Cell.* **2007**, *28*, 886–898. [CrossRef]
128. Jensen, D.R.; Woytovich, C.; Li, M.; Duvnjak, P.; Cassidy, M.S.; Frederick, R.O.; Bergeman, L.F.; Peterson, F.C.; Volkman, B.F. Rapid, robotic, small-scale protein production for NMR screening and structure determination. *Protein Sci.* **2010**, *19*, 570–578. [CrossRef]
129. Renschler, F.A.; Bruekner, S.R.; Salomon, P.L.; Mukherjee, A.; Kullmann, L.; Schutz-Stoffregen, M.C.; Henzler, C.; Pawson, T.; Krahn, M.P.; Wiesner, S. Structural basis for the interaction between the cell polarity proteins Par3 and Par6. *Sci. Signal.* **2018**, *11*. [CrossRef]
130. Poliak, S.; Matlis, S.; Ullmer, C.; Scherer, S.S.; Peles, E. Distinct claudins and associated PDZ proteins form different autotypic tight junctions in myelinating Schwann cells. *J. Cell Biol.* **2002**, *159*, 361–372. [CrossRef]
131. Adachi, M.; Hamazaki, Y.; Kobayashi, Y.; Itoh, M.; Tsukita, S.; Furuse, M.; Tsukita, S. Similar and distinct properties of MUPP1 and Patj, two homologous PDZ domain-containing tight-junction proteins. *Mol. Cell Biol.* **2009**, *29*, 2372–2389. [CrossRef] [PubMed]
132. Elkins, J.M.; Papagrigoriou, E.; Berridge, G.; Yang, X.; Phillips, C.; Gileadi, C.; Savitsky, P.; Doyle, D.A. Structure of PICK1 and other PDZ domains obtained with the help of self-binding C-terminal extensions. *Protein Sci.* **2007**, *16*, 683–694. [CrossRef] [PubMed]
133. Markley, J.L.; Aceti, D.J.; Bingman, C.A.; Fox, B.G.; Frederick, R.O.; Makino, S.; Nichols, K.W.; Phillips, G.N., Jr.; Primm, J.G.; Sahu, S.C.; et al. The Center for Eukaryotic Structural Genomics. *J. Struct. Funct. Genom.* **2009**, *10*, 165–179. [CrossRef] [PubMed]
134. Moleirinho, S.; Guerrant, W.; Kissil, J.L. The Angiomotins–from discovery to function. *Febs. Lett.* **2014**, *588*, 2693–2703. [CrossRef] [PubMed]
135. Sugihara-Mizuno, Y.; Adachi, M.; Kobayashi, Y.; Hamazaki, Y.; Nishimura, M.; Imai, T.; Furuse, M.; Tsukita, S. Molecular characterization of angiomotin/JEAP family proteins: Interaction with MUPP1/Patj and their endogenous properties. *Genes Cells* **2007**, *12*, 473–486. [CrossRef]
136. Citi, S.; Sabanay, H.; Jakes, R.; Geiger, B.; Kendrick-Jones, J. Cingulin, a new peripheral component of tight junctions. *Nature* **1988**, *333*, 272–276. [CrossRef]
137. D'Atri, F.; Nadalutti, F.; Citi, S. Evidence for a functional interaction between cingulin and ZO-1 in cultured cells. *J. Biol. Chem.* **2002**, *277*, 27757–27764. [CrossRef]

138. Guillemot, L.; Schneider, Y.; Brun, P.; Castagliuolo, I.; Pizzuti, D.; Martines, D.; Jond, L.; Bongiovanni, M.; Citi, S. Cingulin is dispensable for epithelial barrier function and tight junction structure, and plays a role in the control of claudin-2 expression and response to duodenal mucosa injury. *J. Cell Sci.* **2012**, *125*, 5005–5014. [CrossRef]
139. Balda, M.S.; Garrett, M.D.; Matter, K. The ZO-1-associated Y-box factor ZONAB regulates epithelial cell proliferation and cell density. *J. Cell Biol.* **2003**, *160*, 423–432. [CrossRef]
140. Lima, W.R.; Parreira, K.S.; Devuyst, O.; Caplanusi, A.; N'Kuli, F.; Marien, B.; Van Der Smissen, P.; Alves, P.M.; Verroust, P.; Christensen, E.I.; et al. ZONAB promotes proliferation and represses differentiation of proximal tubule epithelial cells. *J. Am. Soc. Nephrol.* **2010**, *21*, 478–488. [CrossRef]
141. Schindelin, H.; Marahiel, M.A.; Heinemann, U. Universal nucleic acid-binding domain revealed by crystal structure of the B. subtilis major cold-shock protein. *Nature* **1993**, *364*, 164–168. [CrossRef] [PubMed]
142. Yang, X.J.; Zhu, H.; Mu, S.R.; Wei, W.J.; Yuan, X.; Wang, M.; Liu, Y.; Hui, J.; Huang, Y. Crystal structure of a Y-box binding protein 1 (YB-1)-RNA complex reveals key features and residues interacting with RNA. *J. Biol. Chem.* **2019**, *294*, 10998–11010. [CrossRef] [PubMed]
143. Krug, S.M.; Schulzke, J.D.; Fromm, M. Tight junction, selective permeability, and related diseases. *Semin. Cell Dev. Biol.* **2014**, *36*, 166–176. [CrossRef] [PubMed]
144. Boivin, F.J.; Schmidt-Ott, K.M. Transcriptional mechanisms coordinating tight junction assembly during epithelial differentiation. *Ann. N. Y. Acad. Sci.* **2017**, *1397*, 80–99. [CrossRef]
145. Werth, M.; Walentin, K.; Aue, A.; Schonheit, J.; Wuebken, A.; Pode-Shakked, N.; Vilianovitch, L.; Erdmann, B.; Dekel, B.; Bader, M.; et al. The transcription factor grainyhead-like 2 regulates the molecular composition of the epithelial apical junctional complex. *Development* **2010**, *137*, 3835–3845. [CrossRef]
146. Kojima, T.; Kohno, T.; Kubo, T.; Kaneko, Y.; Kakuki, T.; Kakiuchi, A.; Kurose, M.; Takano, K.I.; Ogasawara, N.; Obata, K.; et al. Regulation of claudin-4 via p63 in human epithelial cells. *Ann. N. Y. Acad. Sci.* **2017**, *1405*, 25–31. [CrossRef]
147. Kaneko, Y.; Kohno, T.; Kakuki, T.; Takano, K.I.; Ogasawara, N.; Miyata, R.; Kikuchi, S.; Konno, T.; Ohkuni, T.; Yajima, R.; et al. The role of transcriptional factor p63 in regulation of epithelial barrier and ciliogenesis of human nasal epithelial cells. *Sci. Rep.* **2017**, *7*, 10935. [CrossRef]
148. Renaud, J.P.; Rochel, N.; Ruff, M.; Vivat, V.; Chambon, P.; Gronemeyer, H.; Moras, D. Crystal structure of the RAR-gamma ligand-binding domain bound to all-trans retinoic acid. *Nature* **1995**, *378*, 681–689. [CrossRef]
149. Matias, P.M.; Donner, P.; Coelho, R.; Thomaz, M.; Peixoto, C.; Macedo, S.; Otto, N.; Joschko, S.; Scholz, P.; Wegg, A.; et al. Structural evidence for ligand specificity in the binding domain of the human androgen receptor. Implications for pathogenic gene mutations. *J. Biol. Chem.* **2000**, *275*, 26164–26171. [CrossRef]
150. Chen, C.; Gorlatova, N.; Kelman, Z.; Herzberg, O. Structures of p63 DNA binding domain in complexes with half-site and with spacer-containing full response elements. *Proc. Natl. Acad. Sci. USA* **2011**, *108*, 6456–6461. [CrossRef]
151. Ming, Q.; Roske, Y.; Schuetz, A.; Walentin, K.; Ibraimi, I.; Schmidt-Ott, K.M.; Heinemann, U. Structural basis of gene regulation by the Grainyhead/CP2 transcription factor family. *Nucleic Acids Res.* **2018**, *46*, 2082–2095. [CrossRef] [PubMed]
152. Fernandez-Leiro, R.; Scheres, S.H. Unravelling biological macromolecules with cryo-electron microscopy. *Nature* **2016**, *537*, 339–346. [CrossRef] [PubMed]
153. Punjani, A.; Rubinstein, J.L.; Fleet, D.J.; Brubaker, M.A. cryoSPARC: Algorithms for rapid unsupervised cryo-EM structure determination. *Nat. Methods* **2017**, *14*, 290–296. [CrossRef] [PubMed]
154. Cheng, Y. Single-particle cryo-EM-How did it get here and where will it go. *Science* **2018**, *361*, 876–880. [CrossRef] [PubMed]
155. Terwilliger, T.C.; Adams, P.D.; Afonine, P.V.; Sobolev, O.V. A fully automatic method yielding initial models from high-resolution cryo-electron microscopy maps. *Nat. Methods* **2018**, *15*, 905–908. [CrossRef]
156. Beck, M.; Lucic, V.; Forster, F.; Baumeister, W.; Medalia, O. Snapshots of nuclear pore complexes in action captured by cryo-electron tomography. *Nature* **2007**, *449*, 611–615. [CrossRef]
157. Schaffer, M.; Pfeffer, S.; Mahamid, J.; Kleindiek, S.; Laugks, T.; Albert, S.; Engel, B.D.; Rummel, A.; Smith, A.J.; Baumeister, W.; et al. A cryo-FIB lift-out technique enables molecular-resolution cryo-ET within native Caenorhabditis elegans tissue. *Nat. Methods* **2019**, *16*, 757–762. [CrossRef]
158. O'Reilly, F.J.; Rappsilber, J. Cross-linking mass spectrometry: Methods and applications in structural, molecular and systems biology. *Nat. Struct. Mol. Biol.* **2018**, *25*, 1000–1008. [CrossRef]

159. Yu, C.; Huang, L. Cross-Linking Mass Spectrometry: An Emerging Technology for Interactomics and Structural Biology. *Anal. Chem.* **2018**, *90*, 144–165. [CrossRef]
160. Blanchet, C.E.; Svergun, D.I. Small-angle X-ray scattering on biological macromolecules and nanocomposites in solution. *Annu. Rev. Phys. Chem.* **2013**, *64*, 37–54. [CrossRef]
161. Tuukkanen, A.T.; Spilotros, A.; Svergun, D.I. Progress in small-angle scattering from biological solutions at high-brilliance synchrotrons. *IUCrJ* **2017**, *4*, 518–528. [CrossRef] [PubMed]
162. Nannenga, B.L.; Gonen, T. The cryo-EM method microcrystal electron diffraction (MicroED). *Nat. Methods* **2019**, *16*, 369–379. [CrossRef] [PubMed]

Review

Computational Modeling of Claudin Structure and Function

Shadi Fuladi [1], Ridaka-Wal Jannat [1], Le Shen [2,3], Christopher R. Weber [2,*] and Fatemeh Khalili-Araghi [1,*]

1 Department of Physics, University of Illinois at Chicago, Chicago, IL 60607, USA; sfulad2@uic.edu (S.F.); rjanna2@uic.edu (R.-W.J.)
2 Department of Pathology, University of Chicago, Chicago, IL 60637, USA; leshen@uchicago.edu
3 Department of Surgery, University of Chicago, Chicago, IL 60637, USA
* Correspondence: cweber@bsd.uchicago.edu (C.R.W.); akhalili@uic.edu (F.K.-A.)

Received: 15 December 2019; Accepted: 16 January 2020; Published: 23 January 2020

Abstract: Tight junctions form a barrier to control passive transport of ions and small molecules across epithelia and endothelia. In addition to forming a barrier, some of claudins control transport properties of tight junctions by forming charge- and size-selective ion channels. It has been suggested claudin monomers can form or incorporate into tight junction strands to form channels. Resolving the crystallographic structure of several claudins in recent years has provided an opportunity to examine structural basis of claudins in tight junctions. Computational and theoretical modeling relying on atomic description of the pore have contributed significantly to our understanding of claudin pores and paracellular transport. In this paper, we review recent computational and mathematical modeling of claudin barrier function. We focus on dynamic modeling of global epithelial barrier function as a function of claudin pores and molecular dynamics studies of claudins leading to a functional model of claudin channels.

Keywords: claudin; molecular dynamics; tight junction; ion transport; ion channel

---

1. Introduction to Tight Junctions

Tight junctions form paracellular barriers between cells. The barriers are established by interactions between transmembrane tight junction proteins located within the apical junctional complex. The regions of interaction were first appreciated ultrastructurally and it was determined that they form a barrier to paracellular flux [1]. In the years since these early observations, we now understand that members of the claudin family of transmembrane proteins are major components of these strands and are important contributors to paracellular charge- and size-selectivity. Extensive data supports that interactions between claudin proteins on adjacent cells are important for tight junction permeability. Understanding how claudins create such charge- and size-selective barriers has been a major research focus. Experiments in combination with mathematical and computational modeling have contributed to understanding the role of claudins in defining and regulating this paracellular barrier. Here, we will provide an overview of the theoretical and computational modeling of claudin pores in recent years and discuss how they have contributed to our understanding of tight junction function. Specifically, we will address what is understood about the dynamic claudin barrier, structural characteristics of claudin pores and the nature of ion selectivity and the ability to self-assemble into multimers (strands).

2. Resistive Models of Paracellular Flux through the Tight Junctions

Using transmission electron microscopy, tight junctions (TJ) are appreciated as a region of close apposition of the apical intercellular membranes [1]. Within this region, the membranes appear to be fused at multiple points. Freeze fracture electron microscopy (FFEM), a technique which allows the membrane to be fractured between the two leaflets of the lipid bilayers, permits visualization of the detailed structures within the tight junction. On the outer leaflet, the tight junction appears to contain multiple interconnecting ridges and, on the inner leaflet, the tight junction appears to contain multiple interconnecting grooves [2,3]. These results lead to the interpretation that the tight junction contains multiple interconnecting strands. Within different epithelia, strand numbers vary. Some epithelia have one or two strands, whereas other epithelia have eight strands on average [3]. Claude et al. hypothesized that the strand number may be related to the tightness of the tight junction [4]. If TJ strands act as series of resistors, then there would be a linear relationship between number of strands and transepithelial junctional resistance (since resistors sum in series). However, Claude observed an exponential relationship between strand number and barrier function. To explain this nonlinear behavior of resistive barrier generated by multiple strands, Claude developed a more complex model and hypothesized that strands are populated by "pores" that can open or close to regulate TJ ion conductance [4]. This is the first model to predict the tight junction barrier as a dynamic structure, however we now know such a model is an oversimplification since, even in epithelia with the same number of tight junction strands, the expression of certain claudin proteins can dramatically influence transepithelial electrical resistance (TER) [5,6]. Thus, tight junction barrier is far more complex than originally predicted by Claude et al. [4].

3. Dynamic Models of Claudin Function

We now understand certain claudin proteins (e.g., claudin-2, -15, -10a, -10b) can selectively increase paracellular permeability of small molecules ($<\sim 6\,\text{Å}$ diameter) in a charge-selective manner [7–11]. Since such charge- and size-selectivity could theoretically be explained by insertion of ion channels, this has been termed the "pore" pathway, which is in contrast to large pathways of non-selective conductance termed the "leak" pathway. Because of the hypothesis that tight junctions are populated by dynamic pores, Weber et al. adapted patch clamp technique to study individual claudin-2-dependent pores (Figure 1A) [12]. Comparison of patch clamp recordings of low versus high expressing claudin-2 monolayers revealed a distinct population of dynamic ion channel openings over tight junctions, which were absent from recordings made on the plasma membrane away from tight junctions. These channel openings demonstrated charge- and size-selectivity which reflected globally measured barrier function. The openings occurred during burst-like intervals in which the channels flickered between open and closed rapidly (sub millisecond flickering) with prolonged periods of closure between bursts (Figure 1B). These opening and closing events were strictly due to claudin-2 since mutating one residue within the claudin-2 channel (I66) to a cysteine, allows linkage of the I66C to a bulky molecule (MTSET), through sulfhydryl modification [13]. This modification rapidly and specifically blocked the opening and closing events indicating that the channels can be blocked like conventional transmembrane ion channels [12].

To further understand such dynamic claudin-2 opening behavior, an in silico resistive multistrand model was developed. Claudin-2 channels were arranged within a network of three branching tight junction strands (as observed by FFEM). The channels were allowed to randomly vary between high resistance closed states (10 pS) and low resistance open states (222 pS) with fixed probabilities defined by the patch clamp recordings. Increasing the density of claudin pores from 6 to 36 (closed and open) per strand micron in each of the parallel strands recapitulated tight junction patch clamp recordings in MDCK monolayers with low and high claudin expression (Figure 1B). However, on its own, the dynamic pore pathway could not account for all of the claudin-2 dependent barrier function, as measured by global TER measurements in epithelia monolayers. To make the data fit globally determined TER values, it was necessary to add a parallel leak component. Such a leak component could be due to a steady state conductive component, rare highly conductive openings or high conductance at tricellular junctions which are not readily detectable using tight junction patch clamp recordings. Thus, while the model explains local pore function well, it cannot fully explain leak pathway conductance [12,14]. More complex models to describe leak pathway flux have been postulated. These include both leak through a tricellulin-dependent pathway and leak through breaks in tight junction strands [15–18]. These studies are beyond the scope of the present review. Thus, patch clamp recordings and modeling suggest the presence of gated ion selective claudin ion channels. However, these studies fail to define the atomic details of the pore, mechanisms of channel selectivity or mechanisms of gating. More detailed studies and atomic-level models are needed.

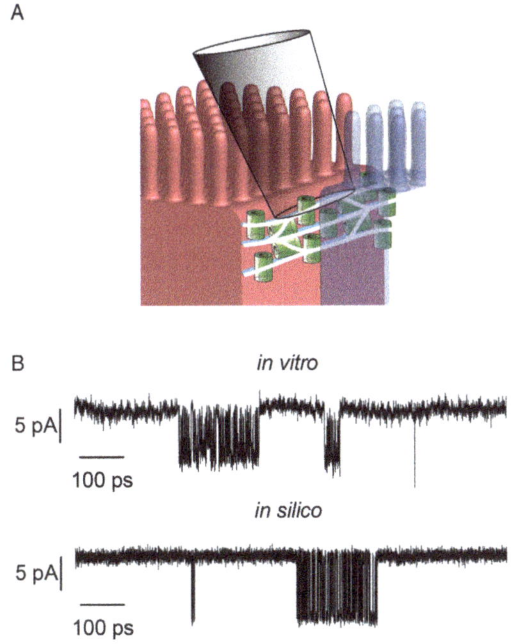

Figure 1. Mathematical modeling of tight junction permeability. (**A**) Electrodes were sealed over the intracellular space between MDCK I induced to express claudin-2 [12]. Each claudin-2 pore was defined by two closed states (C_1, C_2) and a single open state (O) according to the equation in [14] (**B**) In silico patch clamp recordings resembled in vitro tight junction patch clamp recordings from MDCK I monolayers expressing claudin-2 (recording from dataset of Weber et al. [12]). For the simulation, each green pore shown in panel A was modeled as 36 resistors per micron with defined opening and closing probabilities [14].

4. A Structural Model of Claudin Pores

A recently resolved crystal structure of mouse claudin-15 provides the first snapshot of a claudin monomer with atomic resolution [19]. The claudin-15 monomer is made of four transmembrane (TM) helices and two extracellular segments ECS1 and ECS2. The two extracellular segments as well as TM helices are known to mediate polymerization of claudins through side-to-side (*cis*-) interactions as well as head-to-head (*trans*-) interactions of claudins in plasma membrane. However, the crystal structure does not contain key portions of the extracellular segments and does not consider interactions between adjacent cells, making it impossible to define the channel structure.

Soon after resolving the crystal structure of claudin-15, Suzuki et al. [20] proposed a model of paracellular pores based on the crystallographic arrangement of claudin-15 as well as cysteine cross-linking experiments obtained from mutants of claudin-15 and claudin-2 [13,21,22] and strand dimensions observed in freeze-fracture electron microscopy images [20]. In this model, claudin pores are formed by association of two anti-parallel double rows of claudins in the membrane of two adjacent cells, forming β-barrel like channels parallel to plasma membrane (Figure 2A,B).

The stability of this model was investigated in all-atom molecular dynamics (MD) simulations of claudin-15 molecules in their natural environment of lipid bilayers. Simulations of a single- [23] and a double-pore model [23,24] and a periodic strand model of three pores [25] proved that the proposed architecture of Suzuki et al. [20] creates a stable arrangement of claudin channels. During the 200–300 ns simulation trajectories, *cis*- and *trans*-interactions between claudins in the lipid bilayers are maintained and the β-barrel like scaffold of the pore is preserved. However, simulations of a four-pore model of claudin-15 indicate that there might be multiple and alternative interfaces not captured in this arrangement [24].

In this model, the linear arrangement of claudins in the membrane is maintained through side-by-side (*cis*-) interactions between adjacent claudins involving residues S67 and M68 on extracellular helix (ECH) and residues F146 and F147 on TM3 and E157 and L158 on ECS2 domains. Moreover, the double-row arrangement of claudins in the membrane, as was suggested for the first time by Suzuki et al. [20], was maintained through hydrogen bonds between β-sheets of adjacent claudins. In addition, simulations identified two distinct head-to-head (*trans*-) interactions between claudins in two opposing membranes, residues 39–42 on two opposing ECS1 loops and 146–155 on the ECS2 of two claudins form strong hydrophobic interactions that were maintained throughout the simulations [23,25]. Simulation trajectories show that claudins form well-defined pores in paracellular space that are filled with water and ions. Each pore runs parallel to the membranes and is made of eight claudin monomers. The pores are approximately 50 Å long with a minimum diameter of 5–6 Å [23,25]. The claudin-15 pore is anisotropic in the two directions orthogonal to permeation pathway. Spatial distribution of claudin-15 pores in these simulations suggests that this model corresponds to a density of 300 pores per μm in TJ strands as shown in Figure 2C [25].

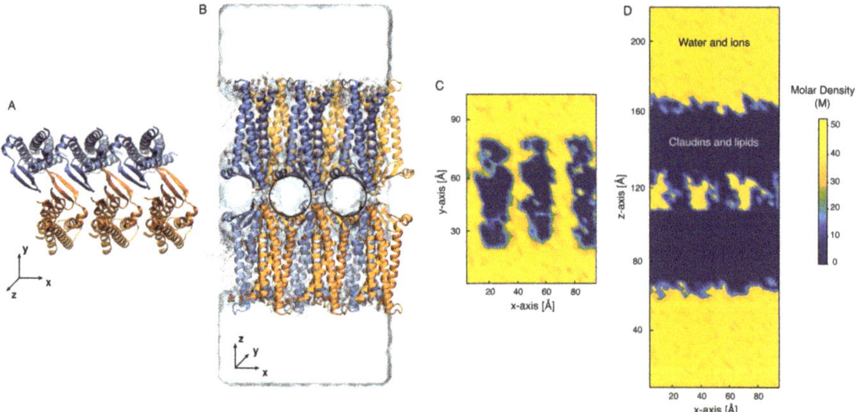

Figure 2. A refined model of claudin-15 paracellular channels [25]. Claudin-15 monomers assemble into double-row strands and form paracellular channels. (**A**) Top view of six claudin-15 monomers assembling into a double-row strand. (**B**) Snapshot of the simulation system showing claudin-15 monomers assembled between adjacent lipid bilayers. The pore regions are marked with circles. The water density across the simulation system and averaged over ~250 ns of MD simulations, is shown: (**C**) parallel to the two membranes crossing the pores in the middle and (**D**) normal to the two membranes and crossing the pores in the middle. Figure is taken from [25] with slight modification.

Analysis of the water density from simulations of Samanta et al. [25] indicates that water molecules, and thus ions, are confined to well-defined pathways within claudin-15 pores (Figure 2C,D).

Water molecules occupy the vestibules formed by claudin pores with a density very close to the bulk density of water. The paracellular space between claudin pores is completely sealed to water molecules. Simulations show that *trans*-interactions between opposing residues (39–42 in claudin-15) on ECS1 (hydrophobic patches) are essential to seal the paracellular space *between* neighboring claudin pores. This is consistent with previous studies suggesting that claudin-15 form water channels [26] and that water and ions are transported through a similar pathway in claudin-2 and claudin-15 pores [26,27].

Based on these refined models of claudin paracellular channels, the underlying mechanism for ion transport through the pores has been studied further.

5. Ion Transport Simulations

All-atom molecular dynamics (MD) simulations studies have been very successful in describing functional mechanisms of membrane proteins, particularly ion channels [28–31]. Having an atomic model of claudin pores provides a unique opportunity to test hypotheses about the structural basis of charge- and size-selectivity in claudin pores. Using MD simulations Samanta et al. [25] investigated the transport properties of claudin-15 pores in a refined model based on the proposed architecture of Suzuki et al. [20]. Transport of ions through the pores was simulated using all-atom MD simulations of three claudin pores in two parallel lipid bilayers. Ionic currents under an external voltage-bias [32] were calculated from the displacement charge associated with all ionic species over the course of simulations resulting in current-voltage relationships and enabling them to estimate the permeability of claudin pores to each ion independently.

Permeability of the pore to monovalent cations of different sizes, Na^+, methylammonium (MA^+), ethylammonium (EA^+), tetramethylammonium (TMA^+) and tetraethylammonium (TEA^+) as well as Cl^- was calculated. The simulations indicate that claudin-15 pores are cation-selective, with a relative permeability of 4.2 for Na^+ with respect to Cl^-. Moreover, simulations suggest that the pore permeability to cations decreases as their size increases. While the claudin-15 pore is permeable to small cations such as MA^+, EA^+ or TMA^+, transport of larger cations such as TEA^+ (radius 3.5 Å) through the pore is significantly hampered by their interaction with the pore surface. This is consistent with previous experiments and Brownian dynamics simulations of the cation-selective claudin-2 pore, which suggest a minimum pore radius of 3.25 Å [9].

Trajectories of ion transport obtained from the simulations showed that transport pathways of cations and anions in claudin-15 are different. Cl^- ions interact minimally with the pore surface and pass through the middle of the pore, while cations such as Na^+ slide along the inner surface of the pore and interact strongly with negatively charged amino acids on the pore surface. Simulations identified four binding sites for cations inside the pore. These binding sites are located in the middle of the pore and are mainly formed by sidechains of D55. One, two or three Na^+ ions were observed to bind to D55 residues simultaneously, while larger cations such as TEA^+ bind in the middle of four D55 residues. To further quantify the charge-selectivity mechanism of claudin-15, Samanta et al. [25] calculated the contact time of the permeating ions with the amino acids lining the pore surface. The peak interacting site for cations was at D55, while negatively charged residues within the pore (E46, D64, D145 and E157) showed significant but much shorter interaction time with cations. Analysis of the contact map of claudin-15 suggests that D55 is the key residue controlling the charge-selectivity of claudin-15 in this model.

To define the role of D55 and other residues of claudin-15, Samanta et al. simulated conductance of several claudin-15 pore mutants [25]. Mutation of D55 to a neutral amino acid (D55N) decreased the charge- (cation) selectivity of the pore and mutation of D55 to a positively charged amino acid (D55K) reversed the charge-selectivity of claudin-15. Moreover, D64K and E46K mutations decreased the cation-selectivity of the pore. However, the triple mutation of E46K/D55K/D64K, similar to D55K mutant reversed the charge-selectivity of claudin-15 resulting in anion-selective pores. These findings were supported by in vitro measurements of transepithelial resistance (TER) of claudin-15 and its mutants in MDCK I monolayers expressing claudin-15. However, one feature that differed from the modeling is that the triple mutant (E46K/D55K/D64K) was not strongly anion-selective in vitro. We speculate that this may relate to the expression of other tight junction proteins in vitro.

In a more recent study, Alberini et al. [33] calculated the potential of mean force (PMF) for the ion permeation through claudin-15. The PMF along the ion permeation pathway provides a quantitative measure of the pore selectivity by quantifying the free energy barriers/wells that each ion encounters as it passes through the pore. The simulations were performed on an equilibrated configuration of a single-pore model of claudin-15 [23] based on the proposed architecture of Suzuki et al. [20]. The free energy profiles were calculated for Na^+, K^+ and Cl^- along the central line of the pore from 325 ns of equilibrium simulation using the umbrella sampling method [34]. The free energy profiles show a free energy barrier of 8 kcal/mol (13.5 k_BT at room temperature) for Cl^- ions and an attractive well of approximately 4 kcal/mol (6.7 k_BT) for Na^+ and K^+. There was no significant difference between the free energy profiles of Na^+ and K^+ in this case. A barrier of 8 kcal/mol is high enough to prevent passage of Cl^- ions through the pore at room temperature, and an attractive well of 4 kcal/mol is strong enough to attract cations while at the same time allowing them to pass through the pore at room temperature. Interestingly, the position of the free energy barrier/well in the PMF corresponds to the location of D55 residues in claudin-15.

To study the kinetics of ions through the pore, Alberini et al. used the Voronoi-tessellated milestoning method [35,36] to calculate the transport rate of ions through claudin-15 and determined a ratio of 25 to 1 for the transport rates of Na^+ with respect to Cl^- and a ratio of 1 to 1.4 for Na^+ with respect to K^+. It has to be noted that transport rates calculated from free energy profiles usually have large error bars due to the limited accuracy of the force fields and the inherent errors in the free energy calculations. However, in a recent experimental study, the relative permeability of claudin-15 for Na^+ and K^+ was estimated to be close to 1 [26]. Simulations of Alberini et al. [33] showed that cations become partially dehydrated as they pass through claudin pores. The narrowest region of the pore is approximately 6 Å in diameter, while the diameter of a fully hydrated Na^+ ion is $\sim$7Å. Further analysis of the hydration state of Na^+ ions in these simulations showed that Na^+ ions are coordinated by approximately 5.7 oxygen atoms throughout the simulation. At the entrance of the pore all these oxygens are contributed by water molecules. However, as the cation moves along the pore, interaction with protein side chains temporarily replaces the interaction with water molecules. In particular, as the ions approach D55 in the middle of the pore, one or two of the oxygen atoms within water molecules are replaced by oxygen atoms of the D55 residues, consistent with favorable interactions between protein and cations at this location.

Simulations of ion transport show that the putative architecture of Suzuki et al. [20] is a viable model of claudin pores that is energetically stable and describes the nature of ion selectivity in TJs. However, this model of claudin-15 assembly does not conclusively explain all of the experimentally obtained results across various claudins. In particular, this model is not compatible with studies suggesting the role of TM helices in claudin assembly and strand formation through *cis*-interactions [37–41]. Based on these interactions, alternative pore models have been proposed [37,42,43]. With recent advances in computational techniques, molecular dynamics simulations can reach time scales and sizes relevant to study assembly and initial stages of protein aggregation [44–49] and can provide an insight into alternative conformations of claudins in the TJ strands.

6. Claudin Polymerization in Lipid Bilayers

Coarse-graining (CG) methods complement atomistic MD simulations and enable researchers to study larger biological systems over longer time scales. In particular, CG modeling is used to study membrane protein self-assembly [47–49]. These approaches have been used to understand how claudin monomers may interact within TJ strands (reviewed in [50]).

Using CG methodologies [51], Irudayanathan et al. [52] studied the self-assembly process of a homology model of claudin-5 in a lipid bilayer. The authors studied two systems of uniformly distributed claudin-5 monomers and simulated self-assembly of these monomers into multimeric units. These simulations revealed rapid association of claudin-5 into dimeric structures, which later aggregated into strands of 16 to 36 claudin-5 molecules within 10 μs. The CG simulations predicted four prevalent interfaces between claudin-5 monomers within a lipid bilayer. Two of these dimeric structures are compatible with the *cis*-interactions proposed in earlier models [20,53], one between adjacent claudin monomers in a row (side-to-side) as observed in the crystal structure of claudin-15 and the other, between β-sheets of claudin molecules from opposing rows (face-to-face). These two interaction surfaces are consistent with the proposed model of Suzuki et al. [20]. A third dimeric interaction was between TM helices (TM2–TM3) of two claudins. This claudin-5 dimeric interface is consistent with previous studies suggesting the close vicinity of TM2 and TM3 helices between claudins [38,39]. A fourth dimeric structure involves interactions between TM helices and ECSs (TM3–ECS2–TM4) of two adjacent monomers. The role of TM3 and TM4 was shown previously using alanine-insertion mutagenesis (AIM) analysis of claudin-16 and claudin-19 heterodimers [41]. These CG simulations were extended to investigate assembly of homology models of other claudins, claudin-1, -2, -4, -15, and -19 in single lipid bilayers and the same four dimeric interfaces were observed [54].

To assess the relative stability of the observed dimers, Iruyadanathan et al. constructed atomic models of the four dimers of claudin-5 observed in CG simulations and calculated the binding free energy (PMF) of claudin-5 monomers in each configuration [52]. Their results show that dimeric conformations that involve more interactions sites through TM helices are more stable compared to dimers involving ECS interfaces. While these results are intuitive, the accuracy of the PMF calculations could be improved. Simulations of longer time scales will be needed to fully sample positional, orientational and conformational changes of proteins in all-atom simulations [55–57] in contrast to CG simulations [58–63].

Experimental studies support that protein-protein interactions may not be the same for all claudins. For example, there are significant differences in how claudin-5 and claudin-3 oligomerize and this is thought to be due to differences in key residues in TM3 [39]. To investigate the role of TM3 helices in the self-assembly of claudins in the membrane, Irudayanathan et al. [43] simulated the assembly of homology models of claudin-3 and claudin-5 in lipid bilayers at CG resolution. The self-assembly trajectory of claudin-3 monomers showed that the dimer with TM2–TM3 interface, detected in claudin-5 trajectories, was absent in claudin-3, while other dimer interfaces were detected. Mutating two key residues on the TM3 of claudin-5 as proposed by Rossa et al. [39] to those in claudin-3 (F139S/I142T) decreased the likelihood of TM2–TM3 dimerization in claudin-5 and vice versa. Mutating the same residues in claudin-3 to those in claudin-5 (S138F/T141I) resulted in a slight increase in dimerization probability of conformations similar to those that involve TM2–TM3. These simulations suggest that distortion of TM3 in claudin-3 prevents dimerization in the self-assembly process. While no significant changes were observed in the shape of strands in wild-type claudins compared to mutants, it is not true that all claudin protein-protein interactions result in the same strand structure. For example, recent X-ray crystallography of claudin-3 has shown that the TM3 helix is bent, in contrast to a straighter helical structure in claudin-4, -15 and -19 [64]. The bend in this helix is shown to affect the morphology of claudin strands.

CG simulations have also been used to study pore formation through *trans*-interactions between claudins in two parallel lipid bilayers. Irudayanathan et al. [43] performed CG simulations of 128 claudin-5 monomers in two adjacent lipid bilayers. 20 μs of self-assembly simulation showed instances of *trans*-interactions between the smaller ECS2 loop of monomers in adjacent membranes. Although these interactions were proposed previously [20,65], formation of pore-like structures requires more involved interactions, requiring longer simulation times.

Lipid environment is important in the regulation of claudin function [66,67] and it has been hypothesized to influence claudin oligomerization. Evidence suggests that removal of cholesterol can disrupt barrier function by altering the stability of tight junction complexes [68,69]. To investigate the influence of membrane composition on claudin oligomerization, Irudayanathan et al. [52] carried out simulations of claudin-5 monomers in lipid bilayers with different lengths of hydrophobic moieties (DLPC, DYPC, DPPC and POPC) and different cholesterol containing membranes (PC, PCS, DPC). Simulations suggest that the presence of cholesterol reduces the frequency of dimerization of claudins and, thus, the likelihood of strand formation in the lipid bilayers. In particular, the presence of cholesterol in the lipid bilayer prevented formation of dimers that involve interaction of TM helices [52]. The prevalent dimers observed in these simulations were those that involved limited interactions between ECS segments of claudin-5. The effect of claudin palmitoylation has also been studied using CG simulations but the study revealed no effect on strand formation [70]. However, cholesterol binding prevents dimerization of palmitoylated claudins, too.

The above CG studies have provided insight into claudin-claudin interactions and this type of model may improve our understanding of the factors which govern strand shape. However, time scales of computations are still too short to simulate dynamics of strand formation, even in CG simulations. Moreover, the limited resolution of CG simulations limits their ability to describe functional properties of proteins, such as ion transport, where channel function is controlled by specific and finely tuned interactions amongst proteins, ions and water.

7. Conclusions

Computational modeling and simulations have proven to be a very useful tool that have enabled researchers to combine much of what we have learned from numerous in vitro and in vivo studies of tight junctions in the past several decades. Recent advances in computational hardware and methodologies have permitted us to reach time scales and resolutions relevant for understanding claudin organization and function, ranging from atomistic simulations of ion permeation to CG simulations of claudin assembly to dynamic simulations of transport across multiple strands and Brownian dynamics simulations of ions across many channels. All of the models presented in this review have limitations and will need to be continually refined as more data becomes available. It will be an iterative process where modeling is continually compared with experimental data. As computing power increases and models become more accurate, it will become possible to expand simulations spatially and temporally to better understand the intersecting pathways which dynamically regulate tight junction structure and function. Such models will help us form and test new hypotheses, understand mechanisms of barrier dysfunction, discover means to increase or decrease tight junction barrier function and gain insight into approaches to promote trans-tight junction drug absorption.

Author Contributions: All authors participated in writing and editing of this article. All authors have read and agreed to the published version of the manuscript.

Funding: S.F., R.J., and F.K.-A. were supported by a grant from National Science Foundation (award number: 186021 to F.K.-A.), and C.R.W. and F.K.-A. were supported by a grant from Crohn's and Colitis Foundation of America (award number: 608820 to C.R.W.).

Conflicts of Interest: The authors declare no conflict of interest.

Abbreviations

The following abbreviations are used in this manuscript:

MD	Molecular Dynamics
CG	Coarse-grained
PMF	Potential of Mean Force
TER	Transepithelial Electrical Resistance
TJ	Tight Junction
TM	Transmembrane
ECS	Extracellular Segment
ECH	Extracellular Helix
FFEM	Freeze Fracture Electron Microscopy

References

1. Farquhar, M.G.; Palade, G.E. Junctional complexes in various epithelia. *J. Cell Biol.* **1963**, *17*, 375–412. [CrossRef] [PubMed]
2. Staehelin, L.A.; Mukherjee, T.; Williams, A.W. Freeze-etch appearance of the tight junctions in the epithelium of small and large intestine of mice. *Protoplasma* **1969**, *67*, 165–184. [CrossRef] [PubMed]
3. Claude, P.; Goodenough, D.A. Fracture faces of zonulae occludentes from "tight" and "leaky" epithelia. *J. Cell Biol.* **1973**, *58*, 390–400. [CrossRef] [PubMed]
4. Claude, P. Morphological factors influencing transepithelial permeability: A model for the resistance of theZonula Occludens. *J. Membr. Biol.* **1978**, *39*, 219–232. [CrossRef]
5. Weber, C.R.; Raleigh, D.R.; Su, L.; Shen, L.; Sullivan, E.A.; Wang, Y.; Turner, J.R. Epithelial myosin light chain kinase activation induces mucosal interleukin-13 expression to alter tight junction ion selectivity. *J. Biol. Chem.* **2010**, *285*, 12037–12046. [CrossRef]
6. Furuse, M.; Furuse, K.; Sasaki, H.; Tsukita, S. Conversion of zonulae occludentes from tight to leaky strand type by introducing claudin-2 into Madin-Darby canine kidney I cells. *J. Cell Biol.* **2001**, *153*, 263–272. [CrossRef]
7. Van Itallie, C.M.; Rogan, S.; Yu, A.S.; Vidal, L.S.; Holmes, J.; Anderson, J.M. Two splice variants of claudin-10 in the kidney create paracellular pores with different ion selectivities. *Am. J. Physiol.-Ren. Physiol.* **2006**, *291*, F1288–F1299. [CrossRef]
8. Van Itallie, C.M.; Holmes, J.; Bridges, A.; Gookin, J.L.; Coccaro, M.R.; Proctor, W.; Colegio, O.R.; Anderson, J.M. The density of small tight junction pores varies among cell types and is increased by expression of claudin-2. *J. Cell Sci.* **2008**, *121*, 298–305. [CrossRef]
9. Yu, A.S.; Cheng, M.H.; Angelow, S.; Günzel, D.; Kanzawa, S.A.; Schneeberger, E.E.; Fromm, M.; Coalson, R.D. Molecular basis for cation selectivity in claudin-2–based paracellular pores: Identification of an electrostatic interaction site. *J. Gen. Physiol.* **2009**, *133*, 111–127. [CrossRef]
10. Watson, C.; Rowland, M.; Warhurst, G. Functional modeling of tight junctions in intestinal cell monolayers using polyethylene glycol oligomers. *Am. J. Physiol.-Cell Physiol.* **2001**, *281*, C388–C397. [CrossRef]
11. Krug, S.M.; Günzel, D.; Conrad, M.P.; Lee, I.F.M.; Amasheh, S.; Fromm, M.; Yu, A.S. Charge-selective claudin channels. *Ann. N. Y. Acad. Sci.* **2012**, *1257*, 20–28. [CrossRef] [PubMed]
12. Weber, C.R.; Liang, G.H.; Wang, Y.; Das, S.; Shen, L.; Yu, A.S.; Nelson, D.J.; Turner, J.R. Claudin-2-dependent paracellular channels are dynamically gated. *Elife* **2015**, *4*, e09906. [CrossRef] [PubMed]
13. Li, J.; Angelow, S.; Linge, A.; Zhuo, M.; Yu, A.S. Claudin-2 pore function requires an intramolecular disulfide bond between two conserved extracellular cysteines. *Am. J. Physiol.-Cell Physiol.* **2013**, *305*, C190–C196. [CrossRef] [PubMed]
14. Weber, C.R.; Turner, J.R. Dynamic modeling of the tight junction pore pathway. *Ann. N. Y. Acad. Sci.* **2017**, *1397*, 209. [CrossRef]
15. Tervonen, A.; Ihalainen, T.O.; Nymark, S.; Hyttinen, J. Structural dynamics of tight junctions modulate the properties of the epithelial barrier. *PLoS ONE* **2019**, *14*, e0214876. [CrossRef]
16. Buschmann, M.M.; Shen, L.; Rajapakse, H.; Raleigh, D.R.; Wang, Y.; Wang, Y.; Lingaraju, A.; Zha, J.; Abbott, E.; McAuley, E.M.; et al. Occludin OCEL-domain interactions are required for maintenance and regulation of the tight junction barrier to macromolecular flux. *Mol. Biol. Cell* **2013**, *24*, 3056–3068. [CrossRef]
17. Krug, S.M.; Amasheh, S.; Richter, J.F.; Milatz, S.; Günzel, D.; Westphal, J.K.; Huber, O.; Schulzke, J.D.; Fromm, M. Tricellulin forms a barrier to macromolecules in tricellular tight junctions without affecting ion permeability. *Mol. Biol. Cell* **2009**, *20*, 3713–3724. [CrossRef]
18. Marcial, M.; Madara, J. Analysis of absorptive cell occluding junction structure-function relationships in a state of enhanced junctional permeability. *Lab. Investig. A J. Tech. Methods Pathol.* **1987**, *56*, 424–434.
19. Suzuki, H.; Nishizawa, T.; Tani, K.; Yamazaki, Y.; Tamura, A.; Ishitani, R.; Dohmae, N.; Tsukita, S.; Nureki, O.; Fujiyoshi, Y. Crystal structure of a claudin provides insight into the architecture of tight junctions. *Science* **2014**, *344*, 304–307. [CrossRef]
20. Suzuki, H.; Tani, K.; Tamura, A.; Tsukita, S.; Fujiyoshi, Y. Model for the architecture of claudin-based paracellular ion channels through tight junctions. *J. Mol. Biol.* **2015**, *427*, 291–297. [CrossRef]
21. Angelow, S.; Yu, A.S. Structure-function studies of claudin extracellular domains by cysteine-scanning mutagenesis. *J. Biol. Chem.* **2009**, *284*, 29205–29217. [CrossRef]

22. Krause, G.; Protze, J.; Piontek, J. Assembly and function of claudins: Structure–function relationships based on homology models and crystal structures. *Semin. Cell Dev. Biol.* **2015**, *42*, 3–12.
23. Alberini, G.; Benfenati, F.; Maragniano, L. A refined model of claudin-15 tight junction paracellular architecture by molecular dynamics simulations. *PLoS ONE* **2017**, *12*, e0184190. [CrossRef]
24. Zhao, J.; Krystofiak, E.S.; Ballesteros, A.; Cui, R.; Van Itallie, C.M.; Anderson, J.M.; Fenollar-Ferrer, C.; Kachar, B. Multiple claudin–claudin cis interfaces are required for tight junction strand formation and inherent flexibility. *Commun. Biol.* **2018**, *1*, 50. [CrossRef]
25. Samanta, P.; Wang, Y.; Fuladi, S.; Zou, J.; Li, Y.; Shen, L.; Weber, C.; Khalili-Araghi, F. Molecular determination of claudin-15 organization and channel selectivity. *J. Gen. Physiol.* **2018**, *150*, 949–968. [CrossRef]
26. Rosenthal, R.; Günzel, D.; Piontek, J.; Krug, S.M.; Ayala-Torres, C.; Hempel, C.; Theune, D.; Fromm, M. Claudin-15 forms a water channel through the tight junction with distinct function compared to claudin-2. *Acta Physiol.* **2019**, e13334. [CrossRef]
27. Rosenthal, R.; Günzel, D.; Krug, S.M.; Schulzke, J.D.; Fromm, M.; Yu, A.S. Claudin-2-mediated cation and water transport share a common pore. *Acta Physiol.* **2017**, *219*, 521–536. [CrossRef]
28. Song, X.; Jensen, M.Ø.; Jogini, V.; Stein, R.A.; Lee, C.H.; Mchaourab, H.S.; Shaw, D.E.; Gouaux, E. Mechanism of NMDA receptor channel block by MK-801 and memantine. *Nature* **2018**, *556*, 515. [CrossRef]
29. Jensen, M.Ø.; Jogini, V.; Borhani, D.W.; Leffler, A.E.; Dror, R.O.; Shaw, D.E. Mechanism of voltage gating in potassium channels. *Science* **2012**, *336*, 229–233. [CrossRef]
30. Ostmeyer, J.; Chakrapani, S.; Pan, A.C.; Perozo, E.; Roux, B. Recovery from slow inactivation in K+ channels is controlled by water molecules. *Nature* **2013**, *501*, 121. [CrossRef]
31. Li, Q.; Wanderling, S.; Paduch, M.; Medovoy, D.; Singharoy, A.; McGreevy, R.; Villalba-Galea, C.A.; Hulse, R.E.; Roux, B.; Schulten, K.; et al. Structural mechanism of voltage-dependent gating in an isolated voltage-sensing domain. *Nat. Struct. Mol. Biol.* **2014**, *21*, 244. [CrossRef]
32. Gumbart, J.; Khalili-Araghi, F.; Sotomayor, M.; Roux, B. Constant electric field simulations of the membrane potential illustrated with simple systems. *Biochim. Biophys. Acta (BBA)-Biomembr.* **2012**, *1818*, 294–302. [CrossRef]
33. Alberini, G.; Benfenati, F.; Maragniano, L. Molecular dynamics simulations of ion selectivity in a claudin-15 paracellular channel. *J. Phys. Chem. B* **2018**, *122*, 10783–10792. [CrossRef]
34. Torrie, G.M.; Valleau, J.P. Nonphysical sampling distributions in Monte Carlo free-energy estimation: Umbrella sampling. *J. Comput. Phys.* **1977**, *23*, 187–199. [CrossRef]
35. Maragliano, L.; Vanden-Eijnden, E.; Roux, B. Free energy and kinetics of conformational transitions from Voronoi tessellated milestoning with restraining potentials. *J. Chem. Theory Comput.* **2009**, *5*, 2589–2594. [CrossRef]
36. Faradjian, A.K.; Elber, R. Computing time scales from reaction coordinates by milestoning. *J. Chem. Phys.* **2004**, *120*, 10880–10889. [CrossRef]
37. Conrad, M.P.; Piontek, J.; Günzel, D.; Fromm, M.; Krug, S.M. Molecular basis of claudin-17 anion selectivity. *Cell. Mol. Life Sci.* **2016**, *73*, 185–200. [CrossRef]
38. Van Itallie, C.M.; Mitic, L.L.; Anderson, J.M. Claudin-2 forms homodimers and is a component of a high molecular weight protein complex. *J. Biol. Chem.* **2011**, *286*, 3442–3450. [CrossRef]
39. Rossa, J.; Ploeger, C.; Vorreiter, F.; Saleh, T.; Protze, J.; Günzel, D.; Wolburg, H.; Krause, G.; Piontek, J. Claudin-3 and claudin-5 protein folding and assembly into the tight junction are controlled by non-conserved residues in the transmembrane 3 (TM3) and extracellular loop 2 (ECL2) segments. *J. Biol. Chem.* **2014**, *289*, 7641–7653. [CrossRef]
40. Milatz, S.; Piontek, J.; Hempel, C.; Meoli, L.; Grohe, C.; Fromm, A.; Lee, I.F.M.; El-Athman, R.; Günzel, D. Tight junction strand formation by claudin-10 isoforms and claudin-10a/-10b chimeras. *Ann. N. Y. Acad. Sci.* **2017**, *1405*, 102–115. [CrossRef]
41. Gong, Y.; Renigunta, V.; Zhou, Y.; Sunq, A.; Wang, J.; Yang, J.; Renigunta, A.; Baker, L.A.; Hou, J. Biochemical and biophysical analyses of tight junction permeability made of claudin-16 and claudin-19 dimerization. *Mol. Biol. Cell* **2015**, *26*, 4333–4346. [CrossRef]
42. Piontek, A.; Rossa, J.; Protze, J.; Wolburg, H.; Hempel, C.; Günzel, D.; Krause, G.; Piontek, J. Polar and charged extracellular residues conserved among barrier-forming claudins contribute to tight junction strand formation. *Ann. N. Y. Acad. Sci.* **2017**, *1397*, 143–156. [CrossRef]

43. Irudayanathan, F.J.; Wang, N.; Wang, X.; Nangia, S. Architecture of the paracellular channels formed by claudins of the blood–brain barrier tight junctions. *Ann. N. Y. Acad. Sci.* **2017**, *1405*, 131–146. [CrossRef]
44. Ingólfsson, H.I.; Melo, M.N.; Van Eerden, F.J.; Arnarez, C.; Lopez, C.A.; Wassenaar, T.A.; Periole, X.; De Vries, A.H.; Tieleman, D.P.; Marrink, S.J. Lipid organization of the plasma membrane. *J. Am. Chem. Soc.* **2014**, *136*, 14554–14559. [CrossRef]
45. Zhao, G.; Perilla, J.R.; Yufenyuy, E.L.; Meng, X.; Chen, B.; Ning, J.; Ahn, J.; Gronenborn, A.M.; Schulten, K.; Aiken, C.; et al. Mature HIV-1 capsid structure by cryo-electron microscopy and all-atom molecular dynamics. *Nature* **2013**, *497*, 643. [CrossRef]
46. Grime, J.M.; Dama, J.F.; Ganser-Pornillos, B.K.; Woodward, C.L.; Jensen, G.J.; Yeager, M.; Voth, G.A. Coarse-grained simulation reveals key features of HIV-1 capsid self-assembly. *Nat. Commun.* **2016**, *7*, 11568. [CrossRef]
47. Yu, H.; Schulten, K. Membrane sculpting by F-BAR domains studied by molecular dynamics simulations. *PLoS Comput. Biol.* **2013**, *9*, e1002892. [CrossRef]
48. Koldsø, H.; Sansom, M.S. Organization and dynamics of receptor proteins in a plasma membrane. *J. Am. Chem. Soc.* **2015**, *137*, 14694–14704. [CrossRef]
49. Simunovic, M.; Evergren, E.; Golushko, I.; Prévost, C.; Renard, H.F.; Johannes, L.; McMahon, H.T.; Lorman, V.; Voth, G.A.; Bassereau, P. How curvature-generating proteins build scaffolds on membrane nanotubes. *Proc. Natl. Acad. Sci. USA* **2016**, *113*, 11226–11231. [CrossRef]
50. Rajagopal, N.; Irudayanathan, F.J.; Nangia, S. Computational Nanoscopy of Tight Junctions at the Blood–Brain Barrier Interface. *Int. J. Mol. Sci.* **2019**, *20*, 5583. [CrossRef]
51. Marrink, S.J.; Risselada, H.J.; Yefimov, S.; Tieleman, D.P.; De Vries, A.H. The MARTINI force field: Coarse grained model for biomolecular simulations. *J. Phys. Chem. B* **2007**, *111*, 7812–7824. [CrossRef]
52. Irudayanathan, F.J.; Trasatti, J.P.; Karande, P.; Nangia, S. Molecular architecture of the blood brain barrier tight junction proteins—A synergistic computational and in vitro approach. *J. Phys. Chem. B* **2015**, *120*, 77–88. [CrossRef]
53. Milatz, S.; Piontek, J.; Schulzke, J.D.; Blasig, I.E.; Fromm, M.; Günzel, D. Probing the cis-arrangement of prototype tight junction proteins claudin-1 and claudin-3. *Biochem. J.* **2015**, *468*, 449–458. [CrossRef]
54. Irudayanathan, F.J.; Wang, X.; Wang, N.; Willsey, S.R.; Seddon, I.A.; Nangia, S. Self-Assembly Simulations of Classic Claudins—Insights into the Pore Structure, Selectivity, and Higher Order Complexes. *J. Phys. Chem. B* **2018**, *122*, 7463–7474. [CrossRef]
55. Gumbart, J.C.; Roux, B.; Chipot, C. Standard binding free energies from computer simulations: What is the best strategy? *J. Chem. Theory Comput.* **2012**, *9*, 794–802. [CrossRef]
56. Gumbart, J.C.; Roux, B.; Chipot, C. Efficient determination of protein–protein standard binding free energies from first principles. *J. Chem. Theory Comput.* **2013**, *9*, 3789–3798. [CrossRef]
57. Woo, H.J.; Roux, B. Calculation of absolute protein—ligand binding free energy from computer simulations. *Proc. Natl. Acad. Sci. USA* **2005**, *102*, 6825–6830. [CrossRef]
58. Periole, X.; Knepp, A.M.; Sakmar, T.P.; Marrink, S.J.; Huber, T. Structural determinants of the supramolecular organization of G protein-coupled receptors in bilayers. *J. Am. Chem. Soc.* **2012**, *134*, 10959–10965. [CrossRef]
59. Provasi, D.; Johnston, J.M.; Filizola, M. Lessons from free energy simulations of δ-opioid receptor homodimers involving the fourth transmembrane helix. *Biochemistry* **2010**, *49*, 6771–6776. [CrossRef]
60. Johnston, J.M.; Filizola, M. Differential stability of the crystallographic interfaces of mu-and kappa-opioid receptors. *PLoS ONE* **2014**, *9*, e90694. [CrossRef]
61. Castillo, N.; Monticelli, L.; Barnoud, J.; Tieleman, D.P. Free energy of WALP23 dimer association in DMPC, DPPC, and DOPC bilayers. *Chem. Phys. Lipids* **2013**, *169*, 95–105. [CrossRef]
62. Baaden, M.; Marrink, S.J. Coarse-grain modelling of protein–protein interactions. *Curr. Opin. Struct. Biol.* **2013**, *23*, 878–886. [CrossRef]
63. Periole, X. Interplay of G protein-coupled receptors with the membrane: Insights from supra-atomic coarse grain molecular dynamics simulations. *Chem. Rev.* **2016**, *117*, 156–185. [CrossRef] [PubMed]
64. Nakamura, S.; Irie, K.; Tanaka, H.; Nishikawa, K.; Suzuki, H.; Saitoh, Y.; Tamura, A.; Tsukita, S.; Fujiyoshi, Y. Morphologic determinant of tight junctions revealed by claudin-3 structures. *Nat. Commun.* **2019**, *10*, 816. [CrossRef]

65. Piontek, J.; Winkler, L.; Wolburg, H.; Muller, S.L.; Zuleger, N.; Piehl, C.; Wiesner, B.; Krause, G.; Blasig, I.E. Formation of tight junction: Determinants of homophilic interaction between classic claudins. *FASEB J.* **2008**, *22*, 146–158. [CrossRef] [PubMed]
66. Li, Q.; Zhang, Q.; Wang, M.; Zhao, S.; Ma, J.; Luo, N.; Li, N.; Li, Y.; Xu, G.; Li, J. Interferon-γ and tumor necrosis factor-α disrupt epithelial barrier function by altering lipid composition in membrane microdomains of tight junction. *Clin. Immunol.* **2008**, *126*, 67–80. [CrossRef]
67. Chen-Quay, S.C.; Eiting, K.T.; Li, A.W.A.; Lamharzi, N.; Quay, S.C. Identification of tight junction modulating lipids. *J. Pharm. Sci.* **2009**, *98*, 606–619. [CrossRef]
68. Lambert, D.; O'NEILL, C.A.; Padfield, P.J. Depletion of Caco-2 cell cholesterol disrupts barrier function by altering the detergent solubility and distribution of specific tight-junction proteins. *Biochem. J.* **2005**, *387*, 553–560. [CrossRef]
69. Nusrat, A.; Parkos, C.; Verkade, P.; Foley, C.; Liang, T.; Innis-Whitehouse, W.; Eastburn, K.; Madara, J. Tight junctions are membrane microdomains. *J. Cell Sci.* **2000**, *113*, 1771–1781.
70. Rajagopal, N.; Irudayanathan, F.J.; Nangia, S. Palmitoylation of Claudin-5 Proteins Influences Their Lipid Domain Affinity and Tight Junction Assembly at the Blood–Brain Barrier Interface. *J. Phys. Chem. B* **2019**, *123*, 983–993. [CrossRef]

International Journal of
Molecular Sciences

Review

Computational Nanoscopy of Tight Junctions at the Blood–Brain Barrier Interface

Nandhini Rajagopal, Flaviyan Jerome Irudayanathan and Shikha Nangia *

Department of Biomedical and Chemical Engineering, Syracuse University, Syracuse, NY 13244, USA; nrajagop@syr.edu (N.R.); firudaya@syr.edu (F.J.I.)
* Correspondence: snangia@syr.edu; Tel.: +1-315-443-0571

Received: 15 October 2019; Accepted: 6 November 2019; Published: 8 November 2019

Abstract: The selectivity of the blood–brain barrier (BBB) is primarily maintained by tight junctions (TJs), which act as gatekeepers of the paracellular space by blocking blood-borne toxins, drugs, and pathogens from entering the brain. The BBB presents a significant challenge in designing neurotherapeutics, so a comprehensive understanding of the TJ architecture can aid in the design of novel therapeutics. Unraveling the intricacies of TJs with conventional experimental techniques alone is challenging, but recently developed computational tools can provide a valuable molecular-level understanding of TJ architecture. We employed the computational methods toolkit to investigate claudin-5, a highly expressed TJ protein at the BBB interface. Our approach started with the prediction of claudin-5 structure, evaluation of stable dimer conformations and nanoscale assemblies, followed by the impact of lipid environments, and posttranslational modifications on these claudin-5 assemblies. These led to the study of TJ pores and barriers and finally understanding of ion and small molecule transport through the TJs. Some of these in silico, molecular-level findings, will need to be corroborated by future experiments. The resulting understanding can be advantageous towards the eventual goal of drug delivery across the BBB. This review provides key insights gleaned from a series of state-of-the-art nanoscale simulations (or computational nanoscopy studies) performed on the TJ architecture.

Keywords: claudin; tight junctions; blood–brain barrier; in silico; drug discovery; membrane proteins; protein interactions; molecular dynamics

---

1. The Blood–Brain Barrier

The blood–brain barrier (BBB) is a selective permeability barrier that impedes the influx of potentially harmful blood-borne chemicals into the brain [1–4]. Although discovered by Ehrlich in 1885, the term blood–brain barrier was coined almost four decades later in the 1920s, after a series of reports by Lewandowsky, Goldmann, Stern, and Gautier [5–22]. Their collective work in vertebrate brain showed that tracers and toxins were unable to permeate the blood–brain interface; for example, trypan blue dye injected into the brain of a mouse remained confined to the brain, whereas the same dye injected outside the brain stained the entire body except the brain (Figure 1a). This permeability barrier is considered an evolutionary adaptation in vertebrates, and is vital for maintaining homeostasis in the CNS [1,5,21–23]. Unlike other organs, the brain lacks energy reserves [24–26]; therefore, for normal function, it requires a constant supply of nutrients from the bloodstream [27,28]. The brain's microvasculature regulates the transfer of essential nutrients while protecting the brain against small chemical imbalances, which can lead to serious disease pathologies [29–33].

A critical component of the BBB is a specialized physical barrier comprised of tight junctions (TJs) present along the apical side of endothelial cells [1,23,34]. These cell-to-cell adhesion structures act as gatekeepers of the paracellular pathway (Figure 1b) by regulating the passive diffusion of molecules and ions into the brain [23,35,36]. The establishment of TJs at the BBB leads to a high electric resistance

(>1000–3000 Ω cm^2) across the interface [37,38]. The presence of TJs also leads to cell polarization whereby the apical and basolateral sides of the endothelia are isolated [2,35]. The polarized endothelia control transcellular permeability by specific expression of transporters and receptor proteins in the apical side of the membrane (Figure 1b). In the event of disease pathologies, TJs are compromised, leading to the loss of barrier selectivity in both the transcellular and paracellular pathways [33,35].

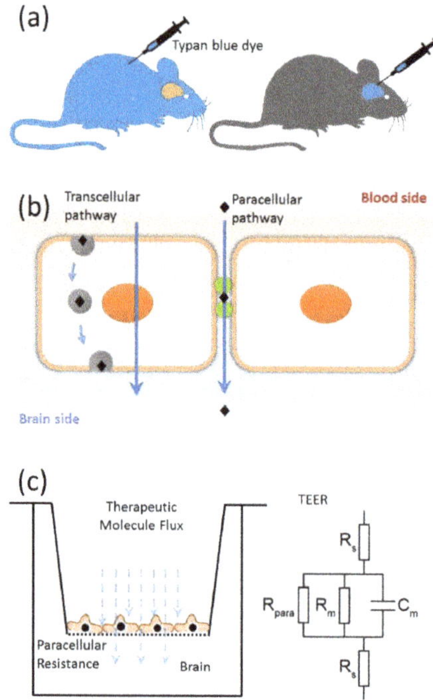

Figure 1. The blood–brain barrier. (**a**) Discovery of the blood–brain barrier (BBB), (**b**) pathways of molecular transport across the neurovascular endothelial barrier: transcellular (arrow through the cell) and paracellular (arrow through space between the two cells) and (**c**) in vitro characterization of the paracellular barrier strength of movement of therapeutic molecules (downward arrows) using transepithelial electric resistance (TEER) model.

Although the BBB is vital for neurological functions, its selective permeability is a deterrent to drug delivery into the brain [31,39–41]. The BBB obstructs effective treatment of neurodegenerative diseases such as Alzheimer's and Parkinson's, because therapeutics are unable to cross the barrier [39]. To develop new therapeutics, in vivo animal models provide an experimental environment that mimics the intricacies of human physiology. However, high cost and lack of complete translation of disease pathologies from animal models to humans make in vivo studies less effective for drug development. In vitro models are therefore being used early in a drug's development to investigate its permeability across the BBB [42–44]. Drug permeability is typically studied using cell cultures grown on permeable membranes in a transwell, where the transendothelial/epithelial electrical resistance (TEER) is used to measure the change in the electrical resistance across the cellular monolayer (Figure 1c) [37,45]. The TEER measurements evaluate the permeability and integrity of the endothelial cell layer [45,46], but the static cell culture cannot mimic the fluid characteristics of the BBB. More recently, microfluidic organ-on-chip in vitro models have been employed for the assessment of TJ barrier properties. Efforts are underway to develop BBB-on-a-chip that can recreate the human BBB using cells derived from the brain endothelial cells [47,48]. Additionally, in silico models of the BBB TJs have been developed to reliably compute the permeability of ions and small molecules across the paracellular barrier [49–52]. With advances in computer algorithms and hardware, strategies are being developed to design novel therapeutics that can cross the BBB. A concerted in vivo, in vitro, and in silico effort is underway to unravel the complexities of the BBB.

2. Tight Junctions: Role of Claudin-5 in the BBB

Claudin family of membrane proteins establishes TJs in endothelial and epithelial cells [53–55]. Claudins are expressed in vascular endothelial cells and all known epithelial cells throughout the body [56–58]. To date, 27 members of the claudin family are known to be functionally expressed in mammals [59,60]. They are classified into classic claudins (1–10, 14, 15, 17, and 19) and non-classic claudins (11–13, 16, 18, and 20–27) based on their sequence similarity and structural homology [59,61,62]. Most tissues express multiple members of the claudin family at the TJs [56,63–67]. Claudin proteins (~20–30 kDa) fold into a four-transmembrane helix bundle (TM1–4) with two extracellular loops (ECL1–2) and an intracellular loop (ICL; Figure 2) [60,62,68]. The lateral interactions mediated by TM and ECL domains of claudins within the membrane of the same cell are termed cis interactions. Cis interactions between multiple claudins occur in membranes of expressing cells. Subsequently, the claudins undergo trans assembly via head-on interactions of their ECL loops and form the macromolecular TJ assembly with adjacent claudin expressing cells. The TJ strands (often observable under freeze-fracture microscopy) are multi-component structures with claudins as the primary component alongside other TJ-associated proteins such as occludin, tricellulin, junctional adhesion molecules (JAMs), and zonula occludens (ZO-1,2) [53,60,61].

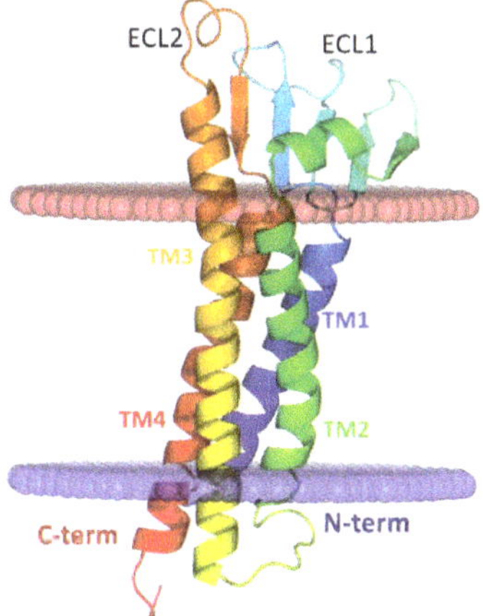

Figure 2. Homology modeled claudin-5 structure shows four transmembrane helices (TM1–4), extracellular loops (ECL1 and ECL2), and intracellular loop (ICL) in cartoon representation colored from N-terminal (blue) to truncated C-terminal (red). The membrane spanning region lies within the horizontal (red and blue) planes.

Claudin-5 is the key TJ protein at the BBB with an expression level that is much higher than claudin-1, -3, and -12 [69,70]. Knockout experiments have shown that claudin-5 is responsible for controlling the paracellular permeability of molecules up to ~800 Da [37,38,71]. In vitro biochemical analyses have suggested that claudin-5's ECL2 loop strongly contributes to the trans interactions [72,73]. Mutations to conserved residues in ECL2 lead to a marked increase in paracellular permeability because of compromised size selectivity [59,74]. Studies also show that claudin-5 knockout mice die within 10 h of birth, although their TJs remain impermeable to molecules >800 Da in size [71]. In contrast a double knockdown study of both claudin-5 and occludin showed enhanced permeability of 3–10 kDa tracers indicating that occludin serves a structural role in the BBB TJs [75,76].

3. Need for Computational Modeling

In silico models complement in vitro and in vivo approaches used to investigate biological systems. Computer models augment our ability to analyze aspects of biological systems that are too complex to be investigated by conventional experiments alone [77–80]. Computational approaches provide a molecular-level basis for experimentally observed biomacromolecular assembly, selectivity, dynamics, and macroscopic properties.

For example, X-ray crystallography of a membrane protein provides essential structural information about its conformation at low temperatures [81,82]. However, proteins participate in dynamic processes and have thermal motions at physiological temperatures that cannot be resolved by current experimental techniques [83,84]. In silico methods can readily build upon the crystal structure data to provide information about molecular displacements, conformational dynamics, and reorientation frequencies of proteins [80]. The resulting models can also provide insights into the intramolecular and intermolecular interactions of a protein in its native environment, including other proteins, lipids, ions, and solvents. Another area where computational methods have been transformative is in predicting the three-dimensional structure of membrane proteins. The hydrophobic nature of membrane proteins makes them experimentally unwieldy for crystallization and X-ray structure determination [83–85]. In recent years, advances in protein sequencing have enabled the utilization of sequence co-evolution information to infer and predict the native fold(s) of a protein to a high-level of accuracy that can be validated by experimental biochemical assays [86–90].

Advances in computational hardware also have facilitated the study of complex biological systems. In the last decade, molecular simulations of biological systems with millions of interacting particles over microsecond to millisecond timescales have been reported [64,91–116]. These simulations provide unprecedented insights into the system under investigation and bridge nanoscale data to macroscale experimental results.

4. Computational Toolkit

Computational methods, in the form of in-house computer programs, freely-distributed and commercial software suites, and online servers, offer an extensive toolkit that can be used to investigate different aspects of a biological system. Based on the desired qualitative or quantitative output, an array of computational models can be employed. Figure 3 summarizes a hierarchical computational approach relevant to the study of membrane proteins, using claudin-5 as an example.

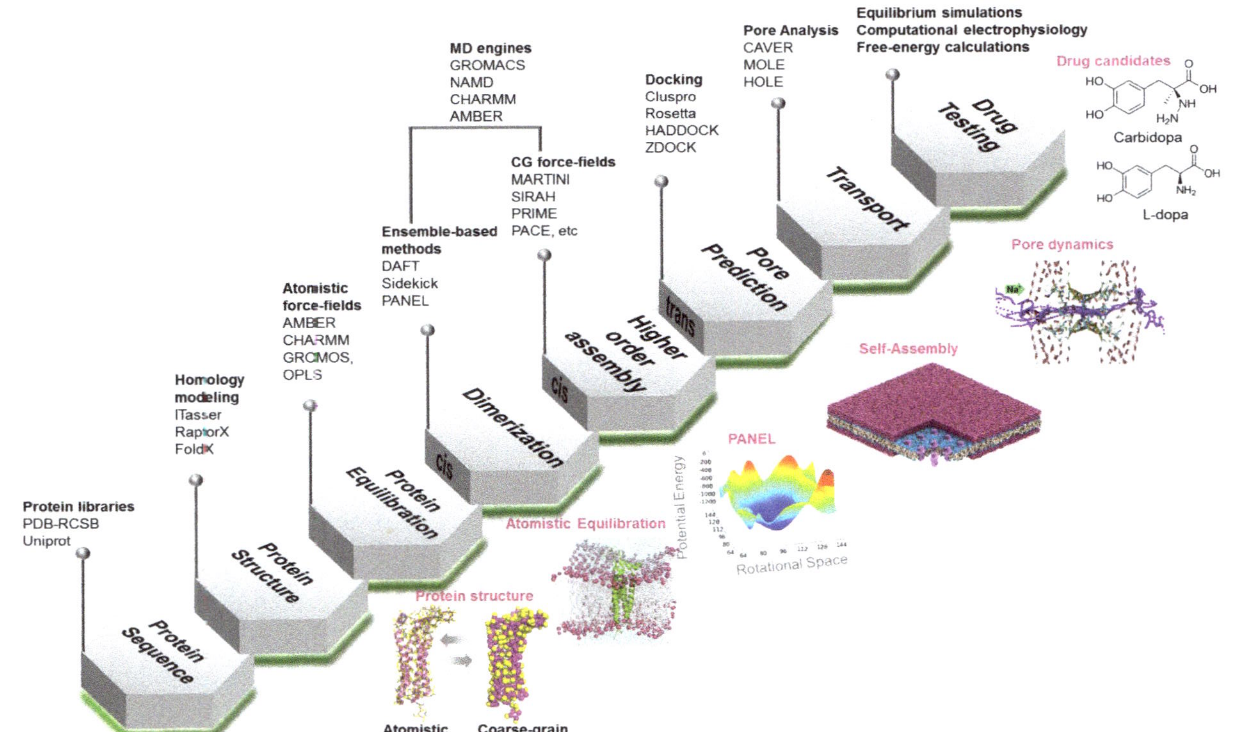

Figure 3. An overview of computational approaches and tools used to study the tight junctions at the blood–brain barrier. The stepwise research strategy included: obtaining protein sequence from online libraries; determining protein structure; equilibrating the protein in membrane lipids; investigating cis assembly (dimer and higher-order); investigating trans assembly for predicting tight junction barrier and pores; evaluating transport properties on ions and molecules; and testing tight junction modulators for enhanced drug delivery. The computational approaches (homology modeling, atomistic and coarse-grained molecular dynamics self-assembly simulations, PANEL, etc.) and tools (online libraries, webservers, force-fields, and software suites) associated with each step of our research strategy are also provided.

4.1. Protein Structure Prediction

Crystal structures of murine and human claudins mClaudin-15 (4P79) [68], mClaudin-19 (3X29) [117], hClaudin-4 (5B2G) [118], and mClaudin-3 (6AKE) [119] are available. Most of the currently available claudin structures have missing loop and c-terminal domains, due to crystallographic artifacts. To study a claudin that has not yet been structurally resolved, the first step is to construct its homology model [120–122]. Reliable homology models can be derived if the target and template structures have at least 30% residue identity spanning the entire length of the domain of interest [120,123]. Since most claudins share ~30% sequence identity, this technique can reliably model structures of the other members of the claudin protein family [59,62,124]. Multiple tools are available online that can build homology models; [125,126] among them the I-TASSER [127–129] and RaptorX [130,131] servers are well established in the structural biology field (Figure 3). Both servers take the primary sequence of the target protein as input and give as output the lowest energy 3D-modeled structure of the protein. In building homology models care should be taken in modeling the nonconserved domains. For example, the flexible c-terminal domains of claudin family of proteins are challenging to model because they are intrinsically disordered and adopt nonspecific secondary structures [132]. Homology modeling methods are designed to provide a correct global structure but do not optimize side-chain orientations or flexible-loop conformations [121,132–134]. Such details can be obtained by loop modeling techniques such as ModLoop and FoldX [135,136]. Once the 3D-structure has been computationally determined, the next step is to embed the claudin in the desired lipid membrane and optimize the structure using molecular dynamics (MD) [137] simulation at atomistic resolution.

4.2. Protein Dynamics

Simulations of biological systems are performed at multiple resolutions—atomistic, coarse-grained, and mesoscale [137–139]. In atomistic resolution, the motion of each atom is evaluated as a function of time, whereas, in the coarse-grain (CG) resolution, a group of atoms in an amino acid is mapped to a bead to evaluate its dynamics. Mesoscale models take coarse graining up another notch; clusters of amino acids are mapped to a superbead, and simple harmonic oscillators represent the interaction between those superbeads [140–142]. Since each resolution has merits and pitfalls, multiscale modeling approaches are increasingly being used to take advantage of the benefits available at each resolution [143]. Multiscale simulations are vital to interpreting experimental results and guiding new lines of experimental investigation [80,144–148].

The classical atomistic MD approach is to simulate each atom of the system using pairwise interatomic potentials to estimate the interaction strength of each atom with its remaining atoms in the system. The interaction energy of each atom in a system is modeled using a molecular force field, which comprises of both bonded (bond stretch, angle bend, dihedral rotation, and torsional) and non-bonded (electrostatic and van der Waal's) interaction terms (Figure 4). There are many force fields available in the literature for proteins, such as CHARMM [137], AMBER [149], GROMOS [150], and OPLS [151] among others. Each force field has a unique set of parameters that are optimized to reproduce experimental reference data. Therefore, selection of a force field is an important decision before starting an MD simulation for any system.

The all-atom simulations are restricted to small time-steps (1–4 fs) to capture the intra- and intermolecular motion of atoms over time. Atomistic simulations result in detailed structural, thermodynamic, and kinetic data that can be compared to experimental data. A typical biomolecular system comprises of millions of atoms, embedded in ionic media to mimic the physiological conditions, simulating biological processes over micro- or millisecond timescales. Although atomic- level molecular modeling provides specific interatomic interactions, its practical application is limited by computational resources.

Setting up membrane protein systems for molecular simulations is challenging when it comes to embedding a protein in multicomponent lipid membranes. To streamline the process, the recent development of publicly hosted interactive web-based servers (such as charmm-gui) [152] is enabling the generation of input files required for membrane protein simulations. Among the existing simulation packages are popular MD engines such as GROMACS [153,154], NAMD [155], AMBER [156,157], CHARMM [152,158,159], and visualization tools such as VMD [160], Pymol [161], and YASARA (Figure 3) [162–164].

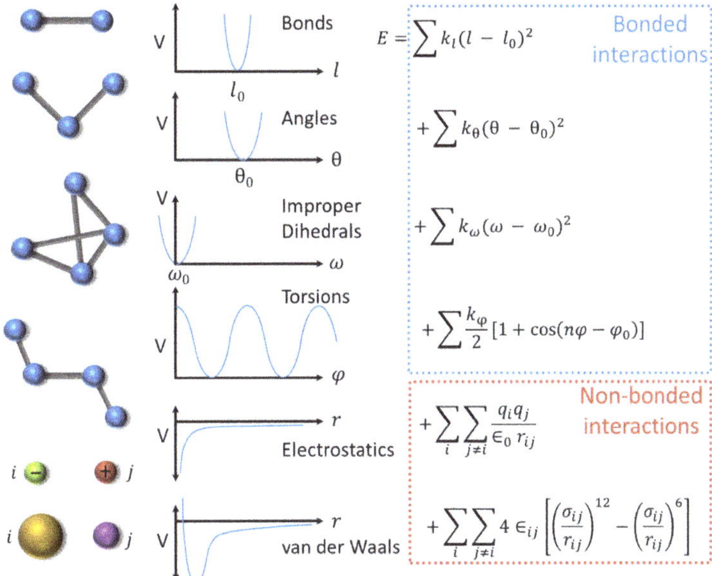

Figure 4. Typical form of a molecular force-field constituted by bonded (bond, angle, dihedral, and torsion) interactions and nonbonded (electrostatic and van der Waal's) interaction energy terms. The force filed parameter set include: equilibrium bond length (l_0), bond angle (θ_0), dihedral angle (ω_0), torsion angle (φ_0), and their respective force constants, k_l, k_θ, k_ω, and k_φ, as well as charge (q_i) on each atom i, dielectric constant (ϵ_0), strength of dispersion interactions (ϵ_{ij}), and contact distance (σ_{ij}) between the atoms i and j. The interatomic distance is represented by r_{ij}.

The use of CG simulations has increased in the last 10 years with the development of several force fields for a variety of biomolecules. Popular CG force fields include MARTINI [165–169], SIRAH [170,171], PRIME [172,173], and PACE [174,175]. Among these, MARTINI is the most popular for membrane protein simulations [165,166,176]. The MARTINI force field maps ~4 nonhydrogenic atoms into one bead; for example, a glycine amino acid residue is mapped as one bead [165,166,168]. By mapping a small number of atoms into a bead, CG resolution reduces the number of particles in a simulation and allows the system to be sampled less frequently with a larger time-step (20–40 fs). This approach enables simulation timescales that are 2 to 3 orders of magnitude larger than atomistic simulations. Thus, the advantages of CG simulations have spurred interest in simulations of larger sizes (hundreds of nanometers) and longer timescales (tens of microseconds) [177–180]. Others and we have rigorously demonstrated the ability of MD simulations to capture residue-level precision in membrane protein interactions [91,94,97,108,110,115].

4.3. Protein–Protein Interactions

4.3.1. Self-Assembly Simulations

Molecular self-assembly simulation is a powerful technique that mimics interactions among molecules over time [94,98,145,181]. These simulations capture the spontaneous organization of molecules dictated by intra- and intermolecular nonbonded interactions (van der Waals forces, electrostatics, hydrogen bonds, hydrophobic, and hydrophilic interactions) under near-equilibrium conditions. Self-assembly simulations are routinely used to study complex membrane protein–protein interactions. These simulations, however, have several limitations including high computational cost. Membrane protein assembly is particularly computationally expensive in part because protein diffusion requires much longer equilibration time than the relaxation of membrane lipids. Moreover, the simulations need to be performed over a sufficiently long time to obtain reliable data, although ensuring convergence of such simulations is a challenge. There is no guarantee that a 10–100 μs timescale is long enough for proteins to assemble [182,183]. Employing multiscale simulations mitigates the challenges of long timescale simulation.

4.3.2. Molecular Docking

Molecular docking, another popular method of studying protein associations, is a highly effective method to examine and rank interactions based on rigorous algorithms that evaluate empirical energetics data. Docking approaches are simple geometry-based methods where tens of thousands of protein–protein interfaces are generated via geometric rigid body motions followed by a scoring scheme where interfaces are ranked based on the energetics [184,185]. The energy function that is used in the scoring algorithm is often empirically modeled or derived from a knowledge-based method. Docking approaches tend to be fast because generating geometries of proteins is a simple task. Depending upon the type of protein–protein interaction, docking methods can give realistic results. Popular docking methods include Cluspro [186,187], Rosetta [188], HADDOCK [189,190], and ZDOCK [191] among others [192–195]. Cluspro has robust reproducibility, and it is the current state of the art in protein–protein docking. Rosetta is a more advanced software suite of protein modeling and analysis tools. The Rosetta docking protocol has been optimized for membrane proteins, and one can define membrane domain boundaries to guide the docking. However, despite the ease and speed of docking methods, static docking is not optimal for macromolecular assemblies that involve multiple proteins [196]. Moreover, for membrane proteins, the presence of the lipid environment significantly influences proteins' association patterns, but none of the currently available docking methods offers a realistic lipid environment for protein docking [197–202].

4.3.3. Protein Association Energy Landscape

An alternate strategy to study membrane protein association is to employ ensemble-based sampling methods. These methods use hundreds to a few thousand starting structures that each run independently to yield an unbiased sampling of the protein–protein association landscape. One such ensemble-based method, docking assay for transmembrane (DAFT) components [203] uses the population density of the resulting protein association conformations as a proxy for stability. Other ensemble-based methods use amino acid contact-based stability as a metric [204,205]; however, none of the available methods provides the association energies needed to determine the relative stability of protein–protein conformations obtained in the ensemble.

In our work, we used a combined stochastic sampling and MD approach to sample pairwise protein–protein interactions [206]. Taking advantage of in-plane diffusion of membrane proteins relative to each other in the lipid bilayer, we defined a two-dimensional rotational space of the interacting proteins. We adopted a stochastic sampling technique to generate thousands of seed geometries that uniformly sample the rotational space around a pair of interacting proteins. These seed geometries were then propagated independently to equilibrium using standard MD. The interaction energy associated with each conformation visited in the dynamics was computed to generate the protein association energy landscape (PANEL) [206]. By placing the proteins within the interacting distance of the seed geometries, we eliminated the need for protein diffusion, thereby reducing the simulation times for each system and boosting the PANEL performance by two orders of magnitude over classic self-assembly simulations. The PANEL method also provides an absolute measure of interaction energy for any pair of interacting proteins in dimeric or even in trimeric association [206]. We used the PANEL method to report the landscape of claudin-5 dimers (Figure 3), corroborating our previous self-assembly simulation results [115]. Furthermore, the robustness, computational affordability, and user-friendly implementation of the PANEL method makes it attractive for application to any membrane protein.

4.4. Pore Characterization and Paracellular Transport

The molecular architecture of the TJ pores and barriers are the determinants of the charge and size-selective transport of ions and molecules. Analysis of the pore's structural features is important to understand TJ selectivity. The structural characteristics of TJ pores (such as length, diameter, volume, and cross-sectional area) provide valuable insight into the size selectivity of the pore. Additionally, identification of pore-lining residues in the channel walls and the bottleneck regions of the pore offer significant clues about their charge selectivity. Several software tools, such as CAVER [207], MOLE [208], HOLE [209], MOLAXIS [210,211], and POREWALKER [212], (each based on slightly different algorithms) provide structural analysis of protein pores (Figure 3).

Paracellular transport involves calculation of the free energy profiles of the translocation of a substrate through the pore. This calculated energy directly correlates to the permeability of the substrate across the pore [213,214]. The free energy profile can be obtained from the advanced sampling techniques of umbrella sampling [215] or metadynamics [216–218]. In the umbrella sampling technique, where multiple frames or conformations of a solute are generated along the length of the pore, an external force-constant is applied on the solute to restrain its position. Based on the amount of force applied, the free energy profile along the transport path can be evaluated [215,219]. An alternative method to determine the free energy profile is the recently introduced metadynamics approach where the system is biased gradually such that the ion translocation is made feasible within the affordable timescale of MD simulation. Metadynamics is available in GROMACS through the PLUMED plugin [220].

A simplistic equilibrium MD simulation can also be set up with different ion concentrations to observe the number of ion translocation events that occur during the simulation. This method may provide insights about preferred ion transport under unbiased, equilibrium conditions; however, the limitation of the method is the timesale in which an ion transport event will occur may not be within the simulaion time frame. To accelerate the ion transport event, computational electrophysiology has emerged as a method that involves biasing the molecular system by creating localized charge imbalances in system, which induces ion transport [221,222]. The small charge imbalances provide biologically realistic electrochemical gradients across the pore and facilitate ion transport.

5. Computational Nanoscopy of Claudin-5 in BBB Tight Junction Architecture

Claudin proteins embedded in the lipid bilayer undergo cis assembly via their transmembrane helical domains, or by the extracellular loops, or by both [94,115]. In general, protein–protein cis interactions are mediated by a multicomponent mixture of saturated and unsaturated lipids with variety of charged head groups and lipid tail lengths [91,106,201,202,223]. The helical bundle of membrane proteins has a rigid tertiary structure that provides stability to the protein to reside in the membrane environment. This rigidity in turn affects the local lipid environment and manifests as a hydrophobic mismatch between the protein and the lipids [198,224,225]. Hydrophobic mismatch is energetically unfavorable and serves as a driving force for the embedded protein to find interactions that would minimize its hydrophobic mismatch with the lipids and undergo cis assembly [103,226–230].

There are multiple aspects of claudin cis and trans assembly that influence the TJ architecture. The cis-assembled claudins can provide the template for the trans interactions [97,231,232]; therefore, the study of cis interfaces is vital to understand the TJ architecture. Besides, claudins are reported to be post-translationally modified (PTM) with phosphorylation and palmitoylation. Of these PTMs, studies suggest that palmitoylation alters recruitment of claudin proteins in the TJs and TJ permeability [91,233,234]. Since most aspartate–histidine–histidine–cysteine (DHHC) enzymes that assist in palmitoylation [235] are thought to be localized to the endoplasmic reticulum (ER) [198,236], it is likely that claudin cis interactions that are influenced by palmitoylation are localized in the ER. However, DHHC enzymes have been reported in other organelles and in the plasma membrane [237], which makes both the location of claudin-5 palmitoylation in the secretory pathway and the cis dimerization ambiguous. It is not surprising that the complexity of the BBB TJ architecture has been a deterrent in unraveling the mystery of the size and charge selectivity of the TJ. A list of basic questions that have confounded experts in the TJ research field include:

(1) How do claudin proteins form contiguous strands? Do the claudin proteins have preferred cis conformations?
(2) Where along the secretory pathway do claudin-5 proteins undergo cis assembly?
(3) How do posttranslational modifications of claudin-5 influence the cis interactions?
(4) How are TJ pores and barriers formed?
(5) How do BBB TJs get their size and change selectivity?

To provide a new perspective to these long-standing challenges, we developed tools for in silico investigation of the TJ claudin-5 proteins. In a series of publications over the past few years [91,94,97,115,206], we have reported on the various aspects of claudin-5 BBB TJ architecture. Here, we reviewed the outcomes of these five lines of inquiry (Figure 5).

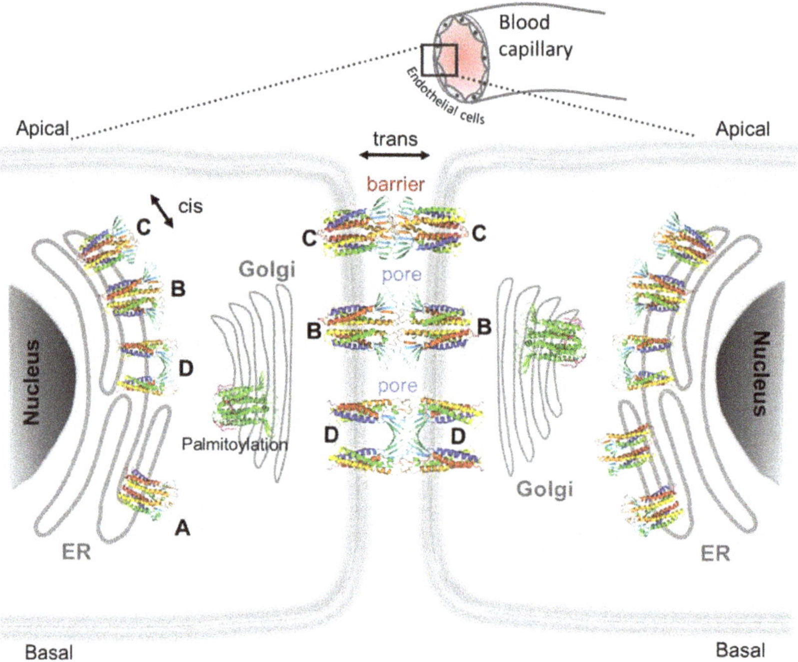

Figure 5. A schematic view of claudin-5's cis and trans assembly along the secretory pathway. The cis dimerization of claudin-5 proteins in endoplasmic membrane (ER) and Golgi complex membrane can result in multiple conformations that are categorized as dimers A, B, C, and D [94,97,115]; posttranslational modification can influence the relative stability of these dimers [91]; the trans interaction of C dimer pair results in a paracellular barrier, whereas trans interaction of B–B or D–D dimer pairs result in paracellular pore.

5.1. Claudin-5 Forms Symmetric and Asymmetric Dimer Conformations that Form Contiguous Strands

The cis interactions among claudins, determined by computational methods, corroborate results of biochemical assays [115]. The claudin-5 cis interactions were studied using both the self-assembly and the PANEL approaches [115,206].

The self-assembly simulations, performed in CG resolution, suggest that claudin-5 monomers embedded in a simple mixture of saturated and unsaturated lipids readily form dimers (within 400–500 ns) [115]. These dimers subsequently (in 5–10 μs) form contiguous strands and high-order assemblies (Figure 6a). For example, in a system composed of 64 claudin-5 monomers, a contiguous strand of 36 monomers was observed with a ring-like pentamer (Figure 6b), which match the size description of ~10 nm particles observed under freeze fracture [238–240]. The claudin strands present at the end of the 10 μs simulation were extracted and reverse mapped to atomistic resolution for conformational and structural analysis (Figure 6c). A pair-wise conformational analysis of the self-assembled claudins revealed four, most prevalent conformations (labeled dimer A–D) that concurrently exist to form the strands. Several repeat simulations have been performed that all show formation of dimers A–D although in varying populations. Since it is practically impossible to achieve convergence in self-assembly simulations due to finite timescales, we were unable to provide converged relative populations of each of the four dimers. However, these simulations indicate that a dimer is the smallest stable unit of the self-assembled strand and that the cis assembly can occur in the absence of trans interacting partners from the adjacent cell. In an experimental study, claudin-2 cis dimerization was also observed independent of cell-surface expression and intercellular cross-linking [241].

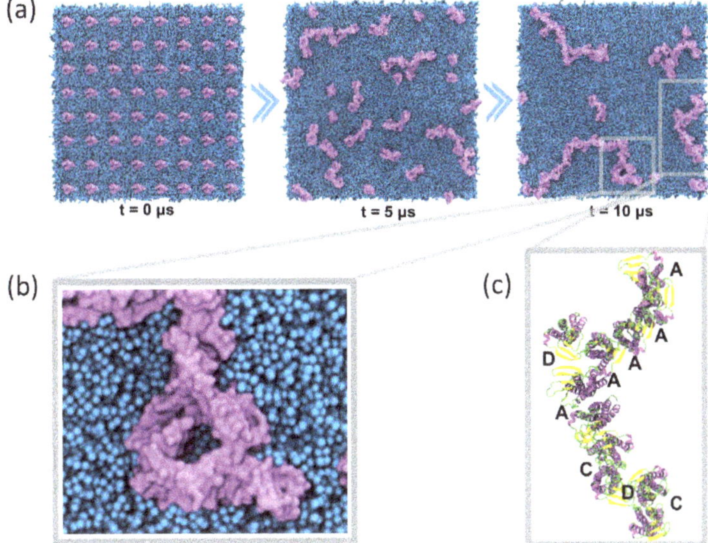

Figure 6. Claudin-5 strand formation in coarse-grain molecular dynamics (CG MD) self-assembly simulations. (**a**) Snapshots (top view) of claudin-5 (purple beads) assembly in lipid (cyan beads) membrane at $t = 0$, 5, and 10 μs, (**b**) close up view of a self-assembled claudin-5 pentamer, and (**c**) reversed-mapped strand showing dimer A, C, and D conformations.

To delve deeper into the stability of the claudin-5 dimers, we developed the PANEL approach [206]. The PANEL plot for claudin-5 provided a comprehensive sampling of dimer conformations and a clear visualization of the entire energy landscape as a function of the orientation angles (Figure 7). The claudin-5 dimer interaction energies range from 0 (least stable) to approximately −1500 kJ mol^{-1} (most stable) [206]. The PANEL contour plot shows multiple low energy conformations. Conformations along the diagonal (Figure 7b) represent symmetric dimers, while those off the diagonal are asymmetric dimers. It was reassuring that the key symmetric Dimers B, C, and D, also observed in self-assembly and in experiments, are shown in the minimum energy PANEL plot. Dimer B is a TM3:TM3-mediated dimer, while Dimer C is mediated by TM4:TM4, and Dimer D, the symmetric ECL1-mediated interaction, also proposed by Suzuki et al. [242]. There are also several asymmetric dimers (Dimer A), which are marked as Dimers

A1, A2, and A3 in the PANEL plot. Dimer A1 corresponds to the linear arrangement observed in the claudin-15 crystal structure [242]. The relative orientation of the claudins in each of the Dimers A to D (Figure 7b) is described in our earlier work [115,206]. Although the presence of claudin dimers and higher order complexes has been reported in several experimental studies [73,240,241,243,244], isolating intact dimers or higher order complexes still remains a challenge. Use of computational approach provides a direct look into the structural details of these claudin interfaces.

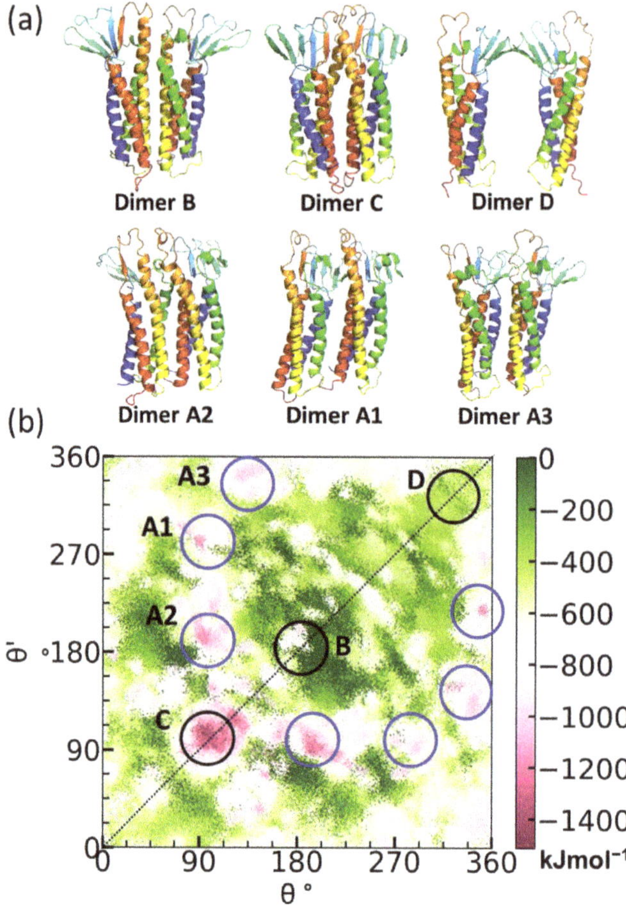

Figure 7. Claudin-5 cis dimerization. (**a**) Key cis dimers formed in claudin-5, reverse mapped to atomistic resolutions, and (**b**) PANEL plot showing location of low energy dimers A1, A2, and A3 (blue circles); and dimers B, C, and D (black circles).

5.2. Primary Cis Interactions Occur Early in the Secretory Pathway

Like other TJ proteins, claudins are secreted in the endoplasmic reticulum (ER) and reach the plasma membrane via the Golgi complex [231] Along this secretory pathway, the membrane composition changes drastically [245–247] To mimic the complexity of cell organelles encountered by a typical membrane protein during its secretory pathway, we performed claudin-5 self-assembly in three model membrane compositions: ER, cholesterol-enriched endoplasmic reticulum (ERc), and plasma membrane (PM) [91]. The ER membrane model constituted a binary mixture of saturated and unsaturated lipids in 1:1 ratio; the ERc was a 2:2:1 ternary mixture of saturated lipid, unsaturated lipid, and cholesterol; and the PM model comprised ten lipid components including gangliosides, sphingomyelin, and cholesterol [91].

Our findings indicate that claudin-5 monomers are more likely to form dimers in simple membranes like the ER where saturated and unsaturated phospholipids are the principal components. As claudins move to the Golgi complex, trans-Golgi vesicles, and the plasma membrane, the higher cholesterol concentrations in these membranes limit claudin-5 assembly. We observed that an increase in membrane complexity was negatively correlated with the length of claudin-5 strands (Figure 8a) [91].

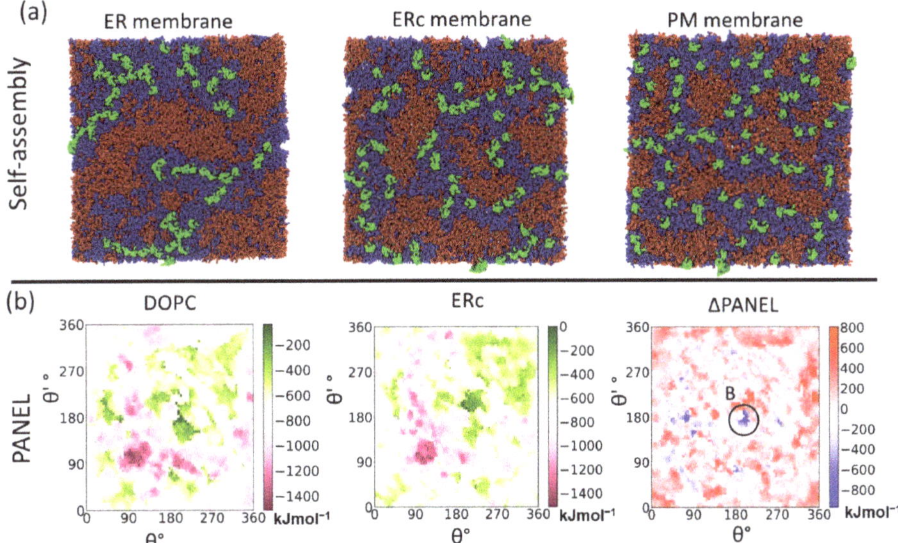

Figure 8. Effects of lipid complexity on claudin-5 dimerization. (**a**) Snapshots (top view) of self-assembled claudins (green) embedded in ER, cholesterol-enriched endoplasmic reticulum (ERc), and plasma membrane (PM) membranes with saturated lipids (red), unsaturated (blue), and cholesterol (white). (**b**) PANEL plots generated for claudin-5 dimers embedded in symmetric DOPC bilayer and ERc membrane. The ΔPANEL plot shows stabilization (blue) and destabilization (red) of the dimer geometries due to change in lipid environment DOPC to ERc. The stabilized dimer B location is marked (circle, black) on the ΔPANEL plot.

A comparison of PANEL plots for claudin-5 dimers in a simple unsaturated di-oleoyl-phosphatidylcholine (DOPC) lipid membrane and ERc membrane shows the impact of the lipid composition on dimer stability (Figure 8a). The effect is quantified by taking the difference in interaction energies (ΔPANEL) between corresponding geometries (θ, θ') on the two plots. A vast majority of the dimer interactions were destabilized in ERc membrane, although a few conformations did show more favorable interactions. In particular, Dimer B conformation (mediated via TM3:TM3 interaction) is stabilized. Overall, our results suggest that dimerization is more favorable in conditions existing in the early stages of the secretory pathway.

5.3. *Posttranslational Modifications Alter Relative Stability of cis Dimers*

Several studies have indicated the significance of palmitoylation in the recruitment of specific claudins to TJ and the consequent effect on the TJ permeability [233,234]. Since palmitoylation and depalmitoylation are highly labile reactions, the evaluation of these modifications on claudin assembly is experimentally challenging [237]. However, in silico methods provide the necessary spatial and temporal resolutions required to study these posttranslational modifications.

The presence of palmitoyl chains at the protein–lipid interface introduces variations in interfacial properties, effecting both the extent of dimerization as well as the distribution of specific dimer interfaces. Our results show that palmitoylation increases cholesterol and saturated lipid accumulation around the claudins and enhances cis interactions to the Dimer B interface while disrupting Dimer C largely due to steric factors (Figure 9a,b) [91]. The impact of palmitoylation on claudin-5 cis dimers is evident from the ΔPANEL landscape plotted as the difference between palmitoylated claudin (claudin-5P) and non-palmitoylated claudin-5 PANEL plots (Figure 9c). The ΔPANEL plot shows the stabilization of Dimer B and the destabilization of Dimer C regions. The change in dimer stability upon palmitoylation indicates that posttranslational modification can be a potential strategy to control paracellular permeability [91,97,233,234,237,248,249].

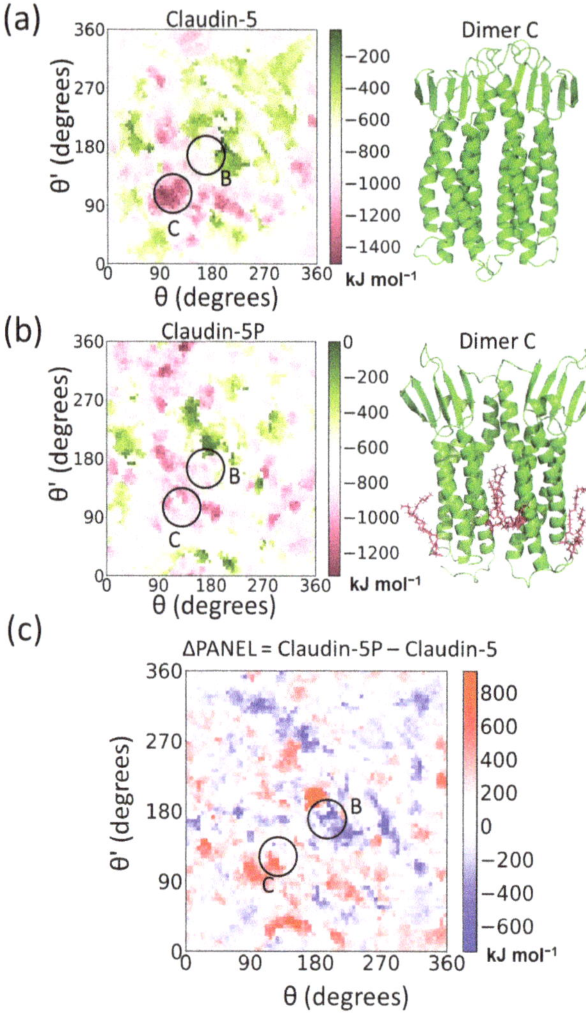

Figure 9. Effects of palmitoylation on claudin-5 cis dimerization. (**a**) PANEL plot Figure claudin. and dimer C conformation (green ribbon), (**b**) PANEL plot for claudin-5P and dimer C conformation (green ribbon) with palmitoylation (pink sticks), and (**c**) ΔPANEL showing the difference in stability upon palmitoylation. The location of dimers B and C are marked by circles (black) on all plots.

5.4. Cis Dimers Act as Precursors to Form Trans Interfaces

Trans interactions are formed when claudins on adjacent cells interact head-on to form the macromolecular TJ assembly. We adopted both CG- and atomistic-level simulations to decipher the nature of trans interfaces. Numerous challenges were posed both on conceptual and computational resource fronts while designing the trans assembly simulations.

To mimic paracellular space, our initial studies of trans interactions used two membrane patches with multiple (64) equidistant claudin-5 proteins [97]. They were placed head-on, such that the ECL loops from either membrane could interact within the gap between the two membranes (Figure 10). Sampling 20 μs of trans self-assembly took about 12 days on a supercomputer using 20 high performance computational nodes. Despite the intensive computation, the simulations represented only a small membrane patch that contained 128 claudin monomers across the two membranes. In contrast to a single membrane simulation, trans assembly simulations were unable to capture a few cis dimeric interactions that were prevalent in the single membrane self-assembly simulations. The simulations were intentionally designed so that a diffusing claudin-5 monomer would encounter a trans interacting partner before encountering a cis interacting partner; this ensured generation of more trans interaction variants. The absence of cis dimers rendered this a monomer-driven, trans interaction assembly [97]. Multiple pore-like assemblies were observed in these simulations showing ECL contacts stabilized by hydrogen-bonds and salt bridges. These trans pairs also strongly resembled pore models suggested in previous studies [72–74,250]. However, this model represented a very low fraction of the actual trans interactions, which span the entire perimeter of a cell and involve tens of millions of protein monomers.

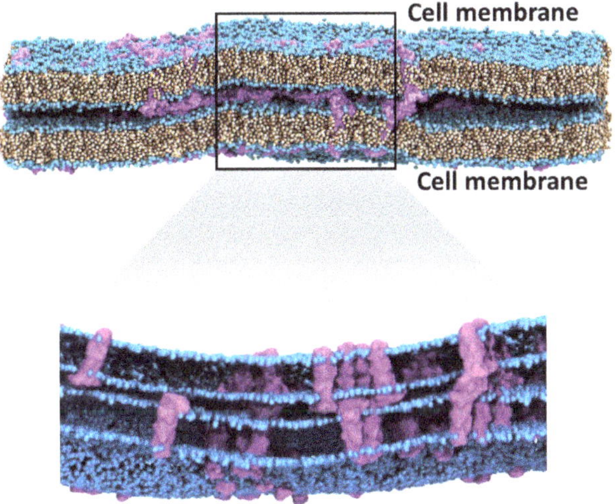

Figure 10. Claudin-5 assembly in two adjacent membranes. A 20 μs snapshot of claudin-5 (purple) trans and cis assembly in a coarse gained MD simulation. The zoomed-in view shows the interacting claudins embedded in the membrane (lipid head groups (blue); lipid tails (light brown). Water and ions in the system are not shown for clarity.

From the insights obtained from the cis and trans self-assembly simulations a dimer-guided trans interaction study was carried out using molecular docking approaches. This method yielded highly promising pore-forming trans interfaces. Moreover, the formation of cis dimers before trans interactions has been suggested in several findings [231,241,251,252]. Based on our work, we suggest that the cis dimers act as precursors for forming trans interfaces [91,97,115]. The fact that cis assembly simulation also self-assembles into a dimer in shorter timescales, lends support to that cis interactions may act as the template to guide the trans assembly.

The findings from these simulations were able to fill significant gaps in our understanding of the structure–function relationships between claudin monomers. Although several questions still need to be answered, and further investigations need to be performed to achieve a comprehensive understanding of the TJ architecture, our current understanding has been made possible by the an array of techniques in the computational toolkit.

5.5. Sub-Nanometer Wide Claudin-5 TJ Pores, Lined by Charged Residue Determine Pore Selectivity

For claudin-5, two pore models, mediated by different cis dimers, have been reported based on trans interaction simulations and mutation analyses of cis dimers [97]. The pore I model is a face-to-face stacking of the dimer D interface. This pore resembles the proposed claudin-15 pore [242]. The pore II model was based on the Dimer B interactions observed in claudin-5. These models were generated using docking methods in combination with symmetric refinement to arrive at the pore structures based on Dimers D and B (Figure 11). Fundamental pore features, such as diameter, length of the pore, pore bottleneck diameter, and the pore lining residues, were characterized.

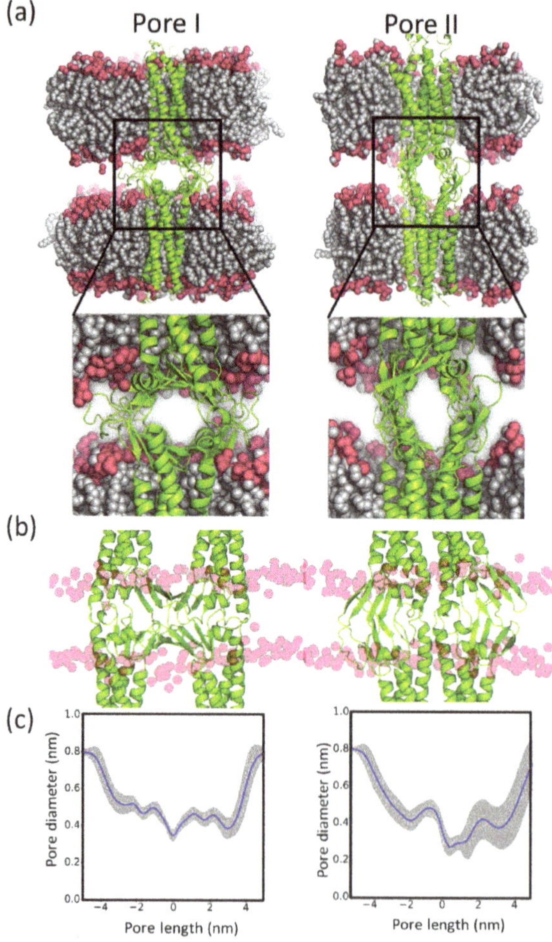

Figure 11. Model for claudin-5 pore I (left) and pore II (right). (**a**) Front-view, (**b**) side-view, and (**c**) pore diameter along the length of the pore.

Pore characteristics provide a general idea of the nature of the pore. However, evaluating the energetics of molecular transport along these pores is essential to further enhance our understanding of paracellular selectivity of the BBB's TJs. Using advanced sampling methods, translocation of water and glucose was reported across claudin-5 pore model formed by Dimer B. These simulations showed that the claudin-5 Dimer B pore allows water to translocate without energy barriers but had a higher energy barrier for glucose translocation compared with the GLUT1 glucose transporter [97,253,254]. Further extension of this work revealed that claudin-2 is selective to sodium ions with multiple observations of ion translocation under equilibrium conditions. Similar approaches are gaining attention within the field (e.g., claudin-15 pores probed using MD simulations) [255–257]. The more comprehensive undertaking of looking at ion transport in pores formed by different claudins is still in its infancy, but several computational groups are developing the essential methods needed to pursue this line of inquiry. In the meantime, the proposed pore structures present opportunities for further refinement to obtain better insights about their structural as well as functional aspects.

6. Additional Factors that Influence Tight Junctions

Computational methods can efficiently study cis and trans interactions of claudins at the molecular level. Claudins and other TJ proteins exist in a chemical environment that varies in type and amount of proteins and lipids, hydrophobicity, protein lipidation states, and other aspects [63,200,225,231,246,247]. These physiological factors contribute to the final TJ architecture. Moreover, TJs are known to be dynamic entities that undergo constant structural change [258,259]. Thus, to understand paracellular transport in its entirety, other factors need to be considered. A few key aspects important to TJ formation are discussed below.

6.1. Higher Order Claudin Aggregates

Several studies have reported the presence of claudin aggregates (in addition to strands) [61,118], suggesting that higher-order aggregates such as trimers, tetramers, or hexamers may be formed. *Clostridium perfringens* enterotoxin, which causes food poisoning, has been reported to form higher-order aggregates upon interacting with claudins. We proposed that the trimeric form of the enterotoxin could potentially bind to the trimeric form of claudins in the intestinal epithelium [94,117]. In another recent report of claudin-11 TJs, the carbon replica method revealed an interesting view of consistent double-stranded structures [260]. These reports emphasize that in order to fully understand the architecture of the TJ strands, it is important to consider the higher-order interfaces in addition to dimers.

6.2. Involvement of Other TJ Proteins: Key to TJ Structural Stability

In addition to claudins, proteins from approximately 80 other families are known to be involved in the TJ protein complex assembly including occludins and JAMs [53,54,60]. Occludins and JAMs contribute to the structural stability of TJs. Although these proteins are not known to affect directly TJs' selective permeability or barrier function, their role in maintaining the structural integrity of the TJ strongly suggests they form interfaces with claudins in order to support the strand. This may lead to masking of some claudins' interfacing surfaces while binding to others, thereby redistributing the relative populations of dimer interfaces. The study of occludins and JAMs is crucial to provide a complete perspective of TJs due to their significant presence within the TJ assembly.

6.3. Cytoskeletal Involvement

Claudins and other TJ membrane proteins are dynamically regulated by scaffolding proteins (zonula occludens-1/2/3) and actin cytoskeleton [53,231]. Although studying claudins individually provides insights about their interface forming behavior, thereby helping us to predict the architecture and nature of the corresponding TJ strands, continuous regulation by the cytoskeleton may induce several factors such as localized redistribution of claudin populations, or regulation of biophysical properties such as protein tilting, membrane anchoring, and so on. Since the cytoskeletal regulation tends to be highly dynamic and specific to the needs of the cell, deciphering their roles remains one of the greatest challenges.

7. Future Work

Future work on the BBB research will be directed towards the development of strategies to enable and enhance drug delivery to the brain. A fundamental understanding of the molecular construct that guards the blood–brain interface has been obtained through computational tools, which have enabled the design of in silico models to explore and the understanding of the architecture of this interface. One future research goal is to identify small molecules that can act as TJ modulators for use in screening claudin-5 trans interactions, reducing specificity, increasing pore size, and enhancing pore permeability in a controlled and reversible manner. Having the molecular-level pore structure is foundational to this endeavor. We envisioned that in vitro studies would be designed in the future to validate the predictions of the computational models. Other future work includes pursuing an integrated computational and in vitro/in vivo study of multiple drug pathways for neurotherapeutics, evaluating molecular drug interactions, obtaining the dynamics of transport of potential drug molecules with the vision of quantifying drug permeability through the BBB.

Author Contributions: N.R., F.J.I., and S.N.; conceptualization, S.N., N.R., and F.J.I.; methodology, S.N., N.R., and F.J.I.; software, N.R. and F.J.I.; validation, N.R. and F.J.I.; formal analysis, N.R. and F.J.I.; investigation, N.R. and F.J.I.; data curation, N.R. and F.J.I.; writing—original draft preparation, N.R. and F.J.I.; writing—review and editing, S.N., N.R., and F.J.I.; visualization, S.N., N.R., and F.J.I.; supervision, S.N.; project administration, S.N.; funding acquisition, S.N.

Funding: This research was funded by the CAREER CBET-1453312 grant from the National Science Foundation. Computational resources for the work were provided by the following groups—Information and Technology Services at Syracuse University; Anton 2, provided by the Pittsburgh Supercomputing Center (PSC) through Grant R01GM116961 from the National Institutes of Health (NIH); Extreme Science and Engineering Discovery Environment (XSEDE), which is supported by National Science Foundation grant number ACI-1053575.

Conflicts of Interest: The authors declare no conflict of interest.

Abbreviations

BBB	Blood–brain barrier
CG	Coarse grained
Claudin-5P	Palmitoylated claudin
DAFT	Docking assay for transmembrane
DHHC	Aspartate-Histidine-Histidine-Cysteine
DOPC	1,2-dioleoyl-sn-glycero-3-phosphocholine
ECL	Extracellular loop
ER	Endoplasmic Reticulum
ERc	Cholesterol enriched Endoplasmic Reticulum
ICL	Intracellular loop
JAMS	Junctional adhesion molecules
MD	Molecular dynamics
PANEL	Protein Association Energy Landscape
TEER	Transendothelial/epithelial electrical resistance
TJ	Tight junction
TM	Transmembrane loop
ZO	Zonula Occuldens

References

1. Abbott, N.J. Comparative Physiology of the Blood-Brain Barrier. In *Physiology and Pharmacology of the Blood-Brain Barrier*; Bradbury, M.W.B., Ed.; Springer: Berlin/Heidelberg, Germany, 1992; pp. 371–396.
2. Bauer, H.C.; Krizbai, I.A.; Bauer, H.; Traweger, A. "You Shall Not Pass"-tight junctions of the blood brain barrier. *Front. Neurosci.* **2014**, *8*, 392. [CrossRef] [PubMed]

3. Begley, D.J.; Brightman, M.W. Structural and functional aspects of the blood-brain barrier. In *Peptide Transport and Delivery into the Central Nervous System*; Prokai, L., Prokai-Tatrai, K., Eds.; Birkhäuser Basel: Basel, Switzerland, 2003; pp. 39–78.
4. Daneman, R.; Prat, A. The blood-brain barrier. *Cold Spring Harb. Perspect. Biol.* **2015**, *7*, a020412. [CrossRef] [PubMed]
5. Saunders, N.R.; Dreifuss, J.-J.; Dziegielewska, K.M.; Johansson, P.A.; Habgood, M.D.; Møllgård, K.; Bauer, H.-C. The rights and wrongs of blood-brain barrier permeability studies: A walk through 100 years of history. *Front. Neurosci.* **2014**, *8*, 404. [CrossRef] [PubMed]
6. Lewandowsky, M. Zur Lehre von der Cerebrospinalflüssgkeit. *Z. Clin. Med.* **1900**, *40*, 480–494.
7. Goldmann, E.E. Die äussere und innere Sekretion des gesunden und kranken Organismus im Lichte der 'vitalen Färbung. *Beiträg Klin. Chir.* **1909**, *64*, 192–265.
8. Goldmann, E.E. Vitalfärbung am Zentralnervensyatem. Beitrag zur Physio-Pathologie des plexus chorioideus und der Hirnhäute. *Abh. Preuss Akad. Wiss. Phys. Math. Kl.* **1913**, *1*, 1–60.
9. Ehrlich, P. *The Relations Existing between Chemical Constitution, Distribution and Pharmacological Action*; Collected Studies on Immunity New York: John Wiley & Sons, translated by C Bolduana from Ch XXXIV of Gesammelte Arbeiten zur Immunitätsforschung; Ehrlich, P., Ed.; Hirschwald: Berlin, Germany, 1906.
10. Ehrlich, P. Das Sauerstoffbedürfnis des Organismus. In *Eine Farbenanalytische Studie*; Hirschwald: Berlin, Germany, 1885.
11. Stern, L.; Gautier, R. II.—Les Rapports Entre Le Liquide Céphalo-Rachidien Et Les éléments Nerveux De L'axe Cerebrospinal. *Arch. Int. Physiol.* **1922**, *17*, 391–448. [CrossRef]
12. Stern, L.; Gautier, R. III.—Rapports Entre Le Liquide Céphalo-Rachidien Des Espaces Ventriculaires Et Celui Des Espaces Sous-Arachnoïdiens. *Arch. Int. Physiol.* **1923**, *20*, 403–436. [CrossRef]
13. Stern, L.; Gautier, R. Passage simultané des substances dans le liquide céphalo-rachidien et dans les centres nerveux. *Rcr D. la Soc. De Phys. Et D'hist. Natur. De Genève* **1918**, *35*, 58–60.
14. Stern, L.; Gautier, R. Le passage dans le liquide céphalo-rachidien de substances introduites dans la circulation et leur action sur le système nerveux central chez les différentes espèces animales. *Rcr D. la Soc. De Phys. Et D'hist. Natur. De Genève* **1918**, *35*, 91–94.
15. Stern, L.; Gautier, R. Recherches sur le liquide céphalo-rachidien. 1. Les rapports entre le liquide céphalo-rachidien et la circulation sanguine. *Arch. Int. Physiol.* **1921**, *17*, 138–192. [CrossRef]
16. Roux, E.; Borrel, A. Tétanos cérébral et immunité contre le tétanus. *Ann. Inst. Pasteur* **1898**, *12*, 225–239.
17. Bouffard, G. Injection des couleurs de benzidine aux animaux normaux. *Ann. D. L'lnst. Pasteur. Paris* **1906**, *20*, 539–548.
18. Weed, L.H. An anatomical consideration of the cerebro-spinal fluid. *Anat. Res.* **1917**, 461–496. [CrossRef]
19. Weed, L.H. The development of the cerebrospinal fluid spaces in pig and in man. *Contrib. Embryol. Carnegie Inst.* **1917**, *5*, 3–116.
20. Wislocki, G.B. Experimental studies on fetal absorption. I. The vitality stained fetus. *Embryol. Carnegie Inst.* **1920**, *11*, 45–60.
21. Cohen, H.; Davies, S. The Morphology and Permeability of the Roof of the Fourth Ventricle in some Mammalian Embryos. *J. Anat.* **1938**, *72*, 430–455.
22. Friedemann, U. Blood-brain barrier. *Physiol. Rev.* **1942**, *22*, 125–145. [CrossRef]
23. Wolburg, H.; Lippoldt, A. Tight junctions of the blood–brain barrier: Development, composition and regulation. *Vasc. Pharmacol.* **2002**, *38*, 323–337. [CrossRef]
24. Duelli, R.; Kuschinsky, W. Brain Glucose Transporters: Relationship to Local Energy Demand. *Physiology* **2001**, *16*, 71–76. [CrossRef]
25. Howarth, C.; Gleeson, P.; Attwell, D. Updated energy budgets for neural computation in the neocortex and cerebellum. *J. Cereb. Blood Flow Metab.* **2012**, *32*, 1222–1232. [CrossRef] [PubMed]
26. Harris, J.; Jolivet, R.; Attwell, D. Synaptic Energy Use and Supply. *Neuron* **2012**, *75*, 762–777. [CrossRef] [PubMed]
27. Mergenthaler, P.; Lindauer, U.; Dienel, G.A.; Meisel, A. Sugar for the brain: The role of glucose in physiological and pathological brain function. *Trends Neurosci.* **2013**, *36*, 587–597. [CrossRef] [PubMed]
28. Patching, S.G. Glucose Transporters at the Blood-Brain Barrier: Function, Regulation and Gateways for Drug Delivery. *Mol. Neurobiol.* **2017**, *54*, 1046–1077. [CrossRef]

29. Abbott, N.J. Inflammatory Mediators and Modulation of Blood–Brain Barrier Permeability. *Cell. Mol. Neurobiol.* **2000**, *20*, 131–147. [CrossRef]
30. Huber, J.D.; Egleton, R.D.; Davis, T.P. Molecular physiology and pathophysiology of tight junctions in the blood–brain barrier. *Trends Neurosci.* **2001**, *24*, 719–725. [CrossRef]
31. Pardridge, W.M. Drug and Gene Delivery to the Brain: The Vascular Route. *Neuron* **2002**, *36*, 555–558. [CrossRef]
32. Ballabh, P.; Braun, A.; Nedergaard, M. The blood–brain barrier: An overview: Structure, regulation, and clinical implications. *Neurobiol. Dis.* **2004**, *16*, 1–13. [CrossRef]
33. Hawkins, B.T.; Davis, T.P. The Blood-Brain Barrier/Neurovascular Unit in Health and Disease. *Pharmacol. Rev.* **2005**, *57*, 173. [CrossRef]
34. Kniesel, U.; Wolburg, H. Tight Junctions of the Blood–Brain Barrier. *Cell. Mol. Neurobiol.* **2000**, *20*, 57–76. [CrossRef]
35. Luissint, A.-C.; Artus, C.; Glacial, F.; Ganeshamoorthy, K.; Couraud, P.-O. Tight junctions at the blood brain barrier: Physiological architecture and disease-associated dysregulation. *Fluids Barriers CNS* **2012**, *9*, 23. [CrossRef] [PubMed]
36. Greene, C.; Campbell, M. Tight junction modulation of the blood brain barrier: CNS delivery of small molecules. *Tissue Barriers* **2016**, *4*, e1138017. [CrossRef] [PubMed]
37. Butt, A.M.; Jones, H.C.; Abbott, N.J. Electrical resistance across the blood-brain barrier in anaesthetized rats: A developmental study. *J. Physiol.* **1990**, *429*, 47–62. [CrossRef] [PubMed]
38. Abbott, N.J.; Patabendige, A.A.K.; Dolman, D.E.M.; Yusof, S.R.; Begley, D.J. Structure and function of the blood–brain barrier. *Neurobiol. Dis.* **2010**, *37*, 13–25. [CrossRef] [PubMed]
39. Pardridge, W.M. The blood-brain barrier: Bottleneck in brain drug development. *NeuroRX* **2005**, *2*, 3–14. [CrossRef]
40. Pardridge, W.M. Blood–brain barrier delivery. *Drug Discov. Today* **2007**, *12*, 54–61. [CrossRef]
41. Pardridge, W.M. Drug Transport across the Blood–Brain Barrier. *J. Cereb. Blood Flow Metab.* **2012**, *32*, 1959–1972. [CrossRef]
42. Cucullo, L.; Bennani-Baiti, B.; Rapp, E.; Janigro, D. Drug delivery and in vitro models of the blood–brain barrier. *Curr. Opin. Drug Discov. Dev.* **2005**, *8*, 89–99.
43. Wilhelm, I.; Krizbai, I. In Vitro Models of the Blood-Brain Barrier for the Study of Drug Delivery to the Brain. *Mol. Pharm.* **2014**, *11*, 1949–1963. [CrossRef]
44. Heymans, M.; Sevin, E.; Gosselet, F.; Lundquist, S.; Culot, M. Mimicking brain tissue binding in an in vitro model of the blood-brain barrier illustrates differences between in vitro and in vivo methods for assessing the rate of brain penetration. *Eur. J. Pharm. Biopharm.* **2018**, *127*, 453–461. [CrossRef]
45. Srinivasan, B.; Kolli, A.R.; Esch, M.B.; Abaci, H.E.; Shuler, M.L.; Hickman, J.J. TEER Measurement Techniques for In Vitro Barrier Model Systems. *J. Lab. Autom.* **2015**, *20*, 107–126. [CrossRef] [PubMed]
46. Santaguida, S.; Janigro, D.; Hossain, M.; Oby, E.; Rapp, E.; Cucullo, L. Side by side comparison between dynamic versus static models of blood–brain barrier in vitro: A permeability study. *Brain Res.* **2006**, *1109*, 1–13. [CrossRef] [PubMed]
47. Achyuta, A.K.H.; Conway, A.J.; Crouse, R.B.; Bannister, E.C.; Lee, R.N.; Katnik, C.P.; Behensky, A.A.; Cuevas, J.; Sundaram, S.S. A modular approach to create a neurovascular unit-on-a-chip. *Lab. A Chip* **2013**, *13*, 542–553. [CrossRef] [PubMed]
48. Wang, Y.I.; Abaci, H.E.; Shuler, M.L. Microfluidic blood-brain barrier model provides in vivo-like barrier properties for drug permeability screening. *Biotechnol. Bioeng.* **2017**, *114*, 184–194. [CrossRef] [PubMed]
49. Bennion, B.J.; Be, N.A.; McNerney, M.W.; Lao, V.; Carlson, E.M.; Valdez, C.A.; Malfatti, M.A.; Enright, H.A.; Nguyen, T.H.; Lightstone, F.C.; et al. Predicting a Drug's Membrane Permeability: A Computational Model Validated With in Vitro Permeability Assay Data. *J. Phys. Chem. B* **2017**, *121*, 5228–5237. [CrossRef] [PubMed]
50. Dickson, C.J.; Hornak, V.; Pearlstein, R.A.; Duca, J.S. Structure–Kinetic Relationships of Passive Membrane Permeation from Multiscale Modeling. *J. Am. Chem. Soc.* **2017**, *139*, 442–452. [CrossRef]
51. Carpenter, T.S.; Kirshner, D.A.; Lau, E.Y.; Wong, S.E.; Nilmeier, J.P.; Lightstone, F.C. A method to predict blood-brain barrier permeability of drug-like compounds using molecular dynamics simulations. *Biophys. J.* **2014**, *107*, 630–641. [CrossRef]
52. Geldenhuys, W.J.; Mohammad, A.S.; Adkins, C.E.; Lockman, P.R. Molecular determinants of blood-brain barrier permeation. *Ther. Deliv.* **2015**, *6*, 961–971. [CrossRef]

53. González-Mariscal, L.; Betanzos, A.; Nava, P.; Jaramillo, B.E. Tight junction proteins. *Prog. Biophys. Mol. Biol.* **2003**, *81*, 1–44. [CrossRef]
54. Tsukita, S.; Furuse, M.; Itoh, M. Multifunctional strands in tight junctions. *Nat. Rev. Mol. Cell Biol.* **2001**, *2*, 285–293. [CrossRef]
55. Morita, K.; Furuse, M.; Fujimoto, K.; Tsukita, S. Claudin multigene family encoding four-transmembrane domain protein components of tight junction strands. *Proc. Natl. Acad. Sci. USA* **1999**, *96*, 511. [CrossRef] [PubMed]
56. Schulzke, J.D.; Ploeger, S.; Amasheh, M.; Fromm, A.; Zeissig, S.; Troeger, H.; Richter, J.; Bojarski, C.; Schumann, M.; Fromm, M. Epithelial Tight Junctions in Intestinal Inflammation. *Ann. N. Y. Acad. Sci.* **2009**, *1165*, 294–300. [CrossRef] [PubMed]
57. Morita, K.; Sasaki, H.; Furuse, M.; Tsukita, S. Endothelial Claudin. *J. Cell Biol.* **1999**, *147*, 185. [CrossRef] [PubMed]
58. Furuse, M.; Hata, M.; Furuse, K.; Yoshida, Y.; Haratake, A.; Sugitani, Y.; Noda, T.; Kubo, A.; Tsukita, S. Claudin-based tight junctions are crucial for the mammalian epidermal barrier. *J. Cell Biol.* **2002**, *156*, 1099. [CrossRef]
59. Günzel, D.; Yu, A.S.L. Claudins and the Modulation of Tight Junction Permeability. *Physiol. Rev.* **2013**, *93*, 525–569. [CrossRef]
60. Günzel, D.; Fromm, M. Claudins and Other Tight Junction Proteins. *Compr. Physiol.* **2012**, *2*, 1819–1852.
61. Van Itallie, C.M.; Anderson, J.M. Claudin interactions in and out of the tight junction. *Tissue Barriers* **2013**, *1*, e25247. [CrossRef]
62. Krause, G.; Winkler, L.; Mueller, S.L.; Haseloff, R.F.; Piontek, J.; Blasig, I.E. Structure and function of claudins. *Biochim. Biophys. Acta (BBA) Biomembr.* **2008**, *1778*, 631–645. [CrossRef]
63. Buckley, A.; Turner, J.R. Cell Biology of Tight Junction Barrier Regulation and Mucosal Disease. *Cold Spring Harb. Perspect. Biol.* **2018**, *10*, a029314. [CrossRef]
64. Garcia-Hernandez, V.; Quiros, M.; Nusrat, A. Intestinal epithelial claudins: Expression and regulation in homeostasis and inflammation. *Ann. N. Y. Acad. Sci.* **2017**, *1397*, 66–79. [CrossRef]
65. Hou, J.; Rajagopal, M.; Yu, A.S.L. Claudins and the kidney. *Annu. Rev. Physiol.* **2013**, *75*, 479–501. [CrossRef] [PubMed]
66. Hewitt, K.J.; Agarwal, R.; Morin, P.J. The claudin gene family: Expression in normal and neoplastic tissues. *BMC Cancer* **2006**, *6*, 186. [CrossRef] [PubMed]
67. Wang, H.; Yang, X. The expression patterns of tight junction protein claudin-1, -3, and -4 in human gastric neoplasms and adjacent non-neoplastic tissues. *Int. J. Clin. Exp. Pathol.* **2015**, *8*, 881–887. [PubMed]
68. Suzuki, H.; Nishizawa, T.; Tani, K.; Yamazaki, Y.; Tamura, A.; Ishitani, R.; Dohmae, N.; Tsukita, S.; Nureki, O.; Fujiyoshi, Y. Crystal Structure of a Claudin Provides Insight into the Architecture of Tight Junctions. *Science* **2014**, *344*, 304. [CrossRef] [PubMed]
69. Ohtsuki, S.; Yamaguchi, H.; Katsukura, Y.; Asashima, T.; Terasaki, T. mRNA expression levels of tight junction protein genes in mouse brain capillary endothelial cells highly purified by magnetic cell sorting. *J. Neurochem.* **2008**, *104*, 147–154. [CrossRef]
70. Daneman, R.; Zhou, L.; Agalliu, D.; Cahoy, J.D.; Kaushal, A.; Barres, B.A. The Mouse Blood-Brain Barrier Transcriptome: A New Resource for Understanding the Development and Function of Brain Endothelial Cells. *PLoS ONE* **2010**, *5*, e13741. [CrossRef]
71. Nitta, T.; Hata, M.; Gotoh, S.; Seo, Y.; Sasaki, H.; Hashimoto, N.; Furuse, M.; Tsukita, S. Size-selective loosening of the blood-brain barrier in claudin-5-deficient mice. *J. Cell Biol.* **2003**, *161*, 653–660. [CrossRef]
72. Krause, G.; Winkler, L.; Piehl, C.; Blasig, I.; Piontek, J.; Müller, S.L. Structure and Function of Extracellular Claudin Domains. *Mol. Biol. Cell* **2009**, *1165*, 4333–4346. [CrossRef]
73. Rossa, J.; Ploeger, C.; Vorreiter, F.; Saleh, T.; Protze, J.; Günzel, D.; Wolburg, H.; Krause, G.; Piontek, J. Claudin-3 and Claudin-5 Protein Folding and Assembly into the Tight Junction Are Controlled by Non-conserved Residues in the Transmembrane 3 (TM3) and Extracellular Loop 2 (ECL2) Segments. *J. Biol. Chem.* **2014**, *289*, 7641–7653. [CrossRef]
74. Piehl, C.; Piontek, J.; Cording, J.; Wolburg, H.; Blasig, I.E. Participation of the second extracellular loop of claudin-5 in paracellular tightening against ions, small and large molecules. *Cell. Mol. Life Sci.* **2010**, *67*, 2131–2140. [CrossRef]

75. Keaney, J.; Walsh, D.M.; O'Malley, T.; Hudson, N.; Crosbie, D.E.; Loftus, T.; Sheehan, F.; McDaid, J.; Humphries, M.M.; Callanan, J.J.; et al. Autoregulated paracellular clearance of amyloid-β across the blood-brain barrier. *Sci. Adv.* **2015**, *1*, e1500472. [CrossRef] [PubMed]
76. Greene, C.; Hanley, N.; Campbell, M. Claudin-5: Gatekeeper of neurological function. *Fluids Barriers CNS* **2019**, *16*, 3. [CrossRef] [PubMed]
77. Di Ventura, B.; Lemerle, C.; Michalodimitrakis, K.; Serrano, L. From in vivo to in silico biology and back. *Nature* **2006**, *443*, 527–533. [CrossRef] [PubMed]
78. Ekins, S.; Mestres, J.; Testa, B. In silico pharmacology for drug discovery: Methods for virtual ligand screening and profiling. *Br. J. Pharmacol.* **2007**, *152*, 9–20. [CrossRef]
79. McInnes, C. Virtual screening strategies in drug discovery. *Curr. Opin. Chem. Biol.* **2007**, *11*, 494–502. [CrossRef]
80. Huggins, D.J.; Biggin, P.C.; Dämgen, M.A.; Essex, J.W.; Harris, S.A.; Henchman, R.H.; Khalid, S.; Kuzmanic, A.; Laughton, C.A.; Michel, J.; et al. Biomolecular simulations: From dynamics and mechanisms to computational assays of biological activity. *Wiley Interdiscip. Rev. Comput. Mol. Sci.* **2019**, *9*, e1393. [CrossRef]
81. Hope, H. Crystallography of Biological Macromolecules at Ultra-Low Temperature. *Annu. Rev. Biophys. Biophys. Chem.* **1990**, *19*, 107–126. [CrossRef]
82. Smyth, M.S.; Martin, J.H. x ray crystallography. *Mol. Pathol.* **2000**, *53*, 8–14. [CrossRef]
83. Carpenter, E.P.; Beis, K.; Cameron, A.D.; Iwata, S. Overcoming the challenges of membrane protein crystallography. *Curr. Opin. Struct. Biol.* **2008**, *18*, 581–586. [CrossRef]
84. Caffrey, M. Crystallizing Membrane Proteins for Structure Determination: Use of Lipidic Mesophases. *Annu. Rev. Biophys.* **2009**, *38*, 29–51. [CrossRef]
85. Seddon, A.M.; Curnow, P.; Booth, P.J. Membrane proteins, lipids and detergents: Not just a soap opera. *Biochim. Biophys. Acta (BBA) Biomembr.* **2004**, *1666*, 105–117. [CrossRef] [PubMed]
86. Dimmic, M.W.; Hubisz, M.J.; Bustamante, C.D.; Nielsen, R. Detecting coevolving amino acid sites using Bayesian mutational mapping. *Bioinformatics* **2005**, *21*, i126–i135. [CrossRef] [PubMed]
87. Lui, S.; Tiana, G. The network of stabilizing contacts in proteins studied by coevolutionary data. *J. Chem. Phys.* **2013**, *139*, 155103. [CrossRef] [PubMed]
88. Jana, B.; Morcos, F.; Onuchic, J.N. From structure to function: The convergence of structure based models and co-evolutionary information. *Phys. Chem. Chem. Phys.* **2014**, *16*, 6496–6507. [CrossRef] [PubMed]
89. Wang, Y.; Barth, P. Evolutionary-guided de novo structure prediction of self-associated transmembrane helical proteins with near-atomic accuracy. *Nat. Commun.* **2015**, *6*, 7196. [CrossRef] [PubMed]
90. Teixeira, P.L.; Mendenhall, J.L.; Heinze, S.; Weiner, B.; Skwark, M.J.; Meiler, J. Membrane protein contact and structure prediction using co-evolution in conjunction with machine learning. *PLoS ONE* **2017**, *12*, e0177866. [CrossRef]
91. Rajagopal, N.; Irudayanathan, F.J.; Nangia, S. Palmitoylation of Claudin-5 Proteins Influences Their Lipid Domain Affinity and Tight Junction Assembly at the Blood–Brain Barrier Interface. *J. Phys. Chem. B* **2019**, *123*, 983–993. [CrossRef]
92. Murthy, A.C.; Dignon, G.L.; Kan, Y.; Zerze, G.H.; Parekh, S.H.; Mittal, J.; Fawzi, N.L. Molecular interactions underlying liquid–liquid phase separation of the FUS low-complexity domain. *Nat. Struct. Mol. Biol.* **2019**, *26*, 637–648. [CrossRef]
93. Zheng, M.; Zhao, J.; Cui, C.; Fu, Z.; Li, X.; Liu, X.; Ding, X.; Tan, X.; Li, F.; Luo, X.; et al. Computational chemical biology and drug design: Facilitating protein structure, function, and modulation studies. *Med. Res. Rev.* **2018**, *38*, 914–950. [CrossRef]
94. Irudayanathan, F.J.; Wang, X.; Wang, N.; Willsey, S.R.; Seddon, I.A.; Nangia, S. Self-Assembly Simulations of Classic Claudins—Insights into the Pore Structure, Selectivity, and Higher Order Complexes. *J. Phys. Chem. B* **2018**, *122*, 7463–7474. [CrossRef]
95. Ibsen, K.N.; Ma, H.; Banerjee, A.; Tanner, E.E.L.; Nangia, S.; Mitragotri, S. Mechanism of Antibacterial Activity of Choline-Based Ionic Liquids (CAGE). *ACS Biomater. Sci. Eng.* **2018**, *4*, 2370–2379. [CrossRef]
96. Sengupta, D.; Prasanna, X.; Mohole, M.; Chattopadhyay, A. Exploring GPCR–Lipid Interactions by Molecular Dynamics Simulations: Excitements, Challenges, and the Way Forward. *J. Phys. Chem. B* **2018**, *122*, 5727–5737. [CrossRef] [PubMed]

97. Irudayanathan, F.J.; Wang, N.; Wang, X.; Nangia, S. Architecture of the paracellular channels formed by claudins of the blood–brain barrier tight junctions. *Ann. N. Y. Acad. Sci.* **2017**, *1405*, 131–146. [CrossRef] [PubMed]
98. Grime, J.M.A.; Dama, J.F.; Ganser-Pornillos, B.K.; Woodward, C.L.; Jensen, G.J.; Yeager, M.; Voth, G.A. Coarse-grained simulation reveals key features of HIV-1 capsid self-assembly. *Nat. Commun.* **2016**, *7*, 11568. [CrossRef] [PubMed]
99. Ingólfsson, H.I.; Arnarez, C.; Periole, X.; Marrink, S.J. Computational 'microscopy' of cellular membranes. *J. Cell Sci.* **2016**, *129*, 257. [CrossRef] [PubMed]
100. Reddy, T.; Sansom, M.S. The Role of the Membrane in the Structure and Biophysical Robustness of the Dengue Virion Envelope. *Structure* **2016**, *24*, 375–382. [CrossRef]
101. Arnarez, C.; Marrink, S.J.; Periole, X. Molecular mechanism of cardiolipin-mediated assembly of respiratory chain supercomplexes. *Chem. Sci.* **2016**, *7*, 4435–4443. [CrossRef]
102. Holdbrook, D.A.; Huber, R.G.; Piggot, T.J.; Bond, P.J.; Khalid, S. Dynamics of Crowded Vesicles: Local and Global Responses to Membrane Composition. *PLoS ONE* **2016**, *11*, e0156963. [CrossRef]
103. Castillo, N.; Monticelli, L.; Barnoud, J.; Tieleman, D.P. Free energy of WALP23 dimer association in DMPC, DPPC, and DOPC bilayers. *Chem. Phys. Lipids* **2013**, *169*, 95–105. [CrossRef]
104. Bennett, W.F.D.; MacCallum, J.L.; Hinner, M.J.; Marrink, S.J.; Tieleman, D.P. Molecular View of Cholesterol Flip-Flop and Chemical Potential in Different Membrane Environments. *J. Am. Chem. Soc.* **2009**, *131*, 12714–12720. [CrossRef]
105. Periole, X.; Knepp, A.M.; Sakmar, T.P.; Marrink, S.J.; Huber, T. Structural Determinants of the Supramolecular Organization of G Protein-Coupled Receptors in Bilayers. *J. Am. Chem. Soc.* **2012**, *134*, 10959–10965. [CrossRef] [PubMed]
106. Schäfer, L.V.; de Jong, D.H.; Holt, A.; Rzepiela, A.J.; de Vries, A.H.; Poolman, B.; Killian, J.A.; Marrink, S.J. Lipid packing drives the segregation of transmembrane helices into disordered lipid domains in model membranes. *Proc. Natl. Acad. Sci. USA* **2011**, *108*, 1343. [CrossRef] [PubMed]
107. Duncan, A.L.; Reddy, T.; Koldsø, H.; Hélie, J.; Fowler, P.W.; Chavent, M.; Sansom, M.S.P. Protein crowding and lipid complexity influence the nanoscale dynamic organization of ion channels in cell membranes. *Sci. Rep.* **2017**, *7*, 16647. [CrossRef] [PubMed]
108. Koldsø, H.; Sansom, M.S.P. Organization and Dynamics of Receptor Proteins in a Plasma Membrane. *J. Am. Chem. Soc.* **2015**, *137*, 14694–14704. [CrossRef]
109. Shorthouse, D.; Hedger, G.; Koldsø, H.; Sansom, M.S.P. Molecular simulations of glycolipids: Towards mammalian cell membrane models. *Biochimie* **2016**, *120*, 105–109. [CrossRef]
110. Koldsø, H.; Reddy, T.; Fowler, P.W.; Duncan, A.L.; Sansom, M.S.P. Membrane Compartmentalization Reducing the Mobility of Lipids and Proteins within a Model Plasma Membrane. *J. Phys. Chem. B* **2016**, *120*, 8873–8881. [CrossRef]
111. Duncan, A.L.; Song, W.; Sansom, M.S.P. Lipid-Dependent Regulation of Ion Channels and G Protein–Coupled Receptors: Insights from Structures and Simulations. *Annu. Rev. Pharmacol. Toxicol.* **2019**. [CrossRef]
112. Koldsø, H.; Shorthouse, D.; Hélie, J.; Sansom, M.S.P. Lipid Clustering Correlates with Membrane Curvature as Revealed by Molecular Simulations of Complex Lipid Bilayers. *PLoS Comput. Biol.* **2014**, *10*, e1003911. [CrossRef]
113. Duncan, A.L.; Bandurka, M.A.R.; Chavent, M.G.; Rassam, P.; Song, W.; Birkholz, O.; Helie, J.; Reddy, T.; Beliaev, D.; Hambly, B.; et al. How Nanoscale Protein Interactions Determine the Mesoscale Dynamic Organisation of Membrane Proteins. *Biophys. J.* **2019**, *116*, 365a. [CrossRef]
114. Lin, X.; Gorfe, A.A.; Levental, I. Protein Partitioning into Ordered Membrane Domains: Insights from Simulations. *Biophys. J.* **2018**, *114*, 1936–1944. [CrossRef]
115. Irudayanathan, F.J.; Trasatti, J.P.; Karande, P.; Nangia, S. Molecular Architecture of the Blood Brain Barrier Tight Junction Proteins–A Synergistic Computational and In Vitro Approach. *J. Phys. Chem. B* **2016**, *120*, 77–88. [CrossRef] [PubMed]
116. Ma, H.; Irudayanathan, F.J.; Jiang, W.; Nangia, S. Simulating Gram-Negative Bacterial Outer Membrane: A Coarse Grain Model. *J. Phys. Chem. B* **2015**, *119*, 14668–14682. [CrossRef] [PubMed]
117. Saitoh, Y.; Suzuki, H.; Tani, K.; Nishikawa, K.; Irie, K.; Ogura, Y.; Tamura, A.; Tsukita, S.; Fujiyoshi, Y. Structural insight into tight junction disassembly by Clostridium perfringens enterotoxin. *Science* **2015**, *347*, 775. [CrossRef] [PubMed]

118. Shinoda, T.; Shinya, N.; Ito, K.; Ohsawa, N.; Terada, T.; Hirata, K.; Kawano, Y.; Yamamoto, M.; Kimura-Someya, T.; Yokoyama, S.; et al. Structural basis for disruption of claudin assembly in tight junctions by an enterotoxin. *Sci. Rep.* **2016**, *6*, 33632. [CrossRef] [PubMed]
119. Nakamura, S.; Irie, K.; Tanaka, H.; Nishikawa, K.; Suzuki, H.; Saitoh, Y.; Tamura, A.; Tsukita, S.; Fujiyoshi, Y. Morphologic determinant of tight junctions revealed by claudin-3 structures. *Nat. Commun.* **2019**, *10*, 816. [CrossRef] [PubMed]
120. Xiang, Z. Advances in homology protein structure modeling. *Curr. Protein Pept. Sci.* **2006**, *7*, 217–227. [CrossRef]
121. Shen, M.-y.; Sali, A. Statistical potential for assessment and prediction of protein structures. *Protein Sci.* **2006**, *15*, 2507–2524. [CrossRef]
122. Kopp, J.; Schwede, T. Automated protein structure homology modeling: A progress report. *Pharmacogenomics* **2004**, *5*, 405–416. [CrossRef]
123. Fiser, A. Template-based protein structure modeling. *Methods Mol. Biol.* **2010**, *673*, 73–94.
124. Mineta, K.; Yamamoto, Y.; Yamazaki, Y.; Tanaka, H.; Tada, Y.; Saito, K.; Tamura, A.; Igarashi, M.; Endo, T.; Takeuchi, K.; et al. Predicted expansion of the claudin multigene family. *FEBS Lett.* **2011**, *585*, 606–612. [CrossRef]
125. Kelley, L.A.; Mezulis, S.; Yates, C.M.; Wass, M.N.; Sternberg, M.J.E. The Phyre2 web portal for protein modeling, prediction and analysis. *Nat. Protoc.* **2015**, *10*, 845. [CrossRef] [PubMed]
126. Schwede, T.; Kopp, J.; Guex, N.; Peitsch, M.C. SWISS-MODEL: An automated protein homology-modeling server. *Nucleic Acids Res.* **2003**, *31*, 3381–3385. [CrossRef] [PubMed]
127. Roy, A.; Kucukural, A.; Zhang, Y. I-TASSER: A unified platform for automated protein structure and function prediction. *Nat. Protoc.* **2010**, *5*, 725–738. [CrossRef] [PubMed]
128. Yang, J.; Yan, R.; Roy, A.; Xu, D.; Poisson, J.; Zhang, Y. The I-TASSER Suite: Protein structure and function prediction. *Nat. Methods* **2014**, *12*, 7. [CrossRef] [PubMed]
129. Yang, J.; Zhang, Y. I-TASSER server: New development for protein structure and function predictions. *Nucleic Acids Res.* **2015**, *43*, W174–W181. [CrossRef]
130. Källberg, M.; Wang, H.; Wang, S.; Peng, J.; Wang, Z.; Lu, H.; Xu, J. Template-based protein structure modeling using the RaptorX web server. *Nat. Protoc.* **2012**, *7*, 1511–1522. [CrossRef]
131. Peng, J.; Xu, J. RaptorX: Exploiting structure information for protein alignment by statistical inference. *Proteins* **2011**, *79* (Suppl. 10), 161–171. [CrossRef]
132. Fiser, A.; Do, R.K.G.; Šali, A. Modeling of loops in protein structures. *Protein Sci.* **2000**, *9*, 1753–1773. [CrossRef]
133. Baker, D.; Sali, A. Protein Structure Prediction and Structural Genomics. *Science* **2001**, *294*, 93. [CrossRef]
134. Wiltgen, M. Algorithms for Structure Comparison and Analysis: Homology Modelling of Proteins. In *Encyclopedia of Bioinformatics and Computational Biology*; Ranganathan, S., Gribskov, M., Nakai, K., Schönbach, C., Eds.; Academic Press: Oxford, UK, 2019; pp. 38–61. [CrossRef]
135. Schymkowitz, J.; Borg, J.; Stricher, F.; Nys, R.; Rousseau, F.; Serrano, L. The FoldX web server: An online force field. *Nucleic Acids Res.* **2005**, *33*, W382–W388. [CrossRef]
136. Fiser, A.; Sali, A. ModLoop: Automated modeling of loops in protein structures. *Bioinformatics* **2003**, *19*, 2500–2501. [CrossRef] [PubMed]
137. MacKerell, A.D.; Bashford, D.; Bellott, M.; Dunbrack, R.L.; Evanseck, J.D.; Field, M.J.; Fischer, S.; Gao, J.; Guo, H.; Ha, S.; et al. All-Atom Empirical Potential for Molecular Modeling and Dynamics Studies of Proteins. *J. Phys. Chem. B* **1998**, *102*, 3586–3616. [CrossRef] [PubMed]
138. Wang, H.; Junghans, C.; Kremer, K. Comparative atomistic and coarse-grained study of water: What do we lose by coarse-graining? *Eur. Phys. J. E* **2009**, *28*, 221–229. [CrossRef] [PubMed]
139. Kmiecik, S.; Gront, D.; Kolinski, M.; Wieteska, L.; Dawid, A.E.; Kolinski, A. Coarse-Grained Protein Models and Their Applications. *Chem. Rev.* **2016**, *116*, 7898–7936. [CrossRef] [PubMed]
140. Arkhipov, A.; Freddolino, P.L.; Imada, K.; Namba, K.; Schulten, K. Coarse-Grained Molecular Dynamics Simulations of a Rotating Bacterial Flagellum. *Biophys. J.* **2006**, *91*, 4589–4597. [CrossRef]
141. Shih, A.Y.; Arkhipov, A.; Freddolino, P.L.; Schulten, K. Coarse Grained Protein−Lipid Model with Application to Lipoprotein Particles. *J. Phys. Chem. B* **2006**, *110*, 3674–3684. [CrossRef]
142. Akhmatskaya, E.; Reich, S. Meso-GSHMC: A stochastic algorithm for meso-scale constant temperature simulations. *Procedia Comput. Sci.* **2011**, *4*, 1353–1362. [CrossRef]

143. de Jong, D.H.; Periole, X.; Marrink, S.J. Dimerization of Amino Acid Side Chains: Lessons from the Comparison of Different Force Fields. *J. Chem. Theory Comput.* **2012**, *8*, 1003–1014. [CrossRef]
144. Ayton, G.S.; Voth, G.A. Systematic multiscale simulation of membrane protein systems. *Curr. Opin. Struct. Biol.* **2009**, *19*, 138–144. [CrossRef]
145. Yuan, C.; Li, S.; Zou, Q.; Ren, Y.; Yan, X. Multiscale simulations for understanding the evolution and mechanism of hierarchical peptide self-assembly. *Phys. Chem. Chem. Phys.* **2017**, *19*, 23614–23631. [CrossRef]
146. Singh, N.; Li, W. Recent Advances in Coarse-Grained Models for Biomolecules and Their Applications. *Int. J. Mol. Sci.* **2019**, *20*, 3774. [CrossRef] [PubMed]
147. Jiang, W.; Luo, J.; Nangia, S. Multiscale Approach to Investigate Self-Assembly of Telodendrimer Based Nanocarriers for Anticancer Drug Delivery. *Langmuir* **2015**, *31*, 4270–4280. [CrossRef] [PubMed]
148. Jiang, W.; Wang, X.; Guo, D.; Luo, J.; Nangia, S. Drug-Specific Design of Telodendrimer Architecture for Effective Doxorubicin Encapsulation. *J. Phys. Chem. B* **2016**, *120*, 9766–9777. [CrossRef] [PubMed]
149. Cornell, W.D.; Cieplak, P.; Bayly, C.I.; Gould, I.R.; Merz, K.M.; Ferguson, D.M.; Spellmeyer, D.C.; Fox, T.; Caldwell, J.W.; Kollman, P.A. A Second Generation Force Field for the Simulation of Proteins, Nucleic Acids, and Organic Molecules. *J. Am. Chem. Soc.* **1995**, *117*, 5179–5197. [CrossRef]
150. Oostenbrink, C.; Villa, A.; Mark, A.E.; Van Gunsteren, W.F. A biomolecular force field based on the free enthalpy of hydration and solvation: The GROMOS force-field parameter sets 53A5 and 53A6. *J. Comput. Chem.* **2004**, *25*, 1656–1676. [CrossRef] [PubMed]
151. Jorgensen, W.L.; Maxwell, D.S.; Tirado-Rives, J. Development and Testing of the OPLS All-Atom Force Field on Conformational Energetics and Properties of Organic Liquids. *J. Am. Chem. Soc.* **1996**, *118*, 11225–11236. [CrossRef]
152. Jo, S.; Kim, T.; Iyer, V.G.; Im, W. CHARMM-GUI: A web-based graphical user interface for CHARMM. *J. Comput. Chem.* **2008**, *29*, 1859–1865. [CrossRef]
153. Abraham, M.J.; Murtola, T.; Schulz, R.; Páll, S.; Smith, J.C.; Hess, B.; Lindahl, E. GROMACS: High performance molecular simulations through multi-level parallelism from laptops to supercomputers. *SoftwareX* **2015**, *1–2*, 19–25. [CrossRef]
154. Van Der Spoel, D.; Lindahl, E.; Hess, B.; Groenhof, G.; Mark, A.E.; Berendsen, H.J.C. GROMACS: Fast, flexible, and free. *J. Comput. Chem.* **2005**, *26*, 1701–1718. [CrossRef]
155. Phillips, J.C.; Braun, R.; Wang, W.; Gumbart, J.; Tajkhorshid, E.; Villa, E.; Chipot, C.; Skeel, R.D.; Kalé, L.; Schulten, K. Scalable molecular dynamics with NAMD. *J. Comput. Chem.* **2005**, *26*, 1781–1802. [CrossRef]
156. Thompson, J.M.T.; Grindon, C.; Harris, S.; Evans, T.; Novik, K.; Coveney, P.; Laughton, C. Large-scale molecular dynamics simulation of DNA: Implementation and validation of the AMBER98 force field in LAMMPS. *Philos. Trans. R. Soc. Lond. Ser. A Math. Phys. Eng. Sci.* **2004**, *362*, 1373–1386.
157. Case, D.A.; Cheatham Iii, T.E.; Darden, T.; Gohlke, H.; Luo, R.; Merz, K.M., Jr.; Onufriev, A.; Simmerling, C.; Wang, B.; Woods, R.J. The Amber biomolecular simulation programs. *J. Comput. Chem.* **2005**, *26*, 1668–1688. [CrossRef] [PubMed]
158. Wu, E.L.; Cheng, X.; Jo, S.; Rui, H.; Song, K.C.; Dávila-Contreras, E.M.; Qi, Y.; Lee, J.; Monje-Galvan, V.; Venable, R.M.; et al. CHARMM-GUI Membrane Builder toward realistic biological membrane simulations. *J. Comput. Chem.* **2014**, *35*, 1997–2004. [CrossRef] [PubMed]
159. Brooks, B.R.; Brooks Iii, C.L.; Mackerell Jr, A.D.; Nilsson, L.; Petrella, R.J.; Roux, B.; Won, Y.; Archontis, G.; Bartels, C.; Boresch, S.; et al. CHARMM: The biomolecular simulation program. *J. Comput. Chem.* **2009**, *30*, 1545–1614. [CrossRef] [PubMed]
160. Humphrey, W.; Dalke, A.; Schulten, K. VMD: Visual molecular dynamics. *J. Mol. Graph.* **1996**, *14*, 33–38. [CrossRef]
161. *The PyMOL Molecular Graphics System*; Version 2.0; Schrödinger, LLC: New York, NY, USA, 2015.
162. Land, H.; Humble, M.S. YASARA: A Tool to Obtain Structural Guidance in Biocatalytic Investigations. In *Protein Engineering: Methods and Protocols*; Bornscheuer, U.T., Höhne, M., Eds.; Springer New York: New York, NY, USA, 2018; pp. 43–67. [CrossRef]
163. Krieger, E.; Vriend, G. YASARA View—molecular graphics for all devices—From smartphones to workstations. *Bioinformatics* **2014**, *30*, 2981–2982. [CrossRef] [PubMed]
164. Krieger, E.; Koraimann, G.; Vriend, G. Increasing the precision of comparative models with YASARA NOVA—A self-parameterizing force field. *Proteins Struct. Funct. Bioinform.* **2002**, *47*, 393–402. [CrossRef]

165. Marrink, S.J.; Risselada, H.J.; Yefimov, S.; Tieleman, D.P.; de Vries, A.H. The MARTINI Force Field: Coarse Grained Model for Biomolecular Simulations. *J. Phys. Chem. B* **2007**, *111*, 7812–7824. [CrossRef]
166. Monticelli, L.; Kandasamy, S.K.; Periole, X.; Larson, R.G.; Tieleman, D.P.; Marrink, S.-J. The MARTINI Coarse-Grained Force Field: Extension to Proteins. *J. Chem. Theory Comput.* **2008**, *4*, 819–834. [CrossRef]
167. Marrink, S.J.; Tieleman, D.P. Perspective on the Martini model. *Chem. Soc. Rev.* **2013**, *42*, 6801–6822. [CrossRef]
168. Periole, X.; Marrink, S.-J. The Martini Coarse-Grained Force Field. In *Biomolecular Simulations: Methods and Protocols*; Monticelli, L., Salonen, E., Eds.; Humana Press: Totowa, NJ, USA, 2013; pp. 533–565.
169. de Jong, D.H.; Singh, G.; Bennett, W.F.D.; Arnarez, C.; Wassenaar, T.A.; Schäfer, L.V.; Periole, X.; Tieleman, D.P.; Marrink, S.J. Improved Parameters for the Martini Coarse-Grained Protein Force Field. *J. Chem. Theory Comput.* **2013**, *9*, 687–697. [CrossRef] [PubMed]
170. Darré, L.; Machado, M.R.; Brandner, A.F.; González, H.C.; Ferreira, S.; Pantano, S. SIRAH: A Structurally Unbiased Coarse-Grained Force Field for Proteins with Aqueous Solvation and Long-Range Electrostatics. *J. Chem. Theory Comput.* **2015**, *11*, 723–739. [CrossRef] [PubMed]
171. Machado, M.R.; Pantano, S. SIRAH tools: Mapping, backmapping and visualization of coarse-grained models. *Bioinformatics* **2016**, *32*, 1568–1570. [CrossRef] [PubMed]
172. Kar, P.; Gopal, S.M.; Cheng, Y.-M.; Predeus, A.; Feig, M. PRIMO: A Transferable Coarse-grained Force Field for Proteins. *J. Chem. Theory Comput.* **2013**, *9*, 3769–3788. [CrossRef] [PubMed]
173. Cheon, M.; Chang, I.; Hall, C.K. Extending the PRIME model for protein aggregation to all 20 amino acids. *Proteins* **2010**, *78*, 2950–2960. [CrossRef] [PubMed]
174. Han, W.; Wan, C.-K.; Wu, Y.-D. PACE Force Field for Protein Simulations. 2. Folding Simulations of Peptides. *J. Chem. Theory Comput.* **2010**, *6*, 3390–3402. [CrossRef]
175. Han, W.; Wan, C.-K.; Jiang, F.; Wu, Y.-D. PACE Force Field for Protein Simulations. 1. Full Parameterization of Version 1 and Verification. *J. Chem. Theory Comput.* **2010**, *6*, 3373–3389. [CrossRef]
176. Uusitalo, J.J.; Ingólfsson, H.I.; Marrink, S.J.; Faustino, I. Martini Coarse-Grained Force Field: Extension to RNA. *Biophys. J.* **2017**, *113*, 246–256. [CrossRef]
177. Bernardi, R.C.; Melo, M.C.R.; Schulten, K. Enhanced sampling techniques in molecular dynamics simulations of biological systems. *Biochim. Et Biophys. Acta (BBA) Gen. Subj.* **2015**, *1850*, 872–877. [CrossRef]
178. Perilla, J.R.; Goh, B.C.; Cassidy, C.K.; Liu, B.; Bernardi, R.C.; Rudack, T.; Yu, H.; Wu, Z.; Schulten, K. Molecular dynamics simulations of large macromolecular complexes. *Curr. Opin. Struct. Biol.* **2015**, *31*, 64–74. [CrossRef]
179. Perilla, J.R.; Schulten, K. Physical properties of the HIV-1 capsid from all-atom molecular dynamics simulations. *Nat. Commun.* **2017**, *8*, 15959. [CrossRef] [PubMed]
180. Xiao, G.; Ren, M.; Hong, H. 50 million atoms scale molecular dynamics modelling on a single consumer graphics card. *Adv. Eng. Softw.* **2018**, *124*, 66–72. [CrossRef]
181. Orsi, M. 15-Molecular simulation of self-assembly. In *Self-Assembling Biomaterials*; Azevedo, H.S., da Silva, R.M.P., Eds.; Woodhead Publishing: Cambridge, MA, USA, 2018; pp. 305–318. [CrossRef]
182. Ben-Nissan, G.; Sharon, M. Capturing protein structural kinetics by mass spectrometry. *Chem. Soc. Rev.* **2011**, *40*, 3627–3637. [CrossRef] [PubMed]
183. Sekhar, A.; Vallurupalli, P.; Kay, L.E. Defining a length scale for millisecond-timescale protein conformational exchange. *Proc. Natl. Acad. Sci. USA* **2013**, *110*, 11391. [CrossRef]
184. Agrawal, P.; Singh, H.; Srivastava, H.K.; Singh, S.; Kishore, G.; Raghava, G.P.S. Benchmarking of different molecular docking methods for protein-peptide docking. *BMC Bioinform.* **2019**, *19*, 426. [CrossRef]
185. Pagadala, N.S.; Syed, K.; Tuszynski, J. Software for molecular docking: A review. *Biophys. Rev.* **2017**, *9*, 91–102. [CrossRef]
186. Kozakov, D.; Hall, D.R.; Xia, B.; Porter, K.A.; Padhorny, D.; Yueh, C.; Beglov, D.; Vajda, S. The ClusPro web server for protein–protein docking. *Nat. Protoc.* **2017**, *12*, 255. [CrossRef]
187. Comeau, S.R.; Gatchell, D.W.; Vajda, S.; Camacho, C.J. ClusPro: A fully automated algorithm for protein–protein docking. *Nucleic Acids Res.* **2004**, *32*, W96–W99. [CrossRef]
188. Lyskov, S.; Gray, J.J. The RosettaDock server for local protein-protein docking. *Nucleic Acids Res.* **2008**, *36*, W233–W238. [CrossRef]
189. de Vries, S.J.; van Dijk, M.; Bonvin, A.M.J.J. The HADDOCK web server for data-driven biomolecular docking. *Nat. Protoc.* **2010**, *5*, 883. [CrossRef]

190. Dominguez, C.; Boelens, R.; Bonvin, A.M.J.J. HADDOCK: A Protein−Protein Docking Approach Based on Biochemical or Biophysical Information. *J. Am. Chem. Soc.* **2003**, *125*, 1731–1737. [CrossRef] [PubMed]
191. Chen, R.; Li, L.; Weng, Z. ZDOCK: An initial-stage protein-docking algorithm. *Proteins Struct. Funct. Bioinform.* **2003**, *52*, 80–87. [CrossRef] [PubMed]
192. Liu, Y.; Zhao, L.; Li, W.; Zhao, D.; Song, M.; Yang, Y. FIPSDock: A new molecular docking technique driven by fully informed swarm optimization algorithm. *J. Comput. Chem.* **2013**, *34*, 67–75. [CrossRef] [PubMed]
193. Gagnon, J.K.; Law, S.M.; Brooks Iii, C.L. Flexible CDOCKER: Development and application of a pseudo-explicit structure-based docking method within CHARMM. *J. Comput. Chem.* **2016**, *37*, 753–762. [CrossRef] [PubMed]
194. Pons, C.; Garzon, J.I.; Lopéz-Blanco, J.R.; Fernandez-Recio, J.; Kovacs, J.; Chacon, P.; Abagyan, R. FRODOCK: A new approach for fast rotational protein–protein docking. *Bioinformatics* **2009**, *25*, 2544–2551.
195. Liu, M.; Wang, S. MCDOCK: A Monte Carlo simulation approach to the molecular docking problem. *J. Comput. Aided Mol. Des.* **1999**, *13*, 435–451. [CrossRef]
196. Yellapu, N.K. Molecular Modelling, Dynamics, and Docking of Membrane Proteins: Still a Challenge. In *Applied Case Studies and Solutions in Molecular Docking-Based Drug Design*; IGI Global: Hershey, PA, USA, 2016; pp. 186–208.
197. Burke, K.A.; Yates, E.A.; Legleiter, J. Biophysical insights into how surfaces, including lipid membranes, modulate protein aggregation related to neurodegeneration. *Front. Neurol* **2013**, *4*, 17. [CrossRef]
198. Bogdanov, M.; Dowhan, W.; Vitrac, H. Lipids and topological rules governing membrane protein assembly. *Biochim. Biophys. Acta* **2014**, *1843*, 1475–1488. [CrossRef]
199. Bogdanov, M.; Mileykovskaya, E.; Dowhan, W. Lipids in the assembly of membrane proteins and organization of protein supercomplexes: Implications for lipid-linked disorders. *Subcell. Biochem.* **2008**, *49*, 197–239.
200. Ho, C.-C. Chapter 7—Membranes for Bioseparations. In *Bioprocessing for Value-Added Products from Renewable Resources*; Yang, S.-T., Ed.; Elsevier: Amsterdam, The Netherlands, 2007; pp. 163–183. [CrossRef]
201. Page, R.C.; Li, C.; Hu, J.; Gao, F.P.; Cross, T.A. Lipid bilayers: An essential environment for the understanding of membrane proteins. *Magn. Reson. Chem.* **2007**, *45*, S2–S11. [CrossRef]
202. Sengupta, D.; Marrink, S.J. Lipid-mediated interactions tune the association of glycophorin A helix and its disruptive mutants in membranes. *Phys. Chem. Chem. Phys.* **2010**, *12*, 12987–12996. [CrossRef] [PubMed]
203. Wassenaar, T.A.; Pluhackova, K.; Moussatova, A.; Sengupta, D.; Marrink, S.J.; Tieleman, D.P.; Böckmann, R.A. High-Throughput Simulations of Dimer and Trimer Assembly of Membrane Proteins. The DAFT Approach. *J. Chem. Theory Comput.* **2015**, *11*, 2278–2291. [CrossRef] [PubMed]
204. Altwaijry, N.A.; Baron, M.; Wright, D.W.; Coveney, P.V.; Townsend-Nicholson, A. An Ensemble-Based Protocol for the Computational Prediction of Helix–Helix Interactions in G Protein-Coupled Receptors using Coarse-Grained Molecular Dynamics. *J. Chem. Theory Comput.* **2017**, *13*, 2254–2270. [CrossRef] [PubMed]
205. Hall, B.A.; Halim, K.B.A.; Buyan, A.; Emmanouil, B.; Sansom, M.S.P. Sidekick for Membrane Simulations: Automated Ensemble Molecular Dynamics Simulations of Transmembrane Helices. *J. Chem. Theory Comput.* **2014**, *10*, 2165–2175. [CrossRef] [PubMed]
206. Rajagopal, N.; Nangia, S. Obtaining Protein Association Energy Landscape (PANEL) for Integral Membrane Proteins. *J. Chem. Theory Comput.* **2019**. [CrossRef] [PubMed]
207. Petrek, M.; Otyepka, M.; Banás, P.; Kosinová, P.; Koca, J.; Damborský, J. CAVER: A new tool to explore routes from protein clefts, pockets and cavities. *BMC Bioinform.* **2006**, *7*, 316. [CrossRef]
208. Petřek, M.; Košinová, P.; Koča, J.; Otyepka, M. MOLE: A Voronoi Diagram-Based Explorer of Molecular Channels, Pores, and Tunnels. *Structure* **2007**, *15*, 1357–1363. [CrossRef]
209. Smart, O.S.; Neduvelil, J.G.; Wang, X.; Wallace, B.A.; Sansom, M.S. HOLE: A program for the analysis of the pore dimensions of ion channel structural models. *J. Mol Graph.* **1996**, *14*, 354–360. [CrossRef]
210. Yaffe, E.; Fishelovitch, D.; Wolfson, H.J.; Halperin, D.; Nussinov, R. MolAxis: A server for identification of channels in macromolecules. *Nucleic Acids Res.* **2008**, *36*, W210–W215. [CrossRef]
211. Yaffe, E.; Fishelovitch, D.; Wolfson, H.J.; Halperin, D.; Nussinov, R. MolAxis: Efficient and accurate identification of channels in macromolecules. *Proteins* **2008**, *73*, 72–86. [CrossRef]
212. Pellegrini-Calace, M.; Maiwald, T.; Thornton, J.M. PoreWalker: A Novel Tool for the Identification and Characterization of Channels in Transmembrane Proteins from Their Three-Dimensional Structure. *PLoS Comput. Biol.* **2009**, *5*, e1000440. [CrossRef] [PubMed]

213. Peter, C.; Hummer, G. Ion Transport through Membrane-Spanning Nanopores Studied by Molecular Dynamics Simulations and Continuum Electrostatics Calculations. *Biophys. J.* **2005**, *89*, 2222–2234. [CrossRef] [PubMed]
214. Hummer, G.; Pratt, L.R.; García, A.E. Molecular Theories and Simulation of Ions and Polar Molecules in Water. *J. Phys. Chem. A* **1998**, *102*, 7885–7895. [CrossRef]
215. Torrie, G.M.; Valleau, J.P. Nonphysical sampling distributions in Monte Carlo free-energy estimation: Umbrella sampling. *J. Comput. Phys.* **1977**, *23*, 187–199. [CrossRef]
216. Barducci, A.; Bonomi, M.; Parrinello, M. Metadynamics. *Wiley Interdiscip. Rev. Comput. Mol. Sci.* **2011**, *1*, 826–843. [CrossRef]
217. Laio, A.; Gervasio, F.L. Metadynamics: A method to simulate rare events and reconstruct the free energy in biophysics, chemistry and material science. *Rep. Prog. Phys.* **2008**, *71*, 126601. [CrossRef]
218. Barducci, A.; Bussi, G.; Parrinello, M. Well-Tempered Metadynamics: A Smoothly Converging and Tunable Free-Energy Method. *Phys. Rev. Lett.* **2008**, *100*, 020603. [CrossRef]
219. Kumar, S.; Rosenberg, J.M.; Bouzida, D.; Swendsen, R.H.; Kollman, P.A. THE weighted histogram analysis method for free-energy calculations on biomolecules. I. The method. *J. Comput. Chem.* **1992**, *13*, 1011–1021. [CrossRef]
220. Bonomi, M.; Branduardi, D.; Bussi, G.; Camilloni, C.; Provasi, D.; Raiteri, P.; Donadio, D.; Marinelli, F.; Pietrucci, F.; Broglia, R.A.; et al. PLUMED: A portable plugin for free-energy calculations with molecular dynamics. *Comput. Phys. Commun.* **2009**, *180*, 1961–1972. [CrossRef]
221. Kutzner, C.; Grubmüller, H.; de Groot, B.L.; Zachariae, U. Computational electrophysiology: The molecular dynamics of ion channel permeation and selectivity in atomistic detail. *Biophys. J.* **2011**, *101*, 809–817. [CrossRef]
222. Kutzner, C.; Köpfer, D.A.; Machtens, J.-P.; de Groot, B.L.; Song, C.; Zachariae, U. Insights into the function of ion channels by computational electrophysiology simulations. *Biochim. Biophys. Acta (BBA) Biomembr.* **2016**, *1858*, 1741–1752. [CrossRef]
223. Epand, R.M. Lipid polymorphism and protein–lipid interactions. *Biochim. Biophys. Acta (BBA) Rev. Biomembr.* **1998**, *1376*, 353–368. [CrossRef]
224. Killian, J.A. Hydrophobic mismatch between proteins and lipids in membranes. *Biochim. Biophys. Acta (BBA) Rev. Biomembr.* **1998**, *1376*, 401–416. [CrossRef]
225. de Jesus, A.J.; Allen, T.W. The determinants of hydrophobic mismatch response for transmembrane helices. *Biochim. Biophys. Acta (BBA) Biomembr.* **2013**, *1828*, 851–863. [CrossRef] [PubMed]
226. Parton, D.L.; Klingelhoefer, J.W.; Sansom, M.S.P. Aggregation of model membrane proteins, modulated by hydrophobic mismatch, membrane curvature, and protein class. *Biophys. J.* **2011**, *101*, 691–699. [CrossRef] [PubMed]
227. Bavi, O.; Vossoughi, M.; Naghdabadi, R.; Jamali, Y. The Combined Effect of Hydrophobic Mismatch and Bilayer Local Bending on the Regulation of Mechanosensitive Ion Channels. *PLoS ONE* **2016**, *11*, e0150578. [CrossRef] [PubMed]
228. Webb, R.J.; East, J.M.; Sharma, R.P.; Lee, A.G. Hydrophobic Mismatch and the Incorporation of Peptides into Lipid Bilayers: A Possible Mechanism for Retention in the Golgi. *Biochemistry* **1998**, *37*, 673–679. [CrossRef]
229. Milovanovic, D.; Honigmann, A.; Koike, S.; Göttfert, F.; Pähler, G.; Junius, M.; Müllar, S.; Diederichsen, U.; Janshoff, A.; Grubmüller, H.; et al. Hydrophobic mismatch sorts SNARE proteins into distinct membrane domains. *Nat. Commun.* **2015**, *6*, 5984. [CrossRef]
230. Fowler, P.W.; Williamson, J.J.; Sansom, M.S.P.; Olmsted, P.D. Roles of Interleaflet Coupling and Hydrophobic Mismatch in Lipid Membrane Phase-Separation Kinetics. *J. Am. Chem. Soc.* **2016**, *138*, 11633–11642. [CrossRef]
231. Koval, M. Differential pathways of claudin oligomerization and integration into tight junctions. *Tissue Barriers* **2013**, *1*, e24518. [CrossRef]
232. Koval, M. Claudins—Key Pieces in the Tight Junction Puzzle. *Cell Commun. Adhes.* **2006**, *13*, 127–138. [CrossRef] [PubMed]
233. Heiler, S.; Mu, W.; Zöller, M.; Thuma, F. The importance of claudin-7 palmitoylation on membrane subdomain localization and metastasis-promoting activities. *Cell Commun. Signal.* **2015**, 13. [CrossRef] [PubMed]
234. Van Itallie, C.M.; Gambling, T.M.; Carson, J.L.; Anderson, J.M. Palmitoylation of claudins is required for efficient tight-junction localization. *J. Cell Sci.* **2005**, *118*, 1427–1436. [CrossRef]

235. Fukata, Y.; Bredt, D.S.; Fukata, M. *Protein Palmitoylation by DHHC Protein Family, the Dynamic Synapse: Molecular Methods in Ionotropic Receptor Biology*; CRC Press/Taylor & Francis: Boca Raton, FL, USA, 2006.
236. Gorleku, O.A.; Barns, A.-M.; Prescott, G.R.; Greaves, J.; Chamberlain, L.H. Endoplasmic Reticulum Localization of DHHC Palmitoyltransferases Mediated by Lysine-based Sorting Signals. *J. Biol. Chem.* **2011**, *286*, 39573–39584. [CrossRef] [PubMed]
237. Chamberlain, L.H.; Shipston, M.J. The Physiology of Protein S-acylation. *Physiol. Rev.* **2015**, *95*, 341–376. [CrossRef]
238. Hou, J. Chapter 2—Paracellular Channel Formation. In *The Paracellular Channel*; Hou, J., Ed.; Academic Press: Cambridge, MA, USA, 2019; pp. 9–27. [CrossRef]
239. Hou, J.; Renigunta, A.; Konrad, M.; Gomes, A.S.; Schneeberger, E.E.; Paul, D.L.; Waldegger, S.; Goodenough, D.A. Claudin-16 and claudin-19 interact and form a cation-selective tight junction complex. *J. Clin. Investig.* **2008**, *118*, 619–628. [CrossRef]
240. Mitic, L.L.; Unger, V.M.; Anderson, J.M. Expression, solubilization, and biochemical characterization of the tight junction transmembrane protein claudin-4. *Protein Sci.* **2003**, *12*, 218–227. [CrossRef]
241. Van Itallie, C.M.; Mitic, L.L.; Anderson, J.M. Claudin-2 Forms Homodimers and Is a Component of a High Molecular Weight Protein Complex. *J. Biol. Chem.* **2011**, *286*, 3442–3450. [CrossRef]
242. Suzuki, H.; Tani, K.; Tamura, A.; Tsukita, S.; Fujiyoshi, Y. Model for the architecture of claudin-based paracellular ion channels through tight junctions. *J. Mol. Biol.* **2015**, *427*, 291–297. [CrossRef]
243. Coyne, C.B.; Gambling, T.M.; Boucher, R.C.; Carson, J.L.; Johnson, L.G. Role of claudin interactions in airway tight junctional permeability. *Am. J. Physiol. Lung Cell. Mol. Physiol.* **2003**, *285*, L1166–L1178. [CrossRef]
244. Angelow, S.; Yu, A.S.L. Structure-Function Studies of Claudin Extracellular Domains by Cysteine-scanning Mutagenesis. *J. Biol. Chem.* **2009**, *284*, 29205–29217. [CrossRef] [PubMed]
245. Anbazhagan, V.; Schneider, D. The membrane environment modulates self-association of the human GpA TM domain—Implications for membrane protein folding and transmembrane signaling. *Biochim. Biophys. Acta (BBA) Biomembr.* **2010**, *1798*, 1899–1907. [CrossRef] [PubMed]
246. van Meer, G.; Voelker, D.R.; Feigenson, G.W. Membrane lipids: Where they are and how they behave. *Nat. Rev. Mol. Cell Biol.* **2008**, *9*, 112. [CrossRef] [PubMed]
247. Von Heijne, G. The membrane protein universe: What's out there and why bother? *J. Intern. Med.* **2007**, *261*, 543–557. [CrossRef]
248. Blaskovic, S.; Blanc, M.; Goot, F.G. What does S-palmitoylation do to membrane proteins? *FEBS J.* **2013**, *280*, 2766–2774. [CrossRef]
249. Linder, M.E.; Deschenes, R.J. Palmitoylation: Policing protein stability and traffic. *Nat. Rev. Mol. Cell Biol.* **2007**, *8*, 74–84. [CrossRef]
250. Rossa, J.; Lorenz, D.; Ringling, M.; Veshnyakova, A.; Piontek, J. Overexpression of claudin-5 but not claudin-3 induces formation of trans-interaction–dependent multilamellar bodies. *Ann. N. Y. Acad. Sci.* **2012**, *1257*, 59–66. [CrossRef]
251. Piontek, J.; Winkler, L.; Wolburg, H.; Müller, S.L.; Zuleger, N.; Piehl, C.; Wiesner, B.; Krause, G.; Blasig, I.E. Formation of tight junction: Determinants of homophilic interaction between classic claudins. *FASEB J.* **2007**, *22*, 146–158. [CrossRef]
252. Kausalya, P.J.; Amasheh, S.; Günzel, D.; Wurps, H.; Müller, D.; Fromm, M.; Hunziker, W. Disease-associated mutations affect intracellular traffic and paracellular Mg^{2+} transport function of Claudin-16. *J. Clin Investig.* **2006**, *116*, 878–891. [CrossRef]
253. Galochkina, T.; Ng Fuk Chong, M.; Challali, L.; Abbar, S.; Etchebest, C. New insights into GluT1 mechanics during glucose transfer. *Sci. Rep.* **2019**, *9*, 998. [CrossRef]
254. Pan, Y.; Zhang, Y.; Gongpan, P.; Zhang, Q.; Huang, S.; Wang, B.; Xu, B.; Shan, Y.; Xiong, W.; Li, G.; et al. Single glucose molecule transport process revealed by force tracing and molecular dynamics simulations. *Nanoscale Horiz.* **2018**, *3*, 517–524. [CrossRef]
255. Samanta, P.; Wang, Y.; Fuladi, S.; Zou, J.; Li, Y.; Shen, L.; Weber, C.; Khalili-Araghi, F. Molecular determination of claudin-15 organization and channel selectivity. *J. Gen. Physiol.* **2018**, *150*, 949. [CrossRef] [PubMed]
256. Alberini, G.; Benfenati, F.; Maragliano, L. Molecular Dynamics Simulations of Ion Selectivity in a Claudin-15 Paracellular Channel. *J. Phys. Chem. B* **2018**, *122*, 10783–10792. [CrossRef] [PubMed]
257. Alberini, G.; Benfenati, F.; Maragliano, L. A refined model of claudin-15 tight junction paracellular architecture by molecular dynamics simulations. *PLoS ONE* **2017**, *12*, e0184190. [CrossRef] [PubMed]

258. Weber, C.R.; Turner, J.R. Dynamic modeling of the tight junction pore pathway. *Ann. N. Y. Acad. Sci.* **2017**, *1397*, 209–218. [CrossRef]
259. Shen, L.; Weber, C.R.; Turner, J.R. The tight junction protein complex undergoes rapid and continuous molecular remodeling at steady state. *J. Cell Biol.* **2008**, *181*, 683. [CrossRef]
260. Krystofiak, E.S.; Heymann, J.B.; Kachar, B. Carbon replicas reveal double stranded structure of tight junctions in phase-contrast electron microscopy. *Commun. Biol.* **2019**, *2*, 98. [CrossRef]

International Journal of
Molecular Sciences

Article

Phosphatidylcholine Passes by Paracellular Transport to the Apical Side of the Polarized Biliary Tumor Cell Line Mz-ChA-1

Wolfgang Stremmel [1,*], Simone Staffer [2] and Ralf Weiskirchen [3]

1 Institute of Pharmacy and Molecular Biotechnology, University of Heidelberg, D-69120 Heidelberg, Germany
2 University Clinics of Heidelberg, D-69120 Heidelberg, Germany
3 Institute of Molecular Pathobiochemistry, Experimental Gene Therapy and Clinical Chemistry, RWTH University Hospital Aachen, D-52074 Aachen, Germany
* Correspondence: wolfgangstremmel@aol.com

Received: 6 July 2019; Accepted: 16 August 2019; Published: 19 August 2019

Abstract: Phosphatidylcholine (PC) translocation into mucus of the intestine was shown to occur via a paracellular transport across the apical/lateral tight junction (TJ) barrier. In case this could also be operative in biliary epithelial cells, this may have implication for the pathogenesis of primary sclerosing cholangitis (PSC). We here evaluated the transport of PC across polarized cholangiocytes. Therefore, the biliary tumor cell line Mz-ChA-1 was grown to confluency. In transwell culture systems the translocation of PC to the apical compartment was analyzed. After 21 days in culture, polarized Mz-ChA-1 cells revealed a predominant apical translocation of choline containing phospholipids including PC with minimal intracellular accumulation. Transport was suppressed by TJ destruction employing chemical inhibitors and pretreatment with siRNA to TJ forming proteins as well as the apical transmembrane mucin 3 as PC acceptor. Apical translocation was dependent on a negative apical electrical potential created by the cystic fibrosis transmembrane conductance regulator (CFTR) and the anion exchange protein 2 (AE2). It was stimulated by apical application of secretory mucins. The results indicated the existence of a paracellular PC passage across apical/lateral TJ of the polarized biliary epithelial tumor cell line Mz-ChA-1. This has implication for the generation of a protective mucus barrier in the biliary tree.

Keywords: Mz-ChA-1 cells; biliary epithelial cells; phosphatidylcholine; mucus; tight junctions; paracellular transport

---

1. Introduction

It has been known for a long time that tight junctions (TJ) are responsible for sealing epithelial cells at their apical–lateral side. This enables stable boundaries with their respective functional implication, e.g., at the blood–brain barrier. In intestinal mucosa, they prevent the attack of microbiota. Additionally, they serve the environmental control by allowing water and electrolyte exchange. Macromolecules were not known to pass this barrier. However, in recent studies the novel pathway of phosphatidylcholine (PC) transport across lateral TJ to the luminal side of mucosal cells was described [1].

After apical translocation, PC from systemic sources (lipoproteins) is enriched within the mucus by binding to mucin 2 to establish a hydrophobic barrier against the colonic lumen. Indeed, genetic mouse models with intestinal specific TJ deletion revealed an impaired mucus barrier with lack of PC [2]. This caused an ulcerative colitis (UC) phenotype [2]. The intrinsic lack of mucus PC, the disposition for microbiota invasion and the consequent inflammation matches the situation in human UC [3,4]. The disease is often associated with primary sclerosing cholangitis (PSC), the pathogenesis of which remains obscure [5]. It was indeed postulated that the luminal lack of PC may be of etiological relevance.

When canalicular secretion of PC through the multiple drug resistance gene 2 (MDR2) was deleted in genetically modified mice, a severe cholangitis occurred [6,7]. It is believed that PC is secreted to neutralize the aggressiveness of bile acids by packing them into micelles. Thus, it prevents the attack of bile acids against the plasma membrane phospholipid bilayer of biliary epithelial cells. However, analysis in patients with PSC did not show a lack of PC in bile [8].

These studies neglected the fact that between lumen and the epithelial cells, there is a mucus layer containing secretory mucins. They bind PC to constitute a protective shield towards bile. It is unlikely that mucus PC originates from bile, containing PC-bile acid micelles with high mutual affinity. Therefore, we assume that the biliary mucus compartment is fed with PC from systemic sources as it is the case in intestinal mucosa [2]. We further assume that PC passes through lateral TJ to the apical side of the biliary epithelial cells. Indeed, a common genetic defect in intestinal and biliary epithelial cells could explain the high association of UC and PSC. To prove this concept, we first analyzed the pathway of PC from basal to the apical side employing the biliary tumor cell line Mz-ChA-1. These cells share the characteristics of physiologic biliary epithelial cells [9].

2. Results

According to our hypothesis, the transport of PC to the apical side of biliary epithelial cells requires the presence of TJ. We utilized the well-characterized biliary epithelial tumor cell line Mz-ChA-1 [9] for our experiments and grew them in transwell tissue culture systems for up to 21 days. This cell line has a highly differentiated phenotype producing large quantities of mucus. In comparison to many other human biliary tract carcinoma cell lines such as TGBC-1, TBC-51, Mz-ChA-2 and SK-ChA-1, the cell line Mz-ChA-1 is 10–1000 times less invasive through both the collagen and the basement membrane [10]. The cell line Mz-ChA-1 was originally described as an adenocarcinoma cell line. However, many subsequent studies showed that the cell line has many features of human cholangiocyte cells including TJ complexes rendering these cells as appropriate tool for respective studies [11–14].

While at day three only marginal transepithelial resistance (TER) was detectable with $120 \pm 40\ \Omega$ (non-polarized cells), it increased to $220 \pm 30\ \Omega$ at day nine and remained stable after 21 days (polarized cells) with $400 \pm 80\ \Omega$. Western blot analysis of representative TJ proteins confirmed their presence after 21 days in culture (Figure 1).

Then apical and basal PC transport was evaluated by a respective translocation of PC, provided to the opposite compartment together with taurocholate (TC) at indicated concentrations. For comparison inulin as a non-cell permeable compound was analyzed. Indeed, polarized Mz-ChA-1 cells had the capacity to translocate PC to the apical compartment, whereas basal transport diminished (Figure 2). Inulin, instead, revealed preferential transport to the basal side. These observations were in contrast to non-polarized Mz-ChA-1 cells which resulted in equilibrated PC and inulin concentrations between apical and basal compartments (Figure 2).

The intracellular accumulation was tested in non-polarized cells incubated with 100 μM for 1 h and accounted for <5% of the incubated PC, whether it was provided together with TC or albumin. Only when PC was provided with apolipoprotein B (ApoB) an uptake rate of 12.84 ± 5 nmol·mg protein^{-1}·h^{-1} was observed indicative of lipoprotein-mediated endocytosis of PC. In comparison, 1 h uptake rates for radiolabeled TC and oleate, the later complexed with TC or albumin, were significantly higher ($p < 0.01$) (Figure 3).

Apical transport of PC increased linearly with time of exposure and incubated PC concentrations, was temperature dependent with the highest rates between 25–37 °C, and a pH optimum between pH 7.0 to pH 8.0 (Figure 4).

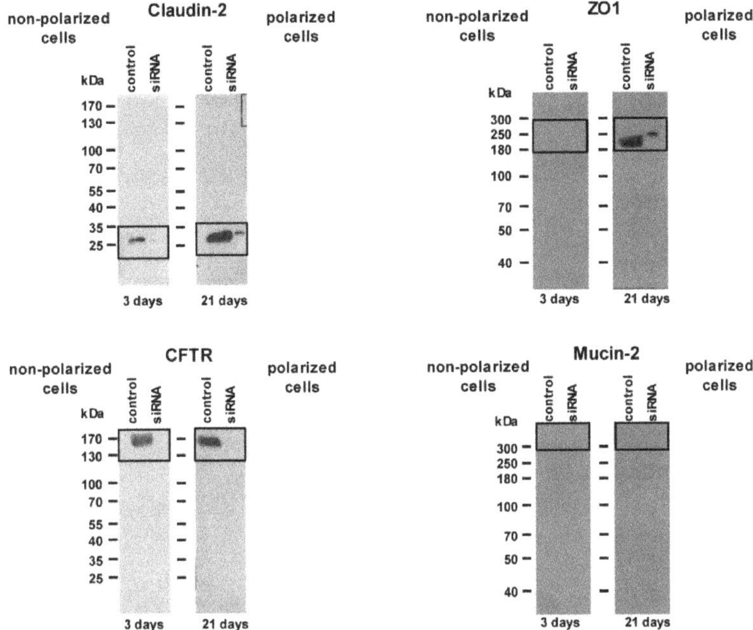

Figure 1. Western blot with non-polarized (three days cultured) and polarized (21 days cultured) MzChA-1 cells with representative proteins of the tight junction (TJ) complex (claudin-2; Zonula Occludens-1, ZO1). Cystic fibrosis transmembrane conductance regulator (CFTR) is a constitutive, non-TJ protein used as housekeeping gene, and mucin 2 as a protein that is not expressed in cholangiocytes. Applied were 30 µg of cell homogenates. The data indicate the enhancement of appearance of the TJ proteins claudin-2 and ZO1 with the polarization of the cells.

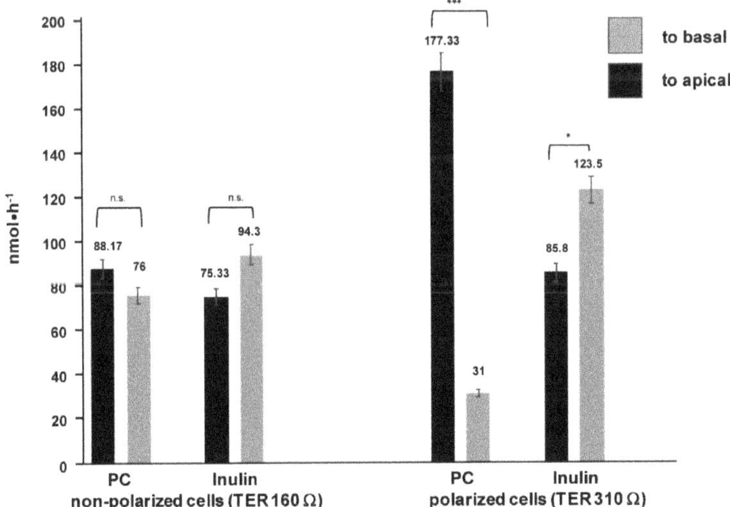

Figure 2. Apical vs. basal transport of phosphatidylcholine (PC) and inulin applied to a transwell culture system of non-polarized (three days cultured) and polarized (21 days cultured) Mz-ChA-1 cells. Means ± SD of $n = 6$; n.s. = not significant, * $p < 0.05$, *** $p < 0.001$. TER, transepithelial resistance.

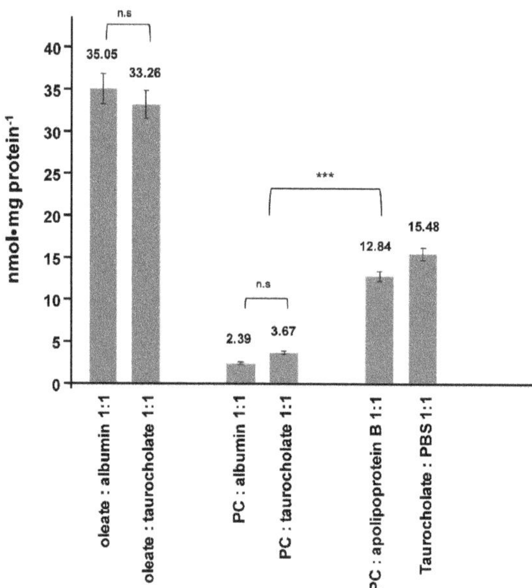

Figure 3. Intracellular accumulation of phosphatidylcholine (PC), fatty acids and taurocholate (TC) in non-polarized Mz-ChA-1 cells. After 1 h incubation of the different substrates with various binding molecules at 37 °C, the amount taken up by the cells was determined after washing the cells three times with 10 mM TC in phosphate-buffered saline (PBS). Means ± SD of $n = 6$; n.s. = not significantly different, *** $p < 0.001$.

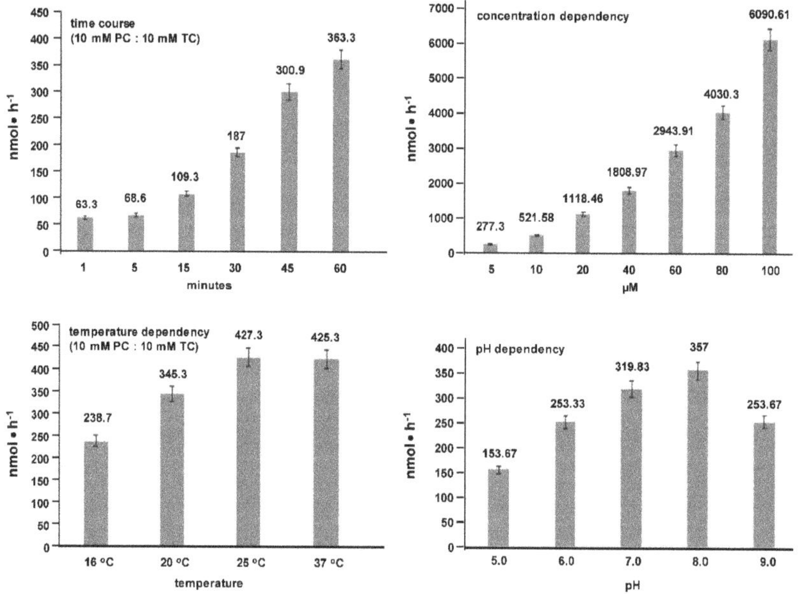

Figure 4. Transport characteristics of phosphatidylcholine (PC) to the apical side of polarized Mz-ChA-1 cells. Apical transport of PC was evaluated after exposure to the basal side. Transport was linear in regard to time and concentration (1 h incubation) with a temperature optimum between 25–37 °C and a pH optimum in the range of pH 7.0 to pH 8.0. Means ± SD of $n = 6$.

Transport was specific for the choline containing phospholipids PC, lysophosphatidylcholine (LPC) and sphingomyelin (SM) but not for other phospholipids (Figure 5).

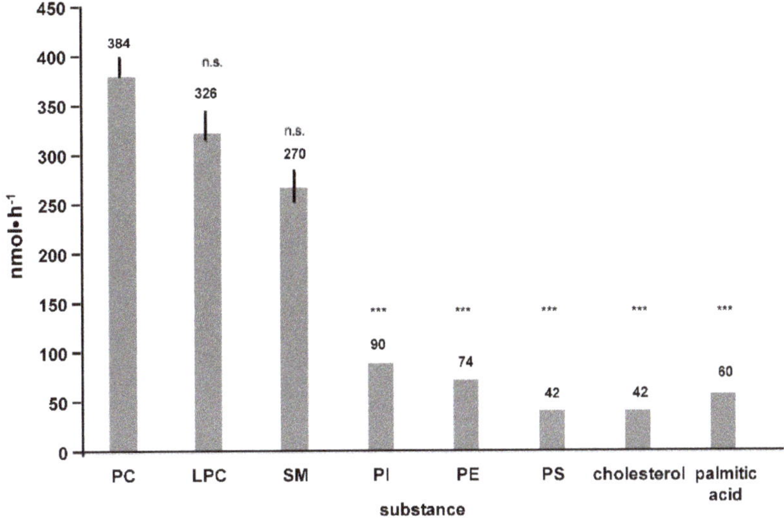

Figure 5. Specificity of apical transport for choline-containing phospholipids in polarized, nine-day cultured Mz-ChA-1 cells. Apical transport was examined for 1 h after basal application of the different substrates at 10 mM phosphatidylcholine (PC) together with 10 mM taurocholate (TC). Means ± SD of $n = 6$. Significances were calculated in relation to PC; n.s. = not significant, *** $p < 0.001$. Abbreviations used are: LPC, lysophosphatidylcholine; SM, sphingomyelin; PI, phosphatidylinositol; PE, phosphatidylethanolamine; PS, phosphatidylserine.

Apical transport was inhibited when a positive charge was generated by application of ammonium chloride (NH_4Cl), sodium thiocyanate (NaSCN) or urea, but remained stable by application of negative charge to the apical surface (Figure 6) [1]. Negative charge in vivo is generated by the cystic fibrosis transmembrane conductance regulator (CFTR) or the anion exchange protein 2 (AE2) which are both present in cholangiocytes [15].

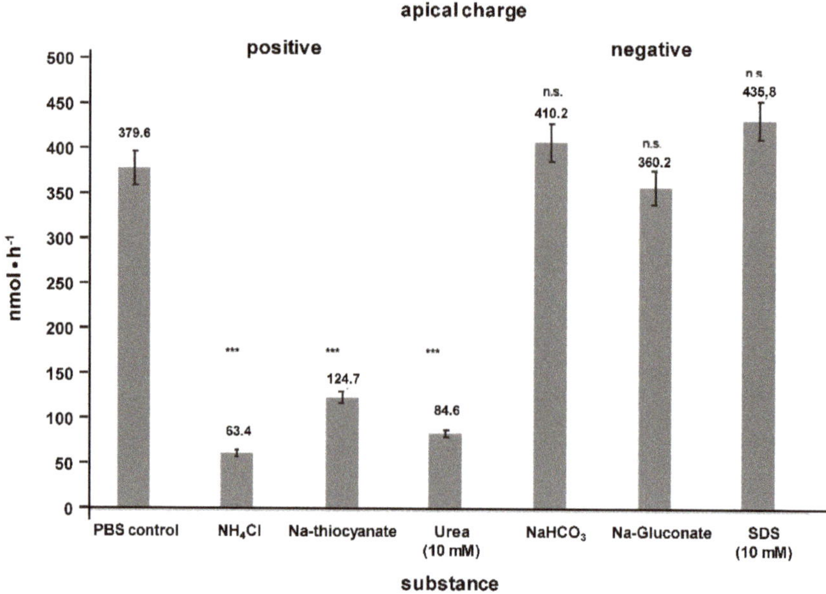

Figure 6. Ionic driving forces for apical translocation of phosphatidylcholine (PC) in polarized Mz-ChA-1 cells. Apical transport of basally applied 10 mM PC: 10 mM taurocholate was examined in polarized (21 days cultured) Mz-ChA-1 cells apically equilibrated for 1 h with indicated salts and substrates (130 mM) generating apical positive or negative charge. Means ± SD of $n = 6$. Significances were calculated in relation to the phosphate-buffered saline (PBS) control; n.s. = not significant, *** $p < 0.001$.

When CFTR or AE2 were reduced by siRNA pretreatment, PC transport was strongly inhibited (Figure 7).

The hypothesis of a TJ-mediated translocation process was supported by the observation of its inhibition by exposure of polarized Mz-ChA-1 cells to acetaldehyde (ACA) vapor or the selective peroxisome proliferator-activated receptor γ (PPARγ) inhibitors T0070907 or GW9962 (Figure 8).

Moreover, apical transport was diminished by pretreatment of polarized Mz-ChA-1 cells with siRNAs against members of the TJ complex or their upstream control proteins kindlin-1 and -2 (Figure 7). Although cholangiocytes do not synthesize mucin 1 and 2, they were shown to have mucin 3 which is a transmembrane protein serving in accepting apically transported PC as it was shown in two previous studies [1,2]. Its inhibition by siRNA reduced PC translocation significantly. From mucin 3 PC is handled to secreted mucin 1 and 2 which originates from goblet cells [16]. When it was applied to the apical surface of polarized Mz-ChA-1 cells transport was dose dependently enhanced as it was also the case when increasing concentrations of TC were apically added which favor the drainage of transported PC to the apical side by incorporation of PC into micelles (Figure 9). However, it has to be considered that Mz-ChA-1 cells as well as biliary epithelial cells in vivo do not exhibit a mucin 2 containing mucus layer.

To document a vectoral transport of PC from the basal to the apical surface, it was tested when increasing, but equal concentrations of PC in both compartments were provided. In contrast to inulin, where basal transport was higher than apical translocation, PC was overwhelmingly transported apically with only marginal appearance at the basal side indicating a predominant paracellular transport across TJ to the luminal surface (Figure 10).

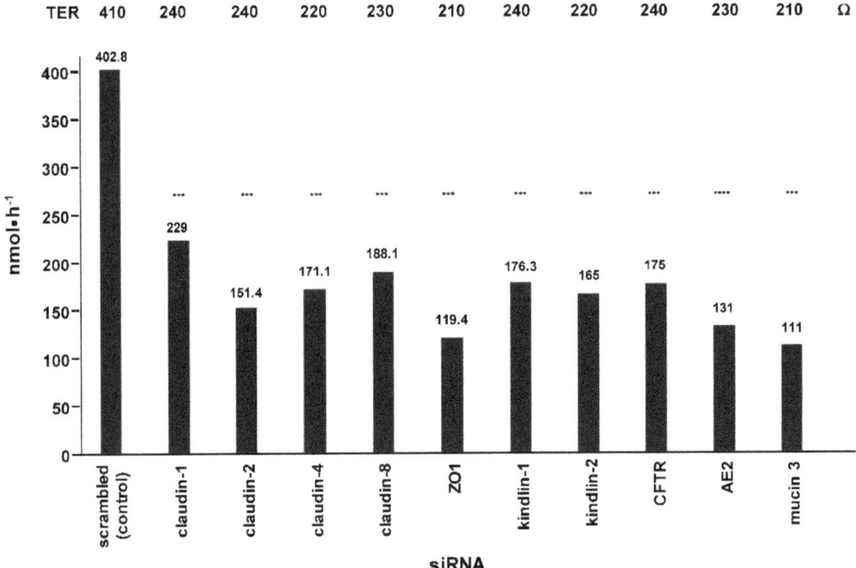

Figure 7. Effect of siRNA suppression of proteins involved in apical translocation of phosphatidylcholine (PC). Apical transport of PC after basal application of 10 mM PC: 10 mM taurocholate was examined after siRNA knockdown of indicated proteins which are all involved in tight junction-mediated luminal secretion of PC. Means ± SD of $n = 6$. Significances were calculated in relation to the control pretreated with scrambled siRNA; n.s. = not significant, *** $p < 0.001$. ZO1, Zonula Occludens-1; CFTR, cystic fibrosis transmembrane conductance regulator; AE2, anion exchange protein 2; TER, transepithelial resistance.

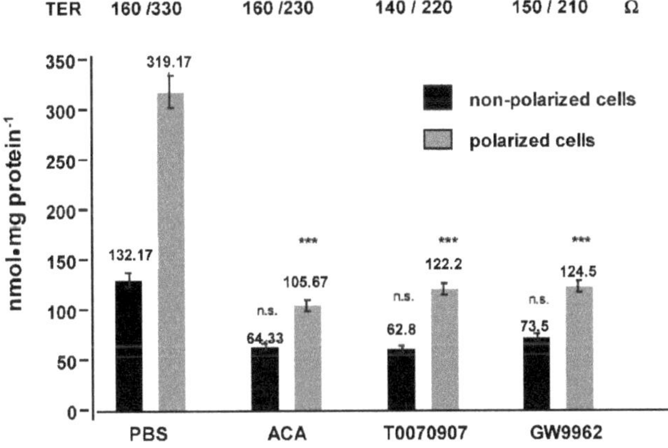

Figure 8. Effect of disruption of tight junctions (TJ) by acetaldehyde (ACA) or the indicated peroxisome proliferator-activated receptor γ (PPARγ) inhibitors on apical transport of phosphatidylcholine (PC). TJ were disrupted by exposure to ACA vapor or incubation of upper transwell chambers with the PPARγ inhibitors. Then apical transport of 10 mM PC: 10 mM taurocholate for 1 h was registered. Means ± SD of $n = 6$. Significances were calculated separately for non-polarized and polarized Mz-ChA-1 cells and compared to phosphate-buffered saline (PBS)-treated controls; n.s. = not significant, *** $p < 0.001$.

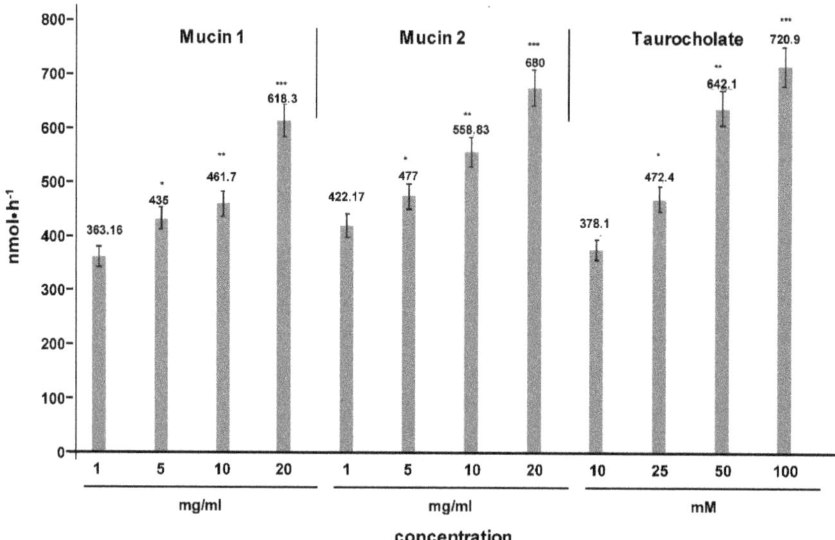

Figure 9. Enhancement of apical phosphatidylcholine (PC) transport by application of increasing concentrations of secretory mucin 1, mucin 2, and taurocholate (TC) as luminal acceptor substances. Apical transport of 10 mM PC: 10 mM TC with polarized Mz-ChA-1 cells was determined as a function of increasing concentrations of mucin 1, mucin 2 and TC in the upper culture system. Means ± SD of $n = 6$. Significances were calculated in relation to the lowest concentration of each substance added. * $p < 0.05$; ** $p < 0.01$; *** $p < 0.001$.

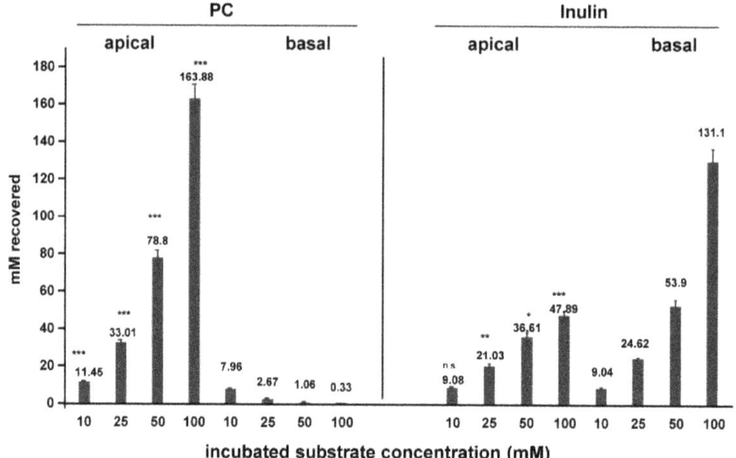

Figure 10. Equilibrium distribution of phosphatidylcholine (PC) and inulin when applied in increasing concentrations to upper and lower compartment of the transwell culture system. Increasing concentrations up to 100 mM of PC and inulin were added to the apical and basal compartments. Over 1 h observation, vectoral transport of PC was completely directed apically with decreasing PC at the luminal side (less than 1%). For inulin there was a significantly higher basal accumulation starting at 25 mM incubated. The process appeared to be less effective, at least at lower concentrations, as for the transport of PC to the apical side. Means ± SD of $n = 6$. Significances were calculated for apical versus basal transport for each in relation to the lowest concentration of each substrate concentration. n.s. = not significant; * $p < 0.05$; ** $p < 0.01$; *** $p < 0.001$.

3. Discussion

Kinetic analyses of the translocation of PC across the polarized biliary epithelial tumor cell line Mz-ChA-1 revealed a TJ-mediated process. This was as unexpected as it was for TJ-mediated PC transport in the intestinal mucosa [1,2]. This transport is strongly unidirectional to the apical side, driven by a negative apical charge generated by CFTR and AE2. The translocated PC is bound to mucin 3 from where it is handled to mucin 1 and 2 in the mucus layer of the biliary tract.

The PC in mucus serves in the biliary tree to protect the epithelial cells from the attack of bile acids, which achieve concentrations in the millimolar range within the small ducts [17]. This can be neutralized by canalicular secretion of PC via MDR2/MDR3 (ABCB4) through the formation of micelles.

However, the excessive load of bile acids could in vivo take PC from the mucus layer still maintaining a shield towards biliary epithelium, because there is a constant basal supply to the apical surface. In fact, as shown in this study, PC secretion is proportionally increased with TC in the apical (biliary) compartment. Thus, PC in cell membranes is not exposed to the luminal bile acid load. However, when, for example by a genetic defect, the apical PC secretion through the biliary epithelial cell layer is impaired, they are attacked, and a cholangitis occurs. This hypothesis will now be proven in a genetic mouse model where the biliary TJ are deleted. The observation of a disturbed TJ barrier as a potential cause of impaired secretion of PC to biliary mucus could shed light on the pathogenesis of PSC. It associates to UC with a genetically determined disruption of the TJ barrier and diminished secretion of PC to the intestinal mucus [2–4]. When this relation can also be verified in the biliary epithelium, new therapeutic strategies for PSC will be developed.

The shortcoming of this study is the use of a tissue culture model, even with a cholangiocyte-derived tumor cell line. However, biliary epithelial cells are not obtainable in high yield, in reproducible fashion and sufficient functionality. Moreover, isolated cholangiocytes do not exhibit a mucus layer. It can only be pointed out that the employed Mz-ChA-1 cell line exhibits TJ complexes and is more differentiated in comparison to other cell lines obtained from cholangiocellular carcinomas. Another shortcoming is the lack of data on the structure of the TJ under the various experimental conditions. Moreover, imaging of transport employing fluorescent PC could not be examined in these experiments. All of these shortcomings can be tackled, when a genetic mouse model with TJ deletion selective for the biliary epithelium becomes available.

4. Materials and Methods

4.1. Cell Culture Transport Studies

Transwell tissue culture for basal–apical polarization was established for the biliary epithelial tumor cell line Mz-ChA-1 obtained from Dr. Alexander Knuth (Mainz, Germany) in our laboratory [10]. In each well of the 12-well collagen-coated transwell culture dish 7.5×10^4 cells (corresponding to 80 ± 18 µg protein) were placed and cultured in Dulbecco's modified Eagle's medium containing 5% fetal calf serum ((FCS) (Life Technologies, Carlsbad, CA, USA). They were grown for up to 21 days and polarity was examined by TER [1].

Translocation of radiolabeled substrates to the apical or basal side was examined by its application to one side and its recovery in the opposite compartment. For equilibrium distribution studies substrates were located in both compartments and transport to apical and basal side was determined. Standard incubation procedures used 10 mM PC (100,000 cpm) bound to 10 mM TC in phosphate buffered saline (PBS) (pH 7.4) at 37 °C in 1 mL placed at basal side and recovery was analyzed after 1 h in the upper compartment with 1 mL 10 mM TC–PBS. As substrates we used [^{3}H]phosphatidylcholine (PC), [^{14}C]lysophosphatidylcholine (LPC), [^{3}H]sphingomyelin (SM), [^{14}C]phosphatidylethanolamine (PE), [^{3}H]phosphatidylinositol (PI), [^{3}H]palmitate (PA), [^{3}H]oleate (OA), [^{3}H]taurocholate (TC) and [^{14}C]inulin (all brought up with unlabeled substrate to desired concentrations). Transport characteristics included dependency on time, concentration, temperature and pH and was evaluated as described [1]. For driving force analysis, the 10 mM TC in the apical medium was applied in different buffers at

130 mM and pH 7.0 which generated a more positive charge in the medium (NH_4Cl, Na-thiocyanate and 10 mM urea in PBS) or a more negative charge ($NaHCO_3$, Na-gluconate and 10 mM sodium dodecylsulphate (SDS)) in PBS [1].

Radiolabeled compounds were purchased from PerkinElmer (Waltham, MA, USA). For TJ disruption apical application of 150 µM ACA for 3 h [18] or the peroxisome proliferator-activated receptor gamma inhibitors T0070907 (10 µM) or GW9662 (1 µM) for 1 h were used [19].

siRNA knockdown experiments were employed to test the functionality of proteins involved in TJ constitution and proteins of significance for PC translocation. As in earlier studies with CaCo2 cells, we used in each case 78 pmol siRNA scrambled as control and targeted to claudin-1, -2, -4 and -8, Zonula Occludens-1 (ZO1), kindlin-1 and -2, CFTR, AE2 and mucin 1, mucin 2, and mucin 3 [1]. siRNAs were apically applied for 16 h at 37 °C. After washing transport studies were initiated.

4.2. siRNA Knockdown Experiments

All sense and antisense probes except for kindlin-1 and kindlin-2 were obtained from Sigma (St. Louis, MO, USA) and are depicted in Table 1. The probes for kindlin-1 and kindlin-2 were a kind gift of Reinhard Faessler, Max Planck Institute of Biochemistry, Munich, Germany. The efficiency of siRNA knockdown was confirmed by Western blotting. Transport characteristics included dependency on time, concentration, temperature and pH and was evaluated as described [1].

Table 1. siRNAs used in knockdown Experiments.

siRNA	Sense Oligo	Antisense Oligo
scrambled	5′-gaugggaccuggccaguga-3′[dT][dT]	5′-ucacuggccagguccaauc-3′[dT][dT]
claudin-1	5′-cagucaaugccagguacga-3′[dT][dT]	5′-ucguaccuggcauugacug-3′[dT][dT]
claudin-2	5′-gacacuaccacuggaucgu-3′[dT][dT]	5′-acgauccagugguaguguc-3′[dT][dT]
claudin-4	5′-gaccaucugggagggccua-3′[dT][dT]	5′-uaggcccucccagauggguc-3′[dT][dT]
claudin-8	5′gguucaagcaucuacucuu-3′[dT][dT]	5′-aagaguagaugcuugaacc-3′[dT][dT]
mucin 2	5′-gcaacauuaccgucugcaa-3′[dT][dT]	5′-uugcagacgguaauguugc-3′[dT][dT]
mucin 3	5′-ccaaacuacucuuacuaca-3′[dT][dT]	5′-uguaguaagaguaguuugg-3′[dT][dT]
CFTR	5′-gaacacauaccuucgauau-3′[dT][dT]	5′-auaucgaagguauguguuc-3′[dT][dT]
AE2	5′-gagaucuucgccuucuuga-3′[dT][dT]	5′-ucaagaaggcgaagaucuc-3′[dT][dT]
ZO1	5′-gagaugaacgggcuacgcu-3′[dT][dT]	5′-agcguagcccguucaucuc-3′[dT][dT]
kindlin-1	5′-ggacauuacugauaucccu-3′[dT][dT]	5′-agggauaucaguaauguccc-3′[dT][dT]
kindlin-2	5′-gugugaauagaaauacugu-3′[dT][dT]	5′-acaguauuucuauucacac-3′[dT][dT]

4.3. Immunoblot Analysis

Cell homogenate samples with 30 µg protein were applied to the gel slots for electrophoretic separation and immunoblotting using a standard protocol [1]. Primary antibodies against the following human proteins were used: claudin 2 (cat. no.: TA347352; 1:500) (Acris Antibodies, Herford, Germany); ZO1 (cat. no.: ab96587; 1:200; Abcam, Cambridge, MA, USA); kindlin-2 (1:500; gift from Reinhard Faessler, Max Planck Institute of Biochemistry, Munich, Germany); mucin 2 (cat. no.: sc-59859; 1:500; Santa Cruz Biotech., Santa Cruz, CA, USA) and CFTR (sc-376683; 1:200; Santa Cruz Biotech.

4.4. Statistical Analysis

Statistical analysis was performed using Prism 4.0 software (GraphPad Software Inc., LaJolla, CA, USA). Differences between groups were evaluated using the Mann–Whitney U test. Data are presented as means ± SD and $p < 0.05$ was considered statistically significant.

5. Conclusions

The manuscript shows for the first time that phosphatidylcholine can pass by paracellular transport from systemic sources to the apical side of biliary epithelial cells. It passes the TJ barrier driven by an apical negative electrical potential, generated by CFTR and AE2. Mucin 3 is the initial apical acceptor molecule from where PC is handled to secretory mucin 1 and 2 to establish the barrier against the biliary lumen. It protects from high bile acid concentration within the biliary lumen (Figure 11).

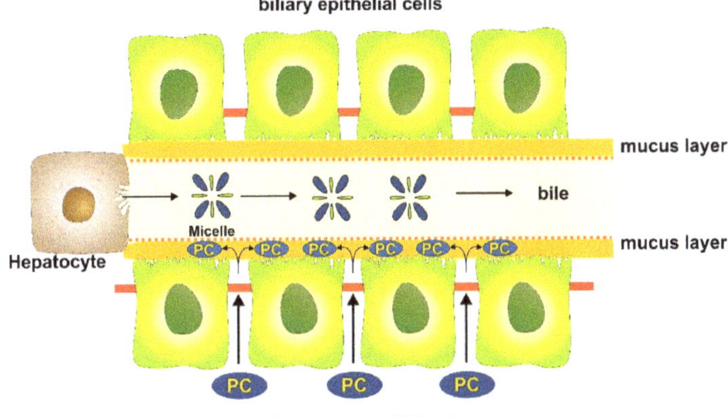

Figure 11. Scheme illustrating the paracellular transport of phosphatidylcholine (PC) across the tight junction barrier to the mucus layer at the apical side of biliary epithelium. The transport is driven by a negative electrical gradient with consequent binding to membrane-localized mucin 3 and an equilibrated shift to secretory mucin 2.

Author Contributions: Conceptualization, W.S.; formal analysis, W.S.; investigation, S.S.; validation, W.S.; visualization, W.S. and R.W.; writing—original draft preparation, W.S.; writing—review and editing, W.S. and R.W.

Funding: R.W. is supported by the German Research Foundation (SFB/TRR57, projects P13 and Q03). The funder had no role in the design of the study, data collection, analyses or interpretation of data, writing of the manuscript, or in the decision to publish the results.

Conflicts of Interest: The authors declare no conflict of interest.

Abbreviations

ACA	acetaldehyde
AE2	anion exchange protein 2
ApoB	apolipoprotein B
CFTR	cystic fibrosis transmembrane conductance regulator
LPC	lysophosphatidylcholine
PBS	phosphate-buffered saline
PC	phosphatidylcholine
PSC	primary sclerosing cholangitis
SM	sphingomyelin
TER	transepithelial resistance
TC	taurocholate
TJ	tight junction(s)
UC	ulcerative colitis
ZO1	Zonula Occludens-1

References

1. Stremmel, W.; Staffer, S.; Gan-Schreier, H.; Wannhoff, A.; Bach, M.; Gauss, A. Phosphatidylcholine passes through lateral tight junctions for paracellular transport to the apical side of the polarized intestinal tumor cell-line CaCo2. *Biochim. Biophys. Acta* **2016**, *1861 Pt A*, 1161–1169. [CrossRef]

2. Stremmel, W.; Staffer, S.; Schneider, M.J.; Gan-Schreier, H.; Wannhoff, A.; Stuhrmann, N.; Gauss, A.; Wolburg, H.; Mahringer, A.; Swidsinski, A.; et al. Genetic mouse models with intestinal-specific tight junction deletion resemble an ulcerative colitis phenotype. *J. Crohns. Colitis* **2017**, *11*, 1247–1257. [CrossRef] [PubMed]
3. Ehehalt, R.; Wagenblast, J.; Erben, G.; Lehmann, W.D.; Hinz, U.; Merle, U.; Stremmel, W. Phosphatidylcholine and lysophosphatidylcholine in intestinal mucus of ulcerative colitis patients. A quantitative approach by nanoElectrospray-tandem mass spectrometry. *Scand. J. Gastroenterol.* **2004**, *39*, 737–742. [CrossRef] [PubMed]
4. Braun, A.; Schönfeld, U.; Welsch, T.; Kadmon, M.; Funke, B.; Gotthardt, D.; Zahn, A.; Autschbach, F.; Kienle, P.; Zharnikov, M.; et al. Reduced hydrophobicity of the colonic mucosal surface in ulcerative colitis as a hint at a physicochemical barrier defect. *Int. J. Colorectal. Dis.* **2011**, *26*, 989–998. [CrossRef] [PubMed]
5. Saich, R.; Chapman, R. Primary sclerosing cholangitis, autoimmune hepatitis and overlap syndromes in inflammatory bowel disease. *World J. Gastroenterol.* **2008**, *14*, 331–337. [CrossRef] [PubMed]
6. Smit, J.J.; Schinkel, A.H.; Oude Elferink, R.P.; Groen, A.K.; Wagenaar, E.; van Deemter, L.; Mol, C.A.; Ottenhoff, R.; van der Lugt, N.M.; van Roon, M.A.; et al. Homozygous disruption of the murine mdr2 P-glycoprotein gene leads to a complete absence of phospholipid from bile and to liver disease. *Cell* **1993**, *75*, 451–462. [CrossRef]
7. Oude Elferink, R.P.; Paulusma, C.C. Function and pathophysiological importance of ABCB4 (MDR3 P-glycoprotein). *Pflug. Arch.* **2007**, *453*, 601–610. [CrossRef] [PubMed]
8. Gauss, A.; Ehehalt, R.; Lehmann, W.D.; Erben, G.; Weiss, K.H.; Schaefer, Y.; Kloeters-Plachky, P.; Stiehl, A.; Stremmel, W.; Sauer, P.; et al. Biliary phosphatidylcholine and lysophosphatidylcholine profiles in sclerosing cholangitis. *World J. Gastroenterol.* **2013**, *19*, 5454–5463. [CrossRef] [PubMed]
9. Knuth, A.; Gabbert, H.; Dippold, W.; Klein, O.; Sachsse, W.; Bitter-Suermann, D.; Prellwitz, W.; Meyer zum Buschenfelde, K.H. Biliary adenocarcinoma. Characterisation of three new human tumor cell lines. *J. Hepatol.* **1985**, *1*, 579–596. [CrossRef]
10. Koike, N.; Todoroki, T.; Kawamoto, T.; Yoshida, S.; Kashiwagi, H.; Fukao, K.; Ohno, T.; Watanabe, T. The invasion potentials of human biliary tract carcinoma cell lines: Correlation between invasiveness and morphologic characteristics. *Int. J. Oncol.* **1998**, *13*, 1269–1274. [CrossRef] [PubMed]
11. Braconi, C.; Swenson, E.; Kogure, T.; Huang, N.; Patel, T. Targeting the IL-6 dependent phenotype can identify novel therapies for cholangiocarcinoma. *PLoS ONE* **2010**, *5*, e15195. [CrossRef] [PubMed]
12. Schrumpf, E.; Tan, C.; Karlsen, T.H.; Sponheim, J.; Björkström, N.K.; Sundnes, O.; Alfsnes, K.; Kaser, A.; Jefferson, D.M.; Ueno, Y.; et al. The biliary epithelium presents antigens to and activates natural killer T cells. *Hepatology* **2015**, *62*, 1249–1259. [CrossRef] [PubMed]
13. Onori, P.; Wise, C.; Gaudio, E.; Franchitto, A.; Francis, H.; Carpino, G.; Lee, V.; Lam, I.; Miller, T.; Dostal, D.E.; et al. Secretin inhibits cholangiocarcinoma growth via dysregulation of the cAMP-dependent signaling mechanisms of secretin receptor. *Int. J. Cancer* **2010**, *127*, 43–54. [CrossRef] [PubMed]
14. Zach, S.; Birgin, E.; Rückert, F. Primary cholangiocellular carcinoma cell lines. *J. Stem Cell Res. Transpl.* **2015**, *2*, 1013.
15. Tietz, P.S.; Marinelli, R.A.; Chen, X.M.; Huang, B.; Cohn, J.; Kole, J.; McNiven, M.A.; Alper, S.; LaRusso, N.F. Agonist-induced coordinated trafficking of functionally related transport proteins for water and ions in cholangiocytes. *J. Biol. Chem.* **2003**, *278*, 20413–20419. [CrossRef] [PubMed]
16. Perez-Vilar, J.; Hill, R.L. The structure and assembly of secreted mucins. *J. Biol. Chem.* **1999**, *274*, 31751–31754. [CrossRef] [PubMed]
17. Dawson, P.A.; Lan, T.; Rao, A. Bile acid transporters. *J. Lipid Res.* **2009**, *50*, 2340–2357. [CrossRef] [PubMed]
18. Dunagan, M.; Chaudhry, K.; Samak, G.; Rao, R.K. Acetaldehyde disrupts tight junctions in Caco-2 cell monolayers by a protein phosphatase 2A-dependent mechanism. *Am. J. Physiol. Gastrointest. Liver Physiol.* **2012**, *303*, G1356–G1364. [CrossRef] [PubMed]
19. Ogasawara, N.; Kojima, T.; Go, M.; Ohkuni, T.; Koizumi, J.; Kamekura, R.; Masaki, T.; Murata, M.; Tanaka, S.; Fuchimoto, J.; et al. PPARγ agonists upregulate the barrier function of tight junctions via a PKC pathway in human nasal epithelial cells. *Pharm. Res.* **2010**, *61*, 489–498. [CrossRef] [PubMed]

International Journal of
Molecular Sciences

Review

Reciprocal Association between the Apical Junctional Complex and AMPK: A Promising Therapeutic Target for Epithelial/Endothelial Barrier Function?

Kazuto Tsukita [1,2], Tomoki Yano [3], Atsushi Tamura [1,4] and Sachiko Tsukita [1,4,*]

1 Laboratory of Biological Science, Graduate School of Frontier Biosciences, Osaka University, Suita, Osaka 565-0811, Japan; kazusan@kuhp.kyoto-u.ac.jp (K.T.); atamura@biosci.med.osaka-u.ac.jp (A.T.)
2 Department of Neurology, Graduate School of Medicine, Kyoto University, Kyoto 606-8507, Japan
3 Laboratory of Biological Science, Graduate School of Medicine, Osaka University, Suita, Osaka 565-0811, Japan; t-yano@biosci.med.osaka-u.ac.jp
4 Strategic Innovation and Research Center, Teikyo University, Tokyo 173-8605, Japan
* Correspondence: atsukita@biosci.med.osaka-u.ac.jp; Tel.: +81-6-6879-3321; Fax: +81-6-6879-3329

Received: 8 November 2019; Accepted: 26 November 2019; Published: 29 November 2019

Abstract: Epithelial/endothelial cells adhere to each other via cell–cell junctions including tight junctions (TJs) and adherens junctions (AJs). TJs and AJs are spatiotemporally and functionally integrated, and are thus often collectively defined as apical junctional complexes (AJCs), regulating a number of spatiotemporal events including paracellular barrier, selective permeability, apicobasal cell polarity, mechano-sensing, intracellular signaling cascades, and epithelial morphogenesis. Over the past 15 years, it has been acknowledged that adenosine monophosphate (AMP)-activated protein kinase (AMPK), a well-known central regulator of energy metabolism, has a reciprocal association with AJCs. Here, we review the current knowledge of this association and show the following evidences: (1) as an upstream regulator, AJs activate the liver kinase B1 (LKB1)–AMPK axis particularly in response to applied junctional tension, and (2) TJ function and apicobasal cell polarization are downstream targets of AMPK and are promoted by AMPK activation. Although molecular mechanisms underlying these phenomena have not yet been completely elucidated, identifications of novel AMPK effectors in AJCs and AMPK-driven epithelial transcription factors have enhanced our knowledge. More intensive studies along this line would eventually lead to the development of AMPK-based therapies, enabling us to manipulate epithelial/endothelial barrier function.

Keywords: tight junction; adherens junction; apical junctional complex; AMP-activated protein kinase (AMPK); paracellular barrier

---

1. Introduction

Epithelial/endothelial cells adhere to each other via cell–cell junctions, thereby collectively functioning as a permselective barrier between the internal and external environments of the body to define the framework of every biological compartment of our body [1–6]. Because the size and function of each compartment and their inter-relationships are critical for biological systems, the specific and general principles regulating the permselective barrier function of epithelial/endothelial cells are of considerable interest. Additionally, it is attractive from a therapeutic point of view to investigate methods to manipulate barrier function, given that its disruption can lead to various diseases [7–17]. For example, changes in paracellular barrier function can initiate and accelerate inflammation (both locally and systemically) [10–12,15–17], contribute to tumor formation [14,18], initiate infection [19], and disrupt ion homeostasis [9,13,20], and have even been suggested as being potentially involved in the initiation and progression of psychiatric and neurodegenerative diseases [21–26].

Tight junctions (TJs) are vertebrate-specific junctions that are located most apically in epithelial cell–cell junctions. It is generally accepted that adherens junctions (AJs), which are located immediately at the basal side of TJs, are essential for epithelial cell sheet formation, on the basis of which TJs create a paracellular permselective barrier [6,27,28]. Therefore, TJs and AJs are considered to be spatiotemporally and functionally integrated, and are thus often collectively known as apical junctional complexes (AJCs). In addition to creating a paracellular barrier, AJCs play essential roles in apicobasal cell polarity, mechano-sensing, intracellular signaling cascades, and epithelial morphogenesis [1,6,28–32]. A number of signaling networks are involved in AJC-related events and, over the past 15 years, adenosine monophosphate (AMP)-activated protein kinase (AMPK), a well-known central regulator of energy metabolism signaling pathway, has been shown to have a reciprocal association with AJCs [33–40]. More specifically, AJs activate AMPK particularly in response to applied junctional tension, whereas TJ barrier function and apicobasal cell polarization are promoted by AMPK activation. Herein, we review the critical studies regarding this reciprocal association and discuss the therapeutic potential of AMPK to manipulate epithelial/endothelial barrier function.

2. Overview of Function and Regulation of AMPK

In the 1970s and 1980s, a series of biochemical experiments identified the presence of a 3-hydroxy-3-methylglutaryl-coenzyme A (HMG-CoA) reductase kinase, which was found to be activated by AMP [41–43]. Meanwhile, an acetyl-coenzyme A (acetyl-CoA) carboxylase kinase was purified from rat liver in the mid-1980s and was also found to be activated by AMP [44]. Later, the independently identified HMG-CoA reductase kinase and acetyl-CoA carboxylase kinase were shown to be the same protein, and this protein was subsequently given the name "AMPK" after its characteristic feature of activation by AMP [45,46].

AMPK is a heterotrimeric serine/threonine kinase composed of one catalytic α-subunit and two regulatory β- and γ-subunits, and is highly conserved in all eukaryotes from protozoa to vertebrates (Figure 1) [47–56].

AMPK functions as a central regulator of energy metabolism and, once activated, it turns on catabolic processes as well as turns off anabolic processes, leading to upregulation of adenosine triphosphate (ATP) concentrations (Figure 2). Activation of AMPK is determined by the balance of the following three factors (Figure 2) [48,49,57]: (1) the AMP/ATP ratio [48,57], (2) the activity of upstream kinases to phosphorylate α-subunit Thr-172 including liver kinase B1 (LKB1) and Ca^{2+}/calmodulin-dependent protein kinase kinase β (CaMKKβ) [58,59], and (3) the activity of upstream protein phosphatases that dephosphorylate α-subunit Thr-172 such as protein phosphatase 2A and 2C [49,60].

Notably, binding of AMP to the γ-subunit can activate AMPK by three mechanisms that can be antagonized by ATP (Figure 3): (1) AMP drives direct allosteric activation of AMPK [61], (2) AMP renders AMPK susceptible to α-subunit Thr-172 phosphorylation by upstream kinases [62,63], and (3) AMP protects phosphorylated α-subunit Thr-172 from dephosphorylation by upstream phosphatases [64]. It should also be noted that the effect of direct allosteric activation by AMP on AMPK activity is modest (approximately a fivefold increase in activity) [61] compared with that of α-subunit Thr-172 phosphorylation (more than 100-fold increase in activity) [64]; therefore, the phosphorylation status of α-subunit Thr-172 is essentially a molecular on–off switch for AMPK activity [57] (Figure 3). Because AMPK activation has potential to mediate a variety of different cellular consequences, AMPK activation is controlled, not only temporally by the phosphorylation status of α-subunit Thr-172, but also spatially [47,57]. Spatially, AMPK is mobilized to various subcellular organelles only in response to appropriate signals. As examples, glucose starvation leads to AMPK targeting to lysosomes, increased autophagic influx results in AMPK targeting to autophagosomes, and mitochondrial damage causes AMPK targeting to mitochondria [65].

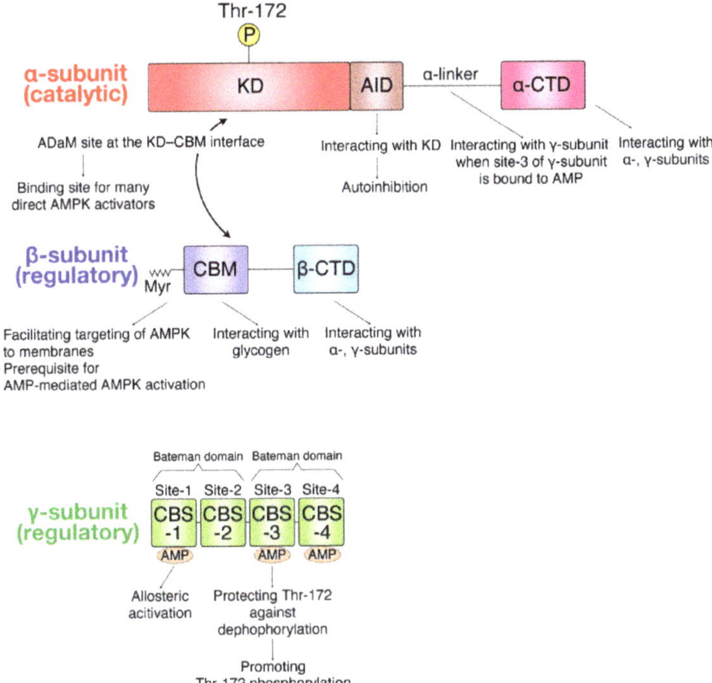

Figure 1. Domains of adenosine monophosphate (AMP)-activated protein kinase (AMPK) subunits. The kinase domain (KD) in the catalytic α-subunit exerts AMPK kinase activity, and Thr-172 phosphorylation of the KD is critical for its activity. The autoinhibitory domain (AID) binds to the backside of KD, constraining its mobility, thereby inhibiting kinase activity [52]. When AMP is bound at site-3 in the γ-subunit, the α-linker binds to the γ-subunit, thereby pulling AID away from the KD [53]. The C-terminal domain of the α-subunit (α-CTD) is responsible for interacting with the β- and γ-subunits. The allosteric drug and metabolite (ADaM) site at the KD–carbohydrate-binding module (CBM) interface is a binding site for many direct AMPK activators that induces allosteric activation, including A-769662 and compound 991 (see Table 1) [56]. The N-terminal of the regulatory β-subunit has been shown to be myristoylated, which facilitates targeting of AMPK to membranes and is prerequisite for AMP-mediated α-subunit Thr-172 phosphorylation [54]. The CBM in the β-subunit interacts with glycogen, which facilitates targeting of AMPK to glycogen for glycogen turnover regulation [55]. The C-terminal domain of the β-subunit (β-CTD) interacts with both α- and γ- subunits and serves as a connection between the α- and γ- subunits [51]. The regulatory γ-subunits contain four cystathionine beta synthase (CBS) domains, packing together to generate two Bateman domains, thereby providing four potential AMP-binding sites (site-1 to site-4) [48]. However, in mammalian AMPK, only site-1, site-3, and site-4 bind to AMP because aspartic acid residues in a conserved potential binding site in site-2 are changed to arginine [48]. Site-4 contains a permanently-bound AMP molecule, whereas AMP transiently binds to site-1 and site-3 (site-1 has a much stronger binding affinity for AMP than site-3). AMP binding to site-1 causes allosteric activation of AMPK, and AMP binding to site-3 promotes α-subunit Thr-172 phosphorylation by protecting AMPK from upstream phosphatases [49]. Myr, myristoylation.

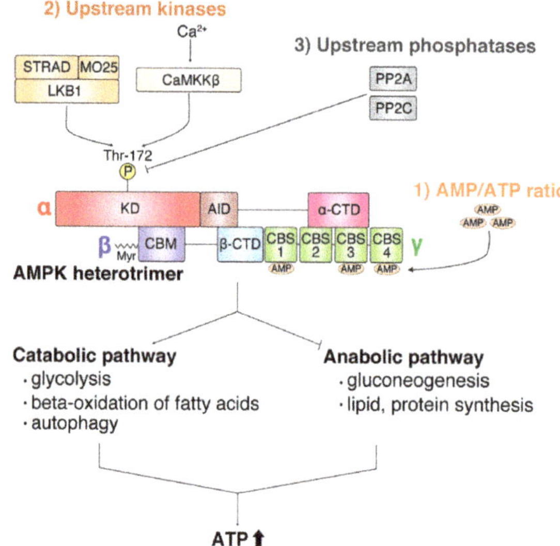

Figure 2. Overview of adenosine monophosphate (AMP)-activated protein kinase (AMPK) activation and the biological consequences. Activation of AMPK is determined by the balance of (1) AMP/adenosine triphosphate (ATP) ratio, (2) upstream kinase activity, and (3) upstream phosphatase activity. Once activated, AMPK activates catabolic pathways such as glycolysis, beta-oxidation of fatty acids, and autophagy, as well as suppresses anabolic pathway such as gluconeogenesis and lipid/protein synthesis to increase ATP levels. LKB1, liver kinase B1; STRAD, ste-20 related adaptor; MO25, mouse protein 25; CaMKKβ, CaMKK Ca^{2+}/calmodulin-dependent protein kinase kinase β; PP2A, protein phosphatase 2A; PP2C, protein phosphatase 2C; KD, kinase domain; AID, auto-inhibitory domain; CTD, C-terminal domain; CBM, carbohydrate-binding module; Myr, myristoylation; CBS, cystathionine beta synthase.

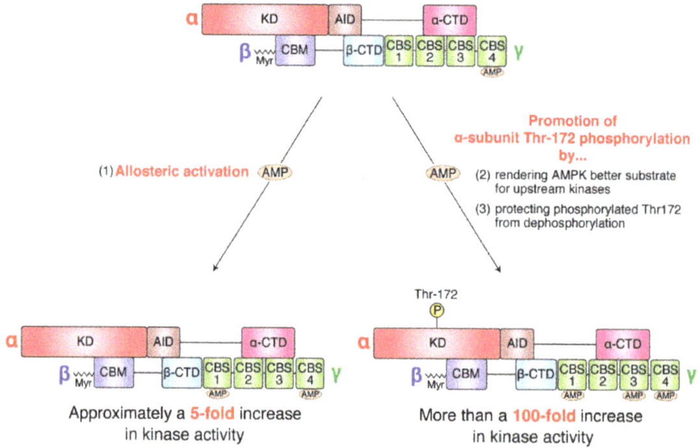

Figure 3. Adenosine monophosphate (AMP) activates AMP-activated protein kinase (AMPK) in several different ways. Allosteric activation results in approximately a fivefold increase in kinase activity, whereas α-subunit Thr-172 phosphorylation causes more than a 100-fold increase in kinase activity; therefore, α-subunit Thr-172 phosphorylation is essentially considered to be a molecular on–off switch for AMPK. KD, kinase domain; AID, auto-inhibitory domain; CTD, C-terminal domain; CBM, carbohydrate-binding module; Myr, myristoylation; CBS, cystathionine beta synthase.

3. Notable Findings in the Mid-2000s Established the Important Role of the LKB1–AMPK Axis in Apicobasal Epithelial Cell Polarity

In 2004, Clevers and colleagues shocked the epithelial cell research community with their seminal paper [33]. Until then, it was believed that apicobasal epithelial cell polarity was established after cells formed cell–cell junctions [66]. However, using intestinal epithelial cell lines with inducible LKB1 activity, they showed that, even in the absence of junctional cell–cell contacts, activation of LKB1 alone can fully establish apicobasal cell polarity in a cell-autonomous manner with the three major aspects of cell polarity all fulfilled, these being the appropriate sorting of apical and basolateral plasma membrane markers, formation of a full brush border, and correct junctional localization of junctional proteins. LKB1 forms a complex with ste-20 related adaptor (STRAD) and mouse protein 25 (MO25) that exerts many functions through the phosphorylation of many substrates [67]. In 2005, the downstream LKB1 effector responsible for establishing apicobasal cell polarity was suggested to be AMPK by investigating the functional consequences of a missense mutation identified in Peutz–Jeghers syndrome (PJS) that is caused by germline mutations of the LKB1 gene [34]. It was found that there is a certain LKB1-C-terminal missense mutation found in PJS patients that abolish the kinase activity responsible for almost all effects of the LKB1/STRAD/MO25 complex but still impair AMPK activity and apicobasal polarity establishment [34]. Subsequently, in 2007, it was found that not only *Drosophila* LKB1-null mutants but also *Drosophila* AMPK-null mutants display severe abnormalities in apicobasal cell polarity [36]. These findings collectively showed the crucial role of the LKB1–AMPK axis in apicobasal epithelial cell polarity establishment and have certainly paved the way for subsequent elucidation of the intimate association between AMPK and AJCs.

4. AJs and Junctional Tension Promote the LKB1–AMPK Axis to Provide Energy to Resist Applied Force

Interestingly, in 2009, it was found that AJs can activate LKB1 complexes [38]. In this study, it was shown that active LKB1 complexes co-localize with E-cadherin in polarized Caco-2 and Madin–Darby canine kidney (MDCK) II cells, that E-cadherin-dependent junctional maturation gradually mobilizes LKB1 complexes to AJs, and that AJ localization of LKB1 complexes is critical for the activation of AMPK. However, the physiological relevance of the LKB1–AMPK axis regulation at AJs remained unknown until 2017 when this regulation was shown to be critical for epithelial cells to resist against applied force. In epithelial cells, mechano-sensing, especially at AJs, triggers robust actin cytoskeletal rearrangements to resist externally applied forces [68–70]. From an energetic point of view, these changes are very costly, consuming about half of the total ATP in a cell. In 2017, it was found that junctional tension directly applied on E-cadherin, as well as shear stress on MDCK II cells, facilitates AJ-mediated LKB1–AMPK activation. Activated AMPK subsequently increases ATP levels and activates RhoA to support peri-junctional actomyosin filament rearrangement, thereby contributing to the cell stiffness critical for resisting against applied force [39] (Figure 4). In vivo importance of this AJ-mediated LKB1–AMPK regulation is suggested by the study of the tumor suppressor gene folliculin (FLCN), the causative gene for a rare autosomal dominant disorder named Birt–Hogg–Dubé (BHD) syndrome. It was shown that FLCN deletion in lung epithelium results in downregulation of E-cadherin and LKB1 expression, which leads to the impairment of AMPK activation and alveolar collapse [71]. Collectively, these findings established the critical role of AJs and junctional tension in promoting the LKB1–AMPK axis to resist applied force.

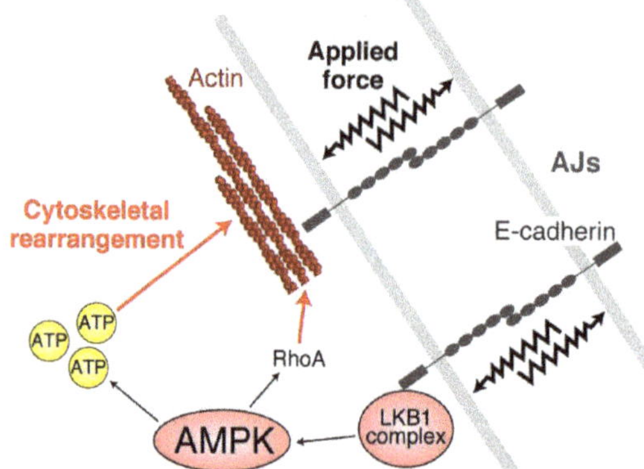

Figure 4. Adherens junctions (AJs) activate the liver kinase B1 (LKB1)–AMP-activated protein kinase (AMPK) axis. AJs can promote LKB1 complex activation [38], which is facilitated by applied force (zigzag arrows) [39]. Activated LKB1 complexes subsequently activate AMPK, which increases ATP levels and promotes RhoA activation to rearrange cytoskeletons, providing stiffness to resist against applied force.

5. TJ Function Is Reinforced by AMPK

The first report that showed the association of AMPK with TJs was published in 2006 [35]. At that time, AMPK was shown to facilitate apicobasal polarity establishment [34,36] and TJs have long been believed to function as a "fence" to prevent the diffusion of proteins and lipids, thereby having a role in apicobasal cell polarity [72] (later, "fence" function of TJs has been basically denied [73,74], although some controversial discussions appeared again recently [75]). Therefore, the role of AMPK in TJ assembly was investigated using a calcium switch experiment, a prevalent experiment to study TJ assembly. In a calcium switch experiment, calcium depletion from the medium of confluent epithelial cells destabilizes AJs, which triggers TJ disassembly with zonula occludens-1 (ZO-1) translocation from cell–cell junctions to the cytoplasm. Calcium restoration causes TJ assembly with ZO-1 redistribution from the cytoplasm to cell–cell junctions to form nascent ZO-1 puncta, which later forms continuous ZO-1 structures. Using this experimental procedure, it was shown that AMPK is activated during TJ assembly in MDCK II cells and that 5-aminoimidazole-4-carboxamide ribonucleoside (AICAR), a potent AMPK activating reagent, facilitates ZO-1 redistribution from the cytoplasm to cell–cell junctions, whereas expression of a dominant-negative AMPK (a kinase-dead form of AMPK) significantly delays ZO-1 redistribution with impaired establishment of the paracellular barrier. It was also noted that AMPK activation alone does not lead to the formation of continuous ZO-1 structures from puncta, suggesting that AMPK is required in the initiation of TJ assembly but other factors are needed to support further TJ maturation [35]. Shortly after this, it was also shown that AMPK activation by AICAR partially protects TJs from disassembly induced by calcium depletion [37]. Altogether, these studies established that TJs are a downstream target of AMPK and are reinforced by its activation.

6. Molecular Basis of the Effects of AMPK on AJCs: Implication of Novel AMPK Effectors in AJCs and AMPK-Driven Epithelial Transcription Factors

Since these early studies, there has been a general consensus that AMPK has a role in reinforcing TJ function and/or establishing apicobasal polarity in a variety of epithelial cells both in vitro and in vivo [76–85]. Although the molecular mechanisms underlying these phenomena are not yet completely elucidated, recent identification of novel AMPK effectors in AJCs and AMPK-driven epithelial transcription factors have improved our knowledge.

To date, myosin regulatory light chain (MRLC), afadin, cingulin, claudins, and G-alpha interacting vesicle associated protein (GIV)/Girdin were identified as AMPK effectors in AJCs. The first one identified was MRLC, which was originally shown in 2007 to be directly phosphorylated at Thr-21 by AMPK using in vitro phosphorylation assays [36]. In that report, it was shown that this phosphorylation of MRLC is critical in the phenotype of AMPK-null mutants in *Drosophila* which displays severe abnormalities in apicobasal cell polarity [36]. However, it should be noted that this finding has since been questioned because this phosphorylation site does not match the optimal AMPK substrate motif found in all other established AMPK substrates [86]. In 2011, afadin was identified as the second AMPK substrate in AJCs by the study investigating adhesion-related molecules, which are important in AMPK-dependent TJ assembly [87]. Prior to that study, afadin, together with another cell adhesion molecule nectin, was known to constitute an important adhesion systems in epithelial cells [88], and nectin–afadin-based adhesion was also known to appear during the earliest stages in the process of epithelial morphogenesis prior to the expression of TJ proteins [89]. In that study, it was found that afadin is a direct substrate for AMPK using in vitro phosphorylation assays. It was also found that AMPK activation by AICAR treatment increases the interaction of afadin with ZO-1 and that knocking down afadin expression prevented AMPK-mediated facilitation of TJ assembly in MDCK II cells, suggesting the possibility that AMPK reinforces nectin–afadin-based adhesion at an early stage of cell–cell junction establishment to facilitate TJ assembly [87]. The third AMPK effector in AJCs was identified in 2013, during the search for microtubule-binding proteins in the AJC fraction, when we found that cingulin is a direct AMPK substrate and AMPK-mediated phosphorylation of cingulin facilitates cingulin-mediated linkages of apical microtubule networks to TJs in epithelial cells [80,90]. Cingulin is a specific TJ protein [91] and has two potential evolutionally conserved AMPK target motifs at residues Ser-132 and Ser-150, and we showed that wild-type cingulin can be directly phosphorylated by AMPK at both sites using in vitro phosphorylation assays. Furthermore, we showed that AMPK-mediated cingulin phosphorylation strengthens the interaction of cingulin with microtubules in mouse mammary gland epithelial Eph4 cells and is important in epithelial three-dimensional morphogenesis [80,92]. The fourth group of AMPK effectors in AJCs is the claudins. In 2014, it was shown that AMPK activation by AICAR increases phosphorylation of claudin-4 at Ser-199 in an extracellular signal-regulated kinase 1/2-dependent manner in SV40 immortalized rat submandibular acinar cell lines [93]. Shortly after this, it was shown that claudin-1 can also be directly phosphorylated at Thr-191 by AMPK using in vitro phosphorylation assays and that this phosphorylation may potentially be important in TJ reinforcement in epithelial Eph4 cells [94]. The last AMPK effector in AJCs discovered so far was GIV/Girdin, which was identified as a direct AMPK substrate in 2016 [95]. Prior to that work, GIV/Girdin was shown to be a guanine-nucleotide-exchange factor (GEF) for Gαi3 in trimeric G proteins and to play an important role in TJ formation and establishment of apicobasal epithelial cell polarity [96]. In that work, it was found that GIV/Girdin has an evolutionally conserved, potential AMPK phosphorylation motif at Ser-245 in the N-terminus of GIV/Girdin and that AMPK directly phosphorylates GIV/Girdin at Ser-245 using in vitro phosphorylation assays. It was also shown that, during apicobasal cell polarity establishment in MDCK II cells, GIV/Girdin becomes phosphorylated by AMPK and subsequently localizes at TJs through its binding affinity for the TJ-linked apical microtubule network. Further, it was shown that this phosphorylation of GIV/Girdin is essential in AMPK-dependent TJ formation and apicobasal epithelial cell polarity establishment, and that an oncogenic mutant of GIV which cannot be phosphorylated by AMPK increases tumor cell growth through the defects in establishing apicobasal cell polarity [95]. These results collectively show that there are several AMPK substrates in AJCs (MRLC, cingulin, claudins, and GIV/Girdin) and that the phosphorylation of these substrates impacts TJ function and apicobasal cell polarity, although how AMPK coordinates these different pathways merits future investigation (Figure 5).

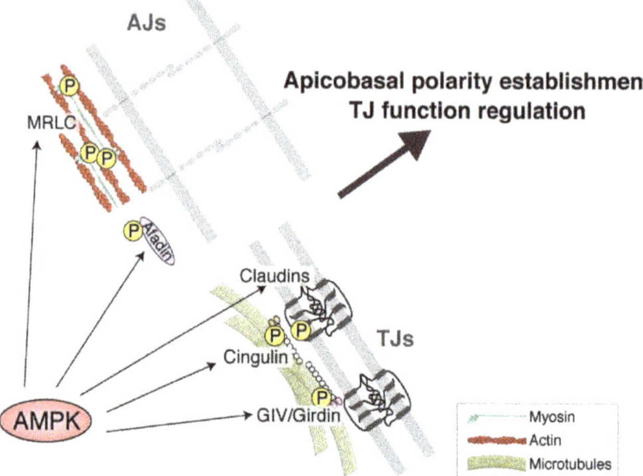

Figure 5. AMP-activated protein kinase (AMPK)-mediated regulation of apicobasal polarity establishment and tight junction (TJ) function are at least partially dependent on its phosphorylation of its substrates in AJCs. AMPK has been shown to phosphorylate myosin regulatory light chain (MRLC), cingulin, G-alpha interacting vesicle associated protein (GIV)/Girdin, claudins, and afadin, which can explain the role of AMPK in apicobasal polarity establishment and tight junction function regulation, at least partially. P, phosphate; AJs, adherens junctions.

Another important mechanism explaining the association between AMPK and TJs was described in 2017 [97]. Prior studies suggested the possibility that AMPK upregulates the expression of TJ proteins, thereby exerting a positive effect on TJ function; however, the underlying mechanism remained unknown [81,84]. In that study, it was found that, in intestinal epithelial Caco-2 cells, AMPK activation by AICAR upregulates expression of caudal type homeobox 2 (CDX2), a key transcription factor governing the differentiation of intestinal epithelial cells. It was also shown that this upregulation is caused by AMPK-dependent recruitment of methylase PRC2 and demethylase LSD1 to induce histone modifications of the CDX2 promoter. Furthermore, it was shown that intestine-conditional AMPK-knockout mice displayed impairment of intestinal epithelial differentiation and barrier function with a significant reduction in CDX2 expression [97]. Subsequent research found that AMPK activation by purple potato extract (PPE) in Caco-2 cells upregulates TJ protein expression in correlation with CDX2 expression and that AMPK-knockout abolishes the positive effects of PPE on CDX2 expression and TJ protein expression [98]. These findings suggest that, in addition to the AMPK substrates in AJCs, an AMPK-driven epithelial transcription factor plays an important role in AMPK-dependent TJ reinforcement via upregulation of TJ protein content (Figure 6).

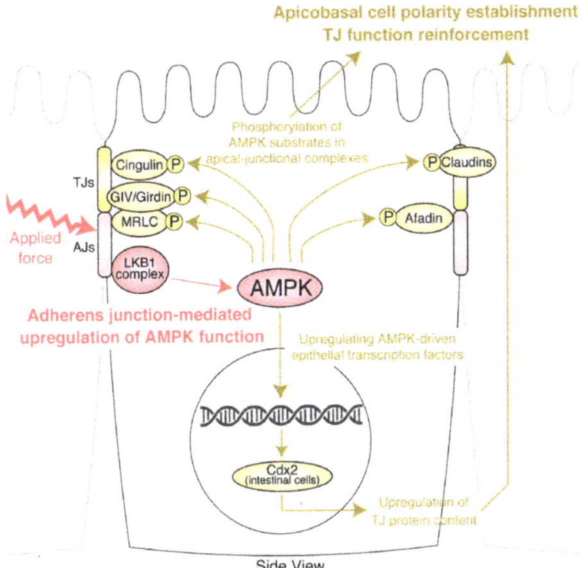

Figure 6. Overview of the reciprocal association between apical junctional complexes and AMP-activated kinase (AMPK). Adherens junctions (AJs) activate the liver kinase B1 (LKB1)–AMPK axis particularly in response to applied junctional tension. AMPK reinforces tight junction (TJ) function and promotes apicobasal cell polarization establishment through AMPK effectors in AJCs and AMPK-driven epithelial transcription factors.

7. Role of AMPK in Endothelial Barrier Function

Thus far, we have mainly focused on the role of AMPK in epithelial AJCs. Another intriguing aspect of AMPK is that it also has been shown to enhance endothelial TJ function [99–102]. An endothelial cell monolayer lines the inside of the intima of all blood and lymphatic vessels, forming a semipermeable barrier between the interstitial fluids within tissues and circulating fluids [103]. In general, TJs together with AJs form pericellular zipper-like structures between endothelial cells to restrict free diffusion of molecules, but the specific architecture and composition varies in different vascular beds. For example, TJs are more sophisticated in small arterioles, especially in highly specialized vascular beds like the blood–brain barrier (BBB) and inner blood–retinal barrier, whereas AJs are more prominent in postcapillary venules [103–105].

The overall molecular composition of endothelial TJs and AJs are similar to those of epithelial cells, with some small differences [103]. For example, the cadherin-family protein in epithelial AJs is E-cadherin, whereas it is vascular endothelial-cadherin (VE-cadherin) in endothelial AJs. It is noteworthy that N-cadherin is another major cadherin-family protein expressed in endothelial cells, but is excluded from endothelial AJs for an unknown reason [106,107]. N-cadherin has been shown to be important in the interactions of endothelial cells with pericytes and smooth muscle cells, which are thought to play an essential role in the maturation and stabilization of endothelial cells [108,109]. Interestingly, in 2011, using a pulmonary microvascular endothelial cell (PMVEC) wound healing model, it was found that AMPK-mediated TJ reinforcement is critical for PMVEC repair and that AMPK colocalizes with N-cadherin in PMVEC [99]. Following this finding, the biological importance of the AMPK–N-cadherin interaction was investigated using the same PMVEC wound healing model, and it was found that a short-hairpin RNA which decreases the expression of N-cadherin causes translocation of AMPK away from the membrane and inhibits AMPK-mediated barrier repair [110].

These findings collectively suggest that the N-cadherin–AMPK axis may be important for endothelial barrier assembly, but further studies are warranted for validation of these intriguing findings.

Table 1. List of commonly-used AMPK activators and inhibitors.

	Compound Name	Mechanism	Isoform Selectivity	Reference
Indirect activators	Metformin (biguanides)	Inhibiting the mitochondrial respiratory chain Complex I → Increase in AMP/ATP ratio	No	[111,112]
	Troglitazone (thiazolidinediones)	Inhibiting the mitochondrial respiratory chain Complex I → Increase in AMP/ATP ratio	No	[112,113]
	Resveratrol (polyphenols)	Inhibiting F1F0–ATPase/ATP synthase → Increase in AMP/ATP ratio	No	[112,114]
	Quercetin (polyphenols)	Inhibiting F1F0–ATPase/ATP synthase → Increase in AMP/ATP ratio	No	[112,115]
	α-Lipoic acid	Increasing calcium level?	No	[116]
Direct activators	5-Aminoimidazole-4-carboxamide ribonucleoside (AICAR)	Prodrug, converted to 5-amino-1-β-d-ribofuranosylimidazole-4-carboxamide-5′-monophosphate (ZMP) → ZMP binds to site-3 in the γ-subunit	No	[112,117]
	Compound-13 (C-13)	Prodrug, converted to Compound-2 (C-2) → C-2 binds to the γ-subunit	α1 > α2	[118]
	A-769662	Binding to the allosteric drug and metabolite (ADaM) site → Allosteric activation	β1 > β2	[56,119]
	Compound 991	Binding to the ADaM site → Allosteric activation	β1 > β2	[56]
	Salicylate	Binding to the ADaM site → Allosteric activation	β1	[120]
	MT63-78	Binding to the ADaM site → Allosteric activation	β1	[121]
Inhibitors	Compound C (dorsomorphin)	Binding to the α-subunit → Competitive inhibition of ATP	No	[111,122]
	MT47-100	Binding to the β2-subunit → Allosteric inhibition	β2 (β1 activation)	[123]
	SBI-0206965	Binding to the α-subunit → Mixed-type inhibition of ATP	No	[65]

8. Therapeutic Potential of Targeting AMPK for Manipulation of Epithelial Barrier Function

One of the fascinating features of AMPK is that it is "druggable". Indeed, we can artificially activate AMPK either directly (direct activator) or indirectly through AMP or calcium accumulation (indirect activator) using preexisting food compounds such as polyphenols and α-lipoic acid, as well as drugs such as metformin, thiazolidinediones, AICAR, and A79662 (summarized in Table 1). This leads to the question of whether these agents have any potential as therapeutics for diseases caused by barrier dysfunction. The answer may be "yes".

By improving intestinal epithelial barrier dysfunction, metformin-induced activation of AMPK can alleviate inflammatory bowel disease in mouse models induced by interleukin-10 deficiency, dextran sulfate sodium, and heat stress [84,85,124]. Metformin-induced intestinal barrier reinforcement was also shown to alleviate fructose-induced non-alcoholic fatty liver disease by blocking translocation of bacterial endotoxin [78]. Moreover, it was shown that metformin-induced AMPK activation may inhibit respiratory infection via reinforcement of airway epithelial barriers [79], attenuate the severity of ischemia-induced renal tubular injury by facilitating TJ assembly in renal epithelial cells [77], and facilitate recovery from acute ischemic stroke by attenuating BBB disruption [102,125] (metformin is an indirect AMPK activator that has other AMPK-independent effects; however, these protective effects of metformin on epithelial barrier function was shown to be dependent on AMPK [77–79,84,85,102,124,125]). Furthermore, AICAR-induced AMPK activation was also shown to alleviate lipopolysaccharide (LPS)-induced lung injury via restoring pulmonary endothelial barrier integrity [110], and LPS-induced cardiac dysfunction by attenuating left ventricular wall edema [100]. Therefore, these results collectively show that AMPK is a promising therapeutic target for epithelial/endothelial barrier manipulation. However, it should be noted that AMPK has wide variety of function in wide variety of organs with its activity being spatiotemporally regulated. Therefore, systemic and uniform AMPK activation would cause too many off-target effects. AMPK subunits (α, β, or γ) occur as multiple isoforms encoded by distinct genes (α1 or α2; β1 or β2; γ1, γ2, or γ3), giving rise to 12 possible AMPK α/β/γ heterotrimers that are differently expressed in different tissues and exert different effects [47]. For example, the α1/β2/γ1 heterotrimer is predominant in intestinal epithelial cells [126] and, while α1/β2/γ1, α2/β2/γ1, and α1/β2/γ3 heterotrimers are present in skeletal muscles [127,128], only the α2/β2/γ3 heterotrimer is activated during high intensity exercise [128,129]. Most AMPK activators/inhibitors that are commonly used today are not isoform-specific; therefore, further elucidation of isoform specific localization and/or function, as well as further development of isoform-specific activators and inhibitors, is important for AMPK-based organ- and/or barrier-specific therapy.

9. Conclusions

AMPK is a well-known central regulator of energy metabolism. As we have reviewed here, it has also been shown over 15 years that AMPK has a reciprocal association with AJCs, and fine-tunes TJ function and apicobasal cell polarity establishment in epithelial/endothelial cells. The complete molecular mechanisms underlying these phenomena have not yet been elucidated; however, identifications of novel AMPK effectors in AJCs and AMPK-driven epithelial transcription factors have enhanced our knowledge, and recent technological advances will contribute to better understanding. For example, recent development of transgenic mice expressing a highly sensitive fluorescence resonance energy transfer-based biosensor would visualize dynamic AMPK responses in vivo [130,131]. The formation and regulation of AJC-mediated epithelial cell–cell adhesion and paracellular barrier are critical for the biological systems with its disruption being implicated in various diseases; therefore, further elucidation of molecular details underlying an association between AMPK and AJCs will contribute to the better understanding of epithelial cell physiology as well as the development of therapies, enabling us to manipulate epithelial/endothelial barrier function.

Author Contributions: K.T., writing—original draft preparation; T.Y., A.T. and S.T., writing—review and editing.

Funding: This work was supported by Core Research for Evolutionary Science and Technology (CREST) of the Japan Science and Technology Agency (JST) (to Sachiko Tsukita), Grant-in-Aid for Specially Promoted Research (JP19H05468, to Sachiko Tsukita), Grant-in-Aid for Early-Career Scientists (JP18K14696, to Tomoki Yano), and Grant-in-Aid for Scientific Research (B) (JP16H05121, to Atsushi Tamura) from Japan Society for the Promotion of Science (JSPS) of Japan.

Acknowledgments: We thank all members of Tsukita Lab for their useful comments and discussions.

Conflicts of Interest: The authors declare no conflict of interest.

Abbreviations

TJs	Tight junctions
AJs	Adherens junctions
AMP	Adenosine monophosphate
AMPK	AMP-activated protein kinase
HMG-CoA	3-hydroxy-3-methylglutaryl-coenzyme A
acetyl-CoA	Acetyl-coenzyme A
ATP	Adenosine triphosphate
LKB1	Liver kinase B1
CaMKKβ	Ca^{2+}/calmodulin-dependent protein kinase kinase β
PJS	Peutz–Jeghers syndrome
MDCK	Madin–Darby canine kidney
FLCN	Folliculin
BHD	Birt–Hogg–Dubé
AICAR	5-Aminoimidazole-4-carboxamide ribonucleoside

MRLC	Myosin regulatory light chain
GIV	G-alpha interacting vesicle associated protein
GEF	Guanine-nucleotide-exchange factor
CDX2	Caudal type homeobox 2
PPE	Purple potato extract
BBB	Blood–brain barrier
VE-cadherin	vascular endothelial-cadherin
PMVEC	Pulmonary microvascular endothelial cell
ZMP	5-amino-1-β-d-ribofuranosylimidazole- 4-carboxamide-5′-monophosphate
LPS	Lipopolysaccharide

References

1. Tsukita, S.; Furuse, M.; Itoh, M. Multifunctional strands in tight junctions. *Nat. Rev. Mol. Cell Biol.* **2001**, *2*, 285–293. [CrossRef] [PubMed]
2. Tamura, A.; Yamazaki, Y.; Hayashi, D.; Suzuki, K.; Sentani, K.; Yasui, W.; Tsukita, S. Claudin-based paracellular proton barrier in the stomach. *Ann. N. Y. Acad. Sci.* **2012**, *1258*, 108–114. [CrossRef] [PubMed]
3. Tamura, A.; Tsukita, S. Paracellular barrier and channel functions of TJ claudins in organizing biological systems: Advances in the field of barriology revealed in knockout mice. *Semin. Cell Dev. Biol.* **2014**, *36*, 177–185. [CrossRef] [PubMed]
4. Suzuki, H.; Tani, K.; Tamura, A.; Tsukita, S.; Fujiyoshi, Y. Model for the architecture of claudin-based paracellular ion channels through tight junctions. *J. Mol. Biol.* **2015**, *427*, 291–297. [CrossRef]
5. Nakamura, S.; Irie, K.; Tanaka, H.; Nishikawa, K.; Suzuki, H.; Saitoh, Y.; Tamura, A.; Tsukita, S.; Fujiyoshi, Y. Morphologic determinant of tight junctions revealed by claudin-3 structures. *Nat. Commun.* **2019**, *10*, 816. [CrossRef]
6. Tsukita, S.; Tanaka, H.; Tamura, A. The Claudins: From Tight Junctions to Biological Systems. *Trends Biochem. Sci.* **2019**, *44*, 141–152. [CrossRef]
7. Tanaka, H.; Tamura, A.; Suzuki, K.; Tsukita, S. Site-specific distribution of claudin-based paracellular channels with roles in biological fluid flow and metabolism. *Ann. N. Y. Acad. Sci.* **2017**, *1405*, 44–52. [CrossRef]
8. Wada, M.; Tamura, A.; Takahashi, N.; Tsukita, S. Loss of claudins 2 and 15 from mice causes defects in paracellular Na+ flow and nutrient transport in gut and leads to death from malnutrition. *Gastroenterology* **2013**, *144*, 369–380. [CrossRef]
9. Matsumoto, K.; Imasato, M.; Yamazaki, Y.; Tanaka, H.; Watanabe, M.; Eguchi, H.; Nagano, H.; Hikita, H.; Tatsumi, T.; Takehara, T.; et al. Claudin 2 deficiency reduces bile flow and increases susceptibility to cholesterol gallstone disease in mice. *Gastroenterology* **2014**, *147*, 1134–1145.e10. [CrossRef]
10. Tanaka, H.; Takechi, M.; Kiyonari, H.; Shioi, G.; Tamura, A.; Tsukita, S. Intestinal deletion of Claudin-7 enhances paracellular organic solute flux and initiates colonic inflammation in mice. *Gut* **2015**, *64*, 1529–1538. [CrossRef]
11. Tokumasu, R.; Yamaga, K.; Yamazaki, Y.; Murota, H.; Suzuki, K.; Tamura, A.; Bando, K.; Furuta, Y.; Katayama, I.; Tsukita, S. Dose-dependent role of claudin-1 in vivo in orchestrating features of atopic dermatitis. *Proc. Natl. Acad. Sci. USA* **2016**, *113*, E4061–E4068. [CrossRef] [PubMed]
12. Citi, S. Intestinal barriers protect against disease. *Science* **2018**, *359*, 1097–1098. [CrossRef] [PubMed]
13. Tanaka, H.; Imasato, M.; Yamazaki, Y.; Matsumoto, K.; Kunimoto, K.; Delpierre, J.; Meyer, K.; Zerial, M.; Kitamura, N.; Watanabe, M.; et al. Claudin-3 regulates bile canalicular paracellular barrier and cholesterol gallstone core formation in mice. *J. Hepatol.* **2018**, *69*, 1308–1316. [CrossRef] [PubMed]
14. Suzuki, K.; Sentani, K.; Tanaka, H.; Yano, T.; Suzuki, K.; Oshima, M.; Yasui, W.; Tamura, A.; Tsukita, S. Deficiency of Stomach-Type Claudin-18 in Mice Induces Gastric Tumor Formation Independent of H pylori Infection. *Cell. Mol. Gastroenterol. Hepatol.* **2019**, *8*, 119–142. [CrossRef]
15. Nalle, S.C.; Zuo, L.; Ong, M.L.D.M.; Singh, G.; Worthylake, A.M.; Choi, W.; Manresa, M.C.; Southworth, A.P.; Edelblum, K.L.; Baker, G.J.; et al. Graft-versus-host disease propagation depends on increased intestinal epithelial tight junction permeability. *J. Clin. Investig.* **2019**, *129*, 902–914. [CrossRef]

16. Graham, W.V.; He, W.; Marchiando, A.M.; Zha, J.; Singh, G.; Li, H.-S.; Biswas, A.; Ong, M.L.D.M.; Jiang, Z.-H.; Choi, W.; et al. Intracellular MLCK1 diversion reverses barrier loss to restore mucosal homeostasis. *Nat. Med.* **2019**, *25*, 690–700. [CrossRef]
17. Kuo, W.-T.; Shen, L.; Zuo, L.; Shashikanth, N.; Ong, M.L.D.M.; Wu, L.; Zha, J.; Edelblum, K.L.; Wang, Y.; Wang, Y.; et al. Inflammation-induced Occludin Downregulation Limits Epithelial Apoptosis by Suppressing Caspase-3 Expression. *Gastroenterology* **2019**, *157*, 1323–1337. [CrossRef]
18. Hagen, S.J.; Ang, L.-H.; Zheng, Y.; Karahan, S.N.; Wu, J.; Wang, Y.E.; Caron, T.J.; Gad, A.P.; Muthupalani, S.; Fox, J.G. Loss of Tight Junction Protein Claudin 18 Promotes Progressive Neoplasia Development in Mouse Stomach. *Gastroenterology* **2018**, *155*, 1852–1867. [CrossRef]
19. Shah, J.; Rouaud, F.; Guerrera, D.; Vasileva, E.; Popov, L.M.; Kelley, W.L.; Rubinstein, E.; Carette, J.E.; Amieva, M.R.; Citi, S. A Dock-and-Lock Mechanism Clusters ADAM10 at Cell-Cell Junctions to Promote α-Toxin Cytotoxicity. *Cell Rep.* **2018**, *25*, 2132–2147.e7. [CrossRef]
20. King, A.J.; Siegel, M.; He, Y.; Nie, B.; Wang, J.; Koo-McCoy, S.; Minassian, N.A.; Jafri, Q.; Pan, D.; Kohler, J.; et al. Inhibition of sodium/hydrogen exchanger 3 in the gastrointestinal tract by tenapanor reduces paracellular phosphate permeability. *Sci. Transl. Med.* **2018**, *10*, eaam6474. [CrossRef]
21. Clairembault, T.; Leclair-Visonneau, L.; Coron, E.; Bourreille, A.; Le Dily, S.; Vavasseur, F.; Heymann, M.-F.; Neunlist, M.; Derkinderen, P. Structural alterations of the intestinal epithelial barrier in Parkinson's disease. *Acta Neuropathol. Commun.* **2015**, *3*, 12. [CrossRef]
22. Montagne, A.; Zhao, Z.; Zlokovic, B.V. Alzheimer's disease: A matter of blood-brain barrier dysfunction? *J. Exp. Med.* **2017**, *214*, 3151–3169. [CrossRef] [PubMed]
23. Greene, C.; Kealy, J.; Humphries, M.M.; Gong, Y.; Hou, J.; Hudson, N.; Cassidy, L.M.; Martiniano, R.; Shashi, V.; Hooper, S.R.; et al. Dose-dependent expression of claudin-5 is a modifying factor in schizophrenia. *Mol. Psychiatry* **2018**, *23*, 2156–2166. [CrossRef] [PubMed]
24. Tsukita, K.; Taguchi, T.; Sakamaki-Tsukita, H.; Tanaka, K.; Suenaga, T. The vagus nerve becomes smaller in patients with Parkinson's disease: A preliminary cross-sectional study using ultrasonography. *Parkinsonism Relat. Disord.* **2018**, *55*, 148–149. [CrossRef] [PubMed]
25. Tsukita, K.; Sakamaki-Tsukita, H.; Tanaka, K.; Suenaga, T.; Takahashi, R. Value of in vivo α-synuclein deposits in Parkinson's disease: A systematic review and meta-analysis. *Mov. Disord.* **2019**, *34*, 1452–1463. [CrossRef] [PubMed]
26. Van IJzendoorn, S.C.D.; Derkinderen, P. The Intestinal Barrier in Parkinson's Disease: Current State of Knowledge. *J. Parkinsons Dis.* **2019**, *9*, S323–S329. [CrossRef] [PubMed]
27. Rosenthal, R.; Günzel, D.; Theune, D.; Czichos, C.; Schulzke, J.-D.; Fromm, M. Water channels and barriers formed by claudins. *Ann. N. Y. Acad. Sci.* **2017**, *1397*, 100–109. [CrossRef] [PubMed]
28. Konishi, S.; Yano, T.; Tanaka, H.; Mizuno, T.; Kanoh, H.; Tsukita, K.; Namba, T.; Tamura, A.; Yonemura, S.; Gotoh, S.; et al. Vinculin is critical for the robustness of the epithelial cell sheet paracellular barrier for ions. *Life Sci. Alliance* **2019**, *2*, e201900414. [CrossRef]
29. Yonemura, S.; Wada, Y.; Watanabe, T.; Nagafuchi, A.; Shibata, M. alpha-Catenin as a tension transducer that induces adherens junction development. *Nat. Cell Biol.* **2010**, *12*, 533–542. [CrossRef]
30. Zihni, C.; Mills, C.; Matter, K.; Balda, M.S. Tight junctions: From simple barriers to multifunctional molecular gates. *Nat. Rev. Mol. Cell Biol.* **2016**, *17*, 564–580. [CrossRef]
31. Yano, T.; Kanoh, H.; Tamura, A.; Tsukita, S. Apical cytoskeletons and junctional complexes as a combined system in epithelial cell sheets. *Ann. N. Y. Acad. Sci.* **2017**, *1405*, 32–43. [CrossRef] [PubMed]
32. Citi, S. The mechanobiology of tight junctions. *Biophys. Rev.* In press. [CrossRef] [PubMed]
33. Baas, A.F.; Kuipers, J.; van der Wel, N.N.; Batlle, E.; Koerten, H.K.; Peters, P.J.; Clevers, H.C. Complete polarization of single intestinal epithelial cells upon activation of LKB1 by STRAD. *Cell* **2004**, *116*, 457–466. [CrossRef]
34. Forcet, C.; Etienne-Manneville, S.; Gaude, H.; Fournier, L.; Debilly, S.; Salmi, M.; Baas, A.; Olschwang, S.; Clevers, H.; Billaud, M. Functional analysis of Peutz-Jeghers mutations reveals that the LKB1 C-terminal region exerts a crucial role in regulating both the AMPK pathway and the cell polarity. *Hum. Mol. Genet.* **2005**, *14*, 1283–1292. [CrossRef] [PubMed]
35. Zhang, L.; Li, J.; Young, L.H.; Caplan, M.J. AMP-activated protein kinase regulates the assembly of epithelial tight junctions. *Proc. Natl. Acad. Sci. USA* **2006**, *103*, 17272–17277. [CrossRef] [PubMed]

36. Lee, J.H.; Koh, H.; Kim, M.; Kim, Y.; Lee, S.Y.; Karess, R.E.; Lee, S.-H.; Shong, M.; Kim, J.-M.; Kim, J.; et al. Energy-dependent regulation of cell structure by AMP-activated protein kinase. *Nature* **2007**, *447*, 1017–1020. [CrossRef]
37. Zheng, B.; Cantley, L.C. Regulation of epithelial tight junction assembly and disassembly by AMP-activated protein kinase. *Proc. Natl. Acad. Sci. USA* **2007**, *104*, 819–822. [CrossRef]
38. Sebbagh, M.; Santoni, M.-J.; Hall, B.; Borg, J.-P.; Schwartz, M.A. Regulation of LKB1/STRAD localization and function by E-cadherin. *Curr. Biol.* **2009**, *19*, 37–42. [CrossRef]
39. Bays, J.L.; Campbell, H.K.; Heidema, C.; Sebbagh, M.; DeMali, K.A. Linking E-cadherin mechanotransduction to cell metabolism through force-mediated activation of AMPK. *Nat. Cell Biol.* **2017**, *19*, 724–731. [CrossRef]
40. Rowart, P.; Wu, J.; Caplan, M.J.; Jouret, F. Implications of AMPK in the Formation of Epithelial Tight Junctions. *Int. J. Mol. Sci.* **2018**, *19*, 2040. [CrossRef]
41. Beg, Z.H.; Allmann, D.W.; Gibson, D.M. Modulation of 3-hydroxy-3-methylglutaryl coenzyme A reductase activity with cAMP and wth protein fractions of rat liver cytosol. *Biochem. Biophys. Res. Commun.* **1973**, *54*, 1362–1369. [CrossRef]
42. Carlson, C.A.; Kim, K.H. Regulation of hepatic acetyl coenzyme A carboxylase by phosphorylation and dephosphorylation. *J. Biol. Chem.* **1973**, *248*, 378–380. [CrossRef]
43. Ferrer, A.; Caelles, C.; Massot, N.; Hegardt, F.G. Activation of rat liver cytosolic 3-hydroxy-3-methylglutaryl coenzyme A reductase kinase by adenosine 5′-monophosphate. *Biochem. Biophys. Res. Commun.* **1985**, *132*, 497–504. [CrossRef]
44. Carling, D.; Zammit, V.A.; Hardie, D.G. A common bicyclic protein kinase cascade inactivates the regulatory enzymes of fatty acid and cholesterol biosynthesis. *FEBS Lett.* **1987**, *223*, 217–222. [CrossRef]
45. Sim, A.T.; Hardie, D.G. The low activity of acetyl-CoA carboxylase in basal and glucagon-stimulated hepatocytes is due to phosphorylation by the AMP-activated protein kinase and not cyclic AMP-dependent protein kinase. *FEBS Lett.* **1988**, *233*, 294–298. [CrossRef]
46. Grahame Hardie, D.; Carling, D.; Sim, A.T.R. The AMP-activated protein kinase: A multisubstrate regulator of lipid metabolism. *Trends Biochem. Sci.* **1989**, *14*, 20–23. [CrossRef]
47. Carling, D. The AMP-activated protein kinase cascade—A unifying system for energy control. *Trends Biochem. Sci.* **2004**, *29*, 18–24. [CrossRef]
48. Xiao, B.; Heath, R.; Saiu, P.; Leiper, F.C.; Leone, P.; Jing, C.; Walker, P.A.; Haire, L.; Eccleston, J.F.; Davis, C.T.; et al. Structural basis for AMP binding to mammalian AMP-activated protein kinase. *Nature* **2007**, *449*, 496–500. [CrossRef]
49. Xiao, B.; Sanders, M.J.; Underwood, E.; Heath, R.; Mayer, F.V.; Carmena, D.; Jing, C.; Walker, P.A.; Eccleston, J.F.; Haire, L.F.; et al. Structure of mammalian AMPK and its regulation by ADP. *Nature* **2011**, *472*, 230–233. [CrossRef]
50. Langendorf, C.G.; Ngoei, K.R.W.; Scott, J.W.; Ling, N.X.Y.; Issa, S.M.A.; Gorman, M.A.; Parker, M.W.; Sakamoto, K.; Oakhill, J.S.; Kemp, B.E. Structural basis of allosteric and synergistic activation of AMPK by furan-2-phosphonic derivative C2 binding. *Nat. Commun.* **2016**, *7*, 10912. [CrossRef]
51. Salt, I.P.; Hardie, D.G. AMP-Activated Protein Kinase: An Ubiquitous Signaling Pathway With Key Roles in the Cardiovascular System. *Circ. Res.* **2017**, *120*, 1825–1841. [CrossRef] [PubMed]
52. Chen, L.; Jiao, Z.-H.; Zheng, L.-S.; Zhang, Y.-Y.; Xie, S.-T.; Wang, Z.-X.; Wu, J.-W. Structural insight into the autoinhibition mechanism of AMP-activated protein kinase. *Nature* **2009**, *459*, 1146–1149. [CrossRef] [PubMed]
53. Li, X.; Wang, L.; Zhou, X.E.; Ke, J.; de Waal, P.W.; Gu, X.; Tan, M.H.E.; Wang, D.; Wu, D.; Xu, H.E.; et al. Structural basis of AMPK regulation by adenine nucleotides and glycogen. *Cell Res.* **2015**, *25*, 50–66. [CrossRef] [PubMed]
54. Oakhill, J.S.; Chen, Z.-P.; Scott, J.W.; Steel, R.; Castelli, L.A.; Ling, N.; Macaulay, S.L.; Kemp, B.E. β-Subunit myristoylation is the gatekeeper for initiating metabolic stress sensing by AMP-activated protein kinase (AMPK). *Proc. Natl. Acad. Sci. USA* **2010**, *107*, 19237–19241. [CrossRef]
55. Oligschlaeger, Y.; Miglianico, M.; Chanda, D.; Scholz, R.; Thali, R.F.; Tuerk, R.; Stapleton, D.I.; Gooley, P.R.; Neumann, D. The recruitment of AMP-activated protein kinase to glycogen is regulated by autophosphorylation. *J. Biol. Chem.* **2015**, *290*, 11715–11728. [CrossRef]

56. Xiao, B.; Sanders, M.J.; Carmena, D.; Bright, N.J.; Haire, L.F.; Underwood, E.; Patel, B.R.; Heath, R.B.; Walker, P.A.; Hallen, S.; et al. Structural basis of AMPK regulation by small molecule activators. *Nat. Commun.* **2013**, *4*, 3017. [CrossRef]
57. Carling, D.; Thornton, C.; Woods, A.; Sanders, M.J. AMP-activated protein kinase: New regulation, new roles? *Biochem. J.* **2012**, *445*, 11–27. [CrossRef]
58. Hawley, S.A.; Boudeau, J.; Reid, J.L.; Mustard, K.J.; Udd, L.; Mäkelä, T.P.; Alessi, D.R.; Hardie, D.G. Complexes between the LKB1 tumor suppressor, STRAD alpha/beta and MO25 alpha/beta are upstream kinases in the AMP-activated protein kinase cascade. *J. Biol.* **2003**, *2*, 28. [CrossRef]
59. Hawley, S.A.; Pan, D.A.; Mustard, K.J.; Ross, L.; Bain, J.; Edelman, A.M.; Frenguelli, B.G.; Hardie, D.G. Calmodulin-dependent protein kinase kinase-beta is an alternative upstream kinase for AMP-activated protein kinase. *Cell Metab.* **2005**, *2*, 9–19. [CrossRef]
60. Davies, S.P.; Helps, N.R.; Cohen, P.T.; Hardie, D.G. 5′-AMP inhibits dephosphorylation, as well as promoting phosphorylation, of the AMP-activated protein kinase. Studies using bacterially expressed human protein phosphatase-2C alpha and native bovine protein phosphatase-2AC. *FEBS Lett.* **1995**, *377*, 421–425.
61. Cheung, P.C.; Salt, I.P.; Davies, S.P.; Hardie, D.G.; Carling, D. Characterization of AMP-activated protein kinase gamma-subunit isoforms and their role in AMP binding. *Biochem. J.* **2000**, *346*, 659–669. [CrossRef] [PubMed]
62. Hong, S.-P.; Leiper, F.C.; Woods, A.; Carling, D.; Carlson, M. Activation of yeast Snf1 and mammalian AMP-activated protein kinase by upstream kinases. *Proc. Natl. Acad. Sci. USA* **2003**, *100*, 8839–8843. [CrossRef] [PubMed]
63. Oakhill, J.S.; Steel, R.; Chen, Z.-P.; Scott, J.W.; Ling, N.; Tam, S.; Kemp, B.E. AMPK is a direct adenylate charge-regulated protein kinase. *Science* **2011**, *332*, 1433–1435. [CrossRef] [PubMed]
64. Sanders, M.J.; Grondin, P.O.; Hegarty, B.D.; Snowden, M.A.; Carling, D. Investigating the mechanism for AMP activation of the AMP-activated protein kinase cascade. *Biochem. J.* **2007**, *403*, 139–148. [CrossRef] [PubMed]
65. Dite, T.A.; Langendorf, C.G.; Hoque, A.; Galic, S.; Rebello, R.J.; Ovens, A.J.; Lindqvist, L.M.; Ngoei, K.R.W.; Ling, N.X.Y.; Furic, L.; et al. AMP-activated protein kinase selectively inhibited by the type II inhibitor SBI-0206965. *J. Biol. Chem.* **2018**, *293*, 8874–8885. [CrossRef] [PubMed]
66. Wodarz, A. Establishing cell polarity in development. *Nat. Cell Biol.* **2002**, *4*, E39–E44. [CrossRef] [PubMed]
67. Lizcano, J.M.; Göransson, O.; Toth, R.; Deak, M.; Morrice, N.A.; Boudeau, J.; Hawley, S.A.; Udd, L.; Mäkelä, T.P.; Hardie, D.G.; et al. LKB1 is a master kinase that activates 13 kinases of the AMPK subfamily, including MARK/PAR-1. *EMBO J.* **2004**, *23*, 833–843. [CrossRef]
68. Daniel, J.L.; Molish, I.R.; Robkin, L.; Holmsen, H. Nucleotide exchange between cytosolic ATP and F-actin-bound ADP may be a major energy-utilizing process in unstimulated platelets. *Eur. J. Biochem.* **1986**, *156*, 677–684. [CrossRef]
69. Bernstein, B.W.; Bamburg, J.R. Actin-ATP hydrolysis is a major energy drain for neurons. *J. Neurosci.* **2003**, *23*, 1–6. [CrossRef]
70. Borghi, N.; Sorokina, M.; Shcherbakova, O.G.; Weis, W.I.; Pruitt, B.L.; Nelson, W.J.; Dunn, A.R. E-cadherin is under constitutive actomyosin-generated tension that is increased at cell-cell contacts upon externally applied stretch. *Proc. Natl. Acad. Sci. USA* **2012**, *109*, 12568–12573. [CrossRef]
71. Goncharova, E.A.; Goncharov, D.A.; James, M.L.; Atochina-Vasserman, E.N.; Stepanova, V.; Hong, S.-B.; Li, H.; Gonzales, L.; Baba, M.; Linehan, W.M.; et al. Folliculin controls lung alveolar enlargement and epithelial cell survival through E-cadherin, LKB1, and AMPK. *Cell Rep.* **2014**, *7*, 412–423. [CrossRef] [PubMed]
72. Diamond, J.M. Twenty-first Bowditch lecture. The epithelial junction: Bridge, gate, and fence. *Physiologist* **1977**, *20*, 10–18. [PubMed]
73. Umeda, K.; Ikenouchi, J.; Katahira-Tayama, S.; Furuse, K.; Sasaki, H.; Nakayama, M.; Matsui, T.; Tsukita, S.; Furuse, M.; Tsukita, S. ZO-1 and ZO-2 independently determine where claudins are polymerized in tight-junction strand formation. *Cell* **2006**, *126*, 741–754. [CrossRef] [PubMed]
74. Ikenouchi, J.; Umeda, K.; Tsukita, S.; Furuse, M.; Tsukita, S. Requirement of ZO-1 for the formation of belt-like adherens junctions during epithelial cell polarization. *J. Cell Biol.* **2007**, *176*, 779–786. [CrossRef] [PubMed]
75. Otani, T.; Nguyen, T.P.; Tokuda, S.; Sugihara, K.; Sugawara, T.; Furuse, K.; Miura, T.; Ebnet, K.; Furuse, M. Claudins and JAM-A coordinately regulate tight junction formation and epithelial polarity. *J. Cell Biol.* **2019**, *218*, 3372–3396. [CrossRef]

76. Peng, L.; Li, Z.-R.; Green, R.S.; Holzman, I.R.; Lin, J. Butyrate enhances the intestinal barrier by facilitating tight junction assembly via activation of AMP-activated protein kinase in Caco-2 cell monolayers. *J. Nutr.* **2009**, *139*, 1619–1625. [CrossRef]
77. Seo-Mayer, P.W.; Thulin, G.; Zhang, L.; Alves, D.S.; Ardito, T.; Kashgarian, M.; Caplan, M.J. Preactivation of AMPK by metformin may ameliorate the epithelial cell damage caused by renal ischemia. *Am. J. Physiol. Renal Physiol.* **2011**, *301*, F1346–F1357. [CrossRef]
78. Spruss, A.; Kanuri, G.; Stahl, C.; Bischoff, S.C.; Bergheim, I. Metformin protects against the development of fructose-induced steatosis in mice: Role of the intestinal barrier function. *Lab. Investig.* **2012**, *92*, 1020–1032. [CrossRef]
79. Garnett, J.P.; Baker, E.H.; Naik, S.; Lindsay, J.A.; Knight, G.M.; Gill, S.; Tregoning, J.S.; Baines, D.L. Metformin reduces airway glucose permeability and hyperglycaemia-induced Staphylococcus aureus load independently of effects on blood glucose. *Thorax* **2013**, *68*, 835–845. [CrossRef]
80. Yano, T.; Matsui, T.; Tamura, A.; Uji, M.; Tsukita, S. The association of microtubules with tight junctions is promoted by cingulin phosphorylation by AMPK. *J. Cell Biol.* **2013**, *203*, 605–614. [CrossRef]
81. Park, H.-Y.; Kunitake, Y.; Hirasaki, N.; Tanaka, M.; Matsui, T. Theaflavins enhance intestinal barrier of Caco-2 Cell monolayers through the expression of AMP-activated protein kinase-mediated Occludin, Claudin-1, and ZO-1. *Biosci. Biotechnol. Biochem.* **2015**, *79*, 130–137. [CrossRef] [PubMed]
82. Wang, B.; Wu, Z.; Ji, Y.; Sun, K.; Dai, Z.; Wu, G. L-Glutamine Enhances Tight Junction Integrity by Activating CaMK Kinase 2-AMP-Activated Protein Kinase Signaling in Intestinal Porcine Epithelial Cells. *J. Nutr.* **2016**, *146*, 501–508. [CrossRef] [PubMed]
83. Patkee, W.R.A.; Carr, G.; Baker, E.H.; Baines, D.L.; Garnett, J.P. Metformin prevents the effects of Pseudomonas aeruginosa on airway epithelial tight junctions and restricts hyperglycaemia-induced bacterial growth. *J. Cell. Mol. Med.* **2016**, *20*, 758–764. [CrossRef] [PubMed]
84. Xue, Y.; Zhang, H.; Sun, X.; Zhu, M.-J. Metformin Improves Ileal Epithelial Barrier Function in Interleukin-10 Deficient Mice. *PLoS ONE* **2016**, *11*, e0168670. [CrossRef] [PubMed]
85. Chen, L.; Wang, J.; You, Q.; He, S.; Meng, Q.; Gao, J.; Wu, X.; Shen, Y.; Sun, Y.; Wu, X.; et al. Activating AMPK to Restore Tight Junction Assembly in Intestinal Epithelium and to Attenuate Experimental Colitis by Metformin. *Front. Pharmacol.* **2018**, *9*, 761. [CrossRef] [PubMed]
86. Shackelford, D.B.; Shaw, R.J. The LKB1-AMPK pathway: Metabolism and growth control in tumour suppression. *Nat. Rev. Cancer* **2009**, *9*, 563–575. [CrossRef] [PubMed]
87. Zhang, L.; Jouret, F.; Rinehart, J.; Sfakianos, J.; Mellman, I.; Lifton, R.P.; Young, L.H.; Caplan, M.J. AMP-activated protein kinase (AMPK) activation and glycogen synthase kinase-3β (GSK-3β) inhibition induce Ca2+-independent deposition of tight junction components at the plasma membrane. *J. Biol. Chem.* **2011**, *286*, 16879–16890. [CrossRef]
88. Yamada, A.; Fujita, N.; Sato, T.; Okamoto, R.; Ooshio, T.; Hirota, T.; Morimoto, K.; Irie, K.; Takai, Y. Requirement of nectin, but not cadherin, for formation of claudin-based tight junctions in annexin II-knockdown MDCK cells. *Oncogene* **2006**, *25*, 5085–5102. [CrossRef]
89. Brakeman, P.R.; Liu, K.D.; Shimizu, K.; Takai, Y.; Mostov, K.E. Nectin proteins are expressed at early stages of nephrogenesis and play a role in renal epithelial cell morphogenesis. *Am. J. Physiol. Renal Physiol.* **2009**, *296*, F564–F574. [CrossRef]
90. Yano, T.; Torisawa, T.; Oiwa, K.; Tsukita, S. AMPK-dependent phosphorylation of cingulin reversibly regulates its binding to actin filaments and microtubules. *Sci. Rep.* **2018**, *8*, 15550. [CrossRef]
91. Citi, S.; Sabanay, H.; Jakes, R.; Geiger, B.; Kendrick-Jones, J. Cingulin, a new peripheral component of tight junctions. *Nature* **1988**, *333*, 272–276. [CrossRef] [PubMed]
92. Mangan, A.J.; Sietsema, D.V.; Li, D.; Moore, J.K.; Citi, S.; Prekeris, R. Cingulin and actin mediate midbody-dependent apical lumen formation during polarization of epithelial cells. *Nat. Commun.* **2016**, *7*, 12426. [CrossRef] [PubMed]
93. Xiang, R.-L.; Mei, M.; Cong, X.; Li, J.; Zhang, Y.; Ding, C.; Wu, L.-L.; Yu, G.-Y. Claudin-4 is required for AMPK-modulated paracellular permeability in submandibular gland cells. *J. Mol. Cell Biol.* **2014**, *6*, 486–497. [CrossRef] [PubMed]
94. Shiomi, R.; Shigetomi, K.; Inai, T.; Sakai, M.; Ikenouchi, J. CaMKII regulates the strength of the epithelial barrier. *Sci. Rep.* **2015**, *5*, 13262. [CrossRef] [PubMed]

95. Aznar, N.; Patel, A.; Rohena, C.C.; Dunkel, Y.; Joosen, L.P.; Taupin, V.; Kufareva, I.; Farquhar, M.G.; Ghosh, P. AMP-activated protein kinase fortifies epithelial tight junctions during energetic stress via its effector GIV/Girdin. *Elife* **2016**, *5*, e20795. [CrossRef]
96. Sasaki, K.; Kakuwa, T.; Akimoto, K.; Koga, H.; Ohno, S. Regulation of epithelial cell polarity by PAR-3 depends on Girdin transcription and Girdin-Gαi3 signaling. *J. Cell Sci.* **2015**, *128*, 2244–2258. [CrossRef]
97. Sun, X.; Yang, Q.; Rogers, C.J.; Du, M.; Zhu, M.-J. AMPK improves gut epithelial differentiation and barrier function via regulating Cdx2 expression. *Cell Death Differ.* **2017**, *24*, 819–831. [CrossRef]
98. Sun, X.; Du, M.; Navarre, D.A.; Zhu, M.-J. Purple Potato Extract Promotes Intestinal Epithelial Differentiation and Barrier Function by Activating AMP-Activated Protein Kinase. *Mol. Nutr. Food Res.* **2018**, *62*. [CrossRef]
99. Creighton, J.; Jian, M.; Sayner, S.; Alexeyev, M.; Insel, P.A. Adenosine monophosphate-activated kinase alpha1 promotes endothelial barrier repair. *FASEB J.* **2011**, *25*, 3356–3365. [CrossRef]
100. Castanares-Zapatero, D.; Bouleti, C.; Sommereyns, C.; Gerber, B.; Lecut, C.; Mathivet, T.; Horckmans, M.; Communi, D.; Foretz, M.; Vanoverschelde, J.-L.; et al. Connection between cardiac vascular permeability, myocardial edema, and inflammation during sepsis: Role of the α1AMP-activated protein kinase isoform. *Crit. Care Med.* **2013**, *41*, e411–e422. [CrossRef]
101. Takata, F.; Dohgu, S.; Matsumoto, J.; Machida, T.; Kaneshima, S.; Matsuo, M.; Sakaguchi, S.; Takeshige, Y.; Yamauchi, A.; Kataoka, Y. Metformin induces up-regulation of blood-brain barrier functions by activating AMP-activated protein kinase in rat brain microvascular endothelial cells. *Biochem. Biophys. Res. Commun.* **2013**, *433*, 586–590. [CrossRef] [PubMed]
102. Liu, Y.; Tang, G.; Li, Y.; Wang, Y.; Chen, X.; Gu, X.; Zhang, Z.; Wang, Y.; Yang, G.-Y. Metformin attenuates blood-brain barrier disruption in mice following middle cerebral artery occlusion. *J. Neuroinflammation* **2014**, *11*, 177. [CrossRef] [PubMed]
103. Komarova, Y.A.; Kruse, K.; Mehta, D.; Malik, A.B. Protein Interactions at Endothelial Junctions and Signaling Mechanisms Regulating Endothelial Permeability. *Circ. Res.* **2017**, *120*, 179–206. [CrossRef] [PubMed]
104. Simionescu, M.; Simionescu, N.; Palade, G.E. Segmental differentiations of cell junctions in the vascular endothelium. The microvasculature. *J. Cell Biol.* **1975**, *67*, 863–885. [CrossRef] [PubMed]
105. Schneeberger, E.E. Structure of intercellular junctions in different segments of the intrapulmonary vasculature. *Ann. N. Y. Acad. Sci.* **1982**, *384*, 54–63. [CrossRef] [PubMed]
106. Navarro, P.; Ruco, L.; Dejana, E. Differential localization of VE- and N-cadherins in human endothelial cells: VE-cadherin competes with N-cadherin for junctional localization. *J. Cell Biol.* **1998**, *140*, 1475–1484. [CrossRef] [PubMed]
107. Gentil-dit-Maurin, A.; Oun, S.; Almagro, S.; Bouillot, S.; Courçon, M.; Linnepe, R.; Vestweber, D.; Huber, P.; Tillet, E. Unraveling the distinct distributions of VE- and N-cadherins in endothelial cells: A key role for p120-catenin. *Exp. Cell Res.* **2010**, *316*, 2587–2599. [CrossRef]
108. Kim, J.H.; Kim, J.H.; Yu, Y.S.; Kim, D.H.; Kim, K.-W. Recruitment of pericytes and astrocytes is closely related to the formation of tight junction in developing retinal vessels. *J. Neurosci. Res.* **2009**, *87*, 653–659. [CrossRef]
109. Hellström, M.; Gerhardt, H.; Kalén, M.; Li, X.; Eriksson, U.; Wolburg, H.; Betsholtz, C. Lack of pericytes leads to endothelial hyperplasia and abnormal vascular morphogenesis. *J. Cell Biol.* **2001**, *153*, 543–553. [CrossRef]
110. Jian, M.-Y.; Liu, Y.; Li, Q.; Wolkowicz, P.; Alexeyev, M.; Zmijewski, J.; Creighton, J. N-cadherin coordinates AMP kinase-mediated lung vascular repair. *Am. J. Physiol. Lung Cell Mol. Physiol.* **2016**, *310*, L71–L85. [CrossRef]
111. Zhou, G.; Myers, R.; Li, Y.; Chen, Y.; Shen, X.; Fenyk-Melody, J.; Wu, M.; Ventre, J.; Doebber, T.; Fujii, N.; et al. Role of AMP-activated protein kinase in mechanism of metformin action. *J. Clin. Investig.* **2001**, *108*, 1167–1174. [CrossRef] [PubMed]
112. Hawley, S.A.; Ross, F.A.; Chevtzoff, C.; Green, K.A.; Evans, A.; Fogarty, S.; Towler, M.C.; Brown, L.J.; Ogunbayo, O.A.; Evans, A.M.; et al. Use of cells expressing gamma subunit variants to identify diverse mechanisms of AMPK activation. *Cell Metab.* **2010**, *11*, 554–565. [CrossRef] [PubMed]
113. Fryer, L.G.D.; Parbu-Patel, A.; Carling, D. The Anti-diabetic drugs rosiglitazone and metformin stimulate AMP-activated protein kinase through distinct signaling pathways. *J. Biol. Chem.* **2002**, *277*, 25226–25232. [CrossRef] [PubMed]
114. Ahn, J.; Lee, H.; Kim, S.; Park, J.; Ha, T. The anti-obesity effect of quercetin is mediated by the AMPK and MAPK signaling pathways. *Biochem. Biophys. Res. Commun.* **2008**, *373*, 545–549. [CrossRef] [PubMed]

115. Baur, J.A.; Pearson, K.J.; Price, N.L.; Jamieson, H.A.; Lerin, C.; Kalra, A.; Prabhu, V.V.; Allard, J.S.; Lopez-Lluch, G.; Lewis, K.; et al. Resveratrol improves health and survival of mice on a high-calorie diet. *Nature* **2006**, *444*, 337–342. [CrossRef]
116. Shen, Q.W.; Zhu, M.J.; Tong, J.; Ren, J.; Du, M. Ca $^{2+}$ /calmodulin-dependent protein kinase kinase is involved in AMP-activated protein kinase activation by α-lipoic acid in C2C12 myotubes. *Am. J. Physiol. Cell Physiol.* **2007**, *293*, C1395–C1403. [CrossRef]
117. Corton, J.M.; Gillespie, J.G.; Hawley, S.A.; Hardie, D.G. 5-aminoimidazole-4-carboxamide ribonucleoside. A specific method for activating AMP-activated protein kinase in intact cells? *Eur. J. Biochem.* **1995**, *229*, 558–565. [CrossRef]
118. Gómez-Galeno, J.E.; Dang, Q.; Nguyen, T.H.; Boyer, S.H.; Grote, M.P.; Sun, Z.; Chen, M.; Craigo, W.A.; van Poelje, P.D.; MacKenna, D.A.; et al. A Potent and Selective AMPK Activator That Inhibits de Novo Lipogenesis. *ACS Med. Chem. Lett.* **2010**, *1*, 478–482. [CrossRef]
119. Cool, B.; Zinker, B.; Chiou, W.; Kifle, L.; Cao, N.; Perham, M.; Dickinson, R.; Adler, A.; Gagne, G.; Iyengar, R.; et al. Identification and characterization of a small molecule AMPK activator that treats key components of type 2 diabetes and the metabolic syndrome. *Cell Metab.* **2006**, 3, 403–416. [CrossRef]
120. Calabrese, M.F.; Rajamohan, F.; Harris, M.S.; Caspers, N.L.; Magyar, R.; Withka, J.M.; Wang, H.; Borzilleri, K.A.; Sahasrabudhe, P.V.; Hoth, L.R.; et al. Structural Basis for AMPK Activation: Natural and Synthetic Ligands Regulate Kinase Activity from Opposite Poles by Different Molecular Mechanisms. *Structure* **2014**, *22*, 1161–1172. [CrossRef]
121. Zadra, G.; Photopoulos, C.; Tyekucheva, S.; Heidari, P.; Weng, Q.P.; Fedele, G.; Liu, H.; Scaglia, N.; Priolo, C.; Sicinska, E.; et al. A novel direct activator of AMPK inhibits prostate cancer growth by blocking lipogenesis. *EMBO Mol. Med.* **2014**, *6*, 519–538. [CrossRef] [PubMed]
122. Handa, N.; Takagi, T.; Saijo, S.; Kishishita, S.; Takaya, D.; Toyama, M.; Terada, T.; Shirouzu, M.; Suzuki, A.; Lee, S.; et al. Structural basis for compound C inhibition of the human AMP-activated protein kinase α2 subunit kinase domain. *Acta Crystallogr. D Biol. Crystallogr.* **2011**, *67*, 480–487. [CrossRef] [PubMed]
123. Scott, J.W.; Galic, S.; Graham, K.L.; Foitzik, R.; Ling, N.X.Y.; Dite, T.A.; Issa, S.M.A.; Langendorf, C.G.; Weng, Q.P.; Thomas, H.E.; et al. Inhibition of AMP-Activated Protein Kinase at the Allosteric Drug-Binding Site Promotes Islet Insulin Release. *Chem. Biol.* **2015**, *22*, 705–711. [CrossRef] [PubMed]
124. Xia, Z.; Huang, L.; Yin, P.; Liu, F.; Liu, Y.; Zhang, Z.; Lin, J.; Zou, W.; Li, C. L-Arginine alleviates heat stress-induced intestinal epithelial barrier damage by promoting expression of tight junction proteins via the AMPK pathway. *Mol. Biol. Rep.*. in press. [CrossRef] [PubMed]
125. Farbood, Y.; Sarkaki, A.; Khalaj, L.; Khodagholi, F.; Badavi, M.; Ashabi, G. Targeting Adenosine Monophosphate-Activated Protein Kinase by Metformin Adjusts Post-Ischemic Hyperemia and Extracellular Neuronal Discharge in Transient Global Cerebral Ischemia. *Microcirculation* **2015**, *22*, 534–541. [CrossRef]
126. Harmel, E.; Grenier, E.; Bendjoudi Ouadda, A.; El Chebly, M.; Ziv, E.; Beaulieu, J.F.; Sané, A.; Spahis, S.; Laville, M.; Levy, E. AMPK in the small intestine in normal and pathophysiological conditions. *Endocrinology* **2014**, *155*, 873–888. [CrossRef]
127. Wojtaszewski, J.F.P.; Birk, J.B.; Frøsig, C.; Holten, M.; Pilegaard, H.; Dela, F. 5′AMP activated protein kinase expression in human skeletal muscle: Effects of strength training and type 2 diabetes. *J. Physiol.* **2005**, *564*, 563–573. [CrossRef]
128. Birk, J.B.; Wojtaszewski, J.F.P. Predominant alpha2/beta2/gamma3 AMPK activation during exercise in human skeletal muscle. *J. Physiol.* **2006**, *577*, 1021–1032. [CrossRef]
129. Thomson, D.M. The Role of AMPK in the Regulation of Skeletal Muscle Size, Hypertrophy, and Regeneration. *Int. J. Mol. Sci.* **2018**, *19*, 3125. [CrossRef]
130. Tsou, P.; Zheng, B.; Hsu, C.-H.; Sasaki, A.T.; Cantley, L.C. A Fluorescent Reporter of AMPK Activity and Cellular Energy Stress. *Cell Metab.* **2011**, *13*, 476–486. [CrossRef]
131. Konagaya, Y.; Terai, K.; Hirao, Y.; Takakura, K.; Imajo, M.; Kamioka, Y.; Sasaoka, N.; Kakizuka, A.; Sumiyama, K.; Asano, T.; et al. A Highly Sensitive FRET Biosensor for AMPK Exhibits Heterogeneous AMPK Responses among Cells and Organs. *Cell Rep.* **2017**, *21*, 2628–2638. [CrossRef] [PubMed]

International Journal of
Molecular Sciences

Article

Enteropathogenic *Escherichia coli* (EPEC) Recruitment of PAR Polarity Protein Atypical PKCζ to Pedestals and Cell–Cell Contacts Precedes Disruption of Tight Junctions in Intestinal Epithelial Cells

Rocio Tapia [1,†], Sarah E. Kralicek [1,†] and Gail A. Hecht [1,2,3,*]

1 Department of Medicine, Division of Gastroenterology and Nutrition, Loyola University Chicago, Maywood, IL 60153, USA; rtpastrana@luc.edu (R.T.); sknopf@luc.edu (S.E.K.)
2 Department of Microbiology and Immunology, Loyola University Chicago, Maywood, IL 60153, USA
3 Edward Hines Jr. VA Hospital, Hines, IL 60141, USA
* Correspondence: ghecht@lumc.edu; Tel.: +1-708-216-5965
† These authors contributed equally to this work.

Received: 13 December 2019; Accepted: 10 January 2020; Published: 14 January 2020

Abstract: Enteropathogenic *Escherichia coli* (EPEC) uses a type three secretion system to inject effector proteins into host intestinal epithelial cells, causing diarrhea. EPEC induces the formation of pedestals underlying attached bacteria, disrupts tight junction (TJ) structure and function, and alters apico-basal polarity by redistributing the polarity proteins Crb3 and Pals1, although the mechanisms are unknown. Here we investigate the temporal relationship of PAR polarity complex and TJ disruption following EPEC infection. EPEC recruits active aPKCζ, a PAR polarity protein, to actin within pedestals and at the plasma membrane prior to disrupting TJ. The EPEC effector EspF binds the endocytic protein sorting nexin 9 (SNX9). This interaction impacts actin pedestal organization, recruitment of active aPKCζ to actin at cell–cell borders, endocytosis of JAM-A S285 and occludin, and TJ barrier function. Collectively, data presented herein support the hypothesis that EPEC-induced perturbation of TJ is a downstream effect of disruption of the PAR complex and that EspF binding to SNX9 contributes to this phenotype. aPKCζ phosphorylates polarity and TJ proteins and participates in actin dynamics. Therefore, the early recruitment of aPKCζ to EPEC pedestals and increased interaction with actin at the membrane may destabilize polarity complexes ultimately resulting in perturbation of TJ.

Keywords: enteropathogenic *E. coli* (EPEC); tight junctions (TJ); polarity; atypical aPKCζ; transepithelial electrical resistance (TER); sorting nexin 9 (SNX9); EspF

1. Introduction

Enteropathogenic *Escherichia coli* (EPEC) delivers bacterial effector proteins into host intestinal epithelial cells (IECs) through a type III secretion system (TTSS), inducing actin pedestal formation, attaching and effacing lesions, and physiological changes in IECs that contribute to diarrhea [1]. EPEC alters the architecture and barrier function of tight junctions (TJ) [2,3] although the mechanisms are not well understood.

TJ are localized at the most apical region of the lateral membrane and constitute a paracellular diffusion barrier modulating the flow of ions and solutes. These structures consist of integral membrane proteins (claudin family, occludin, tricellulin, MarvelD3, and JAM-A) that interact with adhesion molecules of adjacent cells and with intracellular domains that associate with cytoplasmic adaptor proteins (MAGUK family, cingulin, paracingulin, MAGI-1-3, and MUPP-1) [4,5]. TJ also constitute a fence contributing to the maintenance of apico-basal polarity by restricting the intermixing of apical and lateral plasma membrane components. Three main protein complexes control epithelial polarity,

Crumbs (Crb3/Pals1/Patj), PAR (Par3/Par6/aPKCζ/Cdc42), and Scribble (Scrib/Lgl/Dlg). Apico-basal polarity contributes to cell morphology, directional vesicle transportation, ion and solute transport, and specific localization of proteins and lipids to different membrane domains [6,7].

The interdependence between apico-basal polarity complexes and TJ is well established. Alterations in Crb3 expression or reduced expression of Patj/Pals1 impair apical polarity and TJ development [8–12]. Inhibition of aPKCζ activity, impaired phosphorylation of Par3, as well as deletion of the aPKCζ binding domain of Par6, delays TJ assembly [13–16]. aPKC activity also maintains TJ integrity and membrane localization of occludin and ZO-1 [17]. Downregulation of Scrib or Dlg compromises TJ establishment [18–20]. In contrast, increased expression of Scrib in MCF10A cells promotes the formation of functional TJ [21]. These data demonstrate that polarity complexes are crucial to TJ assembly, maintenance, and function.

EPEC effectors perturb TJ structure and function and alter apico-basal polarity of IECs. EspF perturbs barrier function in vivo and in vitro by redistributing TJ proteins from the cell–cell contacts, decreasing transepithelial electrical resistance (TER), and increasing paracellular permeability [2,22–24]. Map increases permeability to charged and non-charged molecules, indicating a failure in gate function [24,25]. NleA mislocalizes occludin and ZO-1 from the cell–cell contacts leading to barrier dysfunction [26]. EspG also contributes to leaky barrier, perturbs microtubule networks, and induces cytoplasmic accumulation of occludin and delays TJ recovery [27–29].

EPEC infection causes progressive redistribution of the basolateral proteins, β1-integrin and Na^+/K^+ ATPase, to the apical compartment and the mislocalization of TJ proteins, occludin, claudin-1, and ZO-1 from cell–cell contacts to the lateral membrane and cytoplasm [22,23,26,30–32], suggesting that cell polarity is altered. We recently reported that EPEC drives Crb3 and Pals1 away from the apical membrane, and cell–cell contacts into the cytoplasm of IECs and EspF is crucial for this phenotype [32]. EspF is a multifunctional molecule that interacts with several host proteins including actin, profilin, Arp2, N-WASP, SNX9, Abcf2, cytokeratin 18, 14-3-3, WIPF1, SNX18, and SNX33 [33–38]. EspF interacts with the SH3 domain of sorting nexin 9 (SNX9) through its RxAPxxP motif [33,35]. The interaction of EspF with SNX9 promotes the formation of elongated plasma membrane tubules, as well as the internalization of EPEC into IECs [39]. EspF/SNX9 complex is required for impairment of both cell polarity and altered TJ structure and function [32,33,40].

Despite extensive investigation into the mechanisms by which EPEC effectors directly perturb TJ, no such evidence has been reported. In view of the interdependence of polarity and TJ complexes, we hypothesized that EPEC-induced disruption of intestinal epithelial cell TJ structure and function stems from the initial targeting of polarity complexes. This study examines the effect of EPEC on the PAR complex with particular focus on aPKCζ, which phosphorylates several targets crucial for the establishment and maintenance of apico–basal polarity and TJ function. The data presented herein support the notion that EPEC-induced perturbation of TJ is a downstream consequence of EspF-induced disruption of the PAR polarity complex, particularly the recruitment of aPKCζ to actin-rich pedestals, and its increased co-localization with actin at the membrane.

2. Results

2.1. EPEC Disrupts PAR Polarity Complexes In Vivo and In Vitro

EPEC alters the localization of Crb complex proteins resulting in perturbed cell polarity [32]. Here, we investigate the effect of EPEC on PAR polarity components. EPEC does not change the localization of Par3 in murine colonocytes as compared with uninfected (UI) tissues (Figure 1A). In contrast, colonocytes of EPEC-infected mice show that Par6 is redistributed from the apical membrane to the cytoplasm (Figure 1A). Similarly, phosphorylated aPKCζ–T410, aPKCζ–T560, and total aPKCζ in the apical membrane of colonocytes of EPEC-infected mice appear diffuse and shift from the apical to the lateral membrane and into the cytoplasm compared to its restricted localization to the apical membrane in UI cells (Figure 1A). To begin to explore the mechanisms involved in PAR complex perturbation caused by EPEC, in vitro models were used. SKCO-15 monolayers were infected for 1-2 h and the localization and protein levels of PAR complex proteins were examined. In accordance with in vivo data, Par3 localization in cultured IECs was unchanged by EPEC infection (Figure 1B). In contrast, EPEC displaced Par6 from cell–cell contacts into the cytoplasm (Figure 1B). Interestingly, aPKCζ displayed a ringlet pattern around attached bacteria (Figure 1B, zoom). Expression levels of PAR complex proteins are not altered by EPEC infection (Figure 1C). These data demonstrate that EPEC alters the distribution of the PAR complex proteins, Par6 and aPKCζ without changing their expression levels.

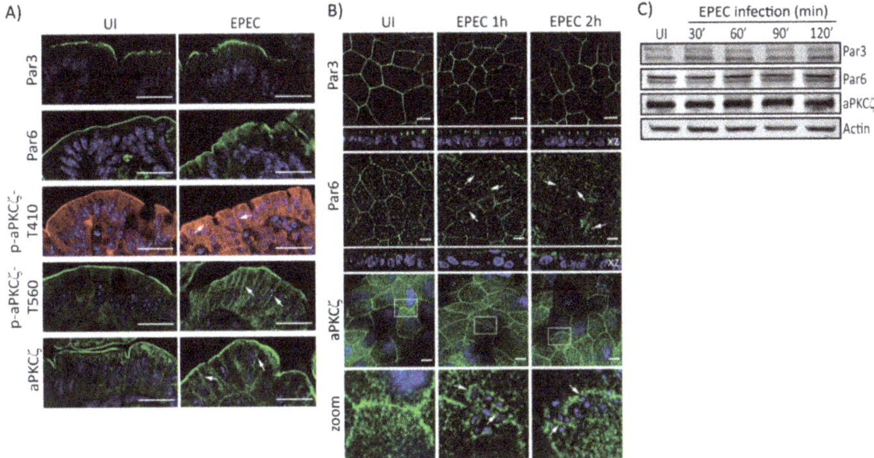

Figure 1. EPEC alters PAR complex localization. (**A**) Immunofluorescence microscopy of murine colonic epithelial tissue 3 days post-infection with EPEC or not (uninfected—UI). Par3 localization is unchanged. Par6 accumulates within the cytoplasm. Phosphorylated aPKCζ–T410, aPKCζ–T560, and total aPKCζ mislocalize from the apical membrane and accumulate within the cytoplasm and lateral membrane (arrows). Scale bar: 40 µm. (**B** and **C**) SKCO-15 monolayers were infected, or not (UI), with EPEC at indicated times, then the localization and protein levels of PAR complex proteins were assessed by immunofluorescence and western blot, respectively. (**B**) Par3 localization is unaltered by EPEC infection, and Par6 redistributes from cell–cell contacts to the cytoplasm (arrows). aPKCζ localizes under attached bacteria (arrows, inset zoom). Hoechst was used to stain host and bacteria nuclei (blue). Scale bar: 10 µm. (**C**) EPEC does not change the expression levels of PAR complex proteins; β-actin was used as a loading control.

2.2. EspF, Via Its SNX9-Binding Domain, Impacts the Structural Organization of aPKCζ and F-actin at Pedestals

EPEC induces the translocation of aPKCζ from the cytoplasm to the plasma membrane [41,42]. aPKCζ localization under attached EPEC in SKCO-15 cells suggests it is recruited to actin-rich pedestals. Indeed, aPKCζ co-localizes with filamentous actin (F-actin) around attached bacteria as early as 30 min and becomes more evident at 60 min post-infection (Figure 2A). The pattern around attached bacteria is highly organized in which actin surrounds bacteria, then a ring of aPKCζ and F-actin co-localization, and finally aPKCζ alone as seen in Figure 2A zoom, and depicted in the schematic in Figure 2B. Confocal microscopy confirms pedestal organization at 60 min post-infection consisting of attached bacteria atop actin, an interface of actin and aPKCζ co-localization, and then a region of aPKCζ alone (Figure 2A z-stack and Figure 2B). This structural organization is maintained at 2 h post-infection (Figure 2C zoom and z-stack, Figure 2B,D). Interestingly, aPKCζ is also seen within large filopodia induced by attached EPEC co-localizing with actin (Figure 2E).

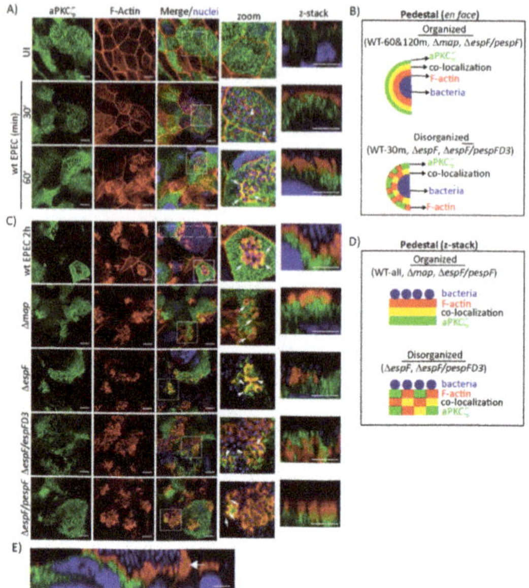

Figure 2. EspF and its SNX9-binding domain are crucial for the structural organization of F-actin and aPKCζ within EPEC pedestals. SKCO-15 monolayers were infected with wt EPEC, Δ*map*, Δ*espF*, Δ*espF/pespF*D3, or Δ*espF/pespF* to assess F-actin and aPKCζ co-localization. (**A**) F-actin and aPKCζ are recruited to and co-localize around attached bacteria at 30 and 60 min post-infection with wt EPEC. Zoom and z-stack images at 60 min display EPEC pedestals with attached bacteria surrounded by F-actin, an interface of co-localized F-actin and aPKCζ, then aPKCζ alone. (**B**) Schematic of F-actin (red), aPKCζ (green), and co-localizing interface (yellow) organization around bacteria after infection with wt EPEC and mutant strains. (**C**) aPKCζ recruitment to pedestals is maintained at 2 h post-infection with wt EPEC. Zoom and z-stacks reveal the robust organization of F-actin and aPKCζ co-localization. Deletion of *map* (Δ*map*) does not alter aPKCζ recruitment to pedestals. Infection with Δ*espF* and Δ*espF/pespF*D3 causes disorganization of co-localized F-actin and aPKCζ within pedestals. Complementation of *espF* (Δ*espF/pespF*) restores the organized phenotype. Squares correspond to zoom and z-stack areas in all except wt EPEC, in which a large rectangle corresponds to z-stack image. Arrows and arrowheads indicate organized and disorganized pedestals, respectively. (**D**) Schematic representation of pedestals showing the organized and disorganized localization of F-actin (red), aPKCζ (green), and co-localizing interface (yellow) after infection with wt EPEC and mutant strains. (**E**) aPKCζ also co-localizes with F-actin in filopodia (arrow) 2 h post-infection with wt EPEC. Scale bars: 10 μm (en face); 5 μm (z-stack).

We previously demonstrated that the effectors Map and EspF, through its interaction with SNX9, disrupt the Crb polarity complex [32]. Therefore, we investigated the contribution of Map and EspF to the recruitment of aPKCζ to pedestals. aPKCζ recruitment to EPEC pedestals is not diminished by deletion of *map* (Δ*map*) or *espF* (Δ*espF*), or by mutation of the SNX9-binding domain of EspF (Δ*espF/pespF*D3) (Figure 2C, green channel). However, the localization of F-actin and aPKCζ within pedestals is disorganized when infected with Δ*espF* and Δ*espF/pespF*D3, but not Δ*map*, compared to infection with wild-type (wt) EPEC, seen both en face and within z-stack images (Figure 2B–D). Complementation of *espF* (Δ*espF/pespF*) restores the organization of actin and aPKCζ within pedestals (Figure 2C, zoom and z-stack, Figure 2B,D).

Interestingly, infection of T84 monolayers with Δ*espF/pespF*D3 produces aPKCζ aggregates under attached bacteria in actin pedestals similar to wt EPEC and Δ*espF/pespF*, while only infection with Δ*espF* induces less co-localization of aPKCζ and actin (Supplemental Figure S1A,B). These data indicate that aPKCζ recruitment to pedestals is not dependent on EspF, however, the SNX9-binding domain of EspF is required for the structural organization of actin and aPKCζ within pedestals in a cell-specific manner.

2.3. EspF and Its SNX9-Binding Domain Contribute to the Recruitment of Phosphorylated aPKCζ–T560 to Pedestals without Altering Kinase Activity

To determine if EPEC infection alters kinase activity, total PKC activity was measured in T84 and SKCO-15 monolayers following infection with wt EPEC or EspF mutant strains for 1–2 h. In agreement with our previous observations, a significant increase in the kinase activity of total PKC in T84 cells is observed following infection with EPEC and EspF mutant strains (Supplementary Figure S1C) [41,42]. Interestingly, EPEC failed to alter the kinase activity of total PKC in SKCO-15 monolayers (Figure 3A). aPKCζ autophosphorylation at Thr560 (p-aPKCζ–T560) activates kinase activity [43,44]. We questioned whether active aPKCζ is recruited to pedestals. Figure 3B shows the co-localization of p-aPKCζ-T560 with F-actin in EPEC at the cell–cell borders and pedestals. Heat maps were generated by merging the green and red channels depicted in Figure 3B in order to visually quantitate the co-localization of p-aPKCζ–T560 with F-actin; white indicates the most abundant co-localization and blue-black the least to none (Figure 3C). To more specifically quantify F-actin and p-aPKCζ–T560 co-localization, immunofluorescence intensity was measured in regions of interest within cell–cell contacts, the cytoplasm, or pedestals (Figure 3D). In uninfected cells, there is no significant association of F-actin and p-aPKCζ–T560 at cell membranes or within the cytoplasm (Figure 3D), despite their apparent co-localization (Figure 3B,C). However, co-localization within the membrane intensifies as early as 15 min post-infection with wt EPEC (Figure 3B,C) and significant correlation between F-actin and p-aPKCζ–T560 is present at 30 min post-infection (Figure 3D). Interestingly, at both 15 and 30 min post-infection with wt EPEC, there is significant association between p-aPKCζ–T560 and F-actin within pedestals (Figure 3B–D) and confocal microscopy confirms this co-localization (Figure 3E). Interestingly, both Δ*espF*and Δ*espF/pespF*D3 diminish p-aPKCζ–T560 and F-actin co-localization at both the plasma membrane and within pedestals 30 min post-infection (Figure 3B–E). Complementation of Δ*espF*with wt EspF (Δ*espF/pespF*) restores the interaction of F-actin and p-aPKCζ–T560 at cell–cell borders and pedestals (Figure 3B–E). Together, these results suggest that although PKC activity is unchanged by EPEC infection in SKCO-15 cells, active aPKCζ localization within the cell is altered and is dependent on EspF and its SNX9-binding domain for the recruitment of p-aPKCζ–T560 to actin at cell–cell borders and EPEC-induced pedestals.

2.4. SNX9-Binding Domain of EspF Is Essential to Disrupt TJ Structure and Function in SKCO-15 Monolayers

EPEC effectors contribute to loss of intestinal epithelial TJ structure and function, and EspF is largely responsible for this phenotype [2]. To determine the individual contribution of EspF to barrier disruption, TER was measured in SKCO-15 monolayers harboring pRetroX–Tight–Pur–EspF-HA with and without inducible expression. EspF-HA expression levels were determined by western blot of control and Tet–On cells incubated with increasing amounts of doxycycline (dox) (Figure 4A). The basal expression of EspF–HA is observed in the absence of dox compared to control cells and increases at 500 ng/mL of dox (Figure 4A). Control and EspF–HA cells were grown in Transwells with and without dox (500 ng/mL) for three days, then TER was determined. Interestingly, uninduced (-dox) cells transfected with EspF–HA show a significant reduction in TER values as compared to control monolayers (647 ± 44 ohms.cm^2 vs. 1545 ± 22 ohms.cm^2) (Figure 4B), likely due to basal expression of EspF (Figure 4A) as previously reported [32]. Induced expression of EspF–HA (+dox) significantly decreases TER compared to uninduced (-dox) (308 ± 86 ohms.cm^2 vs. 647 ± 44 ohms.cm^2) (Figure 4B) indicating that EspF alone significantly reduces TER in SKCO-15 monolayers.

To assess in detail the EspF motifs involved in perturbation of TJ barrier function, SKCO-15 monolayers were infected with wt EPEC, Δ*espF*, Δ*espF/pespF*, or Δ*espF* complemented to express specific site-directed EspF mutations (Δ*espF/pespF*D3, Δ*espF/pSer47A*, and Δ*espF/pSer47/50A*). In addition to binding SNX9, EspF binds to 14-3-3 family proteins [36]. EPEC strains expressing EspF harboring 14-3-3ζ binding motif mutations (Δ*espF/pSer47A* and Δ*espF/pSer47/50A)* were used as controls because of their inability to perturb the Crb polarity complex [32]. Infection with Δ*espF/pSer47A* and Δ*espF/pSer47/50A* perturbed barrier function to a similar degree as wt EPEC (–41.0 ± 6.0% and –40.0 ± 5.0% vs. –38.0 ± 7.0%, change from baseline, respectively) (Figure 4C). In contrast, Δ*espF* and Δ*espF*/*pespF*D3 mutants induced minimal change in TER at 2 h compared to wt EPEC (–9.0 ± 7.0% and –7.0 ± 6.0% vs. –38.0 ± 7.0%, change from baseline, respectively) (Figure 4C). As reported previously in T84 cells [33], Δ*espF/pespF*D3 decreased TER to the same extent as wt EPEC (–35.7 ± 5.8% vs. –32.3 ± 5.1%, change from baseline) (Supplemental Figure S1D). Infection with Δ*espF* attenuates the decrease in resistance (–18.3 ± 5.1%, change from baseline) and complementation of *espF* (Δ*espF/pespF*) restores this phenotype (–38.6 ± 7.8%, change from baseline) (Supplemental Figure S1D), indicating the cell-specific responses to the EspF SNX9-binding domain mutant.

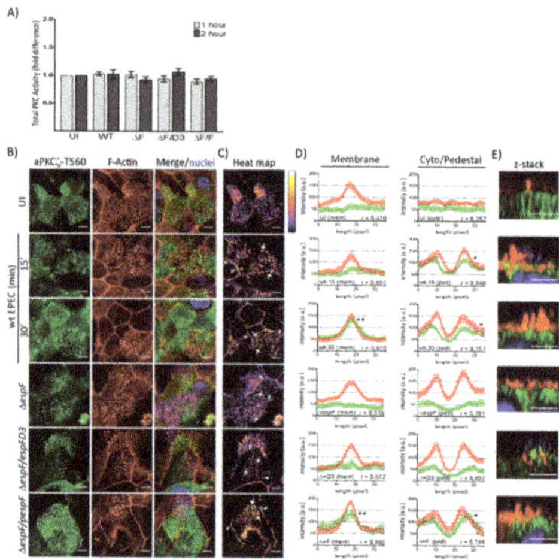

Figure 3. EspF and its SNX9-binding domain contribute to the co-localization of phosphorylated aPKCζ–T560 and F-actin at the plasma membrane and within pedestals. SKCO-15 monolayers plated on Transwells were infected with wt EPEC or EspF mutant strains, then PKC kinase activity, and F-actin and p-aPKCζ-T560 co-localization were assessed. (**A**) Fold change in PKC activity does not change at 1 or 2 h post-infection with EPEC (WT), Δ*espF* (ΔF), Δ*espF* complemented with mutant *espF* (ΔF/D3) or wt *espF* (ΔF/F) compared to UI monolayers. (**B–E**) p-aPKCζ–T560 is recruited to and co-localizes with F-actin at the cell–cell contacts and within pedestals 15 and 30 min post-infection with wt EPEC. p-aPKCζ–T560 recruitment is reduced following infection with Δ*espF* or Δ*espF/pespF*D3. Complementation of *espF* (Δ*espF/pespF*) restores F-actin and p-aPKCζ–T560 co-localization levels similar to wt. (**C**) Heat map generated from merged images in B indicating the intensity of F-actin and p-aPKCζ–T560 co-localization at the plasma membrane (arrowheads) and pedestals (arrows). (**D**) Immunofluorescence (IF) quantification of images represented in B. IF intensity in arbitrary units (a.u.) plotted against length (pixel) of p-aPKCζ–T560 (green) and F-actin (red) at the membrane and within the cell. Graphs represent data from >15 regions of interest from three biological replicates. Average peak intensity for p-aPKCζ–T560 reveals significantly higher levels at the membrane with wt and Δ*espF/pespF* at 30 min post-infection and significantly higher levels within pedestals with wt-15, wt-30, and Δ*espF/pespF* 30 min post-infection compared to uninfected controls; one-way ANOVA, * $p < 0.01$ and ** $p < 0.001$. r = Pearson correlation coefficient indicating an association between F-actin and p-aPKCζ–T560 localization at the membrane and within pedestals in an EspF and SNX9-binding domain dependent manner. (**E**) Z-stack images reveal F-actin and p-aPKCζ–T560 co-localization within pedestals. Scale bars: 10 μm (en face); 5 μm (z-stack).

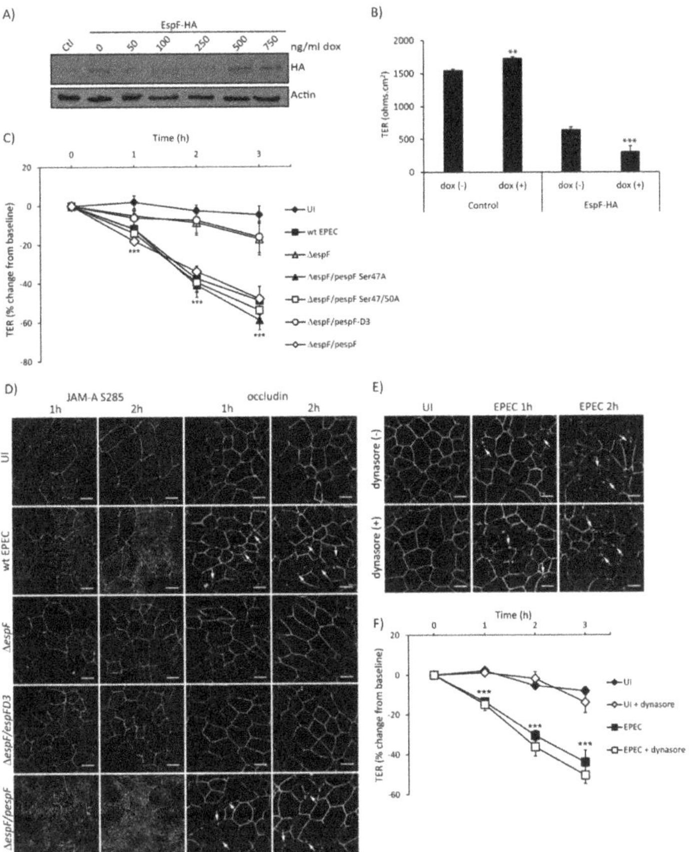

Figure 4. The SNX9 binding domain of EspF is crucial for disruption of TJ structure and function. (**A** and **B**) To generate Tet–On cells, EspF–HA was cloned into the doxycycline-inducible p–RetroX–Tight–Pur vector and transfected or not (control – Ctl) into wild-type SKCO-15 cells. Tet–On SKCO-15 cells were plated in absence of doxycycline (dox). (**A**) EspF–HA expression was induced with increasing concentrations of dox for three days. Cells were processed for western blot and detection of HA was performed. (**B**) Tet–On SKCO-15 cells were cultured on Transwells for 1 week (-dox) then expression of pRetroX–Tight–Pur–EspF-HA was induced (+dox) for three days. Basal expression of EspF–HA induces a significant reduction in TER (-dox) as compared to controls. Induction of EspF–HA (+dox) significantly reduces TER even further. (**C** and **D**) SKCO-15 monolayers were infected with wt EPEC, Δ*espF*, or Δ*espF*complemented with wt *espF* or site-directed *espF* mutants (Δ*espF/pespF*, Δ*espF/pespF*D3, Δ*espF/pSer47A*, or Δ*espF/pSer47/50A*). TER and localization of JAM-A S285 and occludin were determined. (**C**) Infection with Δ*espF/pespF*, Δ*espF/pSer47A* and Δ*espF/pSer47/50A*mutant strains reduce TER to levels similar to those induced by wt EPEC infection. Infection with Δ*espF*or Δ*espF/pespF*D3 significantly blunt the decrease in TER compared to wt EPEC. ***$p < 0.001$. (**D**) Infection with wt EPEC and Δ*espF/pespF*redistributes JAM-A S285 and occludin from intercellular junctions to the cytoplasm. Infection with Δ*espF*or Δ*espF/pespFD3*does not alter the localization of JAM-A S285 and occludin. Scale bar, 10 µm. (**E** and **F**) SKCO-15 monolayers were treated with or without dynasore 1 h prior to EPEC infection, then occludin localization and TER were assessed. (**E**) Blocking dynamin-dependent endocytosis with dynasore fails to prevent occludin internalization induced by wt EPEC infection. Arrows indicate the endocytosis of occludin triggered by EPEC infection. Scale bar, 10 µm. (**F**) Dynasore does not prevent the drop in TER induced by EPEC. TER reported as percent change from baseline. *$p < 0.01$, ***$p < 0.001$.

We next questioned whether EspF and the SNX9-binding domain play a role in the maintenance of TJ structure and function. We evaluated the distribution of phosphorylated JAM-A at Ser285 and occludin, which are both involved in the assembly and maintenance of TJ. Interestingly, Δ*espF/pespF*D3 ablates the redistribution of JAM-A S285 and occludin from the membrane to the cytoplasm caused by wt EPEC in a similar manner to those cells infected with Δ*espF* (Figure 4D). We previously demonstrated that the EspF/SNX9 interaction is crucial for the endocytosis of Crb3 in a clathrin-dependent manner [32]. We therefore investigated whether the redistribution of occludin caused by EPEC is mediated by a similar endocytic pathway. SKCO-15 monolayers were incubated, or not, with dynasore, a dynamin inhibitor that plays an important role in clathrin-dependent endocytosis, and infected, or not, with EPEC for 1–2 h. Occludin localization and TER measurements were followed. Treatment with dynasore did not prevent the endocytosis of occludin (Figure 4E) or the decrease in TER caused by EPEC at 1–2 h post-infection (–15.0 ± 3.0% vs. –14.0 ± 2.0%; –36.0 ± 5.0% vs. –30.0 ± 2.0%, with and without dynasore at 1 and 2 h post-infection, respectively) (Figure 4F). These results indicate that occludin endocytosis and barrier dysfunction mediated by EPEC occurs via a dynamin-independent pathway.

2.5. Temporal Disruption of TJ Proteins Mediated by EPEC Infection

The interplay between cell polarity and TJ establishment and function is well defined. Active aPKCζ targets TJ proteins and participates in the assembly and maintenance of barrier function. EPEC impairs both cell polarity and intestinal epithelial TJ barrier function. To determine if the temporal sequence of TJ disruption following EPEC infection correlates with aPKCζ redistribution, SKCO-15 monolayers were infected with EPEC and the localization of TJ proteins and TER were analyzed over time. EPEC does not alter the localization of total JAM-A from the cell–cell contacts (Figure 5A). In contrast, JAM-A S285 is internalized by 30 min and TER significantly decreases 45 min post-infection (Figure 5A,C). Phosphorylation of JAM-A at tyrosine 280 (JAM-A Y280) is related to loss of barrier function [45]. Interestingly, we found that JAM-A Y280 is detectable only after 1–2 h EPEC infection, corresponding to leaky TJ (Figure 5A). Occludin disruption was seen at 1 h post-infection (Figure 5B) corresponding with a more significant drop in TER (–14.0 ± 8.0%, change from baseline) (Figure 5C). ZO-1 was the last TJ protein altered by EPEC moving into the cytoplasm at 2 h post-infection (Figure 5B) and corresponding to a profound drop in TER (–29.0 ± 5.0%, change from baseline) (Figure 5C).

To further delineate the temporal relationship of EPEC-induced recruitment of p-aPKCζ-T560 to pedestals and the redistribution of occludin from the cell–cell borders, localization of F-actin, p-aPKCζ-T560, and occludin, was assessed from 5–120 min post-infection. EPEC induces the recruitment of F-actin almost immediately upon bacterial attachment (5–15 min) (Figure 6A). Recruitment of p-aPKCζ–T560 to pedestals occurs as early as 5–15 min, and its co-localization with F-actin is seen by 15–30 min post-infection (Figure 6A). Consistent with our previous results, occludin remains at the cell borders at early time points after infection being displaced from the membrane at 60–120 min post-infection (Figure 6A). To examine if inhibition of aPKCζ activity affects its recruitment to EPEC pedestals and subsequent impairment in barrier function, SKCO-15 monolayers were incubated or not (-) with aPKCζ pseudosubstrate (PS) prior to EPEC infection, then localization of p-aPKCζ–T410 and TER were determined. Monolayers treated with PS alone show the redistribution of p-aPKCζ–T410 from the apical membrane and cell–cell contacts to the lateral membrane domain and cytoplasm, and cell death is apparent (Figure 6B). Interestingly, PS treatment does not alter the recruitment of p-aPKCζ–T410 to EPEC pedestals (Figure 6B). In agreement with previous reports [17], inhibition of aPKCζ with PS disrupts barrier function (–28.0% ± 4.0 and –73.0% ± 3.0 change from baseline, PS 5 μM and PS 10 μM, respectively) (Figure 6C). EPEC infection does not alter the drop in TER caused by PS (Figure 6C). Interestingly, monolayers incubated with PS alone (5μM) begin to recover TER (Figure 6C). In contrast, EPEC infection (2 h) in the presence of 5 μM PS causes a progressive drop in TER (–46.0% ± 4.0 change from baseline) similar to infection with EPEC alone (–45.0% ± 2.0 change from baseline) (Figure 6C), suggesting that the displacement of p-aPKCζ–T410 away from TJ results in a perturbation of barrier function. Together these results demonstrate the progressive dismantling of TJ by

EPEC and the corresponding impact on barrier function, events that occur after the recruitment of active aPKCζ to actin pedestals.

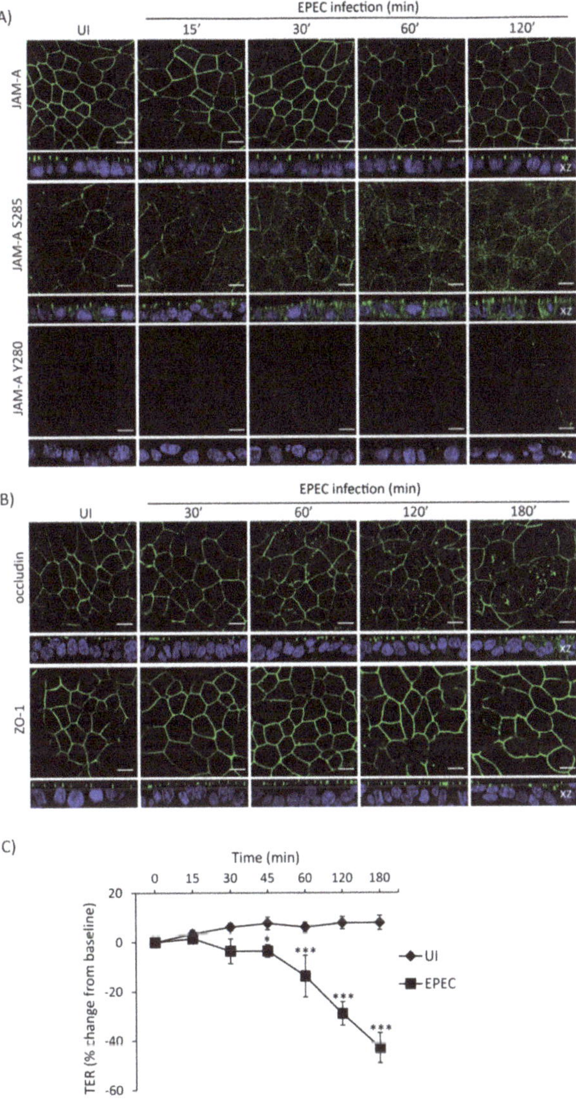

Figure 5. Temporal redistribution of TJ proteins and barrier dysfunction caused by EPEC. (**A–C**) SKCO-15 cells were plated on Transwells and infected or not (UI) with EPEC. Total JAM-A, JAM-A S285, JAM-A Y280, occludin, and ZO-1 localization and TER were determined. (**A**) EPEC does not alter the distribution of total JAM-A. In contrast, JAM-A S285 is displaced from the cell–cell contacts to the cytoplasm at 30 min post-infection. Tyrosine phosphorylation of JAM-A Y280 is apparent at 60–120 min post-infection. Scale bars, 10 μm. (**B**) EPEC induces the endocytosis of occludin and ZO-1 at 1 and 2 h post-infection, respectively. Scale bars, 10 μm. (**C**) TER drops significantly as early as 45 min post-infection and progressively decreases over time as more TJ proteins are displaced. TER reported as percent change from baseline. * $p < 0.01$, *** $p < 0.001$.

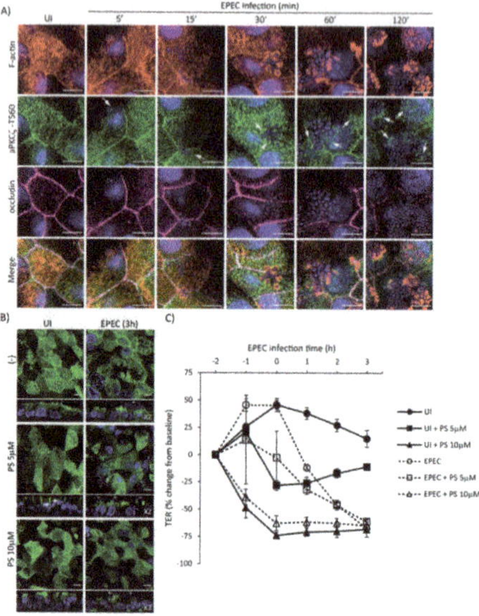

Figure 6. EPEC induces the recruitment of p-aPKCζ–T560 to actin pedestals prior to redistribution of occludin from cell borders. (**A**) SKCO-15 cells were plated on Transwells and infected or not (UI) with EPEC for 5–120 min. Immunofluorescence microscopy for F-actin, p-aPKCζ–T560, and occludin was performed. p-aPKCζ–T560 is recruited to pedestals co-localizing with actin consistently as early as 15 min and increases as infection progresses. Occludin localization is not displaced from cell borders until 1–2 h post-infection. Arrows indicate the presence of p-aPKCζ–T560 within EPEC pedestals. Scale bars, 10 µm. (**B** and **C**) SKCO-15 cells were plated on Transwells and pre-incubated or not with aPKCζ pseudosubstrate (PS) for 1 h prior to EPEC infection. Immunofluorescence of p-aPKCζ–T410 was performed and TER was measured. (**B**) Treatment with PS redistributes p-aPKCζ–T410 from the apical membrane and intercellular junctions to the lateral membrane (arrows) and cytoplasm without altering recruitment to EPEC pedestals (stars). Scale bars, 10 µm. (**C**) Inhibition of aPKCζ results in rapid disruption of barrier function in both UI and EPEC-infected monolayers.

3. Discussion

The present study provides evidence that EPEC perturbs PAR polarity complex integrity and induces the recruitment of aPKCζ to actin pedestals almost immediately following EPEC attachment. The SNX9-binding domain of EspF is important for the recruitment and organization of p-aPKCζ–T560 organization within EPEC pedestals, and the endocytosis of TJ proteins. We speculate that the very early recruitment of active aPKCζ to actin within pedestals and cell–cell contacts triggers downstream signalling events that ultimately disrupt intestinal epithelial TJ structure and function.

In view of the well-established interdependence between polarity and TJ complexes, we questioned if the apico-basal polarity defects caused by EPEC might be upstream of events that lead to TJ barrier perturbation. We provide evidence that EPEC perturbs PAR polarity complex integrity, mislocalizes active aPKCζ to the lateral membrane in vivo, and induces aPKCζ recruitment to pedestals and co-localization with actin at cell–cell borders almost immediately following EPEC attachment in vitro. These events precede the progressive dismantling of TJ proteins from the cell–cell contacts and barrier dysfunction. This is the first report correlating the temporal disruption of the PAR polarity complex with TJ disassembly by EPEC. Phosphorylated JAM-A S285, which is involved in TJ assembly [46], is displaced from cell–cell contacts at early times following EPEC infection. Subsequently, occludin then ZO-1 are displaced from TJ temporally correlating with the progressive loss of TER. Furthermore, pedestal formation and recruitment of active aPKCζ to pedestals occurs prior to occludin mislocalization. Interestingly, we found that EPEC induces the presence of phosphorylated JAM-A Y280 at intercellular contacts at times that correspond to barrier loss. Inflammatory stimuli increase JAM-A Y280 phosphorylation by balancing activity of the Src kinase, YES-1, and the phosphatase PTPN13, ultimately leading to disruption of barrier function [45]. EPEC induces a pro-inflammatory response through several signaling pathways resulting in the activation of NF-κB, ERK1/2, p38, JNK and PKCζ, and the upregulation of IL-8 expression, which contribute to intestinal barrier dysfunction [42,47–49]. EPEC also regulates the activity of multiple kinases, including the Src family, that contribute to actin polymerization [50,51]. Interestingly, Src family kinases contribute to Tir phosphorylation and actin pedestal formation [52,53]. Further analysis is required to understand how these pathways are involved in aPKCζ activity and relocalization.

EspF and Map are major EPEC effectors that disrupt the Crb polarity complex and TJ, however, the mechanisms are not known. Here we demonstrate that EspF likely through binding SNX9 recruits active aPKCζ to actin within pedestals and at the plasma membrane. Map does not have a role in aPKCζ recruitment, but as aPKCζ also associates with actin in filopodia, we cannot discard the notion that Map may regulate aPKCζ activity during filopodia dynamics [33,54,55]. The EspF/SNX9/N-WASP complex participates in F-actin polymerization, membrane remodeling during EPEC pathogenesis, impairment of TJ structure and function, and recruitment of ZO-1 and ZO-2 to pedestals, but this complex does not affect EPEC pedestal formation [33–35,39,40]. This is in agreement with our data in which clear pedestals form after infection with Δ*espF*, however, the organization of total aPKCζ within pedestals and the recruitment of active aPKCζ to actin within pedestals and at cell–cell borders are severely altered following infection with Δ*espF* and Δ*espF/pespFD3*. Interestingly, EspF of rabbit EPEC (REPEC) and EPEC is involved in pedestal maturation [34,56], suggesting that EspF coordination of aPKCζ and actin within pedestals and at the membrane influences downstream signaling events that lead to TJ disruption.

EspF interacts with 14-3-3ζ [36], a protein that binds Par3 to regulate cell polarity [57]. This interaction is not involved in TJ barrier disruption by EPEC as determined by infection with Δ*espF/pSer47A* and Δ*espF/pSer47/50A* and is consistent with our finding that Par3 localization remains unchanged after EPEC infection. EspF also forms a complex with SNX9 and N-WASP. This binding is required to disrupt Crb3, ZO-1, and E-cadherin from cell–cell contacts increasing their cytoplasmic accumulation, thus leading to impaired cell polarity and barrier function [32,40]. We found that EspF is essential for the displacement of JAM-A S285 and occludin from cell–cell contacts, as well as disruption of barrier function in SKCO-15 cells, as occurs in Caco-2 and T84 cells [33,40]. While the EspF–SNX9 interaction is important for barrier disruption in SKCO-15 and Caco-2, it does not play a role in T84 cells, as infection with the EspF-D3 mutant failed to protect against a loss of TER and the redistribution of occludin in polarized T84 cells [33]. Furthermore, examination of pedestals in EPEC-infected T84 cells reveals that aPKCζ and actin co-localization is not impacted by the EspF-D3 mutant. PKC activity post-EPEC-infection also differs between T84 and SKCO-15 cells [41,42]. Although all derived from colonic cancer cell lines, Caco-2 when grown for extended periods differentiate into small intestine-like cells, whereas SKCO-15 and T84 cells are more colonic in nature. Several studies have focused on the biochemical and structural differences between these colonic cell lines [58–60]. In addition, SNX9 expression levels differ between colon cancer cell lines, as well as having other varying redundant sorting nexin proteins [61–63]. These data highlight the cell-specific aPKCζ responses and different mechanisms that contribute to TJ barrier dysfunction and lends support to the notion that mislocalization of aPKCζ activity contributes to TJ disruption in SKCO-15 cells.

EPEC modulates clathrin-mediated endocytosis, thus it has been suggested that EPEC-induced TJ disassembly may occur via this mechanism. For instance, EspF recruits clathrin, AP2, early (Rab5a and EEA1) and recycling (Rab4a, Rab11a, Rab11b, FIP2, Myo5b) endocytic proteins to sites of infection [38]. The EspF binding partner SNX9 is recruited to clathrin-coated pits and associates with N-WASP, dynamin, Arp2/3 and other associated proteins to promote endocytosis of plasma membrane receptors [64–69]. In addition, EPEC mediates Crb3 endocytosis in a dynamin-dependent manner [32].

However, data presented herein indicate that EPEC-induced endocytosis of occludin and disruption of barrier function occur via a dynamin-independent mechanism indicating that an alternative endocytic pathway is responsible and supports aPKCζ involvement in TJ disassembly.

Interestingly, aPKCζ has direct and indirect roles in the formation and maintenance of polarity and TJ structure and function. aPKCζ phosphorylates the TJ proteins occludin, JAM-A, claudin-4, and ZO-1 to establish and maintain TJ structure and barrier function [17,46,70,71]. Silencing of aPKCζ or inhibition of its kinase activity with PS are associated with dephosphorylation of occludin and ZO-1, delayed assembly and perturbed maintenance of barrier function [17]. We found that PS treatment mislocalizes p-aPKCζ–T410 to the lateral membrane and cytoplasm but does not affect its recruitment to EPEC pedestals. In addition, when aPKCζ is absent from the lateral membrane, whether due to high concentrations of PS or recruitment to EPEC pedestals, barrier function is severely altered. aPKCζ also phosphorylates polarity proteins including Crb3, Par3, Lgl, and Par1b, and is dependent on interactions with Par6 and Cdc42 for its activity at the apical domain [13,72–75]. aPKC activation during endothelial morphogenesis is determined by the adaptor protein Nck [76]. After Nck is recruited to actin pedestals [77,78], it could serve as a hub for aPKCζ pedestal activity. Therefore, one could speculate that the early recruitment of aPKCζ to pedestals redirects its kinase activity away from TJ and polarity complexes, thus disrupting apico-basal polarity and barrier function.

Besides phosphorylation of polarity and TJ proteins, aPKCζ plays a role in actin dynamics. In *Drosophila*, Cdc42 and active aPKC cause the interaction of DSH3PX1/Dock/Wasp/Arp2/3, homologues of human SNX9/Nck/N-WASP/Arp2/3. This complex is involved in actin rearrangement and adherens junction stability [79–81]. Interestingly, we found that the EspF–SNX9 interaction was essential not only for the highly organized structure of aPKCζ and actin within pedestals but also for increased aPKCζ and actin co-localization at the membrane. aPKC activates ezrin, a plasma membrane protein that links and stabilizes the actin cytoskeleton to the membrane facilitating microvilli formation and endocytosis [82–84]. Ezrin and SNX9 are enriched on curved membranes; ezrin may interact with SNX9 via its curvature-sensing I-BAR domain facilitating ezrin to tether and close membranes during membrane tubule formation [62,85]. EspF/SNX9 binding is involved in F-actin polymerization and promotes the formation of tubular vesicles in EPEC-infected cells [33,35,39]. We previously determined that ezrin recruitment to the cytoskeleton is dependent on EspF, and ezrin activation contributes to disruption of TJ barrier function [86]. EspF, through its interaction with several host proteins including actin, profilin, Arp2, N-WASP, Abcf2, cytokeratin 18, WIPF1, SNX9/18/33, regulates cytoskeletal dynamics [33,34,36–38]. Together, these data indicate that through EspF interactions with actin-binding proteins and its structured localization of aPKCζ with actin at the membrane, aPKCζ activity likely participates in actin dynamics controlling endocytic pathways and TJ maintenance following EPEC infection.

The recruitment of cytoskeletal, adaptor, and signaling proteins into pedestals, as well as rearrangement of the actin cytoskeleton in host cells during EPEC infection are crucial steps in EPEC pathogenesis [77,87–90]. Together the results reported herein suggest a mechanism by which EPEC perturbs intestinal epithelial TJ (Figure 7). EPEC recruits active aPKCζ to interact with actin within pedestals and at the membrane of cell–cell contacts immediately following bacterial attachment. The recruitment of active aPKCζ away from polarity complexes and into highly organized actin pedestals via an EspF–SNX9 dependent process may be the initial upstream event that triggers downstream pathways that alter actin dynamics and disrupt intestinal epithelial TJ structure and function thus contributing to EPEC pathogenesis.

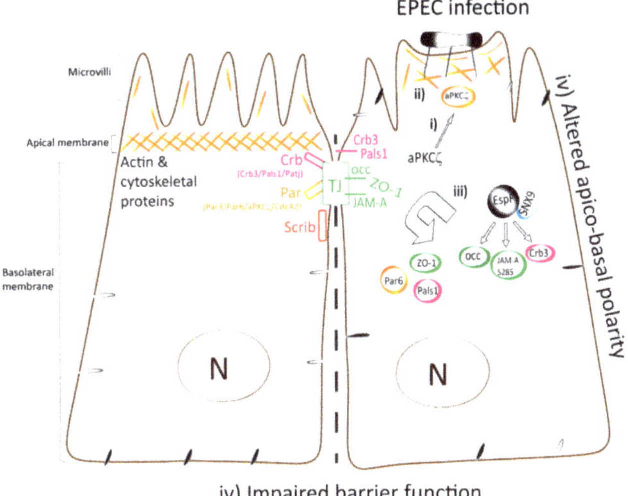

Figure 7. Model displaying the temporal mislocalization of aPKCζ leading to the dismantling of polarity and TJ complexes. Polarized epithelial cells consist of the apical membrane facing the lumen and the basolateral domain in contact with the underlying basement membrane. TJ, localized at the most apical area of the lateral membrane, contribute to the maintenance of apico-basal polarity. Polarity complexes (Crb, PAR, and Scrib) localize along the lateral membrane and modulate TJ assembly and function. (**i**) EPEC infection induces the recruitment of active aPKCζ to actin pedestals and at plasma membrane almost immediately upon bacterial attachment. (**ii**) Actin and aPKCζ are highly organized within EPEC pedestals via a mechanism that is dependent on the EspF/SNX9 binding domain. (**iii**) TJ (JAM-A S285, occludin, and ZO-1) and polarity (Par6, Crb3, and Pals1) proteins are displaced from cell–cell contacts and internalized into the cytoplasm. (**iv**) Apico-basal polarity, evidenced by redistribution of basolateral proteins to the apical membrane [32], and TJ are perturbed. These events are largely dependent on EspF–SNX9 binding domain, although other effectors may contribute.

4. Materials and Methods

4.1. Tissue Culture

SKCO-15 cell line derived from human adenocarcinoma of the colon displays typical intestinal epithelial cell features with adhesion complexes and microvilli [59,60]. SKCO-15 cultures were grown in Dulbecco's Modified Eagle Medium (DMEM) complemented with 10% fetal bovine serum with antibiotics as previously described [32]. T84 cells were grown in low glucose DMEM/Ham's F12 (Gibco) medium complemented with 10% newborn calf serum (Gibco, Life Technologies, Carlsbad, CA, USA) and antibiotics as previously reported [91]. Cells were grown at 37 °C in 5% CO_2, once monolayers were confluent (7–10 days), media was replaced with bacterial growth medium 24 h prior to infection.

4.2. Antibodies and Reagents

Par3 (07-330, EMD Millipore), Par6 (ab49776 and ab6022, Abcam, Cambridge, MA, USA), aPKCζ (sc-17781, Santa Cruz Biotechnology, Dallas, TX, USA), p-aPKCζ–T560 (ab62372, Abcam, Cambridge, MA, USA), p-aPKCζ–T410 (sc-12894R, Santa Cruz Biotechnology, Dallas, TX, USA), actin (A2066, Sigma-Aldrich, St. Louis, MO, USA), F-actin BODIPY 558/568 Phalloidin (B3475, Invitrogen, Life Technologies Carlsbad, CA, USA), occludin (33-1500, Invitrogen, Life Technologies, Carlsbad, CA, USA), JAM-A S285 (sc-17430, Santa Cruz Biotechnology, Dallas, TX, USA), JAM-A Y280 (600-401-GN5, Rockland Immunochemicals Inc, Limerick, PA), JAM-A (361700, Invitrogen, Life Technologies, Carlsbad, CA, USA), and ZO-1 (61-7300, Invitrogen, Life Technologies,

Carlsbad, CA, USA). Secondary antibodies used for immunofluorescence were Alexa Fluor (Life Technologies, Carlsbad, CA, USA). Dynasore (Sigma-Aldrich, St. Louis, MO, USA) was used at 80 μM. aPKCζ pseudosubstrate (539624, Millipore-Sigma, Burlington, MA, USA) was used at 5 and 10 μM.

4.3. Bacterial Culture

The following EPEC strains were used: wt EPEC 0127:H6 E2348/69, Δ*espF* [92], Δ*espF*/*pespF*, Δ*map* [93], Δ*espF/pespF*D3 [33], Δ*espF/pSer47A* and Δ*espF/pSer47/50A* [Hecht unpublished]. Bacterial cultures were grown overnight in Luria-Bertani broth with appropriate selective antibiotics. For infections, bacterial cultures were processed as previously reported [32,91]. Monolayers plated on Transwells (Costar #3740), or six-well plates were infected with bacterial strains at a multiplicity of infection of 50. Infected cells were incubated at 37 °C in 5% CO2 for indicated times.

4.4. Murine Infection

Male mice C57BL/6J from 8 to 10 weeks old were used (Jackson Laboratory, Bar Harbor, ME, USA), and housed in a specific pathogen-free facility at Loyola University Chicago (LUC) Medical Campus. LUC Animal Care and Use Committee approved all animal protocols. Animals were infected with EPEC by oral gavage and sacrificed on day three post-infection [94]. Intestinal tissues were processed for immunofluorescence as reported [91].

4.5. Immunofluorescence

Cells were plated on coverslips or Transwells and fixed with cold methanol at –20 °C for 10 min or with 4% PFA for 15–30 min and permeabilized with 0.1% Triton X-100 (5–15 min). Samples were blocked and incubated with primary antibodies overnight at 4 °C with Invitrogen blocking solution, washed with PBS, then processed for immunofluorescence. Paraffin embedded intestinal sections of infected mice were processed as previously described [91].

4.6. Imaging

Slides were analyzed using a confocal Leica TCS SPE DMI 4000B (LAS X software, Leica, Wetzlar, Germany) microscope. Z-stack images were acquired in 0.33 μm sections and processed using the 3-D volume setting of LasX software. Images were processed using Adobe Photoshop CC 2018 and FIJI-ImageJ-64 software. Heat maps were generated from red and green channels using ImageJ software (Analyze> Colocalization> colocalization threshold > show colocalized pixel map), then channels were split and co-localization channel assigned LUT > Fire and auto-contrasted. Quantitation of p-PKCζ–T560 and F-actin were accomplished with ImageJ. Green and red channels were analyzed separately. A 35-pixel length line was drawn and centered over cell–cell contacts or pedestals, and the line duplicated in each channel (Edit > selection > restore selection). Pixel intensity was measured over the length of the line (Analyze > plot profile), recorded for at least 15 regions of interest from membranes and pedestals of three biological replicates and statistical analysis was performed as described below.

4.7. Western Blot Analysis

SKCO-15 monolayers were rinsed twice with cold PBS, protein extraction was achieved with RIPA buffer containing protease inhibitors as previously reported [32]. Cell lysates were processed by electrophoresis (SDS-PAGE), transferred to Immobilon membrane (IPFL00010, Millipore, Burlington, MA, USA). Membranes were incubated with primary antibodies overnight at 4 °C and LI-COR secondary antibodies for 1 h at room temperature. Immunoblotting was performed using LI-COR Odyssey system (LI-COR Bioscience, Lincoln, NE, USA)

4.8. Generation of Tet-On System

SKCO-15 cells were co-transfected with p-RetroX–Tet–On Advanced and pRetroX–Tight–Pur–EspF–HA [32]. To generate stable cell lines, transfected SKCO-15 cells were selected with G418 and puromycin for 2–3 weeks. Expression of the transgene pRetroX–Tight–Pur–EspF–HA was induced in presence of doxycycline (500 ng/mL) for three days [32].

4.9. Measurement of Transepithelial Electrical Resistance

Wild type and Tet–On SKCO-15 cells (3×10^5 cells) were plated in triplicate on Transwell filters in DMEM. TER was measured using cellZscope (nanoAnalytics, Munster, Germany) for 5–7 days. Cells were infected with wt EPEC or mutant strains and TER measurements were taken at indicated intervals. Expression of pRetroX–Tight–Pur–EspF–HA was induced with doxycycline (+dox) and measurements were followed for 3–6 days. TER values were normalized to UI monolayers.

4.10. PKC activity Assays

SKCO-15 and T84 cells were grown in 6-well plates and infected for the indicated time points. PKC Kinase Activity Assay Kit (ab139437, Abcam, Cambridge, MA) was used according to the manufacturer's instructions.

4.11. Statistical Analysis

All experiments were performed in triplicate. TER results are the mean ± SEM of three independent experiments performed in triplicate. *P* values were calculated by ANOVA Tukey's Multiple Comparison Test and Pearson correlation co-efficient using GraphPad Prism v7.

Supplementary Materials: Supplementary materials can be found at http://www.mdpi.com/1422-0067/21/2/527/s1.

Author Contributions: Conceptualization, R.T., S.E.K. and G.A.H.; methodology, R.T., S.E.K. and G.A.H.; validation, R.T., S.E.K.; formal analysis, R.T., S.E.K.; investigation, R.T., S.E.K.; resources, G.A.H.; data curation, R.T., S.E.K.; writing—original draft preparation, review and editing, R.T., S.E.K. and G.A.H.; visualization, R.T., S.E.K. and G.A.H.; supervision, G.A.H.; project administration, G.A.H.; funding acquisition, G.A.H. All authors have read and agreed to the published version of the manuscript.

Funding: This research was funded by National Institutes of Health grant (DK097043 to G.A.H.) and Edward Hines JR VA Hospital grant (BX002687 to G.A.H.).

Conflicts of Interest: The authors declare no conflict of interest.

Abbreviations

aPKCζ	Atypical protein kinase C zeta
EPEC	Enteropathogenic *Escherichia coli*
F-actin	Filamentous actin
JAM-A S285	Junctional adhesion molecule-A phosphorylated at Serine 285
JAM-A Y280	Junctional adhesion molecule-A phosphorylated at Tyrosine 280
p-aPKCζ-T410	aPKCζ phosphorylated at Threonine 410
p-aPKCζ-T560	aPKCζ phosphorylated at Threonine 560
PS	aPKCζ pseudosubstrate inhibitor
SNX9	Sorting nexin 9
TJ	Tight junctions
TER	Transepithelial electrical resistance
ZO-1	Zonula Occludens -1

References

1. Pearson, J.S.; Giogha, C.; Wong Fok Lung, T.; Hartland, E.L. The genetics of enteropathogenic escherichia coli virulence. *Annu. Rev. Genet.* **2016**, *50*, 493–513. [CrossRef]
2. McNamara, B.P.; Koutsouris, A.; O'Connell, C.B.; Nougayrede, J.P.; Donnenberg, M.S.; Hecht, G. Translocated EspF protein from enteropathogenic escherichia coli disrupts host intestinal barrier function. *J. Clin. Invest.* **2001**, *107*, 621–629. [CrossRef]
3. Muza-Moons, M.M.; Schneeberger, E.E.; Hecht, G.A. Enteropathogenic escherichia coli infection leads to appearance of aberrant tight junctions strands in the lateral membrane of intestinal epithelial cells. *Cell. Microbiol.* **2004**, *6*, 783–793. [CrossRef]
4. Tsukita, S.; Tanaka, H.; Tamura, A. The claudins: From Tight junctions to biological systems. *Trends Biochem. Sci.* **2019**, *44*, 141–152. [CrossRef]
5. Van Itallie, C.M.; Anderson, J.M. Architecture of tight junctions and principles of molecular composition. *Semin. Cell Dev. Biol.* **2014**, *36*, 157–165. [CrossRef]
6. Wen, W.; Zhang, M. Protein complex assemblies in epithelial cell polarity and asymmetric cell division. *J. Mol. Biol.* **2018**, *430*, 3504–3520. [CrossRef]
7. Rodriguez-Boulan, E.; Macara, I.G. Organization and execution of the epithelial polarity programme. *Nat. Rev. Mol. Cell Biol.* **2014**, *15*, 225–242. [CrossRef]
8. Hurd, T.W.; Gao, L.; Roh, M.H.; Macara, I.G.; Margolis, B. Direct interaction of two polarity complexes implicated in epithelial tight junction assembly. *Nat. Cell Biol.* **2003**, *5*, 137–142. [CrossRef] [PubMed]
9. Roh, M.H.; Fan, S.; Liu, C.J.; Margolis, B. The Crumbs3-Pals1 complex participates in the establishment of polarity in mammalian epithelial cells. *J. Cell. Sci.* **2003**, *116*, 2895–2906. [CrossRef] [PubMed]
10. Fogg, V.C.; Liu, C.J.; Margolis, B. Multiple regions of Crumbs3 are required for tight junction formation in MCF10A Cells. *J. Cell. Sci.* **2005**, *118*, 2859–2869. [CrossRef] [PubMed]
11. Straight, S.W.; Shin, K.; Fogg, V.C.; Fan, S.; Liu, C.J.; Roh, M.; Margolis, B. Loss of PALS1 expression leads to tight junction and polarity defects. *Mol. Biol. Cell* **2004**, *15*, 1981–1990. [CrossRef] [PubMed]
12. Tilston-Lunel, A.M.; Haley, K.E.; Schlecht, N.F.; Wang, Y.; Chatterton, A.L.; Moleirinho, S.; Watson, A.; Hundal, H.S.; Prystowsky, M.B.; Gunn-Moore, F.J.; et al. Crumbs 3b promotes tight junctions in an Ezrin-dependent manner in mammalian cells. *J. Mol. Cell. Biol.* **2016**, *8*, 439–455. [CrossRef] [PubMed]
13. Nagai-Tamai, Y.; Mizuno, K.; Hirose, T.; Suzuki, A.; Ohno, S. Regulated protein-protein interaction between aPKC and PAR-3 plays an essential role in the polarization of epithelial cells. *Genes Cells* **2002**, *7*, 1161–1171. [CrossRef] [PubMed]
14. Suzuki, A.; Yamanaka, T.; Hirose, T.; Manabe, N.; Mizuno, K.; Shimizu, M.; Akimoto, K.; Izumi, Y.; Ohnishi, T.; Ohno, S. Atypical protein kinase c is involved in the evolutionarily conserved par protein complex and plays a critical role in establishing epithelia-specific junctional structures. *J. Cell Biol.* **2001**, *152*, 1183–1196. [CrossRef] [PubMed]
15. Chen, X.; Macara, I.G. Par-3 controls tight junction assembly through the Rac exchange factor Tiam1. *Nat. Cell Biol.* **2005**, *7*, 262–269. [CrossRef] [PubMed]
16. Yamanaka, T.; Horikoshi, Y.; Suzuki, A.; Sugiyama, Y.; Kitamura, K.; Maniwa, R.; Nagai, Y.; Yamashita, A.; Hirose, T.; Ishikawa, H.; et al. PAR-6 regulates aPKC activity in a novel way and mediates cell-cell contact-induced formation of the epithelial junctional complex. *Genes Cells* **2001**, *6*, 721–731. [CrossRef]
17. Jain, S.; Suzuki, T.; Seth, A.; Samak, G.; Rao, R. Protein kinase czeta phosphorylates occludin and promotes assembly of epithelial tight junctions. *Biochem. J.* **2011**, *437*, 289–299. [CrossRef]
18. Qin, Y.; Capaldo, C.; Gumbiner, B.M.; Macara, I.G. The mammalian scribble polarity protein regulates epithelial cell adhesion and migration through E-cadherin. *J. Cell Biol.* **2005**, *171*, 1061–1071. [CrossRef]
19. Stucke, V.M.; Timmerman, E.; Vandekerckhove, J.; Gevaert, K.; Hall, A. The MAGUK protein MPP7 binds to the polarity protein hDlg1 and facilitates epithelial tight junction formation. *Mol. Biol. Cell* **2007**, *18*, 1744–1755. [CrossRef]
20. Ivanov, A.I.; Young, C.; Den Beste, K.; Capaldo, C.T.; Humbert, P.O.; Brennwald, P.; Parkos, C.A.; Nusrat, A. Tumor suppressor scribble regulates assembly of tight junctions in the intestinal epithelium. *Am. J. Pathol.* **2010**, *176*, 134–145. [CrossRef]
21. Elsum, I.A.; Martin, C.; Humbert, P.O. Scribble regulates an EMT polarity pathway through modulation of MAPK-ERK signaling to mediate junction formation. *J. Cell. Sci.* **2013**, *126*, 3990–3999. [CrossRef] [PubMed]

22. Shifflett, D.E.; Clayburgh, D.R.; Koutsouris, A.; Turner, J.R.; Hecht, G.A. Enteropathogenic E. Coli disrupts tight junction barrier function and structure in Vivo. *Lab. Invest.* **2005**, *85*, 1308–1324. [CrossRef] [PubMed]
23. Guttman, J.A.; Li, Y.; Wickham, M.E.; Deng, W.; Vogl, A.W.; Finlay, B.B. Attaching and effacing pathogen-induced tight junction disruption in Vivo. *Cell. Microbiol.* **2006**, *8*, 634–645. [CrossRef] [PubMed]
24. Dean, P.; Kenny, B. Intestinal barrier dysfunction by Enteropathogenic Escherichia Coli is mediated by two effector molecules and a bacterial surface protein. *Mol. Microbiol.* **2004**, *54*, 665–675. [CrossRef]
25. Singh, A.P.; Aijaz, S. Generation of a MDCK cell line with constitutive expression of the Enteropathogenic E. Coli effector protein map as an in Vitro model of pathogenesis. *Bioengineered* **2015**, *6*, 335–341. [CrossRef]
26. Thanabalasuriar, A.; Koutsouris, A.; Weflen, A.; Mimee, M.; Hecht, G.; Gruenheid, S. The bacterial virulence factor NleA is required for the disruption of intestinal tight junctions by Enteropathogenic Escherichia Coli. *Cell. Microbiol.* **2010**, *12*, 31–41. [CrossRef]
27. Tomson, F.L.; Viswanathan, V.K.; Kanack, K.J.; Kanteti, R.P.; Straub, K.V.; Menet, M.; Kaper, J.B.; Hecht, G. Enteropathogenic Escherichia Coli EspG disrupts microtubules and in conjunction with Orf3 enhances perturbation of the tight junction barrier. *Mol. Microbiol.* **2005**, *56*, 447–464. [CrossRef]
28. Matsuzawa, T.; Kuwae, A.; Abe, A. Enteropathogenic Escherichia Coli Type III effectors EspG and EspG2 alter epithelial paracellular permeability. *Infect. Immun.* **2005**, *73*, 6283–6289. [CrossRef]
29. Glotfelty, L.G.; Zahs, A.; Hodges, K.; Shan, K.; Alto, N.M.; Hecht, G.A. Enteropathogenic E. Coli effectors EspG1/G2 disrupt microtubules, contribute to tight junction perturbation and inhibit restoration. *Cell. Microbiol.* **2014**, *16*, 1767–1783. [CrossRef]
30. Thanabalasuriar, A.; Kim, J.; Gruenheid, S. The inhibition of COPII trafficking is important for intestinal epithelial tight junction disruption during Enteropathogenic Escherichia Coli and Citrobacter Rodentium infection. *Microbes Infect.* **2013**, *15*, 738–744. [CrossRef]
31. Muza-Moons, M.M.; Koutsouris, A.; Hecht, G. Disruption of cell polarity by Enteropathogenic Escherichia Coli enables basolateral membrane proteins to migrate apically and to potentiate physiological consequences. *Infect. Immun.* **2003**, *71*, 7069–7078. [CrossRef] [PubMed]
32. Tapia, R.; Kralicek, S.E.; Hecht, G.A. EPEC Effector EspF Promotes Crumbs3 Endocytosis and Disrupts Epithelial Cell Polarity. *Cell. Microbiol.* **2017**, *19*, e12757. [CrossRef] [PubMed]
33. Alto, N.M.; Weflen, A.W.; Rardin, M.J.; Yarar, D.; Lazar, C.S.; Tonikian, R.; Koller, A.; Taylor, S.S.; Boone, C.; Sidhu, S.S.; et al. The type III effector EspF coordinates membrane trafficking by the spatiotemporal activation of two eukaryotic signaling pathways. *J. Cell Biol.* **2007**, *178*, 1265–1278. [CrossRef] [PubMed]
34. Peralta-Ramirez, J.; Hernandez, J.M.; Manning-Cela, R.; Luna-Munoz, J.; Garcia-Tovar, C.; Nougayrede, J.P.; Oswald, E.; Navarro-Garcia, F. EspF interacts with nucleation-promoting factors to recruit junctional proteins into pedestals for pedestal maturation and disruption of paracellular permeability. *Infect. Immun.* **2008**, *76*, 3854–3868. [CrossRef] [PubMed]
35. Marches, O.; Batchelor, M.; Shaw, R.K.; Patel, A.; Cummings, N.; Nagai, T.; Sasakawa, C.; Carlsson, S.R.; Lundmark, R.; Cougoule, C.; et al. EspF of Enteropathogenic Escherichia Coli binds sorting nexin 9. *J. Bacteriol.* **2006**, *188*, 3110–3115. [CrossRef]
36. Viswanathan, V.K.; Lukic, S.; Koutsouris, A.; Miao, R.; Muza, M.M.; Hecht, G. Cytokeratin 18 interacts with the Enteropathogenic Escherichia Coli Secreted Protein F (EspF) and is redistributed after infection. *Cell. Microbiol.* **2004**, *6*, 987–997. [CrossRef]
37. Nougayrede, J.P.; Foster, G.H.; Donnenberg, M.S. Enteropathogenic Escherichia Coli effector EspF Interacts with host protein Abcf2. *Cell. Microbiol.* **2007**, *9*, 680–693. [CrossRef]
38. Kassa, E.G.; Zlotkin-Rivkin, E.; Friedman, G.; Ramachandran, R.P.; Melamed-Book, N.; Weiss, A.M.; Belenky, M.; Reichmann, D.; Breuer, W.; Pal, R.R.; et al. Enteropathogenic Escherichia Coli remodels host endosomes to promote endocytic turnover and breakdown of surface polarity. *PLoS Pathog.* **2019**, *15*, e1007851. [CrossRef]
39. Weflen, A.W.; Alto, N.M.; Viswanathan, V.K.; Hecht, G.E. Coli secreted protein f promotes EPEC invasion of intestinal epithelial cells via an SNX9-Dependent mechanism. *Cell. Microbiol.* **2010**, *12*, 919–929. [CrossRef]
40. Garber, J.J.; Mallick, E.M.; Scanlon, K.M.; Turner, J.R.; Donnenberg, M.S.; Leong, J.M.; Snapper, S.B. Attaching-and-Effacing pathogens exploit junction regulatory activities of N-WASP and SNX9 to disrupt the intestinal barrier. *Cell. Mol. Gastroenterol. Hepatol.* **2017**, *5*, 273–288. [CrossRef]

41. Tomson, F.L.; Koutsouris, A.; Viswanathan, V.K.; Turner, J.R.; Savkovic, S.D.; Hecht, G. Differing roles of protein kinase C-Zeta in disruption of tight junction barrier by Enteropathogenic and Enterohemorrhagic Escherichia Coli. *Gastroenterology* **2004**, *127*, 859–869. [CrossRef] [PubMed]
42. Savkovic, S.D.; Koutsouris, A.; Hecht, G. PKC zeta participates in activation of inflammatory response induced by Enteropathogenic E. Coli. *Am. J. Physiol. Cell. Physiol.* **2003**, *285*, C512–C521. [CrossRef] [PubMed]
43. Keranen, L.M.; Dutil, E.M.; Newton, A.C. Protein Kinase C is regulated in Vivo by Three functionally distinct phosphorylations. *Curr. Biol.* **1995**, *5*, 1394–1403. [CrossRef]
44. Tsutakawa, S.E.; Medzihradszky, K.F.; Flint, A.J.; Burlingame, A.L.; Koshland, D.E., Jr. Determination of in Vivo phosphorylation sites in Protein Kinase C. *J. Biol. Chem.* **1995**, *270*, 26807–26812. [CrossRef] [PubMed]
45. Fan, S.; Weight, C.M.; Luissint, A.C.; Hilgarth, R.S.; Brazil, J.C.; Ettel, M.; Nusrat, A.; Parkos, C.A. Role of JAM-A tyrosine phosphorylation in epithelial barrier dysfunction during intestinal inflammation. *Mol. Biol. Cell* **2019**, *30*, 566–578. [CrossRef] [PubMed]
46. Iden, S.; Misselwitz, S.; Peddibhotla, S.S.; Tuncay, H.; Rehder, D.; Gerke, V.; Robenek, H.; Suzuki, A.; Ebnet, K. aPKC Phosphorylates JAM-A at Ser285 to promote cell contact maturation and tight junction formation. *J. Cell Biol.* **2012**, *196*, 623–639. [CrossRef]
47. Savkovic, S.D.; Koutsouris, A.; Hecht, G. Attachment of a noninvasive enteric pathogen, Enteropathogenic Escherichia Coli, to cultured human intestinal epithelial monolayers induces transmigration of neutrophils. *Infect. Immun.* **1996**, *64*, 4480–4487. [CrossRef]
48. Savkovic, S.D.; Koutsouris, A.; Hecht, G. Activation of NF-kappaB in intestinal epithelial cells by Enteropathogenic Escherichia Coli. *Am. J. Physiol.* **1997**, *273*, C1160–C1167. [CrossRef]
49. Czerucka, D.; Dahan, S.; Mograbi, B.; Rossi, B.; Rampal, P. Implication of mitogen-activated protein kinases in T84 cell responses to Enteropathogenic Escherichia Coli infection. *Infect. Immun.* **2001**, *69*, 1298–1305. [CrossRef]
50. Young, J.C.; Clements, A.; Lang, A.E.; Garnett, J.A.; Munera, D.; Arbeloa, A.; Pearson, J.; Hartland, E.L.; Matthews, S.J.; Mousnier, A.; et al. The Escherichia Coli Effector EspJ Blocks Src Kinase activity Via Amidation and ADP Ribosylation. *Nat. Commun.* **2014**, *5*, 5887. [CrossRef]
51. Pollard, D.J.; Berger, C.N.; So, E.C.; Yu, L.; Hadavizadeh, K.; Jennings, P.; Tate, E.W.; Choudhary, J.S.; Frankel, G. Broad-spectrum regulation of nonreceptor tyrosine kinases by the bacterial ADP-Ribosyltransferase EspJ. *MBio* **2018**, *9*. [CrossRef] [PubMed]
52. Phillips, N.; Hayward, R.D.; Koronakis, V. Phosphorylation of the Enteropathogenic E. Coli receptor by the Src-Family Kinase C-Fyn triggers actin pedestal formation. *Nat. Cell Biol.* **2004**, *6*, 618–625. [CrossRef] [PubMed]
53. Hayward, R.D.; Hume, P.J.; Humphreys, D.; Phillips, N.; Smith, K.; Koronakis, V. Clustering transfers the translocated escherichia coli receptor into lipid rafts to stimulate reversible activation of C-Fyn. *Cell. Microbiol.* **2009**, *11*, 433–441. [CrossRef] [PubMed]
54. Huang, Z.; Sutton, S.E.; Wallenfang, A.J.; Orchard, R.C.; Wu, X.; Feng, Y.; Chai, J.; Alto, N.M. Structural insights into Host GTPase isoform selection by a Family of bacterial GEF mimics. *Nat. Struct. Mol. Biol.* **2009**, *16*, 853–860. [CrossRef] [PubMed]
55. Berger, C.N.; Crepin, V.F.; Jepson, M.A.; Arbeloa, A.; Frankel, G. The Mechanisms used by Enteropathogenic Escherichia Coli to control filopodia dynamics. *Cell. Microbiol.* **2009**, *11*, 309–322. [CrossRef] [PubMed]
56. Ugalde-Silva, P.; Navarro-Garcia, F. Coordinated transient interaction of ZO-1 and Afadin is required for pedestal maturation induced by EspF from Enteropathogenic Escherichia Coli. *Microbiologyopen* **2019**, *8*, e931. [CrossRef] [PubMed]
57. Hurd, T.W.; Fan, S.; Liu, C.J.; Kweon, H.K.; Hakansson, K.; Margolis, B. Phosphorylation-dependent binding of 14-3-3 to the polarity protein Par3 regulates cell polarity in Mammalian Epithelia. *Curr. Biol.* **2003**, *13*, 2082–2090. [CrossRef]
58. Devriese, S.; Van den Bossche, L.; Van Welden, S.; Holvoet, T.; Pinheiro, I.; Hindryckx, P.; De Vos, M.; Laukens, D. T84 monolayers are superior to Caco-2 as a model system of Colonocytes. Histochem. *Cell Biol.* **2017**, *148*, 85–93.
59. Le Bivic, A.; Real, F.X.; Rodriguez-Boulan, E. Vectorial targeting of apical and basolateral plasma membrane proteins in a human adenocarcinoma epithelial cell line. *Proc. Natl. Acad. Sci. USA* **1989**, *86*, 9313–9317. [CrossRef]

60. Yoo, B.K.; Yanda, M.K.; No, Y.R.; Yun, C.C. Human intestinal epithelial cell line SK-CO15 is a new model system to study Na(+)/H(+) exchanger 3. *Am. J. Physiol. Gastrointest. Liver Physiol.* **2012**, *303*, G180–G188. [CrossRef]
61. Zhang, J.; Zhang, X.; Guo, Y.; Xu, L.; Pei, D. Sorting nexin 33 induces mammalian cell micronucleated phenotype and actin polymerization by interacting with Wiskott-Aldrich syndrome protein. *J. Biol. Chem.* **2009**, *284*, 21659–21669. [CrossRef] [PubMed]
62. Park, J.; Kim, Y.; Lee, S.; Park, J.J.; Park, Z.Y.; Sun, W.; Kim, H.; Chang, S. SNX18 shares a redundant role with SNX9 and modulates endocytic trafficking at the plasma membrane. *J. Cell. Sci.* **2010**, *123*, 1742–1750. [CrossRef] [PubMed]
63. Haberg, K.; Lundmark, R.; Carlsson, S.R. SNX18 is an SNX9 paralog that acts as a membrane Tubulator in AP-1-positive endosomal trafficking. *J. Cell. Sci.* **2008**, *121*, 1495–1505. [CrossRef] [PubMed]
64. Hua, Y.; Yan, K.; Wan, C. Clever cooperation: Interactions between EspF and host proteins. *Front. Microbiol.* **2018**, *9*, 2831. [CrossRef]
65. Badour, K.; McGavin, M.K.; Zhang, J.; Freeman, S.; Vieira, C.; Filipp, D.; Julius, M.; Mills, G.B.; Siminovitch, K.A. Interaction of the Wiskott-Aldrich syndrome protein with sorting Nexin 9 is required for CD28 endocytosis and cosignaling in T cells. *Proc. Natl. Acad. Sci. USA* **2007**, *104*, 1593–1598. [CrossRef]
66. Merrifield, C.J.; Qualmann, B.; Kessels, M.M.; Almers, W. Neural Wiskott Aldrich Syndrome Protein (N-WASP) and the Arp2/3 complex are recruited to sites of clathrin-mediated endocytosis in cultured fibroblasts. *Eur. J. Cell Biol.* **2004**, *83*, 13–18. [CrossRef]
67. Benesch, S.; Polo, S.; Lai, F.P.; Anderson, K.I.; Stradal, T.E.; Wehland, J.; Rottner, K. N-WASP deficiency impairs EGF internalization and Actin assembly at clathrin-coated pits. *J. Cell. Sci.* **2005**, *118*, 3103–3115. [CrossRef]
68. Yarar, D.; Waterman-Storer, C.M.; Schmid, S.L. SNX9 couples actin assembly to phosphoinositide signals and is required for membrane remodeling during endocytosis. *Dev. Cell.* **2007**, *13*, 43–56. [CrossRef]
69. Lundmark, R.; Carlsson, S.R. SNX9—A prelude to vesicle release. *J. Cell. Sci.* **2009**, *122*, 5–11. [CrossRef]
70. Andreeva, A.Y.; Krause, E.; Muller, E.C.; Blasig, I.E.; Utepbergenov, D.I. Protein kinase C regulates the phosphorylation and cellular localization of occludin. *J. Biol. Chem.* **2001**, *276*, 38480–38486. [CrossRef]
71. Aono, S.; Hirai, Y. Phosphorylation of claudin-4 is required for tight junction formation in a human keratinocyte cell line. *Exp. Cell Res.* **2008**, *314*, 3326–3339. [CrossRef] [PubMed]
72. Sotillos, S.; Diaz-Meco, M.T.; Caminero, E.; Moscat, J.; Campuzano, S. DaPKC-dependent phosphorylation of crumbs is required for epithelial cell polarity in drosophila. *J. Cell Biol.* **2004**, *166*, 549–557. [CrossRef] [PubMed]
73. Hurov, J.B.; Watkins, J.L.; Piwnica-Worms, H. Atypical PKC phosphorylates PAR-1 kinases to regulate localization and activity. *Curr. Biol.* **2004**, *14*, 736–741. [CrossRef] [PubMed]
74. Plant, P.J.; Fawcett, J.P.; Lin, D.C.; Holdorf, A.D.; Binns, K.; Kulkarni, S.; Pawson, T. A Polarity complex of mPar-6 and atypical PKC binds, phosphorylates and regulates mammalian Lgl. *Nat. Cell Biol.* **2003**, *5*, 301–308. [CrossRef]
75. Tobias, I.S.; Newton, A.C. Protein scaffolds control localized protein kinase Czeta activity. *J. Biol. Chem.* **2016**, *291*, 13809–13822. [CrossRef]
76. Chaki, S.P.; Barhoumi, R.; Rivera, G.M. Actin remodeling by Nck regulates endothelial lumen formation. *Mol. Biol. Cell* **2015**, *26*, 3047–3060. [CrossRef]
77. Gruenheid, S.; DeVinney, R.; Bladt, F.; Goosney, D.; Gelkop, S.; Gish, G.D.; Pawson, T.; Finlay, B.B. Enteropathogenic E. Coli Tir binds Nck to initiate actin pedestal formation in host cells. *Nat. Cell Biol.* **2001**, *3*, 856–859. [CrossRef]
78. Nieto-Pelegrin, E.; Kenny, B.; Martinez-Quiles, N. Nck adaptors, besides promoting N-WASP mediated actin-nucleation activity at pedestals, influence the cellular levels of Enteropathogenic Escherichia Coli Tir effector. *Cell Adhes Migr.* **2014**, *8*, 404–417. [CrossRef]
79. Worby, C.A.; Simonson-Leff, N.; Clemens, J.C.; Huddler, D., Jr.; Muda, M.; Dixon, J.E. Drosophila Ack targets its substrate, the sorting nexin DSH3PX1, to a protein complex involved in axonal guidance. *J. Biol. Chem.* **2002**, *277*, 9422–9428. [CrossRef]
80. Worby, C.A.; Simonson-Leff, N.; Clemens, J.C.; Kruger, R.P.; Muda, M.; Dixon, J.E. The sorting nexin, DSH3PX1, connects the axonal guidance receptor, dscam, to the actin cytoskeleton. *J. Biol. Chem.* **2001**, *276*, 41782–41789. [CrossRef]

81. Georgiou, M.; Marinari, E.; Burden, J.; Baum, B. Cdc42, Par6, and aPKC regulate Arp2/3-mediated endocytosis to control local adherens junction stability. *Curr. Biol.* **2008**, *18*, 1631–1638. [CrossRef] [PubMed]
82. Wald, F.A.; Oriolo, A.S.; Mashukova, A.; Fregien, N.L.; Langshaw, A.H.; Salas, P.J. Atypical protein kinase C (Iota) activates Ezrin in the apical domain of intestinal epithelial cells. *J. Cell. Sci.* **2008**, *121*, 644–654. [CrossRef] [PubMed]
83. Rouven Bruckner, B.; Pietuch, A.; Nehls, S.; Rother, J.; Janshoff, A. Ezrin is a major regulator of membrane tension in epithelial cells. *Sci. Rep.* **2015**, *5*, 14700. [CrossRef] [PubMed]
84. Crawley, S.W.; Mooseker, M.S.; Tyska, M.J. Shaping the intestinal brush border. *J. Cell Biol.* **2014**, *207*, 441–451. [CrossRef] [PubMed]
85. Tsai, F.C.; Bertin, A.; Bousquet, H.; Manzi, J.; Senju, Y.; Tsai, M.C.; Picas, L.; Miserey-Lenkei, S.; Lappalainen, P.; Lemichez, E.; et al. Ezrin enrichment on curved membranes requires a specific conformation or interaction with a curvature-sensitive partner. *Elife* **2018**, *7*, e37262. [CrossRef] [PubMed]
86. Simonovic, I.; Arpin, M.; Koutsouris, A.; Falk-Krzesinski, H.J.; Hecht, G. Enteropathogenic Escherichia Coli activates ezrin, which participates in disruption of tight junction barrier function. *Infect. Immun.* **2001**, *69*, 5679–5688. [CrossRef]
87. Law, H.T.; Chua, M.; Moon, K.M.; Foster, L.J.; Guttman, J.A. Mass spectrometry-based proteomics identification of Enteropathogenic Escherichia Coli pedestal constituents. *J. Proteome Res.* **2015**, *14*, 2520–2527. [CrossRef]
88. Campellone, K.G.; Giese, A.; Tipper, D.J.; Leong, J.M. A tyrosine-phosphorylated 12-amino-acid sequence of Enteropathogenic Escherichia Coli Tir Binds the host adaptor protein Nck and is required for Nck localization to actin pedestals. *Mol. Microbiol.* **2002**, *43*, 1227–1241. [CrossRef]
89. Kalman, D.; Weiner, O.D.; Goosney, D.L.; Sedat, J.W.; Finlay, B.B.; Abo, A.; Bishop, J.M. Enteropathogenic E. Coli acts through WASP and Arp2/3 complex to form actin pedestals. *Nat. Cell Biol.* **1999**, *1*, 389–391. [CrossRef]
90. Lommel, S.; Benesch, S.; Rottner, K.; Franz, T.; Wehland, J.; Kuhn, R. Actin pedestal formation by Enteropathogenic Escherichia Coli and intracellular motility of Shigella Flexneri are abolished in N-WASP-defective cells. *EMBO Rep.* **2001**, *2*, 850–857. [CrossRef]
91. Kralicek, S.E.; Nguyen, M.; Rhee, K.J.; Tapia, R.; Hecht, G. EPEC NleH1 is significantly more effective in reversing colitis and reducing mortality than NleH2 via differential effects on host signaling pathways. *Lab. Invest.* **2018**, *98*, 477–488. [CrossRef] [PubMed]
92. McNamara, B.P.; Donnenberg, M.S. A novel proline-rich protein, EspF, is secreted from Enteropathogenic Escherichia Coli via the Type III export pathway. *FEMS Microbiol. Lett.* **1998**, *166*, 71–78. [CrossRef] [PubMed]
93. Kenny, B.; Jepson, M. Targeting of an Enteropathogenic Escherichia Coli (EPEC) effector protein to host mitochondria. *Cell. Microbiol.* **2000**, *2*, 579–590. [CrossRef] [PubMed]
94. Rhee, K.J.; Cheng, H.; Harris, A.; Morin, C.; Kaper, J.B.; Hecht, G. Determination of spatial and temporal colonization of Enteropathogenic E. Coli and Enterohemorrhagic E. Coli in mice using bioluminescent in Vivo imaging. *Gut Microbes* **2011**, *2*, 34–41. [CrossRef]

International Journal of
Molecular Sciences

MDPI

Review

Tight Junctions in Cell Proliferation

Mónica Díaz-Coránguez, Xuwen Liu and David A. Antonetti *

Department of Ophthalmology and Visual Sciences, University of Michigan, Kellogg Eye Center, Ann Arbor, MI 48105, USA; mdiazcor@med.umich.edu (M.D.-C.); xuwen@med.umich.edu (X.L.)

* Correspondence: dantonet@med.umich.edu; Tel.: +(734)-232-8230; Fax: +(734)-232-8030

Received: 1 November 2019; Accepted: 22 November 2019; Published: 27 November 2019

Abstract: Tight junction (TJ) proteins form a continuous intercellular network creating a barrier with selective regulation of water, ion, and solutes across endothelial, epithelial, and glial tissues. TJ proteins include the claudin family that confers barrier properties, members of the MARVEL family that contribute to barrier regulation, and JAM molecules, which regulate junction organization and diapedesis. In addition, the membrane-associated proteins such as MAGUK family members, i.e., zonula occludens, form the scaffold linking the transmembrane proteins to both cell signaling molecules and the cytoskeleton. Most studies of TJ have focused on the contribution to cell-cell adhesion and tissue barrier properties. However, recent studies reveal that, similar to adherens junction proteins, TJ proteins contribute to the control of cell proliferation. In this review, we will summarize and discuss the specific role of TJ proteins in the control of epithelial and endothelial cell proliferation. In some cases, the TJ proteins act as a reservoir of critical cell cycle modulators, by binding and regulating their nuclear access, while in other cases, junctional proteins are located at cellular organelles, regulating transcription and proliferation. Collectively, these studies reveal that TJ proteins contribute to the control of cell proliferation and differentiation required for forming and maintaining a tissue barrier.

Keywords: tight junctions; cell growth; epithelia; endothelia; proliferation; migration

---

1. TJ Expression in Epithelial Differentiation

The role of TJs in growth, proliferation, and differentiation becomes evident in development during epithelial specialization in eukaryotic organisms. Studies of junction formation in early development reveal the contribution of TJs to the early differentiation process and are largely associated with barrier formation. In mice, influenced by the reproductive niche, the germ stem cells undergo the first cell proliferation cycle. The subsequent transcription and translational activity from the embryonic genome during the 2-cell and 4-cell stages is crucial for the first morphogenetic transition in the embryo that occurs during the 8-cell stage, a process that is known as compaction (Figure 1A). Compaction is characterized by the expression and organization of intercellular adhesion and polarization complexes. At a molecular level, protein kinase C isoform alpha (PKCα) and myosin light-chain kinase signaling become activated and maintain cell contact through the regulation of the adherens junction (AJ) proteins E-cadherin, nectin-2, epithin, vezatin, and β-catenin [1,2]. Meanwhile, the membrane-associated guanylate kinase (MAGUK) homolog family member zonula occludens-1 (ZO-1) α^-, the smaller splice variant of the two ZO-1α isoforms [3], arrives at the blastocyst membrane cell contact where it binds with the Rab-GTPase, Rab13 [4–6]. The junctional adhesion molecule (JAM)-A is also expressed at this stage [7] and is localized at the apical microvillous pole together with the partitioning defective (PAR) complex proteins PAR-3, PAR-6, and the atypical PKC isoform iota (aPKCι) as well as Cdc42 [8]. *Cldn3, 6, 7, 8, 10, 12,* and *15*, and *Ocln* genes [9] are also expressed in this compaction stage. *Cldn3* gene shows a particularly high expression compared to other claudins in the progenitors of the trophectoderm during late compaction [10], suggesting a role of claudin-3 protein in differentiation. The TJ protein content increases as the embryo continues to develop and shows an even distribution at the lateral domain similar to the AJs.

The first epithelial specialization differentiation takes places in cavitation, which is a process of the formation of the first cavity in the embryonic body or blastocoel. Cavitation is characterized by asymmetrical divisions and the coordinated polarization of the trophectoderm epithelial cells, a process regulated by junctional components. At the 16-cell stage, E-cadherin assembles at the membrane at cell-cell contact sites and confers the apical and basolateral cell polarization of blastomeres, which leads to a restricted localization of polarity proteins [11]. The polarity complexes Crumbs, which includes Crumbs3, a protein associated with lin seven 1 (PALS1) and a protein associated with tight junctions (PATJ); PAR, including PAR3, PAR6, and aPKCι; and Scribble, including Scribble, lethal giant larvae (LGL), and disc large (DLG), distribute in three specific domains in a mechanism orchestrated by the recruitment of JAM-A into the junctions. JAM-A is recruited to newly formed cell–cell junctions simultaneously with ZO-1. Then, PAR-3 interacts with JAM-A and also tethers to the junction. Although the mechanism is not completely understood, it has been proposed that the JAM-induced localization of the PAR protein complex controls the asymmetrical divisions of the blastocyst [12,13]. Meanwhile, cingulin and ZO-2 assemble to the TJ for the first time [2,14]. At the late 32-cell stage, ZO-1α^+ [5] is expressed and co-localizes with occludin [15] at the Golgi. Together with claudin-1 and 3, ZO-1α^+ and occludin assemble at the junctions [16], allowing the embryo to generate a barrier between trophectoderm cells and the nascent blastocoel cavity. The specific function of the ZO-1 α-domain is not completely clear, but a recent study showed that ZO-1α^+ expression correlates with the grade of Caco-2 cell differentiation. During the exponential phase of Caco-2 cell growth, cells shift in isoform expression from elevated ZO-1α^- to elevated ZO-1α^+ as cells reach confluence [17]. Further, this shift in expression pattern could be manipulated by the cell substrate and correlated with differentiation of the cells, implicating ZO-1α^+ in differentiation of junctional properties. Little is known about the molecular mechanisms that promote the proper localization of these junctional domains, but PKC isoforms have been implicated through chemical activation of PKC isoform zeta (PKCζ) in trophectoderm cells, which promotes the assembly of ZO-1α^+ [18], while PKCι interacts with PAR3/6 proteins, promoting localization and positioning at the junctions [19].

The mechanisms of epithelial polarization and differentiation during development differ between species. In addition to mammals, this process has been studied in invertebrate organisms including *C. elegans*, *Xenopus*, and *Drosophila* embryos. More detailed reviews of these species may be found in [11,20]. In contrast with mammals, the polarization of *C. elegans* blastomeres is not directly linked to cell fate specialization since at the 4-cell stage the blastomeres are already polarized but do not form junctions. In fact, the first epithelial specialization of *C. elegans* appears later during organogenesis [21]. In *Xenopus* embryos, both polarization and junction formation start together with the first cleavage, but in this case, the epithelial differentiation process occurs independently of cell adhesion [22]. Distinct from these organisms, the *Drosophila* embryo has a unique cleavage mechanism named cellularization. In this process, the embryo undergoes multiple cell divisions at the same time that are mediated through membrane invaginations. The resultant tightly packed epithelium of 13 columnar hexagonal cells, possesses cytoskeleton-based landmarks that act as localized clusters for AJ and septate junction (SJ) recruitment [23,24]. In *Drosophila*, SJ performs a similar role as TJ. These studies highlight important differences in epithelial junction specializations among species and reveal unique evolutionary resolution of epithelial differentiation.

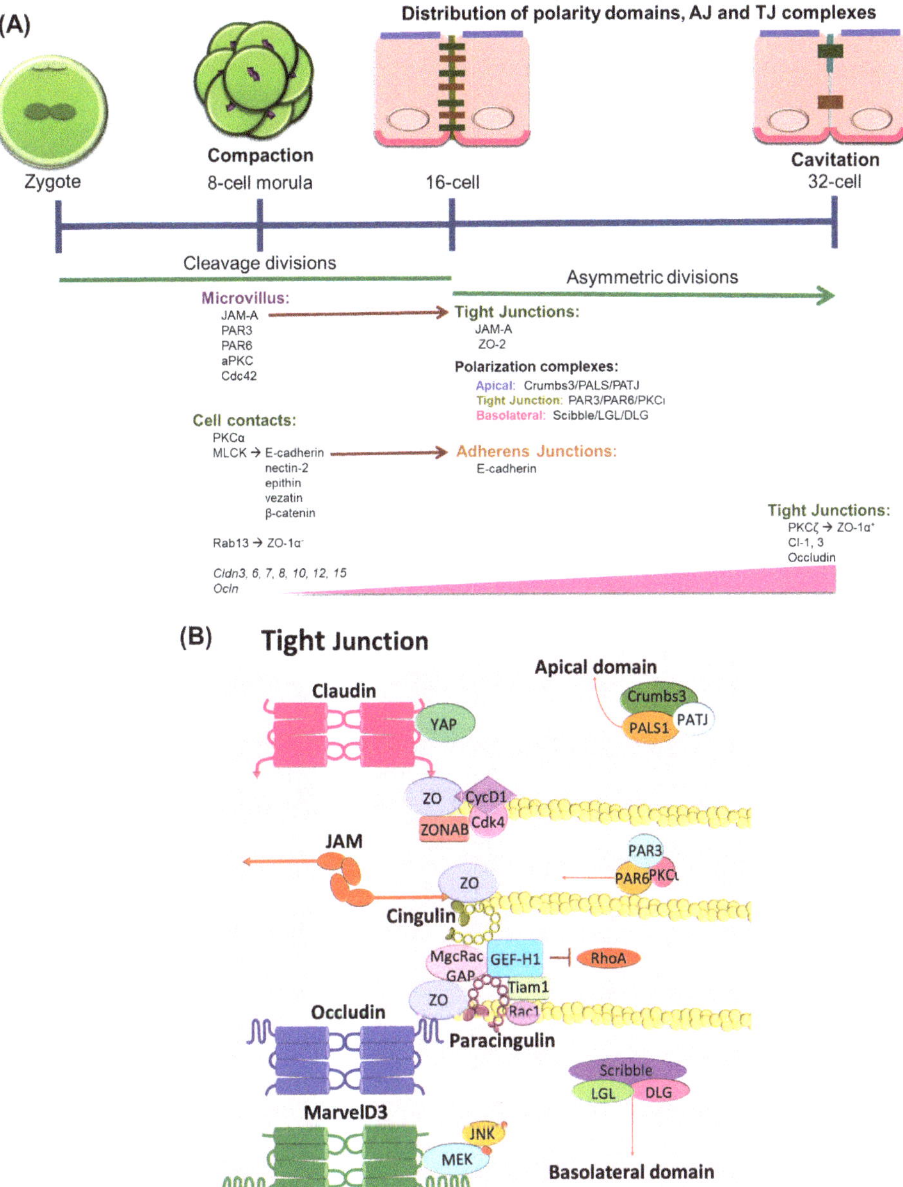

Figure 1. Tight junction proteins in the differentiation of mouse epithelial cells. (**A**) The morula is produced by a series of cleavage divisions of the early embryo, starting with a single cell zygote. Once the embryo has divided into 8 cells, it begins to form a clustered cell mass that expresses AJ and TJ proteins at cell contacts. While JAM and the PAR complex are already expressed at this stage, they localize at the microvillus. As the embryo continues to develop, JAM is recruited into the junctions and polarization complexes are tethered into three specific domains. The subsequent asymmetric divisions initiate the cavitation process at the 32-cell stage. The expression of TJ proteins occludin and claudins

increases and together with ZO-1α^+, they assemble at the junctions, correlating with the differentiation of the first epithelium specialization in the body. (**B**) In fully differentiated tissues, TJ proteins localize at the contact sites of adjacent cells and create a selective paracellular barrier. In addition, the junctional proteins bind and regulate a number of signal transduction factors including those involved in the regulation of cell proliferation. These include sequestering GEF-H1 preventing RhoA activity, the transcription factor ZONAB, and the CycD1/Cdk4 complex, all involved in G1 to S transition. The complexes Crumbs, PAR, and Scribble regulate the proper polarization of apical, TJ, and basolateral domains, respectively.

2. TJ Expression in Endothelial Differentiation

In the central nervous system (CNS), the vasculature has unique anatomic and functional properties that contribute to blood-brain (BBB) and blood-retinal (BRB) barriers. The endothelial cells are in close association with neurons, glia, and pericytes that promote the differentiation to a specialized endothelium with TJ proteins that restrict the transport of molecules into the neural tissue. The development of CNS vascularization and angiogenesis begins around embryonic (E) day 9 in rodents. Starting from the perineural vascular plexus, the endothelial network sprouts and covers the entire surface of the neural tube [25].

Endothelial differentiation and specialization in the CNS barriers result from the polarization of the endothelial network and the localization of junctional components along endothelial cell contacts. For the mouse BBB, the differentiation process takes place between E15.5 and E18.5 [26], while the BRB angiogenesis starts after birth, and the barrier properties are established simultaneously with capillary formation between postnatal day 7 and 15 [27]. In humans, both BBB and BRB develop during gestation, but it is not completely clear if the establishment of barrier properties culminate with birth [28]. Although all endothelial cells from the most primitive linage express the TJ protein claudin-5, a junctional protein required for brain barrier properties [29], the sealing of a functional barrier occurs later.

The Wnt/β-catenin signaling pathway is vital for growth and specialization of the CNS vasculature. During mouse brain development, the neuroepithelium-derived Wnt7a/Wnt7b ligands activate signaling through Frizzled-4 (Fzd4) receptor, low-density lipoprotein receptor-related protein 5/6 (Lrp5/6) co-receptor, G-protein-coupled receptor 124 (Gpr124), and reversion-inducing cysteine-rich protein (Reck) co-activators to promote β-catenin activation and angiogenesis [30–32]. This signaling is essential for neural tube vascularization because its depletion causes CNS vascular malformation and lethality. Moreover, β-catenin signaling contributes to the specialization of the CNS endothelium by inducing the expression of the glucose transporter and the TJ protein claudin-5 [31]. The non-Wnt ligand norrin activates β-catenin signaling though its receptor Fzd4, the Lrp5/6 co-receptor, and the tetraspanin-12 (Tspan12) co-activator, and this signaling is essential for proper retinal angiogenesis and barrier formation as well as cerebellum barrier formation [33]. In the retina, while Wnt3, Wnt7a, and Wnt7b ligands have only a small contribution in barrier formation [34], the Müller glia-derived norrin is the predominant β-catenin activator that regulates BRB formation and maintenance. Downstream of β-catenin, gene expression of the sex-determining region Y-related high-mobility group box factors 7, 17, and 18 (*Sox7*, *Sox17*, and *Sox18*, respectively) promotes BRB formation [35].

3. Epithelial-Mesenchymal Transition (EMT)

As part of normal development, epithelial cells transition to mesenchymal phenotype in a process called epithelial mesenchymal transition (EMT). This occurs, for example, during implantation of embryos, as part of gastrulation and neural crest formation. EMT also contributes to both wound repair and to the pathology of neoplasia as a part of tumor dedifferentiation. Under a mesenchymal state, tumor cells are highly invasive and able to migrate to distant sites, establishing metastases. After invading a new tissue, tumors cells may differentiate towards the recovery of their epithelial characteristics in a process called mesenchymal-epithelial transition (MET) [36].

Changes in TJ protein expression and organization are a fundamental step in EMT. Indeed, the most common markers that indicate the epithelial cell transitioning to mesenchymal phenotype include the suppression of TJ protein transcription and loss of junctional complex organization and polarity. These changes coincide with additional changes such as increased synthesis of cytoskeletal components, actin stress fibers, spindle-shaped appearance, increased movement, and resistance to apoptosis [37,38].

An early step in EMT in cancer cells is loss of gene transcription for junctional components, where the high expression of the transcription regulators Snail, Slug, Twist, zinc finger E-box-binding homeobox 1 (ZEB-1), SMAD interacting protein 1 (Sip-1), and lymphoid enhancer-binding factor 1 (LEF-1) reduces *Ocln* and *Cldn* synthesis [39,40] (Figure 2). With the progression of EMT, the junction complex is disassembled via transforming growth factor beta (TGFβ) signaling. The binding of TGFβ to its receptor TGFβR2 results in its recruitment to the junctional complex where it binds to occludin and promotes phosphorylation of the polarity protein PAR6. Then, the endogenous E3 ubiquitin ligase Smurf1 redistributes to cell junctions and promotes RhoA ubiquitination and degradation, thus leading to cytoskeleton rearrangement and TJ disassembly [41]. Another example is epidermal growth factor (EGF) activation of its receptor (ERBB2), which then interacts with the PAR6-aPKC complex and causes PAR3 dissociation and ultimately TJ breakdown [42]. Other growth factors that promote EMT through their tyrosine kinase receptors include the hepatocyte growth factor (HGF) through its receptor Met; the fibroblast growth factor (FGF); and the bone morphogenetic protein (BMP) [39]. While BMP2 and BMP4 promote EMT [43,44], BMP7 induces MET [45].

While TJ proteins are often reduced in cancers of epithelial origin, in several human cancers, TJ proteins are overexpressed potentially by epigenetic regulation [46]. It is important to note in these cases that the junctional complexes have lost their membrane organization and barrier function. However, their intracellular localization suggests that they might contribute to other cellular functions. Recently, E-cadherin was found to promote metastasis in models of invasive ductal carcinomas [47]. In this work, E-cadherin gene (*Cdh1*) was depleted in MMTV-PyMT invasive ductal carcinoma cells and as expected, this resulted in increased invasion. However, *Cdh1*-null cells also exhibited increased TGFβ/SMAD2/3 and reactive oxygen species signaling, which resulted in reduced cell proliferation, lower survival, and inhibition of metastasis. These studies suggest that E-cadherin acts as a survival factor in invasive ductal carcinomas by limiting reactive oxygen-mediated apoptosis and highlight the non-barrier function of this adherens junction protein.

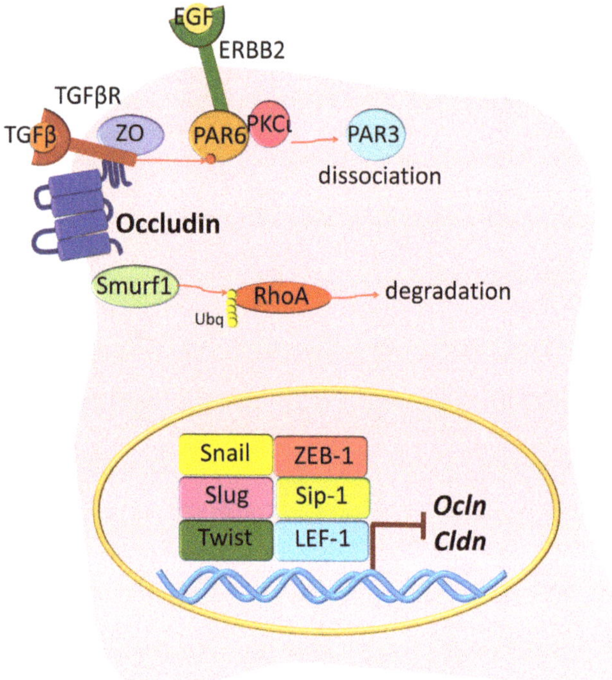

Figure 2. Tight junction proteins in EMT. As an early step in EMT, epithelial cells lose polarity and TJs are disrupted. TGFβ binds its receptor and is recruited to the junction where it interacts with ZO-1 and occludin. TGFβR activation promotes PAR6 phosphorylation. ERBB2 binds to PAR6/PKCι proteins, but PAR3 becomes dissociated from the complex, and this results in overall altered cell polarization. Smurf1 is also recruited into the TJ, where it induces RhoA ubiquitination (Ubq) and degradation. Meanwhile, during EMT, a series of nuclear transcription factors inhibit the expression of TJ genes *Ocln* and *Cldn*. Growth factors including FGF, HGF, and BMPs promote EMT, but the exact mechanism is not completely clear. Together, these molecular mechanisms promote epithelial transformation into a mesenchymal state.

4. Endothelial-Mesenchymal Transition (EndMT)

Similar to epithelial cells, endothelial tissues have the plasticity to transition to a mesenchymal phenotype (EndMT), and this transition is commonly observed in vasculogenesis. EndMT was first described by Leonard M. Eisenberg in 1995 [48], and recently, considerable attention has been paid to EndMT due to the vital role it plays in human diseases, including cerebral cavernous malformation (CCM) [49] ischemic stroke, atherosclerosis, cardiac fibrosis associated with diabetes mellitus, nephrosis, pulmonary arterial hypertension, diabetic nephropathy, alcoholic liver disease, multiple myeloma, and other cancers [50]. CCM is caused by loss of function mutations in *CCM* genes 1, 2 or 3. The *CCM* gene products bind to the endothelial adherens junction complex in the cytoplasm [51]. In CCM, increased TGFβ and BMP signaling and the consequent EndMT in *CCM1*-null endothelial cells are crucial events in the onset and progression of the disease [49]. Further, recent evidence reveals limited *CCM3*-null endothelial cell clonal expansion induces EndMT in wild-type cells [52]. Importantly, while in epithelial cells all three TGFβ isoforms can induce EMT [53,54], EndMT is primarily stimulated by the TGFβ2 isoform [55–57], by BMP2 or by BMP4 [43,44].

In cancer, it has been proposed that EndMT may contribute to metastasis. The cancer microenvironment, which is rich in growth factors secreted by stromal cells, promotes a sustained tip phenotype in most of the endothelial cells undergoing angiogenesis in tumors [58]. Moreover, recent studies suggest that TGFβ induces EndMT in brain endothelial cells [59]. In this study, cells with the mesenchymal phenotype had better adhesion to melanoma cells with increased migration potential, suggesting that EndMT might facilitate brain metastasis.

In some diseases, EndMT remains controversial. Lineage tracing studies employing multiple independent murine Cre lines suggest that fibroblasts do not originate from hematopoietic cells, endothelial cells (through EndMT) or epicardial cells (through EMT) but proliferate from resident fibroblast lineages [60]. Moreover, only few studies have described the presence of EndMT in vascular cells with endothelial barriers. The difficulty in identifying EndMT may be due to the presence of various differentiated states within the growing vessel. Loss of the endothelial markers CD31, von Willebrand factor (vWF), and vascular endothelial cadherin (VE-cadherin) and an increase in the mesenchymal markers alpha smooth muscle actin (α-SMA), fibroblast-specific protein 1 (FSP1), and vimentin have been considered a hallmark of EndMT. However, it has been shown that some level of endothelial markers such as VE-Cadherin, Tie-1, vWF, and cytokeratins is still detected in cells after EndMT. In addition, these markers have heterogeneous expression within the same vessel. In CNS vessels undergoing angiogenesis, the tip cells have a mesenchymal character that allows them to migrate [61,62], suggesting that EndMT may occur at the leading edge of the branching vessels. While tip cells have a fibroblastic phenotype, the intermediate vascular stalk cells proliferate but also have a more differentiated character and contact with neighboring cells. As such, tip cells demonstrate low expression of the TJ protein claudin-5 but high expression of the transcytosis marker PLVAP (plasmalemma vesicle-associated protein). However, the stalk cells simultaneously proliferate and form a barrier through the high expression of claudin-5 [33]. Better knowledge about the mechanisms that control EndMT might help in the development of new therapies in several diseases.

While extensive research has elaborated a role for AJ proteins in the control of epithelial (reviewed in [63]) and endothelial (reviewed in [51]) cell proliferation, growing evidence now implicates the TJ proteins in cell cycle control and regulation of proliferation.

5. Role of Tight Junction Scaffold Proteins in Controlling Cell Proliferation

5.1. Zona Occludens (ZO)

ZO proteins serve as links between the transmembrane TJ proteins and the cytoskeleton. ZO proteins bind to the TJ proteins claudins, occludin, and JAMs (Figure 1B) as well as other ZO proteins, promoting the polymerization of the junctional complex and binding cytoskeletal-associated proteins α-catenin and afadin (AF6) linking the TJ to the cytoskeleton [64–66]. ZO proteins are part of the MAGUK family, and there are three genes that encode these proteins: ZO-1, -2, and -3 [67–69]. Although they have redundant functions, knockout mice studies suggest a role of ZO-1 and -2 in early development. Mice deficient of ZO-1 gene (*Tjp1*) die in E10.5 due to embryonic defects mediated by apoptosis in the neural tube, the notochord, and the allantois areas, as well as extra-embryonic defects in the angiogenesis of the yolk sac, suggesting that ZO-1 regulates tissue organization and remodeling in both epithelial and endothelial tissues [70]. Similarly, mice deficient of ZO-2 gene (*Tjp2*) die shortly after implantation due to an arrest in early gastrulation [71].

Recent publications suggest that ZO proteins are able to transmit information about the degree of cell-cell contacts to the nucleus, thus maintaining a balance between proliferation and differentiation. ZO -1, -2, and -3 sequester key regulators of cell cycle progression at the junction sites [72] (Figure 1B), including ZO-1-associated nucleic acid binding protein (ZONAB). At low-density cell numbers, the Y-box transcription factor ZONAB is located at the nuclear fractions, promoting the expression of G1/S-phase transition through regulation of proliferating cell nuclear antigen (*PCNA*) gene expression (Figure 3). However, at higher density and cell contact, ZO-1 sequesters ZONAB at the TJ, reducing nuclear concentration and thus controlling cell proliferation [73,74]. Cyclin D1 (CycD1) is also sequestrated at the junctions by ZO proteins, together with its associated cyclin-dependent kinase 4

(Cdk4). CycD1 forms a complex with the ZO-3 protein, specifically through its PDZ-binding motif. This interaction regulates epithelial cell proliferation by CycD1 stabilization at the membrane during cell proliferation, since knockdown of ZO-3 gene (*TJP3*) with siRNA results in G0/G1 cell-cycle arrest [75].

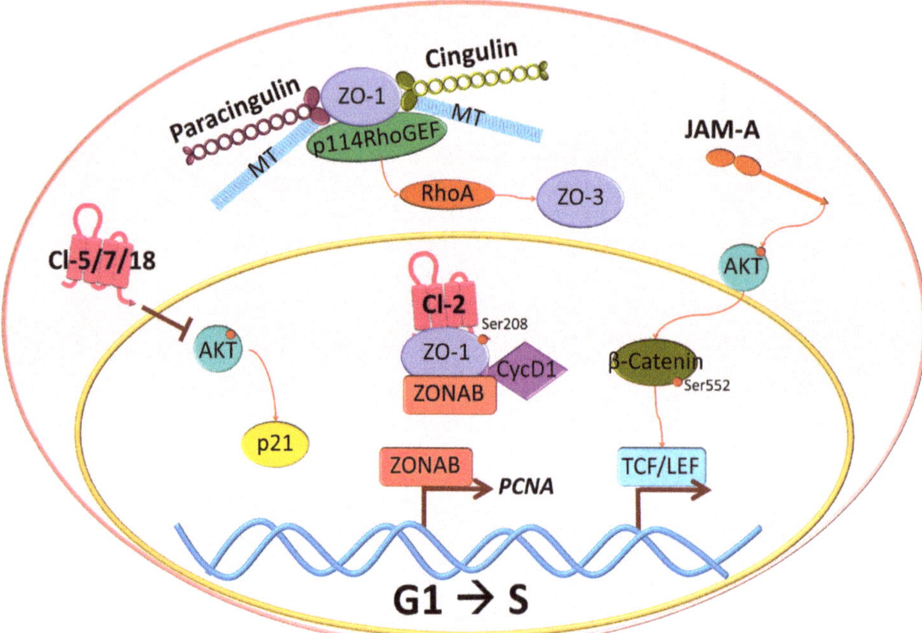

Figure 3. Tight junction proteins in G1/S transition. TJ proteins have been implicated in the control of cell cycle progression. Claudin-5, -7 or -18 (Cl-5/7/18) expression in cancer cells inhibits G1/S transition associated with AKT activation. Alternatively, cell cycle progression can be activated by cingulin and paracingulin dissociation from the junction, which results in a conformational change that allows interaction with microtubules (MT), ZO-1, and p114RhoGEF, promoting RhoA activation and ZO-3 and claudin-2 accumulation that are related to increased G1/S transition. ZO-1 also has been found in nuclear fractions and interacts with claudin-2 (Cl-2) phosphorylated at Ser208, CycD1, and ZONAB. The latter, in sparse cells, can promote *PCNA* gene expression and increase proliferation. In mice deficient of JAM-A gene (*F11r*), AKT becomes activated and phosphorylates β-catenin on Ser552. As a result, β-catenin translocates to the nucleus and activates transcription mediated by TCF/LEF.

Other studies reveal changes in ZO proteins in carcinogenesis. The ZO-1 protein has been found to be highly expressed in adenocarcinoma samples, as compared with healthy tissue [76]. Also, in pancreatic cancer cells, EGFR activation correlates with ZO-1 protein localization at the cytoplasm or at the nucleus, and inhibition of EGFR with AG1478 induces the redistribution of ZO-1 to the junctions [77] (Figure 4). In lung cancer, PKC isoform epsilon (PKCε) activation regulates the interaction between α5-integrin and ZO-1, and this correlates with poor prognosis [78]. Moreover, in EMT, receptor serine/threonine kinase TGFβR type II and I co-localize with ZO-1 at the junctions in a mechanism dependent on TGFβ stimulation and the phosphorylation of PAR6 by TGFβRII [41] (Figure 2). Collectively, these studies suggest that ZO proteins contribute to contact regulated control of cell proliferation.

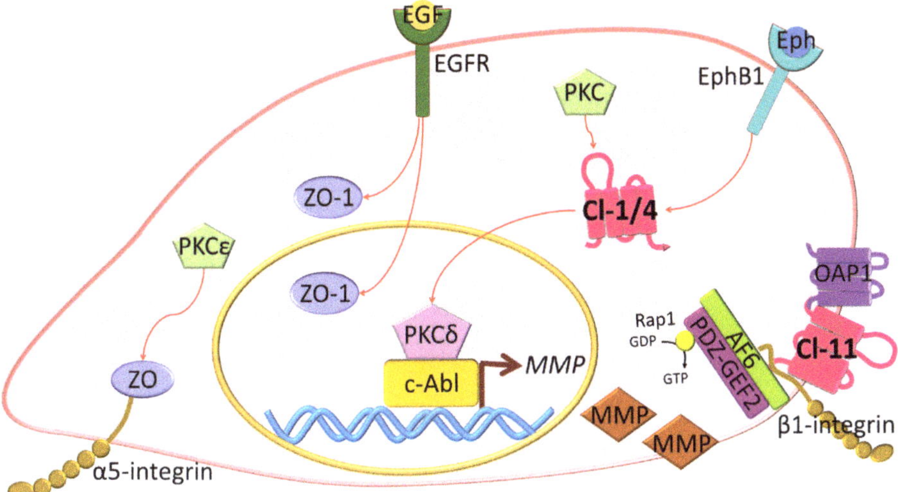

Figure 4. Tight junction proteins in migration and invasion. Overexpression of TJ proteins has been associated with the promotion of cell migration and invasion in cancer cells. EGFR expression has been correlated with ZO-1 localization at nuclear and cytoplasmic fractions, and inhibition of EGFR phosphorylation leads to relocalization of ZO-1 to cell-contacts. PKCε promotes ZO interaction with α5-integrin. Claudin-1 (Cl-1) activates PKCδ, which in turn, binds to c-Abl transcription factors and activates *MMP* transcription. MMPs are secreted and induce basal membrane degradation, increasing the invasive potential of cancer cells. Similarly, EphB1 receptor phosphorylation has been associated with claudin-4 (Cl-4) altered expression promoting MMP expression and secretion. Claudin-11 (Cl-11) interaction with OAP1 and β1-integrin increases cell migration through AF6 and PDZ-GEF2 interaction and Rap1 activation.

5.2. Cingulin

Cingulin is a cytoskeletal adaptor protein that has a crucial role in transducing the mechanical force generated by the contraction of the actin-myosin cytoskeleton into functional regulation of the epithelial and endothelial barriers [79]. Its localization at the junctions is mediated by the interaction with the TJ proteins ZO and JAMs, along with its anchoring to the actin cytoskeleton (Figure 1B).

Recent studies have demonstrated a role of cingulin in cell proliferation and migration through its ability to interact with microtubule (MT)-associated small GTPase activators of RhoA, such as the guanine nucleotide exchange factor H1 (GEF-H1) [80–83]. Knockdown of cingulin gene (*CGN*) in Mardin-Darby Canine Kidney (MDCK) cells results in increased expression of claudin-2, ZO-3, and RhoA activation (Figure 3). While the knockdown did not affect TJ protein localization or its barrier function, loss of *CGN* increased RhoA-induced G1/S phase transition through its interaction with GEF-H1 [84]. During neural tube closure, the pre-migratory neural crest cells initiate EMT by TJ disruption and cingulin-induced delamination in the neuroepithelium. Both depletion and overexpression of *CGN* increase the migratory neural crest cell population, associated with loss of basal lamina and disruption of the neural tube [85]. Moreover, cingulin participates in cell polarization in epithelial cysts by interacting with the Rab11 family interacting protein 5 (FIP5), an effector of Rab11 GTPase. Cingulin serves as the tethering factor to ensure the fidelity of apical endosome targeting the apical membrane initiation sites [86].

5.3. Paracingulin

Paracingulin, cingulin-like 1 or the junction-associated-coiled-coil protein (JACOP), is a scaffold TJ protein that maintains the integrity of the association between the MT cytoskeleton and cell junctions [87,88]. With 39% sequence homology to cingulin, paracingulin also interacts with ZO-1 at the globular head domain, promoting its localization at the junctions [89].

Similar to cingulin, paracingulin activates RhoA GTPases to promote cell proliferation. Through its interaction with Rac1, paracingulin promotes the recruitment of Tiam1 (T-cell lymphoma invasion and metastasis 1) and GEF-H1 activators at the junctions (Figure 1B). Therefore, a reduced paracingulin expression in MDCK cells increases RhoA activity and promotes G1/S phase transition [90]. However, when the Rac1 inhibitor MgcRacGAP is present, this interacts with both cingulin and paracingulin at TJs and reduces cell proliferation [91].

Recent studies on endothelial cells suggest a role of paracingulin in angiogenesis. In a primary culture of human dermal microvascular endothelial cells (HDMEC), ZO-1 promoted the recruitment of paracingulin and the RhoA activator 114RhoGEF into the junctions, and this complex increased the angiogenic potential of HDMEC [92]. Supporting this idea, the silencing of paracingulin gene (*Cgnl1*), greatly impaired tubule structure formation in 3D co-culture assays and diminished the number of vascular structures formed during vascular expansion in the developing retina, suggesting paracingulin as a defining factor in new vessel formation [93]. Furthermore, using baculovirus-infected insect cells, Vasileva et al. provided evidence for the interaction of paracingulin with MTs, which is important for the formation of both strong AJ and focal adhesions to ensure stabilization and further elongation of neovascular tubules [93,94]. Together, these results suggest that paracingulin might control endothelial cell proliferation and angiogenesis through the regulation of MT and RhoA activity.

6. Transmembrane TJ Proteins in the Control of Cell Proliferation

6.1. Claudins

Claudins are transmembrane proteins that contribute to paracellular transport by forming ion selective barriers and pores in a tissue-specific manner. As mentioned previously, a hallmark of EMT includes loss of barrier properties. Indeed, claudin dysregulation is commonly found in human carcinomas, often with decreased *CLDN* gene expression. Further, claudin overexpression can sometimes reverse the malignant phenotype. In lung squamous cell carcinoma, the transfection of claudins -5, -7, and -18 was able to suppress proliferation and inhibit G1/S transition associated with inhibition of AKT phosphorylation [95] (Figure 3).

However, some cancer studies reveal a correlation between claudin overexpression and tumor progression [96–99]. Examples include claudin-3 expression, which has been correlated with increased tubulogenesis and bromo-deoxy uridine (BrdU) incorporation in mouse inner medullary collecting duct cells (mIMCD-3) [100]. Similarly, overexpression of claudin-6, -7, or -9 enhanced invasiveness and proliferation of an adenocarcinoma cell line [96], and *CLDN4* and *18* were upregulated in pancreatic cancer tissues [101,102]. Claudin-18 isoform a2 is highly expressed in gastric, esophageal, pancreatic, lung, and ovarian cancers, and while it is considered as a putative marker, its exact role in tumor progression remains unknown [103]. Importantly, the high expression of claudins in neoplastic tissues does not coincide with border localization or barrier regulation. For example, claudin-4 is highly expressed in poorly differentiated pancreatic cancer cells and is enriched at basolateral membranes rather than the apical junctional complex [104]. Claudin-11 overexpression has been associated with proliferation and migration of oligodendrocytes in a mechanism dependent on its interaction with outer surface protein-associated protein 1 (OAP1) and β1-integrin [105] (Figure 4). Moreover, claudin-2 has been found localized at the nuclear fractions in highly proliferative lung adenocarcinoma cells [106].

The molecular consequences of claudin overexpression in cancer biology have not been clearly defined. In human liver cells, it has been suggested that the activation of the protein kinase complex c-Abl/PKCδ (PKC isoform delta) is critical for the acquisition of a malignant phenotype induced by claudin-1 overexpression (Figure 4) [99]. In this study, the overexpression of claudin-1 in normal liver hepatocytes led to an increased expression of metalloproteinsases (MMPs), promoting an invasive phenotype. Similar results were found in human melanoma cells where PKC activation by phorbol myristic acid (PMA) increased *CLDN1* transcription and contributed to invasion [107]. Recent evidence also indicates that ephrin (Eph), through its EphB1 receptor, can control epithelial transformation through interaction with claudin-1 and -4. This interaction promotes EphB1 tyrosine phosphorylation, which in cancer cells mediates migration and invasion through the downstream exocytosis of MMPs [108]. Importantly, claudins can also be phosphorylated by Eph receptors that regulate cell-contact formation [108,109]. A role of claudin-2 in proliferation has been also suggested in studies with lung adenocarcinoma cells. In this work, it was proposed that claudin-2 could induce proliferation when it is phosphorylated at Ser208 and located at the nuclear fractions, interacting with ZO-1, ZONAB, and CycD1 [106] (Figure 3). Moreover, there was a positive correlation between claudin-4 expression, production of interleukin 8 (IL-8), and increased angiogenesis in mouse xenografts [110].

The *Cldn15* knockout mouse model suggests that alterations in barrier function might lead to homeostatic changes that activate other cellular processes including proliferation. *Cldn15* null mice have an enlarged upper small intestinal phenotype or mega-intestine due to the enhanced proliferation of the crypt cells [111]. In this knockout model, the expression and localization of claudins-1, -2, -3, -4, -7, -12, -18, -20, and -23 were not affected, but barrier properties were clearly lost. The authors concluded that the increased cellular proliferation is a consequence of altered barrier function that changes the intestinal environment [112]. This alteration in cell proliferation associated with an altered microenvironment may also be observed in metastasis studies of pancreatic carcinomas where the primary tissue has low expression of *CLDN3* while the metastatic cells that invade the liver express high levels of *CLDN3* and *4* but show low expression of *CLDN1* and *7* [113,114]. The altered TJ expression might influence the cancer microenvironment, promoting changes in cancer metastasis. Similar to *Cldn15*, *Cldn18* knockout mice show proliferation changes with lung enlargement, parenchymal expansion, and increased abundance and proliferation of known distal lung progenitors, the alveolar epithelial type II (AT2) cells, activation of Yes-associated protein (YAP), and increased organ size and tumor genesis in mice [115]. Importantly, claudin-18 and YAP were found to interact and co-localize at cell contacts in control samples (Figure 1B), and claudin-18 overexpression decreased YAP nuclear localization, suggesting that similar to ZO proteins, claudin-18 restricts cell proliferation by the retention of YAP at cell contacts.

Together, these data suggest that altered claudin expression, organization or function at the barrier may be observed as a hallmark of EMT and in a number of epithelial cancers. However, some cancers involve increased expression of specific claudins despite loss of barrier function, and overexpression of some claudins can alter the cancer cell phenotype. While loss of barrier function clearly alters the microenvironment of the cancer, the precise role of claudin in proliferation and cancer remains an area that requires additional investigation. Although claudins are promising molecular targets for diagnosis and therapy [102,116], a better understanding of the mechanisms regulated by claudins in the control of cell proliferation and metastasis is needed.

6.2. *Junctional Adhesion Molecules (JAM)*

JAM is the most extensively studied single-span TJ protein, and there are three genes described that encode to the proteins: JAM-A, -B, and -C. JAMs belong to the immunoglobulin superfamily because they contain at least one immunoglobulin domain at the extracellular N-terminus, while the cytoplasmic tail contains a PDZ binding sequence that interacts with the PDZ domain of ZO-1.

JAM-A regulates epithelial proliferation via canonical Wnt signaling. In control mice, JAM-A shows a gradient expression along the intestinal crypt-luminal axis, which is increased in non-proliferating luminal epithelial cells [117]. JAM-A gene (*F11r*) deletion in mice results in increased intestinal epithelial cell proliferation and activation of AKT and its downstream target β-catenin, which is phosphorylated at Ser552 (Figure 3). This in turn, promotes β-catenin accumulation and its nuclear localization that is followed by an increase in T-cell factor (TCF)/LEF-induced transcriptional activity [118]. Moreover, a recent study revealed that JAM-A also regulates the cortical localization of dynein to control planar spindle orientation during mitosis via activation of Cdc42 and phosphatidyl inositol 3-kinase (PI3K) [119]. As expected, JAM-A suppression or expression of a dimerization-deficient isoform resulted in aberrant spindle orientation.

JAM-A dimerization promotes epithelial and endothelial migration. Expression of the JAM-A dimerization-defective mutant in 293T cells and the use of JAM-A dimer-disrupting antibodies reduce cell migration. Disruption of JAM-A dimerization also correlates with β1-integrin degradation, decreased GTPase Rap1 activation, and diminished numbers of focal concentrations of phosphorylated paxillin [120]. This process is mediated by the interaction of JAM-A with AF6 and PDZGEF2 [121,122]. Similar to these studies, isolated endothelial cells from mice deficient in *F11r* have enhanced spontaneous and random motility associated with increased numbers of actin-containing protrusions, reduced MT stability, and impaired focal adhesions, which can be reversed by JAM-A expression or by using glycogen synthase kinase 3-beta (GSK3β) inhibitors [123]. Moreover, mice deficient of JAM-C gene (*Jam3*) show increased *F11r* expression and enhanced retinal vascularization [124,125], supporting a role for JAM in angiogenesis.

Together, these data indicate that JAMs contribute to cell proliferation and migration. However, further studies are clearly needed in order to determine the exact role of JAMs in these processes.

6.3. MARVEL Family Proteins

The MARVEL (for MAL and related proteins for vesicle trafficking and membrane link) family members include occludin (MarvelD1), tricellulin (MarvelD2), and MarvelD3 [126–128]. MARVEL proteins possess a putative protein-lipid-interacting motif containing four intramembrane helices creating the MARVEL domain (Figure 5A). Its function remains unknown, but MARVEL domains of non-TJ proteins, such as myelin and lymphocyte protein (MAL), have been associated with the generation and stabilization of functional membrane domains via the propensity for homo-oligomerization and the ability to attract apical membrane lipids [129]. Knockout studies of TJ MARVEL genes reveal a complex phenotype in a variety of tissues (reviewed in [130]). Many tissues are able to form morphological and functionally normal TJs, suggesting that they are not individually required for TJ formation although duplication of function has yet to be fully explored. Indeed, in vitro experiments demonstrate depletion of occludin gene (*Ocln*) resulting in the redistribution of tricellulin [131]. *Ocln* knockout mice develop hyperplasia of the gastric epithelium and testicular atrophy, while deletion of *MarvelD2* or *Ocln* in mice present degeneration of cochlear hair cells that leads to progressive hearing loss [132,133]. Moreover, Tricellulin is essential for barrier formation and the maintenance of the TJ structure in the inner ear as determined by gene deletion studies [126,134]. Additionally, MARVEL proteins have been found in membrane raft domains with a high cholesterol content, suggesting that they might participate in bending during the formation of endocytic caveolin transport vesicles similar to other proteins with MARVEL domains [135]. To our knowledge, there is no evidence for tricellulin in the control of cell proliferation, and we will focus on occludin and MarvelD3.

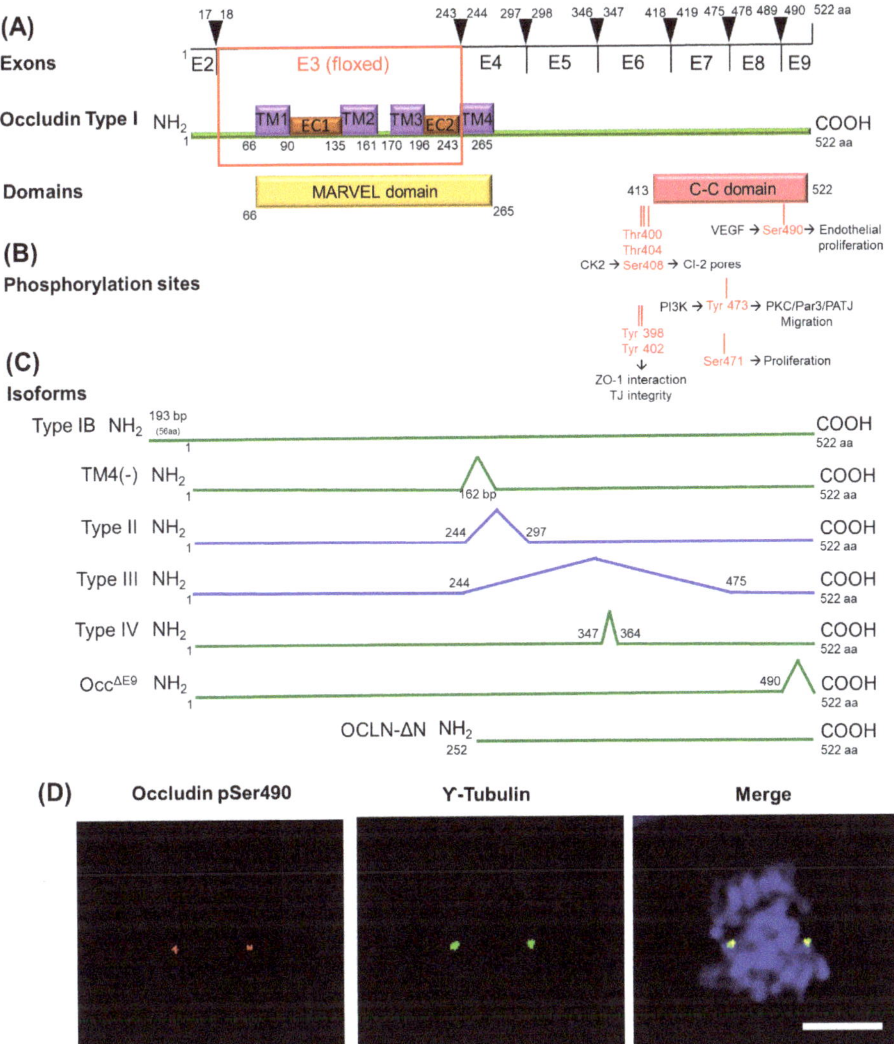

Figure 5. Occludin structure and isoforms. Occludin is a 522 amino acids protein encoded by 9 exons (**A**). Occludin full length (type I) possesses four transmembrane (TM) domains and two extracellular (EC) loops with the MARVEL domain as homology on the cytoplasmic side after each TM region. At the C-terminus (COOH) of occludin, a coiled-coiled (C-C) domain can be phosphorylated at multiple sites (**B**). Phosphorylation of occludin identified and known or implied functions. Occludin can mediate proliferation though the phosphorylation of two sites: Ser471, which regulates post contact proliferation in epithelial cells, and Ser490, which is promoted by VEGF-induced PKCβ activation and regulates both endothelial permeability and neovascularization. To date, several occludin isoforms have been described (**C**). The function of each isoform has not been fully elucidated, but most isoforms localize to the junctions except type II and III (blue lines). Interestingly, occludin deleted in exon 9 ($Occ^{\Delta E9}$) restricts cell migration. (**D**) In bovine retinal endothelial cells, occludin stained with a pS490-specific antibody (red) shows co-localization of phospho-occludin with the centrosome marker γ-tubulin (green) in pro-metaphase. Hoechst dye (blue). Scale bar = 5 μm.

Occludin was the first transmembrane protein discovered at the TJ [136]. While the *Ocln* -null mouse forms intact TJs, the animals have phenotypic alterations including growth retardation, thinning of compact bone, testicular atrophy, male infertility, loss of cytoplasmic granules in salivary epithelial cells, females are not able to lactate, brain calcification [132], and hyper proliferation of mucous epithelial cells in the intestinal lining [137]. In vitro, the silencing of *OCLN* gene has a limited effect on barrier properties, with increases in permeability to divalent organic cations and also to small molecules under hydrostatic pressure [138,139]. In ARPE-19 cells, a human retinal pigmented epithelial cell line, loss of *OCLN* increases the DNA synthesis rate and cell proliferation [139]. Moreover, during neurogenesis, occludin loss has been found in neural tubes of chicken and mouse embryos at E9 [140].

Occludin function is regulated by phosphorylation. Specially, the C-terminal region has been studied extensively (Figure 5B). The 3 sites Thr400, Thr404, and Ser408 that lie just prior to the coiled-coil (C-C) domain regulate its interaction with ZO proteins and TJ integrity [141]. While occludin phosphorylation at Thr404 regulates its localization at the junctions [142], phosphorylation of Ser408 promotes ion flux through control of claudin-2 dimerization. The complex formed by occludin/ZO-1/claudin-2 is dissociated when occludin is phosphorylated by casein kinase 2 (CK2) at Ser408. As a result, claudin-2 interacts in trans with claudin-2 from neighboring cells, forms a cation pore, and increases ion flux [143]. Other phosphorylation sites nearby including Tyr398 and 402 residues may also alter ZO-1 binding [144]. Ser490 phosphorylation controls vascular endothelial growth factor (VEGF)-induced endothelial permeability with the expression of S490A point mutants preventing VEGF-induced permeability in cell culture. In endothelial cells, this results in increased vascular permeability [145] in a PKC isoform beta (PKCβ)-dependent manner [146].

Collectively, these studies demonstrate that occludin phosphorylation contributes to regulation of barrier properties by promoting its binding to ZO proteins and the recruitment of occludin and claudins into the junctions.

Occludin phosphorylation has also been associated with the regulation of cell proliferation and migration. The phosphorylation at Tyr473 promotes directional migration of epithelial cells by the activation of PI3K signaling and by promoting the organization of the aPKC, PAR3, and PATJ polarity complex [147], suggesting a role of occludin in cell migration. In addition, previous studies in our group have identified five phosphorylation sites on the C-C domain of occludin, including Ser471 and Ser490, mentioned above, on the two turns of the C-C domain, which have been associated with the control of cell proliferation. Expression of the S471A mutant as a stable cell line has no effect on sub-confluent proliferation but inhibits proliferation and cell packing after cell contact in MDCK cells as determined by cell number and DNA synthesis, leading to enlarged cells. Inhibition of proliferation by expressing S471A point mutant occludin, inhibition of cell proliferation in cell contacted immature monolayers or inhibition of the Ser471 kinase G-protein coupled receptor kinase (GRK), all profoundly inhibit TJ formation and epithelial monolayer maturation [148]. The Hippo signaling pathway is an important determinant of cell and organ sizes, and nuclear exclusion of the co-activator YAP accompanies proliferative quiescence. The Hippo pathway elements YAP and TEAD (TEA-dependent) have been found to co-localize with occludin in pancreatic cancer cells and regulate cell proliferation [149].

The second phosphorylation site of occludin that regulates cell proliferation was identified at Ser490. This phosphorylation can be induced by VEGF and promotes occludin ubiquitination and its intracellular trafficking. In endothelial cells, this results in increased vascular permeability [145] in a PKCβ-dependent manner [146]. Moreover, studies on the phosphorylation of this site have revealed a novel function of occludin as a regulator of centrosome separation and mitosis initiation. In MDCK cells, the expression of occludin mutated at Ser490 to Ala slows cell proliferation and hindered mitotic entry due to delayed centrosome separation. Stable expression of aspartic acid phosphomimetic (S490D) in MDCK cells results in centrosomal localization of occludin and cell proliferation. However, expression of the nonphosphorylatable alanine mutation (S490A) of occludin impedes centrosome separation, delays mitotic entry, and reduces proliferation. Collectively, these studies demonstrate a novel location and function for occludin in centrosome separation and mitosis [150]. Similar results were found in endothelial cell culture and in retinal tissue, where the induction of endothelial-specific expression of the occludin S490A mutant through viral delivery completely inhibited neovascularization [151]. In this study, primary bovine retinal endothelial cells in collagen matrices responded to VEGF with increased Ser490 phosphorylation coincident with tube formation. Transfection of occludin with the S490A point mutant inhibited tube formation, proliferation, and migration compared to wild-type occludin. Western blotting revealed increased occludin phosphorylation associated with angiogenesis, and whole-mount immunofluorescent staining of the retinas revealed centrosomal occludin organization in proliferating vessels. Further, mice with doxycycline-inducible *VEGF* expression from photoreceptors were used to study occludin control of angiogenesis in vivo. Expression of the occludin S490A mutant by sub-retinal injection of adeno-associated virus significantly reduced retinal vessel growth in vivo. Importantly, in these studies, occludin was located at the centrosomes (Figure 5D), and increased occludin phosphorylation was found in dividing endothelial cells, mouse retinas with neovascularization, and human surgical samples of retinal neovessels, suggesting a novel role for occludin in regulation of endothelial proliferation and neovascularization in a phosphorylation-dependent manner

Similar to claudins, previous publications have demonstrated that oncogenic transformation of a variety of cell types is associated with altered occludin expression [152]. *OCLN* gene is downregulated in premalignant foci in kidneys from patients with germ line tumor suppressor von Hippel–Lindau (*VHL*) gene mutations [153]. At a molecular level, it has been suggested that *OCLN* is transcriptionally repressed following constitutive Raf-1 expression [154] this is mediated through a direct interaction between activated Slug and the E-box in the *Ocln* promoter [155]. Similarly, in murine melanoma cells, *Ocln* is epigenetically silenced through promoter hyper-methylation, and its forced expression also reduces tumor migration. This is supported by other studies suggesting that occludin possess anti-tumorigenic properties [156,157]. The expression of exogenous occludin suppresses tumor growth in nude mice of Raf1-transformed rat salivary gland epithelial cells [158]. Similarly, stable occludin expression in melanoma and breast cancer cells followed by injection into the craniolateral thorax and mammary fat pad, respectively, reduced the size of lung metastases [157]. In a cell culture model of uveal melanoma, blood vessel epicardial substance (BVES) protein overexpression led to an increase in ZO-1 and occludin, which correlated with decreased cell proliferation [159]. Further, occludin induces premature senescence in breast cancer cells, which can be blocked by chemical inhibition of the mitogen-activated protein kinase (MEK) pathway [160]. Conversely, occludin has been also implicated in EMT, where occludin targets the TGFβ receptor to the junctional complex and promotes efficient epithelial transformation [161] (Figure 2), a mechanism that correlates with the simultaneous repression of the genes encoding E-cadherin, claudins, and occludin [39]. Finally, *OCLN* depletion in MDCK cultures demonstrated impaired mitotic spindle orientation due to a reduced interaction with ZO-1, suggesting an alteration of the cues necessary for polarization in cell division [162].

Other studies support a role of occludin in proliferation and interaction with centrosomal proteins. *OCLN* mutations in human patients can lead to microcephaly and band-like calcifications with polymicrogyria characterized by loss of cortical convolutions, shallow or absent sulci, and multiple small gyri giving the cortex surface a roughened irregular appearance [163–168]. To date, thirteen pathogenic mutations in *OCLN* gene have been identified in thirteen families, and seven mutations are situated in exon 3 [165–168]. Primary microcephaly (MCPH, for microcephaly primary hereditary) is a disorder of brain development that results in a head circumference more than three standard deviations below the mean for age and gender. Notably, many of the causative genes for MCPH encode centrosomal proteins involved in centriole biogenesis [169].

Recently, a new isoform of occludin was discovered by Bendriem et al. with a specific function in the proliferation of human embryonic stem cells (hESCs) and in neural progenitors that alter cortex size in the developing mouse brain (Bendriem et al., accepted for publication in ELife). The original *Ocln* knockout mouse line was generated by excising exon 3 and was believed to be a null model.

While mouse full-length occludin (OCLN-FL) is no longer expressed, a truncated form that lacks its N-terminus and three of its four transmembrane domains (OCLN-ΔN) is still expressed. This is a 32-34 kDa protein that results from a shorter ΔN transcript lacking exons 2 and 3 (Figure 5C). Both OCLN-FL and OCLN-ΔN isoforms localize to the centrosomes; however, in the homozygous mutant mouse line *Ocln*$^{\Delta N/\Delta N}$, OCLN-ΔN localizes to interphase and mitotic centrosomes in the embryonic mouse cortex but not at the plasma membrane, suggesting the C-terminal domain of occludin is important for this centrosomal localization. Consistent with *OCLN* mutation in patients with microcephaly, depletion of *Ocln* in mice led to microcephaly. An increased mitotic index was found in *Ocln*$^{\Delta N/\Delta N}$ mutant mice along with a higher percentage of cells of *Ocln*$^{\Delta N/\Delta N}$ E12.5 cortices in prometaphase and metaphase compared to the wild-type. Moreover, prolonged mitosis led to a higher percentage of activated (cleaved) caspase 3 (CC3)-positive apoptotic cells in mutant embryos compared to controls prior to E14.5.

Occludin centrosomal localization was also confirmed in vitro in two hESCs lines that closely resemble the *Ocln*$^{\Delta N/\Delta N}$ mouse mutant. Mutant hESC-derived organoids displayed pronounced proliferation defects, premature differentiation, and apoptosis. Cells numbers with a reduced ratio of the basal neural progenitor marker HOPX (homeodomain-only protein homeobox) compared to the early neuronal marker NeuroD1 may be responsible for the reduced size of human organoids. Importantly, centrosomal occludin co-localized and immune-precipitated with NuMA (for Nuclear Mitotic Apparatus protein) and the small GTPase RAN (RAs-related Nuclear protein), two important proteins in mitotic spindle assembly and stabilization, and mutant hESCs exhibited impaired mitotic spindles and abnormal morphology at the spindle poles. These studies demonstrated an important role for occludin in neurogenesis through its centrosomal interactions and promotion of proper functioning neural progenitor mitotic spindles.

Along with occludin localization at the centrosomes, studies have also shown that intracellular occludin-containing vesicles move along MTs and contribute to the regulation of cell proliferation [170]. MTs interacting with plasma membranes participate in the preservation of epithelial TJ structure and function. This is regulated by the binding of MT plus end–tracking proteins at the scaffold in the AJs or may be achieved through MT minus end binding of nezha/calmodulin-regulated spectrin-associated protein (CAMSAP) and ninein to the AJs [171–175]. The junctional localization of several MT organizing center proteins indicates the crucial role of junctions as sites that orchestrate MT organization in polarized cells [94]. Glotfelty et al. [170] reported that intracellular occludin-containing vesicles move along MTs and that the rate of movement depends on intact MT networks. This suggests that dynein, a key regulator of MT-vesicle trafficking, may regulate this process in the minus-end direction. Consistent with this hypothesis, the siRNA knockdown of dynein/dynactin induced occludin accumulation in the cytosol, whereas plus-end motor kinesin knockdown did not. This model of MT-dependent TJ trafficking was further supported by the results from studies on dynein and Rab11 [176]. Rab11 utilizes MTs for trafficking and has been shown to participate in occludin trafficking by the regulation of the Rab11 FIPs (Rab11 family interacting protein) [176].

Occludin-containing vesicles may traffic bi-directionally on MTs to regulate cell proliferation. Previous studies suggest that MTs traffic likely contribute to TJ assembly by functioning as tracks for the delivery of TJ proteins through MT-associated vesicles. Particularly occludin-containing granules have been found near the tip of oolemma ingression in dividing *Xenopus* oocytes [177]. Rab13 or junction rab (JRab) is a key mediator of the endocytic recycling of occludin [178] through its binding partners Rab13-binding protein and MICAL-like protein 2 (MICAL-L2) [179,180]. VAP-33 (VAMP associated protein of 33 kDa) is implicated in vesicle docking/fusion and binds to occludin, and its overexpression promotes occludin movement along the lateral edge of the plasma membrane [181]. Identification and characterization of a homolog of VAP-33 in *Drosophila* (DVAP-33A) revealed that DVAP-33A regulates the division of boutons at the synaptic terminals by stabilizing and directing the MT cytoskeleton during budding [182]. Thus, TJ assembly mediated by occludin-MT interaction may play an important role in both cell proliferation and junction organization.

Together, these studies indicate that occludin both regulates barrier function by controlling TJ protein internalization and localizes at centrosomes, contributing to the regulation of cell proliferation.

MarvelD3 is identified as the third member of proteins with a MARVEL domain [128]. MarvelD3 is expressed as two isoforms that show a broad tissue distribution. The two isoforms represent splice variants and share the predicted N-terminal cytoplasmic domain of 198 amino acids, but differ in their C-terminal halves that contain the transmembrane domains. MarvelD3 does not have the long cytoplasmic C-terminus ending in a C-C domain found on occludin and tricellulin. RNA interference experiments in steady-state Caco-2 monolayers indicate that MarvelD3 is required to maintain epithelial integrity under osmotic stress, but this is not essential for TJ formation [183]. However, Raleigh et al. reported that knockdown of *MARVELD3* delays TJs assembly in Caco-2 cells [127]. Further, depletion of *MARVELD3* by siRNAs in the human pancreatic cancer cell line (HPAC) resulted in downregulation of barrier function as shown by decreased electrical resistance and increased permeability to fluorescent dextran tracers, whereas knockdown did not affect the fence function of TJs maintaining apical and basolateral membrane protein restriction [184]. Interestingly, in a genome-wide association study, an intergenic single nucleotide in *MARVELD3* correlated with resistance to severe malaria [185] while its specific role has not yet been discovered.

Recent studies suggest a role of MarvelD3 in EMT, cell proliferation, and migration. In tumor cells, MarvelD3 was downregulated during EMT in human pancreatic cancer cells [184]. Similar effects were also observed when MCF-7 breast cancer cells that are known to express MarvelD3 were compared to MiaPaca-2 pancreatic tumor cells, which do not express detectable MarvelD3. In this study, transfection of MarvelD3 in MiaPaca-2 cells reduced cell migration and proliferation. When they were injected in a xenograft model, MarvelD3 overexpressing MiaPaca-2 cells revealed reduced tumor volume as compared to cells with low MarvelD3 expression. Moreover, siRNA-mediated depletion of *MARVELD3* induced Caco-2 cell migration and proliferation, and the effects could be rescued by expressing mouse MarvelD3. These data indicate that MarvelD3 regulates cell proliferation and cell migration of differentiating and dedifferentiated epithelial model cell lines. The authors suggest that MarvelD3 functions as a signaling transmembrane component of TJs.

MarvelD3 activates MEKK1/JNK (mitogen-activated protein kinase kinase 1/c-Jun NH_2-terminal kinase) signaling through its N-terminal cytoplasmic domain to control cell proliferation. [183]. The interaction of the N-terminal domain of MarvelD3 with upstream component MEKK1 that regulates JNK activation was first demonstrated in glutathione S-transferase (GST) pull-down assays in Caco-2 cells. Further studies in MDCK cell lines expressing MarvelD3 fusion proteins carrying the biotin ligase at both the C-terminus and N-terminus supported this conclusion and corroborated the specific interaction of MEKK1 with the N-terminal domain of MarvelD3. Moreover, in MarvelD3 overexpressing MiaPaca-2 cells, MEKK1 was partially recruited to cell–cell contacts when MarvelD3 was expressed in MiaPaca-2 cells but not in control cells, indicating that re-expression of MarvelD3 is sufficient to stimulate membrane recruitment of MEKK1.

MarvelD3 modulates cell proliferation and migration in early embryogenesis. A recent study with *Xenopus* indicated that MarvelD3 modulates cell proliferation in early eye development and regulates cell survival during eye morphogenesis [186]. In these studies, the JNK pathway was required for proper eye morphogenesis by acting as an inhibitor of the expression of eye-field transcription factors (EFTFs). EFTFs specify a single eye-field in the most anterior region of the neural plate. In this region, inhibition of cell-cycle activators occurs to favor EFTF expression, while duration of the expression of the transcription factors is established by cell-cycle-independent factors. The single eye-field is then divided into two eye primordia under the influence of Sonic hedgehog signaling. Moreover, MarvelD3 is part of a regulatory feedback loop that coordinates JNK activity with neural crest formation. *MarvelD3* depletion enhances JNK signaling, which leads to disruption of neural crest derivative differentiation and neural crest precursor formation, as well as displacement of the neural plate border. Consistent with this model, inhibition of JNK signaling is sufficient to rescue the phenotype induced by *marvelD3* depletion, while constitutively active JNK disrupts neural crest development, supporting the importance of controlled regulation of JNK activity [187]. Together, these data present a novel role of MarvelD3 as an essential regulator of early vertebrate development and neural crest induction that relies on interplay between gene expression, cell proliferation, and cell migration.

7. Conclusions

While numerous studies clearly demonstrate the role of TJ proteins in barrier regulation, more recent research suggests a number of tight junction proteins also contribute to the control of cell proliferation and growth. The junction proteins may contribute to proliferation in a number of mechanisms. In some cases, control of barrier properties alters the microenvironment contributing to growth control. Further, TJ proteins may localize transcription factors and cell cycle control proteins at the plasma membrane, restricting nuclear access or inhibiting function. In the presence of growth factors, cell polarization is disrupted and the TJ proteins release these factors, allowing progression through the cell cycle. In addition, a number of TJ proteins demonstrate non-typical cellular localization associated with cell proliferation, revealing distinct and unique functions from their role in cell-cell contact. TJ proteins have been found localized in the nucleus at G1/S cell cycle transition, associated with integrins at the basal membrane during migration, or in mitosis interacting with microtubules or with centrosome proteins. Together, these studies reveal the coordination of epithelial and endothelial barrier formation with cell proliferation, an essential component of cellular growth and differentiation.

Author Contributions: D.A.A. and M.D.-C. designed the manuscript outline, M.D.-C., X.L., and D.A.A. wrote the manuscript.

Funding: This work was supported by the NIH EY012021 (D.A.A.), the Jules and Doris Stein Professorship from Research to Prevent Blindness (D.A.A.), the Core NIH grants P30EY007003 (Core Center for Vision Research at the Kellogg Eye Center) and DK020572 (Morphology and Image Analysis Core of the Michigan Diabetes Research and Training Center) and the American Diabetes Association Minority Postdoctoral Fellowship Award #4-16-PMF-003 (M.D.-C.).

Conflicts of Interest: The authors declare no conflict of interest.

Abbreviations

α-SMA alpha smooth muscle actin
AF6 afadin
AJ adherens junction
AT2 alveolar epithelial type II cells
BBB blood-brain barrier
BMP bone marrow morphogenetic protein
BRB blood-retinal barrier
BrdU bromo-deoxy uridine
BVES blood vessel epicardial substance protein
CAMSAP calmodulin-regulated spectrin-associated protein
C-C coiled-coil domain
CC3 cleaved caspase 3
CCM cerebral cavernous malformation
Cdk4 cyclin-dependent kinase 4
CK2 casein kinase 2
Cl claudin
α-SMA alpha smooth muscle actin
AF6 afadin
AJ adherens junction
AT2 alveolar epithelial type II cells
BBB blood-brain barrier
BMP bone marrow morphogenetic protein
BRB blood-retinal barrier
BrdU bromo-deoxy uridine

BVES	blood vessel epicardial substance protein
CAMSAP	calmodulin-regulated spectrin-associated protein
C-C	coiled-coil domain
CC3	cleaved caspase 3
CCM	cerebral cavernous malformation
Cdk4	cyclin-dependent kinase 4
CK2	casein kinase 2
Cl	claudin
CNS	central nervous system
CycD1	cyclin-D1
DLG	disc large
EC	extracellular
EFTFs	eye-field transcription factors
EGF	epidermal growth factor
EGFR2 or ERBB2	epidermal growth factor receptor
EMT	epithelial-mesenchymal transition
EndMT	endothelial-mesenchymal transition
Eph	ephrin
EphB1	ephrin B1 receptor
FGF	fibroblast growth factor
FSP1	fibroblast-specific protein 1
FIP	family interacting protein
Fzd4	frizzled-4
GEF-H1	guanine nucleotide exchange factor H1
Gpr124GRK	G-protein-coupled receptor 124
GSK3β	G-protein coupled receptor kinase glycogen synthase kinase 3 beta
GST	glutathione S-transferase
HDMEC	human dermal microvascular endothelial cells
hESCs	human embryonic stem cells
HGF	hepatocyte growth factor
HOPX	homeodomain-only protein homeobox
HPAC	human pancreatic cancer cell line
IL-8	interleukin 8
JACOP	junction-associated-coiled-coil protein
JAM	junctional adhesion molecule
JNK	c-Jun NH_2-terminal kinase
JRAB	junction Rab
LEF	lymphoid enhancer-binding factor
LGL	lethal giant larvae
Lrp5/6	low density lipoprotein receptor-related protein 5/6
MAGUK	membrane-associated guanylate kinase homolog
MAL	myelin and lymphocyte protein
MARVEL	MAL and related proteins for vesicle trafficking and membrane link
MCPH	microcephaly primary hereditary
MDCK	Madin–Darby canine kidney
MEK	mitogen-activated protein kinase
MEKK	mitogen-activated protein kinase kinase
MET	mesenchymal-epithelial transition
MICAL-L2	MICAL-like protein 2
mIMCD-3	mouse inner medullary collecting duct cells
MMP	matrix metalloproteinase
MT	microtubules
NuMA	nuclear mitotic apparatus protein
OAP1	outer surface protein -associated protein 1
$Occ^{\Delta E9}$	occludin deleted in exon 9

OCLN-FL	occludin-full length
OCLN-ΔN	occludin N-terminal deleted
PALS1	protein associated with lin seven 1
PAR	partitioning defective
PATJ	protein associated with tight junctions
PCNA	proliferating cell nuclear antigen
PI3K	phosphatidyl inositol 3-kinase
PKC	protein kinase C
PLVAP	plasmalemma vesicle-associated protein
PMA	phorbol myristic acid
RAN	Ras-related nuclear protein
Reck	reversion-inducing cysteine-rich protein
Sip-1	SMAD interacting protein 1
SJ	septate junction
Sox	sex determining region Y-related high mobility group box factors
TCF	T-cell factor
TEAD	TEA-dependent
TGFβ	transforming growth factor-beta
TGFβR	transforming growth factor-beta receptor
Tiam1	T-cell lymphoma invasion and metastasis 1
TJ	tight junctions
TM	transmembrane
Tspan12	tetraspanin-12
Ubq	ubiquitination
VAMP	vesicle-associated membrane proteins
VAP-33	VAMP associated protein of 33 kDa
VE-cadherin	vascular endothelial cadherin
VEGF	vascular endothelial growth factor
VHL	von Hippel–Lindau
vWF	von Willebrand factor
YAP	c-yes associated protein
ZEB-1	zinc finger E-box-binding homeobox 1
ZO	zona occludens
ZONAB	ZO-1-associated nucleic acid binding protein

References

1. Pauken, C.M.; Capco, D.G. Regulation of cell adhesion during embryonic compaction of mammalian embryos: Roles for PKC and beta-catenin. *Mol. Reprod. Dev.* **1999**, *54*, 135–144. [CrossRef]
2. Fleming, T.P.; Wilkins, A.; Mears, A.; Miller, D.J.; Thomas, F.; Ghassemifar, M.R.; Fesenko, I.; Sheth, B.; Kwong, W.Y.; Eckert, J.J. Society for Reproductive Biology Founders' Lecture 2003. The making of an embryo: Short-term goals and long-term implications. *Reprod. Fertil. Dev.* **2004**, *16*, 325–337. [CrossRef] [PubMed]
3. Balda, M.S.; Anderson, J.M. Two classes of tight junctions are revealed by ZO-1 isoforms. *Am. J. Physiol.* **1993**, *264 Pt 4*, C918–C924. [CrossRef]
4. Fleming, T.P.; McConnell, J.; Johnson, M.H.; Stevenson, B.R. Development of tight junctions de novo in the mouse early embryo: Control of assembly of the tight junction-specific protein, ZO-1. *J. Cell Biol.* **1989**, *108*, 1407–1418. [CrossRef] [PubMed]
5. Sheth, B.; Fesenko, I.; Collins, J.E.; Moran, B.; Wild, A.E.; Anderson, J.M.; Fleming, T.P. Tight junction assembly during mouse blastocyst formation is regulated by late expression of ZO-1 alpha+ isoform. *Development* **1997**, *124*, 2027–2037. [PubMed]
6. Sheth, B.; Fontaine, J.J.; Ponza, E.; McCallum, A.; Page, A.; Citi, S.; Louvard, D.; Zahraoui, A.; Fleming, T.P. Differentiation of the epithelial apical junctional complex during mouse preimplantation development: A role for rab13 in the early maturation of the tight junction. *Mech. Dev.* **2000**, *97*, 93–104. [CrossRef]

7. Thomas, F.C.; Sheth, B.; Eckert, J.J.; Bazzoni, G.; Dejana, E.; Fleming, T.P. Contribution of JAM-1 to epithelial differentiation and tight-junction biogenesis in the mouse preimplantation embryo. *J. Cell Sci.* **2004**, *117 Pt 23*, 5599–5608. [CrossRef]
8. Macara, I.G. Par proteins: Partners in polarization. *Curr. Biol.* **2004**, *14*, R160–R162. [CrossRef]
9. Cui, X.S.; Li, X.Y.; Shen, X.H.; Bae, Y.J.; Kang, J.J.; Kim, N.H. Transcription profile in mouse four-cell, morula, and blastocyst: Genes implicated in compaction and blastocoel formation. *Mol. Reprod. Dev.* **2007**, *74*, 133–143. [CrossRef]
10. Hamatani, T.; Carter, M.G.; Sharov, A.A.; Ko, M.S. Dynamics of global gene expression changes during mouse preimplantation development. *Dev. Cell* **2004**, *6*, 117–131. [CrossRef]
11. St Johnston, D.; Ahringer, J. Cell polarity in eggs and epithelia: Parallels and diversity. *Cell* **2010**, *141*, 757–774. [CrossRef] [PubMed]
12. Ebnet, K.; Suzuki, A.; Horikoshi, Y.; Hirose, T.; Zu Brickwedde, M.K.M.; Ohno, S.; Vestweber, D. The cell polarity protein ASIP/PAR-3 directly associates with junctional adhesion molecule (JAM). *EMBO J.* **2001**, *20*, 3738–3748. [CrossRef] [PubMed]
13. Betschinger, J.; Mechtler, K.; Knoblich, J.A. The Par complex directs asymmetric cell division by phosphorylating the cytoskeletal protein Lgl. *Nature* **2003**, *422*, 326–330. [CrossRef] [PubMed]
14. Javed, Q.; Fleming, T.P.; Hay, M.; Citi, S. Tight junction protein cingulin is expressed by maternal and embryonic genomes during early mouse development. *Development* **1993**, *117*, 1145–1151.
15. Sheth, B.; Moran, B.; Anderson, J.M.; Fleming, T.P. Post-translational control of occludin membrane assembly in mouse trophectoderm: A mechanism to regulate timing of tight junction biogenesis and blastocyst formation. *Development* **2000**, *127*, 831–840.
16. Fleming, T.P.; Sheth, B.; Fesenko, I. Cell adhesion in the preimplantation mammalian embryo and its role in trophectoderm differentiation and blastocyst morphogenesis. *Front. Biosci.* **2001**, *6*, D1000–D1007. [CrossRef]
17. Ciana, A.; Meier, K.; Daum, N.; Gerbes, S.; Veith, M.; Lehr, C.M.; Minetti, G. A dynamic ratio of the alpha+ and alpha− isoforms of the tight junction protein ZO-1 is characteristic of Caco-2 cells and correlates with their degree of differentiation. *Cell Biol. Int.* **2010**, *34*, 669–678. [CrossRef]
18. Eckert, J.J.; McCallum, A.; Mears, A.; Rumsby, M.G.; Cameron, I.T.; Fleming, T.P. Relative contribution of cell contact pattern, specific PKC isoforms and gap junctional communication in tight junction assembly in the mouse early embryo. *Dev. Biol.* **2005**, *288*, 234–247. [CrossRef]
19. Gopalakrishnan, S.; Hallett, M.A.; Atkinson, S.J.; Marrs, J.A. aPKC-PAR complex dysfunction and tight junction disassembly in renal epithelial cells during ATP depletion. *Am. J. Physiol. Cell Physiol.* **2007**, *292*, C1094–C1102. [CrossRef]
20. Nance, J. Getting to know your neighbor: Cell polarization in early embryos. *J. Cell Biol.* **2014**, *206*, 823–832. [CrossRef]
21. Nance, J.; Priess, J.R. Cell polarity and gastrulation in *C. elegans*. *Development* **2002**, *129*, 387–397. [PubMed]
22. Cardellini, P.; Davanzo, G.; Citi, S. Tight junctions in early amphibian development: Detection of junctional cingulin from the 2-cell stage and its localization at the boundary of distinct membrane domains in dividing blastomeres in low calcium. *Dev. Dyn.* **1996**, *207*, 104–113. [CrossRef]
23. Harris, T.J.C.; Sawyer, J.K.; Peifer, M. How the Cytoskeleton Helps Build the Embryonic Body Plan: Models of Morphogenesis from *Drosophila*. *Curr. Top. Dev. Biol.* **2009**, *89*, 55–85. [PubMed]
24. Lecuit, T. Junctions and vesicular trafficking during *Drosophila* cellularization. *J. Cell Sci.* **2004**, *117 Pt 16*, 3427–3433. [CrossRef]
25. Risau, W. Mechanisms of angiogenesis. *Nature* **1997**, *386*, 671–674. [CrossRef]
26. Haddad-Tovolli, R.; Dragano, N.R.V.; Ramalho, A.F.S.; Velloso, L.A. Development and Function of the Blood-Brain Barrier in the Context of Metabolic Control. *Front. Neurosci.* **2017**, *11*, 224. [CrossRef]
27. Diaz-Coranguez, M.; Ramos, C.; Antonetti, D.A. The inner blood-retinal barrier: Cellular basis and development. *Vis. Res.* **2017**, *139*, 123–137. [CrossRef]
28. Saunders, N.R.; Dreifuss, J.J.; Dziegielewska, K.M.; Johansson, P.A.; Habgood, M.D.; Mollgard, K.; Bauer, H.C. The rights and wrongs of blood-brain barrier permeability studies: a walk through 100 years of history. *Front. Neurosci.* **2014**, *8*, 404. [CrossRef]
29. Nitta, T.; Hata, M.; Gotoh, S.; Seo, Y.; Sasaki, H.; Hashimoto, N.; Furuse, M.; Tsukita, S. Size-selective loosening of the blood-brain barrier in claudin-5-deficient mice. *J. Cell Biol.* **2003**, *161*, 653–660. [CrossRef]

30. Stenman, J.M.; Rajagopal, J.; Carroll, T.J.; Ishibashi, M.; McMahon, J.; McMahon, A.P. Canonical Wnt signaling regulates organ-specific assembly and differentiation of CNS vasculature. *Science* **2008**, *322*, 1247–1250. [CrossRef]
31. Liebner, S.; Corada, M.; Bangsow, T.; Babbage, J.; Taddei, A.; Czupalla, C.J.; Reis, M.; Felici, A.; Wolburg, H.; Fruttiger, M.; et al. Wnt/beta-catenin signaling controls development of the blood-brain barrier. *J. Cell Biol.* **2008**, *183*, 409–417. [CrossRef] [PubMed]
32. Daneman, R.; Agalliu, D.; Zhou, L.; Kuhnert, F.; Kuo, C.J.; Barres, B.A. Wnt/beta-catenin signaling is required for CNS, but not non-CNS, angiogenesis. *Proc. Natl. Acad. Sci. USA* **2009**, *106*, 641–646. [CrossRef] [PubMed]
33. Zhou, Y.; Wang, Y.; Tischfield, M.; Williams, J.; Smallwood, P.M.; Rattner, A.; Taketo, M.M.; Nathans, J. Canonical WNT signaling components in vascular development and barrier formation. *J. Clin. Investig.* **2014**, *124*, 3825–3846. [CrossRef] [PubMed]
34. Wang, Y.; Cho, C.; Williams, J.; Smallwood, P.M.; Zhang, C.; Junge, H.J.; Nathans, J. Interplay of the Norrin and Wnt7a/Wnt7b signaling systems in blood-brain barrier and blood-retina barrier development and maintenance. *Proc. Natl. Acad. Sci. USA* **2018**, *115*, E11827–E11836. [CrossRef] [PubMed]
35. Zhou, Y.; Williams, J.; Smallwood, P.M.; Nathans, J. Sox7, Sox17, and Sox18 Cooperatively Regulate Vascular Development in the Mouse Retina. *PLoS ONE* **2015**, *10*, e0143650. [CrossRef] [PubMed]
36. Morris, H.T.; Machesky, L.M. Actin cytoskeletal control during epithelial to mesenchymal transition: Focus on the pancreas and intestinal tract. *Br. J. Cancer* **2015**, *112*, 613–620. [CrossRef] [PubMed]
37. Robson, E.J.; Khaled, W.T.; Abell, K.; Watson, C.J. Epithelial-to-mesenchymal transition confers resistance to apoptosis in three murine mammary epithelial cell lines. *Differentiation* **2006**, *74*, 254–264. [CrossRef]
38. Vega, S.; Morales, A.V.; Ocana, O.H.; Valdes, F.; Fabregat, I.; Nieto, M.A. Snail blocks the cell cycle and confers resistance to cell death. *Genes Dev.* **2004**, *18*, 1131–1143. [CrossRef]
39. Ikenouchi, J.; Matsuda, M.; Furuse, M.; Tsukita, S. Regulation of tight junctions during the epithelium-mesenchyme transition: Direct repression of the gene expression of claudins/occludin by Snail. *J. Cell Sci.* **2003**, *116 Pt 10*, 1959–1967. [CrossRef]
40. Nieto, M.A. The snail superfamily of zinc-finger transcription factors. *Nat. Rev. Mol. Cell Biol.* **2002**, *3*, 155–166. [CrossRef]
41. Ozdamar, B.; Bose, R.; Barrios-Rodiles, M.; Wang, H.R.; Zhang, Y.; Wrana, J.L. Regulation of the polarity protein Par6 by TGFbeta receptors controls epithelial cell plasticity. *Science* **2005**, *307*, 1603–1609. [CrossRef] [PubMed]
42. Aranda, V.; Haire, T.; Nolan, M.E.; Calarco, J.P.; Rosenberg, A.Z.; Fawcett, J.P.; Pawson, T.; Muthuswamy, S.K. Par6-aPKC uncouples ErbB2 induced disruption of polarized epithelial organization from proliferation control. *Nat. Cell Biol.* **2006**, *8*, 1235–1245. [CrossRef] [PubMed]
43. Ma, L.; Lu, M.F.; Schwartz, R.J.; Martin, J.F. Bmp2 is essential for cardiac cushion epithelial-mesenchymal transition and myocardial patterning. *Development* **2005**, *132*, 5601–5611. [CrossRef] [PubMed]
44. McCulley, D.J.; Kang, J.O.; Martin, J.F.; Black, B.L. BMP4 is required in the anterior heart field and its derivatives for endocardial cushion remodeling, outflow tract septation, and semilunar valve development. *Dev. Dyn.* **2008**, *237*, 3200–3209. [CrossRef]
45. Zeisberg, M.; Hanai, J.; Sugimoto, H.; Mammoto, T.; Charytan, D.; Strutz, F.; Kalluri, R. BMP-7 counteracts TGF-beta1-induced epithelial-to-mesenchymal transition and reverses chronic renal injury. *Nat. Med.* **2003**, *9*, 964–968. [CrossRef]
46. Bhat, A.A.; Uppada, S.; Achkar, I.W.; Hashem, S.; Yadav, S.K.; Shanmugakonar, M.; Al-Naemi, H.A.; Haris, M.; Uddin, S. Tight Junction Proteins and Signaling Pathways in Cancer and Inflammation: A Functional Crosstalk. *Front. Physiol.* **2018**, *9*, 1942. [CrossRef]
47. Padmanaban, V.; Krol, I.; Suhail, Y.; Szczerba, B.M.; Aceto, N.; Bader, J.S.; Ewald, A.J. E-cadherin is required for metastasis in multiple models of breast cancer. *Nature* **2019**, *573*, 439–444. [CrossRef]
48. Eisenberg, L.M.; Markwald, R.R. Molecular regulation of atrioventricular valvuloseptal morphogenesis. *Circ. Res.* **1995**, *77*, 1–6. [CrossRef]
49. Maddaluno, L.; Rudini, N.; Cuttano, R.; Bravi, L.; Giampietro, C.; Corada, M.; Ferrarini, L.; Orsenigo, F.; Papa, E.; Boulday, G.; et al. EndMT contributes to the onset and progression of cerebral cavernous malformations. *Nature* **2013**, *498*, 492–496. [CrossRef]
50. Hong, L.; Du, X.; Li, W.; Mao, Y.; Sun, L.; Li, X. EndMT: A promising and controversial field. *Eur. J. Cell Biol.* **2018**, *97*, 493–500. [CrossRef]

51. Lampugnani, M.G.; Dejana, E.; Giampietro, C. Vascular Endothelial (VE)-Cadherin, Endothelial Adherens Junctions, and Vascular Disease. *Cold Spring Harb. Perspect. Biol.* **2018**, *10*, a029322. [CrossRef]
52. Malinverno, M.; Maderna, C.; Abu Taha, A.; Corada, M.; Orsenigo, F.; Valentino, M.; Pisati, F.; Fusco, C.; Graziano, P.; Giannotta, M.; et al. Endothelial cell clonal expansion in the development of cerebral cavernous malformations. *Nat. Commun.* **2019**, *10*, 2761. [CrossRef] [PubMed]
53. Akhurst, R.J.; Derynck, R. TGF-beta signaling in cancer—A double-edged sword. *Trends Cell Biol.* **2001**, *11*, S44–S51. [PubMed]
54. Medici, D.; Hay, E.D.; Olsen, B.R. Snail and Slug promote epithelial-mesenchymal transition through beta-catenin-T-cell factor-4-dependent expression of transforming growth factor-beta3. *Mol. Biol. Cell* **2008**, *19*, 4875–4887. [CrossRef]
55. Liebner, S.; Cattelino, A.; Gallini, R.; Rudini, N.; Iurlaro, M.; Piccolo, S.; Dejana, E. Beta-catenin is required for endothelial-mesenchymal transformation during heart cushion development in the mouse. *J. Cell Biol.* **2004**, *166*, 359–367. [CrossRef]
56. Kokudo, T.; Suzuki, Y.; Yoshimatsu, Y.; Yamazaki, T.; Watabe, T.; Miyazono, K. Snail is required for TGFbeta-induced endothelial-mesenchymal transition of embryonic stem cell-derived endothelial cells. *J. Cell Sci.* **2008**, *121 Pt 20*, 3317–3324. [CrossRef]
57. Medici, D.; Potenta, S.; Kalluri, R. Transforming growth factor-beta2 promotes Snail-mediated endothelial-mesenchymal transition through convergence of Smad-dependent and Smad-independent signalling. *Biochem. J.* **2011**, *437*, 515–520. [CrossRef]
58. Gerhardt, H.; Golding, M.; Fruttiger, M.; Ruhrberg, C.; Lundkvist, A.; Abramsson, A.; Jeltsch, M.; Mitchell, C.; Alitalo, K.; Shima, D.; et al. VEGF guides angiogenic sprouting utilizing endothelial tip cell filopodia. *J. Cell Biol.* **2003**, *161*, 1163–1177. [CrossRef] [PubMed]
59. Krizbai, I.A.; Gasparics, A.; Nagyoszi, P.; Fazakas, C.; Molnar, J.; Wilhelm, I.; Bencs, R.; Rosivall, L.; Sebe, A. Endothelial-mesenchymal transition of brain endothelial cells: Possible role during metastatic extravasation. *PLoS ONE* **2015**, *10*, e0123845. [CrossRef]
60. Moore-Morris, T.; Tallquist, M.D.; Evans, S.M. Sorting out where fibroblasts come from. *Circ. Res.* **2014**, *115*, 602–604. [CrossRef]
61. Davis, G.E.; Senger, D.R. Endothelial extracellular matrix: Biosynthesis, remodeling, and functions during vascular morphogenesis and neovessel stabilization. *Circ. Res.* **2005**, *97*, 1093–1107. [CrossRef]
62. Potenta, S.; Zeisberg, E.; Kalluri, R. The role of endothelial-to-mesenchymal transition in cancer progression. *Br. J. Cancer* **2008**, *99*, 1375–1379. [CrossRef]
63. Pinheiro, D.; Bellaiotache, Y. Mechanical Force-Driven Adherens Junction Remodeling and Epithelial Dynamics. *Dev. Cell* **2018**, *47*, 391. [CrossRef] [PubMed]
64. Itoh, M.; Nagafuchi, A.; Moroi, S.; Tsukita, S. Involvement of ZO-1 in cadherin-based cell adhesion through its direct binding to alpha catenin and actin filaments. *J. Cell Biol.* **1997**, *138*, 181–192. [CrossRef]
65. Wittchen, E.S.; Haskins, J.; Stevenson, B.R. Protein interactions at the tight junction. Actin has multiple binding partners, and ZO-1 forms independent complexes with ZO-2 and ZO-3. *J. Biol. Chem.* **1999**, *274*, 35179–35185. [CrossRef]
66. Yamamoto, T.; Harada, N.; Kawano, Y.; Taya, S.; Kaibuchi, K. In vivo interaction of AF-6 with activated Ras and ZO-1. *Biochem. Biophys. Res. Commun.* **1999**, *259*, 103–107. [CrossRef]
67. Gumbiner, B.; Lowenkopf, T.; Apatira, D. Identification of a 160-kDa polypeptide that binds to the tight junction protein ZO-1. *Proc. Natl. Acad. Sci. USA* **1991**, *88*, 3460–3464. [CrossRef]
68. Haskins, J.; Gu, L.; Wittchen, E.S.; Hibbard, J.; Stevenson, B.R. ZO-3, a novel member of the MAGUK protein family found at the tight junction, interacts with ZO-1 and occludin. *J. Cell Biol.* **1998**, *141*, 199–208. [CrossRef]
69. Stevenson, B.R.; Siliciano, J.D.; Mooseker, M.S.; Goodenough, D.A. Identification of ZO-1: A high molecular weight polypeptide associated with the tight junction (zonula occludens) in a variety of epithelia. *J. Cell Biol.* **1986**, *103*, 755–766. [CrossRef]
70. Katsuno, T.; Umeda, K.; Matsui, T.; Hata, M.; Tamura, A.; Itoh, M.; Takeuchi, K.; Fujimori, T.; Nabeshima, Y.; Noda, T.; et al. Deficiency of zonula occludens-1 causes embryonic lethal phenotype associated with defected yolk sac angiogenesis and apoptosis of embryonic cells. *Mol. Biol. Cell* **2008**, *19*, 2465–2475. [CrossRef]
71. Xu, J.; Kausalya, P.J.; Phua, D.C.; Ali, S.M.; Hossain, Z.; Hunziker, W. Early embryonic lethality of mice lacking ZO-2, but Not ZO-3, reveals critical and nonredundant roles for individual zonula occludens proteins in mammalian development. *Mol. Cell Biol.* **2008**, *28*, 1669–1678. [CrossRef] [PubMed]

72. Balda, M.S.; Matter, K. Tight junctions and the regulation of gene expression. *Biochim. Biophys. Acta* **2009**, *1788*, 761–767. [CrossRef]
73. Sourisseau, T.; Georgiadis, A.; Tsapara, A.; Ali, R.R.; Pestell, R.; Matter, K.; Balda, M.S. Regulation of PCNA and cyclin D1 expression and epithelial morphogenesis by the ZO-1-regulated transcription factor ZONAB/DbpA. *Mol. Cell Biol.* **2006**, *26*, 2387–2398. [CrossRef] [PubMed]
74. Lima, W.R.; Parreira, K.S.; Devuyst, O.; Caplanusi, A.; N'Kuli, F.; Marien, B.; Van Der Smissen, P.; Alves, P.M.; Verroust, P.; Christensen, E.I.; et al. ZONAB promotes proliferation and represses differentiation of proximal tubule epithelial cells. *J. Am. Soc. Nephrol.* **2010**, *21*, 478–488. [CrossRef]
75. Capaldo, C.T.; Koch, S.; Kwon, M.; Laur, O.; Parkos, C.A.; Nusrat, A. Tight function zonula occludens-3 regulates cyclin D1-dependent cell proliferation. *Mol Biol Cell* **2011**, *22*, 1677–1685. [CrossRef]
76. Kleeff, J.; Shi, X.; Bode, H.P.; Hoover, K.; Shrikhande, S.; Bryant, P.J.; Korc, M.; Buchler, M.W.; Friess, H. Altered expression and localization of the tight junction protein ZO-1 in primary and metastatic pancreatic cancer. *Pancreas* **2001**, *23*, 259–265. [CrossRef]
77. Takai, E.; Tan, X.; Tamori, Y.; Hirota, M.; Egami, H.; Ogawa, M. Correlation of translocation of tight junction protein Zonula occludens-1 and activation of epidermal growth factor receptor in the regulation of invasion of pancreatic cancer cells. *Int. J. Oncol.* **2005**, *27*, 645–651.
78. Tuomi, S.; Mai, A.; Nevo, J.; Laine, J.O.; Vilkki, V.; Ohman, T.J.; Gahmberg, C.G.; Parker, P.J.; Ivaska, J. PKCepsilon regulation of an alpha5 integrin-ZO-1 complex controls lamellae formation in migrating cancer cells. *Sci. Signal.* **2009**, *2*, ra32. [CrossRef]
79. Turner, J.R. 'Putting the squeeze' on the tight junction: Understanding cytoskeletal regulation. *Semin. Cell Dev. Biol.* **2000**, *11*, 301–308. [CrossRef]
80. Ren, Y.; Li, R.; Zheng, Y.; Busch, H. Cloning and characterization of GEF-H1, a microtubule-associated guanine nucleotide exchange factor for Rac and Rho GTPases. *J. Biol. Chem.* **1998**, *273*, 34954–34960. [CrossRef]
81. Cordenonsi, M.; D'Atri, F.; Hammar, E.; Parry, D.A.; Kendrick-Jones, J.; Shore, D.; Citi, S. Cingulin contains globular and coiled-coil domains and interacts with ZO-1, ZO-2, ZO-3, and myosin. *J. Cell Biol.* **1999**, *147*, 1569–1582. [CrossRef]
82. Citi, S.; D'Atri, F.; Parry, D.A. Human and Xenopus cingulin share a modular organization of the coiled-coil rod domain: Predictions for intra- and intermolecular assembly. *J. Struct. Biol.* **2000**, *131*, 135–145. [CrossRef]
83. Aijaz, S.; D'Atri, F.; Citi, S.; Balda, M.S.; Matter, K. Binding of GEF-H1 to the tight junction-associated adaptor cingulin results in inhibition of Rho signaling and G1/S phase transition. *Dev. Cell* **2005**, *8*, 777–786. [CrossRef]
84. Guillemot, L.; Citi, S. Cingulin regulates claudin-2 expression and cell proliferation through the small GTPase RhoA. *Mol. Biol. Cell* **2006**, *17*, 3569–3577. [CrossRef] [PubMed]
85. Wu, C.Y.; Jhingory, S.; Taneyhill, L.A. The tight junction scaffolding protein cingulin regulates neural crest cell migration. *Dev. Dyn.* **2011**, *240*, 2309–2323. [CrossRef]
86. Mangan, A.J.; Sietsema, D.V.; Li, D.; Moore, J.K.; Citi, S.; Prekeris, R. Cingulin and actin mediate midbody-dependent apical lumen formation during polarization of epithelial cells. *Nat. Commun.* **2016**, *7*, 12426. [CrossRef]
87. Ohnishi, H.; Nakahara, T.; Furuse, K.; Sasaki, H.; Tsukita, S.; Furuse, M. JACOP, a novel plaque protein localizing at the apical junctional complex with sequence similarity to cingulin. *J. Biol. Chem.* **2004**, *279*, 46014–46022. [CrossRef]
88. Paschoud, S.; Yu, D.; Pulimeno, P.; Jond, L.; Turner, J.R.; Citi, S. Cingulin and paracingulin show similar dynamic behaviour, but are recruited independently to junctions. *Mol. Membr. Biol.* **2011**, *28*, 123–135. [CrossRef]
89. Pulimeno, P.; Paschoud, S.; Citi, S. A role for ZO-1 and PLEKHA7 in recruiting paracingulin to tight and adherens junctions of epithelial cells. *J. Biol. Chem.* **2011**, *286*, 16743–16750. [CrossRef]
90. Guillemot, L.; Paschoud, S.; Jond, L.; Foglia, A.; Citi, S. Paracingulin regulates the activity of Rac1 and RhoA GTPases by recruiting Tiam1 and GEF-H1 to epithelial junctions. *Mol. Biol. Cell* **2008**, *19*, 4442–4453. [CrossRef]
91. Guillemot, L.; Guerrera, D.; Spadaro, D.; Tapia, R.; Jond, L.; Citi, S. MgcRacGAP interacts with cingulin and paracingulin to regulate Rac1 activation and development of the tight junction barrier during epithelial junction assembly. *Mol. Biol. Cell* **2014**, *25*, 1995–2005. [CrossRef]

92. Tornavaca, O.; Chia, M.; Dufton, N.; Almagro, L.O.; Conway, D.E.; Randi, A.M.; Schwartz, M.A.; Matter, K.; Balda, M.S. ZO-1 controls endothelial adherens junctions, cell-cell tension, angiogenesis, and barrier formation. *J. Cell Biol.* **2015**, *208*, 821–838. [CrossRef]
93. Chrifi, I.; Hermkens, D.; Brandt, M.M.; van Dijk, C.G.M.; Burgisser, P.E.; Haasdijk, R.; Pei, J.; van de Kamp, E.H.M.; Zhu, C.; Blonden, L.; et al. Cgnl1, an endothelial junction complex protein, regulates GTPase mediated angiogenesis. *Cardiovasc. Res.* **2017**, *113*, 1776–1788. [CrossRef]
94. Vasileva, E.; Citi, S. The role of microtubules in the regulation of epithelial junctions. *Tissue Barriers* **2018**, *6*, 1539596. [CrossRef]
95. Akizuki, R.; Shimobaba, S.; Matsunaga, T.; Endo, S.; Ikari, A. Claudin-5, -7, and -18 suppress proliferation mediated by inhibition of phosphorylation of Akt in human lung squamous cell carcinoma. *Biochim. Biophys. Acta Mol. Cell Res.* **2017**, *1864*, 293–302. [CrossRef]
96. Zavala-Zendejas, V.E.; Torres-Martinez, A.C.; Salas-Morales, B.; Fortoul, T.I.; Montano, L.F.; Rendon-Huerta, E.P. Claudin-6, 7, or 9 overexpression in the human gastric adenocarcinoma cell line AGS increases its invasiveness, migration, and proliferation rate. *Cancer Investig.* **2011**, *29*, 1–11. [CrossRef]
97. Ikari, A.; Sato, T.; Takiguchi, A.; Atomi, K.; Yamazaki, Y.; Sugatani, J. Claudin-2 knockdown decreases matrix metalloproteinase-9 activity and cell migration via suppression of nuclear Sp1 in A549 cells. *Life Sci.* **2011**, *88*, 628–633. [CrossRef]
98. Takehara, M.; Nishimura, T.; Mima, S.; Hoshino, T.; Mizushima, T. Effect of claudin expression on paracellular permeability, migration and invasion of colonic cancer cells. *Biol. Pharm. Bull.* **2009**, *32*, 825–831. [CrossRef]
99. Yoon, C.H.; Kim, M.J.; Park, M.J.; Park, I.C.; Hwang, S.G.; An, S.; Choi, Y.H.; Yoon, G.; Lee, S.J. Claudin-1 acts through c-Abl-protein kinase Cdelta (PKCdelta) signaling and has a causal role in the acquisition of invasive capacity in human liver cells. *J. Biol. Chem.* **2010**, *285*, 226–233. [CrossRef]
100. Haddad, N.; El Andalousi, J.; Khairallah, H.; Yu, M.; Ryan, A.K.; Gupta, I.R. The tight junction protein claudin-3 shows conserved expression in the nephric duct and ureteric bud and promotes tubulogenesis in vitro. *Am. J. Physiol. Ren. Physiol.* **2011**, *301*, F1057–F1065. [CrossRef]
101. Neesse, A.; Griesmann, H.; Gress, T.M.; Michl, P. Claudin-4 as therapeutic target in cancer. *Arch. Biochem. Biophys.* **2012**, *524*, 64–70. [CrossRef]
102. Karanjawala, Z.E.; Illei, P.B.; Ashfaq, R.; Infante, J.R.; Murphy, K.; Pandey, A.; Schulick, R.; Winter, J.; Sharma, R.; Maitra, A.; et al. New markers of pancreatic cancer identified through differential gene expression analyses: Claudin 18 and annexin A8. *Am. J. Surg. Pathol.* **2008**, *32*, 188–196. [CrossRef]
103. Sahin, U.; Koslowski, M.; Dhaene, K.; Usener, D.; Brandenburg, G.; Seitz, G.; Huber, C.; Tureci, O. Claudin-18 splice variant 2 is a pan-cancer target suitable for therapeutic antibody development. *Clin. Cancer Res.* **2008**, *14*, 7624–7634. [CrossRef]
104. Yamaguchi, H.; Kojima, T.; Ito, T.; Kyuno, D.; Kimura, Y.; Imamura, M.; Hirata, K.; Sawada, N. Effects of Clostridium perfringens enterotoxin via claudin-4 on normal human pancreatic duct epithelial cells and cancer cells. *Cell. Mol. Biol. Lett.* **2011**, *16*, 385–397. [CrossRef]
105. Tiwari-Woodruff, S.K.; Buznikov, A.G.; Vu, T.Q.; Micevych, P.E.; Chen, K.; Kornblum, H.I.; Bronstein, J.M. OSP/claudin-11 forms a complex with a novel member of the tetraspanin super family and beta1 integrin and regulates proliferation and migration of oligodendrocytes. *J. Cell Biol.* **2001**, *153*, 295–305. [CrossRef]
106. Ikari, A.; Watanabe, R.; Sato, T.; Taga, S.; Shimobaba, S.; Yamaguchi, M.; Yamazaki, Y.; Endo, S.; Matsunaga, T.; Sugatani, J. Nuclear distribution of claudin-2 increases cell proliferation in human lung adenocarcinoma cells. *Biochim. Biophys. Acta* **2014**, *1843*, 2079–2088. [CrossRef]
107. Leotlela, P.D.; Wade, M.S.; Duray, P.H.; Rhode, M.J.; Brown, H.F.; Rosenthal, D.T.; Dissanayake, S.K.; Earley, R.; Indig, F.E.; Nickoloff, B.J.; et al. Claudin-1 overexpression in melanoma is regulated by PKC and contributes to melanoma cell motility. *Oncogene* **2007**, *26*, 3846–3856. [CrossRef]
108. Tanaka, M.; Kamata, R.; Sakai, R. Phosphorylation of ephrin-B1 via the interaction with claudin following cell-cell contact formation. *EMBO J.* **2005**, *24*, 3700–3711. [CrossRef]
109. Tanaka, M.; Kamata, R.; Sakai, R. EphA2 phosphorylates the cytoplasmic tail of Claudin-4 and mediates paracellular permeability. *J. Biol. Chem.* **2005**, *280*, 42375–42382. [CrossRef]
110. Li, J.; Chigurupati, S.; Agarwal, R.; Mughal, M.R.; Mattson, M.P.; Becker, K.G.; Wood, W.H., 3rd; Zhang, Y.; Morin, P.J. Possible angiogenic roles for claudin-4 in ovarian cancer. *Cancer Biol. Ther.* **2009**, *8*, 1806–1814. [CrossRef]

111. Tamura, A.; Kitano, Y.; Hata, M.; Katsuno, T.; Moriwaki, K.; Sasaki, H.; Hayashi, H.; Suzuki, Y.; Noda, T.; Furuse, M.; et al. Megaintestine in claudin-15-deficient mice. *Gastroenterology* **2008**, *134*, 523–534. [CrossRef]
112. Tsukita, S.; Yamazaki, Y.; Katsuno, T.; Tamura, A.; Tsukita, S. Tight junction-based epithelial microenvironment and cell proliferation. *Oncogene* **2008**, *27*, 6930–6938. [CrossRef]
113. Borka, K.; Kaliszky, P.; Szabo, E.; Lotz, G.; Kupcsulik, P.; Schaff, Z.; Kiss, A. Claudin expression in pancreatic endocrine tumors as compared with ductal adenocarcinomas. *Virchows Arch.* **2007**, *450*, 549–557. [CrossRef]
114. Holczbauer, A.; Gyongyosi, B.; Lotz, G.; Szijarto, A.; Kupcsulik, P.; Schaff, Z.; Kiss, A. Distinct claudin expression profiles of hepatocellular carcinoma and metastatic colorectal and pancreatic carcinomas. *J. Histochem. Cytochem.* **2013**, *61*, 294–305. [CrossRef]
115. Zhou, B.; Flodby, P.; Luo, J.; Castillo, D.R.; Liu, Y.; Yu, F.X.; McConnell, A.; Varghese, B.; Li, G.; Chimge, N.O.; et al. Claudin-18-mediated YAP activity regulates lung stem and progenitor cell homeostasis and tumorigenesis. *J. Clin. Investig.* **2018**, *128*, 970–984. [CrossRef]
116. Michl, P.; Barth, C.; Buchholz, M.; Lerch, M.M.; Rolke, M.; Holzmann, K.H.; Menke, A.; Fensterer, H.; Giehl, K.; Lohr, M.; et al. Claudin-4 expression decreases invasiveness and metastatic potential of pancreatic cancer. *Cancer Res.* **2003**, *63*, 6265–6271.
117. Nava, P.; Capaldo, C.T.; Koch, S.; Kolegraff, K.; Rankin, C.R.; Farkas, A.E.; Feasel, M.E.; Li, L.; Addis, C.; Parkos, C.A.; et al. JAM-A regulates epithelial proliferation through Akt/beta-catenin signalling. *EMBO Rep.* **2011**, *12*, 314–320. [CrossRef]
118. Perry, J.M.; He, X.C.; Sugimura, R.; Grindley, J.C.; Haug, J.S.; Ding, S.; Li, L. Cooperation between both Wnt/[189]-catenin and PTEN/PI3K/Akt signaling promotes primitive hematopoietic stem cell self-renewal and expansion. *Genes Dev.* **2011**, *25*, 1928–1942. [CrossRef]
119. Tuncay, H.; Brinkmann, B.F.; Steinbacher, T.; Schurmann, A.; Gerke, V.; Iden, S.; Ebnet, K. JAM-A regulates cortical dynein localization through Cdc42 to control planar spindle orientation during mitosis. *Nat. Commun.* **2015**, *6*, 8128. [CrossRef]
120. Severson, E.A.; Jiang, L.; Ivanov, A.I.; Mandell, K.J.; Nusrat, A.; Parkos, C.A. Cis-dimerization mediates function of junctional adhesion molecule A. *Mol. Biol. Cell* **2008**, *19*, 1862–1872. [CrossRef]
121. Mandell, K.J.; Babbin, B.A.; Nusrat, A.; Parkos, C.A. Junctional adhesion molecule 1 regulates epithelial cell morphology through effects on beta1 integrins and Rap1 activity. *J. Biol. Chem.* **2005**, *280*, 11665–11674. [CrossRef]
122. Severson, E.A.; Lee, W.Y.; Capaldo, C.T.; Nusrat, A.; Parkos, C.A. Junctional adhesion molecule A interacts with Afadin and PDZ-GEF2 to activate Rap1A, regulate beta1 integrin levels, and enhance cell migration. *Mol. Biol. Cell* **2009**, *20*, 1916–1925. [CrossRef]
123. Bazzoni, G.; Tonetti, P.; Manzi, L.; Cera, M.R.; Balconi, G.; Dejana, E. Expression of junctional adhesion molecule-A prevents spontaneous and random motility. *J. Cell Sci.* **2005**, *118 Pt 3*, 623–632. [CrossRef]
124. Daniele, L.L.; Adams, R.H.; Durante, D.E.; Pugh, E.N., Jr.; Philp, N.J. Novel distribution of junctional adhesion molecule-C in the neural retina and retinal pigment epithelium. *J. Comp. Neurol.* **2007**, *505*, 166–176. [CrossRef]
125. Economopoulou, M.; Avramovic, N.; Klotzsche-von Ameln, A.; Korovina, I.; Sprott, D.; Samus, M.; Gercken, B.; Troullinaki, M.; Grossklaus, S.; Funk, R.H.; et al. Endothelial-specific deficiency of Junctional Adhesion Molecule-C promotes vessel normalisation in proliferative retinopathy. *Thromb. Haemost.* **2015**, *114*, 1241–1249.
126. Ikenouchi, J.; Furuse, M.; Furuse, K.; Sasaki, H.; Tsukita, S.; Tsukita, S. Tricellulin constitutes a novel barrier at tricellular contacts of epithelial cells. *J. Cell Biol.* **2005**, *171*, 939–945. [CrossRef]
127. Raleigh, D.R.; Marchiando, A.M.; Zhang, Y.; Shen, L.; Sasaki, H.; Wang, Y.; Long, M.; Turner, J.R. Tight junction-associated MARVEL proteins marveld3, tricellulin, and occludin have distinct but overlapping functions. *Mol. Biol. Cell* **2010**, *21*, 1200–1213. [CrossRef]
128. Steed, E.; Rodrigues, N.T.; Balda, M.S.; Matter, K. Identification of MarvelD3 as a tight junction-associated transmembrane protein of the occludin family. *BMC Cell Biol.* **2009**, *10*, 95. [CrossRef]
129. Yaffe, Y.; Shepshelovitch, J.; Nevo-Yassaf, I.; Yeheskel, A.; Shmerling, H.; Kwiatek, J.M.; Gaus, K.; Pasmanik-Chor, M.; Hirschberg, K. The MARVEL transmembrane motif of occludin mediates oligomerization and targeting to the basolateral surface in epithelia. *J. Cell Sci.* **2012**, *125 Pt 15*, 3545–3556. [CrossRef]
130. Mariano, C.; Sasaki, H.; Brites, D.; Brito, M.A. A look at tricellulin and its role in tight junction formation and maintenance. *Eur. J. Cell Biol.* **2011**, *90*, 787–796. [CrossRef]

131. Cording, J.; Berg, J.; Kading, N.; Bellmann, C.; Tscheik, C.; Westphal, J.K.; Milatz, S.; Gunzel, D.; Wolburg, H.; Piontek, J.; et al. In tight junctions, claudins regulate the interactions between occludin, tricellulin and marvelD3, which, inversely, modulate claudin oligomerization. *J. Cell Sci.* **2013**, *126 Pt 2*, 554–564. [CrossRef]
132. Saitou, M.; Furuse, M.; Sasaki, H.; Schulzke, J.D.; Fromm, M.; Takano, H.; Noda, T.; Tsukita, S. Complex phenotype of mice lacking occludin, a component of tight junction strands. *Mol. Biol. Cell* **2000**, *11*, 4131–4142. [CrossRef] [PubMed]
133. Kamitani, T.; Sakaguchi, H.; Tamura, A.; Miyashita, T.; Yamazaki, Y.; Tokumasu, R.; Inamoto, R.; Matsubara, A.; Mori, N.; Hisa, Y.; et al. Deletion of Tricellulin Causes Progressive Hearing Loss Associated with Degeneration of Cochlear Hair Cells. *Sci. Rep.* **2015**, *5*, 18402. [CrossRef] [PubMed]
134. Nayak, G.; Lee, S.I.; Yousaf, R.; Edelmann, S.E.; Trincot, C.; Van Itallie, C.M.; Sinha, G.P.; Rafeeq, M.; Jones, S.M.; Belyantseva, I.A.; et al. Tricellulin deficiency affects tight junction architecture and cochlear hair cells. *J. Clin. Investig.* **2013**, *123*, 4036–4049. [CrossRef] [PubMed]
135. Sanchez-Pulido, L.; Martin-Belmonte, F.; Valencia, A.; Alonso, M.A. MARVEL: A conserved domain involved in membrane apposition events. *Trends Biochem. Sci.* **2002**, *27*, 599–601. [CrossRef]
136. Furuse, M.; Hirase, T.; Itoh, M.; Nagafuchi, A.; Yonemura, S.; Tsukita, S.; Tsukita, S. Occludin: A novel integral membrane protein localizing at tight junctions. *J. Cell Biol.* **1993**, *123 Pt 6*, 1777–1788. [CrossRef]
137. Schulzke, J.D.; Gitter, A.H.; Mankertz, J.; Spiegel, S.; Seidler, U.; Amasheh, S.; Saitou, M.; Tsukita, S.; Fromm, M. Epithelial transport and barrier function in occludin-deficient mice. *Biochim. Biophys. Acta* **2005**, *1669*, 34–42. [CrossRef]
138. Yu, A.S.; McCarthy, K.M.; Francis, S.A.; McCormack, J.M.; Lai, J.; Rogers, R.A.; Lynch, R.D.; Schneeberger, E.E. Knockdown of occludin expression leads to diverse phenotypic alterations in epithelial cells. *Am. J. Physiol. Cell Physiol.* **2005**, *288*, C1231–C1241. [CrossRef]
139. Phillips, B.E.; Cancel, L.; Tarbell, J.M.; Antonetti, D.A. Occludin independently regulates permeability under hydrostatic pressure and cell division in retinal pigment epithelial cells. *Investig. Ophthalmol. Vis. Sci.* **2008**, *49*, 2568–2576. [CrossRef]
140. Aaku-Saraste, E.; Hellwig, A.; Huttner, W.B. Loss of occludin and functional tight junctions, but not ZO-1, during neural tube closure–remodeling of the neuroepithelium prior to neurogenesis. *Dev. Biol.* **1996**, *180*, 664–679. [CrossRef]
141. Dorfel, M.J.; Westphal, J.K.; Bellmann, C.; Krug, S.M.; Cording, J.; Mittag, S.; Tauber, R.; Fromm, M.; Blasig, I.E.; Huber, O. CK2-dependent phosphorylation of occludin regulates the interaction with ZO-proteins and tight junction integrity. *Cell Commun. Signal.* **2013**, *11*, 40. [CrossRef] [PubMed]
142. Suzuki, T.; Elias, B.C.; Seth, A.; Shen, L.; Turner, J.R.; Giorgianni, F.; Desiderio, D.; Guntaka, R.; Rao, R. PKC eta regulates occludin phosphorylation and epithelial tight junction integrity. *Proc. Natl. Acad. Sci. USA* **2009**, *106*, 61–66. [CrossRef] [PubMed]
143. Raleigh, D.R.; Boe, D.M.; Yu, D.; Weber, C.R.; Marchiando, A.M.; Bradford, E.M.; Wang, Y.; Wu, L.; Schneeberger, E.E.; Shen, L.; et al. Occludin S408 phosphorylation regulates tight junction protein interactions and barrier function. *J. Cell Biol.* **2011**, *193*, 565–582. [CrossRef] [PubMed]
144. Elias, B.C.; Suzuki, T.; Seth, A.; Giorgianni, F.; Kale, G.; Shen, L.; Turner, J.R.; Naren, A.; Desiderio, D.M.; Rao, R. Phosphorylation of Tyr-398 and Tyr-402 in occludin prevents its interaction with ZO-1 and destabilizes its assembly at the tight junctions. *J. Biol. Chem.* **2009**, *284*, 1559–1569. [CrossRef]
145. Murakami, T.; Felinski, E.A.; Antonetti, D.A. Occludin phosphorylation and ubiquitination regulate tight junction trafficking and vascular endothelial growth factor-induced permeability. *J. Biol. Chem.* **2009**, *284*, 21036–21046. [CrossRef]
146. Murakami, T.; Frey, T.; Lin, C.; Antonetti, D.A. Protein kinase C beta phosphorylates occludin regulating tight junction trafficking in vascular endothelial growth factor-induced permeability in vivo. *Diabetes* **2012**, *61*, 1573–1583. [CrossRef]
147. Du, D.; Xu, F.; Yu, L.; Zhang, C.; Lu, X.; Yuan, H.; Huang, Q.; Zhang, F.; Bao, H.; Jia, L.; et al. The tight junction protein, occludin, regulates the directional migration of epithelial cells. *Dev. Cell* **2010**, *18*, 52–63. [CrossRef]
148. Bolinger, M.T.; Ramshekar, A.; Waldschmidt, H.V.; Larsen, S.D.; Bewley, M.C.; Flanagan, J.M.; Antonetti, D.A. Occludin S471 Phosphorylation Contributes to Epithelial Monolayer Maturation. *Mol. Cell Biol.* **2016**, *36*, 2051–2066. [CrossRef]

149. Cravo, A.S.; Carter, E.; Erkan, M.; Harvey, E.; Furutani-Seiki, M.; Mrsny, R. Hippo pathway elements Co-localize with Occludin: A possible sensor system in pancreatic epithelial cells. *Tissue Barriers* **2015**, *3*, e1037948. [CrossRef]
150. Runkle, E.A.; Sundstrom, J.M.; Runkle, K.B.; Liu, X.; Antonetti, D.A. Occludin localizes to centrosomes and modifies mitotic entry. *J. Biol. Chem.* **2011**, *286*, 30847–30858. [CrossRef]
151. Liu, X.; Dreffs, A.; Diaz-Coranguez, M.; Runkle, E.A.; Gardner, T.W.; Chiodo, V.A.; Hauswirth, W.W.; Antonetti, D.A. Occludin S490 Phosphorylation Regulates Vascular Endothelial Growth Factor-Induced Retinal Neovascularization. *Am. J. Pathol.* **2016**, *186*, 2486–2499. [CrossRef] [PubMed]
152. Runkle, E.A.; Mu, D. Tight junction proteins: From barrier to tumorigenesis. *Cancer Lett.* **2013**, *337*, 41–48. [CrossRef] [PubMed]
153. Harten, S.K.; Shukla, D.; Barod, R.; Hergovich, A.; Balda, M.S.; Matter, K.; Esteban, M.A.; Maxwell, P.H. Regulation of renal epithelial tight junctions by the von Hippel-Lindau tumor suppressor gene involves occludin and claudin 1 and is independent of E-cadherin. *Mol. Biol. Cell* **2009**, *20*, 1089–1101. [CrossRef] [PubMed]
154. Li, D.; Mrsny, R.J. Oncogenic Raf-1 disrupts epithelial tight junctions via downregulation of occludin. *J. Cell Biol.* **2000**, *148*, 791–800. [CrossRef]
155. Wang, Z.; Wade, P.; Mandell, K.J.; Akyildiz, A.; Parkos, C.A.; Mrsny, R.J.; Nusrat, A. Raf 1 represses expression of the tight junction protein occludin via activation of the zinc-finger transcription factor slug. *Oncogene* **2007**, *26*, 1222–1230. [CrossRef]
156. Martin, T.A.; Mansel, R.E.; Jiang, W.G. Loss of occludin leads to the progression of human breast cancer. *Int. J. Mol. Med.* **2010**, *26*, 723–734. [CrossRef]
157. Osanai, M.; Murata, M.; Nishikiori, N.; Chiba, H.; Kojima, T.; Sawada, N. Epigenetic silencing of occludin promotes tumorigenic and metastatic properties of cancer cells via modulations of unique sets of apoptosis-associated genes. *Cancer Res.* **2006**, *66*, 9125–9133. [CrossRef]
158. Wang, Z.; Mandell, K.J.; Parkos, C.A.; Mrsny, R.J.; Nusrat, A. The second loop of occludin is required for suppression of Raf1-induced tumor growth. *Oncogene* **2005**, *24*, 4412–4420. [CrossRef]
159. Jayagopal, A.; Yang, J.L.; Haselton, F.R.; Chang, M.S. Tight junction-associated signaling pathways modulate cell proliferation in uveal melanoma. *Investig. Ophthalmol. Vis. Sci.* **2011**, *52*, 588–593. [CrossRef]
160. Osanai, M.; Murata, M.; Nishikiori, N.; Chiba, H.; Kojima, T.; Sawada, N. Occludin-mediated premature senescence is a fail-safe mechanism against tumorigenesis in breast carcinoma cells. *Cancer Sci.* **2007**, *98*, 1027–1034. [CrossRef]
161. Barrios-Rodiles, M.; Brown, K.R.; Ozdamar, B.; Bose, R.; Liu, Z.; Donovan, R.S.; Shinjo, F.; Liu, Y.; Dembowy, J.; Taylor, I.W.; et al. High-throughput mapping of a dynamic signaling network in mammalian cells. *Science* **2005**, *307*, 1621–1625. [CrossRef] [PubMed]
162. Odenwald, M.A.; Choi, W.; Buckley, A.; Shashikanth, N.; Joseph, N.E.; Wang, Y.; Warren, M.H.; Buschmann, M.M.; Pavlyuk, R.; Hildebrand, J.; et al. ZO-1 interactions with F-actin and occludin direct epithelial polarization and single lumen specification in 3D culture. *J. Cell Sci.* **2017**, *130*, 243–259. [CrossRef] [PubMed]
163. Jenkinson, E.M.; Livingston, J.H.; O'Driscoll, M.C.; Desguerre, I.; Nabbout, R.; Boddaert, N.; Soares, G.; Goncalves da Rocha, M.; D'Arrigo, S.; Rice, G.I.; et al. Comprehensive molecular screening strategy of OCLN in band-like calcification with simplified gyration and polymicrogyria. *Clin. Genet.* **2018**, *93*, 228–234. [CrossRef] [PubMed]
164. Abdel-Hamid, M.S.; Abdel-Salam, G.M.H.; Issa, M.Y.; Emam, B.A.; Zaki, M.S. Band-like calcification with simplified gyration and polymicrogyria: Report of 10 new families and identification of five novel OCLN mutations. *J. Hum. Genet.* **2017**, *62*, 553–559. [CrossRef]
165. Aggarwal, S.; Bahal, A.; Dalal, A. Renal dysfunction in sibs with band like calcification with simplified gyration and polymicrogyria: Report of a new mutation and review of literature. *Eur. J. Med. Genet.* **2016**, *59*, 5–10. [CrossRef]
166. Elsaid, M.F.; Kamel, H.; Chalhoub, N.; Aziz, N.A.; Ibrahim, K.; Ben-Omran, T.; George, B.; Al-Dous, E.; Mohamoud, Y.; Malek, J.A.; et al. Whole genome sequencing identifies a novel occludin mutation in microcephaly with band-like calcification and polymicrogyria that extends the phenotypic spectrum. *Am. J. Med. Genet. A* **2014**, *164*, 1614–1617. [CrossRef]

167. O'Driscoll, M.C.; Daly, S.B.; Urquhart, J.E.; Black, G.C.; Pilz, D.T.; Brockmann, K.; McEntagart, M.; Abdel-Salam, G.; Zaki, M.; Wolf, N.I.; et al. Recessive mutations in the gene encoding the tight junction protein occludin cause band-like calcification with simplified gyration and polymicrogyria. *Am. J. Hum. Genet.* **2010**, *87*, 354–364. [CrossRef]
168. LeBlanc, M.A.; Penney, L.S.; Gaston, D.; Shi, Y.; Aberg, E.; Nightingale, M.; Jiang, H.; Gillett, R.M.; Fahiminiya, S.; Macgillivray, C.; et al. A novel rearrangement of occludin causes brain calcification and renal dysfunction. *Hum. Genet.* **2013**, *132*, 1223–1234. [CrossRef]
169. Jayaraman, D.; Bae, B.I.; Walsh, C.A. The Genetics of Primary Microcephaly. *Annu. Rev. Genom. Hum. Genet.* **2018**, *19*, 177–200. [CrossRef]
170. Glotfelty, L.G.; Zahs, A.; Iancu, C.; Shen, L.; Hecht, G.A. Microtubules are required for efficient epithelial tight junction homeostasis and restoration. *Am. J. Physiol. Cell Physiol.* **2014**, *307*, C245–C254. [CrossRef]
171. Yano, T.; Matsui, T.; Tamura, A.; Uji, M.; Tsukita, S. The association of microtubules with tight junctions is promoted by cingulin phosphorylation by AMPK. *J. Cell Biol.* **2013**, *203*, 605–614. [CrossRef] [PubMed]
172. Moss, D.K.; Bellett, G.; Carter, J.M.; Liovic, M.; Keynton, J.; Prescott, A.R.; Lane, E.B.; Mogensen, M.M. Ninein is released from the centrosome and moves bi-directionally along microtubules. *J. Cell Sci.* **2007**, *120 Pt 17*, 3064–3074. [CrossRef]
173. Shaw, R.M.; Fay, A.J.; Puthenveedu, M.A.; von Zastrow, M.; Jan, Y.N.; Jan, L.Y. Microtubule plus-end-tracking proteins target gap junctions directly from the cell interior to adherens junctions. *Cell* **2007**, *128*, 547–560. [CrossRef]
174. Meng, W.; Mushika, Y.; Ichii, T.; Takeichi, M. Anchorage of microtubule minus ends to adherens junctions regulates epithelial cell-cell contacts. *Cell* **2008**, *135*, 948–959. [CrossRef] [PubMed]
175. Meng, W.; Takeichi, M. Adherens junction: Molecular architecture and regulation. *Cold Spring Harb. Perspect. Biol.* **2009**, *1*, a002899. [CrossRef] [PubMed]
176. Horgan, C.P.; Hanscom, S.R.; Jolly, R.S.; Futter, C.E.; McCaffrey, M.W. Rab11-FIP3 links the Rab11 GTPase and cytoplasmic dynein to mediate transport to the endosomal-recycling compartment. *J. Cell Sci.* **2010**, *123 Pt 2*, 181–191. [CrossRef]
177. Fesenko, I.; Kurth, T.; Sheth, B.; Fleming, T.P.; Citi, S.; Hausen, P. Tight junction biogenesis in the early Xenopus embryo. *Mech. Dev.* **2000**, *96*, 51–65. [CrossRef]
178. Morimoto, S.; Nishimura, N.; Terai, T.; Manabe, S.; Yamamoto, Y.; Shinahara, W.; Miyake, H.; Tashiro, S.; Shimada, M.; Sasaki, T. Rab13 mediates the continuous endocytic recycling of occludin to the cell surface. *J. Biol. Chem.* **2005**, *280*, 2220–2228. [CrossRef]
179. Terai, T.; Nishimura, N.; Kanda, I.; Yasui, N.; Sasaki, T. JRAB/MICAL-L2 is a junctional Rab13-binding protein mediating the endocytic recycling of occludin. *Mol. Biol. Cell* **2006**, *17*, 2465–2475. [CrossRef]
180. Nishimura, N.; Sasaki, T. Cell-surface biotinylation to study endocytosis and recycling of occludin. *Methods Mol. Biol.* **2008**, *440*, 89–96.
181. Lapierre, L.A.; Tuma, P.L.; Navarre, J.; Goldenring, J.R.; Anderson, J.M. VAP-33 localizes to both an intracellular vesicle population and with occludin at the tight junction. *J. Cell Sci.* **1999**, *112 Pt 21*, 3723–3732.
182. Pennetta, G.; Hiesinger, P.R.; Fabian-Fine, R.; Meinertzhagen, I.A.; Bellen, H.J. Drosophila VAP-33A directs bouton formation at neuromuscular junctions in a dosage-dependent manner. *Neuron* **2002**, *35*, 291–306. [CrossRef]
183. Steed, E.; Elbediwy, A.; Vacca, B.; Dupasquier, S.; Hemkemeyer, S.A.; Suddason, T.; Costa, A.C.; Beaudry, J.B.; Zihni, C.; Gallagher, E.; et al. MarvelD3 couples tight junctions to the MEKK1-JNK pathway to regulate cell behavior and survival. *J. Cell Biol.* **2014**, *204*, 821–838. [CrossRef] [PubMed]
184. Kojima, T.; Takasawa, A.; Kyuno, D.; Ito, T.; Yamaguchi, H.; Hirata, K.; Tsujiwaki, M.; Murata, M.; Tanaka, S.; Sawada, N. Downregulation of tight junction-associated MARVEL protein marvelD3 during epithelial-mesenchymal transition in human pancreatic cancer cells. *Exp. Cell Res.* **2011**, *317*, 2288–2298. [CrossRef] [PubMed]
185. Timmann, C.; Thye, T.; Vens, M.; Evans, J.; May, J.; Ehmen, C.; Sievertsen, J.; Muntau, B.; Ruge, G.; Loag, W.; et al. Genome-wide association study indicates two novel resistance loci for severe malaria. *Nature* **2012**, *489*, 443–446. [CrossRef]

186. Vacca, B.; Sanchez-Heras, E.; Steed, E.; Balda, M.S.; Ohnuma, S.I.; Sasai, N.; Mayor, R.; Matter, K. MarvelD3 regulates the c-Jun N-terminal kinase pathway during eye development in Xenopus. *Biol. Open* **2016**, *5*, 1631–1641. [CrossRef]
187. Vacca, B.; Sanchez-Heras, E.; Steed, E.; Busson, S.L.; Balda, M.S.; Ohnuma, S.I.; Sasai, N.; Mayor, R.; Matter, K. Control of neural crest induction by MarvelD3-mediated attenuation of JNK signalling. *Sci. Rep.* **2018**, *8*, 1204. [CrossRef]

International Journal of
Molecular Sciences

Review

ZO-2 Is a Master Regulator of Gene Expression, Cell Proliferation, Cytoarchitecture, and Cell Size

Lorenza González-Mariscal *, Helios Gallego-Gutiérrez, Laura González-González and Christian Hernández-Guzmán

Center for Research and Advanced Studies (Cinvestav), Department of Physiology, Biophysics and Neuroscience, Mexico City 07360, Mexico
* Correspondence: lorenza@fisio.cinvestav.mx; Tel.: +52-55-5747-3966

Received: 17 July 2019; Accepted: 10 August 2019; Published: 24 August 2019

Abstract: ZO-2 is a cytoplasmic protein of tight junctions (TJs). Here, we describe ZO-2 involvement in the formation of the apical junctional complex during early development and in TJ biogenesis in epithelial cultured cells. ZO-2 acts as a scaffold for the polymerization of claudins at TJs and plays a unique role in the blood–testis barrier, as well as at TJs of the human liver and the inner ear. ZO-2 movement between the cytoplasm and nucleus is regulated by nuclear localization and exportation signals and post-translation modifications, while ZO-2 arrival at the cell border is triggered by activation of calcium sensing receptors and corresponding downstream signaling. Depending on its location, ZO-2 associates with junctional proteins and the actomyosin cytoskeleton or a variety of nuclear proteins, playing a role as a transcriptional repressor that leads to inhibition of cell proliferation and transformation. ZO-2 regulates cell architecture through modulation of Rho proteins and its absence induces hypertrophy due to inactivation of the Hippo pathway and activation of mTOR and S6K. The interaction of ZO-2 with viral oncoproteins and kinases and its silencing in diverse carcinomas reinforce the view of ZO-2 as a tumor regulator protein.

Keywords: tight junctions; ZO-2; cholestasis; gene transcription; hypertrophy; tumor suppressor; NLS; NES; CaSR; RhoA

---

1. Introduction

Zonula occludens proteins ZO-1, ZO-2, and ZO-3 belong to the membrane associated guanylate kinase homologue (MAGUK) protein family and are concentrated at the cytoplasmic face of tight junctions (TJs) in epithelial cells. Although these proteins share a core of common structural and functional domains and interacting partners, they also exhibit particular features that give rise to certain non-redundant functions among these proteins (for review see [[1,2]]). This review concentrates on ZO-2 and its non-canonical role in gene transcription, cell proliferation, and modulation of cell size, cytoarchitecture, and cancer.

ZO-2 is a 160 kDa protein, first identified by its co-immunoprecipitation with ZO-1 in epithelial cells [3]. ZO-2 is encoded by the *TJP2* gene located on human chromosome 9 q21.11 [4]. ZO-2 is present at TJs but in non-epithelial cells like fibroblasts that lack TJs, ZO-2 concentrates at adherens junctions (AJs) [5]. In cardiac muscle cells, the observations are contradictory. Some report the presence of ZO-2 in co-localization with ZO-1 at specialized intercellular junctions, known as fascia adherens or intercalated discs, which connect the opposing ends of cardiac muscle cells [5]. Others indicate that only ZO-1 is present at fascia adherens [6] and that ZO-2 has a diffuse cytoplasmic distribution in myocardium tissue [7].

ZO-2 is a scaffold protein, whose amino segment, containing PDZ1-3-SH3-GuK domains, binds to integral and peripheral proteins of the TJs, including occludin, claudins, JAM-A, cingulin and ZO-1, to proteins of the AJs, like α-catenin and β-catenin, and to gap junction connexins (for review see [8]).

Instead, the carboxyl segment of ZO-2, which exhibits the acidic and proline rich regions and ends with a motif that binds PDZ (PSD95, Dlg1 and ZO-1) domains, distributes when separately introduced into epithelial cells, along actin filaments [5] (Figure 1).

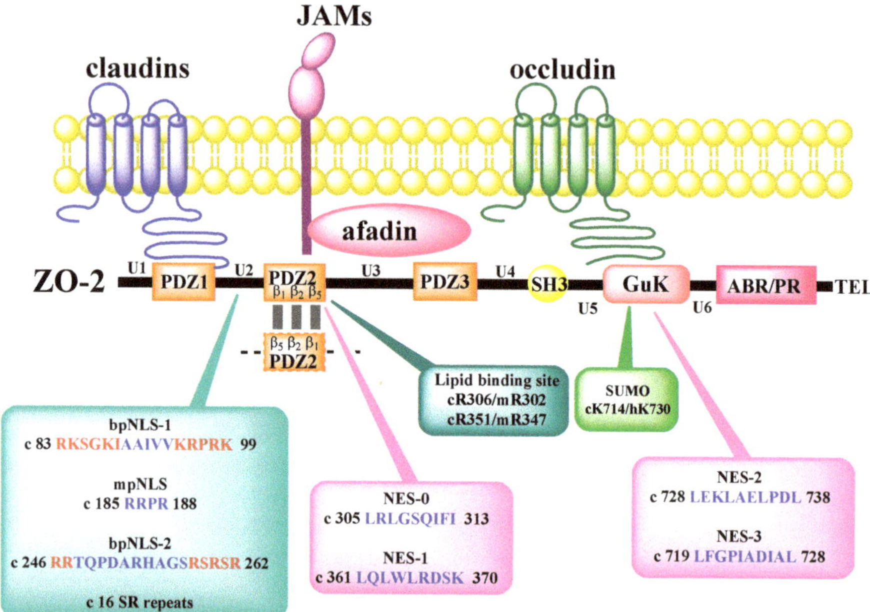

Figure 1. ZO-2 molecular organization and interactions with integral tight junction (TJ) proteins at the plasma membrane. ZO-2 domains (PDZ, SH3, and GuK), regions (U, unique; ABR, actin binding; PR, proline rich), and PDZ-binding motif (TEL) are indicated, as well as the nuclear localization signals (NLS) and exportation signals (NES), SUMOylation (SUMO) and lipid binding sites, and dimerization region. The ZO-2 sequence is identified by letters: c, canine; m, mouse; h, human. Numbers correspond to amino acids. Clusters of basic amino acids (K/R) in the bpNLS are shown in red. bp, bipartite; mp, monopartite.

PDZ1-3 modules, SH3 and GuK domains, and the acidic region of ZO-2 display a high percent of identity and similarity to those in other ZO proteins, with ZO-1 having a higher percent of both than with ZO-3 [9]. In situ ZO-2 is present as a ZO-1/ZO-2 complex, but not in ZO-2/ZO-3 or ZO-1/ZO-2/ZO-3 complexes [10]. The high conservation present between PDZ2 domains in ZO proteins allows these PDZ domains to dimerize via three-dimensional domain swapping, generating heterodimers of ZO-1-PDZ2/ZO-2-PDZ2 and ZO-1-PDZ2/ZO-3-PDZ2 domains only, as well as homodimers of ZO-1-PDZ2 and ZO-2-PDZ2 domains only [11]. Structural analysis of ZO-2 PDZ2 revealed that it has five β sheets and two α helices and that ZO-2-PDZ2 homodimers form by the interaction of three antiparallel β sheets, β1-β5′, β1′-β5, and β2-β2′, due to extensive inter-subunit hydrogen bonds and hydrophobic interactions. In addition, chemical crosslinking and dynamic laser light scatter experiments revealed that ZO-1-PDZ2 and ZO-2-PDZ2 form oligomers in solution. This oligomerization mediated by PDZ2 domains in ZO-1/ZO-2 proteins might provide a scaffold for the assembly of TJs.

Both ZO-1 and ZO-2 independently allow the polymerization of claudins and determine the site of TJ strand formation [12]. Thus, epithelial cells lack TJs when ZO-1 and ZO-2 expression is suppressed, and when either of these proteins is exogenously expressed, claudins polymerize and TJ filaments are again observed in freeze-fracture replicas. However, when a truncated segment of ZO-1 was introduced containing only the PDZ1-3 domains, it localized in the cytoplasm, not in the membrane, and the claudins did not polymerize. However, when a longer construct consisting of PDZ1-3 and SH3–GuK domains was introduced, TJ strands formed. The importance of the SH3–GuK segment is highlighted by the fact that it also plays a role in the dimerization of MAGUK proteins and binds to the AJ proteins afadin and α-catenin, allowing the recruitment of ZO-1 to the proximity of the plasma membrane. If in ZO-1 knock out (KO)/ZO-2 knock down (KD) cells the amino PDZ1-3 segment of ZO-1 was forcibly recruited to the membrane by the addition of a myristoylation signal, no claudin polymerization happened. Instead, if this construct was allowed to dimerize by the introduction of a FK506-binding protein (FKBPv) signal and a homodimerizer, claudin polymerization was abundant throughout the lateral membrane, highlighting the crucial importance of ZO-1 dimerization for claudin polymerization. Although the latter experiments were done with ZO-1, the same results were expected to occur with ZO-2, since their MAGUK domains are so similar. Taken together, these observations suggest that claudin binding through the PDZ1 domain of ZO-1 or ZO-2, as well as the SH3–GuK mediated recruitment of these proteins to the AJ and the dimerization of ZO-1 and ZO-2, allow the polymerization of claudins at the TJ region.

2. ZO-2 During Early Development

In mouse preimplantation embryos, ZO-2 assembles at blastomere junctional sites at the 16-cell stage, at the same time as cingulin but later than ZO-1α^-, rab13, and the Par-3/Par-6/aPKC complex, which reach the junctions after the compaction of 8-cell morulae. At the 16-cell stage, ZO-2 co-localizes with E-cadherin in a transient complex that contains proteins of both the TJ and the AJ. From the 32-cell stage onwards, ZO-2 and the rest of the TJ proteins, also including occludin and claudins, separate from those of the AJ and a blastocoel cavity is formed [13] (for review see [14]). Lack of ZO-2 in the trophectoderm of mouse preimplantation embryos delays the formation of the blastocoel cavity and induces an increased assembly of ZO-1, suggesting a compensatory mechanism of ZO proteins. In accordance, the double KD of ZO-1 and ZO-2 induces a more severe inhibition of blastocoel formation [13]. Early after implantation, KO of ZO-2 is lethal and mouse embryos die, showing decreased proliferation at embryonic day 6.5 (E6.5) and increased apoptosis at E7.5 [15]. However, embryonic lethality of ZO-2 KO mice was rescued by injecting ZO-2 KO embryonic stem cells into wild type blastocysts to generate viable ZO-2 chimeras, hence indicating that ZO-2 is crucial for extraembryonic tissue rather than for the development of the embryo *per se* [16].

3. ZO-2 Presence at TJs Is Crucial in the Mouse Testis and the Human Liver and Inner Ear

ZO-2 is present in the TJs of all epithelial cells and has a role on the barrier function of TJs, as ZO-2 KD cells achieve a much lower peak value of transepithelial electrical resistance (TER) than parental cells [17,18]. Therefore, the observation that the absence of ZO-2 causes severe barrier impairment in mouse testes and human liver and inner ear tissue suggests that compensatory mechanisms, like the over-expression of ZO-1, are not effective in these tissues, where ZO-2 plays a critical and non-redundant role (Table 1).

Table 1. ZO-2 mutations and consequences.

Species	Disease	Alteration	Location within ZO-2	Features of Disease	Reference
Human	ADNSHL	A112T	PDZ1		[19]
		T1188A	TEL		
		G694E	GuK		[20]
		Genomic duplication at chromosome 9p13.3-q21.13			[4]
	Familial hypercholanemia	V48A	PDZ1	Patients with itching, elevated bile acid concentration and fat malabsorption. TJP2 mutation combined with bile acid Coenzyme A (BAAT) mutation	[21]
	Intrahepatic cholestasis	I875T	GuK	71-year-old patient	[22]
		R322W	PDZ2	Disease since early infancy	
		A256T	ZO-2 deficiency		[23]
		S296A			
		Y261S			
		A454G			
		A664S			
		Q318G			
		A632fs			
		S1136A			
		T880S			[24]
		N814Q		Disease since early infancy that leads to hepatocellular carcinoma	[25]
		A273fs			[26]
		Exon 18 deletion			[27]
		T62M	PDZ1	Intrahepatic cholestasis of pregnancy (ICP)	[28]
		T626S	SH3		
Mouse		ZO-2 KO		Reduced male fertility	[16]

ADNSHL: Autosomal dominant non-syndromic hearing loss.

3.1. ZO-2 and the Blood–Testis Barrier

TJs form the blood–testis barrier (BTB) at the lowermost portion of the lateral membrane of Sertoli cells. This barrier divides the paracellular space in two compartments: The basal compartment, which contains diploid spermatogonia, and the adluminal compartment, which contains differentiating spermatocytes and haploid spermatids. The BTB, constituted by occludin, claudin-3 and -11, ZO-1, and ZO-2, is critical for the maintenance of a proper microenvironment for germinal cells and to prevent autoimmunity (for review see [29]). Male ZO-2 chimera mice generated from wild type blastocysts, in which ZO-2 KO embryonic stem cells substituted the normal stem cells, showed reduced fertility, smaller testes, degeneration of seminiferous tubules, and permeable BTB, even when the other TJ proteins were present at the BTB. This suggests that ZO-2 plays a unique and critical role in the BTB [16].

Moreover, the organophosphate pesticide methamidophos (MET), which is widely used in agriculture in developing countries and is known to cause adverse effects in human male reproductive function, opens the BTB of mice and perturbs spermatogenesis. These changes correlate with MET formation of covalent bonds with ZO-2 in serine, tyrosine, and lysine residues. MET bonds formed in ZO-2 are proposed to interfere with ZO-2 degradation and TJ sealing, since cultured Madin-Darby canine kidney (MDCK) cells transfected with ZO-2 mutated at MET target sites developed lower values of TER [30].

3.2. ZO-2 in the Liver

In the liver, TJs present in hepatocytes and cholangiocytes in intrahepatic canaliculi prevent leakage of biliary components through the paracellular space and into the liver parenchyma. Familial hypercholanemia is a rare disease characterized by elevated serum bile acid concentration, pruritus, and fat malabsorption. In Amish individuals with this condition, the missense mutation V48A in the first PDZ domain of TJP2 was identified. This mutation reduces PDZ1 domain stability and binding to the C-terminal domain peptides of claudin-1, -2, -3, -5, and -7. In some individuals, this mutation is associated with an additional missense mutation on the gene for bile acid Coenzyme A (BAAT), a bile-conjugating enzyme. In a patient homozygous for the *TJP2* mutation and heterozygous with respect to the *BAAT* mutation, TJ depth viewed by transmission electron microscopy was greater than in control individuals. Considering that greater TJ depth was observed in liver disease associated with greater TJ permeability, it was concluded that the missense mutation in *TJP2* generated leakier TJs in the liver, which affected bile acid localization [21].

In children with familial intrahepatic cholestasis, a condition where bile does not flow through the intrahepatic ducts of the liver, homozygous protein truncating mutations in *TJP2* leading to a complete absence of the protein were detected. In this case, the absence of ZO-2 is accompanied by a failure of localization of claudin-1, but not of claudin-2, to the canalicular membrane [31]. Another case of compound heterozygous mutations in *TJP2* predicted to abolish TJ protein translation was observed in a young child with familial intrahepatic cholestasis [24]. Additional cases of infants with progressive intrahepatic cholestasis that led to chronic cholestatic hepatitis with cirrhosis and hepatocellular carcinoma were also reported. In these cases, homozygous and heterozygous *TJP2* mutations were present, ZO-2 expression was absent, and claudin-1 expression was decreased [25–27].

Intrahepatic cholestasis of pregnancy, characterized by pruritus, hepatic impairment, and elevated serum bile acids, can lead to spontaneous preterm delivery and stillbirth. In women with this condition, heterozygous mutations in *TJP2* were discovered. These novel T62M and T626S mutations were respectively located in the PDZ1 and the SH3 domains of ZO-2; and no patient harbored more than one mutation [28]. Since only three patients were identified with these mutations, larger cohorts of pregnant patients should be studied to understand why these variants give rise to leaky TJs in the liver during pregnancy (Table 1).

3.3. ZO-2 in the Inner Ear

In the mammalian ear, the hair cells of the cochlea convert the vibrations induced by sound into electrical signals by K^+-mediated depolarization. A K^+-rich fluid, named endolymph and produced by the stria vascularis, bathes the apical surface of hair cells. The basolateral surface of a hair cell is surrounded by the perilymph, which has a low K^+ concentration. To prevent mixing endolymph and perilymph and to maintain a high resting potential in the endolymph (≈90 mV), the cells surrounding the liquid are sealed with TJs, which are very tight in hair cells and supporting cells in the reticular lamina of the organ of Corti and in some cells of the stria vascularis. Accordingly, hearing loss associated with the degeneration of cochlear hair cells was observed in KO mice for claudin-14 [32], occludin [33], tricellulin [34], and ILDR-1/angulin-2 [35], and in a mice carrying a missense mutation of claudin-9 [36]. Claudin-11-null mice lacked TJs in the basal cells of the stria vascularis and their deafness was associated with a low endocochlear potential [37,38].

Autosomal dominant nonsyndromic hearing loss caused by mutations in the *TJP2* gene was observed in a Chinese family involving the missense mutation 2081G>A (G694E) [20], located within the GuK domain of ZO-2, and in the Korean population due to two pathogenic variations: 334G>A (A112T), which is present in the PDZ1 domain of ZO-2, and 3562A>G (T1188A), which is situated within the carboxyl terminal PDZ binding motif TEL of ZO-2 [19]. Since the latter, as well as the PDZ1 and GuK domains of ZO-2, are protein–protein binding regions, it is expected that mutations at these sites would compromise TJ stability and the interaction of ZO-2 with other TJ proteins and signaling molecules (Table 1).

4. Subcellular Localization and Traffic of ZO-2

ZO-2 can be found in the cytoplasm, nuclei, and TJs. In unfertilized mouse eggs, ZO-2 distributes diffusely in the cytoplasm. In the zygote, ZO-2 concentrates in specks around both pronuclei. After the first cell division these specks appear in the cytoplasm close to the plane of cleavage. From late 2-cell embryos until 8-cell compaction, ZO-2 concentrates at the nuclei of the blastomeres, but is excluded from the nucleoli. After 8-cell compaction, nuclear ZO-2 diminishes in intensity and instead concentrates at junctional sites from the 16-cell stage onward [13].

In confluent epithelial cells, ZO-2 concentrates at the TJs, whereas in sparse cultures, ZO-2 is present both at the nucleus and TJs [39], thus revealing that the localization of ZO-2 is regulated by the degree of cell–cell contact (Figure 2A). Nuclear recruitment of ZO-2 can also be induced at sites of monolayer wounding [39] or with environmental stresses, such as heat shock (42 °C) and chemical insult ($CdCl_2$) [40].

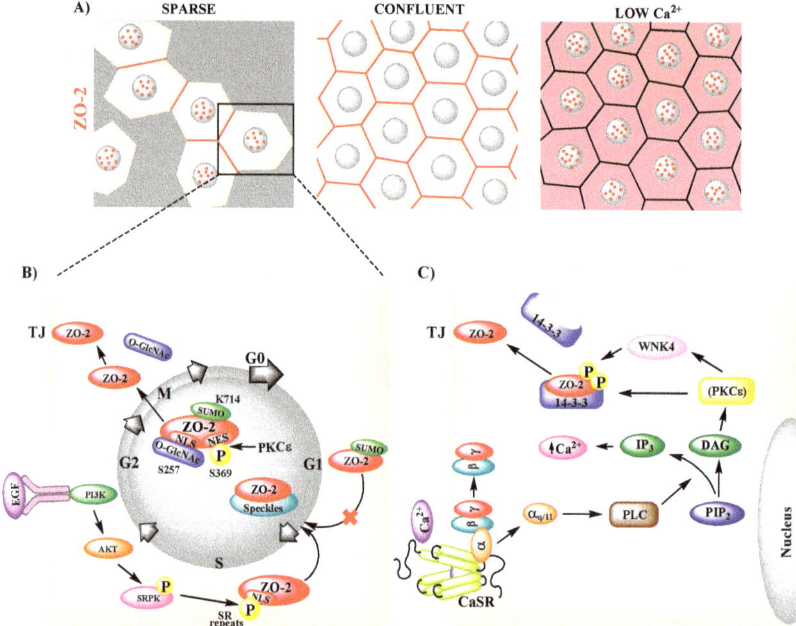

Figure 2. Intracellular movement of ZO-2. (**A**) In sparse cultures, ZO-2 is present in nuclear speckles (red dots) and TJs (red cell borders), whereas in confluent cells ZO-2 concentrates at TJs (red cell borders). In cells cultured with low Ca^{2+} (1–5 μM) ZO-2 distributes diffusely in the cytoplasm (pink cytoplasmic staining) and concentrates in nuclear speckles (red dots). (**B**) In sparse cultures, epidermal growth factor (EGF) induces AKT activation, which triggers ZO-2 phosphorylation of serine-arginine (SR) repeats by SR protein kinase (SRPK). Nuclear localization signals (NLSs) of ZO-2 and phosphorylated SR repeats induce the arrival of ZO-2 to nuclear speckles at late G1. ZO-2 exit from the nucleus requires the phosphorylation of Ser369 within a nuclear exportation signal (NES) by PKCε and the inactivation of bpNLS-2 by *O*-GlcNAc of Ser257, and is facilitated by SUMOylation of Lys714. (**C**) Extracellular Ca^{2+} activates the Ca^{2+}-sensing receptor (CaSR) present in the plasma membrane that signals through the $G\alpha_{q/11}$ subunit, activating PLC, which hydrolyzes PIP2 into DAG and IP3. DAG activates PKC ε, leading to ZO-2 phosphorylation by WNK4. This phosphorylation induces the disassembly of ZO-2 from 14-3-3 proteins in the cytoplasm and triggers the relocation of ZO-2 to the plasma membrane.

ZO-2 stability augments with TJ establishment and maturity. Thus, the half-life of ZO-2 in epithelial MDCK cells goes from 7 h in cells cultured in media with low Ca^{2+} (1–5 μM) that cannot form TJs, to 19.7 h in cultures with 1.8 mM Ca^{2+} [41] and from 8.6 h in sparse cultures to 19.1 h in confluent cultures [3].

In MDCK cells, treatment with L-mimosine, which selectively blocks cell cycle progression in the late G1 phase, revealed that ZO-2 enters the nucleus at the later stages of G1, while treatment with nocodazole, which arrests the cell cycle at the beginning of mitosis, showed that ZO-2 exits the nucleus during mitosis [42]. These observations explain why the nucleus is devoid of ZO-2 in confluent quiescent cells, while proliferating cells have ZO-2 at the nucleus (Figure 2B).

Nuclear importation of ZO-2 relies on nuclear localization signals (NLSs) located within the U2 region of the molecule, between PDZ1 and PDZ2 [43]. The first bipartite (bp) NLS of ZO-2 is conserved in several species, including human, dog, mouse, and chicken, and is functional in a reporter protein nuclear importation assay. The second bpNLS of the molecule, present in dog and mouse, is not active in this assay [44]. The functionality of the second bpNLS is hampered by the phosphorylation of three serine residues (S257, S259, and S261) present within the carboxyl segment of the signal, which neutralizes the positive charges of three contiguous arginine residues (Figure 1).

In addition, the U2 region of ZO-2 has 16, 14, 19, and 16 serine–arginine (SR) motifs in dog, human, mouse, and chicken, respectively. SR motifs are also present in transcription factors and pre-mRNA splicing factors and constitute a signal necessary and sufficient to target proteins for nuclear speckles. ZO-2 also has a docking motif for SR protein kinase 1 (SRPK1) in its U2 domain that specifically phosphorylates serine residues present in SR motifs. Accordingly, ZO-2 associates to SRPK1 and the serine phosphorylation of ZO-2 by SRPK1, which is triggered by the presence of EGF and the subsequent activation of PI3K and AKT, induces the translocation of ZO-2 into the nucleus and its distribution in speckles [44] (Figure 2B). The binding of the PDZ2 domain of ZO-2 to phosphatidylinositol 4,5-bisphosphate (PIP2) regulates ZO-2 recruitment to nuclear speckles, and ZO-2 silencing disrupts speckle morphology, suggesting that ZO-2 plays an active role in the formation and stabilization of nuclear speckles. Residues R302 and R347 in the PDZ2 of mouse ZO-2 are critical for PtdIns binding [45].

ZO-2 exits the nucleus as the monolayer acquires confluence through a process sensitive to leptomycin B, a drug that blocks the interaction between nuclear exportation signals (NES) and exportin1/CRM1 [39]. NES are leucine-rich sequences containing a characteristic spacing of leucine or other hydrophobic residues. Based on the archetypal NES consensus sequence of HIV-1, Rev, PKI-α, IκBα, and zyxin, four NES in ZO-2 canine sequences were identified, three in each of mouse and chicken, and none in human ZO-2. In canine ZO-2, two NES are located at the PDZ-2 domain and the other two at the GuK module (Figure 1). With reporter protein nuclear export assays, the functionality of three of these signals was confirmed, while that of NES-1 proved to be effective only after a phosphomimetic mutation was introduced in a serine residue (Ser369) within the signal that constituted a putative protein kinase C (PKC) phosphorylation site [46] (Figure 2B). Although each of the four NES present in ZO-2 can export a reporter protein like ovalbumin, in full-length ZO-2 the presence of the four intact NES is required to avoid nuclear accumulation of the protein [46]. Since Ser369 is phosphorylated by nPKCε, this phosphorylation constitutes a critical step for the exit of ZO-2 from the nucleus [47].

Two additional post-translational modifications facilitate the nuclear exportation of ZO-2. One is the *O*-linked β-*N*-acetylglucosamination of ZO-2 at Ser257. Since this residue is located within the second bpNLS of ZO-2, it is proposed that this modification inactivates the NLS and consequently facilitates the departure of ZO-2 from the nucleus [44]. The other modification is the SUMOylation at Lys730 of human ZO-2, as mutation of this site to arginine results in prolonged nuclear localization of ZO-2 in nuclear recruitment assays. In addition, a construct mimicking constitutive SUMOylation of ZO-2, preferentially localized in the cytoplasm, suggesting that SUMOylation in ZO-2 occurs in the nucleus and is a signal that inhibits entry to the nucleus and instead facilitates nuclear exportation [48] (Figure 2B).

Newly synthesized ZO-2 first appears at the nucleus of sparse cells and as the monolayer becomes confluent, it relocates from the nucleus to the plasma membrane [39]. Thus, at the earliest times after transfection, around 80% of ZO-2 transfected cells showed ZO-2 at the nucleus; this value diminished to 20% after 18 h. With the help of a nuclear microinjection assay, in which antibodies against ZO-2 were introduced into the nucleus, it was observed that newly synthesized and endogenous ZO-2 made a stopover at the nucleus before reaching the plasma membrane [47]. The nucleus seemed to serve as a cellular reservoir of ZO-2, as the amount of ZO-2 in this compartment was significantly higher than that which eventually reached the plasma membrane. In the latter, the amount of ZO-2 stayed relatively constant while the content of ZO-2 at the nucleus suffered drastic variations with time [47].

In contrast to sparse cultures, in confluent monolayers, blocking the nuclear exportation of ZO-2 with a PKCε permeable inhibitor peptide did not trigger the nuclear accumulation of ZO-2, indicating that in confluent cultures, ZO-2 travels directly to the plasma membrane without stopping at the nucleus [44].

In epithelial monolayers cultured in media with a low concentration of Ca^{2+}, ZO-2 appeared to be diffusely distributed in the cytoplasm and accumulated at the nucleus [39] (Figure 2A). If these cultures were transferred to normal Ca^{2+}-containing medium, TJs assembled and TER developed [49]. However, if the calcium-switch was done in ZO-2 silenced monolayers, a significant delay occurred in the arrival to the cell borders of ZO-1, occludin, and E-cadherin [17]. Integration of ZO-2 to the TJ is triggered by a signaling cascade that starts at the plasma membrane with the activation of the Ca^{2+}-sensing receptor (CaSR) [41], a G-protein coupled receptor [50]. CaSR signals through the $G\alpha_{q/11}$ subunit, which activates phospholipase C (PLC). PLC hydrolyzes PIP2 to diacyl glycerol (DAG) and inositol triphosphate (IP3). The latter induces the release of Ca^{2+} from the endoplasmic reticulum, while DAG leads to the activation of nPKCε. This kinase phosphorylates the kinase WNK4 that binds and phosphorylates ZO-2 in serine residues, triggering its disassembly from 14-3-3 and its relocation to TJs [41].

5. Interaction of ZO-2 with Nuclear Proteins

ZO-2 is present in the nuclear matrix interacting with lamin B1 and actin [51]. Nuclear ZO-2 concentrates at speckles, which are active places that coordinate transcription and splicing. Accordingly, nuclear ZO-2 co-localizes with the essential pre-mRNA splicing protein SC-35 [39], the speckle protein ZASP [52], and the C-terminal portion of scaffold attachment factor-B (SAF-B) [40]. The latter is involved in transcriptional repression (for review on SAFB1/SAFB2 see [53]), associates with E-boxes in DNA [54,55], recruits histone deacetylases allowing histones to wrap the DNA more tightly [55,56], and binds and inhibits SRPK1 [57] (Figure 3).

In addition, a proteomic analysis revealed the presence of ZO-2 in a complex derived from an immunoprecipitation (IP) of SRm160 from a HeLa cell nuclear extract. SRm160 is a serine/arginine repeat nuclear matrix protein that promotes splicing and stimulates 3′-end processing [58].

In HEK293 cells, liquid chromatography-tandem mass spectometry (LC-MS/MS) revealed the presence of ZO-2 in SIRT7 immunoisolates [59]. SIRT7 is a nucleoli enzyme that associates with RNA Pol I machinery and is required for rDNA transcription. This is noteworthy, since ZO-2 does not co-localize at the nucleoli with nucleolin [39]. LC-MS/MS has also shown the presence of ZO-2 in a soluble protein complex with: (a) Ubinuclein [60], a nuclear protein that enhances the stability and function of human histone H3.3 chaperone complex HUCA (HIRA/UBN1/CABIN1/ASF1a) [61] and that in confluent epithelial cells interacts with cingulin and ZO-1 at TJs [62]; (b) WDR36/UTP21 [60], a protein required for the biogenesis of 18S ribosomal RNA [63]; (c) ZMYM3 [60], a chromatin-interacting protein that promotes DNA repair by homologous recombination [64]; (d) the estrogen-related receptor α (ERRα) [60], a nuclear receptor that interacts with estrogen, activates reporters containing steroidogenesis factor 1 response elements [65], and modulates the activity of estrogen receptor α (ERα) in breast, uterus, and bone (for review see [65]); (e) MCM3, a protein involved in the initiation of eukaryotic genome replication [66]; and f) PUS7 [66], a pseudouridine synthase that enhances RNA stability [67].

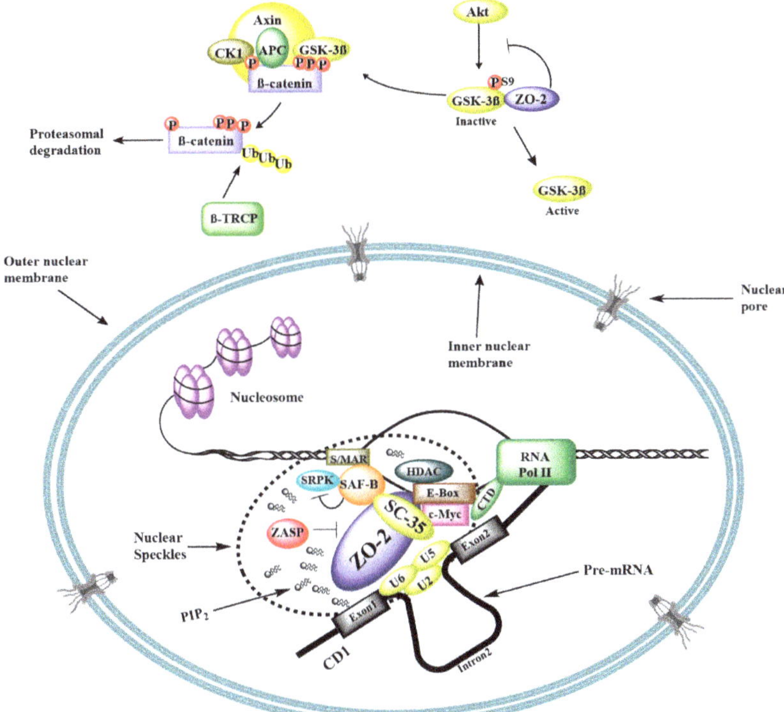

Figure 3. Schematic diagram of ZO-2 as a scaffold for transcription and splicing. ZO-2 expression blocks GSK-3β inhibitory phosphorylation at Ser9. GSK-3β phosphorylates β-catenin and targets it for ubiquitination and proteasomal degradation. Consequently, β-catenin entry into the nucleus and transcriptional activity is inhibited. At the nucleus, ZO-2 localizes in speckles associated to PIP2 and ZASP and inhibits transcription of CD1 through its interaction with an E-box in the promoter region, mediated by c-Myc transcription factor. In addition, ZO-2 is proposed to participate in mRNA splicing due to its association with the essential splicing protein SC-35 and the scaffold attachment factor SAF-B. S/MAR, scaffold/matrix attachment regiŏn; P, phosphate; Ub, ubiquitin; U2, U5, and U6, splicing ribonucleoproteins; PIP2, phosphatidylinositol (4,5)-bisphosphate.

Altogether, these observations suggest that nuclear ZO-2 forms part of a complex that couples chromatin structure to transcription and RNA processing.

6. ZO-2 as Repressor of Gene Transcription

Pull-down and gel shift assays revealed the interaction of ZO-2 with Fos, Jun, and C/EBP transcription factors. In addition, IP and immunolocalization experiments showed that ZO-2 associates with these transcription factors, both at the nucleus and the TJ region [68]. The role of ZO-2 in gene transcription was first studied in reporter gene assays with promoters regulated by AP-1 sites, revealing that ZO-2 downregulates these promoters in a dose-dependent manner [68]. Then, in reporter gene assays, ZO-2 was found to downregulate cyclin D1 (CD1) transcription in a dose-dependent manner, acting upon a region in the promoter that harbored an E-box. IP and chromatin IP (ChIP) assays further revealed that the transcription factor Myc and the histone deacetylase (HDAC) 1 bound to this E-box and formed a complex with ZO-2 to repress CD1 transcription. The over-expression of Myc increased ZO-2-mediated inhibition of CD1 transcription [55], whereas the speckle protein ZASP blocked this transcriptional inhibition [52] (Figure 3).

ZO-2 also represses gene transcription regulated by the Wnt signaling pathway, which leads to cell proliferation and un-differentiation. Reverse transcription quantitative PCR (RT-qPCR) showed that the endogenous expression of axin-2, a Wnt target gene, was repressed upon ZO-2 overexpression [42]. Moreover, with IPs of transfected and endogenous proteins, as well as proximity ligation assays, ZO-2 was found to bind to glycogen synthase kinase 3β (GSK-3β) [48], while ZO-2 overexpression decreased the amount of inactive GSK-3β phosphorylated at Ser9 [42] (Figure 3). This enzyme, which inhibits epithelial to mesenchymal transformation (EMT), phosphorylates slug transcription factor, inducing its subsequent degradation. Accordingly, an increase in GSK-3β phosphorylated at Ser9 was found in tumor cells associated with slug expression. This factor, which represses E-cadherin transcription, belongs to the snail family, and LC-MS/MS revealed that ZO-2 was present in slug IPs [69]. Likewise, affinity purification of epitope tagged proteins followed by MS (AP-MS) identified ZO-2 as an interacting partner of smug, another transcription factor of the snail family [70].

GSK-3β also phosphorylates β-catenin at Ser33, Ser37, and Thr41, leading to its ubiquitination and subsequent proteasomal degradation (Figure 3). Instead, when β-catenin is not degraded, it accumulates in the cytoplasm and is translocated into the nucleus where it forms a complex with T cell factor (TCF)/lymphoid enhancer family (LEF) transcription factor, which induces the transcription of a variety of target genes involved in cell cycle progression and EMT (for reviews see [71,72]). The effect of ZO-2 on the transcription of Wnt target genes was analyzed in two different Wnt reporter constructs, where luciferase expression was regulated either by artificial TCF/LEF binding sites or by the promoter of Siamois, a Wnt target gene. In both cases, dose-dependent repression was exerted by ZO-2. Moreover, the transcriptional activity regulated by artificial TCF/LEF binding sites increased in ZO-2 KD cells in comparison to parental cells, indicating that the absence of ZO-2 facilitated TCF-regulated transcriptional activity [73]. In agreement, in colon carcinoma cells, vitamin D3 promotes ZO-2 expression and inhibits the expression of a variety of β-catenin-TCF4 responsive genes [74].

The capacity of ZO-2 to repress gene transcription regulated by the Wnt pathway was studied in mice *in vivo*, where ZO-2 overexpression in the glomerulus was achieved by hydrodynamic transfection. Treatment with adriamycin, an anthracycline antibiotic, produced a nephrotic syndrome triggered by activation of Wnt signaling, whereas ZO-2 overexpression in these animals controlled proteinuria and podocyte effacement and favored urea and creatinine clearance through a process that inhibited the expression of snail and increased the amount of nephrin and phosphorylated β-catenin in the glomeruli. This indicates that ZO-2 protects against podocyte injury induced by aberrant Wnt signaling activation [75].

The ability of ZO-2 to inhibit gene transcription regulated by the Wnt pathway is controlled by the intracellular localization of ZO-2. Thus, this repressive activity was not observed when a constitutively SUMOylated ZO-2 construct that could not enter the nucleus was employed. Instead, using a ZO-2 SUMOylation mutant (K730R) that exhibited a strong delayed nuclear export increased ZO-2 repressive activity on promoters regulated by TCF/LEF binding. Accordingly, a ZO-2 construct with forced nuclear localization due to fused NLS completely inhibited transcriptional activity mediated by TCF/LEF, even in dense cultures [48].

To address if ZO-2 repression of this transcriptional activity requires the presence of β-catenin, a LEF chimera lacking the β-catenin binding site and instead containing the Herpes virus VP16 transactivation domain (ΔNLEF-1-VP16) was employed. ZO-2 exhibited transcriptional repression with this LEF chimera in reporter assays done either with a synthetic promoter or with the Siamois promoter, thus demonstrating that ZO-2 inhibits TCF/LEF mediated gene transcription in a β-catenin-independent manner [48]. Nevertheless, ZO-2 does form a complex with β-catenin. This was demonstrated in co-IP and proximity ligation assays employing transfected and endogenous proteins. Moreover, pull-down assays with recombinant ZO-2 and β-catenin proteins revealed that ZO-2/β-catenin interaction was direct. These observations hence suggest that ZO-2 repressed TCF/LEF mediated transcriptional activity in both a β-catenin-independent and -dependent manner [48].

Yes-associated protein (YAP), the main downstream effector of the mammalian Hippo pathway that regulates cell size, proliferation, and differentiation (for review see [76]), docks the ubiquitin ligase β-TrCP to the β-catenin destruction complex in cells where the Wnt signaling pathway is inactive, while Wnt ligands dislodge YAP from this complex and trigger the nuclear accumulation of YAP [77]. YAP Ser127 phosphorylation by the large tumor suppressor (LATS) kinase of the Hippo pathway creates a 14-3-3 binding site, which allows the cytoplasmic retention of YAP. However, YAP phosphorylation at Ser381 by LATS primes YAP for a subsequent CK1 phosphorylation that triggers YAP ubiquitination and proteasomal degradation [78]. Instead, the inactivation of the Hippo pathway allows the entry of the un-phosphorylated YAP into the nucleus, where YAP acts as a cofactor for transcriptional enhanced associate domain (TEAD) transcription factors that promote EMT (for review see [76]). MDCK ZO-2 KD cells exhibit a nuclear accumulation of YAP even when grown at high density and display an increased transcriptional activity of a reporter driven by artificial TEAD-binding sites or by the promoter of connective tissue growth factor (CTGF), a YAP target gene, while transfection of ZO-2 reduces promoter activity in both parental and ZO-2 KD cells. Moreover, the absence of ZO-2 increased the amount of CTGF mRNA determined by RT-qPCR, while the overexpression of ZO-2 reduced CTGF mRNA in parental cells. Altogether, these results indicate that ZO-2 repressed the transcriptional activity mediated by TEAD [73].

The involvement of ZO-2 in gene transcription was also studied in human vascular smooth muscle cells (VSMC). ZO-2 was barely observed in isolated control coronary arteries, but was highly expressed after dilation with a balloon catheter in the tunica media, which is composed of SMC, and in the intima, which is made up of endothelial cells. In contrast, the transcription factor stat1 was inversely regulated upon dilation injury. In primary cultures of human VSMC, ZO-2 silencing increased the nuclear and cytosolic expression of stat1 and its mRNA level. When these ZO-2-silenced VSMC cells were transfected with a luciferase reporter plasmid under the control of an interferon stimulated response element (ISRE), to which stat binds, an induction of transcription was observed in comparison to control VSMC. These results suggest that ZO-2 in VSMC repressed stat1 transcriptional activity. The latter in VSMC inhibited the proliferation rate, which was important because an increased proliferation of medial VSMC is one of the hallmarks of negative vascular remodeling [79].

7. ZO-2 and Cell Proliferation

ZO-2 over-expression inhibits cell proliferation blocking, the progression of cells through the G1/S boundary of the cell cycle. This effect is due to the amino terminal segment of ZO-2, which contains the three PDZ domains of ZO-2 [42]. ZO-2 has also been found to block cell proliferation induced by YAP [80].

In MDCK cells, ZO-2 overexpression reduces the level of CD1 mRNA [55], but this effect is not observed when the cultures are synchronized. In these cultures, the level of CD1 mRNA is higher upon arrest after incubation with 0.1% serum than in proliferating cells cultured with 10% serum [42]. ZO-2 also inhibits cell proliferation by augmenting CD1 degradation in the proteasome induced by GSK-3β phosphorylation. Thus, at the S phase, GSK-3β phosphorylates CD1 at Thr286, inducing its exit from the nucleus, ubiquitination, and subsequent degradation in the proteasome [81]. ZO-2 overexpression diminishes GSK-3β inhibitory phosphorylation at Ser-9 and triggers CD1 degradation [42]. Accordingly, LiCl, which inhibits GSK-3β [82,83], blocks the decrease in CD1 protein content induced by ZO-2 overexpression [42].

ZO-2 KD in preimplantation mouse embryos [13] and in monolayers of MDCK cells [73] has no effect on the number of blastocyst cells or proliferating cells, respectively. Instead, ZO-2 silencing in the spontaneously immortalized kidney cell line $mCCD_{d1}$, which displays highly differentiated properties of collecting duct principal cells, decreases cyclin D1 content and blocks cell cycle progression at G0/G1 [84]. This is noteworthy, because cells of the collecting duct show intense proliferation during embryogenesis but very low turnover in the adult kidney, suggesting that ZO-2 might play a regulatory role in this process.

In VSMC, ZO-2 silencing also inhibits cell proliferation, albeit through increased expression of stat1 transcription factor and activation of stat1 specific genes [79].

ZO-2 fused to a strong NLS of SV40, concentrates at the nucleus, increases cell proliferation, and triggers the expression of the M2-type isoenzyme of pyruvate kinase (PKM2) [85]. This enzyme, whose plasma concentration in cancer patients correlates with staging (for review see [86]), is involved in cancerous cell metabolism in a process known as the Warburg effect or aerobic glycolysis, where cancer cells take up high amounts of glucose and produce lactate even in the presence of oxygen. PKM2 catalyzes the last step of glycolysis, the conversion of phosphoenolpyruvate to pyruvate, with the concomitant production of ATP. Generation of ATP by PKM2, in contrast to mitochondrial respiration, is independent of oxygen and hence allows tumor cells to grow in hypoxic conditions. These results therefore suggest that the concentration of ZO-2 in the nuclei through the overexpression of PKM2 might confer a metabolic benefit to cancer cells.

8. ZO-2 in Apoptosis and Cell Degeneration

Apoptosis is a form of programmed cell death present in multicellular organisms. In contrast to necrosis, which is a traumatic cell death after acute cellular injury, apoptosis is a highly regulated and controlled process. Some factors, like Fas receptors and caspases, which are a family of cysteinyl-aspartate proteases that signal through a cascade, promote apoptosis, while proteins of the BCL-2 family inhibit apoptosis.

In human breast epithelial cells, upon apoptosis induction by staurosporine or anti-CD95/Fas antibody treatment, TJs are disrupted and ZO-2 is cleaved by caspases [87]. Caspase-mediated cell death requires cleavage of key proteins involved in cell survival, including PARP-1. This is an enzyme activated by DNA breaks that polyADP-ribosylates proteins. In response to mild DNA damage, PARylation favors DNA repair and cell survival, whereas severe DNA damage leads to apoptosis or necrosis by amplified PARP-1 activity, resulting in high donor nicotinamide adenine dinucleotide (NAD^+) consumption and depletion of ATP pools (for review see [88]). In TK6 lymphoblastoid cells, after induction of apoptosis with hydroquinone, a metabolite of benzene involved in benzene-induced leukemia, ZO-2 is polyADP-ribosylated via interaction with PARP-1 in the nucleus and its expression is upregulated [89].

ZO-2 overexpression does not augment the percentage of early or late apoptotic epithelial cells in culture [42]. Instead, a tandem inverted genomic duplication of the wild type *TJP2* gene in humans leads to progressive non-syndromic hearing loss, accompanied by a change in the expression of BCL2 family genes that favors apoptosis. Although this change in gene expression was studied in the lymphoblast cells of affected individuals, it suggests that ZO-2 overexpression leads to increased susceptibility to apoptosis in inner ear cells and that this mechanism triggers adult-onset hearing loss in this kindred [4].

Ubiquitination regulates protein stability. In particular, lysine 48-linked polyubiquitination chains target proteins for degradation in the proteasome. Protein ubiquitination is counter-regulated by deubiquitinating enzymes, among which the ubiquitin specific proteases (USP) constitute the largest family. Although USP53 has been suggested to be catalytically inactive due to the absence of a critical hystidine residue, mambo mice carrying a point mutation in the catalytic domain of USP53 exhibit rapid progressive hearing loss. USP53 is broadly expressed in the inner ear, where it interacts with ZO-1 and ZO-2. In mambo mice, biotin tracer assays indicated that TJs of the stria vascularis were not affected, while hair cell degeneration preceded the endocochlear potential decline. Moreover, the outer hair cells of these mice evaded degeneration in organ culture, indicating that the unfavorable extracellular conditions that these cells encountered *in vivo* in the mambo mice promoted their degeneration. These observations hence suggest that the inactivation of USP53 promoted the degradation of ZO-1/ZO-2, triggering a disturbance of cochlear homeostasis that resulted in hair cell degeneration and apoptosis [90]. Moreover, in humans, cholestasis and hearing loss was found to correlate with a truncating variant in USP53 [91].

YAP, through its PDZ binding motif, interacts with the first PDZ domain of ZO-2; this interaction facilitates the nuclear localization of YAP. Nuclear YAP exerts a pro-apoptotic function determined by the appearance of a 89 kDa PARP-1 fragment characteristic of caspase-3 and -7 cleavage. The amount of this PARP-1 fragment decreases when ZO-2 expression is silenced, suggesting that ZO-2 facilitates the apoptotic function of nuclear YAP [80,92].

9. ZO-2 Associates with the Cortical Actomyosin Ring and Regulates Cytoarchitecture

In epithelial cells, F-actin associated with the apical junctional complex appears as a thin band that surrounds the apical margins of the cells, also known as the cortical ring of actin. Myosin II staining follows that of actin, but is more punctate and more concentrated at the tricellular junctions [93].

ZO-2 directly interacts with actin, but is not a F-actin cross-linking protein [10]. ZO-2 also binds to 4.1 [94], a cytoskeleton protein that stabilizes spectrin–actin interactions and associates with ankyrin (for review see [95]).

ZO-2 at the cell border is associated with the integral TJ protein JAM-A and the submembranous protein afadin. PDZ-GEF1 binds to this complex and stimulates the activity of the GTPase Rap2c, which inhibits RhoA. Consequently, in JAM-A-deficient cells, the activity of RhoA and the phosphorylation of myosin light chains are enhanced [96]. Moreover, the lack of this complex, which inhibits RhoA signaling, explains the engrossment of the perijunctional ring of actomyosin and punctate distribution of myosin II, which are observed in MDCK cells without ZO-1 and ZO-2 [93].

However, it is noteworthy that, in ZO-1KO/ZO-2KD mammary Eph4 cells, AJs are organized in a fragmented manner containing E-cadherin and actin but lacking myosin-II. In these cells, transfection of either ZO-1 or ZO-2 restores myosin integration and the establishment of regular AJs through a process mediated by Rho activation [97]. The differences in myosin II organization observed in MDCK and Eph4 cells which do not express ZO-1/ZO-2 could be due to the cellular context, since, in contrast to MDCK cells [9], Eph4 cells do not express ZO-3 [98].

In MDCK cells lacking ZO-1 and ZO-2, the change in the appearance of actin and myosin is accompanied by expansion of the apical domain, cell border recruitment of phosphomyosin light chain and Rho kinase 1 (ROCK-I), and contraction of the actomyosin ring [93]. In accordance, atomic force microscopy revealed that the lack of ZO-1 and ZO-2 increased the tension of the apical domain due to elevated myosin II ATPase activity [99]. In addition, in MDCK cells with shRNAs targeting both ZO-1 and ZO-2, the normal convoluted pattern of cell–cell contact becomes much more linear, making the overall cell shape more polygonal [93]. This change does not happen in ZO-2 KD [100] and KO [101] cells, but is clearly observed in ZO-1 KD [100] and KO [101] cells, and can also be attained in control cells after treatment with blebbistatin, an inhibitor of myosin II [101]. Moreover, excessive expression of ZO-1 in ZO-1 KO cells induces an intensive contoured shape of cell–cell junctions. Taken together, these observations indicate that the change from tortuous to linear contacts present in the double ZO-1 and ZO-2 KD is due to the lack of ZO-1 and depends on an actomyosin-generated force.

Structured illumination microscopy of ZO-1 reveals two conformations of the protein, stretched and folded. The latter relies on the interaction between the extreme C-terminal segment and the PDZ3, SH3, U5, and GuK domains of ZO-1. The conformational state of ZO-1 depends on its hetero-dimerization with ZO-2 and the tension of the actomyosin cytoskeleton. Thus, upon ZO-2 depletion, ZO-1 is maintained in the stretched conformation only if the actomyosin tension is high, like when cells are grown on glass coverslips. However, ZO-1 acquires a folded conformation if ZO-2 is absent and actomyosin tension is reduced by treatment with blebbistatin, or in conditions of reduced substrate stiffness (e.g., 3D cultures in Matrigel). In its folded conformation, ZO-1 cannot associate with occludin or recruit the transcription factor zonulin to the membrane, hence, this condition promotes the expression of zonulin target genes like cyclin D1 and proliferating cell nuclear antigen (PCNA) [102]. These observations suggest that, in a condition of low tension due to low substrate stiffness, like the one encountered by epithelial tissues *in vivo* (for review see [103]), the presence of ZO-2 is needed to maintain ZO-1 in a stretched conformation capable of binding to integral TJ proteins and to transcription factors like zonulin, which regulates the expression of genes involved in cell proliferation and differentiation.

ZO-2 alters the cytoarchitecture of epithelial monolayers, as the absence or reduction of ZO-2 content widens the intercellular spaces and detaches regions of the monolayer from the substrate [17,18]. In addition, decreased expression of integrin β1 and claudin-7, which localizes along the basolateral membrane in association to integrin β1, is observed in ZO-2 KD cells. Lack of ZO-2 induces recruitment of vinculin to focal adhesions and proliferation of stress fibers associated with increased activation of RhoA and ROCK-II [18] (Figure 4). Likewise, treatment of scirrhous gastric cancer cells with transforming growth factor β (TGFβ), whose signaling is associated with the invasion of cancer cells, decreases ZO-2 expression and upregulates the activity of RhoA and myosin phosphorylation [104].

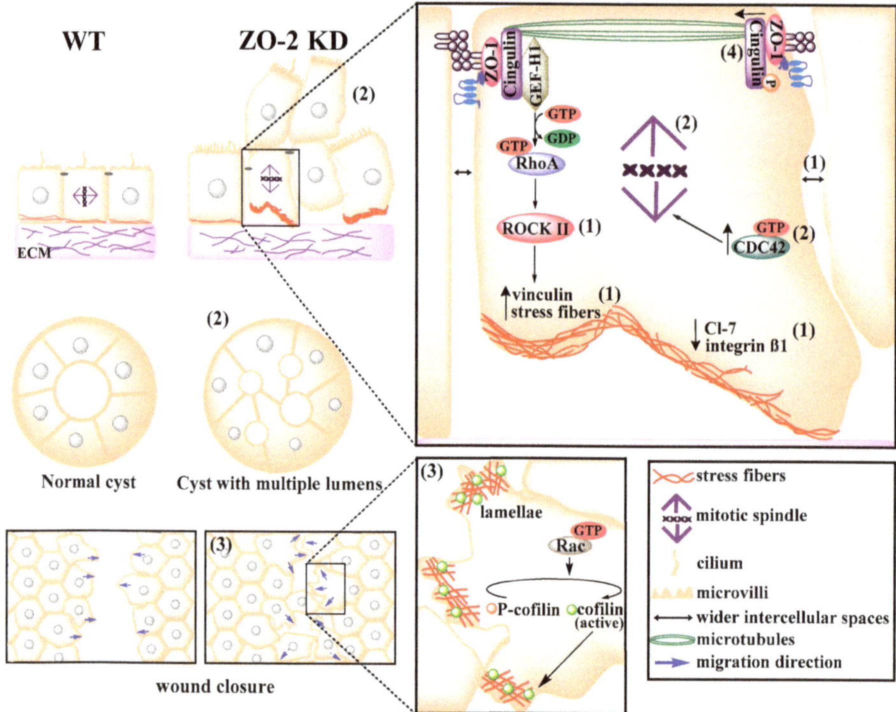

Figure 4. ZO-2 modulates cell size and tissue architecture. Absence of ZO-2 induces cell hypertrophy and development of an atypical cytoarchitecture characterized by: (**1**) Wider intercellular spaces and the detachment of cells from the substrate accompanied by a reduced expression of claudin-7 and integrin β1, an increased recruitment of vinculin to focal adhesions, and the proliferation of stress fibers induced by RhoA/ROCK-II activation; (**2**) an augmented activation of CDC42 that triggers the disorientation of the mitotic spindle and leads to the growth of cells on top of each other in 2D and 3D cultures and then to the formation of multiple lumens per cyst; (**3**) the activation of Rac and cofilin, which promote the formation of multiple lamellipodia, reduce directional persistence, and increase wound closure; and (**4**) the disappearance of a planar network of microtubules at the cell border.

ZO-2 KD cells also exhibit a profusion of lamellae mediated by an increase activity of Rac and cofilin. Rac promotes random cell migration in opposition to directional migration; accordingly, although ZO-2 KD cells move more than parental cells and close wounds more efficiently, they display reduced directional persistence [18] (Figure 4).

ZO-2 KD MDCK cells lose the microtubules that form a non-centrosomal network organized as a planar apical structure and display diminished phosphorylation of cingulin as well as a reduced interaction of cingulin with β-tubulin. ZO-2 favors the phosphorylation of cingulin by the adenosine monophosphate-activated protein kinase (AMPK), which allows microtubules/cingulin interactions at the apical junctional region. In addition, in ZO-2 KD monolayers, some cells grow on top of each other, while cysts of ZO-2 KD cells formed in the extracellular matrix Matrigel display multiple lumens instead of a single lumen per cyst. These aberrations might be due to the disorientation of the mitotic spindle, since in ZO-2 KD cells, the percentage of cells with mitotic spindles parallel to the substrate surface (0°–10°) decreases [18]. ZO-2 KD monolayers also display a reduced number of cilia [73] and more CDC42 activation [18], which controls spindle orientation [105] and centrosome polarization toward the direction of cell migration [106] (Figure 4). Taken together, these observations suggest that ZO-2 affects centrosome function.

The centrosome is a microtubule-organizing center composed of a centriole pair surrounded by pericentriolar material. In addition, centrioles template the formation of primary cilia in non-cycling cells. The mother centriole has sub-distal appendages needed for the organization of the cellular microtubular network and distal appendages involved in docking the centriole to the plasma membrane during ciliogenesis. Recently, using a proximity-dependent biotinylation assay to detect proteins interacting with centrosome and cilium proteins, ZO-2 was found to be associated with [107]: The sub-distal appendage proteins of the mother centriole CEP128, which regulates signaling at the primary cilium [108], and its associated protein cenexin [109] required for centrosome positioning in interphase cells, microtubule stability for proper spindle orientation during mitosis, and the formation of single lumens in cysts [110]; CEP135, a centriole assembly protein that regulates centriole duplication [111]; and centriolin, a protein of maternal centrioles required for the final stages of cytokinesis and entry into S phase [112].

10. ZO-2 as a Modulator of Cell Size

ZO-2 silencing in MDCK cells produces an increase in cell size also known as hypertrophy, where 67% of cells have a cell diameter of 55–60 nm instead of the 35–40 nm present in 73% of parental cells [73]. Hypertrophy has been associated with the activation of cyclin D without the concurrent upregulation of cyclin E, which triggers an increase in the physical growth of cells in early G1 and delayed entry into the S phase [113,114]. In ZO-2 KD cells this appears to be the case, as cyclin D1 content is increased, while entry into the S phase exhibits a delay [73].

In ZO-2 KD cells, hypertrophy is also due to inhibition of the Hippo pathway and activation of mTOR signaling [73]. The absence of ZO-2 induces the concentration of YAP at the nucleus and its transcriptional activity that promotes the expression of the catalytic pik3cb subunit of phosphatidylinositol 3-kinase (PI3K) and, through an miRNA-dependent process, blocks the translation of the phosphate and tensin homologue (PTEN). Consequently, the concentration of phosphatidylinositol(3,4,5)-trisphosphate (PIP3) increases and recruits PDK1 and AKT to the plasma membrane, triggering the phosphorylation and activation of AKT and the subsequent phosphorylation of mTORC1 and its downstream target, the ribosomal protein S6 kinase 1 (S6K1), which increases protein synthesis. Altogether, this leads to an increased protein/DNA ratio that produces hypertrophy [73], in agreement with previous observations in renal cells where hypertrophy appears to be a result of an increase in the protein synthesis/degradation rate [115,116]. Moreover, in renal compensatory hypertrophy induced by the removal of one kidney or uninephrectomy, the size of cells in the proximal tubules of the remaining kidney increased, while the expression of ZO-2 decreased and YAP accumulated in the nucleus. These results hence show that both *in vivo* and *in vitro* ZO-2 silencing correlates with cell hypertrophy.

11. ZO-2 as a Tumor Regulator Protein

Several lines of evidence suggest that ZO-2 is a tumor suppressor protein. First, ZO-2 is highly similar to Disc-Large (DLG) (amino segment with PDZ domains = 63% similarity, SH3 domain = 59% similarity, and GuK domain = 50% similarity), which acts as a tumor suppressor in the larva imaginal discs of Drosophila [6]. Second, some viral oncogenic proteins target ZO-2. Thus, the E4 region-encoded ORF1 (E4-ORF1), the primary oncogenic determinant of adenovirus type 9 that elicits mammary tumors in animals, associates through its PDZ binding motif with the first PDZ domain of ZO-2 and this interaction results in the aberrant sequestration of ZO-2 within the cytoplasm [117]. Likewise, E6 proteins from cancer-causing human papillomavirus (HPV)-16, -18, -31, -51, and -70 have a PDZ binding motif on their extreme carboxy termini that interacts with ZO-2. Instead, E6 proteins from HPV-66 or -40, which are, respectively, rarely or never associated with cancer, do not bind to ZO-2 [118]. E6/ZO-2 interaction appears to stabilize ZO-2, as ablation of E6 expression in HeLa cells reduces ZO-2 levels [118], whereas the expression of E6 from HPV-16 in the skin of transgenic female mice (K14E6) upregulated the expression of ZO-2 in the epidermis; and in monolayers of MDCK cells, transfection of E6 from HPV-16 blocked ZO-2 protein decay [119]. This is noteworthy because E6 proteins from high-risk HPV are known to target degradation of the cell–cell adhesion proteins Dlg [120], MAGI [120], MUPP1 [121], and Scribble [122]. E6 also triggers the mislocalization of ZO-2 as in MDCK cells transfected with E6, ZO-2 disappears from the cell borders and, in some cells, accumulates in the nucleus [119]. In this respect, in testicular carcinoma *in situ* and in bronchopulmonary cancers, ZO-2 disappears from the cell borders and stains the cytoplasm. This aberrant ZO-2 localization correlates with disruption of the blood–testis barrier [123] and the invasive properties of lung cancer cells *in vitro* [124].

c-Src, a non-receptor tyrosine kinase which is the cellular version of v-Src, the oncogene product of Rous sarcoma virus, is frequently overexpressed and active in various carcinomas. c-Src phosphorylates diverse cellular proteins involved in signaling pathways which lead to cancer progression. In MDCK cells, v-Src transfection opens the TJ, which is determined by a profound decrease in TER, and induces tyrosine phosphorylation of ZO-1 and ZO-2 [125]. Moreover, by proteomic analysis using the c-Src SH2 domain as the ligand, ZO-1 and ZO-2 were identified as binding proteins to both c-Src and C-terminal Src kinase (Csk). The latter phosphorylates and negatively regulates c-Src. Therefore ZO-1 and ZO-2 appear to function as scaffolding, regulating c-Src's transformation ability [125].

The *TJP2* gene has six alternatively spliced transcripts encoding various isoforms. In normal epithelia, expression was studied in two isoforms of ZO-2: ZO-2A and ZO-2C, which are respectively transcribed from the downstream promoter P_A and the upstream promoter P_C. The amino terminal region of ZO-2A has a 23-amino acid segment that is absent in ZO-2C [126,127]. In human pancreatic ductal adenocarcinoma and in breast ductal adenocarcinoma, ZO-2A isoform is absent, whereas in colon and acinar prostate adenocarcinomas, ZO-2 isoforms A and C are both expressed [128]. Loss of ZO-2A in pancreas ductal adenocarcinoma is not due to mutation, lack of transcription factors, or methylation of the immediate promoter, although methylation of the P_A promoter inactivates the promoter at the late stages of tumor development [126]. The sequence of promoter P_A reveals that ZO-2A transcription is regulated by numerous Sp1 sites [126].

Among *TJP2* transcripts, one with a longer exon which includes an AluYc sequence (*TJP2*-Alu transcript) was described. Alu proteins are retrotransposons that regulate gene expression patterns. *TJP2*-Alu transcript expression is not tissue specific, although its highest expression is in the testis. *TJP2*-Alu transcript is more highly expressed in normal tissues compared to tumors of the liver, uterus, and breast, however, colon tumor tissue displays a strong expression of *TJP2*-Alu transcript not present in normal tissue, making it a potential biomarker for colorectal cancer diagnosis [129].

The expression of ZO-2 is silenced in several carcinomas, including breast [128,130] and pancreas [127], as well as in a hypoxia-resistant cancer cell lines derived from a scirrhous gastric carcinoma [131]. In patients with incompletely enhancing glioblastoma multiforme (GBM), who have a longer survival rate, ZO-2 is expressed at higher levels than completely enhancing GBM, which is associated with shortened survival [132].

17β-estradiol (E2) and its receptors are risk factors for the initiation and progression of endocrine-related cancers, like those of the breast, prostate, ovarian, and endometrial cells (for reviews see [133,134]). In ovariectomized mice, E2 decreased the expression of ZO-2 in mouse epidermis and, although progesterone induced ZO-2 overexpression, it could not overturn ZO-2 silencing induced by E2 [119].

TGFβ1, which promotes the invasion of cancer cells, decreases the expression of ZO-2 in scirrhous gastric cancer cells [104]. Vascular endothelial growth factor-A (VEGF-A) in pancreas cancer cells increases cell motility and decreases ZO-2 expression [135]. Contrarily, vitamin D3, which promotes the differentiation of colon carcinoma cells, induces the expression of ZO-2 [74].

In summary, ZO-2 could be viewed as a tumor suppressor protein due to its structural similarity with known tumor suppressor proteins and considering that it is a target of oncogenic kinases, is silenced in various carcinomas, and exerts transcriptional repression on genes involved in cell proliferation and EMT. However, since oncogenic HPVs upregulate ZO-2 expression and forced nuclear accumulation of ZO-2 increases cell proliferation and confers metabolic benefit to cancer cells, we think ZO-2 might be considered as a tumor regulator rather than a tumor suppressor.

Author Contributions: L.G.M. wrote the manuscript. H.G.G, L.G.G and C.H.G searched for the information and organized it in different chapters. H.G.G made Figure 3 and Table 1. L.G.G made Figure 4 and C.H.G made Figures 1 and 2.

Funding: This work was supported by a grant from the Miguel Alemán Valdés Foundation to L.G.M. and by a SEP-Cinvestav grant to L.G.M. The APC was funded by Cinvestav.

Acknowledgments: H.G.G.: L.G.G. and C.H.G. were recipients of doctoral fellowship from the Mexican council of Science and Technology (Conacyt) (282075, 340209 and 407499) and H.G.G. was also a recipient of a scholarship from Mexiquense Council of Science and Technology (Comecyt, 2018AD0009-11).

Conflicts of Interest: The authors declare no conflict of interest.

References

1. Gonzalez-Mariscal, L.; Quiros, M.; Diaz-Coranguez, M. ZO proteins and redox-dependent processes. *Antioxid Redox Signal.* **2011**, *15*, 1235–1253. [CrossRef] [PubMed]
2. Gonzalez-Mariscal, L.; Miranda, J.; Ortega-Olvera, J.M.; Gallego-Gutierrez, H.; Raya-Sandino, A.; Vargas-Sierra, O. Zonula Occludens Proteins in Cancer. *Curr. Pathobiol. Rep.* **2016**, *4*, 107–116. [CrossRef]
3. Gumbiner, B.; Lowenkopf, T.; Apatira, D. Identification of a 160-kDa polypeptide that binds to the tight junction protein ZO-1. *Proc. Natl. Acad. Sci. USA* **1991**, *88*, 3460–3464. [CrossRef] [PubMed]
4. Walsh, T.; Pierce, S.B.; Lenz, D.R.; Brownstein, Z.; Dagan-Rosenfeld, O.; Shahin, H.; Roeb, W.; McCarthy, S.; Nord, A.S.; Gordon, C.R.; et al. Genomic duplication and overexpression of TJP2/ZO-2 leads to altered expression of apoptosis genes in progressive nonsyndromic hearing loss DFNA51. *Am. J. Hum. Genet.* **2010**, *87*, 101–109. [CrossRef] [PubMed]
5. Itoh, M.; Morita, K.; Tsukita, S. Characterization of ZO-2 as a MAGUK family member associated with tight as well as adherens junctions with a binding affinity to occludin and alpha catenin. *J. Biol. Chem.* **1999**, *274*, 5981–5986. [CrossRef] [PubMed]
6. Jesaitis, L.A.; Goodenough, D.A. Molecular characterization and tissue distribution of ZO-2, a tight junction protein homologous to ZO-1 and the Drosophila discs-large tumor suppressor protein. *J. Cell Biol.* **1994**, *124*, 949–961. [CrossRef] [PubMed]
7. Jenkins, E.L.; Caputo, M.; Angelini, G.D.; Ghorbel, M.T. Chronic hypoxia down-regulates tight junction protein ZO-2 expression in children with cyanotic congenital heart defect. *Esc Heart Fail.* **2016**, *3*, 131–137. [CrossRef]
8. Gonzalez-Mariscal, L.; Miranda, J.; Raya-Sandino, A.; Dominguez-Calderon, A.; Cuellar-Perez, F. ZO-2, a tight junction protein involved in gene expression, proliferation, apoptosis, and cell size regulation. *Ann. New York Acad. Sci.* **2017**, *1397*, 35–53. [CrossRef]

9. Haskins, J.; Gu, L.; Wittchen, E.S.; Hibbard, J.; Stevenson, B.R. ZO-3, a novel member of the MAGUK protein family found at the tight junction, interacts with ZO-1 and occludin. *J. Cell Biol.* **1998**, *141*, 199–208. [CrossRef]
10. Wittchen, E.S.; Haskins, J.; Stevenson, B.R. Protein interactions at the tight junction. Actin has multiple binding partners, and ZO-1 forms independent complexes with ZO-2 and ZO-3. *J. Biol. Chem.* **1999**, *274*, 35179–35185. [CrossRef]
11. Wu, J.; Yang, Y.; Zhang, J.; Ji, P.; Du, W.; Jiang, P.; Xie, D.; Huang, H.; Wu, M.; Zhang, G.; et al. Domain-swapped dimerization of the second PDZ domain of ZO2 may provide a structural basis for the polymerization of claudins. *J. Biol. Chem.* **2007**, *282*, 35988–35999. [CrossRef] [PubMed]
12. Umeda, K.; Ikenouchi, J.; Katahira-Tayama, S.; Furuse, K.; Sasaki, H.; Nakayama, M.; Matsui, T.; Tsukita, S.; Furuse, M.; Tsukita, S. ZO-1 and ZO-2 independently determine where claudins are polymerized in tight-junction strand formation. *Cell* **2006**, *126*, 741–754. [CrossRef] [PubMed]
13. Sheth, B.; Nowak, R.L.; Anderson, R.; Kwong, W.Y.; Papenbrock, T.; Fleming, T.P. Tight junction protein ZO-2 expression and relative function of ZO-1 and ZO-2 during mouse blastocyst formation. *Exp. Cell Res.* **2008**, *314*, 3356–3368. [CrossRef] [PubMed]
14. Eckert, J.J.; Fleming, T.P. Tight junction biogenesis during early development. *Biochim. Et. Biophys. Acta.* **2008**, *1778*, 717–728. [CrossRef] [PubMed]
15. Xu, J.; Kausalya, P.J.; Phua, D.C.; Ali, S.M.; Hossain, Z.; Hunziker, W. Early embryonic lethality of mice lacking ZO-2, but Not ZO-3, reveals critical and nonredundant roles for individual zonula occludens proteins in mammalian development. *Mol. Cell Biol.* **2008**, *28*, 1669–1678. [CrossRef] [PubMed]
16. Xu, J.; Anuar, F.; Ali, S.M.; Ng, M.Y.; Phua, D.C.; Hunziker, W. Zona occludens-2 is critical for blood-testis barrier integrity and male fertility. *Mol. Biol. Cell* **2009**, *20*, 4268–4277. [CrossRef] [PubMed]
17. Hernandez, S.; Chavez Munguia, B.; Gonzalez-Mariscal, L. ZO-2 silencing in epithelial cells perturbs the gate and fence function of tight junctions and leads to an atypical monolayer architecture. *Exp. Cell Res.* **2007**, *313*, 1533–1547. [CrossRef] [PubMed]
18. Raya-Sandino, A.; Castillo-Kauil, A.; Dominguez-Calderon, A.; Alarcon, L.; Flores-Benitez, D.; Cuellar-Perez, F.; Lopez-Bayghen, B.; Chavez-Munguia, B.; Vazquez-Prado, J.; Gonzalez-Mariscal, L. Zonula occludens-2 regulates Rho proteins activity and the development of epithelial cytoarchitecture and barrier function. *Biochim. Biophys. Acta.* **2017**, *1864*, 1714–1733. [CrossRef]
19. Kim, M.A.; Kim, Y.R.; Sagong, B.; Cho, H.J.; Bae, J.W.; Kim, J.; Lee, J.; Park, H.J.; Choi, J.Y.; Lee, K.Y.; et al. Genetic analysis of genes related to tight junction function in the Korean population with non-syndromic hearing loss. *PLoS ONE* **2014**, *9*, e95646. [CrossRef]
20. Wang, H.Y.; Zhao, Y.L.; Liu, Q.; Yuan, H.; Gao, Y.; Lan, L.; Yu, L.; Wang, D.Y.; Guan, J.; Wang, Q.J. Identification of Two Disease-causing Genes TJP2 and GJB2 in a Chinese Family with Unconditional Autosomal Dominant Nonsyndromic Hereditary Hearing Impairment. *Chin. Med. J.* **2015**, *128*, 3345–3351. [CrossRef]
21. Carlton, V.E.; Harris, B.Z.; Puffenberger, E.G.; Batta, A.K.; Knisely, A.S.; Robinson, D.L.; Strauss, K.A.; Shneider, B.L.; Lim, W.A.; Salen, G.; et al. Complex inheritance of familial hypercholanemia with associated mutations in TJP2 and BAAT. *Nat. Genet.* **2003**, *34*, 91–96. [CrossRef]
22. Vitale, G.; Gitto, S.; Raimondi, F.; Mattiaccio, A.; Mantovani, V.; Vukotic, R.; D'Errico, A.; Seri, M.; Russell, R.B.; Andreone, P. Cryptogenic cholestasis in young and adults: ATP8B1, ABCB11, ABCB4, and TJP2 gene variants analysis by high-throughput sequencing. *J. Gastroenterol.* **2018**, *53*, 945–958. [CrossRef]
23. Sambrotta, M.; Strautnieks, S.; Papouli, E.; Rushton, P.; Clark, B.E.; Parry, D.A.; Logan, C.V.; Newbury, L.J.; Kamath, B.M.; Ling, S.; et al. Mutations in TJP2 cause progressive cholestatic liver disease. *Nat. Genet.* **2014**, *46*, 326–328. [CrossRef]
24. Ge, T.; Zhang, X.; Xiao, Y.; Wang, Y.; Zhang, T. Novel compound heterozygote mutations of TJP2 in a Chinese child with progressive cholestatic liver disease. *Bmc Med. Genet.* **2019**, *20*, 18. [CrossRef]
25. Zhou, S.; Hertel, P.M.; Finegold, M.J.; Wang, L.; Kerkar, N.; Wang, J.; Wong, L.J.; Plon, S.E.; Sambrotta, M.; Foskett, P.; et al. Hepatocellular carcinoma associated with tight-junction protein 2 deficiency. *Hepatology* **2015**, *62*, 1914–1916. [CrossRef]
26. Parsons, D.W.; Roy, A.; Yang, Y.; Wang, T.; Scollon, S.; Bergstrom, K.; Kerstein, R.A.; Gutierrez, S.; Petersen, A.K.; Bavle, A.; et al. Diagnostic Yield of Clinical Tumor and Germline Whole-Exome Sequencing for Children With Solid Tumors. *JAMA Oncol.* **2016**, *2*, 616–624. [CrossRef]
27. Vij, M.; Shanmugam, N.P.; Reddy, M.S.; Sankaranarayanan, S.; Rela, M. Paediatric hepatocellular carcinoma in tight junction protein 2 (TJP2) deficiency. *Virchows. Arch.* **2017**, *471*, 679–683. [CrossRef]

28. Dixon, P.H.; Sambrotta, M.; Chambers, J.; Taylor-Harris, P.; Syngelaki, A.; Nicolaides, K.; Knisely, A.S.; Thompson, R.J.; Williamson, C. An expanded role for heterozygous mutations of ABCB4, ABCB11, ATP8B1, ABCC2 and TJP2 in intrahepatic cholestasis of pregnancy. *Sci. Rep.* **2017**, *7*, 11823. [CrossRef]
29. Stanton, P.G. Regulation of the blood-testis barrier. *Semin. Cell Dev. Biol.* **2016**, *59*, 166–173. [CrossRef]
30. Ortega-Olvera, J.M.; Winkler, R.; Quitanilla-Vega, B.; Shibayama, M.; Chavez-Munguia, B.; Martin-Tapia, D.; Alarcon, L.; Gonzalez-Mariscal, L. The organophosphate pesticide methamidophos opens the blood-testis barrier and covalently binds to ZO-2 in mice. *Toxicol. Appl. Pharmacol.* **2018**, *360*, 257–272. [CrossRef]
31. Sambrotta, M.; Thompson, R.J. Mutations in TJP2, encoding zona occludens 2, and liver disease. *Tissue Barriers* **2015**, *3*, e1026537. [CrossRef]
32. Ben-Yosef, T.; Belyantseva, I.A.; Saunders, T.L.; Hughes, E.D.; Kawamoto, K.; Van Itallie, C.M.; Beyer, L.A.; Halsey, K.; Gardner, D.J.; Wilcox, E.R.; et al. Claudin 14 knockout mice, a model for autosomal recessive deafness DFNB29, are deaf due to cochlear hair cell degeneration. *Hum. Mol. Genet.* **2003**, *12*, 2049–2061. [CrossRef]
33. Kitajiri, S.; Katsuno, T.; Sasaki, H.; Ito, J.; Furuse, M.; Tsukita, S. Deafness in occludin-deficient mice with dislocation of tricellulin and progressive apoptosis of the hair cells. *Biol. Open* **2014**, *3*, 759–766. [CrossRef]
34. Kamitani, T.; Sakaguchi, H.; Tamura, A.; Miyashita, T.; Yamazaki, Y.; Tokumasu, R.; Inamoto, R.; Matsubara, A.; Mori, N.; Hisa, Y.; et al. Deletion of Tricellulin Causes Progressive Hearing Loss Associated with Degeneration of Cochlear Hair Cells. *Sci. Rep.* **2015**, *5*, 18402. [CrossRef]
35. Morozko, E.L.; Nishio, A.; Ingham, N.J.; Chandra, R.; Fitzgerald, T.; Martelletti, E.; Borck, G.; Wilson, E.; Riordan, G.P.; Wangemann, P.; et al. ILDR1 null mice, a model of human deafness DFNB42, show structural aberrations of tricellular tight junctions and degeneration of auditory hair cells. *Hum. Mol. Genet.* **2015**, *24*, 609–624. [CrossRef]
36. Nakano, Y.; Kim, S.H.; Kim, H.M.; Sanneman, J.D.; Zhang, Y.; Smith, R.J.; Marcus, D.C.; Wangemann, P.; Nessler, R.A.; Banfi, B. A claudin-9-based ion permeability barrier is essential for hearing. *Plos Genet.* **2009**, *5*, e1000610. [CrossRef]
37. Gow, A.; Davies, C.; Southwood, C.M.; Frolenkov, G.; Chrustowski, M.; Ng, L.; Yamauchi, D.; Marcus, D.C.; Kachar, B. Deafness in Claudin 11-null mice reveals the critical contribution of basal cell tight junctions to stria vascularis function. *J. Neurosci. Off. J. Soc. Neurosci.* **2004**, *24*, 7051–7062. [CrossRef]
38. Kitajiri, S.; Miyamoto, T.; Mineharu, A.; Sonoda, N.; Furuse, K.; Hata, M.; Sasaki, H.; Mori, Y.; Kubota, T.; Ito, J.; et al. Compartmentalization established by claudin-11-based tight junctions in stria vascularis is required for hearing through generation of endocochlear potential. *J. Cell Sci.* **2004**, *117*, 5087–5096. [CrossRef]
39. Islas, S.; Vega, J.; Ponce, L.; Gonzalez-Mariscal, L. Nuclear localization of the tight junction protein ZO-2 in epithelial cells. *Exp. Cell Res.* **2002**, *274*, 138–148. [CrossRef]
40. Traweger, A.; Fuchs, R.; Krizbai, I.A.; Weiger, T.M.; Bauer, H.C.; Bauer, H. The tight junction protein ZO-2 localizes to the nucleus and interacts with the heterogeneous nuclear ribonucleoprotein scaffold attachment factor-B. *J. Biol Chem.* **2003**, *278*, 2692–2700. [CrossRef]
41. Amaya, E.; Alarcon, L.; Martin-Tapia, D.; Cuellar Perez, F.; Cano-Cortina, M.; Ortega-Olvera, J.M.; Cisneros, B.; Rodriguez, A.; Gamba, G.; Gonzalez-Mariscal, L. Activation of the Ca2+ sensing receptor and the PKC/WNK4 downstream signaling cascade induces arrival of ZO-2 to tight junctions and its separation from 14-3-3. *Mol. Biol. Cell* **2019**. In press. [CrossRef]
42. Tapia, R.; Huerta, M.; Islas, S.; Avila-Flores, A.; Lopez-Bayghen, E.; Weiske, J.; Huber, O.; Gonzalez-Mariscal, L. Zona occludens-2 inhibits cyclin D1 expression and cell proliferation and exhibits changes in localization along the cell cycle. *Mol. Biol. Cell* **2009**, *20*, 1102–1117. [CrossRef]
43. Lopez-Bayghen, E.; Jaramillo, B.E.; Huerta, M.; Betanzos, A.; Gonzalez-Mariscal, L. TJ Proteins That Make Round Trips to the Nucleus. In *Tight Junctions*; Gonzalez-Mariscal, L., Ed.; Springer US: Boston, MA, USA, 2006; pp. 76–100. [CrossRef]
44. Quiros, M.; Alarcon, L.; Ponce, A.; Giannakouros, T.; Gonzalez-Mariscal, L. The intracellular fate of zonula occludens 2 is regulated by the phosphorylation of SR repeats and the phosphorylation/O-GlcNAcylation of S257. *Mol. Biol. Cell* **2013**, *24*, 2528–2543. [CrossRef]
45. Meerschaert, K.; Tun, M.P.; Remue, E.; De Ganck, A.; Boucherie, C.; Vanloo, B.; Degeest, G.; Vandekerckhove, J.; Zimmermann, P.; Bhardwaj, N.; et al. The PDZ2 domain of zonula occludens-1 and -2 is a phosphoinositide binding domain. *Cell Mol. Life Sci.* **2009**, *66*, 3951–3966. [CrossRef]

46. Gonzalez-Mariscal, L.; Ponce, A.; Alarcon, L.; Jaramillo, B.E. The tight junction protein ZO-2 has several functional nuclear export signals. *Exp. Cell Res.* **2006**, *312*, 3323–3335. [CrossRef]
47. Chamorro, D.; Alarcon, L.; Ponce, A.; Tapia, R.; Gonzalez-Aguilar, H.; Robles-Flores, M.; Mejia-Castillo, T.; Segovia, J.; Bandala, Y.; Juaristi, E.; et al. Phosphorylation of zona occludens-2 by protein kinase C epsilon regulates its nuclear exportation. *Mol. Biol. Cell* **2009**, *20*, 4120–4129. [CrossRef]
48. Wetzel, F.; Mittag, S.; Cano-Cortina, M.; Wagner, T.; Kramer, O.H.; Niedenthal, R.; Gonzalez-Mariscal, L.; Huber, O. SUMOylation regulates the intracellular fate of ZO-2. *Cell. Mol. Life Sci. CMLS* **2016**, *74*, 373–392. [CrossRef]
49. Gonzalez-Mariscal, L.; Contreras, R.G.; Bolivar, J.J.; Ponce, A.; Chavez De Ramirez, B.; Cereijido, M. Role of calcium in tight junction formation between epithelial cells. *Am. J. Physiol.* **1990**, *259*, C978–C986. [CrossRef]
50. Jouret, F.; Wu, J.; Hull, M.; Rajendran, V.; Mayr, B.; Schofl, C.; Geibel, J.; Caplan, M.J. Activation of the Ca(2)+-sensing receptor induces deposition of tight junction components to the epithelial cell plasma membrane. *J. Cell Sci.* **2013**, *126*, 5132–5142. [CrossRef]
51. Jaramillo, B.E.; Ponce, A.; Moreno, J.; Betanzos, A.; Huerta, M.; Lopez-Bayghen, E.; Gonzalez-Mariscal, L. Characterization of the tight junction protein ZO-2 localized at the nucleus of epithelial cells. *Exp. Cell Res.* **2004**, *297*, 247–258. [CrossRef]
52. Lechuga, S.; Alarcon, L.; Solano, J.; Huerta, M.; Lopez-Bayghen, E.; Gonzalez-Mariscal, L. Identification of ZASP, a novel protein associated to Zona occludens-2. *Exp. Cell Res.* **2010**, *316*, 3124–3139. [CrossRef]
53. Hong, E.A.; Gautrey, H.L.; Elliott, D.J.; Tyson-Capper, A.J. SAFB1- and SAFB2-mediated transcriptional repression: Relevance to cancer. *Biochem. Soc. Trans.* **2012**, *40*, 826–830. [CrossRef]
54. Lin, J.; Xu, P.; LaVallee, P.; Hoidal, J.R. Identification of proteins binding to E-Box/Ku86 sites and function of the tumor suppressor SAFB1 in transcriptional regulation of the human xanthine oxidoreductase gene. *J. Biol. Chem.* **2008**, *283*, 29681–29689. [CrossRef]
55. Huerta, M.; Munoz, R.; Tapia, R.; Soto-Reyes, E.; Ramirez, L.; Recillas-Targa, F.; Gonzalez-Mariscal, L.; Lopez-Bayghen, E. Cyclin D1 is transcriptionally down-regulated by ZO-2 via an E box and the transcription factor c-Myc. *Mol. Biol. Cell* **2007**, *18*, 4826–4836. [CrossRef]
56. Townson, S.M.; Kang, K.; Lee, A.V.; Oesterreich, S. Structure-function analysis of the estrogen receptor alpha corepressor scaffold attachment factor-B1: Identification of a potent transcriptional repression domain. *J. Biol. Chem.* **2004**, *279*, 26074–26081. [CrossRef]
57. Nikolakaki, E.; Kohen, R.; Hartmann, A.M.; Stamm, S.; Georgatsou, E.; Giannakouros, T. Cloning and characterization of an alternatively spliced form of SR protein kinase 1 that interacts specifically with scaffold attachment factor-B. *J. Biol. Chem.* **2001**, *276*, 40175–40182. [CrossRef]
58. McCracken, S.; Longman, D.; Marcon, E.; Moens, P.; Downey, M.; Nickerson, J.A.; Jessberger, R.; Wilde, A.; Caceres, J.F.; Emili, A.; et al. Proteomic analysis of SRm160-containing complexes reveals a conserved association with cohesin. *J. Biol. Chem.* **2005**, *280*, 42227–42236. [CrossRef]
59. Tsai, Y.C.; Greco, T.M.; Boonmee, A.; Miteva, Y.; Cristea, I.M. Functional proteomics establishes the interaction of SIRT7 with chromatin remodeling complexes and expands its role in regulation of RNA polymerase I transcription. *Mol. Cell. Proteom. MCP* **2012**, *11*, 60–76. [CrossRef]
60. Havugimana, P.C.; Hart, G.T.; Nepusz, T.; Yang, H.; Turinsky, A.L.; Li, Z.; Wang, P.I.; Boutz, D.R.; Fong, V.; Phanse, S.; et al. A census of human soluble protein complexes. *Cell* **2012**, *150*, 1068–1081. [CrossRef]
61. Tang, Y.; Puri, A.; Ricketts, M.D.; Rai, T.S.; Hoffmann, J.; Hoi, E.; Adams, P.D.; Schultz, D.C.; Marmorstein, R. Identification of an ubinuclein 1 region required for stability and function of the human HIRA/UBN1/CABIN1/ASF1a histone H3.3 chaperone complex. *Biochemistry* **2012**, *51*, 2366–2377. [CrossRef]
62. Lupo, J.; Conti, A.; Sueur, C.; Coly, P.A.; Coute, Y.; Hunziker, W.; Burmeister, W.P.; Germi, R.; Manet, E.; Gruffat, H.; et al. Identification of new interacting partners of the shuttling protein ubinuclein (Ubn-1). *Exp. Cell Res.* **2012**, *318*, 509–520. [CrossRef]
63. Bernstein, K.A.; Gallagher, J.E.; Mitchell, B.M.; Granneman, S.; Baserga, S.J. The small-subunit processome is a ribosome assembly intermediate. *Eukaryot Cell* **2004**, *3*, 1619–1626. [CrossRef]
64. Leung, J.W.; Makharashvili, N.; Agarwal, P.; Chiu, L.Y.; Pourpre, R.; Cammarata, M.B.; Cannon, J.R.; Sherker, A.; Durocher, D.; Brodbelt, J.S.; et al. ZMYM3 regulates BRCA1 localization at damaged chromatin to promote DNA repair. *Genes Dev.* **2017**, *31*, 260–274. [CrossRef]

65. Bonnelye, E.; Vanacker, J.M.; Dittmar, T.; Begue, A.; Desbiens, X.; Denhardt, D.T.; Aubin, J.E.; Laudet, V.; Fournier, B. The ERR-1 orphan receptor is a transcriptional activator expressed during bone development. *Mol. Endocrinol.* **1997**, *11*, 905–916. [CrossRef]
66. Wan, C.; Borgeson, B.; Phanse, S.; Tu, F.; Drew, K.; Clark, G.; Xiong, X.; Kagan, O.; Kwan, J.; Bezginov, A.; et al. Panorama of ancient metazoan macromolecular complexes. *Nature* **2015**, *525*, 339–344. [CrossRef]
67. Shaheen, R.; Tasak, M.; Maddirevula, S.; Abdel-Salam, G.M.H.; Sayed, I.S.M.; Alazami, A.M.; Al-Sheddi, T.; Alobeid, E.; Phizicky, E.M.; Alkuraya, F.S. PUS7 mutations impair pseudouridylation in humans and cause intellectual disability and microcephaly. *Hum. Genet.* **2019**, *138*, 231–239. [CrossRef]
68. Betanzos, A.; Huerta, M.; Lopez-Bayghen, E.; Azuara, E.; Amerena, J.; Gonzalez-Mariscal, L. The tight junction protein ZO-2 associates with Jun, Fos and C/EBP transcription factors in epithelial cells. *Exp. Cell Res.* **2004**, *292*, 51–66. [CrossRef]
69. Kao, S.H.; Wang, W.L.; Chen, C.Y.; Chang, Y.L.; Wu, Y.Y.; Wang, Y.T.; Wang, S.P.; Nesvizhskii, A.I.; Chen, Y.J.; Hong, T.M.; et al. GSK3beta controls epithelial-mesenchymal transition and tumor metastasis by CHIP-mediated degradation of Slug. *Oncogene* **2014**, *33*, 3172–3182. [CrossRef]
70. Huttlin, E.L.; Ting, L.; Bruckner, R.J.; Gebreab, F.; Gygi, M.P.; Szpyt, J.; Tam, S.; Zarraga, G.; Colby, G.; Baltier, K.; et al. The BioPlex Network: A Systematic Exploration of the Human Interactome. *Cell* **2015**, *162*, 425–440. [CrossRef]
71. Clevers, H.; Nusse, R. Wnt/beta-catenin signaling and disease. *Cell* **2012**, *149*, 1192–1205. [CrossRef]
72. Nusse, R.; Clevers, H. Wnt/beta-Catenin Signaling, Disease, and Emerging Therapeutic Modalities. *Cell* **2017**, *169*, 985–999. [CrossRef]
73. Dominguez-Calderon, A.; Avila-Flores, A.; Ponce, A.; Lopez-Bayghen, E.; Calderon-Salinas, J.V.; Luis Reyes, J.; Chavez-Munguia, B.; Segovia, J.; Angulo, C.; Ramirez, L.; et al. ZO-2 silencing induces renal hypertrophy through a cell cycle mechanism and the activation of YAP and the mTOR pathway. *Mol. Biol. Cell* **2016**, *27*, 1581–1595. [CrossRef]
74. Palmer, H.G.; Gonzalez-Sancho, J.M.; Espada, J.; Berciano, M.T.; Puig, I.; Baulida, J.; Quintanilla, M.; Cano, A.; de Herreros, A.G.; Lafarga, M.; et al. Vitamin D(3) promotes the differentiation of colon carcinoma cells by the induction of E-cadherin and the inhibition of beta-catenin signaling. *J. Cell Biol.* **2001**, *154*, 369–387. [CrossRef]
75. Bautista-Garcia, P.; Reyes, J.L.; Martin, D.; Namorado, M.C.; Chavez-Munguia, B.; Soria-Castro, E.; Huber, O.; Gonzalez-Mariscal, L. Zona occludens-2 protects against podocyte dysfunction induced by ADR in mice. *Am. J. Physiol. Ren. Physiol.* **2013**, *304*, F77–F87. [CrossRef]
76. Meng, Z.; Moroishi, T.; Guan, K.L. Mechanisms of Hippo pathway regulation. *Genes Dev.* **2016**, *30*, 1–17. [CrossRef]
77. Azzolin, L.; Panciera, T.; Soligo, S.; Enzo, E.; Bicciato, S.; Dupont, S.; Bresolin, S.; Frasson, C.; Basso, G.; Guzzardo, V.; et al. YAP/TAZ incorporation in the beta-catenin destruction complex orchestrates the Wnt response. *Cell* **2014**, *158*, 157–170. [CrossRef]
78. Zhao, B.; Li, L.; Tumaneng, K.; Wang, C.Y.; Guan, K.L. A coordinated phosphorylation by Lats and CK1 regulates YAP stability through SCF(beta-TRCP). *Genes Dev.* **2010**, *24*, 72–85. [CrossRef]
79. Kusch, A.; Tkachuk, S.; Tkachuk, N.; Patecki, M.; Park, J.K.; Dietz, R.; Haller, H.; Dumler, I. The tight junction protein ZO-2 mediates proliferation of vascular smooth muscle cells via regulation of Stat1. *Cardiovasc. Res.* **2009**, *83*, 115–122. [CrossRef]
80. Oka, T.; Remue, E.; Meerschaert, K.; Vanloo, B.; Boucherie, C.; Gfeller, D.; Bader, G.D.; Sidhu, S.S.; Vandekerckhove, J.; Gettemans, J.; et al. Functional complexes between YAP2 and ZO-2 are PDZ domain-dependent, and regulate YAP2 nuclear localization and signalling. *Biochem J.* **2010**, *432*, 461–472. [CrossRef]
81. Diehl, J.A.; Cheng, M.; Roussel, M.F.; Sherr, C.J. Glycogen synthase kinase-3beta regulates cyclin D1 proteolysis and subcellular localization. *Genes Dev.* **1998**, *12*, 3499–3511. [CrossRef]
82. Klein, P.S.; Melton, D.A. A molecular mechanism for the effect of lithium on development. *Proc. Natl. Acad. Sci. USA* **1996**, *93*, 8455–8459. [CrossRef]
83. Hedgepeth, C.M.; Conrad, L.J.; Zhang, J.; Huang, H.C.; Lee, V.M.; Klein, P.S. Activation of the Wnt signaling pathway: A molecular mechanism for lithium action. *Dev. Biol.* **1997**, *185*, 82–91. [CrossRef]
84. Qiao, X.; Roth, I.; Feraille, E.; Hasler, U. Different effects of ZO-1, ZO-2 and ZO-3 silencing on kidney collecting duct principal cell proliferation and adhesion. *Cell Cycle* **2014**, *13*, 3059–3075. [CrossRef]

85. Traweger, A.; Lehner, C.; Farkas, A.; Krizbai, I.A.; Tempfer, H.; Klement, E.; Guenther, B.; Bauer, H.C.; Bauer, H. Nuclear Zonula occludens-2 alters gene expression and junctional stability in epithelial and endothelial cells. *Differ. Res. Biol. Divers.* **2008**, *76*, 99–106. [CrossRef]
86. Iqbal, M.A.; Gupta, V.; Gopinath, P.; Mazurek, S.; Bamezai, R.N. Pyruvate kinase M2 and cancer: An updated assessment. *Febs Lett.* **2014**, *588*, 2685–2692. [CrossRef]
87. Bojarski, C.; Weiske, J.; Schoneberg, T.; Schroder, W.; Mankertz, J.; Schulzke, J.D.; Florian, P.; Fromm, M.; Tauber, R.; Huber, O. The specific fates of tight junction proteins in apoptotic epithelial cells. *J. Cell Sci.* **2004**, *117*, 2097–2107. [CrossRef]
88. Chaitanya, G.V.; Steven, A.J.; Babu, P.P. PARP-1 cleavage fragments: Signatures of cell-death proteases in neurodegeneration. *Cell Commun Signal.* **2010**, *8*, 31. [CrossRef]
89. Liu, J.; Yuan, Q.; Ling, X.; Tan, Q.; Liang, H.; Chen, J.; Lin, L.; Xiao, Y.; Chen, W.; Liu, L.; et al. PARP1 may be involved in hydroquinoneinduced apoptosis by poly ADPribosylation of ZO2. *Mol. Med. Rep.* **2017**, *16*, 8076–8084. [CrossRef]
90. Kazmierczak, M.; Harris, S.L.; Kazmierczak, P.; Shah, P.; Starovoytov, V.; Ohlemiller, K.K.; Schwander, M. Progressive Hearing Loss in Mice Carrying a Mutation in Usp53. *J. Neurosci. : Off. J. Soc. Neurosci.* **2015**, *35*, 15582–15598. [CrossRef]
91. Maddirevula, S.; Alhebbi, H.; Alqahtani, A.; Algoufi, T.; Alsaif, H.S.; Ibrahim, N.; Abdulwahab, F.; Barr, M.; Alzaidan, H.; Almehaideb, A.; et al. Identification of novel loci for pediatric cholestatic liver disease defined by KIF12, PPM1F, USP53, LSR, and WDR83OS pathogenic variants. *Genet. Med. : Off. J. Am. Coll. Med. Genet.* **2019**, *21*, 1164–1172. [CrossRef]
92. Oka, T.; Schmitt, A.P.; Sudol, M. Opposing roles of angiomotin-like-1 and zona occludens-2 on pro-apoptotic function of YAP. *Oncogene* **2012**, *31*, 128–134. [CrossRef]
93. Fanning, A.S.; Van Itallie, C.M.; Anderson, J.M. Zonula occludens-1 and -2 regulate apical cell structure and the zonula adherens cytoskeleton in polarized epithelia. *Mol. Biol. Cell* **2012**, *23*, 577–590. [CrossRef]
94. Mattagajasingh, S.N.; Huang, S.C.; Hartenstein, J.S.; Benz, E.J., Jr. Characterization of the interaction between protein 4.1R and ZO-2. A possible link between the tight junction and the actin cytoskeleton. *J. Biol Chem* **2000**, *275*, 30573–30585. [CrossRef]
95. Baines, A.J.; Lu, H.C.; Bennett, P.M. The Protein 4.1 family: Hub proteins in animals for organizing membrane proteins. *Biochim. Et Biophys. Acta* **2014**, *1838*, 605–619. [CrossRef]
96. Monteiro, A.C.; Sumagin, R.; Rankin, C.R.; Leoni, G.; Mina, M.J.; Reiter, D.M.; Stehle, T.; Dermody, T.S.; Schaefer, S.A.; Hall, R.A.; et al. JAM-A associates with ZO-2, afadin, and PDZ-GEF1 to activate Rap2c and regulate epithelial barrier function. *Mol. Biol. Cell* **2013**, *24*, 2849–2860. [CrossRef]
97. Yamazaki, Y.; Umeda, K.; Wada, M.; Nada, S.; Okada, M.; Tsukita, S.; Tsukita, S. ZO-1- and ZO-2-dependent integration of myosin-2 to epithelial zonula adherens. *Mol. Biol. Cell* **2008**, *19*, 3801–3811. [CrossRef]
98. Umeda, K.; Matsui, T.; Nakayama, M.; Furuse, K.; Sasaki, H.; Furuse, M.; Tsukita, S. Establishment and characterization of cultured epithelial cells lacking expression of ZO-1. *J. Biol Chem* **2004**, *279*, 44785–44794. [CrossRef]
99. Cartagena-Rivera, A.X.; Van Itallie, C.M.; Anderson, J.M.; Chadwick, R.S. Apical surface supracellular mechanical properties in polarized epithelium using noninvasive acoustic force spectroscopy. *Nat. Commun.* **2017**, *8*, 1030. [CrossRef]
100. Van Itallie, C.M.; Fanning, A.S.; Bridges, A.; Anderson, J.M. ZO-1 stabilizes the tight junction solute barrier through coupling to the perijunctional cytoskeleton. *Mol. Biol. Cell* **2009**, *20*, 3930–3940. [CrossRef]
101. Tokuda, S.; Higashi, T.; Furuse, M. ZO-1 knockout by TALEN-mediated gene targeting in MDCK cells: Involvement of ZO-1 in the regulation of cytoskeleton and cell shape. *PLoS ONE* **2014**, *9*, e104994. [CrossRef]
102. Spadaro, D.; Le, S.; Laroche, T.; Mean, I.; Jond, L.; Yan, J.; Citi, S. Tension-Dependent Stretching Activates ZO-1 to Control the Junctional Localization of Its Interactors. *Curr. Biol. : Cb* **2017**, *27*, 3783–3795.e8. [CrossRef]
103. Handorf, A.M.; Zhou, Y.; Halanski, M.A.; Li, W.J. Tissue stiffness dictates development, homeostasis, and disease progression. *Organogenesis* **2015**, *11*, 1–15. [CrossRef]
104. Shinto, O.; Yashiro, M.; Kawajiri, H.; Shimizu, K.; Shimizu, T.; Miwa, A.; Hirakawa, K. Inhibitory effect of a TGFbeta receptor type-I inhibitor, Ki26894, on invasiveness of scirrhous gastric cancer cells. *Br. J. Cancer* **2010**, *102*, 844–851. [CrossRef]
105. Jaffe, A.B.; Kaji, N.; Durgan, J.; Hall, A. Cdc42 controls spindle orientation to position the apical surface during epithelial morphogenesis. *J. Cell Biol.* **2008**, *183*, 625–633. [CrossRef]

106. Etienne-Manneville, S.; Hall, A. Integrin-mediated activation of Cdc42 controls cell polarity in migrating astrocytes through PKCzeta. *Cell* **2001**, *106*, 489–498. [CrossRef]
107. Gupta, G.D.; Coyaud, E.; Goncalves, J.; Mojarad, B.A.; Liu, Y.; Wu, Q.; Gheiratmand, L.; Comartin, D.; Tkach, J.M.; Cheung, S.W.; et al. A Dynamic Protein Interaction Landscape of the Human Centrosome-Cilium Interface. *Cell* **2015**, *163*, 1484–1499. [CrossRef]
108. Monnich, M.; Borgeskov, L.; Breslin, L.; Jakobsen, L.; Rogowski, M.; Doganli, C.; Schroder, J.M.; Mogensen, J.B.; Blinkenkjaer, L.; Harder, L.M.; et al. CEP128 Localizes to the Subdistal Appendages of the Mother Centriole and Regulates TGF-beta/BMP Signaling at the Primary Cilium. *Cell Rep.* **2018**, *22*, 2584–2592. [CrossRef]
109. Kashihara, H.; Chiba, S.; Kanno, S.I.; Suzuki, K.; Yano, T.; Tsukita, S. Cep128 associates with Odf2 to form the subdistal appendage of the centriole. *Genes Cells : Devoted Mol. Cell. Mech.* **2019**, *24*, 231–243. [CrossRef]
110. Hung, H.F.; Hehnly, H.; Doxsey, S. The Mother Centriole Appendage Protein Cenexin Modulates Lumen Formation through Spindle Orientation. *Curr. Biol. : Cb* **2016**, *26*, 793–801. [CrossRef]
111. Ganapathi Sankaran, D.; Stemm-Wolf, A.J.; Pearson, C.G. CEP135 isoform dysregulation promotes centrosome amplification in breast cancer cells. *Mol. Biol. Cell* **2019**, *30*, 1230–1244. [CrossRef]
112. Gromley, A.; Jurczyk, A.; Sillibourne, J.; Halilovic, E.; Mogensen, M.; Groisman, I.; Blomberg, M.; Doxsey, S. A novel human protein of the maternal centriole is required for the final stages of cytokinesis and entry into S phase. *J. Cell Biol.* **2003**, *161*, 535–545. [CrossRef]
113. Liu, B.; Preisig, P. TGF-beta1-mediated hypertrophy involves inhibiting pRB phosphorylation by blocking activation of cyclin E kinase. *Am. J. Physiol.* **1999**, *277*, F186–F194.
114. Liu, B.; Preisig, P.A. Compensatory renal hypertrophy is mediated by a cell cycle-dependent mechanism. *Kidney Int.* **2002**, *62*, 1650–1658. [CrossRef]
115. Jurkovitz, C.T.; England, B.K.; Ebb, R.G.; Mitch, W.E. Influence of ammonia and pH on protein and amino acid metabolism in LLC-PK1 cells. *Kidney Int.* **1992**, *42*, 595–601. [CrossRef]
116. Ling, H.; Vamvakas, S.; Gekle, M.; Schaefer, L.; Teschner, M.; Schaefer, R.M.; Heidland, A. Role of lysosomal cathepsin activities in cell hypertrophy induced by NH4Cl in cultured renal proximal tubule cells. *J. Am. Soc. Nephrol. : Jasn* **1996**, *7*, 73–80.
117. Glaunsinger, B.A.; Weiss, R.S.; Lee, S.S.; Javier, R. Link of the unique oncogenic properties of adenovirus type 9 E4-ORF1 to a select interaction with the candidate tumor suppressor protein ZO-2. *Embo J.* **2001**, *20*, 5578–5586. [CrossRef]
118. Thomas, M.; Myers, M.P.; Massimi, P.; Guarnaccia, C.; Banks, L. Analysis of Multiple HPV E6 PDZ Interactions Defines Type-Specific PDZ Fingerprints That Predict Oncogenic Potential. *Plos Pathog.* **2016**, *12*, e1005766. [CrossRef]
119. Hernandez-Monge, J.; Garay, E.; Raya-Sandino, A.; Vargas-Sierra, O.; Diaz-Chavez, J.; Popoca-Cuaya, M.; Lambert, P.F.; Gonzalez-Mariscal, L.; Gariglio, P. Papillomavirus E6 oncoprotein up-regulates occludin and ZO-2 expression in ovariectomized mice epidermis. *Exp. Cell Res.* **2013**, *319*, 2588–2603. [CrossRef]
120. Grm, H.S.; Banks, L. Degradation of hDlg and MAGIs by human papillomavirus E6 is E6-AP-independent. *J. Gen. Virol.* **2004**, *85*, 2815–2819. [CrossRef]
121. Lee, S.S.; Glaunsinger, B.; Mantovani, F.; Banks, L.; Javier, R.T. Multi-PDZ domain protein MUPP1 is a cellular target for both adenovirus E4-ORF1 and high-risk papillomavirus type 18 E6 oncoproteins. *J. Virol.* **2000**, *74*, 9680–9693. [CrossRef]
122. Nakagawa, S.; Huibregtse, J.M. Human scribble (Vartul) is targeted for ubiquitin-mediated degradation by the high-risk papillomavirus E6 proteins and the E6AP ubiquitin-protein ligase. *Mol. Cell. Biol.* **2000**, *20*, 8244–8253. [CrossRef]
123. Fink, C.; Weigel, R.; Hembes, T.; Lauke-Wettwer, H.; Kliesch, S.; Bergmann, M.; Brehm, R.H. Altered expression of ZO-1 and ZO-2 in Sertoli cells and loss of blood-testis barrier integrity in testicular carcinoma in situ. *Neoplasia* **2006**, *8*, 1019–1027. [CrossRef]
124. Luczka, E.; Syne, L.; Nawrocki-Raby, B.; Kileztky, C.; Hunziker, W.; Birembaut, P.; Gilles, C.; Polette, M. Regulation of membrane-type 1 matrix metalloproteinase expression by zonula occludens-2 in human lung cancer cells. *Clin. Exp. Metastasis* **2013**. [CrossRef]
125. Saito, K.; Enya, K.; Oneyama, C.; Hikita, T.; Okada, M. Proteomic identification of ZO-1/2 as a novel scaffold for Src/Csk regulatory circuit. *Biochem. Biophys. Res. Commun.* **2008**, *366*, 969–975. [CrossRef]

126. Chlenski, A.; Ketels, K.V.; Engeriser, J.L.; Talamonti, M.S.; Tsao, M.S.; Koutnikova, H.; Oyasu, R.; Scarpelli, D.G. zo-2 gene alternative promoters in normal and neoplastic human pancreatic duct cells. *Int J. Cancer* **1999**, *83*, 349–358. [CrossRef]
127. Chlenski, A.; Ketels, K.V.; Tsao, M.S.; Talamonti, M.S.; Anderson, M.R.; Oyasu, R.; Scarpelli, D.G. Tight junction protein ZO-2 is differentially expressed in normal pancreatic ducts compared to human pancreatic adenocarcinoma. *Int. J. Cancer* **1999**, *82*, 137–144. [CrossRef]
128. Chlenski, A.; Ketels, K.V.; Korovaitseva, G.I.; Talamonti, M.S.; Oyasu, R.; Scarpelli, D.G. Organization and expression of the human zo-2 gene (tjp-2) in normal and neoplastic tissues. *Biochim. Et Biophys. Acta* **2000**, *1493*, 319–324. [CrossRef]
129. Kim, Y.J.; Jung, Y.D.; Kim, T.O.; Kim, H.S. Alu-related transcript of TJP2 gene as a marker for colorectal cancer. *Gene* **2013**, *524*, 268–274. [CrossRef]
130. Tokes, A.M.; Szasz, A.M.; Juhasz, E.; Schaff, Z.; Harsanyi, L.; Molnar, I.A.; Baranyai, Z.; Besznyak, I., Jr.; Zarand, A.; Salamon, F.; et al. Expression of tight junction molecules in breast carcinomas analysed by array PCR and immunohistochemistry. *Pathol. Oncol. Res.* **2012**, *18*, 593–606. [CrossRef]
131. Kato, Y.; Yashiro, M.; Noda, S.; Tendo, M.; Kashiwagi, S.; Doi, Y.; Nishii, T.; Matsuoka, J.; Fuyuhiro, Y.; Shinto, O.; et al. Establishment and characterization of a new hypoxia-resistant cancer cell line, OCUM-12/Hypo, derived from a scirrhous gastric carcinoma. *Br. J. Cancer* **2010**, *102*, 898–907. [CrossRef]
132. Pope, W.B.; Chen, J.H.; Dong, J.; Carlson, M.R.; Perlina, A.; Cloughesy, T.F.; Liau, L.M.; Mischel, P.S.; Nghiemphu, P.; Lai, A.; et al. Relationship between gene expression and enhancement in glioblastoma multiforme: Exploratory DNA microarray analysis. *Radiology* **2008**, *249*, 268–277. [CrossRef]
133. Deroo, B.J.; Korach, K.S. Estrogen receptors and human disease. *J. Clin. Investig.* **2006**, *116*, 561–570. [CrossRef]
134. Acconcia, F.; Marino, M. The Effects of 17beta-estradiol in Cancer are Mediated by Estrogen Receptor Signaling at the Plasma Membrane. *Front. Physiol.* **2011**, *2*, 30. [CrossRef]
135. Doi, Y.; Yashiro, M.; Yamada, N.; Amano, R.; Noda, S.; Hirakawa, K. VEGF-A/VEGFR-2 signaling plays an important role for the motility of pancreas cancer cells. *Ann. Surg. Oncol.* **2012**, *19*, 2733–2743. [CrossRef]

International Journal of
Molecular Sciences

MDPI

Article

Tricellulin Effect on Paracellular Water Transport

Carlos Ayala-Torres, Susanne M. Krug, Jörg D. Schulzke, Rita Rosenthal † and Michael Fromm *,†

Institute of Clinical Physiology/Nutritional Medicine, Medical Department, Division of Gastroenterology, Infectiology and Rheumatology, Charité—Universitätsmedizin Berlin, 12203 Berlin, Germany

* Correspondence: michael.fromm@charite.de

† These authors contributed equally to this work.

Received: 10 October 2019; Accepted: 12 November 2019; Published: 14 November 2019

Abstract: In epithelia, large amounts of water pass by transcellular and paracellular pathways, driven by the osmotic gradient built up by the movement of solutes. The transcellular pathway has been molecularly characterized by the discovery of aquaporin membrane channels. Unlike this, the existence of a paracellular pathway for water through the tight junctions (TJ) was discussed controversially for many years until two molecular components of paracellular water transport, claudin-2 and claudin-15, were identified. A main protein of the tricellular TJ (tTJ), tricellulin, was shown to be downregulated in ulcerative colitis leading to increased permeability to macromolecules. Whether or not tricellulin also regulates water transport is unknown yet. To this end, an epithelial cell line featuring properties of a tight epithelium, Madin-Darby canine kidney cells clone 7 (MDCK C7), was stably transfected with small hairpin RNA (shRNA) targeting tricellulin, a protein of the tTJ essential for the barrier against passage of solutes up to 10 kDa. Water flux was induced by osmotic gradients using mannitol or 4 and 40 kDa-dextran. Water flux in tricellulin knockdown (KD) cells was higher compared to that of vector controls, indicating a direct role of tricellulin in regulating water permeability in a tight epithelial cell line. We conclude that tricellulin increases water permeability at reduced expression.

Keywords: tricellulin; tricellular tight junction; paracellular water transport; tight epithelium; MDCK C7 cells

---

1. Introduction

For a long time, the tight junction (TJ) was considered as a more or less perfect barrier to impede the paracellular permeation of solutes and water [1]. A few years after the membrane-spanning proteins occludin and the claudin family were identified to form the TJ, the first claudin forming a charge-selective channel was discovered [2] (for review see [3]), explaining how the TJ also can provide a selective pathway for the passive permeation through the paracellular pathway.

Tight junctions are organized in two ways. (i) The bicellular tight junction (bTJ) connects two neighboring cells and (ii) the tricellular tight junction (tTJ) is formed at the site where three cells meet. Ultrastructurally, at the tTJ the bicellular strands extend in the basal direction and form a central tube of assumedly 1 μm in length and 10 nm in diameter [4]. While the bTJs is comprised of the transmembrane proteins occludin and the family of claudins [5], the predominant proteins making up the tTJ are tricellulin and angulins [6].

Tricellulin is a tetraspan protein that plays a critical role in sealing the tTJ against the passage of solutes. If it is fully knocked out, the TJ structure is altered and barrier function breaks down [7]. At low expression rates as found in MDCK II cells, the tTJ opens a diffusional pathway at the tTJ for macromolecules of up to 10 kDa [8]. It would be reasonable if also small solutes could pass then, but in MDCK II cells this was not detectable. This paradox is explained by the fact that the tTJ is too rare to contribute significantly to ion permeability in low-resistance (leaky) epithelia in which the permeability of the bTJ predominates. In the high-resistance (tight) human colon cell line HT-29/B6 also an effect of the tTJ on ion permeability became significant [9]. In intestinal biopsies from patients suffering from ulcerative colitis, tricellulin expression was decreased and the passage of macromolecules was increased [10].

Angulins are unified names for three single-span proteins located at the tTJ, lipolysis-stimulated lipoprotein receptor (LSR, angulin-1), immunoglobulin-like domain-containing receptor 1 (ILDR1, angulin-2), and ILDR2 (angulin-3). Angulin-1 has been shown to recruit tricellulin to the tTJ [11] and the same effect was observed for angulin-2 and -3 [12]. Angulin-1 and -2 are expressed in a complementary way with angulin-3. In addition, angulin-1 and -2 exert a barrier-maintaining effect, while angulin-3 has a weaker effect in this regard [12].

Whether or not the TJ at all is permeable to water was highly controversial for a long time. The transcellular pathway for water via aquaporin channels was established more than 20 years ago [13] and their presence has been described in a wide range of tissues [14]. Another transcellular pathway for water is provided by the Na^+-glucose cotransporter SGLT1 (and several other carriers) which contributes, especially in small intestinal epithelia, significantly to water transport [15].

Regarding the paracellular pathway through the bTJ, Rosenthal et al. have developed an elaborated though difficult to handle technique to measure water transport on monolayers mounted in Ussing chambers [2]. First it was demonstrated that claudin-2, so far identified as a paracellular cation channel [2], also forms a water channel [16]. Cations and water were then shown to pass through the same pore [17]. Recently, a second water channel through the bTJ was identified to be claudin-15, which also was known so far as a cation channel only [18].

With this in mind, three subtypes of channel-forming claudins can be functionally distinguished regarding their permeability to water: (i) proteins forming channels for anions but not for water (claudin-10a and -17) [19,20], (ii) proteins forming channels for cations but not for water (claudin-10b) [16], and (iii) proteins forming channels for cations and also for water (claudin-2 and -15) [16–18]. After it has been demonstrated that the tTJ can provide another paracellular pathway for solutes [8–10,21], the question arises whether or not it also contributes to water passage.

As for the tTJ, Gong et al. provided compelling evidence that water transport is increased in isolated kidney tubules in the absence of angulin-2, but is normally inhibited in presence of angulin-2 [22]. The role of tricellulin has not been elucidated so far. We hypothesize that tricellulin is a direct negative actuator of tTJ water permeability. Therefore, we aimed to characterize the effect of tricellulin expression on water transport in the tight epithelial cell line MDCK C7 and found that the water passage through the tTJ was indeed enhanced under the influence of reduced tricellulin levels.

2. Results

2.1. Characterization of Tricellulin Knockdown in MDCK C7 Cells

After transfection of Madin-Darby canine kidney cells clone 7 (MDCK C7) with shRNA targeting tricellulin, puromycin-resistant cell clones were screened for tricellulin knockdown by western blotting. In this study, two knockdown clones and two control clones were investigated (KD 23 and its corresponding control 2, and KD 24 and its corresponding control 9). The two KD clones showed high but different reduction in tricellulin expression: clone KD 23 showed a depression of tricellulin by 30% while clone KD 24 exhibited a drop by 40% (Figure 1a and Table 1).

The transepithelial resistance (TER) was reduced in tricellulin knockdown MDCK C7 cells (Figure 1b, Table 1). KD 24 exhibited a stronger reduction of tricellulin expression and stronger decrease in TER compared to the corresponding control than KD 23. This was due to a huge increase in Na^+ and Cl^- permeability without any preference for anions or cations (Figure 1c, Table 1).

Then we examined the permeability of the four clones to fluorescein isothiocyanate-dextran 4 kDa (FITC–dextran, FD4). As a result, FD4 permeability of KD 23 was not significantly different from its control, but that of clone KD 24 exhibited a strong increase (Figure 1d, Table 1). Although clones KD 23 and KD 24 differed only gradually in reduction of tricellulin expression and TER decrease, the two chosen clones differed in macromolecule permeability.

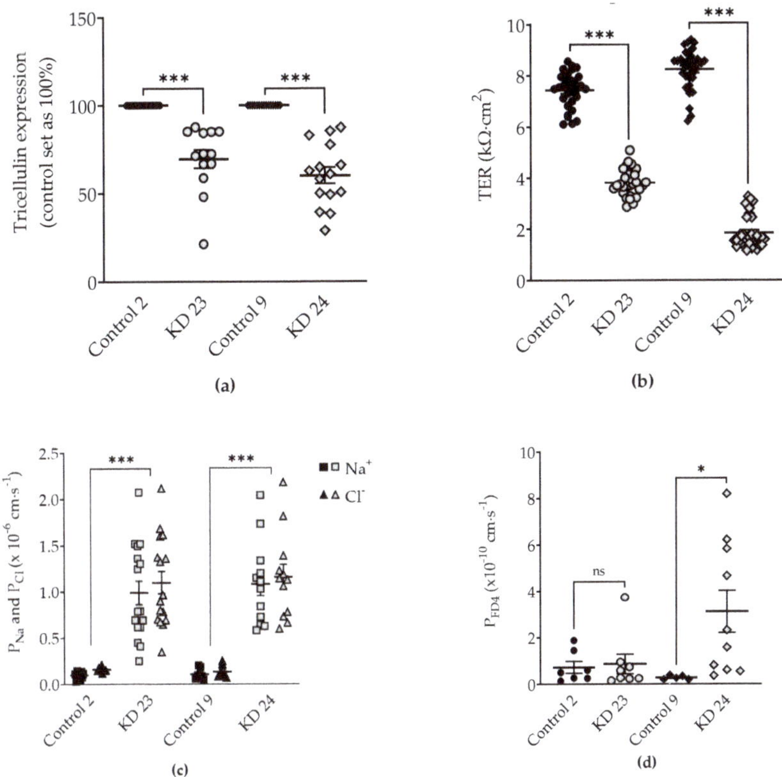

Figure 1. Expression and functional analysis of tricellulin knockdown in MDCK C7 cells. (**a**) Densitometric analysis of tricellulin protein expression levels in stable shTRIC transfectants (KD 23 and KD 24) in comparison to vector-transfected controls (Control 2 and Control 9). shTRIC leads to decreased tricellulin expression (n = 15). (**b**) Effect of tricellulin knockdown on transepithelial resistance (TER). Tricellulin knockdown decreases TER in MDCK C7 cells (n = 24). (**c**) Effect of tricellulin knockdown on ion permeability. Tricellulin knockdown increases Na^+ and Cl^- permeability to the same extent in MDCK C7 cells (n = 7–16). (**d**) Effect of tricellulin knockdown on permeability for 4 kDa-FITC dextran (FD4). Permeability is increased in tricellulin KD 24 cells only (n = 5–10). Statistical analysis was performed using Student's t-test between the KD clone and the corresponding control (KD 23 versus control 2, KD 24 versus control 9). (Symbols of (a), (b) and (d): Control 2: ●; KD 23: ○; Control 9: ◆; KD 24: ◇. * p <0.05, *** p < 0.001, ns = not significant).

2.2. *Effects of Tricellulin Knockdown on Endogenous Proteins of MDCK C7 Cells*

Knockdown of one TJ protein may cause variation in other proteins, which are involved in transepithelial water transport. Therefore, we examined levels of TJ proteins, such as occludin, several claudins, the three angulins, and of aquaporin (AQP) water channels, AQP1, 3, and 4 (Figure 2a). The densitometric analysis revealed some clonal variability in TJ protein expression between the knockdown clones and the controls. In detail, claudin-1 was increased in the tricellulin KD 23, and occludin and claudin-4 and -8 were decreased in the tricellulin KD 24 (Figure 2b). Clonal variability in protein expression was also observed between both control clones and both KD clones. None of the three angulins were significantly altered. Most importantly, the tricellulin knockdown did not affect the expression of the cell membrane water channels expressed in MDCK C7 cells, AQP1, AQP3, and AQP4 (Figure 2b).

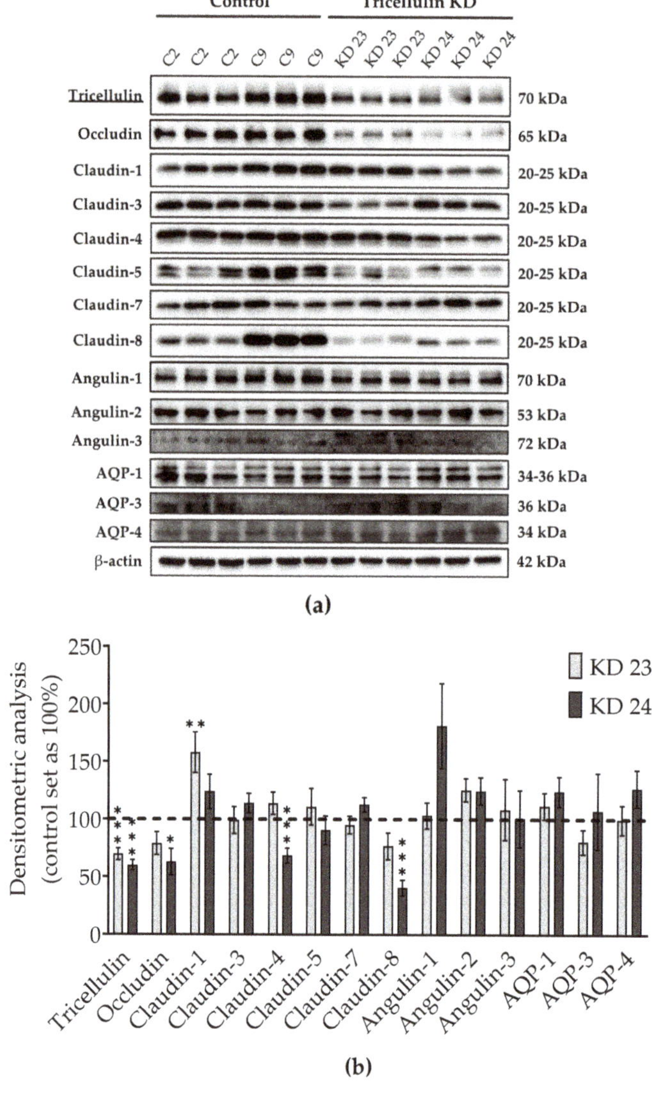

Figure 2. Claudin, occludin, angulin and aquaporin (AQP) expression in tricellulin knockdown MDCK C7 cells. (**a**) Representative western blots. (**b**) Densitometric analysis of protein expression levels in tricellulin knockdown clones KD 23 and KD 24 in comparison to the corresponding vector-transfected controls (n = 9–17, N = 7). β-Actin was used as an internal control for normalization to protein content. Statistical analysis was performed using Student's t-test between the KD clone and the corresponding control (KD 23 versus control 2, KD 24 versus control 9). (* p <0.05, ** p < 0.01, *** p < 0.001).

The proper localization of tricellulin was confirmed for the four clones by immunofluorescence confocal laser-scanning microscopy. Occludin served as a TJ marker (Figure 3).

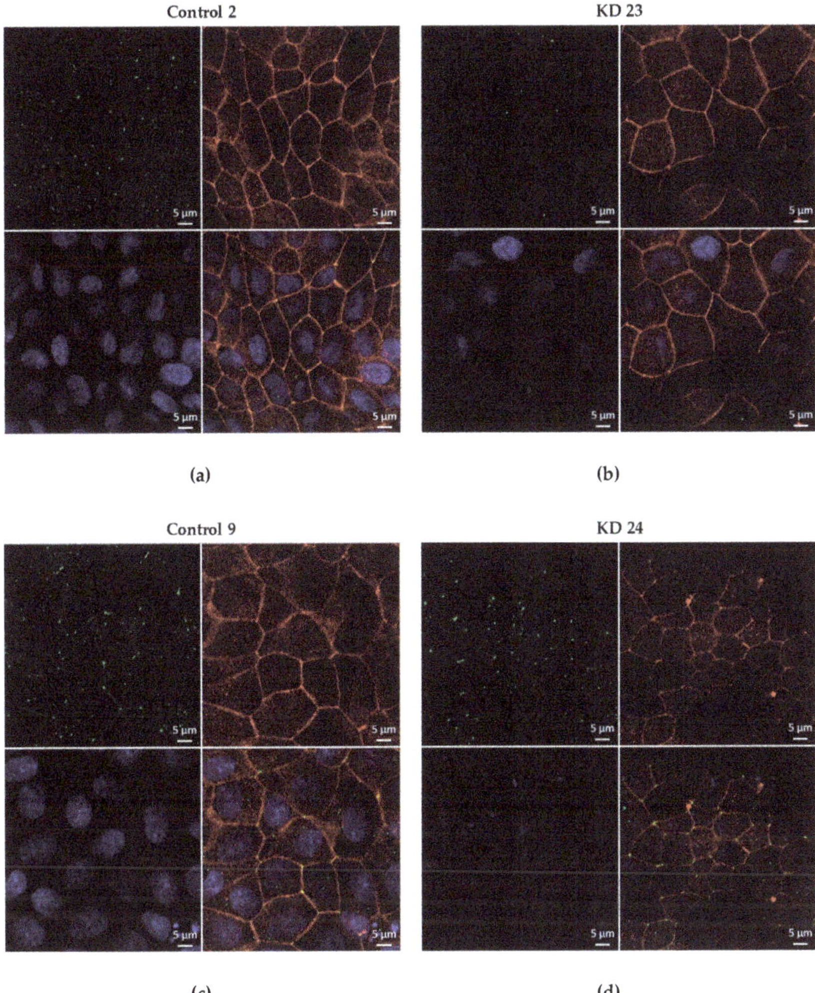

Figure 3. Localization of tricellulin in the four clones used throughout this study. In the KD clones, tricellulin is localized at tricellular tight junction (tTJ) sites. Tricellulin: green; occludin: red; cell nuclei (4′,6-Diamidine-2′-phenylindole dihydrochloride - DAPI): blue. MDCK C7 cells transfected with (**a**,**c**) empty lentiviral vector for shRNA expression (pLKO.1-puro) as a negative control, (**b**) pLKO.1-puro vector containing a sequence for shRNA targeting tricellulin (shRNA 23), and (**d**) pLKO.1-puro vector containing a sequence for shRNA targeting tricellulin (shRNA 24).

2.3. *Effect of Tricellulin Knockdown on Transepithelial Water Transport in MDCK C7 Cells*

To analyze the effect of tricellulin knockdown on water permeability of MDCK C7 cells, water flux was measured after induction with osmotic gradients produced by 100 mM mannitol, by 37 mM 4 kDa-dextran, or by 100 mM 4 kDa-dextran (Figure 4). These concentrations produced measured osmolality gradients of 100 mOsm for 100 mM mannitol and for 37 mM 4 kDa-dextran, and of 900 mOsm for 100 mM 4 kDa-dextran.

Tricellulin KD 23 as well as KD 24 exhibited increased water fluxes compared with their respective vector control clones, however the change obtained by KD 24 was larger under all conditions (Figure 4,

Table 1). In KD 24, the increase under a mannitol-induced osmotic gradient was 2.6-fold (Figure 4a), under a gradient induced by 37 mM 4 kDa-dextran it was 5.2-fold (Figure 4b) and under a gradient induced by 100 mM 4 kDa-dextran the increase was 3.8-fold (Figure 4c), compared to 1.5-fold, 2.1-fold, and 1.4-fold in KD 23, respectively.

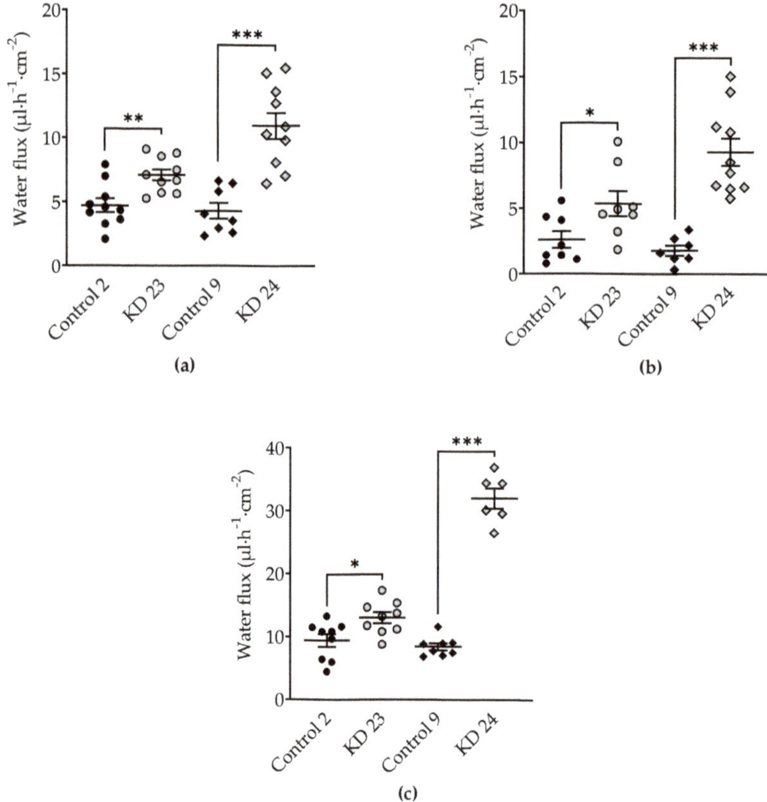

Figure 4. Water flux in tricellulin knockdown MDCK C7 cells stimulated by an osmotic gradient on the apical side of the cell layer. Water flux induced by a gradient of (**a**) 100 mM mannitol (n = 8–10), (**b**) 37 mM 4 kDa-dextran (n = 8–10) and (**c**) 100 mM 4 kDa-dextran (n = 6–9) was increased in both tricellulin knockdown clones, with the stronger increase in clone KD 24 with the higher reduction of tricellulin expression. Statistical analysis was performed using Student's t-test between the KD clone and the corresponding control (KD 23 versus control 2, KD 24 versus control 9). (Control 2: ●; KD 23: ○; Control 9: ◆; KD 24: ◇. * p <0.05, ** p < 0.01, *** p < 0.001).

In a parallel same series of experiments, a 5.5 mM 40 kDa-dextran gradient was applied which produced a measured osmolality gradient of 100 mOsm. In the presence of 100 mOsm 40 kDa-dextran, water flux was lower than in the presence of 100 mOsm mannitol. The 40 kDa-dextran gradient did not significantly change water flux in tricellulin KD 24 compared to control, whereas water flux in KD 23 was increased (Figure 5, Table 1). This finding indicates that the water flux through the tTJ was inhibited in the presence of 40 kDa-dextran in the clone with the strong reduction of tricellulin.

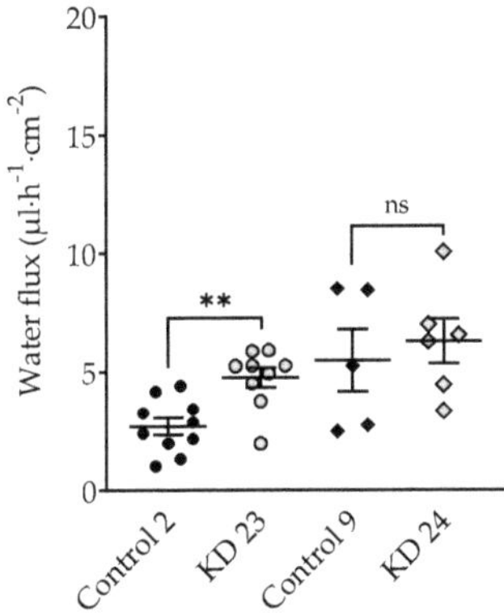

Figure 5. Water flux in tricellulin knockdown MDCK C7 cells stimulated by an osmotic gradient of 5.5 mM 40 kDa-dextran (100 mOsm) on the apical side of the cell layer. No effect could be observed in tricellulin KD 24 cells, whereas in KD 23 with the higher tricellulin expression, the increase was in the same order of magnitude as in the presence of 100 mM (100 mOsm) mannitol. Statistical analysis was performed using Student' *t*-test between the KD clone and the corresponding control (KD 23 versus control 2, KD 24 versus control 9). (Control 2: ●; KD 23: ○; Control 9: ◆; KD 24: ◇. ** $p \leq 0.01$, ns = not significant).

Table 1. Characteristics of MDCK C7 tricellulin knockdown clones and the corresponding controls. Two tricellulin knockdown clones and their corresponding controls were analyzed in this study (Control 2 and KD 23, Control 9 and KD 24). Data of tricellulin expression have been obtained by densitometric analysis of western blots using β-actin for normalization. Data of Na^+/Cl^- permeability and absolute permeabilities for Na^+ and Cl^- (P_{Na}, P_{Cl}) were obtained from dilution potential measurements in the Ussing chamber. Water flux measurements were performed in a modified Ussing chamber with water flux induced by different osmotic gradients. Significances refer to respective controls. *n* number of experiments, * $p \leq 0.05$, ** $p \leq 0.01$, *** $p \leq 0.001$.

		Control 2	KD 23	Control 9	KD 24
Tricellulin Expression (%)		100 (*n* = 15)	69.9 ± 4.8 *** (*n* = 13)	100 (*n* = 15)	60.1 ± 4.6 *** (*n* = 15)
TER (kΩ·cm^2)		7.4 ± 0.2 (*n* = 24)	3.8 ± 0.2 *** (*n* = 24)	8.2 ± 0.2 (*n* = 24)	1.8 ± 0.1 *** (*n* = 24)
P_{FD4} (×10^{-10} cm·s^{-1})		0.72 ± 0.23 (*n* = 7)	0.87 ± 0.39 (*n* = 8)	0.28 ± 0.03 (*n* = 5)	3.11 ± 0.86 * (*n* = 10)
P_{Na} (×10^{-6} cm·s^{-1})		0.10 ± 0.01 (*n* = 9)	0.99 ± 0.13 *** (*n* = 16)	0.11 ± 0.02 (*n* = 7)	1.08 ± 0.13 *** (*n* = 12)
P_{Cl} (×10^{-6} cm·s^{-1})		0.16 ± 0.01 (*n* = 9)	1.10 ± 0.12 *** (*n* = 16)	0.14 ± 0.03 (*n* = 7)	1.16 ± 0.14 *** (*n* = 12)
P_{Na}/P_{Cl}		0.81 ± 0.05 (*n* = 9)	0.94 ± 0.04 (*n* = 16)	0.85 ± 0.07 (*n* = 7)	1.02 ± 0.03 (*n* = 12)
Water flux (µl·h^{-1}·cm^{-2})	100 mM mannitol (100 mOsm)	4.7 ± 0.5 (*n* = 10)	7.1 ± 0.4 ** (*n* = 10)	4.3 ± 0.6 (*n* = 8)	11.0 ± 1.0 *** (*n* = 10)
	37 mM 4 kDa-dextran (100 mOsm)	2.6 ± 0.7 (*n* = 8)	5.4 ± 0.8 * (*n* = 8)	1.8 ± 0.5 (*n* = 7)	9.3 ± 1.0 *** (*n* = 10)
	100 mM 4 kDa-dextran (900 mOsm)	9.4 ± 1.0 (*n* = 9)	13.0 ± 0.8 * (*n* = 9)	8.4 ± 0.5 (*n* = 8)	32.0 ± 1.4 *** (*n* = 6)
	5.5 mM 40 kDa-dextran (100 mOsm)	2.7 ± 0.3 (*n* = 10)	4.7 ± 0.6 ** (*n* = 9)	5.2 ± 1.2 (*n* = 5)	6.3 ± 0.9 (*n* = 6)

3. Discussion

This study shows that tricellulin downregulation in the tight epithelial cell line MDCK C7 resulted in an increased ion permeability, as shown by the reduction of TER and, dependent on the reduction of tricellulin, in an increased macromolecule permeability. Additionally, a decrease in tricellulin expression is associated with an increased transepithelial water flux. Thus, tricellulin is able to regulate osmotically-induced transepithelial water flux in a tight epithelium. This regulation occurs in a reciprocal way, i.e., lowered tricellulin expression causes higher water fluxes.

3.1. Effect of Tricellulin Knockdown on Macromolecule Passage

The MDCK cell line is derived from canine kidney cells and common strains feature properties of either proximal (MDCK II, MDCK C11) or distal nephron (MDCK I, MDCK C7) [23,24].

Transfection of the tight epithelial cell line MDCK C7 with shRNA targeting tricellulin resulted in a reduction of this protein in the tTJ. In agreement with former experiments on cell cultures [10], the reduction in the expression of tricellulin resulted in a lowered transepithelial resistance and the development of a pathway for macromolecules. Complete knockout, however, leads to a disorganization of not only tTJs but also bTJs and complete breakdown of barrier parameters [7]. This is of clinical importance in genetic defects in tricellulin causing hearing loss [25], however, for studies focused selectively on the tTJ, graded knockdown approaches that reduce tricellulin only in part are suitable.

In addition, the permeability to the paracellular marker 4 kDa-dextran was increased by tricellulin knockdown as normally this protein is responsible for blocking the passage of macromolecules with diameters from 1.3 to 4.6 nm (0.9 to 10 kDa) [8]. This is consistent with the present data showing that in tight epithelial cell lines a downregulation of tricellulin strongly regulates the permeability for 4 kDa-macromolecules. It might be surprising that a moderate knockdown of tricellulin by 40% already leads to five times higher permeability for macromolecules. This is explained by the fact that normally the permeability for macromolecules is extremely low (in the order of 10^{-12} cm/s) so that even a slight opening of a pathway for macromolecules causes a significant effect [21]. Furthermore, the reduction by 30% (KD 23) had no significant effect on macromolecule permeability, however, did have an effect on ion permeability, indicating that there is already a significant impairment of the tTJ barrier which however is only affecting small solutes, while 4 kDa-dextran is still not able to pass. This leads to the assumption that the varying range of tricellulin expression may lead to different effects. The limit of macromolecule passage seems to range between a 30% to 40% reduction of tricellulin, while for small solutes the regulation is stricter.

3.2. Effects of Tricellulin Knockdown on Other Proteins of MDCK C7 Cells

Immunoblot analysis with an anti-tricellulin antibody demonstrated a clear reduction of tricellulin expression (Figure 2a). Nevertheless, the generated tricellulin knockdown clones compared with the controls showed clonal variation in other TJ proteins which reached significance for occludin, claudin-1, claudin-4, and claudin-8 (Figure 2a,b). The clonal variation also concerns the two control clones, here the differences are most prominent for claudin-1, claudin-5, and claudin-8. These claudins seem to have no effect on paracellular water transport, since the water transport of the two control clones did not show any difference. In the tricellulin knockdown clones, the expression of occludin, claudin-1, claudin-4, and claudin-8 was changed compared to the corresponding vector control. Considering the functional aspects of the variations it can be said that occludin, which is downregulated only in one of the tricellulin knockdown clones, is assumed to have no effect on TJ barrier function, either genuinely or through compensation of its downregulation by other tight junction proteins [26–28]. All claudins present in MDCK C7 cells are known to behave as barrier formers, acting in concert and compensating their respective up- and down-regulation [29]. Ion and water channel-forming claudins can be disregarded as they are not genuinely present in MDCK C7 cells. A reduced expression of claudin-4 and claudin-8 and an increased expression of claudin-1 was found in one of the knockdown clones, respectively. However, claudin-4 is described to be dispensable for the barrier properties of TJs in wild-type MDCK II cells [30]. Furthermore, claudin-1 and claudin-4 were shown to be dispensable for water barrier formation in human submerged keratinocyte cultures [31]. Therefore, we feel safe to conclude that the observed clonal variation of the claudins in sum is balanced and provides a constant barrier function of the bTJ. Most importantly, the three angulins as well as the aquaporins AQP-1, -3, and -4 did not significantly change. Thus, the changes in ion, macromolecule and water permeability found in the knockdown clones can be assumed to be exclusively due to the reduced tricellulin expression in the tTJ.

That the knockout of a tTJ-related protein, angulin-2, increases water transport was shown in elegant experiments by Gong et al. on isolated mouse kidney tubules [24]. The authors suggest that the barrier function of angulin-2 derives from a trans-interaction between its extracellular immunoglobulin-like domain. All three angulins are known to be able to recruit tricellulin to the tTJ [11]. It is beyond the scope of the present study but will be subject of future investigation, to experimentally clarify whether angulins mediate a direct effect on water permeability independent of tricellulin, or angulins act indirectly via tricellulin regulation.

3.3. Water Transport as Driven by Different Osmotic Gradients

In other studies from this lab using the method of generating water flux by an osmotic gradient e.g., by mannitol, this solute was impermeable and could not pass the TJ by itself [16–20]. In the present study, for the first time this was different and more complicated. In the tricellulin KD clones the tTJ central tube is opened, and through this pathway mannitol can diffuse following its concentration gradient. This has two consequences: First, in the course of the experiment, mannitol may be diluted and its gradient is reduced. However, this will be below significance, because the amount of mannitol in the 9 mL bath solution is high compared to the amount diffusing through the tTJ. Second and more disturbing, mannitol would diffuse through the central tube in the opposite direction to the presumed water flux. As long as a molecule like mannitol is small compared to the diameter of the central tube, this may not have a large effect. As this mechanism may alter water transport rates at least slightly, we decided not to calculate water permeability coefficients in this study and present data as fluxes only.

In order to investigate the side effect of mannitol diffusion, we compared data obtained with 100 mM mannitol with those obtained with larger molecules, 37 mM 4 kDa- and 5.5 mM 40 kDa-dextran. For this, dextran concentrations were chosen which all produced an osmolality of 100 mOsm, as used for mannitol. This resulted in lower nominal concentrations due to a known water-sequestering effect of some macromolecules [32].

It turned out that gradients of 100 mM mannitol and 37 mM 4 kDa-dextran produced comparable water fluxes and that an elevated 4 kDa-dextran concentration (100 mM resulting in roughly 900 mOsm) led to higher water fluxes (see Table 1). Since these numbers satisfactorily fit together this indicates the reliability of the methods used.

The most valuable results should have been obtained with an osmotic gradient obtained by 5.5 mM 40-dextran (measured osmolality 100 mOsm). Surprisingly, this was not the case: In the presence of 40 kDa-dextran, water flux was reduced in the control clones as well as in the knockdown clones compared to water fluxes in the presence of mannitol with the same osmolality. An induction of water flux by a gradient with 40 kDa-dextran did not produce significantly changed water fluxes in the KD 24 clone with the stronger tricellulin reduction, whereas KD 23 showed an increased water flux compared to the control clone. As an explanation for this effect, the sizes of the molecule should be related to that of the tTJ central tube. The hydrodynamic radius of 40 kDa-dextran reportedly amounts 6.6 nm [33] while the diameter of the tTJ central tube was assumed to be 10 nm [4]. We suggest that due to its molecular size the 40 kDa-dextran only barely fits through the central tube or even clogs its entrance so that water is hindered to pass. Thus, water flux was reduced in all clones compared to mannitol-induced fluxes. The strongest reduction was observed in the KD 24 clone with the lowest tricellulin expression and possibly the largest diameter of the tTJ central tube. These assumptions are supported by the findings of Krug et al. that the tTJ central tube at low tricellulin levels allows for passage of molecules of 10 kDa but not of 20 kDa [8]. This reasoning finds its equivalent in experiments dealing with the bTJ, where a NaCl gradient was used to drive water flux through the claudin-15 channel. Under this condition, claudin-15-mediated water flux was inhibited by Na^+ diffusion in the opposite direction [18].

3.4. Effect of Tricellulin Knockdown on Transepithelial Water Transport

For the full line of experiments, two tricellulin KD clones were selected, which differ in the grade of tricellulin reduction, namely by 30% (clone KD 23) and by 40% (clone KD 24) compared to the respective controls. While both KD clones caused a decrease in TER, only KD 24 produced an increase in permeability to a 4 kDa-macromolecule and, most importantly, an increase of water flux driven by different osmotic gradients as discussed above.

The question arises whether this is an all-or-nothing effect. If yes, one would postulate a threshold of tricellulin depression somewhere between 30% and 40%. In order to have closer look, an additional experiment was performed employing a clone with tricellulin reduced by 35% (clone KD 22). Being aware that this is based on three points only, resulting water flux did appear to correlate with the tricellulin expression, suggesting that the effect of tricellulin KD depression is gradual, starting at the level found for the controls (Figure 6).

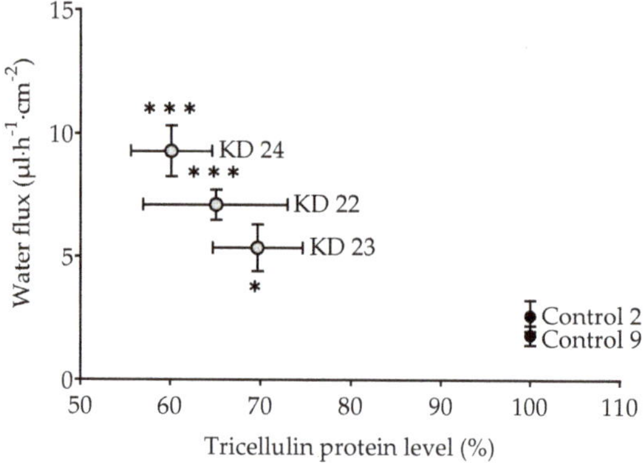

Figure 6. Water flux as a function of the level of tricellulin expression in MDCK C7 cells driven by an osmotic gradient of 37 mM 4 kDa-dextran (100 mOsm). Tricellulin KD 22 ($n = 7$) is an additional KD clone that was tested only for this purpose (shRNA 22: TRCN0000072635, NM_144724.1-989s1c1, Sigma-Aldrich, Schnelldorf, Germany) and compared with Control 9. Statistical analysis was performed using Student's *t*-test between the KD clone and the corresponding control (KD 23 versus control 2, KD 24 versus control 9). In case of the KD 22 and KD 24 and their control 9, the Bonferroni-Holm adjustment for multiple testing was applied. (* $p < 0.05$, *** $p < 0.001$).

4. Materials and Methods

4.1. Cell Culture, Transfection, and TER Measurement

The kidney cell line MDCK C7 (RRID: CVCL_0423) exhibits a high transepithelial resistance and other basic properties making it an excellent model of a tight epithelium. For stable tricellulin knockdown, MDCK C7 cells were transfected with pLKO.1-puro vector containing a sequence for shRNA targeting tricellulin (shRNA 23: TRCN0000072633, NM_144724.1-2011s1c1; and shRNA 24: TRCN0000072634, NM_144724.1-1097s1c1; Sigma-Aldrich, Schnelldorf, Germany) or pLKO.1-puro empty vector as a negative control (Sigma-Aldrich, Schnelldorf, Germany).

The transfected cells were incubated at 37 °C and 5% CO_2 in sterilization incubators held. For the cultivation of the cells sterile culture vessels made of plastic were used. The cells were cultured in a nutrient medium Earl's salts MEM (minimal essential medium) supplemented with 10% FCS as well as 100 U/mL penicillin/100 µg/mL streptomycin and 1.5 µg/mL of puromycin. Every second to third day the medium was changed. Consumption of nutrients was also visible due to a color change of the medium which contained a detecting pH indicator.

For protein quantification, water flux measurements and electrophysiological studies, cell monolayers were cultured on porous culture plate inserts (Millicell PCF filters, pore size 0.4 µm, effective area 0.6 cm^2, Millipore GmbH, Schwalbach, Germany) for 7–9 days before they were used for experiments.

Transepithelial resistance (TER) was measured at 37 °C using chopstick electrodes (STX2, World Precision Instruments, Friedberg, Germany). Electrodes were reproducibly positioned by a semi-automatic motor-driven device and signals were processed by a low-frequency clamp (both own design). The resistances of the bathing solution and the blank filter support were subtracted from measured values, which finally were converted to Ω cm^2.

4.2. Western Blot Analysis

Cells grown on culture-plate inserts were scraped and homogenized in total lysis buffer containing 10 mM of Tris, 150 mM of NaCl, 0.5% of Triton X-100, and 0.1% of sodium dodecyl sulphate (SDS), and protease inhibitors (complete™ ethylenediaminetetraacetic acid (EDTA)-free; Roche, Basel, Switzerland). The samples were incubated 2 h at 4 °C (vortexed every 20 min), and then centrifuged at 11,000× *g* during 20 min. The pellet was discarded and the protein concentration in the supernatant was determined by the BCA (bicinchoninic acid) method (reagents were purchased from Pierce (Perbio Science, Bonn, Germany)) and quantified with a plate reader (Tecan Deutschland, Crailsheim, Germany). Aliquots between 10 and 15 µg protein samples were separated by 12% SDS-polyacrylamide gel electrophoresis and then transferred to a PVDF membrane (Perkin Elmer, Rodgau, Germany) for detection of tight-junction-associated marvel proteins (TAMP), claudins, angulins and AQPs. After blocking for 2 h in 1% PVP-40 and 0.05% Tween-20, membranes were incubated overnight with primary antibodies specific for tricellulin (Abfinity, Invitrogen), occludin (Invitrogen), angulin-1, -2, and -3 (Sigma-Aldrich, Germany), claudin-1, -3, -4, -5, -7, and -8 (Invitrogen), and AQP-1, -3, and -4 (Santa Cruz, Heidelberg, Germany). After removing the first antibody and three washing steps, the membranes were incubated 2 h with the second peroxidase-conjugated antibody (anti-mouse or anti-rabbit) in 2% of milk powder prepared in TBST 1X. For detection of the chemiluminescence signal induced by addition of Lumi-LightPLUS western blotting kit (Roche) a Fusion FX7 (Vilber Lourmat, Eberhardzell, Germany) were used. Densitometric analysis was performed with quantification software (Image Studio™ Lite, LI-COR Biosciences, Lincoln, Nebraska USA). Equal protein loading in each lane was verified by comparison with signals for β-actin (Sigma-Aldrich). For western blot analysis, lysates of at least three individual cell cultures were used and one representative experiment is shown.

4.3. Immunofluorescent Staining

Immunofluorescence studies were performed on culture-plate inserts. Confluent monolayers were rinsed with phosphate-buffered saline (PBS), fixed with 4% paraformaldehyde for 20 min and permeabilized for 10 min with PBS containing 0.5% (*v/v*) Triton X-100. To block non-specific binding sites, cells were then incubated in PBS containing 1% (*w/v*) BSA and 5% (*v/v*) goat serum (blocking solution; Biochrom) for 60 min. All subsequent washing procedures were performed with this blocking solution. After blocking, cells were incubated 60 min at room temperature with primary antibodies for tricellulin (Abfinity, Invitrogen; 1:250) and occludin (Invitrogen, 1:250), followed by washing steps and incubation during 60 min at room temperature with the respective secondary antibodies (Alexa Fluor 488 goat anti-rabbit and Alexa Fluor 594 goat anti-mouse, each 1:500; Molecular Probes MoBiTec) and 4′,6-diamidino-2-phenylindole (1:1000). Images were obtained with a confocal laser-scanning microscope (LSM 780, Zeiss, Jena, Germany) and processed using ZEN software (Zeiss, Oberkochen, Germany).

4.4. Measurement of 4 kDa-FITC-Dextran Flux

Flux studies were performed in conventional Ussing chambers under voltage clamp conditions. Dextran flux was measured in 5 mL circulating Ringer's containing 37 mM unlabeled 4 kDa-dextran on the apical side. After addition of 100 μL of 37 mM FITC-labeled dialyzed dextran (4 kDa-FITC-dextran; Sigma-Aldrich) to the apical bath, basolateral samples (200 μL) were collected at 0, 20, 40, 60, 80, 100, and 120 min. Tracer fluxes were determined from FITC-dextran samples, which were measured with a fluorometer at 520 nm (Spectramax Gemini, Molecular Devices, Ismaning, Germany). Dextran permeability was calculated from $P = J/\Delta c$ with P = permeability (cm s^{-1}), J = flux (mol h^{-1} cm^{-2}) and c = concentration (mol/L).

4.5. Dilution Potential Measurements

Dilution potential measurements for the determination of ion permeabilities were performed in Ussing chambers modified for cell-culture inserts. Water-jacketed gas lifts kept at 37 °C were filled with 10 mL circulating fluid on each side. The bathing solution contained (in mM) 119 NaCl, 21 $NaHCO_3$, 5.4 KCl, 1.2 $CaCl_2$, 1 $MgSO_4$, 3 HEPES, and 10 D(+)-glucose, and was gassed with 95% O_2 and 5% CO_2 to ensure a pH value of 7.4. All experimental data were corrected for the resistance of the empty filter and the bathing solution. Dilution potentials were measured with modified bathing solution on the apical or basolateral side of the epithelial monolayer. In the modified bathing solution, NaCl was iso-osmotically replaced by mannitol. The ratio of P_{Na} and P_{Cl} and the absolute permeabilities for Na^+ and Cl^- were calculated as described before [19].

4.6. Measurement of Transepithelial Water Transport

Water flux measurements were performed using a modified Ussing chamber with two glass tubes instead of the gas lifts as described before [16–18]. Throughout these experiments, transepithelial resistance (TER, $\Omega \cdot cm^2$), short-circuit current (I_{SC}, $\mu A \cdot cm^{-2}$), and transepithelial voltage (mV) were recorded routinely. Resistances of bathing solution and blank filter support were measured prior to each experiment and subtracted. The stability of TER served as an indicator of cell viability.

Cell filters were mounted in Ussing chambers and perfused with HEPES-buffered solution with the following composition (in mM): 144.8 NaCl, 2.4 Na_2HPO_4, 0.6 NaH_2PO_4, 5.4 KCl, 1.2 $MgCl_2$, 1.2 $CaCl_2$, 10.6 HEPES, and 10 D(+)-glucose. The pH value of the perfusion solution was pH 7.4. A rotary pump ensured constant circulation of the perfusion solution (4.0 $mL \cdot min^{-1}$) and thus a fast fluid exchange in both hemichambers (volume 500 μL) to avoid effects of unstirred layers on water permeability. Water flux was induced by a transepithelial osmotic gradient: (i) 100 mM mannitol, (ii), 37 mM 4 kDa-dextran, (iii) 100 mM 4 kDa-dextran, or (iv) 5.5 mM 40 kDa-dextran. The solution was added in the apical compartment of the Ussing chamber.

The osmolality of the perfusion solutions (mosmol/kg, abbreviated mOsm) was determined using a Vapor Pressure Osmometer (5100B, Wescor, Logan, UT). The osmolality of the HEPES-buffered solution for water flux measurements was 287.4 ± 5.7 mOsm (n = 15). The solutions for measuring water flux ((i) 100 mM mannitol, (ii), 37 mM 4 kDa-dextran, (iii) 100 mM 4 kDa-dextran or (iv) 5.5 mM 40 kDa-dextran) increased the osmolality to 388.2 ± 3.2 mOsm (n = 5), 399.4 ± 2.4 mOsm (n = 10), 877.8 ± 13.7 mOsm (n = 5), and 390.2 ± 2.5 mOsm (n = 5), respectively. Molecular diameters of the dextrans were estimated using the equation: $d_{(Å)} = 0.215 \cdot (MW^{0.587})$.

The fluid level in both glass tubes was monitored by a visual system ColorView XS (Olympus Soft Imaging Solutions GmbH, Munster, Germany), at time 0 min and with intervals of 10 min over a period of 120 min. Transepithelial water flux, given as flux per square centimeter and hour, was calculated after special calibration from the difference between the menisci at the registration times. Fluxes directed from the basolateral to the apical compartment were defined as positive flux.

4.7. Statistical Analysis

Data are expressed as mean values ± SEM (standard error of the mean), indicating n as the number of single measurements, and N as the number of independent experiments, which means independent seeding of cells. Statistical analysis was performed using Student's t-test between the KD clone and the corresponding control (KD 23 versus control 2, KD 24 versus control 9). In case of the KD 22 and KD 24 data of Figure 6, the Bonferroni–Holm adjustment for multiple testing was applied. $p < 0.05$ was considered significant (* $p < 0.05$, ** $p < 0.01$, *** $p < 0.001$).

5. Conclusions

This study analyzed the contribution of the tTJ to water permeability as regulated by its prominent protein tricellulin. Our results demonstrate that tricellulin KD in MDCK C7 cells, which lack the bTJ cation and water channels claudin-2 and -15, results in an increase in transepithelial water transport. This water flux appears to be negatively correlated with the tricellulin expression level. We conclude that in a tight epithelial cell line, tricellulin contributes to regulate water flux through the paracellular pathway.

Author Contributions: Conceptualization, R.R., M.F., S.M.K. and C.A.-T.; Funding Acquisition, M.F., J.D.S.; Methodology, R.R. and M.F.; Investigation, C.A.-T.; Supervision and Validation, R.R., M.F. and S.M.K.; Visualization, C.A.-T.; Writing–Original Draft Preparation, C.A.-T.; Writing–Review and Editing, M.F., R.R., S.M.K. and J.D.S.

Funding: This research was funded by the Deutsche Forschungsgemeinschaft (DFG), Research Training Group GRK 2318 A3, and the Open Access Publication Fund of the Charité–Universitätsmedizin Berlin.

Acknowledgments: We would like to thank Britta Jebautzke, In-Fah M. Lee, and Anja Fromm (all three: Institute of Clinical Physiology/Nutritional Medicine, Medical Department, Division of Gastroenterology, Infectiology and Rheumatology, Charité – Universitätsmedizin Berlin, 12203 Berlin, Germany) for their expert technical assistance.

Conflicts of Interest: The authors declare no conflict of interest.

Abbreviations

bTJ	Bicellular tight junction
$d_{(Å)}$	Diameter (Å)
EDTA	Ethylenediaminetetraacetic acid
FD4	4 kDa-FITC-dextran 4 kDa
FITC	fluorescein isothiocyanate-dextran
KD	Knockdown
MDCK C7	Madin-Darby canine kidney cells, clone 7
shRNA	small hairpin RNA
TJ	Tight junction
tTJ	Tricellular tight junction
P	Apparent permeability (cm s^{-1})
PBS	phosphate-buffered saline
SEM	Standard error of the mean
TAMP	tight-junction-associated marvel protein
TER	Transepithelial resistance (Ω cm^2)

References

1. Farquhar, M.G.; Palade, G.E. Junctional complexes in various epithelia. *J. Cell Biol.* **1963**, *17*, 375–412. [CrossRef] [PubMed]
2. Amasheh, S.; Meiri, N.; Gitter, A.H.; Schöneberg, T.; Mankertz, J.; Schulzke, J.D.; Fromm, M. Claudin-2 expression induces cation-selective channels in tight junctions of epithelial cells. *J. Cell Sci.* **2002**, *115*, 4969–4976. [CrossRef] [PubMed]
3. Krug, S.M.; Schulzke, J.D.; Fromm, M. Tight junction, selective permeability, and related diseases. *Semin. Cell Dev. Biol.* **2014**, *36*, 166–176. [CrossRef] [PubMed]

4. Staehelin, L.A. Further observations on the fine structure of freeze-cleaved tight junctions. *J. Cell Sci.* **1973**, *13*, 763–786.
5. Günzel, D.; Fromm, M. Claudins and other tight junction proteins. *Compr. Physiol* **2012**, *2*, 1819–1852. [CrossRef]
6. Higashi, T.; Miller, A.L. Tricellular junctions: How to build junctions at the TRICkiest points of epithelial cells. *Mol. Biol. Cell* **2017**, *28*, 2023–2034. [CrossRef]
7. Ikenouchi, J.; Furuse, M.; Furuse, K.; Sasaki, H.; Tsukita, S.; Tsukita, S. Tricellulin constitutes a novel barrier at tricellular contacts of epithelial cells. *J. Cell Biol.* **2005**, *171*, 939–945. [CrossRef]
8. Krug, S.M.; Amasheh, S.; Richter, J.F.; Milatz, S.; Günzel, D.; Westphal, J.K.; Huber, O.; Schulzke, J.D.; Fromm, M. Tricellulin forms a barrier to macromolecules in tricellular tight junctions without affecting ion permeability. *Mol. Biol. Cell* **2009**, *20*, 3713–3724. [CrossRef]
9. Krug, S.M. Contribution of the tricellular tight junction to paracellular permeability in leaky and tight epithelia. *Ann. N. Y. Acad. Sci.* **2017**, *1397*, 219–230. [CrossRef]
10. Krug, S.M.; Bojarski, C.; Fromm, A.; Lee, I.M.; Dames, P.; Richter, J.F.; Turner, J.R.; Fromm, M.; Schulzke, J.D. Tricellulin is regulated via interleukin-13-receptor alpha2, affects macromolecule uptake, and is decreased in ulcerative colitis. *Mucosal Immunol.* **2018**, *11*, 345–356. [CrossRef]
11. Masuda, S.; Oda, Y.; Sasaki, H.; Ikenouchi, J.; Higashi, T.; Akashi, M.; Nishi, E.; Furuse, M. LSR defines cell corners for tricellular tight junction formation in epithelial cells. *J. Cell Sci.* **2011**, *124*, 548–555. [CrossRef] [PubMed]
12. Higashi, T.; Tokuda, S.; Kitajiri, S.; Masuda, S.; Nakamura, H.; Oda, Y.; Furuse, M. Analysis of the 'angulin' proteins LSR, ILDR1 and ILDR2–tricellulin recruitment, epithelial barrier function and implication in deafness pathogenesis. *J. Cell Sci.* **2013**, *126*, 966–977. [CrossRef] [PubMed]
13. Agre, P.; Preston, G.M.; Smith, B.L.; Jung, J.S.; Raina, S.; Moon, C.; Guggino, W.B.; Nielsen, S. Aquaporin CHIP: The archetypal molecular water channel. *Am. J. Physiol.* **1993**, *265*, F463–F476. [CrossRef] [PubMed]
14. Day, R.E.; Kitchen, P.; Owen, D.S.; Bland, C.; Marshall, L.; Conner, A.C.; Bill, R.M.; Conner, M.T. Human aquaporins: Regulators of transcellular water flow. *Biochim. Biophys. Acta* **2014**, *1840*, 1492–1506. [CrossRef] [PubMed]
15. Zeuthen, T.; Gorraitz, E.; Her, K.; Wright, E.M.; Loo, D.D. Structural and functional significance of water permeation through cotransporters. *Proc. Natl. Acad. Sci. USA* **2016**, *113*, E6887–E6894. [CrossRef] [PubMed]
16. Rosenthal, R.; Milatz, S.; Krug, S.M.; Oelrich, B.; Schulzke, J.D.; Amasheh, S.; Günzel, D.; Fromm, M. Claudin-2, a component of the tight junction, forms a paracellular water channel. *J. Cell Sci.* **2010**, *123*, 1913–1921. [CrossRef]
17. Rosenthal, R.; Günzel, D.; Krug, S.M.; Schulzke, J.D.; Fromm, M.; Yu, A.S. Claudin-2-mediated cation and water transport share a co Günzel (with u-umlaut)mmon pore. *Acta Physiol.* **2017**, *219*, 521–536. [CrossRef]
18. Rosenthal, R.; Günzel, D.; Piontek, J.; Krug, S.M.; Ayala-Torres, C.; Hempel, C.; Theune, D.; Fromm, M. Claudin-15 forms a water channel through the tight junction with distinct function compared to claudin-2. *Acta Physiol.* **2019**, e13334. [CrossRef]
19. Krug, S.M.; Günzel, D.; Conrad, M.P.; Rosenthal, R.; Fromm, A.; Amasheh, S.; Schulzke, J.D.; Fromm, M. Claudin-17 forms tight junction channels with distinct anion selectivity. *Cell. Mol. Life Sci.* **2012**, *69*, 2765–2778. [CrossRef]
20. Rosenthal, R.; Günzel, D.; Theune, D.; Czichos, C.; Schulzke, J.D.; Fromm, M. Water channels and barriers formed by claudins. *Ann. N. Y. Acad. Sci.* **2017**, *1397*, 100–109. [CrossRef]
21. Krug, S.M.; Hayaishi, T.; Iguchi, D.; Watari, A.; Takahashi, A.; Fromm, M.; Nagahama, M.; Takeda, H.; Okada, Y.; Sawasaki, T.; et al. Angubindin-1, a novel paracellular absorption enhancer acting at the tricellular tight junction. *J. Control. Release* **2017**, *260*, 1–11. [CrossRef] [PubMed]
22. Gong, Y.; Himmerkus, N.; Sunq, A.; Milatz, S.; Merkel, C.; Bleich, M.; Hou, J. ILDR1 is important for paracellular water transport and urine concentration mechanism. *Proc. Natl. Acad. Sci. USA* **2017**, *114*, 5271–5276. [CrossRef] [PubMed]
23. Wegmann, M.; Nüsing, R.M. Prostaglandin E2 stimulates sodium reabsorption in MDCK C7 cells, a renal collecting duct principal cell model. *Prostaglandins Leukot Essent Fatty Acids* **2003**, *69*, 315–322. [CrossRef] [PubMed]

24. Zak, J.; Schneider, S.W.; Eue, I.; Ludwig, T.; Oberleithner, H. High-resistance MDCK-C7 monolayers used for measuring invasive potency of tumour cells. *Pflügers Archiv Eur. J. Physiol.* **2000**, *440*, 179–183. [CrossRef] [PubMed]
25. Riazuddin, S.; Ahmed, Z.M.; Fanning, A.S.; Lagziel, A.; Kitajiri, S.; Ramzan, K.; Khan, S.N.; Chattaraj, P.; Friedman, P.L.; Anderson, J.M.; et al. Tricellulin is a tight-junction protein necessary for hearing. *Am. J. Hum. Genet.* **2006**, *79*, 1040–1051. [CrossRef]
26. Saitou, M.; Furuse, M.; Sasaki, H.; Schulzke, J.D.; Fromm, M.; Takano, H.; Noda, T.; Tsukita, S. Complex phenotype of mice lacking occludin, a component of tight junction strands. *Mol. Biol. Cell* **2000**, *11*, 4131–4142. [CrossRef]
27. Schulzke, J.D.; Gitter, A.H.; Mankertz, J.; Spiegel, S.; Seidler, U.; Amasheh, S.; Saitou, M.; Tsukita, S.; Fromm, M. Epithelial transport and barrier function in occludin-deficient mice. *Biochim. Biophys. Acta* **2005**, *1669*, 34–42. [CrossRef]
28. Raleigh, D.R.; Marchiando, A.M.; Zhang, Y.; Shen, L.; Sasaki, H.; Wang, Y.; Long, M.; Turner, J.R. Tight junction-associated MARVEL proteins marveld3, tricellulin, and occludin have distinct but overlapping functions. *Mol. Biol. Cell* **2010**, *21*, 1200–1213. [CrossRef]
29. Günzel, D.; Yu, A.S. Claudins and the modulation of tight junction permeability. *Physiol. Rev.* **2013**, *93*, 525–569. [CrossRef]
30. Weber, C.R.; Tokuda, S.; Hirai, T.; Furuse, M. Claudin-4 knockout by TALEN-mediated gene targeting in MDCK cells: Claudin-4 is dispensable for the permeability properties of tight junctions in wild-type MDCK cells. *PLoS ONE* **2017**, *12*, e0182521. [CrossRef]
31. Kirschner, N.; Rosenthal, R.; Furuse, M.; Moll, I.; Fromm, M.; Brandner, J.M. Contribution of tight junction proteins to ion, macromolecule, and water barrier in keratinocytes. *J. Investig. Dermatol.* **2013**, *133*, 1161–1169. [CrossRef] [PubMed]
32. Rudan-Tasic, D.; Klofutar, C. Osmotic Coefficients of Aqueous Solutions of Some Poly(oxyethylene) Glycols at the Freezing Point of Solution. *Chem. Mon.* **2004**, *135*, 773–784. [CrossRef]
33. Wen, H.; Hao, J.; Li, S.K. Characterization of human sclera barrier properties for transscleral delivery of bevacizumab and ranibizumab. *J. Pharm. Sci.* **2013**, *102*, 892–903. [CrossRef] [PubMed]

International Journal of
Molecular Sciences

MDPI

Article

Apoptotic Fragmentation of Tricellulin

Susanne Janke, Sonnhild Mittag, Juliane Reiche and Otmar Huber *

Department of Biochemistry II, Jena University Hospital, Friedrich Schiller University Jena, 07743 Jena, Germany; susanne.janke@med.uni-jena.de (S.J.); sonnhild.mittag@med.uni-jena.de (S.M.); juliane.reiche@med.uni-jena.de (J.R.)

* Correspondence: otmar.huber@med.uni-jena.de; Tel.: +49-3641-939-6400

Received: 26 July 2019; Accepted: 26 September 2019; Published: 1 October 2019

Abstract: Apoptotic extrusion of cells from epithelial cell layers is of central importance for epithelial homeostasis. As a prerequisite cell–cell contacts between apoptotic cells and their neighbors have to be dissociated. Tricellular tight junctions (tTJs) represent specialized structures that seal polarized epithelial cells at sites where three cells meet and are characterized by the specific expression of tricellulin and angulins. Here, we specifically addressed the fate of tricellulin in apoptotic cells. Methods: Apoptosis was induced by staurosporine or camptothecin in MDCKII and RT-112 cells. The fate of tricellulin was analyzed by Western blotting and immunofluorescence microscopy. Caspase activity was inhibited by Z-VAD-FMK or Z-DEVD-FMK. Results: Induction of apoptosis induces the degradation of tricellulin with time. Aspartate residues 487 and 441 were identified as caspase cleavage-sites in the C-terminal coiled-coil domain of human tricellulin. Fragmentation of tricellulin was inhibited in the presence of caspase inhibitors or when Asp487 or Asp441 were mutated to asparagine. Deletion of the tricellulin C-terminal amino acids prevented binding to lipolysis-stimulated lipoprotein receptor (LSR)/angulin-1 and thus should impair specific localization of tricellulin to tTJs. Conclusions: Tricellulin is a substrate of caspases and its cleavage in consequence contributes to the dissolution of tTJs during apoptosis.

Keywords: tight junction; tricellulin; lipolysis-stimulated lipoprotein receptor (LSR); angulin; epithelial barrier; cell–cell contact; apoptosis; caspase

---

1. Introduction

Tight junctions (TJs) represent the most apical cell–cell contacts in polarized epithelial cell layers. They form essential barriers for solutes including nutrients, metabolites and toxins as well as ions, thereby separating luminal and external compartments from the interior of multicellular bodies. Accordingly, TJs are also important to exclude pathogens such as bacteria and viruses from organisms. Break-down of this barrier function is associated with different diseases and is involved in inflammatory bowel diseases [1] or in organ failure during sepsis [2,3]. However, it turned out that TJ components may also represent targets of specific bacteria and viruses to invade cells or tissues resulting in a deregulation of the cytoskeleton or of signaling pathways required for barrier maintenance [4]. Primarily, claudins and occludin form the apical sealing belt between two opposing cells where they are arranged in a network of TJ strands that restricts and regulates paracellular flux depending on the specific claudin composition. The situation is different at tricellular contacts where apical TJs strands of three neighboring cells meet and thus are assumed to represent weak points for the paracellular TJ network [5]. At these tricellular tight junctions, bicellular TJs converge, and TJ strands turn basally and form the so-called central sealing element [6]. These tricellular TJs (tTJs) differ from bicellular TJs (bTJs) by a specific set of proteins including tricellulin [7] and one of the three angulins [8].

Tricellulin (marvelD2) together with occludin and marvelD3 form the TJ-associated MARVEL protein (TAMP) family of 4-transmembrane-domain proteins with long N- and C-terminal domains both located in the cytosol. The tricellulin C-terminus interacts with the cytosolic adapter protein zona occludens-1 (ZO-1) and is thereby linked to the actin cytoskeleton [9–11]. Knock-down of tricellulin impairs epithelial barrier function and results in disorganized bTJs, indicating an essential role of tricellulin in maintenance of overall TJ structure and function [7]. Tricellulin at tTJs was shown to restrict macromolecular passage and strong overexpression contributes to reduced paracellular permeability [12,13]. Interestingly, in occludin knock-down cells, tricellulin is mislocated to bTJs preferentially at edges of elongating bTJs [14]. How homomeric or heteromeric tricellulin/occludin complexes may be involved in this process or in the transport of tricellulin to the TJs [15] is currently not understood. However, it is obvious that tricellulin and occludin do not form heterophilic trans-interactions within established TJs [11,16].

With the identification of LSR (lipolysis-stimulated lipoprotein receptor)/angulin-1 as a tTJ-located receptor that recruits tricellulin to tTJs, it became clear that a specific anchor–protein is responsible for the defined localization of tricellulin. Knock-down of LSR/angulin-1 impaired barrier function in reducing transepithelial resistance (TER) probably as a consequence of mislocalization of tricellulin to bTJs [17]. In addition, mislocalization of LSR was associated with cancer progression and metastasis [18]. Based on sequence homology, two LSR/angulin-1-related proteins including immunoglobulin-like domain-containing receptor (ILDR) 1/angulin-2 and ILDR 2/angulin-3 were identified. The three members show tissue-specific expression patterns and at least one family member is located at tTJs [8,19]. However, it is still open regarding how angulins are directed to tTJs.

The physiological importance of tricellulin for hair cells is emphasized by mutations in the tricellulin C-terminal domain found in patients with nonsyndromic deafness (DFNB49). All mutations result in tricellulin variants with premature stop codons and, consequently, structurally impaired C-terminal ends and limited binding to ZO-1 [20,21]. Moreover, truncation of the tricellulin C-terminal domain diminishes its correct localization to tricellular contacts [22]. Tricellulin knock-out mice show a progressive hearing loss due to apoptotic death of hair cells [23]. It is currently not clear why tricellulin knock-out mice show this limited and hair cell-specific phenotype although tricellulin was deleted in all other tissues too. Interestingly, ILDR1/angulin-2 deficiency causes a similar phenotype with progressing hearing loss by outer hair cell degeneration [19,24,25]. Whether compensatory effects or regulatory roles of tricellulin and/or angulins may play a role still has to be solved.

Little is known in respect to mechanisms involved in regulation of tricellulin function. Like occludin, tricellulin can be phosphorylated [7]. However, in occludin, a multitude of kinases and corresponding phosphosites as well as functional consequences of these modifications have been studied in detail [26]. A phosphorylation hotspot region [27] that seems to be involved in the regulation of occludin mobility, apical junctional complex dynamics and cell migration [28] has been identified. In contrast, tricellulin phosphorylation-sites have been found by mass spectrometry, but, to our knowledge, none of them has been functionally characterized. However, there is the first evidence that occludin and tricellulin are differentially targeted by specific kinases [29]. Interestingly, JNK1/JNK2-dependent phosphorylation of LSR/angulin-1 was reported to be crucial for its exclusive localization to tTJs [30]. In addition, tricellulin was shown to be targeted by the ubiquitin ligase Itch, which might affect its stability and trafficking [31].

To maintain epithelial functionality, old or damaged cells have to be removed by apoptosis and replaced by new cells. Apoptosis is induced by extrinsic and intrinsic signals. Both pathways in an initial step lead to activation of initiator caspases, which subsequently activate effector caspases. These finally hydrolyze essential structural and housekeeping proteins by cleavage after aspartate residues [32]. Caspases have multiple functions also in inflammation and immunity [33]. Extrusion of apoptotic cells from the cell layer ideally occurs without loss of barrier function and disruption of the cytoarchitecture [34]. In consequence, contacts between the dying and neighboring cells have to be disassembled. Previous studies have analyzed bTJs in this respect [35], but, to our knowledge, tTJs have not been investigated in more detail. Here, we induced apoptosis with staurosporine or camptothecin and observed a caspase-dependent cleavage of the C-terminus of tricellulin. We identified two specific caspase cleavage-sites and mutation of these sites protected tricellulin from fragmentation. In addition, we provide evidence that LSR/angulin-1, which anchors tricellulin to tTJs, is also targeted by caspases.

2. Results

2.1. Caspase-Dependent Cleavage of Tricellulin upon Apoptotic Stimuli

To investigate the fate of tricellulin during apoptosis, MDCKII cells were treated with staurosporine as a known inducer of apoptosis [36]. At different time points after addition of staurosporine, both floating and adherent cells were collected and lysed for subsequent Western blot analysis. Already three hours after induction of apoptosis, a time-dependent decline of endogenous tricellulin was detectable compared to DMSO-treated control cells (Figure 1A,B). Induction of apoptosis was confirmed by detection of cleaved poly(ADP-ribose)-polymerase (PARP), a well-known effector caspase substrate. The typical PARP cleavage fragment was detectable only in staurosporine-treated cells but not in control cells (Figure 1A). Moreover, when camptothecin was used as an alternative apoptosis inducer, the loss of endogenous tricellulin in MDCKII cells was detectable between 8 h and 24 h (Supplementary Figure S1). The produced tricellulin cleavage fragments were not detectable using the anti-tricellulin monoclonal antibody (clone 54H19L38). Immunofluorescence microscopy confirmed degradation of tricellulin. Signals for endogenous tricellulin disappeared with time (Figure 1C).

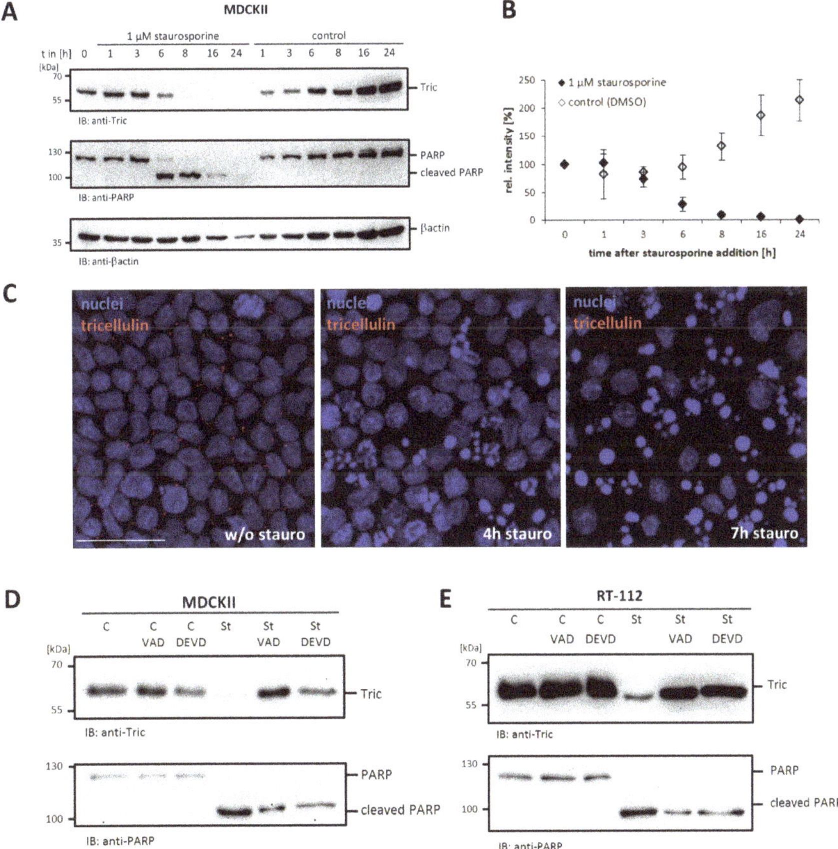

Figure 1. Tricellulin is a target of caspase cleavage in apoptotic cells. (**A**) MDCKII cells were treated with 1 μM staurosporine or DMSO as solvent control for the indicated times. Cell lysates were analyzed by Western blotting using anti-tricellulin, anti-poly[ADP-ribose] polymerase (PARP) and anti-βactin antibodies. (**B**) Quantification of tricellulin degradation by densitometric analysis. The graph represents mean values +/− SD of three independent experiments. (**C**) Loss of endogenous tricellulin detected by immunofluorescence microscopy (scale bar 30 μm). (**D**) Pre-treatment with 10 μM Z-VAD-FMK or 20 μM Z-DEVD-FMK for 1 h before stimulation with 1 μM staurosporine inhibited tricellulin degradation. Lysates were generated 6 h after induction of apoptosis. (**E**) Staurosporine-induced cleavage of tricellulin in RT-112 cells is inhibited by caspase inhibitors (10 μM Z-VAD-FMK or 20 μM Z-DEVD-FMK). All experiments are representatives of at least three independent experiments.

From our observations, we hypothesized that tricellulin like occludin [35] is a caspase target during apoptosis. To verify that the loss of tricellulin is indeed the consequence of caspase cleavage MDCKII cells were pre-treated with the pan-caspase inhibitor Z-VAD-FMK or the caspase-3 inhibitor Z-DEVD-FMK before adding staurosporine. Both caspase inhibitors significantly reduced degradation of endogenous tricellulin (Figure 1D). Inhibition of caspase activity was confirmed by the reduced formation of the PARP cleavage-fragment. A similar effect was obtained in RT-112 cells (bladder carcinoma) (Figure 1E) showing that the observed effect is not restricted to MDCKII cells.

2.2. *Mapping of Potential Caspase Cleavage-Sites in Tricellulin*

For prediction of possible caspase cleavage-sites in human tricellulin, the online software tool CaspDB [37] was applied. Two potential caspase cleavage-sites with the highest scores were predicted in the C-terminal domain of human tricellulin after amino acid aspartate 487 (D487) and/or aspartate 441 (D441) (Figure 2A). To confirm caspase-3-mediated cleavage in the C-terminal domain of tricellulin, we expressed and purified a GST-TricC fusion protein as reported previously [29,31] and applied in vitro digestion using recombinant active caspase-3. In the presence of caspase-3, two additional fragments with a molecular mass of about 40 and 35 kDa were detectable by Western blot analysis, representing potential caspase-3 cleavage-products. The molecular mass of these fragments coincides with the predicted molecular weight of GST-TricC fragments when cleaved at D487 (frag1, ~41 kDa) and D441 (frag2, ~35 kDa) (Figure 2B). Remarkably, fragment 1 is generated earlier and more efficiently than fragment 2, suggesting that D487 is the preferred caspase-3 cleavage-site and D441 appears to be used as a sequential second site after cleavage at D487. Addition of the pan-caspase inhibitor Z-VAD-FMK inhibited the generation of both GST-TricC fragments (Figure 2C). As control, purified recombinant GST alone was analyzed for caspase-3 cleavage, but no fragmentation was detectable (Figure 2D). Finally, we generated GST-TricC-D487N, GST-TricC-D441N and GST-TricC-D441N/D487N double mutant constructs and subjected them to in vitro cleavage-assays with recombinant caspase-3. Consistent with our previous results, caspase cleavage of GST-TricC-D487N only resulted in formation of fragment 2 (frag2), whereas cleavage of GST-TricC-D441N only generated fragment 1 (frag1). The GST-TricC-D441N/D487N double-mutated construct was no longer cleaved by recombinant caspase-3 (Figure 2E).

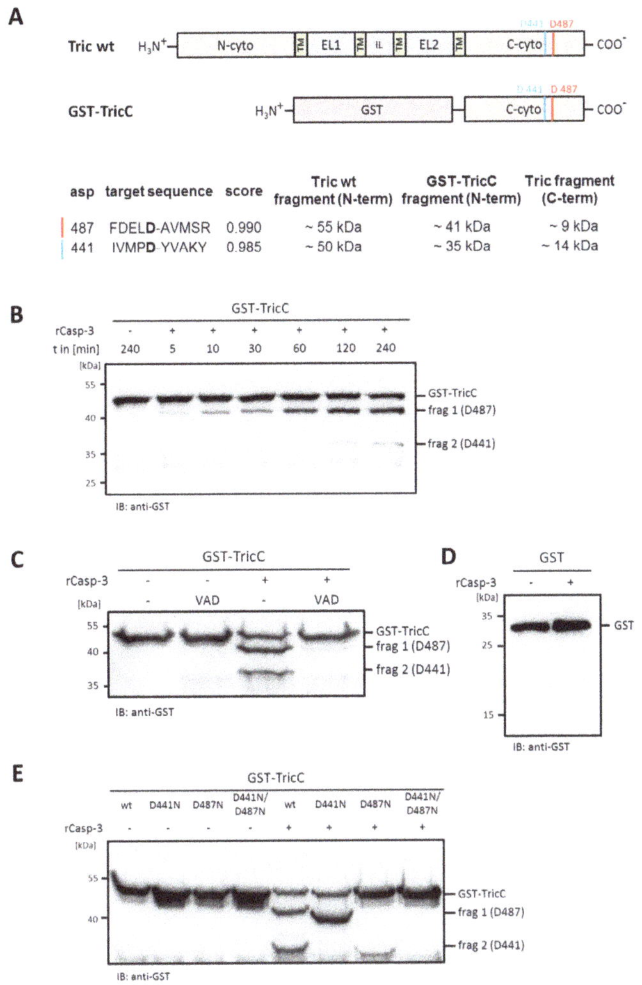

Figure 2. Identification of D487 and D441 as potential caspase-3 cleavage-sites in human tricellulin. (**A**) Schematic overview of the full-length tricellulin structure including the four transmembrane domains (TM), extracellular loops 1 and 2 (EL1, EL2), the intracellular loop (IL) and the potential caspase cleavage-sites D487 and D441. The table summarizes the molecular masses of the corresponding caspase cleavage-products expected for an in vitro caspase assay using recombinant GST-TricC fusion protein as substrate. (**B**) Time-dependent generation of the GST-TricC cleavage-products frag 1 and frag 2 after addition of recombinant caspase-3. The image is a representative of $n = 2$. (**C**) In the presence of pan-caspase inhibitor Z-VAD-FMK, fragmentation of GST-TricC is inhibited. (**D**) GST was used as a control and was not fragmented by caspase-3. (**E**) Mutation of potential caspase-3 cleavage-sites disable cleavage of GST-TricC partly (GST-TricC-D441N, GST-TricC-D487N) or completely (GST-TricC-D441N/D487N).

To validate the in vitro results, N-terminally $FLAG_3$-tagged human tricellulin was transiently transfected into MDCKII cells. Cells were subsequently treated with or without staurosporine for 6 h in the presence or absence of pan-caspase inhibitor Z-VAD-FMK and the caspase-3 inhibitor Z-DEVD-FMK, respectively. After induction of apoptosis, two bands with a molecular weight of about 55 kDa and 65 kDa were detectable (Figure 3A). Fragmentation was abrogated in the presence of each of the caspase inhibitors. In contrast to wildtype $FLAG_3$-Tric, transient transfection of a mutated $FLAG_3$-Tric-D441N construct abolished the generation of caspase-3 cleavage product frag 2 (~ 55 kDa) upon induction of apoptosis with staurosporine. Generation of cleavage-product frag 1 (~ 65 kDa) was not affected. Transfection of mutated $FLAG_3$-Tric-D487N revealed no fragment 1 and only to a very limited amount fragment 2. When the double-mutated $FLAG_3$-Tric-D441N/D487N was transfected, none of the fragments was detectable. These observations suggest that cleavage at D487 supports caspase-3-mediated cleavage at D441 (Figure 3B). Taken together, these results confirm D487 and D441 as potential caspase-sites in human tricellulin that are targeted in apoptotic cells.

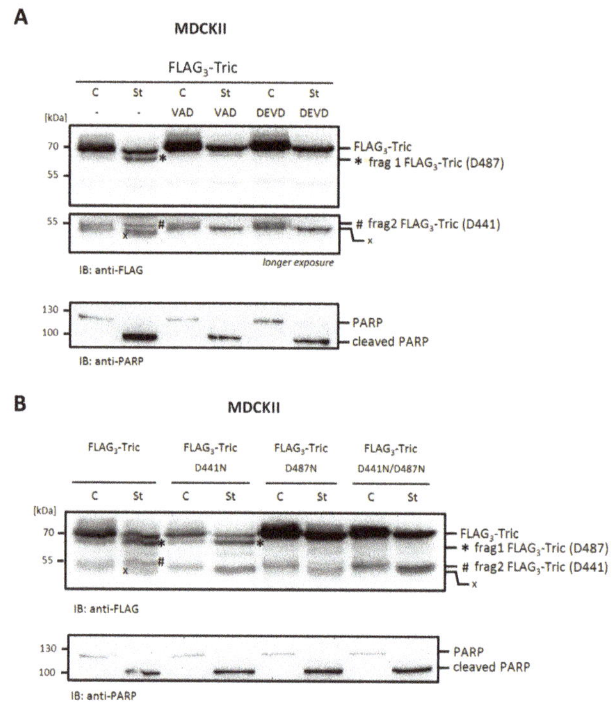

Figure 3. Caspase-3-mediated cleavage of $FLAG_3$-tricellulin in MDCKII upon apoptosis induction. (**A**) MDCKII cells were transiently transfected with p3xFLAG-CMV10-tricellulin, pre-treated with caspase inhibitors Z-VAD-FMK (VAD) or Z-DEVD-FMK (DEVD) for 1 h before induction of apoptosis with 1 μM staurosporine for 6 h. (**B**) MDCKII cells transiently transfected with $FLAG_3$-tricellulin wild-type or caspase-site mutated constructs as indicated were treated with 1 μM staurosporine for 6 h. The lower panels in (**A**) and (**B**) show Western blot detection of the typical PARP fragment generated by caspases confirming induction of apoptosis. Representative images of at least three independent experiments are shown. Bands marked with * represent caspase-dependent cleavage product frag 1 (~65 kDa) and # represents frag 2 (~55 kDa). The other bands (x) represent undefined or at least caspase-independent fragments.

2.3. *The Functional Interaction of Tricellulin and LSR Is Disrupted during Apoptosis*

Tricellulin is recruited to tTJs by lipolysis-stimulated lipoprotein receptor (LSR/angulin-1) in epithelial and endothelial cells [38,39]. This interaction is mediated by the cytosolic C-terminus of tricellulin [17]. In this context, the question arises if caspase-mediated cleavage within the cytosolic C-terminus of tricellulin affects its interaction with LSR. Therefore, co-immunoprecipitation experiments were performed using cell lysates obtained from HEK-293 cells transiently transfected with LSR together with either full-length tricellulin or deletion constructs lacking amino acids 487–558 ($FLAG_3$-TricΔ487–558), amino acid 441–558 ($FLAG_3$-TricΔ441–558) or complete cytosolic C-terminus ($FLAG_3$-TricΔC) (Figure 4A). Confirming literature, co-transfection of $FLAG_3$-Tric and green fluorescent protein GFP-tagged LSR in HEK-293 cells revealed an interaction of both proteins in co-immunoprecipitation experiments, whereas $FLAG_3$-TricΔC did only show a weak signal for GFP-LSR (Figure 4B). Similar to $FLAG_3$-TricΔC, an interaction between GFP-LSR and $FLAG_3$-TricΔ487–558 or $FLAG_3$-TricΔ441–558 protein lacking the cytosolic C-terminal parts released by caspases was not detectable (Figure 4B). In this context, it is interesting to note that, in a Western blot experiment, using the anti-tricellulin (clone 54H19L38) ABfinity^{TM} rabbit monoclonal antibody generated against amino acids 369–558 of human tricellulin did no longer detect neither the $FLAG_3$-TricΔ487–558 nor the $FLAG_3$-TricΔ441–558 protein in transiently transfected cells, thus suggesting that the epitope of this antibody is located between amino acids 487–558 (Supplementary Figure S2).

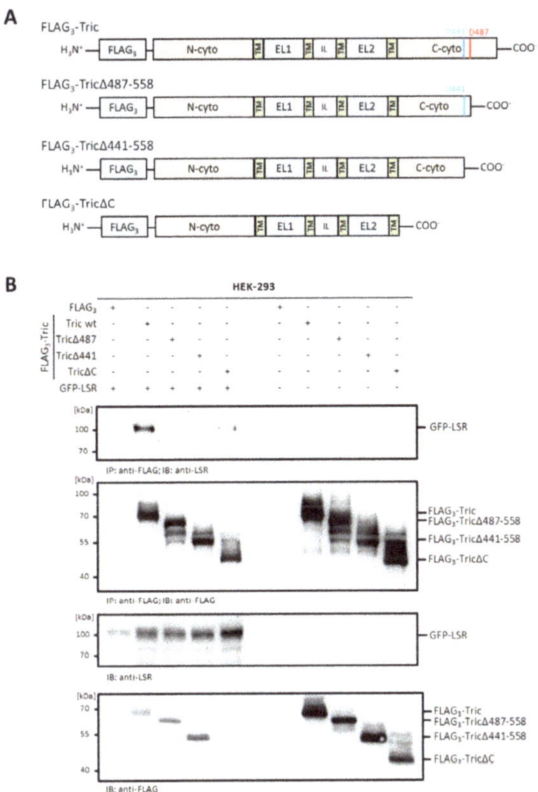

Figure 4. Deletion of the tricellulin C-terminus impairs binding to LSR/angulin-1. (**A**) Schematic representation of the $FLAG_3$-tricellulin constructs used in the transient transfection experiments. TM, transmembrane domain; EL, extracellular loop; IL, intracellular loop. (**B**) HEK-293 cells were transiently transfected with $FLAG_3$-tricellulin constructs missing the C-terminal tails corresponding to a cleavage by caspases together with GFP-LSR as indicated. Cells were lysed 48 h after transfection. Immunoprecipitaion (IP) was performed with monoclonal anti-FLAG antibody. Protein complexes from IP (upper two blots) and cell lysates (lower two blots) were analyzed by Western blotting with anti-LSR and anti-FLAG antibodies. The images are representatives of at least three independent experiments.

These results indicate that caspase-mediated cleavage of tricellulin liberates it from angulin and thus from tTJs. However, this does not exclude that LSR/angulin-1 itself is targeted during apoptosis.

Therefore, we next analyzed the fate of LSR/angulin-1 after induction of apoptosis in MDCKII cells. Indeed, staurosporine treatment led to a fragmentation of endogenous LSR/angulin-1 with time. Already 4 h after induction of apoptosis, four fragments of LSR were detectable using an anti-LSR antibody targeting the cytosolic part of LSR (Figure 5A). LSR/angulin-1 cleavage during apoptosis was inhibited by both caspase inhibitors Z-VAD-FMK and Z-DEVD-FMK. This indicates that LSR fragmentation is a consequence of caspase activation during apoptosis (Figure 5B).

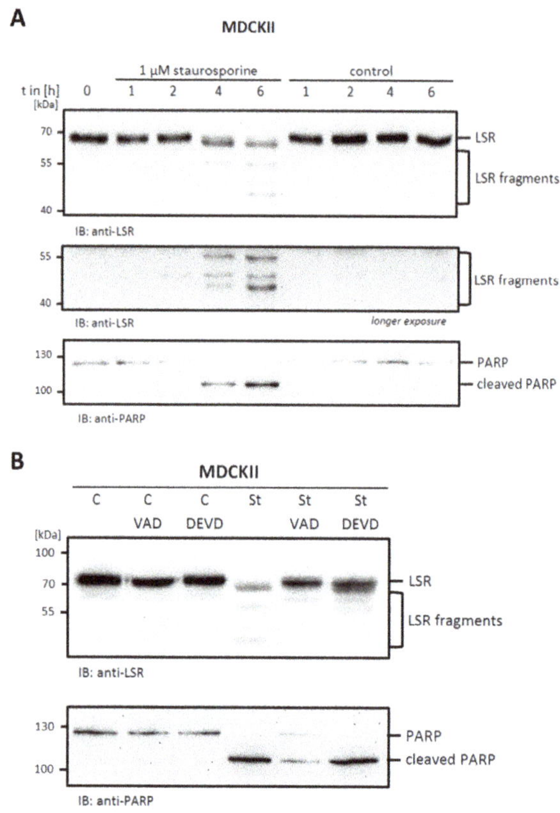

Figure 5. LSR in apoptotic cells. (**A**) Treatment of MDCKII cells with 1 µM staurosporine or solvent control for different times as indicated induced fragmentation of LSR. (**B**) Pre-treatment with caspase inhibitors (10 µM Z-VAD-FMK or 20 µM Z-DEVD-FMK for 1 h) inhibited fragmentation induced by staurosporine (1 µM for 6 h). Cell lysates were analyzed by Western blotting with anti-LSR and anti-PARP antibodies. Representative images of three independent experiments are shown.

3. Discussion

Release of apoptotic cells from epithelial and endothelial layers is critical for the maintenance of the barrier function and is often accompanied with a corresponding local decrease of transepithelial resistance [40]. In the gut, it is of importance that apoptotic cells, which are continuously extruded from the epithelium, do not generate microbial entry sites. Thus, it was not surprising that, in epithelial cell layers, there are mechanisms that efficiently break-down cell–cell contacts between apoptotic and neighboring cells. In previous studies, it was shown that components of the cadherin-catenin adhesion complex are targets of caspases. The adherens junction proteins E- and VE-cadherin are both cleaved by metalloproteinases, resulting in a release of the extracellular domain and thereby disrupting their trans-interactions with cadherin molecules on the surface of neighboring cells. Caspases are responsible for the cleavage of the cytosolic domains, thus disconnecting the interaction with the actin cytoskeleton [36,41]. Moreover, β-catenin as a linker between E-cadherin and the actin cytoskeleton is a caspase target itself [42]. Along with the adherens junctions, desmosomal cadherins and components of the cytosolic desmosomal plaque are inactivated similarly during apoptosis [43]. In addition, occludin, ZO-1 and ZO-2 were identified as caspase substrates [35]. Here, we extended this study to tricellulin and show that the C-terminal cytosolic tail of tricellulin is also targeted by caspases independent from whether apoptosis is induced by staurosporine or camptothecin and also independent from the cell system, in that case MDCKII or RT-112 cells. We identified two cleavage sites in tricellulin and mapped them C-terminal to Asp441 and Asp487. Specificity of cleavage was verified by treatment with caspase inhibitors and by mutation of the corresponding aspartate to asparagine, preventing the formation of cleavage fragments both in apoptotic cells as well as in in vitro fragmentation assays.

From our experiments, we conclude that D487 is the preferred cleavage-site. This is based on the observation that the corresponding cleavage-fragment was detectable far earlier as compared to the product generated by cleavage at D441 in in vitro cleavage-assays (Figure 2B). Moreover, it seems that initial cleavage at D487 is prerequisite for efficient cleavage at D441. This was confirmed by analyzing FLAG-tagged tricellulin where caspase cleavage-sites were mutated. The induction of apoptosis in cells transfected with wildtype $FLAG_3$-tricellulin revealed two cleavage fragments. However, in cells transfected with $FLAG_3$-tricellulin-D487N, a fragment generated by cleavage at D441 was more or less not detectable, most likely due to the lack of the initial cleavage at D487 (Figure 3B).

Caspase cleavage of tricellulin liberates a fragment of 71 C-terminal amino acids containing a region known to form a coiled-coil dimer [10]. This fragment was not detectable, even not when using high percentage SDS-PAGE gels, either because this small fragment escaped detection on Western blots or due to further degradation. Based on our observations, the antibody used in this study appears to bind to this fragment since truncated tricellulin where the C-terminus is deleted after D487 was no longer detectable. Definitively, it is worth testing by other methods if this fragment is detectable and of functional relevance. Not only due to its size can it easily translocate to the nucleus where it might be involved in transcriptional regulation, as shown for the amyloid precursor protein intracellular domain (AICD) [44] or the Notch intracellular domain (NICD) [45]. It has also been reported that nuclear localization of full-length tricellulin in pancreatic cancer promotes cell proliferation and invasiveness [46].

Based on our studies, we cannot exclude that the remaining tricellulin N-terminal part undergoes further processing steps as observed for occludin. There, the induction of apoptosis leads to caspase cleavage in the cytosolic C-terminus and to further fragmentation by metalloproteinases probably in the first extracellular loop [35]. First evidence for a potential contribution of metalloproteinases is provided by a study in Caco2 cells, where MMP-2- and MMP-3-catalyzed degradation of tricellulin is induced by N-3-(oxododecanoyl)-homoserine lactone (C12-HSL) [47].

A consequence of caspase cleavage and the concomitant release of the tricellulin C-terminal fragment is the disruption of the ZO-protein-mediated linkage of tricellulin to the actin cytoskeleton. Moreover, loss of the C-terminal amino acids might impair localization of tricellulin to tTJs by LSR/angulin-1. However, we observed that also LSR/angulin-1 is fragmented after induction of apoptosis. We conclude that, during apoptosis, tTJs are targeted at more than one level.

Taken together, here we showed that tricellulin is efficiently targeted by caspases in apoptotic cells within its C-terminal coiled-coil domain at D487 and D441. This together with concomitant cleavage of LSR/angulin-1 may contribute to a coordinated release of apoptotic cells from cell layers as it occurs continuously in the gastrointestinal tract at high frequency.

4. Materials and Methods

4.1. Cell Culture

Madine–Darby canine kidney II (MDCKII) cells were cultured in MEM with 10% (*v/v*) FBS and 1% (*v/v*) penicillin/streptomycin. Human bladder carcinoma cells (RT-112) as well as human embryonal kidney-293 cells (HEK-293) were cultivated in DMEM with 10% (*v/v*) FBS and 1% (*v/v*) penicillin/streptomycin using standard procedures as described elsewhere [31].

4.2. Reagents and Antibodies

Staurosporine was obtained from AppliChem GmbH (Darmstadt, Germany), caspase inhibitors Z-VAD-FMK and Z-DEVD-FMK as well as recombinant active caspase-3 were obtained from BD Biosciences (Heidelberg, Germany). Monoclonal anti-FLAG® M2 antibody was purchased from Sigma-Aldrich (Taufkirchen, Germany), anti-tricellulin (clone 54H19L38) ABfinity™ rabbit monoclonal antibody was from ThermoFisher Scientific (Darmstadt, Germany), anti-LSR (D3E3N) and anti-βactin (8H10D10) antibodies were obtained from Cell Signaling Technology (Frankfurt am Main, Germany) and anti-PARP (Ab-2) antibody was from Calbiochem (#AM30) (Merck KGaA, Darmstadt, Germany). Rabbit anti-GST antibody was provided by Jürgen Wienands. Goat anti-mouse-HRP and goat anti-rabbit-HRP antibodies were purchased from Sigma-Aldrich (Taufkirchen, Germany) and Alexa-Fluor™594-labeled secondary antibody was obtained from Molecular Probes (ThermoFisherScientific, Darmstadt, Germany). Dilution of the antibodies is summarized in Table 1.

Table 1. Antibodies used in the presented study including dilutions for Western blotting, immunofluorescence microscopy and immunoprecipitation.

Antibody	Dilution	Target	Species
anti-FLAG® M2	WB (1:5.000); IP 2 μg	FLAG-tag	mouse mab
anti-GST	WB (1:20.000)	glutathione-S-transferase	rabbit pab
anti-LSR (D3E3N) XP®	WB (1:1.000)	LSR	rabbit mab
anti-PARP-1 (Ab-2)	WB (1:1.000)	PARP	mouse mab
anti-tricellulin (54H19L38) ABfinity™	WB (1:1.000) IF (1 μg/mL)	tricellulin	rabbit mab
anti-βactin	WB (1:1000)	βactin	mouse
goat anti-mouse-HRP	WB (1:50.000–1:100.000)	mouse IgG	goat
goat anti-rabbit-HRP	WB (1:50.000–1:100.000)	rabbit IgG	goat
donkey anti-rabbit-Alexa594	IF (1: 1.000)	rabbit IgG	donkey

4.3. Plasmids, Site-Directed Mutagenesis and Transient Transfections

Plasmids were cloned by standard procedures and verified by sequencing. The cloning of p3xFLAG-CMV10-Tric and p3xFLAG-CMV10-TricΔC constructs was described previously [15]. To generate caspase cleavage-site mutated tricellulin variants, site-directed mutagenesis of the potential caspase cleavage-sites at positions D441 and D487 to asparagine was performed by SBOE (splicing by overlap extension)-PCR. Oligonucleotides used for the SBOE-PCR reactions are summarized in Table 2. The mutated tricellulin cDNAs were cloned into the vector p3xFLAG-CMV10 (Sigma-Aldrich, Taufkirchen, Germany) using *BamHI* restriction site. The C-terminal tricellulin deletion constructs TricΔ441–558 and TricΔ487–558 were generated by PCR using the oligonucleotides 5′-CGG ATC CTC AAA TGA TGG AAG ATC CAG-3′ as forward primer and 5′-CGG ATC CTT AGG GCA TCA CGA TAG GTT TAG-3′ or 5′-GGA TCC TTA CAG CTC ATC AAA CTT CCT CA-3′ as reverse primers, respectively, and cloned into the *BamHI* restriction site of p3xFLAG-CMV10 (Sigma-Aldrich, Taufkirchen, Germany). For recombinant expression of the tricellulin C-terminal cytosolic domain, pGEX4T1-TricC described previously [31] was used. The caspase-site mutated variants of GST-TricC were constructed by PCR employing oligonucleotides 5′-GCG GGA TCC ATG TGG AGG CAT GAG GCA GCT C-3′ and 5′-GCG GGT ACC GGA TCC TTA AGA ATA ACC TTG TAC ATC-3′ using p3xFLAG-CMV10-Tric-D441N, -D487N or -D441N/D487N as templates. After digestion with *BamHI*, the amplified products were ligated into *BamHI*-cleaved and dephosphorylated pGEX4T1 (GE Healthcare, Freiburg, Germany). pCAGGS-LSR for expression of GFP-tagged LSR was obtained from Mikio Furuse and described in [17].

MDCKII and HEK-293 cells were transiently transfected with the indicated plasmids using a DNA:PEI (poyethylenimine) ratio of 1:4 as described previously [48]. Cells were seeded in 6-well plates (2 × 10^5 cells for MDCKII and 5 × 10^5 HEK-293 cells per well one day before transfection. For transiently transfected HEK-293 or MDCKII cells, one or two 6-wells were lysed per sample. Empty vectors were transfected as a control.

Table 2. Sequences of oligonucleotides used to generate mutated tricellulin constructs. The bases marked in red represent those bases that differ from the wild-type sequence to generate the indicated mutations.

Mutation	Product	Sequence Oligonucleotides
Tric-D441N	1	5′-GCG GGT ACC GGA TCC GCC GCC ATG TCA AAT GAT GGA AGA TCC-3′ 5′-CAC ATA GTT GGG CAT CAC GAT-3′
	2	5′-ATC GTG ATG CCC AAC TAT GTG-3′ 5′-GCG GGT ACC GGA TCC TTA AGA ATA ACC TTG TAC ATC-3′
	1 + 2 (SBOE)	5′-GCG GGT ACC GGA TCC GCC GCC ATG TCA AAT GAT GGA AGA TCC-3′ 5′-GCG GGT ACC GGA TCC TTA AGA ATA ACC TTG TAC ATC-3′
Tric-D487N	1	5′-GCG GGT ACC GGA TCC GCC GCC ATG TCA AAT GAT GGA AGA TCC-3′ 5′-CAC TGC ATT CAG CTC ATC AAA-3′
	2	5′-TTT GAT GAG CTG AAT GCA GTG-3′ 5′-GCG GGT ACC GGA TCC TTA AGA ATA ACC TTG TAC ATC-3′
	1 + 2 (SBOE)	5′-GCG GGT ACC GGA TCC GCC GCC ATG TCA AAT GAT GGA AGA TCC-3′ 5′-GCG GGT ACC GGA TCC TTA AGA ATA ACC TTG TAC ATC-3′

4.4. Induction of Apoptosis and Preparation of Cell Lysates

MDCKII cells were plated on 6-well plates (2 × 10^5 cells per well for transfection; 8 × 10^5 cells per well without transfection) or 12-well plates (4 × 10^5 cells per well) and optionally transfected as described above. One day after seeding or transfection, apoptosis was induced by treatment with 1 μM staurosporine. Cells were lysed at the indicated time points. For inhibition of caspase activity, 10 μM Z-VAD-FMK or 20 μM Z-DEVD-FMK were added 1 h before induction of apoptosis for 6 h. Both floating and adherent cells from one well were harvested and pooled. Cells were lysed with 80 μL (12-well) or 100/150 μL (6-well) ice-cold modified RIPA-buffer (50 mM Tris/HCl pH 7.5, 150 mM NaCl, 0.5% (*w/v*) sodium-deoxycholate, 0.1% (*v/v*) SDS, 1% (*w/v*) Nonidet P-40 including Complete™ protease inhibitor mix (Roche Life Science, Mannheim, Germany)) for at least 15 min on ice. After sonication (15 pulses; cycle 0.5; 70% amplitude; UP100H ultrasound processor, Hilscher Ultrasound Technology, Teltow, Germany) the lysates were centrifuged (10 min, 20,800× *g*, 4 °C) to remove insoluble cell components.

4.5. In Vitro Caspase Cleavage

Recombinant glutathione S-transferase (GST)-tagged c-cytosolic (amino acids 366–558) domain of tricellulin wild type and mutant (D441N; D487N; D441N/D487N) proteins were expressed and purified as reported elsewhere [49]. In total, 4 μg of recombinant protein was digested with 150 ng of recombinant, active caspase-3 (BD Biosciences, Heidelberg, Germany) in 20 mM Pipes pH 7.2, 100 mM NaCl, 1 mM EDTA, 10% (*w/v*) sucrose, 10 mM DTT and 0,1% (*w/v*) CHAPS at 37 °C for the indicated incubation times. For inhibition of the in vitro caspase cleavage, 10 μM of pan-caspase inhibitor Z-VAD-FMK or DMSO as a control was added. Finally, samples were analyzed by SDS-PAGE and Western blot analysis.

4.6. Cell Lysis and Co-Immunoprecipitation

HEK-293 cells were seeded in 6-well plates and transfected (1µg DNA each construct) with PEI as described above. Cells (two 6-wells) were harvested 48 h after transfection and lysed in 300 µL ice-cold modified RIPA-buffer including Complete™ protease inhibitor mix (Roche Life Science, Mannheim, Germany) for 20 min on ice. After lysis, samples were centrifuged (10 min, 20,800× *g*, 4 °C) to remove insoluble material. For co-immunoprecipitation experiments, 200 µL lysate was incubated with 2 µg of anti-FLAG® M2 antibody for 1 h at 4 °C under constant rotation. Subsequently, 30 µL of equilibrated Protein A Sepharose CL-4B beads (GE Healthcare, Freiburg, Germany) were added and incubated for further 30 min. Finally, beads were washed for three times with lysis buffer and protein complexes were eluted using 2 × SDS-loading buffer at 95 °C for 5 min. Samples were separated by SDS-PAGE and subsequently analyzed by Western blotting. For each experiment, 10 µL of cell lysate (5% of input) was loaded as a control.

4.7. Western Blot Analysis

Cell lysates were separated by SDS-PAGE and transferred to polyvinylidene difluoride (PVDF) membrane by tank blotting. After blocking with 5% (*w/v*) dry milk in Tris-buffered saline/Tween20 (TBS-T) for 1 h, membranes were incubated with the primary antibodies overnight at 4 °C and with horseradish peroxidase (HRP)-labeled secondary antibody at least for 1 h at room temperature (Table 1).

4.8. Immunofluorescence Microscopy

For immunofluorescence microscopy, MDCKII cells were seeded on transwell filter supports (Millicell Cell Culture Insert, 12 mm, 0.4 µm, Merck Millipore, Darmstadt, Germany) with a density of 150,000 cells/cm^2. At day 7, cells were treated with 1 µM staurosporine or DMSO for 4 h and 7 h. Subsequently, the cells were fixed with 2% paraformaledhyde in PBS and stained with 1 µg/mL anti-tricellulin (clone 54H19L38) antibody for 18 h at 4 °C followed by secondary antibody incubation (donkey anti-rabbit, Alexa-594 conjugated) for 1 h at room temperature. Nuclei were labeled using DAPI (Sigma-Aldrich, Taufkirchen, Germany). Images (z-stacks) were acquired with a laser scanning system TC/SP5 (Leica Microsystems, Wetzlar, Germany) and were maximum intensity projected.

Supplementary Materials: Supplementary materials can be found at http://www.mdpi.com/1422-0067/20/19/4882/s1.

Author Contributions: S.J., S.M. and J.R. performed experiments, analyzed and validated the data and edited the manuscript; O.H. conducted conceptual planning, discussed the data, and wrote and edited the manuscript.

Funding: This research received no external funding.

Acknowledgments: We thank Mikio Furuse for the GFP-tagged LSR expression vectors. Expert technical assistance by Manuela Neumann is gratefully acknowledged.

Conflicts of Interest: The authors declare no conflict of interest.

Abbreviations

bTJ	Bicellular tight junction
CHAPS	3-[(3-cholamidopropyl)dimethylammonio]-1-propanesulfonate
DAPI	4,6-diamidino-2-phenylindole
DMEM	Dulbecco's Modified Eagle's Medium
DMSO	Dimethyl sulfoxide
EDTA	Ethylenediaminetetraacetic acid
FBS	Fetal bovine serum
GFP	Green-fluorescent protein
GST	Glutathione S-transferase
HRP	Horseradish peroxidase

LSR	Lipolysis-stimulated lipoprotein receptor
mab	Monoclonal antibody
MDCK	Madin-Darby canine kidney cells
pab	Polyclonal antibody
PARP	Poly [ADP-ribose] polymerase
PEI	Polyethlenimine
RIPA	Radioimmunoprecipitation Assay
SBOE-PCR	Splicing by overlap extension PCR
SDS	Sodium dodecyl sulfate
SDS-PAGE	Sodium dodecyl sulfate polyacrylamide gel electrophoresis
TBS-T	Tris-buffered saline/Tween 20
TER	Transepithelial resistance
TJ	Tight junction
tTJ	Tricellular tight junction
ZO-1	Zona occludens-1
Z-DEVD-FMK	N-Benzyloxycarbonyl-Asp-Glu-Val-Asp(OMe) fluoromethylketone
Z-VAD-FMK	N-Benzyloxycarbonyl-Val-Ala-Asp(OMe) fluoromethylketone

References

1. Barmeyer, C.; Schulzke, J.D.; Fromm, M. Claudin-related intestinal diseases. *Semin. Cell Dev. Biol.* **2015**, *42*, 30–38. [CrossRef] [PubMed]
2. Chawla, L.S.; Fink, M.; Goldstein, S.L.; Opal, S.; Gomez, A.; Murray, P.; Gomez, H.; Kellum, J.A.; Workgrp, A.X. The epithelium as a target in sepsis. *Shock* **2016**, *45*, 249–258. [CrossRef] [PubMed]
3. Haussner, F.; Chakraborty, S.; Halbgebauer, R.; Huber-Lang, M. Challenge to the intestinal mucosa during sepsis. *Front. Immunol.* **2019**, *10*, 891. [CrossRef] [PubMed]
4. Zihni, C.; Balda, M.S.; Matter, K. Signalling at tight junctions during epithelial differentiation and microbial pathogenesis. *J. Cell Sci.* **2014**, *127*, 3401–3413. [CrossRef] [PubMed]
5. Higashi, T.; Miller, A.L. Tricellular junctions: How to build junctions at the TRICkiest points of epithelial cells. *Mol. Biol. Cell* **2017**, *28*, 2023–2034. [CrossRef]
6. Staehelin, L.A. Further observation on the fine structure of freeze-cleaved tight junctions. *J. Cell Sci.* **1973**, *13*, 763–786. [PubMed]
7. Ikenouchi, J.; Furuse, M.; Furuse, K.; Sasaki, H.; Tsukita, S.; Tsukita, S. Tricellulin constitutes a novel barrier at tricellular contacts of epithelial cells. *J. Cell Biol.* **2005**, *171*, 939–945. [CrossRef]
8. Higashi, T.; Tokuda, S.; Kitajiri, S.; Masuda, S.; Nakamura, H.; Oda, Y.; Furuse, M. Analysis of the 'angulin' proteins LSR, ILDR1 and ILDR2-tricellulin recruitment, epithelial barrier function and implication in deafness pathogenesis. *J. Cell Sci.* **2013**, *126*, 966–977. [CrossRef]
9. Furuse, M.; Itoh, M.; Hirase, T.; Nagafuchi, A.; Yonemura, S.; Tsukita, S.; Tsukita, S. Direct association of occludin with ZO-1 and its possible involvement in the localization of occludin at tight junctions. *J. Cell Biol.* **1994**, *127*, 1617–1626. [CrossRef]
10. Schuetz, A.; Radusheva, V.; Krug, S.M.; Heinemann, U. Crystal structure of the tricellulin C-terminal coiled-coil domain reveals a unique mode of dimerization. *Ann. N. Y. Acad. Sci.* **2017**, *1405*, 147–159. [CrossRef]
11. Raleigh, D.R.; Marchiando, A.M.; Zhang, Y.; Shen, L.; Sasaki, H.; Wang, Y.; Long, M.; Turner, J.R. Tight junction-associated MARVEL proteins marveld3, tricellulin, and occludin have distinct but overlapping functions. *Mol. Biol. Cell* **2010**, *21*, 1200–1213. [CrossRef] [PubMed]
12. Krug, S.M.; Amasheh, M.; Dittmann, I.; Christoffel, I.; Fromm, M.; Amasheh, S. Sodium caprate as an enhancer of macromolecule permeation across tricellular tight junctions of intestinal cells. *Biomaterials* **2013**, *34*, 275–282. [CrossRef] [PubMed]
13. Krug, S.M.; Amasheh, S.; Richter, J.F.; Milatz, S.; Günzel, D.; Westphal, J.K.; Huber, O.; Schulzke, J.D.; Fromm, M. Tricellulin forms a barrier to macromolecules in tricellular tight junctions without affecting ion permeability. *Mol. Biol. Cell* **2009**, *20*, 3713–3724. [CrossRef] [PubMed]
14. Ikenouchi, J.; Sasaki, H.; Tsukita, S.; Furuse, M.; Tsukita, S. Loss of occludin affects tricellular localization of tricellulin. *Mol. Biol. Cell* **2008**, *19*, 4687–4693. [CrossRef] [PubMed]

15. Westphal, J.K.; Dörfel, M.J.; Krug, S.M.; Cording, J.D.; Piontek, J.; Blasig, I.E.; Tauber, R.; Fromm, M.; Huber, O. Tricellulin forms homomeric and heteromeric tight junctional complexes. *Cell. Mol. Life Sci.* **2010**, *67*, 2057–2068. [CrossRef] [PubMed]
16. Cording, J.; Berg, J.; Kading, N.; Bellmann, C.; Tscheik, C.; Westphal, J.K.; Milatz, S.; Günzel, D.; Wolburg, H.; Piontek, J.; et al. In tight junctions, claudins regulate the interactions between occludin, tricellulin and marvelD3, which, inversely, modulate claudin oligomerization. *J. Cell Sci.* **2013**, *126*, 554–564. [CrossRef]
17. Masuda, S.; Oda, Y.; Sasaki, H.; Ikenouchi, J.; Higashi, T.; Akashi, M.; Nishi, E.; Furuse, M. LSR defines cell corners for tricellular tight junction formation in epithelial cells. *J. Cell Sci.* **2011**, *124*, 548–555. [CrossRef]
18. Kohno, T.; Konno, T.; Kojima, T. Role of tricellular tight junction protein lipolysis-stimulated lipoprotein receptor (LSR) in cancer cells. *Int. J. Mol. Sci.* **2019**, *20*, 3555. [CrossRef]
19. Higashi, T.; Katsuno, T.; Kitajiri, S.; Furuse, M. Deficiency of angulin-2/ILDR1, a tricellular tight junction-associated membrane protein, causes deafness with cochlear hair cell degeneration in mice. *PLoS ONE* **2015**, *10*, e0120674. [CrossRef]
20. Riazuddin, S.; Ahmed, Z.M.; Fanning, A.S.; Lagziel, A.; Kitajiri, S.; Ramzan, K.; Khan, S.N.; Chattaraj, P.; Friedman, P.L.; Anderson, J.M.; et al. Tricellulin is a tight-junction protein necessary for hearing. *Am. J. Hum. Genet.* **2006**, *79*, 1040–1051. [CrossRef]
21. Ramzan, K.; Shaikh, R.S.; Ahmad, J.; Khan, S.N.; Riazuddin, S.; Ahmed, Z.M.; Friedman, T.B.; Wilcox, E.R.; Riazuddin, S. A new locus for nonsyndromic deafness DFNB49 maps to chromosome 5q12.3-q14.1. *Hum. Genet.* **2005**, *116*, 407–412. [CrossRef] [PubMed]
22. Nayak, G.; Lee, S.I.; Yousaf, R.; Edelmann, S.E.; Trincot, C.; Van Itallie, C.M.; Sinha, G.P.; Rafeeq, M.; Jones, S.M.; Belyantseva, I.A.; et al. Tricellulin deficiency affects tight junction architecture and cochlear hair cells. *J. Clin. Investig.* **2013**, *123*, 4036–4049. [CrossRef] [PubMed]
23. Kamitani, T.; Sakaguchi, H.; Tamura, A.; Miyashita, T.; Yamazaki, Y.; Tokumasu, R.; Inamoto, R.; Matsubara, A.; Mori, N.; Hisa, Y.; et al. Deletion of tricellulin causes progressive hearing loss associated with degeneration of cochlear hair cells. *Sci. Rep.* **2015**, *5*, 18402. [CrossRef] [PubMed]
24. Morozko, E.L.; Nishio, A.; Ingham, N.J.; Chandra, R.; Fitzgerald, T.; Martelletti, E.; Borck, G.; Wilson, E.; Riordan, G.P.; Wangemann, P.; et al. ILDR1 null mice, a model of human deafness DFNB42, show structural aberrations of tricellular tight junctions and degeneration of auditory hair cells. *Hum. Mol. Genet.* **2015**, *24*, 609–624. [CrossRef] [PubMed]
25. Sang, Q.; Li, W.; Xu, Y.; Qu, R.G.; Xu, Z.G.; Feng, R.Z.; Jin, L.; He, L.; Li, H.W.; Wang, L. ILDR1 deficiency causes degeneration of cochlear outer hair cells and disrupts the structure of the organ of Corti: A mouse model for human DFNB42. *Biol. Open* **2015**, *4*, 411–418. [CrossRef]
26. Dörfel, M.J.; Huber, O. Modulation of tight junction structure and function by kinases and phosphatases targeting occludin. *J. Biomed. Biotechnol.* **2012**, 807356. [CrossRef] [PubMed]
27. Dörfel, M.J.; Huber, O. A phosphorylation hotspot within the occludin C-terminal domain. *Ann. N. Y. Acad. Sci.* **2012**, *1257*, 38–44. [CrossRef]
28. Manda, B.; Mir, H.; Gangwar, R.; Meena, A.S.; Amin, S.; Shukla, P.K.; Dalal, K.; Suzuki, T.; Rao, R. Phosphorylation hotspot in the C-terminal domain of occludin regulates the dynamics of epithelial junctional complexes. *J. Cell Sci.* **2018**, *131*. [CrossRef]
29. Dörfel, M.J.; Westphal, J.K.; Huber, O. Differential phosphorylation of occludin and tricellulin by CK2 and CK1. *Ann. N. Y. Acad. Sci.* **2009**, *1165*, 69–73. [CrossRef]
30. Nakatsu, D.; Kano, F.; Taguchi, Y.; Sugawara, T.; Nishizono, T.; Nishikawa, K.; Oda, Y.; Furuse, M.; Murata, M. JNK1/2-dependent phosphorylation of angulin-1/LSR is required for the exclusive localization of angulin-1/LSR and tricellulin at tricellular contacts in EpH4 epithelial sheet. *Genes Cells* **2014**, *19*, 565–581. [CrossRef]
31. Jennek, S.; Mittag, S.; Reiche, J.; Westphal, J.K.; Seelk, S.; Dorfel, M.J.; Pfirrmann, T.; Friedrich, K.; Schutz, A.; Heinemann, U.; et al. Tricellulin is a target of the ubiquitin ligase Itch. *Ann. N. Y. Acad. Sci.* **2017**, *1397*, 157–168. [CrossRef] [PubMed]
32. Ramirez, M.L.G.; Salvesen, G.S. A primer on caspase mechanisms. *Semin. Cell Dev. Biol.* **2018**, *82*, 79–85. [CrossRef] [PubMed]
33. Songane, M.; Khair, M.; Saleh, M. An updated view on the functions of caspases in inflammation and immunity. *Semin. Cell Dev. Biol.* **2018**, *82*, 137–149. [CrossRef] [PubMed]

34. Gagliardi, P.A.; Primo, L. Death for life: A path from apoptotic signaling to tissue-scale effects of apoptotic epithelial extrusion. *Cell. Mol. Life Sci.* **2019**, *76*, 3571–3581. [CrossRef] [PubMed]
35. Bojarski, C.; Weiske, J.; Schöneberg, T.; Schröder, W.; Mankertz, J.; Schulzke, J.D.; Florian, P.; Fromm, M.; Tauber, R.; Huber, O. The specific fates of tight junction proteins in apoptotic epithelial cells. *J. Cell Sci.* **2004**, *117*, 2097–2107. [CrossRef] [PubMed]
36. Steinhusen, U.; Weiske, J.; Badock, V.; Tauber, R.; Bommert, K.; Huber, O. Cleavage and shedding of E-cadherin after induction of apoptosis. *J. Biol. Chem.* **2001**, *276*, 4972–4980. [CrossRef] [PubMed]
37. Kumar, S.; van Raam, B.J.; Salvesen, G.S.; Cieplak, P. Caspase cleavage sites in the human proteome: CaspDB, a database of predicted substrates. *PLoS ONE* **2014**, *9*, e110539. [CrossRef] [PubMed]
38. Iwamoto, N.; Higashi, T.; Furuse, M. Localization of angulin-1/LSR and tricellulin at tricellular contacts of brain and retinal endothelial cells in vivo. *Cell. Struct. Funct.* **2014**, *39*, 1–8. [CrossRef] [PubMed]
39. Sohet, F.; Lin, C.; Munji, R.N.; Lee, S.Y.; Ruderisch, N.; Soung, A.; Arnold, T.D.; Derugin, N.; Vexler, Z.S.; Yen, F.T.; et al. LSR/angulin-1 is a tricellular tight junction protein involved in blood-brain barrier formation. *J. Cell Biol.* **2015**, *208*, 703–711. [CrossRef] [PubMed]
40. Bojarski, C.; Gitter, A.H.; Bendfeldt, K.; Mankertz, J.; Schmitz, H.; Wagner, S.; Fromm, M.; Schulzke, J.D. Permeability of human HT-29/B6 colonic epithlium as a function of apoptosis. *J. Physiol.* **2001**, *535*, 541–552. [CrossRef] [PubMed]
41. Herren, B.; Levkau, B.; Raines, E.W.; Ross, R. Cleavage of β-catenin and plakoglobin and shedding of VE-cadherin during endothelial apoptosis: Evidence for a role for caspases and metalloproteinases. *Mol. Biol. Cell* **1998**, *9*, 1589–1601. [CrossRef] [PubMed]
42. Steinhusen, U.; Badock, V.; Bauer, A.; Behrens, J.; Wittmann-Liebold, B.; Dörken, B.; Bommert, K. Apoptosis induced cleavage of β-catenin by caspase-3 results in proteolytic fragments with reduced transactivation potential. *J. Biol. Chem.* **2000**, *275*, 16345–16353. [CrossRef] [PubMed]
43. Weiske, J.; Schöneberg, T.; Schröder, W.; Hatzfeld, M.; Tauber, R.; Huber, O. The fate of desmosomal proteins in apoptotic cells. *J. Biol. Chem.* **2001**, *276*, 41175–41181. [CrossRef] [PubMed]
44. Multhaup, G.; Huber, O.; Buee, L.; Galas, M.C. Amyloid Precursor Protein (APP) Metabolites APP intracellular fragment (AICD), Aβ42, and Tau in nuclear roles. *J. Biol. Chem.* **2015**, *290*, 23515–23522. [CrossRef] [PubMed]
45. Bray, S.J.; Gomez-Lamarca, M. Notch after cleavage. *Curr. Opin. Cell Biol.* **2018**, *51*, 103–109. [CrossRef]
46. Takasawa, A.; Murata, M.; Takasawa, K.; Ono, Y.; Osanai, M.; Tanaka, S.; Nojima, M.; Kono, T.; Hirata, K.; Kojima, T.; et al. Nuclear localization of tricellulin promotes the oncogenic property of pancreatic cancer. *Sci. Rep.* **2016**, *6*, 33582. [CrossRef]
47. Eum, S.Y.; Jaraki, D.; Bertrand, L.; Andras, I.E.; Toborek, M. Disruption of epithelial barrier by quorum-sensing N-3-(oxododecanoyl)-homoserine lactone is mediated by matrix metalloproteinases. *Am. J. Physiol. Gastrointest. Liver Physiol.* **2014**, *306*, G992–G1001. [CrossRef] [PubMed]
48. Wetzel, F.; Mittag, S.; Cano-Cortina, M.; Wagner, T.; Kramer, O.H.; Niedenthal, R.; Gonzalez-Mariscal, L.; Huber, O. SUMOylation regulates the intracellular fate of ZO-2. *Cell. Mol. Life Sci.* **2017**, *74*, 373–392. [CrossRef]
49. Mittag, S.; Valenta, T.; Weiske, J.; Bloch, L.; Klingel, S.; Gradl, D.; Wetzel, F.; Chen, Y.; Petersen, I.; Basler, K.; et al. A novel role for the tumour suppressor nitrilase1 modulating the Wnt/β-catenin signalling pathway. *Cell Discov.* **2016**, *2*. [CrossRef]

MDPI
St. Alban-Anlage 66
4052 Basel
Switzerland
Tel. +41 61 683 77 34
Fax +41 61 302 89 18
www.mdpi.com

International Journal of Molecular Sciences Editorial Office
E-mail: ijms@mdpi.com
www.mdpi.com/journal/ijms

www.ingramcontent.com/pod-product-compliance
Lightning Source LLC
LaVergne TN
LVHW071613170726
843515LV00009B/2387

* 9 7 8 3 0 3 9 4 3 2 2 4 0 *